FOURIER ANALYSIS

FOURIER ANALYSIS

T.W. KÖRNER

Trinity Hall, Cambridge

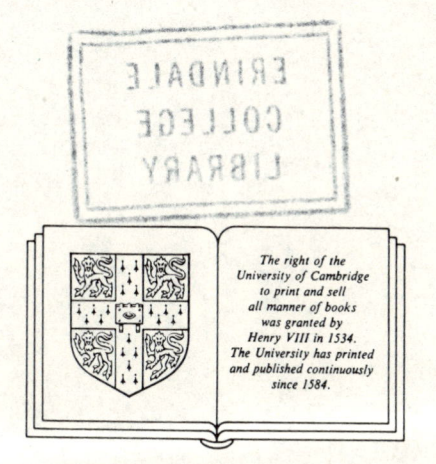

The right of the
University of Cambridge
to print and sell
all manner of books
was granted by
Henry VIII in 1534.
The University has printed
and published continuously
since 1584.

CAMBRIDGE UNIVERSITY PRESS

CAMBRIDGE

NEW YORK   PORT CHESTER

MELBOURNE   SYDNEY

Published by the Press Syndicate of the University of Cambridge
The Pitt Building, Trumpington Street, Cambridge CB2 1RP
40 West 20th Street, New York, NY 10011, USA
10 Stamford Road, Oakleigh, Melbourne 3166, Australia

© Cambridge University Press 1988

First published 1988
First paperback edition (with corrections) 1989

Printed in Great Britain at the University Press, Cambridge

British Library cataloguing in publication data
Körner, T.W.
Fourier analysis.
1. Fourier analysis
I. Title
515′.2433   QA403.5

Library of Congress cataloguing in publication data
Körner, T.W. (Thomas William), 1946–
Fourier analysis.
Includes index.
1. Fourier analysis.   I. Title.
QA403.5.K67   1986   515′.2433   85-17410

ISBN 0 521 25120 6 hard covers
ISBN 0 521 38991 7 paperback

TM

…mathematical ideas originate in empirics, although the genealogy is sometimes long and obscure. But, once they are so conceived, the subject begins to live a peculiar life of its own and is better compared to a creative one, governed by almost entirely aesthetical motivations, than to anything else and, in particular, to an empirical science. There is, however, a further point which, I believe, needs stressing. As a mathematical discipline travels far from its empirical source, or still more, if it is a second and third generation only indirectly inspired by ideas coming from 'reality', it is beset with very grave dangers. It becomes more and more purely aestheticising, more and more purely l'art pour l'art. This need not be bad, if the field is surrounded by correlated subjects, which still have closer empirical connections, or if the discipline is under the influence of men with an exceptionally well-developed taste. But there is a grave danger that the subject will develop along the line of least resistance, that the stream, so far from its source, will separate into a multitude of insignificant branches, and that the discipline will become a disorganised mass of details and complexities. In other words, at a great distance from its empirical source, or after much 'abstract' inbreeding, a mathematical subject is in danger of degeneration.

von Neumann (from the first paper in his collected works)

Some calculus tricks are quite easy. Some are enormously difficult. The fools who write the text books of advanced mathematics – and they are mostly clever fools – seldom take the trouble to show you how easy the easy calculations are. On the contrary, they seem to desire to impress you with their tremendous cleverness by going about it in the most difficult way.

Being myself a remarkably stupid fellow, I have had to unteach myself the difficulties, and now beg to present to my fellow fools the parts that are not hard. Master these thoroughly, and the rest will follow. What one fool can do, another can.

(from *Calculus Made Easy* by Sylvanus P. Thompson)

'Now,' Herbie says, 'wait a minute. A story goes with it.'

(from *A Story Goes With It* by Damon Runyon)

CONTENTS

PREFACE

This book is meant neither as a drill book for the successful nor as a lifebelt for the unsuccessful student. Rather, it is intended as a shop window for some of the ideas, techniques and elegant results of Fourier analysis.

I have tried to write a series of interlinked essays accessible to a student with a good general background in mathematics such as an undergraduate at a British university is supposed to have after two years of study. If the reader has not covered the relevant topic, say contour integration or probability, then she can usually omit, or better, skim through any chapters which involve this topic without impairing her ability to cope with subsequent chapters.

It is a consequence of the plan of this book that nothing is done in great depth or generality. If the reader wants to acquire facility with the Laplace transform or to study the L^2 convergence of the Fourier series of an L^2 function she must look elsewhere. It is very much easier to acquire a skill or to generalise a theorem when one is under the pressure of immediate necessity than when one is told that such a skill or generalisation might just possibly come in useful some day.

Another consequence is that, although anything specifically presented as a proof or statement of a result is intended to meet the pure mathematician's criteria for accuracy, the rigour of the accompanying discussion will vary according to the subject discussed. (Compare Chapter 3, Chapter 8 and Chapter 14.) For this I make no apology. 'It is the mark of the educated mind to use for each subject the degree of exactness which it admits' (Aristotle).

I must however apologise for a major, though perhaps unavoidable, fault in this book. The historical remarks which I make in connection with certain problems are brief and, if only for that reason, paint only a small part of a very complicated picture. Moreover, a glance at the average history of mathematics shows that mathematicians are remarkably incompetent historians. I make no claim to superiority and can only advise that the reader consults the original sources before accepting the truth of any historical sketch drawn in this book.

Any textbook owes more to the books and lectures of others than to the nominal author. At one stage I had a list of over 25 names of unwitting contributors to

this one. However, a long list prompts more interest in its omissions than its inclusions so I shall simply record my immense debt to the inspiring lectures of G. Friedlander, J.P. Kahane, H.P. Swinnerton-Dyer and H. Shapiro.

This book would never have seen the light of day without the labours of several generations of Cambridge typists. I should like to thank in particular Robyn Bringan, Debbie McCleland and Betty Sharples. It would have contained many more great and small mathematical errors without the careful scrutiny of Jonathan Partington, Richard Hildich, Chris Budd and an anonymous referee.

I close the preface by dedicating this book to my parents with love and respect.

T.W. Körner

PART 1
FOURIER SERIES

1
INTRODUCTION

We work on the circle $\mathbb{T} = \mathbb{R}/2\pi\mathbb{Z}$ i.e. the real line mod 2π. Thus if $\theta \in \mathbb{T}$ then $\theta + 2\pi = \theta$. (Or see, if you must, Appendix A.)

Suppose f is a Riemann integrable function from $\mathbb{T}$ to $\mathbb{C}$ or to $\mathbb{R}$. Then we define the Fourier coefficients of f by

$$\hat{f}(r) = (2\pi)^{-1} \int_0^{2\pi} f(t) \exp(-irt) dt = (2\pi)^{-1} \int_{\mathbb{T}} f(t) \exp(-irt) dt.$$

Mathematicians of the eighteenth century (Bernoulli, Euler, Lagrange, etc.) knew 'experimentally' that for some simple functions

$$S_n(f, t) = \sum_{-n}^{n} \hat{f}(r) \exp irt \to f(t) \quad \text{as } n \to \infty.$$

Fourier claimed that this was always true and in a book of outstanding importance in the history of physics (*Théorie Analytique de la Chaleur*) showed how formulae of the kind $\sum_{-\infty}^{\infty} \hat{f}(r) \exp irt$ could be used to solve linear partial differential equations of the kind which dominated 19th century physics.

After several mathematicians (including Cauchy) had produced more or less fallacious proofs of convergence, Dirichlet took up the problem. In a paper which set up new and previously undreamed of standards of rigour and clarity in analysis, he was able to prove convergence under quite general conditions. For example, the following theorem is a consequence of his results.

Theorem 1.1. *If f is continuous and has a bounded continuous derivative except, possibly, at a finite number of points then $S_n(f, t) = \sum_{-n}^{n} \hat{f}(r) \exp irt \to f(t)$ as $n \to \infty$ at all points t where f is continuous.*

However, it turned out that the conditions on f could not be relaxed indefinitely for Du Bois-Reymond constructed the following counter example.

Example 1.2. *There exists a continuous function f such that $\limsup_{n \to \infty} S_n(f, 0) = \infty$.*

(Theorem 1.1 will be proved in Chapter 15 in the case f everywhere continuous

3

and in Chapter 16 in the general case, whilst Example 1.2 will be constructed in Chapter 18.)

A new question was thus posed.

Question 1.3. *If f is a continuous function from $\mathbb{T}$ to $\mathbb{C}$ then, given the Fourier coefficients $\hat{f}(r)[r \in \mathbb{Z}]$ of f, can we find $f(t)$ for $t \in \mathbb{T}$?*

To the surprise of everybody Fejér (then aged only 19) showed that the answer is yes. He started from the observation that if a sequence $s_0, s_1, \ldots$ is not terribly well behaved, its behaviour may be improved by considering averages s_0, $(s_0 + s_1)/2$, $(s_0 + s_1 + s_2)/3, \ldots$ This had, of course, been known since the time of Euler, but the first person to study the phenomenon in detail was Cesàro about ten years before Fejér's discovery.

Lemma 1.4. (i) *If $s_n \to s$ then $(n+1)^{-1} \sum_{j=0}^{n} s_j \to s$.*
(ii) *There exist sequences s_n such that s_n does not tend to a limit but $(n+1)^{-1} \sum_{j=0}^{n} s_j$ does.*

(Thus the 'Cesàro limit' exists, and is equal to the usual limit, whenever the usual limit exists and may exist even if the usual limit does not.)
Proof. (i) Let $\varepsilon > 0$ be given. Since $s_n \to s$ we can find an $N(\varepsilon)$ such that $|s_n - s| \leqslant \varepsilon/2$ for $n \geqslant N(\varepsilon)$. Set $A = \sum_{j=1}^{N(\varepsilon)} |s_j - s|$ and choose $M(\varepsilon) \geqslant N(\varepsilon)$ such that $M(\varepsilon) \geqslant 2A\varepsilon^{-1}$. Then if $n \geqslant M(\varepsilon)$,

$$\left| (n+1)^{-1} \sum_{j=0}^{n} s_j - s \right| = (n+1)^{-1} \left| \sum_{j=0}^{n} (s_j - s) \right|$$

$$\leqslant (n+1)^{-1} \sum_{j=0}^{n} |s_j - s| = (n+1)^{-1} \left(\sum_{j=0}^{N(\varepsilon)} |s_j - s| + \sum_{j=N(\varepsilon)+1}^{n} |s_j - s| \right)$$

$$\leqslant (n+1)^{-1}(A + (n - N(\varepsilon))\varepsilon/2) \leqslant (n+1)^{-1}((n+1)\varepsilon/2 + (n+1)\varepsilon/2) = \varepsilon.$$

(ii) Let $s_n = (-1)^n$ so that s_n fails to converge as $n \to \infty$. Then

$$\left| (n+1)^{-1} \sum_{j=0}^{n} s_j \right| = (n+1)^{-1} \left| \sum_{j=0}^{n} s_j \right| \leqslant (n+1)^{-1} \to 0$$

as $n \to \infty$, so $(n+1)^{-1} \sum_{j=0}^{n} s_j \to 0$ as $n \to \infty$. ∎

Fejér saw that although partial sums $S_n(f, t) = \sum_{r=-n}^{n} \hat{f}(r) \exp irt$ could fail to converge their averages

$$\sigma_n(f, t) = \frac{1}{n+1} \sum_{j=0}^{n} S_j(f, t) = \sum_{r=-n}^{n} \frac{n+1-|r|}{n+1} \hat{f}(r) \exp irt$$

might behave rather better and that a Cesàro limit could take the place of the usual limit.

Theorem 1.5. (i) *If* $f: \mathbb{T} \to \mathbb{C}$ *is Riemann integrable then, if* f *is continuous at* t,

$$\sigma_n(f, t) = \sum_{-n}^{n} \frac{n + 1 - |r|}{n + 1} \hat{f}(r) \exp irt \to f(t).$$

(ii) *If* $f: \mathbb{T} \to \mathbb{C}$ *is continuous then*

$$\sigma_n(f, t) = \sum_{-n}^{n} \frac{n + 1 - |r|}{n + 1} \hat{f}(r) \exp irt \to f(t)$$

uniformly on $\mathbb{T}$.

(Any reader discouraged by Fejér's precocity should note that a few years earlier his school considered him so weak in mathematics as to require extra tuition.)

2

PROOF OF FEJÉR'S THEOREM

We start with the following remark. Suppose $f: \mathbb{T} \to \mathbb{C}$ is Riemann integrable. Then

$$\sigma_n(f, t) = \sum_{r=-n}^{n} \frac{n+1-|r|}{n+1} \hat{f}(r) \exp irt$$

$$= \sum_{r=-n}^{n} \frac{n+1-|r|}{n+1} \frac{1}{2\pi} \left(\int_{\mathbb{T}} f(x) \exp(-irx) dx \right) \exp irt$$

$$= \frac{1}{2\pi} \int_{\mathbb{T}} f(x) \sum_{r=-n}^{n} \frac{n+1-|r|}{n+1} \exp(ir(t-x)) dx = \frac{1}{2\pi} \int_{\mathbb{T}} f(x) K_n(t-x) dx,$$

where

$$K_n(s) = \sum_{r=-n}^{n} \frac{n+1-|r|}{n+1} \exp irs.$$

Further, making the substitution $y = t - x$, we have

$$\sigma_n(f, t) = \frac{1}{2\pi} \int_{-\pi}^{\pi} f(x) K_n(t-x) dx = -\frac{1}{2\pi} \int_{t+\pi}^{t-\pi} f(t-y) K_n(y) dy$$

$$= \frac{1}{2\pi} \int_{t-\pi}^{t+\pi} f(t-y) K_n(y) dy = \frac{1}{2\pi} \int_{\mathbb{T}} f(t-y) K_n(y) dy.$$

We are therefore led to examine the structure of K_n in some detail.

Lemma 2.1.

$$K_n(s) = \frac{1}{n+1} \left(\frac{\sin \dfrac{(n+1)s}{2}}{\sin \dfrac{s}{2}} \right)^2 \qquad [s \neq 0].$$

$$K_n(0) = n + 1.$$

6

Proof. If $s \neq 0$ then

$$\sum_{r=-n}^{n} (n+1-|r|)\exp irs = \left(\sum_{k=0}^{n} \exp i\left(k-\frac{n}{2}\right)s \right)^2 = \left(\exp\left(-\frac{ins}{2}\right) \sum_{k=0}^{n} \exp iks \right)^2$$

$$= \left(\exp\left(-\frac{ins}{2}\right) \frac{1-\exp i(n+1)s}{1-\exp is} \right)^2 = \left(\frac{\exp\left(-\frac{i(n+1)s}{2}\right) - \exp\left(\frac{i(n+1)s}{2}\right)}{\exp\left(-\frac{is}{2}\right) - \exp\frac{is}{2}} \right)^2$$

$$= \left(\frac{\sin\dfrac{(n+1)s}{2}}{\sin\dfrac{s}{2}} \right)^2 .$$

If $s = 0$ then $K_n(0) = n+1$ by direct computation. ∎

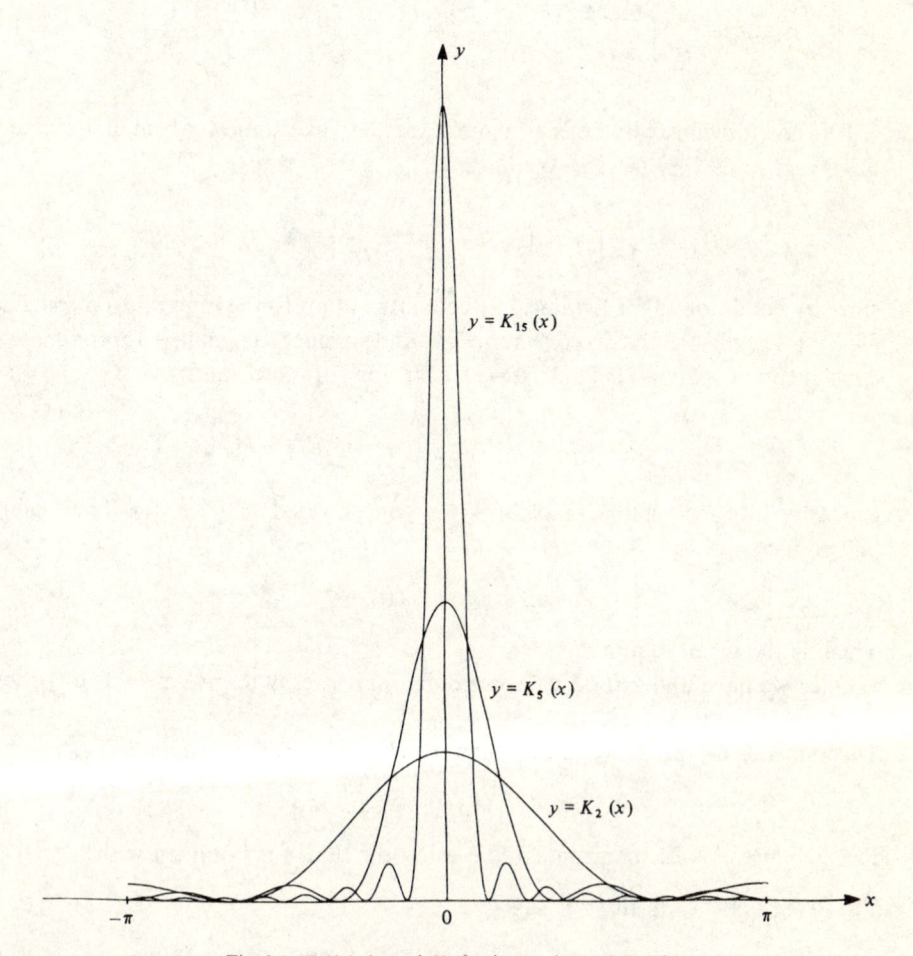

Fig. 2.1. Fejér's kernel K_n for increasing values of n.

We sketch K_n for a few values of n in Figure 2.1. The function has three vital properties.

Lemma 2.2. (i) $K_n(s) \geqslant 0$ *for all* $s \in \mathbb{T}$,

(ii) $K_n(s) \to 0$ *uniformly outside* $[-\delta, \delta]$ *for all* $\delta > 0$,

(iii) $(2\pi)^{-1} \int_{\mathbb{T}} K_n(s) ds = 1$.

Proof. Property (i) is obvious since the square of a real number is positive. Property (ii) follows on observing that

$$K_n(s) \leqslant \frac{1}{n+1} \left(\frac{1}{\sin s/2} \right)^2 \leqslant \frac{1}{n+1} \left(\frac{1}{\sin \delta/2} \right)^2 \to 0 \quad \text{as } n \to \infty$$

for all $\delta \leqslant |s| \leqslant \pi$. To see (iii) note that, returning to our original definition,

$$\frac{1}{2\pi} \int_{-\pi}^{\pi} K_n(s) ds = \frac{1}{2\pi} \int_{-\pi}^{\pi} \sum_{r=-n}^{n} \frac{n+1-|r|}{n+1} e^{irs} ds = \sum_{r=-n}^{n} \frac{n+1-|r|}{n+1} \frac{1}{2\pi} \int_{-\pi}^{\pi} e^{irs} ds = 1.$$

∎

It is now obvious why Fejér's theorem is true. Take some very small, but fixed, $\delta > 0$ and some very large n. We have

$$\sigma_n(f, t) = \frac{1}{2\pi} \int_{-\pi}^{\pi} K_n(s) f(t-s) ds \approx \frac{1}{2\pi} \int_{-\delta}^{\delta} K_n(s) f(t-s) ds,$$

since by condition (ii) of Lemma 2.2 the contribution to the integral from outside $[-\delta, \delta]$ is negligible. But f is continuous at t and so, since δ is small, is approximately constant in $[t-\delta, t+\delta]$. Thus $f(t-s) \approx f(t)$ for $s \in [-\delta, \delta]$ and

$$\sigma_n(f, t) \approx \frac{1}{2\pi} \int_{-\delta}^{\delta} K_n(s) f(t) ds = f(t) \left(\frac{1}{2\pi} \int_{-\delta}^{\delta} K_n(s) ds \right) \approx f(t) \left(\frac{1}{2\pi} \int_{-\pi}^{\pi} K_n(s) ds \right)$$

(since by (ii) most of the mass of K_n is concentrated in $[-\delta, \delta]$). Thus using (iii) we have

$$\sigma_n(f, t) \approx f(t),$$

which is the result required.

Once we have understood what is going on, it is easy to give a rigorous proof.

Theorem 2.3. (i) *If* $f : \mathbb{T} \to \mathbb{C}$ *is Riemann integrable then if* f *is continuous at* t,

$$\sigma_n(f, t) \to f(t) \quad \text{as } n \to \infty.$$

Proof. Since f is Riemann integrable we know that f is bounded with

(a) $|f(s)| \leqslant M$ for all $s \in \mathbb{T}$, say.

Since f is continuous at t we know that given any $\varepsilon > 0$ we can find a

$\delta(t, \varepsilon) > 0$ such that

(b) $|f(s) - f(t)| \leqslant \varepsilon/2$ for $|t - s| < \delta$.

Once $\delta(t, \varepsilon)$ is fixed it follows from property (ii) in Lemma 2.2 that we can find an N, depending on $\delta(t, \varepsilon)$ and so on t and ε, such that

(c) $|K_n(s)| \leqslant \varepsilon/(4M)$ for all $s \in [-\delta, \delta]$ and $n \geqslant N(t, \varepsilon)$.

Thus, using property (iii) of Lemma 2.2 in the first line of calculation and property (i) in the last line but one (to show that

$$\int_{s \in [-\delta, \delta]} |K_n(s)| \, ds = \int_{s \in [-\delta, \delta]} K_n(s) \, ds \leqslant \int_{\mathbb{T}} K_n(s) \, ds),$$

we have

$$|\sigma_n(f, t) - f(t)|$$

$$= \left| \frac{1}{2\pi} \int_{\mathbb{T}} f(t - s) K_n(s) \, ds - \frac{1}{2\pi} \int_{\mathbb{T}} K_n(s) f(t) \, ds \right| = \left| \frac{1}{2\pi} \int_{\mathbb{T}} (f(t - s) - f(t)) K_n(s) \, ds \right|$$

$$\leqslant \left| \frac{1}{2\pi} \int_{s \in [-\delta, \delta]} (f(t - s) - f(t)) K_n(s) \, ds \right| + \left| \frac{1}{2\pi} \int_{s \notin [-\delta, \delta]} (f(t - s) - f(t)) K_n(s) \, ds \right|$$

$$\leqslant \frac{\varepsilon}{2} \cdot \frac{1}{2\pi} \int_{s \in [-\delta, \delta]} |K_n(s)| \, ds + \frac{2M}{2\pi} \int_{s \notin [-\delta, \delta]} |K_n(s)| \, ds$$

$$\leqslant \frac{\varepsilon}{2} \cdot \frac{1}{2\pi} \int_{\mathbb{T}} K_n(s) \, ds + \frac{2M}{2\pi} \int_{s \notin [-\delta, \delta]} \frac{\varepsilon}{4M} \, ds \leqslant \frac{\varepsilon}{2} + \frac{\varepsilon}{2} = \varepsilon.$$

for all $n \geqslant N(t, \varepsilon)$ and the theorem is proved.   ∎

Since $\mathbb{T}$ is compact, any continuous function on $\mathbb{T}$ is uniformly continuous and the same arguments will prove a second version of Fejér's theorem.

Theorem 2.3. (ii) *If $f : \mathbb{T} \to \mathbb{C}$ is continuous then $\sigma_n(f, \ ) \to f$ uniformly on $\mathbb{T}$.*
Proof. Replace $\delta(t, \varepsilon)$ and $N(t, \varepsilon)$ by $\delta(\varepsilon)$ and $N(\varepsilon)$ in the proof of Theorem 2.3 (i).   ∎

Fejér's theorem has two consequences which may not appear very important at first sight but which form the key to the approach adopted in this book.

Theorem 2.4. *If $f, g : \mathbb{T} \to \mathbb{C}$ are continuous and $\hat{f}(r) = \hat{g}(r)$ for all $r \in \mathbb{Z}$ then $f = g$.*
Proof. We have

$$0 = \sigma_n(f, t) - \sigma_n(g, t) \to f(t) - g(t)$$

as $n \to \infty$, so $f(t) = g(t)$ for all $t \in \mathbb{T}$, i.e. $f = g$.   ∎

Theorem 2.4 says that a continuous function in *uniquely determined* by its Fourier coefficients.

Theorem 2.5. *If $f: \mathbb{T} \to \mathbb{C}$ is continuous and $\varepsilon > 0$ then we can find a trigonometric polynomial P with $\sup_{t \in \mathbb{T}} |P(t) - f(t)| \leqslant \varepsilon$.*

Proof. Recall that a trigonometric polynomial Q is a function of the form $Q(t) = \sum_{r=-n}^{n} a_r \exp irt$. Thus $\sigma_n(f, \ )$ is a trigonometric polynomial and the result follows from Theorem 2.3. (ii).   ∎

In the language of analytic topology Theorem 2.5 says that the trigonometric polynomials are *uniformly dense* (i.e. dense in the uniform metric) in the continuous functions.

Note that in the results above we work with functions $f: \mathbb{T} \to \mathbb{C}$ and this will be the case throughout most of the book. In the one or two chapters where we specifically consider functions $f: \mathbb{T} \to \mathbb{R}$ we shall feel free to make use of results like the following without further comment.

Lemma 2.6. (i) *If $f: \mathbb{T} \to \mathbb{R}$ is Riemann integrable then $\hat{f}(-r) = \hat{f}(r)^*$ (where * denotes complex conjugation).*
(ii) *If $a_{-r} = a_r^* [|r| \leqslant n]$ then we can find real numbers A_r and $B_r [|r| \leqslant n]$ such that*

$$\sum_{r=-n}^{n} a_r \exp irt = \tfrac{1}{2} A_0 + \sum_{r=1}^{n} A_r \cos rt + \sum_{r=1}^{n} B_r \sin rt [t \in \mathbb{T}],$$

and so in particular $\sum_{r=-n}^{n} a_r \exp irt$ is real valued.
(iii) *If, in the statement of Theorem 2.5, we add the condition that f is real valued, we may also demand that P be real valued.*

Proof. (i) $\hat{f}(r)^* = \left(\dfrac{1}{2\pi} \int_{\mathbb{T}} \exp(-irt) f(t) dt \right)^* = \dfrac{1}{2\pi} \int_{\mathbb{T}} (\exp(-irt) f(t))^* dt$

$$= \dfrac{1}{2\pi} \int_{\mathbb{T}} (\exp(-irt))^* (f(t))^* dt = \dfrac{1}{2\pi} \int_{\mathbb{T}} \exp(irt) f(t) dt = \hat{f}(-r).$$

(ii) Set $A_r = 2 \operatorname{Re} a_r, B_r = -2 \operatorname{Im} a_r$.
(iii) By part (i) $\hat{f}(r)^* = \hat{f}(-r)$

and so

$$\left(\frac{n+1-|r|}{n+1} \hat{f}(r) \right)^* = \left(\frac{n+1-|r|}{n+1} \hat{f}(-r) \right) [|r| \leqslant n].$$

It follows, by part (ii), that $\sigma_n(f, \ )$ is real valued and the method of Theorem 2.5 now gives the result.   ∎

As Theorem 2.4 and 2.5 and, to a lesser extent, the method of proof of Fejér's theorem form the central core of this book time spent understanding them will not be wasted. Once the reader has absorbed the contents of this section then, with a little good will on her part, she should be able to browse freely through the remainder of the book.

3

WEYL'S EQUIDISTRIBUTION THEOREM

Only time and experience will convince the reader that simple results like those of Theorem 2.4 and 2.5 are often more important and useful than any amount of difficult, ponderous and condition beset theorems on inversion and pointwise convergence. To start the process, we shall prove a very pretty result of Weyl.

For the purposes of this chapter (and this chapter only) let $\langle x \rangle$ denote the fractional part of $x \in \mathbb{R}$ (so that $0 \leqslant \langle x \rangle < 1$ and $x - \langle x \rangle \in \mathbb{Z}$). Weyl's equidistribution theorem states that if γ is irrational then for large n the fractional parts $\langle \gamma \rangle, \langle 2\gamma \rangle, \langle 3\gamma \rangle, \ldots, \langle n\gamma \rangle$ are more or less uniformly scattered over the unit interval. A precise statement of the result runs as follows.

Theorem 3.1 (Weyl). *If γ is irrational then for every $0 \leqslant a \leqslant b \leqslant 1$ we have*

$$n^{-1} \operatorname{card} \{1 \leqslant r \leqslant n : a \leqslant \langle r\gamma \rangle \leqslant b\} \to b - a \quad as \ n \to \infty.$$

The *reviews* of papers springing from Weyl's proof of his theorem fill over 100 pages in Chapter K of *Mathematics Reviews In Number Theory* 1940–72. Theorem 3.1 is a simple restatement of part (ii) of the following result for $\mathbb{T}$ which we shall now prove.

Theorem 3.1′ (Weyl). (i) *Suppose γ is irrational. Then if $f : \mathbb{T} \to \mathbb{C}$ is continuous we have*

$$n^{-1} \sum_{r=1}^{n} f(2\pi r \gamma) \to \frac{1}{2\pi} \int_{\mathbb{T}} f(t)dt \quad as \ n \to \infty.$$

(ii) *Suppose γ is irrational. Then if $0 \leqslant a \leqslant b \leqslant 1$,*

$$n^{-1} \operatorname{card} \{1 \leqslant r \leqslant n : 2\pi r \gamma \in [2\pi a, 2\pi b]\} \to (b - a) \quad as \ n \to \infty.$$

Proof. Write

$$G_n(f) = n^{-1} \sum_{r=1}^{n} f(2\pi r \gamma) - \frac{1}{2\pi} \int_{\mathbb{T}} f(t)dt.$$

11

If we can show that $G_n(f) \to 0$ as $n \to \infty$ we shall have proved (i). We divide the proof into short steps.

Step 1

$$G_n(1) = n^{-1} \sum_{r=1}^{n} 1 - \frac{1}{2\pi} \int_{\mathbb{T}} 1 dt = 1 - 1 = 0.$$

Step 2. If $e_s(t) = \exp ist [t \in \mathbb{T}, s \in \mathbb{Z}]$ then (for $s \neq 0$)

$$|G_n(e_s)| = \left| n^{-1} \sum_{r=1}^{n} \exp(2\pi i r s \gamma) - \frac{1}{2\pi} \int_{\mathbb{T}} \exp(ist) dt \right|$$

$$= \left| n^{-1} \exp(2\pi i s \gamma) \sum_{r=0}^{n-1} \exp(2\pi i r s \gamma) - 0 \right|$$

$$= \left| \frac{1}{n} \cdot \exp(2\pi i s \gamma) \cdot \frac{1 - \exp(2\pi i n s \gamma)}{1 - \exp 2\pi i s \gamma} \right| = \frac{1}{n} \left| \frac{1 - \exp(2\pi i n s \gamma)}{1 - \exp 2\pi i s \gamma} \right|$$

$$\leqslant \frac{1}{n} \frac{2}{|1 - \exp 2\pi i s \gamma|} \to 0 \quad \text{as } n \to \infty.$$

Step 3. If $P = \sum_{s=-m}^{m} a_s e_s$ (i.e. if P is a trigonometric polynomial) then, using linearity and the result of Steps 1 and 2,

$$G_n(P) = \sum_{s=-m}^{m} a_s G_n(e_s) \to 0 \quad \text{as } n \to \infty.$$

Step 4. If $f, g : \mathbb{T} \to \mathbb{C}$ are continuous functions and

$$|f(t) - g(t)| \leqslant \varepsilon \text{ for all } t \in \mathbb{T},$$

then

$$|G_n(f) - G_n(g)| \leqslant n^{-1} \sum_{r=1}^{n} |f(2\pi r \gamma) - g(2\pi r \gamma)| + \frac{1}{2\pi} \int_0^{2\pi} |f(t) - g(t)| dt$$

$$\leqslant \varepsilon + \varepsilon = 2\varepsilon \quad \text{for all } n \geqslant 0.$$

Step 5. If $f : \mathbb{T} \to \mathbb{C}$ is continuous and $\varepsilon > 0$ then by Theorem 2.5 we can find a trigonometric polynomial P with $|P(t) - f(t)| \leqslant \varepsilon/3$ for all $t \in \mathbb{T}$. By the result of Step 3 we can now find an n_0 such that $|G_n(P)| \leqslant \varepsilon/3$ for all $n \geqslant n_0$. But by the result of Step 4, $|G_n(f) - G_n(P)| \leqslant 2\varepsilon/3$ and so

$$|G_n(f)| \leqslant |G_n(P)| + |G_n(f) - G_n(P)| \leqslant \varepsilon \quad \text{for all } n \geqslant n_0.$$

It follows that $G_n(f) \to 0$ as $n \to \infty$ and so part (i) is proved. All that remains is to prove (ii).

Step 6. It is an easy matter to find, for each $\varepsilon > 0$, continuous functions $f_+, f_- : \mathbb{T} \to \mathbb{R}$ such that

(a) $f_+(t) \geqslant 1 \geqslant f_-(t)$ for all $t \in [2\pi a, 2\pi b]$,

(b)$_1$ $f_+(t) \geqslant 0$ for all $t \in \mathbb{T}$,

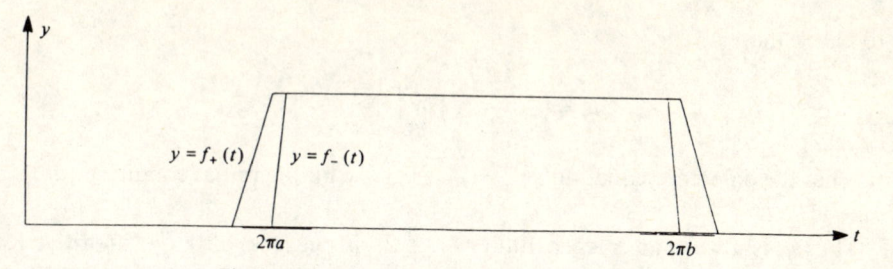

Fig. 3.1. The approximating functions f_+ and f_-.

(b)$_2$ $f_-(t) = 0$ for all $t \notin [2\pi a, 2\pi b]$,
(c)$_1$ $(b - a) + \varepsilon \geqslant (2\pi)^{-1} \int_\mathbb{T} f_+(t) dt$,
(c)$_2$ $(2\pi)^{-1} \int_\mathbb{T} f_-(t) dt \geqslant (b - a) - \varepsilon$.

(The functions f_+ and f_- are sketched in Figure 3.1.) From (a), (b)$_1$ and (b)$_2$ it follows at once that

$$\sum_{r=1}^{n} f_+(2\pi r\gamma) \geqslant \text{card } \{n \geqslant r \geqslant 1 : 2\pi r\gamma \in [2\pi a, 2\pi b]\} \geqslant \sum_{r=1}^{n} f_-(2\pi r\gamma).$$

On the other hand we know from part (i) that there exists an $n_0(\varepsilon)$ such that, whenever $n \geqslant n_0(\varepsilon)$, $|G_n(f_+)|, |G_n(f_-)| \leqslant \varepsilon$ and so

$$\frac{1}{2\pi} \int_\mathbb{T} f_+(t) dt + \varepsilon \geqslant n^{-1} \text{card } \{n \geqslant r \geqslant 1 : 2\pi r\gamma \in [2\pi a, 2\pi b]\} \geqslant \frac{1}{2\pi} \int_\mathbb{T} f_-(t) dt - \varepsilon.$$

Thus using (c)$_1$ and (c)$_2$ we obtain

$$(b - a) + 2\varepsilon \geqslant n^{-1} \text{card } \{n \geqslant r \geqslant 1 : 2\pi r\gamma \in [2\pi a, 2\pi b]\} \geqslant (b - a) - 2\varepsilon,$$

so, since $\varepsilon > 0$ was arbitrary,

$$n^{-1} \text{card } \{n \geqslant r \geqslant 1 : 2\pi r\gamma \in [2\pi a, 2\pi b]\} \to (b - a) \quad \text{as } n \to \infty,$$

as stated. ∎

The results of Theorem 3.1 and 3.1' are trivially false if γ is rational, so the problem of characterising those γ with $\langle r\gamma \rangle$ equidistributed (i.e. with

$$n^{-1} \text{card } \{n \geqslant r \geqslant 1 : \langle r\gamma \rangle \in [a, b]\} \to b - a$$

whenever $0 \leqslant a \leqslant b \leqslant 1$) is solved completely by the condition γ irrational.

On the other hand the problem of characterising those α with $\langle \alpha^r \rangle$ equidistributed is still not completely solved. We conclude this chapter with an example of the kind of idea involved.

Lemma 3.2. $\langle ((1 + \sqrt{5})/2)^r \rangle$ *is not equidistributed.*
Proof. By solving the difference equation or by straightforward induction it is easy

to check that

$$u_r = \left(\frac{1+\sqrt{5}}{2}\right)^r + \left(\frac{1-\sqrt{5}}{2}\right)^r$$

satisfies the difference equation $u_{r+1} = u_r + u_{r-1}$, with the initial conditions $u_0 = 2$, $u_1 = 1$.

Thus u_r is always an integer. But $(1 - \sqrt{5}/2)^r$ is negative for r odd, positive for r even and, more importantly, tends to 0 as $r \to \infty$ (since $0 > (1 - \sqrt{5})/2 > -1$). Thus

$$\langle ((1 + \sqrt{5})/2)^{2r+1} \rangle \to 0 \quad \text{as } r \to \infty,$$
$$\langle ((1 + \sqrt{5})/2)^{2r} \rangle \to 1 \quad \text{as } r \to \infty,$$

and   card $n^{-1}\{n \geqslant r \geqslant 1 : \langle ((1 + \sqrt{5})/2)^r \rangle \in [\frac{1}{4}, \frac{3}{4}]\} \to 0$ as $n \to \infty$. ∎

4
THE WEIERSTRASS POLYNOMIAL
APPROXIMATION THEOREM

Is it always possible to express a smooth function as a Taylor expansion in the neighbourhood of a given point? We shall show that the answer is no.

We start with a preliminary remark.

Lemma 4.1. *Suppose* $f(x) = \sum_{r=0}^{\infty} a_r x^r$ *for all* $|x| < \varepsilon$ *(where* $\varepsilon > 0$ *is fixed). Then*

$$a_n = f^{(n)}(0)/n! \quad [n \geqslant 0].$$

Proof. By definition $\sum_{r=1}^{\infty} a_r x^r$ has radius of convergence at least ε. But we know that we can differentiate term by term as often as we want within the radius of convergence to obtain

$$f^{(n)}(x) = \sum_{r=n}^{\infty} a_r r(r-1)\ldots(r-n+1)x^{r-n} \quad \text{for all } |x| < \varepsilon.$$

Setting $x = 0$ we obtain $f^{(n)}(0) = a_n n!$ which is the required result. ∎

We can now give our counter example.

Example 4.2. *Define* $h: \mathbb{R} \to \mathbb{R}$ *by*

$$h(x) = \exp(-1/x^2) \quad \text{for } x \neq 0,$$
$$h(0) = 0.$$

Then h is infinitely differentiable everywhere yet h cannot be written in the form

$$h(x) = \sum_{r=0}^{\infty} a_r x^r \text{ in } |x| < \varepsilon \text{ for any } \varepsilon > 0.$$

Proof. We divide the proof into steps.
Step 1. We start by showing that h is infinitely differentiable at all $x \neq 0$ and

$$h^{(r)}(x) = Q_r(x^{-1})\exp(-x^{-2}), \tag{1}_r$$

where Q_r is a polynomial. The proof is inductive. We observe that $(1)_0$ is trivially

15

true with $Q_0(t) = 1 [t \in \mathbb{R}]$ and that if $(1)_r$ is true then h is $r + 1$ times differentiable with

$$h^{(r+1)}(x) = -x^{-2}Q_r'(x^{-1})\exp(-x^{-2}) + 2x^{-3}Q_r(x^{-1})\exp(-x^{-2})$$
$$= Q_{r+1}(x^{-1})\exp(-x^{-2}),$$

where $\qquad\qquad Q_{r+1}(t) = -t^2 Q_r'(t) + 2t^3 Q_r(t) \quad [t \in \mathbb{R}],$

so Q_{r+1} is a polynomial and $(1)_{r+1}$ is true.

Step 2. We can now show that h is infinitely differentiable at 0 with

$$h^{(r)}(0) = 0 \tag{2}_r$$

Once again the proof is inductive. We observe that $(2)_0$ is true by definition and that if $(2)_r$ is true then using $(1)_r$

$$\frac{h^{(r)}(x) - h^{(r)}(0)}{x} = x^{-1}Q_r(x^{-1})\exp(-x^{-2}) \to 0 \quad \text{as } x \to 0$$

(since $\exp(t^2) \to \infty$ faster than any polynomial as $|t| \to \infty$) and so $(2)_{r+1}$ is true.

Step 3. We have now shown that h is infinitely differentiable everywhere and that $h^{(r)}(0) = 0$ for all $r \geq 0$. Suppose

$$h(x) = \sum_{r=0}^{\infty} a_r x^r \quad \text{for all } |x| < \varepsilon.$$

By Lemma 4.1,

$$a_r = h^{(r)}(0)/r! = 0$$

and so

$$\sum_{r=0}^{\infty} a_r x^r = 0 \quad \text{for all } |x| < \varepsilon.$$

But $h(x) = \exp(-x^{-2}) \neq 0$ for all $x \neq 0$ so we have a contradiction. Thus h cannot be expanded as a power series in the interval $(-\varepsilon, \varepsilon)$. ∎

The reader will recall that the versions of Taylor's theorem for the real variable case given in (reputable) text books have the form

$$f(a+t) = f(a) + f^{(1)}(a)t + f^{(2)}(a)\frac{t^2}{2!} + \cdots + f^{(n)}(a)\frac{t^n}{n!} + R_n(a, t),$$

where the $R_n(a, t)$ is a remainder term. In the case of the function h described in Lemma 4.2 $R_n(0, t) = h(t)$ and the only contributing term on the right hand side is the remainder. The point of the various versions of Taylor's theorem is that they give various useful formulae for R_n with the aid of which it is sometimes possible to show that $R_n(a, t) \to 0$ as $n \to \infty$ and so, indeed,

$$f(a+t) = \sum_{r=0}^{\infty} f^{(r)}(a)t^r/r!$$

(In the complex variable case Taylor's theorem states that an *analytic* function

can be expanded in power series about any point. This only shows once again how restrictive the condition of analyticity is.)

Power series of the form $\sum_{r=0}^{\infty} a_r x^r$ are rather easy to work with, partly because they are limits of polynomials $\sum_{r=0}^{n} a_r x^r$. The failure of the general form of Taylor's theorem is thus perhaps a little disappointing. However Weierstrass proved the following theorem which is often an acceptable substitute.

Theorem 4.3 (Weierstrass). *If $f:[a,b] \to \mathbb{C}$ is continuous and $\varepsilon > 0$ we can find a polynomial P with $\sup_{t \in [a,b]} |P(t) - f(t)| < \varepsilon$.*

Thus in the language of analytic topology Theorem 4.3 says that the polynomials are uniformly dense (i.e. dense in the uniform metric) in the continuous functions. An obvious use for Weierstrass's theorem occurs in numerical analysis. Here it gives us confidence when applying to more general continuous functions methods like Gaussian quadrature which were obtained by studying polynomials.

Weierstrass's theorem antedates Fejér's by 15 years and in Chapter 59 we shall give Weierstrass's original proof. However, it is clear that Theorem 2.5 and 4.3 are closely related and we shall now show how to deduce Theorem 4.3 from Theorem 2.5.

In fact we shall give two methods: the first quick and obvious, the second longer and more subtle but also perhaps more illuminating. For both proofs it is convenient to make a preliminary simplification by pointing out that it suffices to prove the following simpler version of Theorem 4.3.

Theorem 4.3′. *If $f:[0,1] \to \mathbb{C}$ is continuous and $\varepsilon > 0$ we can find a polynomial P with $\sup_{t \in [0,1]} |P(t) - f(t)| < \varepsilon$.*

Proof of Theorem 4.3 from Theorem 4.3′. Suppose $f:[a,b] \to \mathbb{C}$ is continuous and $\varepsilon > 0$. Define $g:[0,1] \to \mathbb{C}$ by $g(s) = f(a + s(b-a))$. Then g is continuous and so by Theorem 4.3′ we can find a polynomial Q with $\sup_{s \in [0,1]} |Q(s) - g(s)| < \varepsilon$. Set $P(t) = Q((t-a)/(b-a))$. Then P is also a polynomial and

$$\sup_{t \in [a,b]} |f(t) - P(t)| = \sup_{s \in [0,1]} |f(a+s(b-a)) - P(a+s(b-a))| = \sup_{s \in [0,1]} |g(s) - Q(s)| < \varepsilon,$$

as required. ∎

We now give our first proof of Theorem 4.3′ and thus of Theorem 4.3.

First proof of Theorem 4.3′. Let $f:[0,1] \to \mathbb{C}$ be continuous and let $\varepsilon > 0$. Define $g: \mathbb{T} \to \mathbb{C}$ by

$$g(t) = f(|t|) \quad \text{for } |t| \leq 1,$$
$$g(t) = f(1) \quad \text{for } |t| > 1.$$

Then g is continuous and so by Theorem 2.5 we can find an $n \geq 1$ and a_{-n},

$a_{-n+1}, \ldots, a_n \in \mathbb{C}$ such that

$$\left| g(t) - \sum_{r=-n}^{n} a_r \exp irt \right| < \varepsilon/2 \quad \text{for all } t.$$

But we know that $\sum_{k=0}^{m}(is)^k/k!$ converges uniformly to $\exp is$ in any bounded interval $[-R, R]$. Thus we can find for each $-n \leqslant r \leqslant n$ an $m(r)$ such that

$$\left| \sum_{k=0}^{m(r)} (irt)^k/k! - \exp irt \right| \leqslant \varepsilon/((4n+2)|a_r|+1) \quad \text{for all } |t| \leqslant 1.$$

Set
$$P(t) = \sum_{r=-n}^{n} a_r \sum_{k=0}^{m(r)} (irt)^k/k!.$$

Then P is clearly a polynomial and

$$|P(t) - f(t)| = |P(t) - g(t)| \leqslant \left| P(t) - \sum_{r=-n}^{n} a_r \exp irt \right| + \left| \sum_{r=-n}^{n} a_r \exp irt - g(t) \right|$$

$$< \sum_{r=-n}^{n} |a_r| \left| \sum_{k=0}^{m(r)} (irt)^k/k! - \exp irt \right| + \varepsilon/2 < \sum_{r=-n}^{n} \varepsilon/(4n+2) + \varepsilon/2 = \varepsilon$$

for all $0 \leqslant t \leqslant 1$, as required.　∎

We give our second proof in the next chapter.

5

A SECOND PROOF OF WEIERSTRASS'S THEOREM

In this chapter we give a second proof of the following theorem (Theorem 4.3' of Chapter 4).

Theorem 5.1. *If $f:[0,1] \to \mathbb{C}$ is continuous and $\varepsilon > 0$ we can find a polynomial P with*

$$\sup_{t \in [0,1]} |P(t) - f(t)| < \varepsilon.$$

We need two preliminary lemmas.

Lemma 5.2. *Suppose $f:\mathbb{T} \to \mathbb{C}$ is continuous and that $f(t) = f(-t)$ for all $t \in \mathbb{T}$. Then given any $\varepsilon > 0$ we can find $n \geq 1$ and $a_0, a_1, \ldots, a_n \in \mathbb{C}$ with $\sup_{t \in \mathbb{T}} |f(t) - \sum_{r=0}^{n} a_r \cos rt| < \varepsilon$.*

Proof. By making the substitution $s = -t$ we see that

$$\hat{f}(-r) = \frac{1}{2\pi} \int_{-\pi}^{\pi} f(t) \exp irt \, dt = \frac{-1}{2\pi} \int_{\pi}^{-\pi} f(-s) \exp(-irs) \, ds$$

$$= \frac{1}{2\pi} \int_{-\pi}^{\pi} f(-s) \exp(-irs) \, ds = \frac{1}{2\pi} \int_{-\pi}^{\pi} f(s) \exp(-irs) \, ds = \hat{f}(r).$$

Thus

$$\sigma_n(f,t) = \sum_{r=-n}^{n} \frac{n+1-|r|}{n+1} \hat{f}(r) \exp irt$$

$$= \hat{f}(0) + \sum_{r=1}^{n} \frac{n+1-r}{n+1} \hat{f}(r)(\exp irt + \exp - irt) = \hat{f}(0) + \sum_{r=1}^{n} 2\frac{n+1-r}{n+1} \hat{f}(r) \cos rt,$$

and since by Fejér's theorem (Theorem 2.3) $\sigma_n(f,t) \to f(t)$ uniformly, the result follows. ∎

The second lemma is more interesting.

Lemma 5.3. *There exists a real polynomial T_n (of degree exactly n and leading*

19

coefficient 1 *if* $n = 0$, 2^{n-1} *if* $n \geqslant 1$) *such that*

$$\cos n\theta = T_n(\cos \theta) \quad [\theta \in \mathbb{T}].$$

Proof. We prove the result by induction. The result is certainly true for $n = 0$ and $n = 1$ (take $T_0(t) = 1$, $T_1(t) = t$). Suppose it is true for all $n < m$, where $m \geqslant 2$. Then

$$\begin{aligned} \cos m\theta &= \cos m\theta + \cos(m-2)\theta - \cos(m-2)\theta \\ &= \cos((m-1)\theta + \theta) + \cos((m-1)\theta - \theta) - \cos(m-2)\theta \\ &= 2\cos\theta\cos(m-1)\theta - \cos(m-2)\theta \\ &= T_m(\cos\theta), \end{aligned}$$

where $$T_m(t) = 2tT_{m-1}(t) - T_{m-2}(t),$$

and so T_m is a polynomial of the right form. Thus the result is true for $n = m$. ∎

We can now prove Theorem 5.1.
Proof of Theorem 5.1. Let $f:[0,1] \to \mathbb{C}$ a continuous function and $\varepsilon > 0$ be given. Define $g: \mathbb{T} \to \mathbb{C}$ by

$$g(t) = f(|\cos t|) \quad \text{for all } t \in \mathbb{T}.$$

Then g is continuous and $g(t) = g(-t)$ for all $t \in \mathbb{T}$. It follows from Lemma 5.2 that we can find an $n \geqslant 1$ and $a_0, a_1, \ldots, a_n \in \mathbb{C}$ with $\sup_{t \in \mathbb{T}} |g(t) - \sum_{r=0}^{n} a_r \cos rt| < \varepsilon$. But, using Lemma 5.3, we have

$$g(t) - \sum_{r=0}^{n} a_r \cos rt = f(|\cos t|) - \sum_{r=0}^{n} a_r T_r(\cos t)$$

for all $t \in \mathbb{T}$, and so

$$\sup_{x \in [0,1]} \left| f(x) - \sum_{r=0}^{n} a_r T_r(x) \right| \leqslant \sup_{0 \leqslant t \leqslant \pi} \left| f(|\cos t|) - \sum_{r=0}^{n} a_r T_r(\cos t) \right| < \varepsilon.$$

Writing $P(x) = \sum_{r=0}^{n} a_r T_r(x)$ we see that P is a polynomial and $\sup_{x \in [0,1]} |f(x) - P(x)| < \varepsilon$, as required. ∎

    We have deduced Theorem 5.1 from Lemma 5.2 but the reader should have no difficulty in reversing the argument so as to deduce Lemma 5.2 from Theorem 5.1. With a little ingenuity, she should be able, if she wants, to show that Lemma 5.2 can be deduced directly from Theorem 2.5 and vice versa so that Theorems 2.5 and 4.3 are completely equivalent.
    The polynomials T_n which link trigonometric with polynomial approximation are called the Tchebychev (or Chebychev) polynomials. We shall discuss one of their remarkable properties in Chapter 45.

6

HAUSDORFF'S MOMENT PROBLEM

Let $f:\mathbb{R}\to\mathbb{R}$ be a continuous function with $f(x)\geq 0$ for all $x\in\mathbb{R}$. In mechanics we call $\int_{-\infty}^{\infty}f(x)\,dx$ the mass M associated with the mass density f. We call $M^{-1}\int_{-\infty}^{\infty}xf(x)\,dx$ the centre of gravity and $\int_{-\infty}^{\infty}x^2f(x)\,dx$ the moment of inertia about 0. In probability theory if $\int_{-\infty}^{\infty}f(x)\,dx=1$, we say that f is a probability density associated with a random variable, X, say. We call $\mathbb{E}X=\int_{-\infty}^{\infty}xf(x)\,dx$ the mean of the random variable and $\mathbb{E}X^2=\int_{-\infty}^{\infty}x^2f(x)\,dx$ the second moment. If $\mathbb{E}X=0$, we call $\mathbb{E}X^2$ the variance. (Note, however, that it is possible for some f that the integrals do not converge in which case the concepts are not defined.)

Let us call $\int_{-\infty}^{\infty}x^rf(x)\,dx$ (if it converges) the rth moment of f. It is clear that the more moments we know the more information we have concerning the rough shape of f. We are thus led to ask whether (if all the moments exist) the moments by themselves determine f.

Hausdorff showed that the answer is yes provided f is zero outside some fixed interval. More precisely we have the following theorem (which applies to complex valued functions as well).

Theorem 6.1 (Hausdorff). *Let $[a,b]$ be an interval and let $f,g:[a,b]\to\mathbb{C}$ be continuous. Then if*

$$\int_a^b x^r f(x)\,dx = \int_a^b x^r g(x)\,dx$$

for all $r\geq 0$ it follows that $f(x)=g(x)$ for all $x\in[a,b]$.

Proof. Set $h(x)=f(x)-g(x)$ for all $x\in[a,b]$. Then the theorem reduces to the statement that if $h:[a,b]\to\mathbb{C}$ is continuous and $\int_a^b x^r h(x)\,dx=0$ for all r then $h(x)=0$ for all $x\in[a,b]$. We shall prove the theorem in this form.

Observe first that if $P(x)=\sum_{k=0}^{n}a_kx^k$ is any polynomial then automatically

$$\int_a^b P(x)h(x)\,dx = \sum_{k=0}^{n}a_k\int_a^b x^k h(x)\,dx = 0.$$

21

But by Weierstrass's theorem (Theorem 4.3) we can find a sequence of polynomials P_n such that $\sup_{x\in[a,b]} |P_n(x) - h(x)^*| < 1/n$ (here z^* is the complex conjugate of z). Thus $P_n(x) \to h(x)^*$ uniformly on $[a, b]$ so (since h is continuous on a closed interval and thus bounded) $P_n(x)h(x) \to h(x)h(x)^* = |h(x)|^2$ uniformly on $[a, b]$. It follows that $\int_a^b P_n(x)h(x)\,dx \to \int_a^b |h(x)|^2\,dx$ as $n \to \infty$. But we saw at the beginning of the paragraph that $\int_a^b P_n(x)h(x)\,dx = 0$. Thus $\int_a^b |h(x)|^2\,dx = 0$ and so (since the integral of a positive continuous function can only be zero if the function is identically zero) $|h(x)|^2 = 0$ for all $x\in[a, b]$. Hence $h(x) = 0$ for all $x\in[a, b]$ and we are done. ∎

Hausdorff's result has obvious importance in probability theory where we often prefer to work with moments rather than with density functions, since it tells us that (in the bounded case at least) we do not lose information thereby.

(It is of less importance in mechanics since the three-dimensional analogues of the zeroth moment (a scalar giving the mass), the first moment (a vector giving the centre of gravity) and the second moment (a tensor called the inertia tensor determined by the principal axes and moments of inertia) suffice to determine the motion of a rigid body in a uniform gravitational field.)

Unfortunately Hausdorff's result breaks down if we replace $[a, b]$ by $\mathbb{R}$ or even by $[0, \infty)$.

Example 6.2. *Let $h(x) = \exp(-x^{\frac{1}{4}})\sin x^{\frac{1}{4}}$ for $x \geqslant 0$. Then $h:[0, \infty] \to \mathbb{R}$ is continuous and not identically zero yet $\int x^r h(x)\,dx = 0$ for all integers $r \geqslant 0$.*

Proof. Let $\omega = \exp(\pi i/4)$. Then by integration by parts we have

$$\int_0^\infty y^n e^{-\omega y}\,dy = \omega^{-1}[-y^n e^{-\omega y}]_0^\infty + \omega^{-1}n\int_0^\infty y^{n-1}e^{-\omega y}\,dy$$

$$= \omega^{-1}\left[-y^n \exp\left(\frac{-y}{\sqrt{2}}\right)\exp\left(\frac{-iy}{\sqrt{2}}\right)\right]_0^\infty + \omega^{-1}n\int_0^\infty y^{n-1}e^{-\omega y}\,dy$$

$$= \omega^{-1}n\int_0^\infty y^{n-1}e^{-\omega y}\,dy \quad [n \geqslant 1].$$

Thus, by induction,

$$\int_0^\infty y^n e^{-\omega y}\,dy = \omega^{-n}n!\int_0^\infty e^{-\omega y}\,dy = \omega^{-n-1}n!.$$

In particular, $\int_0^\infty y^{4n+3}e^{-\omega y}\,dy$ is real and so, by considering the imaginary part, we have

$$\int_0^\infty y^{4n+3}\exp(-2^{-\frac{1}{2}}y)\sin(2^{-\frac{1}{2}}y)\,dy = 0 \quad \text{for all } n \geqslant 0.$$

We now make the substitution $x = y^4/4$ to obtain

$$\int_0^\infty x^n \exp(-x^{\frac{1}{4}}) \sin(x^{\frac{1}{4}}) dx = 0,$$

which is the result required. ∎

As it stands, the example does not apply to probability theory since h is not positive but it is easy to modify it so that it does.

Example 6.3. *There exist random variables* X_1, X_2 *taking values in* $\mathbb{R}$ *which have the same moments but different probability densities.*
Proof. Set $h_1(x) = 0$ for $x < 0$, $h_1(x) = \max(h(x), 0)$ for $x \geqslant 0$ where h is the function of Example 6.2. We observe that h_1 is a continuous positive function. Further, since

$$0 \leqslant x^n h_1(x) \leqslant x^n \exp(-x^{\frac{1}{4}}) \quad \text{for } x \geqslant 0$$

and

$$\int_0^\infty x^n \exp(-x^{\frac{1}{4}}) dx = 4 \int_0^\infty y^{4n+3} \exp(-y) dy$$

converges, it follows that $\int_0^\infty x^n h_1(x) dx$ exists for all $n \geqslant 0$. Set $A = \int_0^\infty h_1(x) dx$ and set $f_1(x) = A^{-1} h_1(x)$ for all x, $f_2(x) = 0$ for $x < 0$ and $f_2(x) = A^{-1}(h_1(x) - h(x))$ for $x \geqslant 0$.

It is now easy to see that f_1, f_2 are continuous positive functions and (since $f_1(x) - f_2(x) = A^{-1} h(x)$ for $x > 0$) that $\int_{-\infty}^\infty x^n f_1(x) dx$ and $\int_{-\infty}^\infty x^n f_2(x) dx$ exist and are equal for all $n \geqslant 0$ with, in particular, $\int_{-\infty}^\infty f_1(x) dx = \int_{-\infty}^\infty f_2(x) dx = 1$. Thus we can find random variables X_1 and X_2 with distinct probability densities f_1 and f_2 yet with

$$\mathbb{E}X_1^n = \int_0^\infty x^n f_1(x) dx = \int_0^\infty x^n f_2(x) dx = \mathbb{E}X_2^n \quad \text{for all integers } n \geqslant 0. \quad \blacksquare$$

Postscript

I must confess that I have always considered Example 6.2 as a simple mathematical curiosity. However whilst reading the proofs for this book I had the pleasure of listening to a talk by Y. Meyer on ondelettes (i.e. 'wavelets'). These wavelets form the basis for a new type of Fourier analysis which seems extremely promising. The idea originated in signal processing for the oil industry and has been developed using ideas from quantum physics and advanced Fourier analysis. To construct his wavelets Meyer has recourse to functions all of whose moments vanish!

7

THE IMPORTANCE
OF LINEARITY

Much of the progress of nineteenth century science was due to the realisation and systematic exploitation of the fact that many of the laws of nature were linear. That is to say that, given two solutions ϕ_1 and ϕ_2 of a problem, they could be combined linearly to give further solutions $\lambda_1\phi_1 + \lambda_2\phi_2$. Thus, for example, if ϕ_1 is the solution of Laplace's equation $\nabla^2\phi = 0$ in a volume V satisfying the condition $\phi = G_1$ on the boundary ∂V, and if ϕ_2 is the solution with $\phi = G_2$ on ∂V, then $\phi = \lambda_1\phi_1 + \lambda_2\phi_2$ is the solution with $\phi = \lambda_1 G_1 + \lambda_2 G_2$ on ∂V.

The idea can be traced at least as far back as Newton who used it in his explanation as to why certain ports (like Southampton) have four and not two tides per day; but the work of Fourier and his successors saw its full flowering. The reader will readily trace the essential importance of linearity in most of the methods (pure and applied) presented in this book.

The impact of the idea of linearity is memorably expressed in a famous passage of Helmholtz (*On the Sensations of Tone*, Chapter 2, translated by A.J. Ellis).

Indeed it is seldom possible to survey a large surface of water from a high point of sight, without perceiving a great multitude of different systems of waves mutually overtopping and crossing each other. This is best seen on the surface of the sea, viewed from a lofty cliff, when there is a lull after a stiff breeze. We see first the great waves, advancing in far-stretching ranks from the blue distance, here and there more clearly marked out by their white foaming crests, and following one another at regular intervals towards the shore. From the shore they rebound in different directions according to its sinuosities, and cut obliquely across the advancing waves. A passing steamboat forms its own wedge-shaped wake of waves, or a bird, darting on a fish, excites a small circular system. The eye of the spectator is easily able to pursue each one of these different trains of waves, great and small, wide and narrow, straight and curved, and observe how each passes over the surface, as undisturbedly as if the water over which it flits were not agitated at the same time by other forces. I must own that whenever I attentively observe this spectacle it awakens in me a peculiar kind of intellectual

pleasure, because it bares to the bodily eye, what the mind's eye grasps only by the help of a long series of complicated conclusions for the waves of the invisible atmospheric ocean.

These ideas exerted a profound influence even outside the circle of professional mathematicians and physicists. Alexander Bell's initial experiments were intended to find a method for simultaneously conveying several Morse code messages along the same telegraph wire by using different frequencies. In this, as in his final invention of the telephone, he was greatly helped by his reading of Helmholtz.

Let us close this chapter with an example taken from the original work of Fourier. In this he obtains for the first time the equation of heat conduction

$$\frac{\partial \theta}{\partial t}(\mathbf{x}, t) = K \nabla^2 \theta(\mathbf{x}, t),$$

where $\theta(\mathbf{x}, t)$ is the temperature at a point $\mathbf{x}$ at a time t, K depends only on the material involved and

$$\nabla^2 \theta = \sum_{i=1}^{3} \partial^2 \theta / \partial x_i^2 \text{ is the Laplacian.}$$

As an application he considers the temperature θ of the ground at depth y due to the sun's heating. For the purposes of a 'back of an envelope calculation' θ depends only on y and t, and K is fixed. The equation of heat conduction thus becomes

$$\frac{\partial \theta}{\partial t}(y, t) = \frac{K \partial^2 \theta}{\partial y^2}(y, t) \quad [y > 0], \tag{1}$$

where $\theta(y, t)$ is the temperature of the ground at depth y.

The surface temperature clearly depends on the time of day and on the time of year. Since we are only interested in obtaining a qualitative picture of what happens, this suggests setting

$$\theta(0, t) = A \cos 2\pi D^{-1} t + B \cos 2\pi Y^{-1} t + C, \tag{2}$$

where D is the length of a day (in appropriate units), Y the length of a year and A, B and C are appropriate constants. It turns out in the course of calculation that we need some condition on the behaviour of $\theta(y, t)$ for large y. Since we do not believe that the temperature fluctuations at the surface will be magnified at considerable depths, we suppose that

$$|\theta(y, t)| < C_1 \quad \text{for all } y > 0 \text{ and } t, \tag{3}$$

where C_1 is some constant.

We must now find a suitable solution for (1) subject to (2) and (3). Linearity suggests looking at a simpler subproblem.

Lemma 7.1. *A solution of the equation*

$$\frac{\partial \theta}{\partial t}(y, t) = \frac{K \partial^2 \theta}{\partial y^2}(y, t) \quad [y > 0], \tag{1}$$

subject to the boundary conditions

$$\theta(0, t) = \exp i\omega t \tag{2}'$$

and $|\theta(y, t)| < C_2$ *for all* $y > 0$ *and* t *and some* C_2, *(where* ω *is real) is given by*

$$\theta(y, t) = \exp(-(\omega/2K)^{1/2} y) \exp(i(\omega t - (\omega/2K)^{\frac{1}{2}} y)).$$

Proof. The nature of the problem leads us to seek solutions of the form

$$\theta(y, t) = f(y) \exp i\omega t.$$

Substituting in (1) we get

$$i\omega f(y) \exp i\omega t = K f''(y) \exp i\omega t,$$

so that $f''(y) - (i\omega/K)f(y) = 0$ and we obtain

$$f(y) = \alpha \exp((1 + i)(\omega/2K)^{\frac{1}{2}} y) + \beta \exp(-(1 + i)(\omega/2K)^{\frac{1}{2}} y), \quad \text{for some } \alpha, \beta \in \mathbb{C}.$$

Condition (3) tells us that $f(y)$ is bounded for all $y > 0$ and so $\alpha = 0$ whilst condition (2)' tells us that $f(0) = 1$. Thus $\beta = 1$,

$$f(y) = \exp(-(1 + i)(\omega/2K)^{\frac{1}{2}} y)$$

and

$$\theta(y, t) = \exp(-(1 + i)(\omega/2K)^{\frac{1}{2}} y) \exp i\omega t,$$

giving the stated answer. (Alternatively, we could simply have verified the lemma by substitution.) ∎

Now it is easy to check that if $\theta_1, \theta_2, \theta_3, \theta_4$ satisfy

$$\frac{\partial \theta_j}{\partial t}(y, t) = K \frac{\partial^2 \theta_j}{\partial y^2}(y, t) \quad [y > 0], \tag{1_j}$$

$$\theta_j(0, t) = \exp(i\omega_j t), \tag{2'_j}$$

$$|\theta_j(y, t)| < C_j \text{ for all } y > 0 \text{ and } t \text{ and some } C_j, \tag{3_j}$$

then if $\theta = \sum_{j=1}^4 A_j \theta_j$ we have

$$\frac{\partial \theta}{\partial t} = \frac{K \partial^2 \theta}{\partial y^2} \quad [y > 0] \tag{1*}$$

$$\theta(0, t) = \sum_{j=1}^4 A_j \exp i\omega_j t, \tag{2*}$$

$$|\theta(y, t)| < \sum_{j=1}^4 |A_j| C_j. \tag{3*}$$

Thus taking $\omega_1 = 2\pi D^{-1}$, $\omega_2 = -2\pi D^{-1}$, $A_1 = A_2 = A/2$, $\omega_3 = 2\pi Y^{-1}$, $\omega_4 = -2\pi Y^{-1}$ and $A_3 = A_4 = B/2$ we see that our original equations (1), (2) and (3) have a possible solution

$$\theta(y, t) = A \exp\left(-(\pi/KD)^{\frac{1}{2}}y\right) \cos\left(2\pi D^{-1}t - (\pi/KD)^{\frac{1}{2}}y\right)$$
$$+ B \exp\left(-(\pi/KY)^{\frac{1}{2}}y\right) \cos\left(2\pi Y^{-1}t - (\pi/KY)^{\frac{1}{2}}y\right) + C.$$

We assume, on physical grounds, that any reasonable solution is the unique reasonable solution and proceed to examine the consequences.

We notice that $\theta = \theta_D + \theta_Y + C$, where

$$\theta_D(y, t) = A \exp\left(-(\pi/KD)^{\frac{1}{2}}y\right) \cos\left(2\pi D^{-1}t - (\pi/KD)^{\frac{1}{2}}y\right)$$

is the 'effect of daily heating' and $\theta_Y(y, t)$ is the corresponding 'effect of yearly heating'. We notice that both effects die off exponentially with depth but that the high frequency daily effect dies off much more rapidly than the low frequency yearly effect. We also notice that there is a phase lag $(\pi/KD)^{\frac{1}{2}}y$ for the daily (and $(\pi/KY)^{\frac{1}{2}}y$ for the yearly) effect so that at certain depths the temperature will be completely out of step with the surface temperature.

All these predictions are confirmed by measurements which show that annual variations in temperature are imperceptible at quite small depths (this accounts for the permafrost, i.e. permanently frozen subsoil, at high latitudes) and that daily variations are imperceptible at depths measured in tens of centimetres. Choosing a reasonable value of K leads to a prediction that annual temperature changes will lag by six months (i.e. the ground temperature will be hotter in winter and cooler in summer) at about 2–3 metres depth. Again this is confirmed by observation and, as Fourier remarks, gives a good depth for the construction of cellars.

8

COMPASS AND TIDES

Some useful applications of Fourier's ideas are very simple indeed. Consider the problem of correcting a magnetic compass mounted in a ship containing a great deal of iron or steel. Then if the compass indicates an angle of θ to north, the true angle is $f(\theta) = \theta + g(\theta)$ where $g(\theta)$ is the error.

The main contributions to g are the permanent magnetism of the ship (for example, acquired by steel plates subjected to heating and vibration whilst the ship is being built and thus lying in a fixed direction) and induced (or transient) magnetism due to the immediate effect of the earth's magnetic field. It would thus appear extremely hard to compute and tabulate g.

Although g can be quite large for an iron ship (of the order of $20°$) we do not need its value to an accuracy of more than $2°$ or $3°$ (since other errors in setting a course due to unknown currents and so on have a similar order of magnitude). It thus makes sense to try and approximate to g by a trigonometric polynomial g_1 of low degree, say

$$g_1(\theta) = A_0 + A_1 \cos \theta + A_2 \cos 2\theta + B_1 \sin \theta + B_2 \sin 2\theta.$$

Simple theoretical models and practical experiment confirm that this is indeed a reasonable approximation. The values of A_0, A_1, A_2, B_1 and B_2 may be calculated by swinging the ship round in port so that its direction θ is given accurately by reference to known landmarks and observing $f(\theta)$ for various θ.

But rather more can be done than this. In 1871 Kelvin was asked to write a popular article on the compass for a new magazine. 'But', Kelvin wrote, 'when I tried to write on the mariner's compass I found that I did not know nearly enough about it. So I had to learn my subject.' It took three years to write the first part of the article and five more to write the second.

During that time Kelvin completely redesigned the compass, making it small enough to correct the deviation directly by the careful positioning of permanent magnets and spheres of soft iron close to the compass itself. After the customary Victorian battle with the Admiralty, the Kelvin compass was adopted and with

one major improvement (the use of liquid filled rather than dry compasses) remained standard up to the Second World War.

Kelvin worked on several problems associated with navigation – sounding machines, lighthouse lights and so on – but from our point of view his most interesting work was that on the prediction of tides. Tides are primarily due to the gravitational effects of the moon, sun and earth on the oceans but their theoretical investigation, even in the simplest case of a single ocean covering a rigid earth to a uniform depth, is very hard. Even today, the study of only slightly more realistic models is only possible by numerical computer modelling.

How then could Kelvin hope to predict the tides of the real world? We observe first that the height $h(t)$ of the tide at a certain point at time t is influenced by forces which change periodically and whose effect is additive. Thus we might expect

$$h(t) = h_1(t) + h_2(t) + h_3(t) + \cdots,$$

where the period of h_1 is $2\pi\lambda_1^{-1}$ the period of the rotation of the earth with respect to the moon, the period of h_2 is $2\pi\lambda_2^{-1}$ the period of the rotation of the earth with respect to the sun and so on.

But we know that each h_j may be closely approximated by a trigonometric polynomial

$$a_{0j} + \sum_{m=1}^{n} \alpha_{mj} \cos m\lambda_j t + \sum_{m=1}^{n} \beta_{mj} \sin m\lambda_j t.$$

It follows that we may hope to have

$$h(t) \approx A_0 + \sum_{r=1}^{N} A_r \cos \omega_r t + \sum_{r=1}^{N} B_r \sin \omega_r t, \tag{*}$$

where the ω_r are selected from the frequencies of the form $m\lambda_j$.

We can now make use of a simple idea.

Lemma 8.1. *Suppose* $h(t) = A_0 + \sum_{r=1}^{N} A_r \cos \omega_r t + \sum_{r=1}^{N} B_r \sin \omega_r t,$

where $\omega_1, \omega_2, \ldots, \omega_N$ *are distinct strictly positive numbers. Then for any* $S,$

$$\frac{2}{T} \int_S^{S+T} h(t) \cos \omega_r t \, dt \to A_r \quad \text{as } T \to \infty,$$

and similarly

$$\frac{2}{T} \int_S^{S+T} h(t) \sin \omega_r t \, dt \to B_r,$$

$$\frac{1}{T} \int_S^{S+T} h(t) \, dt \to A_0 \quad \text{as } T \to \infty \ [1 \leqslant r \leqslant N].$$

Proof. Observe that if

$$f(t) = \sum_{j=1}^{m} a_j \exp i\lambda_j t,$$

with $\lambda_1, \lambda_2, \ldots, \lambda_m$ distinct real numbers

$$\int_{S}^{S+T} f(t) \exp(-i\lambda_k t) dt = \sum_{j=1}^{m} a_j \int_{S}^{S+T} \exp i(\lambda_j - \lambda_k) t \, dt$$

$$= a_k T + \sum_{j \neq k} \frac{a_j}{i(\lambda_j - \lambda_k)} [\exp i(\lambda_j - \lambda_k) t]_S^{S+T},$$

so that

$$\left| \frac{1}{T} \int_{S}^{S+T} f(t) \exp(-i\lambda_k t) dt - a_k \right| \leqslant \frac{1}{T} \sum_{j \neq k} \frac{2|a_j|}{|\lambda_j - \lambda_k|} \to 0 \quad \text{as } T \to \infty$$

and

$$\frac{1}{T} \int_{S}^{S+T} f(t) \exp(-i\lambda_k t) dt \to a_k \quad \text{for all } 1 \leqslant k \leqslant m.$$

The result now follows by using simple equalities like

$$A_r \cos \omega_r t = (A_r/2) \exp(i\omega_r t) + (A_r/2) \exp(-i\omega_r t). \qquad \blacksquare$$

Thus if we have a record of $h(t)$ over a long period of time $[S, S+T]$ we can compute the A_r and B_r of formula (*) approximately by numerical integrations of the form

$$A_r = \frac{2}{T} \int_{S}^{S+T} h(t) \cos \omega_r t \, dt,$$

and then use (*) to predict the height of the tide at future times.

Unfortunately T has to be taken quite large since there is a substantial term with period a (lunar) fortnight (and indeed a not entirely negligible term with period half a solar year). If the reader contemplates the task of performing such calculations by hand, say by counting squares under a graph, she will see why the computation of a single coefficient '...required not less than twenty hours of calculation by skilled arithmeticians. (Kelvin and Tait, *Treatise On Natural Philosophy*, Appendix B).'

To reduce this labour Kelvin not only designed and built a machine which, given the A_r and B_r, would trace out the predicted height

$$h(t) = A_0 + \sum_{r=1}^{n} A_r \cos \omega_r t + \sum_{r=1}^{n} B_r \sin \omega_r t$$

for a year in a few minutes, but also another machine (the harmonic analyser) to perform the task 'which seemed to the Astronomer Royal so complicated and difficult that no machine could master it' of computing the coefficients A_r and B_r from the record of the past height $h(t)$.

Kelvin's harmonic analyser has a good claim to be the grandfather of today's computers not only because he obtained government money to build it but also because it represents the first major victory in the struggle 'to substitute brass for brain' in calculation. It is pleasant to record that Kelvin's instruments were so well adapted to their purpose that it took electronic computers 20 years to replace them.

9

THE SIMPLEST CONVERGENCE THEOREM

I hope that by now the reader is convinced that many interesting problems in Fourier analysis do not depend on the pointwise convergence of Fourier series. However, it is convenient to have criteria for convergence and in this chapter we give a very simple (and therefore useful) one.

Theorem 9.1. *Let $f:\mathbb{C}\to\mathbb{T}$ be continuous. Then if $\sum_{r=-\infty}^{\infty}|\hat{f}(r)|$ converges it follows that $S_n(f,t)\to f(t)$ uniformly on $\mathbb{T}$ as $n\to\infty$.*

We deduce this from another simple and useful result.

Theorem 9.2. *Suppose $\sum_{r=-n}^{n}|a_r|$ converges as $n\to\infty$. Then $\sum_{n=-n}^{n}a_r\exp irt$ converges uniformly on $\mathbb{T}$ as $n\to\infty$ to some $g(t)$ where $g:\mathbb{T}\to\mathbb{C}$ is continuous and $\hat{g}(r)=a_r$ for all $r\in\mathbb{Z}$.*

Proof. Since $\sum_{r=-n}^{n}|a_r|$ converges, it follows by the general principle of convergence that given any $\varepsilon>0$ we can find an $n_0(\varepsilon)$ such that $\sum_{n\leqslant|r|\leqslant m}|a_r|<\varepsilon$ whenever $m\geqslant n\geqslant n_0(\varepsilon)$. It follows that

$$\left|\sum_{n\leqslant|r|\leqslant m}a_r\exp irt\right|\leqslant\sum_{n\leqslant|r|\leqslant m}|a_r\exp irt|\leqslant\sum_{n\leqslant|r|\leqslant m}|a_r|<\varepsilon$$

for all $t\in\mathbb{T}$ and $m\geqslant n\geqslant n_0(\varepsilon)$, so, by the general principle of uniform convergence, $\sum_{r=-n}^{n}a_r\exp irt$ tends uniformly on $\mathbb{T}$ to some $g(t)$ as $n\to\infty$. (The reader who knows it will recognise the Weierstrass M test.) Since the uniform limit of continuous functions is continuous, it follows that $g:\mathbb{T}\to\mathbb{C}$ is continuous.

Since $\sum_{r=-n}^{n}a_r\exp irt\to g(t)$ uniformly, it follows that $(\sum_{r=-n}^{n}a_r\exp irt)\times\exp(-ikt)\to g(t)\exp(-ikt)$ uniformly and so

$$a_k=\sum_{r=-n}^{n}a_r\frac{1}{2\pi}\int_{\mathbb{T}}\exp(i(r-k)t)\,dt=\frac{1}{2\pi}\int_{\mathbb{T}}\left(\sum_{r=-n}^{n}a_r\exp irt\right)\exp(-ikt)\,dt$$

$$\to\frac{1}{2\pi}\int_{\mathbb{T}}g(t)\exp(-ikt)\,dt=\hat{g}(k)\quad\text{as }n\to\infty.$$

Thus $\hat{g}(k)=a_k$ for all $k\in\mathbb{Z}$. ∎

The proof of Theorem 9.2 did not use any of our previous work on Fourier series, but to complete the proof of Theorem 9.1 we use Theorem 2.4 on the uniqueness of Fourier series.

Proof of Theorem 9.1. Since $\sum_{r=-\infty}^{\infty}|\hat{f}(r)|$ converges, it follows from Theorem 9.2 that

$$S_n(f, t) = \sum_{r=-n}^{n} \hat{f}(r)\exp irt$$

converges uniformly to some $g(t)$ where $g: \mathbb{T} \to \mathbb{C}$ is continuous and $\hat{g}(r) = \hat{f}(r)$ for all $r \in \mathbb{Z}$. But by Theorem 2.4 this last statement implies that $f = g$, so $S_n(f,) \to f$ uniformly and we are done. ∎

Theorem 9.1 is a good criterion not only because it is simple but because the convergence it implies is uniform (indeed uniformly absolute) convergence which is analytically the kind of convergence which gives rise to the fewest problems.

Continuing along this line of thought the reader will recall the following theorem. (If the reader has forgotten the proof she will find it in Chapter 53 (Lemma 53.2).)

Theorem 9.3. *Let $f_1, f_2, \ldots : \mathbb{T} \to \mathbb{C}$ be once continuously differentiable functions with derivatives $f'_1, f'_2, \ldots$. Suppose that $f_n \to f$ and $f'_n \to g$ uniformly on $\mathbb{T}$ as $n \to \infty$. Then f is once continuously differentiable with derivative g.*

As a direct consequence we obtain the following simple theorem on term by term differentiation of Fourier series.

Theorem 9.4. *Let $f: \mathbb{T} \to \mathbb{C}$ be continuous. Then if $\sum_{r=-\infty}^{\infty}|r||\hat{f}(r)|$ converges, it follows that f is once continuously differentiable and that $\sum_{r=-n}^{n} ir\hat{f}(r)\exp irt \to f'(t)$ uniformly as $n \to \infty$.*

Proof. Write $f_n = S_n(f,)$. Since $|\hat{f}(r)| \leqslant |r||\hat{f}(r)| [r \neq 0]$, it follows by the comparison test that $\sum_{-n}^{n}|\hat{f}(r)|$ converges and so $f_n \to f$ uniformly. On the other hand $f'_n(t) = \sum_{r=-n}^{n} ir\hat{f}(r)\exp irt$ so, by Theorem 9.2, f'_n converges uniformly to g say. The result now follows from Theorem 9.3. ∎

Theorem 9.1 and 9.4 are often used in conjunction with the following result which the reader may already know.

Lemma 9.5. *Suppose $f: \mathbb{T} \to \mathbb{C}$ is $n-1$ times continuously differentiable and that $f^{(n-1)}$ is differentiable with continuous derivative except at a finite number of points $x_1, x_2, \ldots, x_n$ say. Then if $|f^{(n)}(t)| \leqslant M$ for all $t \neq x_1, x_2, \ldots, x_n$ it follows that $|\hat{f}(r)| \leqslant Mr^{-n}$ for all $r \neq 0$.*

Proof. Integrating by parts and noting that $\pi = -\pi$ we have

$$\hat{f}(r) = \frac{1}{2\pi} \int_{-\pi}^{\pi} f(t)\exp(-irt)\,dt$$

$$= \frac{1}{2\pi}\left[f(t)\frac{\exp(-irt)}{-ir} \right]_{-\pi}^{\pi} + \frac{1}{2\pi ir}\int_{-\pi}^{\pi} f'(t)\exp(-irt)\,dt$$

$$= \frac{1}{2\pi ir}\int_{-\pi}^{\pi} f'(t)\exp(-irt)\,dt.$$

Repeating this process n times we obtain

$$\hat{f}(r) = \frac{(ir)^{-n}}{2\pi}\int_{-\pi}^{\pi} f^{(n)}(t)\exp(-irt)\,dt,$$

so that $\qquad\qquad |\hat{f}(r)| \leqslant \left|\frac{(ir)^{-n}}{2\pi}\right| \int_{-\pi}^{\pi} |f^{(n)}(t)\exp(-irt)|\,dt.$ ∎

Thus we have the following immediate corollary. (Contrast this result with the situation for Taylor series revealed by Example 4.2.)

Theorem 9.6. *If $f:\mathbb{T}\to\mathbb{C}$ is twice continuously differentiable, then $S_n(f,\ )\to f$ uniformly.*

Proof. By Lemma 9.5, $|\hat{f}(r)| \leqslant Mr^{-2}\ [r\neq 0]$, where $M = \sup_{t\in\mathbb{T}}|f^{(2)}(t)|$. But $\sum_{r\neq 0} r^{-2}$ is convergent so, by the comparison test, $\sum_{r=-n}^{n}|\hat{f}(r)|$ converges as $n\to\infty$. Thus, by Theorem 9.1, $S_n(f,\ )\to f$ uniformly. ∎

The results presented in this chapter are by no means the best possible and we shall prove better ones in the next few chapters. But we shall also see (particularly in Chapter 17) that these more sophisticated results are surrounded by traps for the unwary and have to be used with a caution that the results of this chapter do not demand.

10

THE RATE OF CONVERGENCE

In Chapter 8 we saw how approximation by a trigonometric polynomial of very low degree was sufficiently accurate to help solve a practical problem. The designers of electronic organs use the fact that seven or eight harmonics (corresponding to a trigonometric polynomial of that degree) are sufficient to imitate another musical instrument. (Here sufficient means 'sufficient to produce a saleable product' rather than 'sufficient to fool a musical ear'.) In general it often turns out that $S_n(f)$ is a surprisingly good approximation to some practically given f even for quite low n. Encouraged by this we may then ask how large n must be to get a really good approximation.

Let us take as an example the function $h:\mathbb{T}\to\mathbb{R}$ given by $h(x)=\pi/2-|x|$ $[0\leqslant|x|\leqslant\pi]$. This function described as the 'plucked string' and shown in Figure 10.1 is often given as the first example in books on Fourier series. It is easy to see that

$$\hat{h}(r)=\frac{1}{2\pi}\int_{-\pi}^{\pi}h(t)\exp(-irt)\,dt=\frac{1}{2\pi}\int_{0}^{\pi}h(t)(\exp(irt)+\exp(-irt))\,dt$$

$$=\frac{1}{\pi}\int_{0}^{\pi}h(t)\cos rt\,dt=\frac{1}{\pi}\int_{0}^{\pi}(\pi/2-t)\cos rt\,dt$$

$$=\frac{1}{\pi}\left[(\pi/2-t)\frac{\sin rt}{r}\right]_{0}^{\pi}+\frac{1}{\pi}\int_{0}^{\pi}\frac{\sin rt}{r}\,dt=\frac{1}{\pi r}\int_{0}^{\pi}\sin rt\,dt$$

$$=\frac{1}{\pi r^{2}}[-\cos rt]_{0}^{\pi}=\begin{cases}0 & \text{for } r \text{ even } [r\neq 0]\\ 2/(\pi r^{2}) & \text{for } r \text{ odd}\end{cases}$$

and that $\hat{h}(0)=0$. Thus $\sum_{r=-n}^{n}|\hat{h}(r)|$ converges and, by Theorem 9.1, $S_n(h,x)\to h(x)$ uniformly on $\mathbb{T}$. We turn now to the rate of convergence.

Example 10.1. Let $h(x)=\pi/2-|x|[0\leqslant|x|\leqslant\pi]$. *Then*

(i) $|h(x)-S_n(h,x)|\leqslant 2\pi^{-1}/(n-1)$,

(ii) $h(0)-S_n(h,0)\geqslant 2\pi^{-1}/(n+2)$.

35

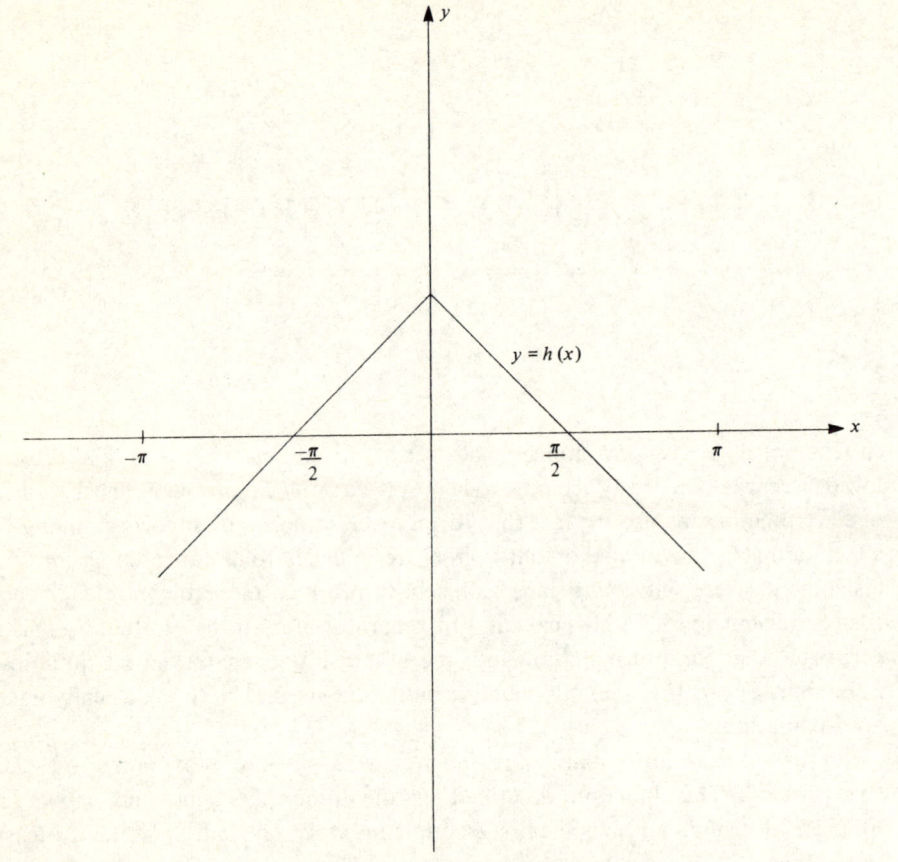

Fig. 10.1. The plucked string function h.

Proof. (i) Observe that, since $S_n(h, x) \rightarrow h(x)$,

$$|h(x) - S_n(h, x)| = \left| \sum_{|r| \geqslant n+1} \hat{h}(r) \exp irx \right| \leqslant \sum_{|r| \geqslant n+1} |\hat{h}(r)| = \sum_{|r| \geqslant n+1, r \, odd} 2/(\pi r^2).$$

Now

$$\sum_{|r| \geqslant n+1, r \, odd} r^{-2} = \sum_{r \geqslant n+1, r \, odd} 2r^{-2} \leqslant \sum_{r \geqslant n} r^{-2} \leqslant \sum_{r \geqslant n} (r(r-1))^{-1}$$

$$= \sum_{r \geqslant n} ((r-1)^{-1} - r^{-1}) = (n-1)^{-1} \text{ so } |h(x) - S_n(h, x)| \leqslant 2\pi^{-1}/(n-1).$$

(ii) Observe that $S_n(h, 0) = \sum_{|r| \leqslant n, r \, odd} 2/(\pi r^2)$, where all the terms are positive, so, since $S_n(h, 0) \rightarrow h(0)$, $h(0) - S_n(h, 0) = \sum_{|r| \geqslant n+1, r \, odd} 2/(\pi r^2)$.
Now

$$\sum_{|r| \geqslant n+1, r \, odd} r^{-2} = \sum_{r \geqslant n+1, r \, odd} 2r^{-2} \geqslant \sum_{r \geqslant n+2} r^{-2} \geqslant \sum_{r \geqslant n+2} (r(r+1))^{-1}$$

$$= \sum_{r \geqslant n+2} (r^{-1} - (r+1)^{-1}) = (n+2)^{-1},$$

so $h(0) - S_n(h, 0) \geqslant 2\pi^{-1}/(n+2)$.   ∎

Thus, to get reasonable accuracy (say to within 10% of the true answer), we need only take $n = 6$ (so that $S_n(h, x) = 4\pi^{-1}(\cos x + 9^{-1}\cos 3x + 25^{-1}\cos 5x)$), but to get six-figure accuracy we need to take hundreds of thousands of terms.

Obviously I have taken an example favourable to my moral, but many common Fourier expansions exhibit (though usually to a lesser degree) the same phenomena of reasonable approximation for small n yet only slow convergence for large n. Moreover, much worse behaviour is possible.

Lemma 10.2. *Given a decreasing sequence* $\delta_0, \delta_1, \ldots$, *with* $\delta_n \to 0$ *as* $n \to \infty$ *we can find a continuous function* $g : \mathbb{T} \to \mathbb{C}$ *with* $S_n(g, t) \to g(t)$ *uniformly, but* $\sup_{t \in \mathbb{T}} |g_n(t) - S_n(g, t)| \geqslant \delta_n$ *for all* $n \geqslant 0$.

Proof. Set $a_r = 0$ for $r \leqslant -1$, $a_0 = \delta_0$ and $a_r = \delta_r - \delta_{r-1}$ for $r \geqslant 1$. Then

$$\sum_{r=-n}^{n} |a_r| = \delta_0 + \sum_{r=1}^{n} (\delta_{r-1} - \delta_r) = 2\delta_0 - \delta_n \leqslant 2\delta_0$$

for all $n \geqslant 1$, so $\sum_{r=-n}^{n} |a_r|$ converges as $n \to \infty$. It follows, by Theorem 9.2, that $\sum_{r=-n}^{n} a_r \exp irt$ converges uniformly to a continuous function $g(t)$ with $\hat{g}(r) = a_r$. Thus $S_n(g, \ ) \to g$ uniformly, but

$$S_n(g, 0) - g(0) = -\sum_{r=n+1}^{\infty} a_r = \sum_{r=n+1}^{\infty} (\delta_{r-1} - \delta_r) = \delta_n.$$ ∎

With the advent of the high speed computer there is now no absolute bar on the use of quite slowly convergent series in attempts to obtain (apparently) very accurate results. The apprentice sorcerer should pause to ask what physical meaning can be attached to, for example, the high frequency terms in a trigonometric polynomial with degree about a thousand.

Finally the reader, whilst admitting that Example 10.1 shows that h is not very well uniformly approximated by $S_n(h)$, may ask if we could not approximate h very well by some other trigonometric polynomial of degree n (for example, $\sigma_n(h)$). If she performs the calculation she will see that $\sigma_n(h)$ does slightly worse and in Chapter 33 (Example 33.4) we shall show that no trigonometric polynomial of degree n, of any form, can be a very good uniform approximation to h.

11

A NOWHERE DIFFERENTIABLE FUNCTION

In Theorem 9.4 we gave a safe but unexciting condition which allowed us to differentiate a Fourier series term by term. What happens if we go ahead and differentiate formally term by term without imposing conditions? Then, *formally*,

$$\text{if } `f(x) \sim \Sigma a_r \exp irx` \quad \text{then} \quad `f'(x) \sim \Sigma ira_r \exp irx`.$$

But it is clear that we can have $\Sigma |a_r|$ convergent so that, by Theorem 9.2, the convergence of $\Sigma a_r \exp irx$ is trouble-free and yet $\lim\sup_{r \to \infty} |ra_r| = \infty$ so that the sum $\Sigma ira_r \exp irx$ cannot possibly converge. (For example, we could take $a_r = r^{-\frac{1}{2}}$ whenever $r = 2^{2n}[n \geqslant 1]$, $a_r = 0$ otherwise.)

And that, apparently, is the end of the story. However, Weierstrass was able to see rather more in the remarks above and use them as a hint for his famous construction of a continuous function which was nowhere differentiable.

To understand the stir that this example caused, the reader must remember that though it was understood that a continuous function could fail to be differentiable at a point (look at $f(x) = |x|$), it was also generally believed that a continuous function must be differentiable at some point. Indeed, many advanced calculus texts carried 'proofs' of this fact (and even Galois seems to have thought himself in possession of such a proof).

(In passing let me add that the proofs were not necessarily worthless. For example some authors, essentially, proved the true, and interesting, theorem that a continuous function with only a finite number of maxima and minima must be differentiable at a dense set of points and then stated mistakenly that a continuous function can only have a finite number of maxima and minima.)

In this chapter we shall construct a version of Weierstrass's continuous nowhere differentiable function.

Example 11.1. $\Sigma_{r=0}^{n}(r!)^{-1} \sin((r!)^2 t)$ *converges uniformly on* $\mathbb{T}$ *as* $n \to \infty$ *to* $h(t)$ *where* $h:\mathbb{T} \to \mathbb{R}$ *is a continuous nowhere differentiable function.*

In order to shorten the main proof we make some preliminary computations.

Lemma 11.2. (i) $\sum_{r=n}^{\infty} (r!)^{-1} \leqslant 2(n!)^{-1}$ *for* $n \geqslant 1$,

(ii) $|\sin x - \sin y| \leqslant |x - y|$ *for all* x, $y \in \mathbb{T}$,

(iii) *If* $K \geqslant 3$ *is an integer and* $x \in \mathbb{T}$ *then we can find* $y \in \mathbb{T}$ *such that* $K^{-1}\pi < |x - y|$ $\leqslant 3K^{-1}\pi$ *and yet* $|\sin Kx - \sin Ky| \geqslant 1$.

Proof. (i) $\sum_{r=n}^{\infty} (r!)^{-1} \leqslant \sum_{r=n}^{\infty} (n!)^{-1}(n+1)^{-(r-n)} \leqslant \sum_{r=n}^{\infty} (n!)^{-1} 2^{-(r-n)} = 2(n!)^{-1}$.

(ii) Use the mean value theorem. (iii) Observe that $\sin Kt$ takes all values between 1 and -1 in the range $(x + K^{-1}\pi, x + 3K^{-1}\pi]$. If we take $y_1, y_2 \in (x + K^{-1}\pi, x + 3K^{-1}\pi]$ with $\sin Ky_1 = 1$ and $\sin Ky_2 = -1$ then at least one of y_1 and y_2 must satisfy the conditions of (iii). ∎

Proof of Example 11.1. The fact that $\sum_{r=0}^{\infty} (r!)^{-1} \sin((r!)^2 t)$ converges uniformly on $\mathbb{T}$ to a continuous function $h(t)$ follows from Theorem 9.2 (or directly from the Weierstrass M test). All that remains therefore is to prove that h is nowhere differentiable and to show this it suffices to show that h is not differentiable at any fixed point x.

Let us write

$$h_n(t) = \sum_{r=0}^{n-1} (r!)^{-1} \sin((r!)^2 t),$$

$$k_n(t) = (n!)^{-1} \sin((n!)^2 t),$$

$$l_n(t) = \sum_{r=n+1}^{\infty} (r!)^{-1} \sin((r!)^2 t),$$

so that $h_n + k_n + l_n = h$. This decomposition is the key to the construction. Now consider any integer $n \geqslant 3$. By Lemma 11.2(iii) we can find an $x_n \in \mathbb{T}$ such that

$$\text{(A)}_n \quad (n!)^{-2}\pi < |x - x_n| \leqslant 3(n!)^{-2}\pi,$$

yet $\quad$ $\text{(B)}_n \quad |k_n(x) - k_n(x_n)| = (n!)^{-1} |\sin((n!)^2 x) - \sin((n!)^2 x_n)| \geqslant (n!)^{-1}$.

Using the inequality $|x - x_n| \leqslant 3(n!)^{-2}\pi$ together with Lemma 11.2 (ii) we have

$\text{(C)}_n \quad |h_n(x) - h_n(x_n)|$

$$\leqslant \sum_{r=0}^{n-1} (r!)^{-1} |\sin((r!)^2 x) - \sin((r!)^2 x_n)| \leqslant \sum_{r=0}^{n-1} (r!)^{-1} |(r!)^2 x - (r!)^2 x_n|$$

$$= \sum_{r=0}^{n-1} r! |x - x_n| = \left((n-1)! + \sum_{r=0}^{n-2} r! \right)|x - x_n| \leqslant \left((n-1)! + \sum_{r=0}^{n-2} (n-2)! \right)|x - x_n|$$

$$= 2(n-1)! |x - x_n| \leqslant 6\pi n^{-1}(n!)^{-1}.$$

On the other hand Lemma 11.2(i) shows that

$$\text{(D)}_n \quad |l_n(t)| \leqslant \sum_{r=n+1}^{\infty} (r!)^{-1} \leqslant 2((n+1)!)^{-1}$$

for all $t \in \mathbb{T}$ and so

$$\text{(E)}_n \quad |l_n(x) - l_n(x_n)| \leqslant |l_n(x)| + |l_n(x_n)| \leqslant 4((n+1)!)^{-1}.$$

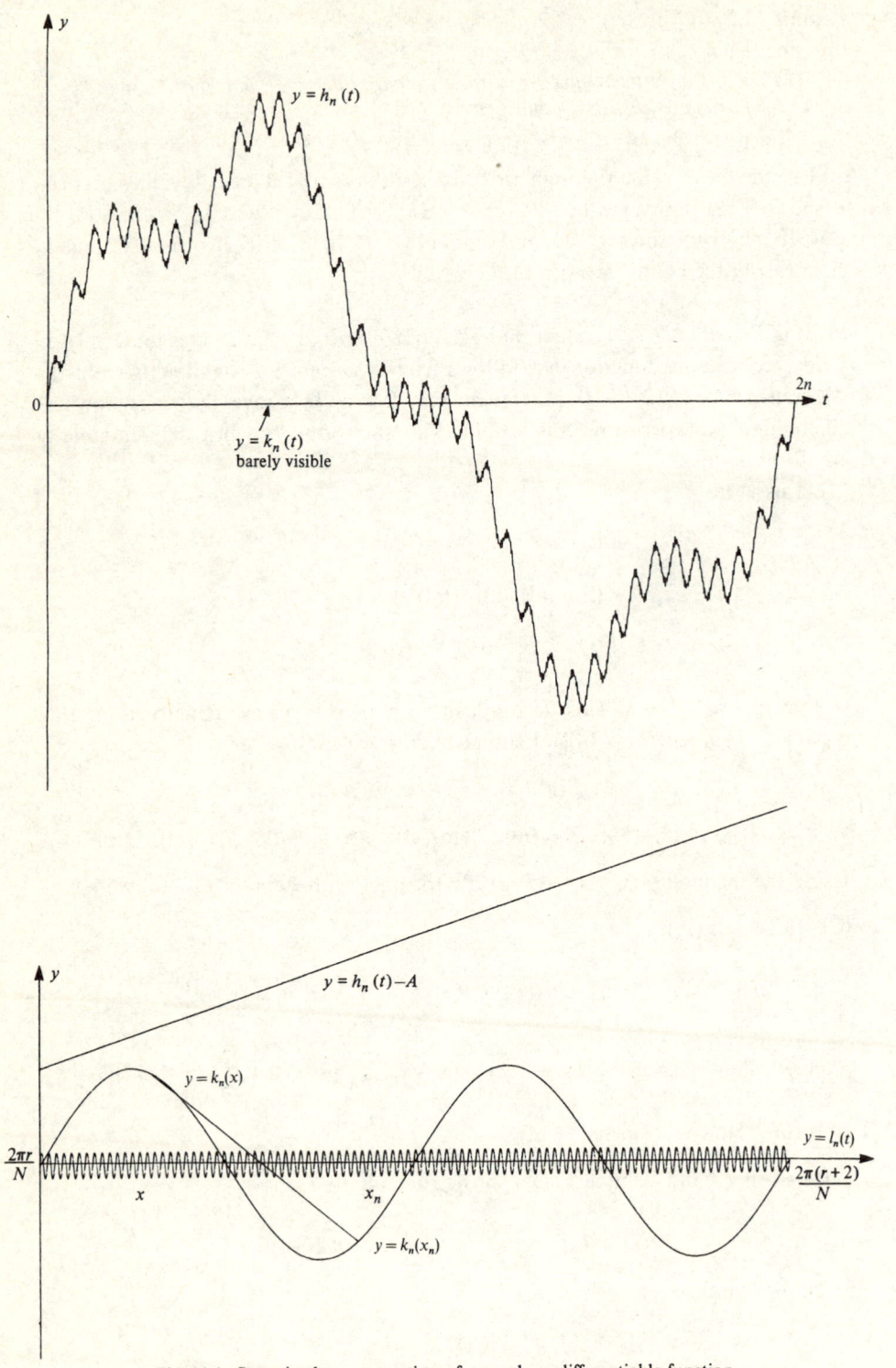

Fig. 11.1. Steps in the construction of a nowhere differentiable function.

Let us sum up what we have done so far. We have chosen an x_n whose distance from x has the same order of magnitude as the period of k_n (this is formula $(A)_n$) but for which $k_n(x) - k(x_n)$ is relatively large and so the chord joining $(x, k(x))$ to $(x_n, k_n(x_n))$ has a steep slope (this is formula $(B)_n$). Since the period of k_n is so small compared with that of h_n we know that, although h_n may be large relative to k_n, the difference $h_n(x) - h_n(x_n)$ is relatively small (this is formula $(C)_n$). Finally, since l_n is small compared to h_n, it follows that $l_n(x) - l_n(x_n)$ is small. (A look at Figure 11.1 may help a bit, reflection and the passage of time will help a great deal.)

Using the inequalities $(B)_n$, $(C)_n$ and $(E)_n$ we obtain

$(F)_n$ $|h(x) - h(x_n)| \geqslant |k_n(x) - k_n(x_n)| - |h_n(x) - h_n(x_n)| - |l_n(x) - l_n(x_n)|$

$$\geqslant (n!)^{-1} - 6n^{-1}\pi(n!)^{-1} - 4((n+1)!)^{-1}$$

$$= (n!)^{-1}(1 - 6\pi n^{-1} - 4(n+1)^{-1}) \geqslant (n!)^{-1}(1 - 30n^{-1}) \geqslant (n!)^{-1}/2$$

whenever $n \geqslant 60$. In other words the value of

$$h(x) - h(x_n) = (k_n(x) - k_n(x_n)) + (h_n(x) - h_n(x_n)) + (l_n(x) - l_n(x_n))$$

is dominated by the $k_n(x) - k_n(x_n)$ term. Thus, since the chord joining $(x, k_n(x))$ to $(x_n, k_n(x_n))$ was steep, the chord joining $(x, h(x))$ to $(x, h(x_n))$ remains so. Using $(A)_n$ and $(F)_n$ we have, in fact,

$(H)_n$ $$\left| \frac{h(x) - h(x_n)}{x - x_n} \right| \geqslant \frac{(n!)^{-1}}{2} \cdot \frac{1}{3\pi(n!)^{-2}} = \frac{n!}{6\pi}$$

for $n \geqslant 60$. Thus

$$\left| \frac{h(x) - h(x_n)}{x - x_n} \right| \to \infty \text{ whilst } x_n \to x,$$

so h cannot be differentiable at x. ∎

Although the construction and proof above may seem hard to grasp at first, the reader who studies them carefully will find that the underlying idea is very simple. She can check her understanding by constructing a proof for Weierstrass's original function $\sum_{n=-\infty}^{\infty} a^{-n} \sin b^n x$ where b is an integer and b/a and a are sufficiently large. (The same function was discovered independently but not published by Cellérier. Many years earlier Bolzano seems to have been close to a continuous nowhere differentiable function constructed along different lines but he too did not publish.)

12
REACTIONS

Weierstrass's nowhere differentiable continuous function was the first in a series of examples of functions exhibiting hitherto unexpected behaviour. (One which shook the mathematical community particularly was Peano's space filling curve, a function $f:[0,1] \to [0,1]^2$ which is both continuous and surjective.)

Many mathematicians held up their hands in (more or less) genuine horror at 'this dreadful plague of continuous nowhere differentiable functions' (Hermite), with which nothing mathematical could be done. They complained that the extra hypotheses required to avoid what they called 'pathological functions' spoilt the elegance of classical analysis and that concentration on such functions would spoil the geometric intuition which is at the heart of analysis.

Poincaré wrote that in the last half century

> we have seen a rabble of functions arise whose only job, it seems, is to look as little as possible like decent and useful functions. No more continuity, or perhaps continuity but no derivatives. Moreover, from the point of view of logic, it is these strange functions which are the most general; whilst those one meets unsearched for and which follow simple laws are seen just as very special cases to be given their own tiny corner.
>
> Yesterday, if a new function was invented it was to serve some practical end; today they are specially invented only to show up the arguments of our fathers, and they will never have any other use.
>
> (*Collected Works*, Vol. 11, p. 130)

Too much can be read into remarks in an article intended to dissuade over-enthusiastic teachers from starting a first course in analysis with a full discussion of nowhere differentiable functions. However, Lebesgue experienced some difficulty in publishing an article which contained a nowhere differentiable function. More-over, he writes, for some time afterwards

> Whenever I tried to take part in a mathematical conversation there would always be an analyst to say 'This could not possibly interest you, it is a question about

42

differentiable functions' and a geometer to repeat in his own language 'We are dealing with surfaces which have a tangent plane.'

(The category- and topos-theorist of the present day will sympathise.)

Others were more enthusiastic. Paul Lévy, at the age of 16, on hearing from his father that Weierstrass had constructed such a thing, produced his own example of a continuous curve (function from $\mathbb{R}$ to $\mathbb{R}^2$) without a tangent. His pride in this discovery still shines through in autobiographical notes written 60 years later.

Another person impressed was the physicist Jean Perrin who investigated Brownian motion in order to confirm the molecular hypothesis. As the reader probably knows, the name of Brownian motion is given to the ceaseless irregular motion of tiny particles suspended in a fluid such as water, first investigated by the Scottish botanist Brown. (Brown was not only a great botanist but also, according to Darwin, the kindest man he had ever met.) The proponents of the view that liquids and gases are composed of molecules in constant motion explained this by the buffeting the small particles are exposed to as they are hit first from one direction and then another by the still smaller molecules. Both the hypothesis and the explanation are generally accepted.

Perrin begins by noting that early workers had attempted to measure the velocities of particles in Brownian motion. 'But', says Perrin, 'such evaluations are *absolutely wrong*'. If we take the position of the particle as $\mathbf{x}(t)$ and try to estimate the velocity at t_0 in the usual way as $(\mathbf{x}(t_0 + \delta t) - \mathbf{x}(t_0))/\delta t$ then, as we make δt smaller,

> the apparent mean velocity of a grain... varies in the *wildest way* in magnitude and direction, and does not tend to a limit as the time taken for an observation decreases, as may easily be seen... by noting the positions occupied by a grain from minute to minute and then every five seconds or, better still, ...every twentieth of a second.
>
> Obviously, it is impossible to fix a tangent, even approximately, at any point on the trajectory. It is one of those cases when we are forced to think of those continuous nowhere differentiable functions which have been wrongly regarded as just mathematical curiosities – wrongly since nature contains suggestions of 'non differentiable' as well as 'differentiable' processes.

Later he gives diagrams (Figure 12.1) in which the position of a particle was marked every 30 seconds but notes that these diagrams

> give only a meagre idea of the extraordinary irregularity of the actual trajectory. If, for example, we marked the position each second, each straight line segment would be replaced by a polygonal path composed of thirty lines which would be relatively just as complicated as the initial diagram and so on. We see by such examples how close the mathematicians stayed to reality when they refused by logical instinct to accept so called geometrical proofs, where the existence of a tangent at each point of a curve is taken as intuitively obvious.

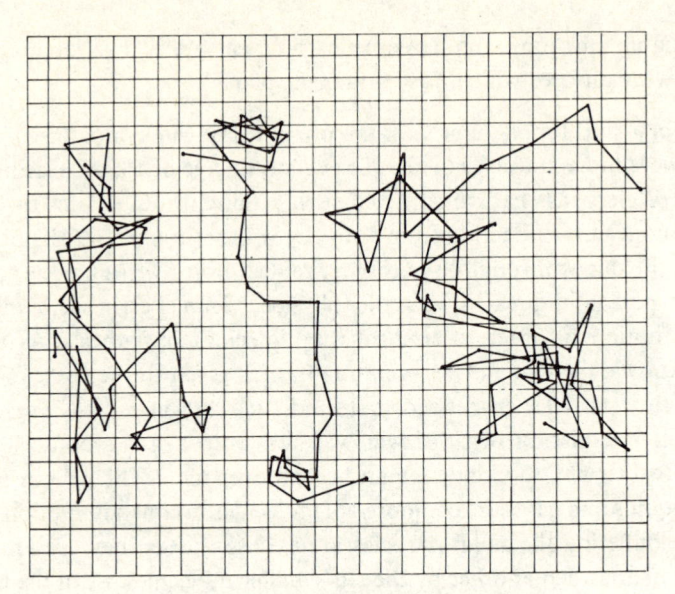

Fig. 12.1. Examples of Brownian motion.

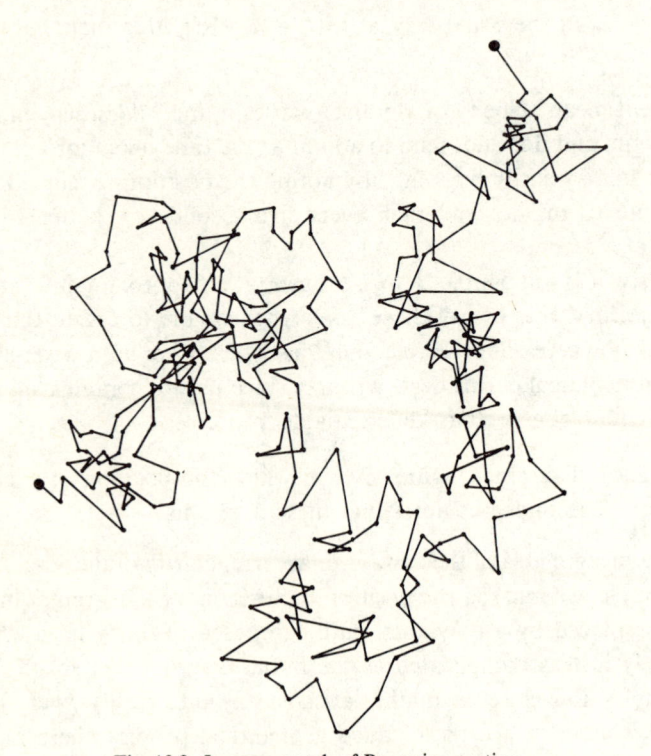

Fig. 12.2. Longer stretch of Brownian motion.

(The two quotations are taken from a paper in *Annales de Chimie et de Physique*, 1909. Later Perrin expanded the paper into a book whose English translation called *Atoms* appeared in 1922. Figure 12.2 is taken from the book.)

The study of Brownian motion had been put on a reasonable (if not a rigorous) basis by Einstein in one of his great 1905 papers. However, his work did not require a study of the random trajectories themselves and a successful attack on the problem of random curves and functions began with the work of Wiener and Lévy in the 1920s. Wiener (who often cited the inspiration of Perrin) was able to give a rigorous construction for a family of random functions which formed a good model for the curves of Brownian motion. In accordance with Perrin's prophetic remarks they turned out to be continuous and nowhere differentiable (with probability one).

These ideas turned out to have practical applications. For example, in any electrical communications circuit, however well constructed, there is always a residual hiss. Thus the actual signal is $g(t) = f(t) + \varepsilon(t)$ where f is the desired signal and ε is the error term. Wiener used his random functions as a model for the 'white noise' $\varepsilon(t)$ and showed how to mitigate the effects of this term.

Finally, the study of random functions turned out to be mathematically extremely rich; so rich indeed that results on these nondifferentiable random functions are now used to prove results about infinitely differentiable and analytic functions.

With hindsight we can see Weierstrass's result not as a counter example marking the boundary of a desert of useless generality but, as an example pointing the way to fruitful areas of study.

13

MONTE CARLO METHODS

Suppose we wish to evaluate numerically the integral

$$I = \int_0^1 \int_0^1 \cdots \int_0^1 f(x_1, x_2, \ldots, x_m)\, dx_1\, dx_2 \ldots dx_m,$$

or, more compactly,

$$I = \int_K f(\mathbf{x})\, d\mathbf{x} \quad (\text{with } K = [0, 1]^m),$$

where, to fix ideas, we shall suppose $f:[0, 1]^m \to \mathbb{R}$ to be continuous with $|f(\mathbf{x})| \leqslant 1$ for all $\mathbf{x} \in [0, 1]^m$. If m is small (for example if $m = 2$) then we could try repeated numerical integration using, for example, Simpson's rule. But such a procedure will demand at least 3^m evaluations of $f(\mathbf{x})$ and if $m = 20$, say, this may be too expensive in time and money.

Such integrals occur in atomic engineering and, just after the Second World War, Fermi, von Neumann and Ulam developed a different method called (for reasons which will become obvious) the Monte Carlo method.

Suppose we have a computer equipped with a source of random numbers Y_1, $Y_2, \ldots$ all independent and uniformly distributed over $[0, 1]$. The associated vectors $\mathbf{X}_1 = (Y_1, Y_2, \ldots, Y_m)$, $\mathbf{X}_2 = (Y_{m+1}, Y_{m+2}, \ldots, Y_{2m}), \ldots$ will be independent and uniformly distributed over the m dimensional cube $[0, 1]^m$. We get the machine to form

$$S_1 = f(\mathbf{X}_1),$$
$$S_2 = f(\mathbf{X}_1) + f(\mathbf{X}_2),$$
$$S_n = S_{n-1} + f(\mathbf{X}_n) = \sum_{r=1}^{n} f(\mathbf{X}_r).$$

Lemma 13.1. *With the notation above*

$$Pr(|n^{-1}S_n - I| \geqslant 2\varepsilon^{-\frac{1}{2}}n^{-\frac{1}{2}}) \leqslant \varepsilon.$$

Thus, after n goes, the value of $n^{-1}S_n$ will differ from the true value of I by less than $(2\varepsilon^{-\frac{1}{2}})n^{-\frac{1}{2}}$ with probability at least $1 - \varepsilon$.

Proof. Observe that with the usual notation of probability theory

$$\mathbb{E}f(\mathbf{X}_r) = \int_K f(\mathbf{x})\,d\mathbf{x} = I,$$

$$\operatorname{var} f(\mathbf{X}_r) = \int_K |f(\mathbf{x}) - I|^2\,d\mathbf{x} \leqslant \int_K 2^2\,d\mathbf{x} = 4.$$

Thus $\mathbb{E}(n^{-1}S_n) = n^{-1}\sum_{r=1}^n \mathbb{E}f(\mathbf{X}_r) = I$ and since $f(\mathbf{X}_1), f(\mathbf{X}_2),\ldots$ are independent,

$$\operatorname{var}(n^{-1}S_n) = n^{-2}\sum_{r=1}^n \operatorname{var} f(\mathbf{X}_r) \leqslant 4n^{-1}.$$

It follows by the Tchebychev inequality (Theorem 50.4(i)) that

$$Pr(|n^{-1}S_n - I| \geqslant 2\varepsilon^{-\frac{1}{2}}n^{-\frac{1}{2}}) = Pr(|n^{-1}S_n - \mathbb{E}(n^{-1}S_n)| \geqslant 2\varepsilon^{-\frac{1}{2}}n^{-\frac{1}{2}}) \leqslant \frac{\operatorname{var} n^{-1}S_n}{(2\varepsilon^{-\frac{1}{2}}n^{-\frac{1}{2}})^2} = \varepsilon,$$

as stated. ∎

This method has various advantages.

(1) It produces reasonable accuracy in a reasonable time (for example, we can get 2% accuracy with probability 99/100 with 10^6 calculations) independent of m. For many engineering applications this is sufficient. (And in any case a reasonable answer in 10^6 steps is better than no answer until 3^{20} steps have been performed.)

(2) The calculation can be halted, S_n and n recorded, and restarted after some more urgent calculation has been performed.

It has several disadvantages. For example,

(1) The accuracy only increases as $n^{-\frac{1}{2}}$. To increase the order of accuracy by 10, we must perform 100 times as many calculations.

(2) The production of random numbers presents problems both philosophical and practical. (An early machine in Manchester had a cosmic ray counter attached to provide random numbers. It is said that there was an annoying tendency for random numbers to 'leak' from the counter into the main machine.) The alert reader may notice a vague connection with Chapter 3 on equidistribution.

However it is not our purpose here to discuss Monte Carlo methods in practice but to use them to introduce the mathematical concept corresponding to Brownian motion. (In particular the reader is warned that the discussion that follows is entirely theoretical.)

Consider the problem of solving Laplace's equation $\nabla^2\phi = 0$ in the bounded region $\Omega \subseteq \mathbb{R}^2$ subject to the boundary condition $\phi(\mathbf{x}) = G(\mathbf{x})$ for all $\mathbf{x}$ on the boundary $\partial\Omega$. We assume that G is continuous and that Ω is not too oddly shaped.

Initially we follow the standard approach to computing such a ϕ. We cover Ω

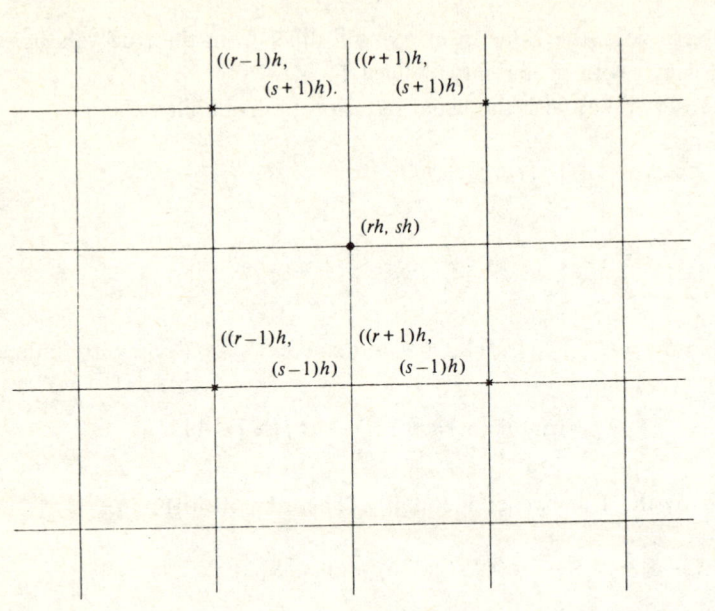

Fig. 13.1. Grid for the Monte Carlo method.

with a mesh of squares having vertices (rh, sh), as in Figure 13.1. The equation

$$\nabla^2 \phi = 0 \left(\text{i.e.} \ \frac{\partial^2 \phi}{\partial x^2} + \frac{\partial^2 \phi}{\partial y^2} = 0 \right)$$

may be replaced, provided h is small enough, by

$$\phi((r+1)h, (s+1)h) + \phi((r-1)h, (s+1)h) + \phi((r+1)h, (s-1)h)$$
$$+ \phi((r-1)h, (s-1)h) - 4\phi(rh, sh) \approx 0,$$

or abbreviating

$$\Sigma a_{uv} \phi((r+u)h, (s+v)h) \approx 0,$$

where $a_{11} = a_{-11} = a_{1-1} = a_{-1-1} = \frac{1}{4}$, $a_{00} = -1$ and $a_{uv} = 0$ otherwise. (The slightly non-standard approximation will make some of the calculations of Chapter 14 easier.)

We therefore try to solve the equations $\Sigma a_{uv} e_{r+u, s+v} = 0$, subject to

$$e_{r,s} = G(rh, sh) \text{ where } (rh, sh) \in \partial\Omega. \qquad (*)$$

We hope that provided h is small enough $e_{r,s}$ will be close to $\phi(rh, sh)$.

(There is an obvious problem here if the boundary does not contain the appropriate grid points but it is easily solved. If the reader is worried she may take Ω to be the square with vertices $(1, 1)$, $(1, -1)$, $(-1, -1)$, $(-1, 1)$ and $h = 1/N$ with N an integer.)

It is only in attacking the equations $(*)$ that our method begins to diverge from

the usual ones. What we do is to let the machine trace the path of a 'particle' jumping from vertex (rh, sh) to vertex $(r'h, s'h)$ according to the following rules.

(1) Start at the point (Rh, Sh).

(2) If at the end of the nth step the particle is at $(rh, sh) \notin \partial\Omega$ then choose one of the four adjoining vertices $((r + 1)h, (s + 1)h)$, $((r - 1)h, (s + 1)h)$, $((r + 1)h, (s - 1)h)$, $((r - 1)h, (s - 1)h)$ at random (each vertex being equally likely) and jump to that one.

The 'particle' will now jump around from vertex to vertex at random until it hits a boundary point $(rh, sh) \in \partial\Omega$. The machine now obeys:

(3) If at the end of the nth stage the particle is at $(rh, sh) \in \partial\Omega$ record the value of $X_{R,S} = G(rh, sh)$ and stop.

Lemma 13.2. (i) *For the process described above* $\mathbb{E}X_{R,S} = e_{R,S}$.

(ii) *If* $|G(rh, sh)| \leqslant 1$ *for all* $(rh, sh) \in \partial\Omega$ *then* var $X_{R,S} \leqslant 4$.

Proof. (i) Observe that $\mathbb{E}(X_{R,S}) = \Sigma Pr(\text{jump from } (Rh, Sh) \text{ to } (R'h, S'h))\mathbb{E}X_{R',S'}$, and so $\Sigma a_{uv}\mathbb{E}(X_{R+u,S+v}) = 0$. On the other hand $\mathbb{E}X_{R,S} = G(Rh, Sh)$ whenever $(Rh, Sh) \in \partial\Omega$. Thus $\mathbb{E}X_{r,s}$ and $e_{r,s}$ satisfy the same equations (*) and so are equal as stated.

(ii) Since $X_{R,S} = G(rh, sh)$ for some $(rh, sh) \in \partial\Omega$ it follows that $|X_{R,S}| \leqslant 1$ so $|\mathbb{E}X_{R,S}| \leqslant 1$ and var $X_{R,S} = \mathbb{E}(X_{R,S} - \mathbb{E}X_{R,S})^2 \leqslant 4$.

Now let the machine run the process over and over again (with different random choices but the same (R, S)) recording the results as $X_{R,S}^{(1)}, X_{R,S}^{(2)}, \ldots$ and forming the sums

$$S_n = \sum_{k=1}^{n} X_{R,S}^{(k)}.$$

Lemma 13.3. *With the notation above,* $Pr(|n^{-1}S_n - G(Rh, Sh)| \geqslant 2\varepsilon^{-\frac{1}{2}}n^{-\frac{1}{2}}) \leqslant \varepsilon$.

Proof. As for Lemma 13.1 ∎

Thus we have indeed got a method for solving the set of equations (*).

14

MATHEMATICAL BROWNIAN MOTION

The process described in the second half of Chapter 13 is a good candidate for a model of Brownian motion. Let us ignore the boundary $\partial\Omega$ for the moment and just consider a particle on a grid $\{(rh, sh) : r, s \in \mathbb{Z}\}$ moving according to the rule:

(2)′ If the position $\mathbf{Y}_n$ of the particle at the end of the nth step is given by $\mathbf{Y}_n = (rh, sh)$, then $\mathbf{Y}_{n+1} = ((r+u)h, (s+v)h)$ with probability $\frac{1}{4}$ if $(u, v) \in \{(1, 1), (-1, 1), (1, -1), (-1, -1)\}$ and probability 0 otherwise.

We have the following simple two dimensional extension of de Moivre's theorem. (We shall discuss the generalisation of de Moivre's theorem called the central limit theorem in Chapters 69–71. A proof of de Moivre's theorem itself is given as Theorem 71.1.)

Theorem 14.1. *For the process above*

$$Pr\{(\mathbf{Y}_{m+n} - \mathbf{Y}_n)/(m^{\frac{1}{2}}h) \in [a, b] \times [c, d]\} \to \frac{1}{2\pi} \int_a^b \int_c^d \exp(-(x^2 + y^2)/2) \, dx \, dy$$

$$as\ m \to \infty.$$

Proof. Write $\mathbf{Y}_{n+r} - \mathbf{Y}_{n+r-1} = (U_r, V_r)$. Then $U_1, U_2, \ldots, U_m$ are independent identically distributed random variables with $Pr(U_r = h) = Pr(U_r = -h) = \frac{1}{2}$. Thus by de Moivre's theorem

$$Pr\left(\sum_{r=1}^m U_r/(m^{\frac{1}{2}}h) \in [a, b]\right) \to \frac{1}{\sqrt{(2\pi)}} \int_a^b \exp(-x^2/2) \, dx.$$

Similarly, $\qquad Pr\left(\sum_{r=1}^m V_r/(m^{\frac{1}{2}}h) \in [c, d]\right) \to \frac{1}{\sqrt{(2\pi)}} \int_c^d \exp(-y^2/2) \, dy.$

But U_r and V_r are independent (since $Pr(U_r = u, V_r = v) = Pr(U_r = u)Pr(V_r = v)$ for all $u, v \in \{-1, 1\}$) so all the U_r and V_s are independent. Thus $\sum_{r=1}^m U_r/(m^{\frac{1}{2}}h)$ and $\sum_{r=1}^m V_r/(m^{\frac{1}{2}}h)$ are independent and

$$Pr\{(\mathbf{Y}_{m+n} - \mathbf{Y}_n)/(m^{\frac{1}{2}}h) \in [a, b] \times [c, d]\}$$

50

$$= Pr\left(\sum_{r=1}^{m} U_r/(m^{\frac{1}{2}}h) \in [a,b], \sum_{r=1}^{m} V_r/(m^{\frac{1}{2}}h) \in [c,d] \right)$$

$$= Pr\left(\sum_{r=1}^{m} U_r/(m^{\frac{1}{2}}h) \in [a,b] \right) Pr\left(\sum_{r=1}^{m} V_r/(m^{\frac{1}{2}}h) \in [c,d] \right)$$

$$\to \frac{1}{\sqrt{(2\pi)}} \int_a^b \exp(-x^2/2)dx \, \frac{1}{\sqrt{(2\pi)}} \int_c^d \exp(-y^2/2)dy$$

$$= \frac{1}{2\pi} \int_a^b \int_c^d \exp(-(x^2+y^2)/2)dx\,dy.$$ ∎

Thus if A is any reasonably shaped area

$$Pr\{(\mathbf{Y}_{m+n} - \mathbf{Y}_n)/(m^{\frac{1}{2}}h) \in A\} \to \frac{1}{2\pi} \int\!\!\int_A \exp(-(x^2+y^2)/2)dx\,dy$$

$$= \frac{1}{2\pi} \int\!\!\int_A \exp(-r^2/2)r\,dr\,d\theta$$

(rewriting the integral in polars).

Suppose now we run the process so that each step takes time τ (the grid length being h). Let us write $\mathbf{Y}_{(\tau,h)}(t) = \mathbf{Y}_{t\tau^{-1}}$ (assuming $t\tau^{-1}$ an integer). What can we say about the quantities

$$R_{(\tau,h)}(\delta t) = |\mathbf{Y}_{(\tau,h)}(t_0 + \delta t) - \mathbf{Y}_{(\tau,h)}(t_0)|$$

and $$V_{\tau,h}(\delta t) = (\mathbf{Y}_{(\tau,h)}(t_0 + \delta t) - \mathbf{Y}_{(\tau,h)}(t_0))/\delta t?$$

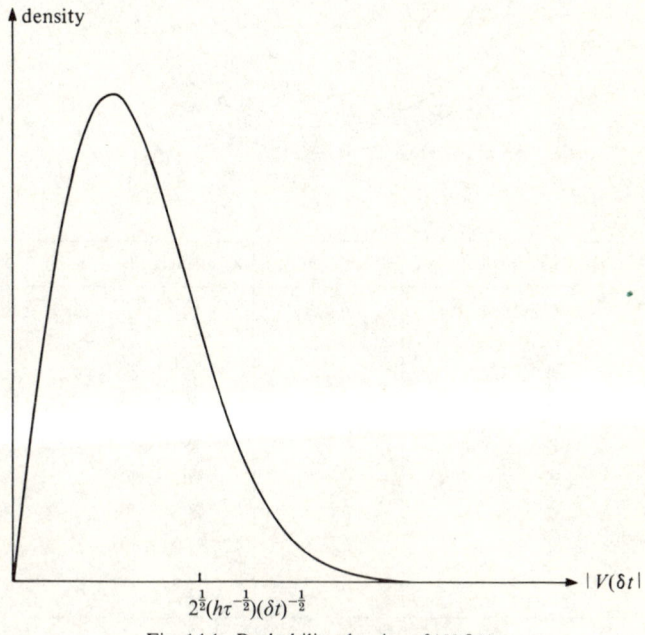

Fig. 14.1. Probability density of $|V(\delta t)|$.

From the remarks above we see that if δt is large compared with τ and so the number $m = \tau^{-1}\delta t$ of steps taken is large, it will be approximately two dimensionally normal with mean 0 and variance $(2h^2m)/(\delta t^2) = 2\tau^{-1}h^2(\delta t)^{-1}$. Thus $\mathbf{V}(\delta t)$ is approximately uniformly distributed in direction and its norm is distributed roughly as in Figure 14.1. (Compare also Figure 14.2.) Thus a 'typical' value of $|\mathbf{V}(\delta t)|$ is about $(h\tau^{-\frac{1}{2}})(\delta t)^{-\frac{1}{2}}$. Naturally, $R(\delta t) = |\mathbf{V}(\delta t)|\delta t$ is similarly distributed with typical value about $(h\tau^{-\frac{1}{2}})(\delta t)^{\frac{1}{2}}$.

Now suppose not only that δt is large compared with τ but that $10^{-3}\delta t$ is large compared with τ. The same arguments as above apply but this time a typical value of $|\mathbf{V}(10^{-3}\delta t)|$ is $(10^{\frac{3}{2}}(h\tau)^{-\frac{1}{2}})(\delta t)^{\frac{1}{2}}$ which is at least 30 times as large as before (though, of course $R(10^{-3}\delta t)$ has a typical value at least 30 times as small as $R(\delta t)$). In Figure 14.2 Perrin plots the horizontal displacement during a 30 second period due to

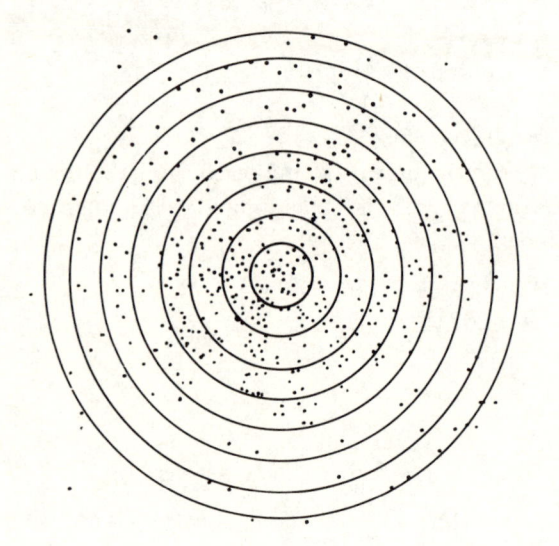

Displacement between	P for each ring	n calculated	n found
0 and $\frac{e}{4}$	0.063	32	34
$\frac{e}{4}$ $2\frac{e}{4}$	0.167	83	78
$2\frac{e}{4}$ $3\frac{e}{4}$	0.214	107	106
$3\frac{e}{4}$ $4\frac{e}{4}$	0.210	107	103
$4\frac{e}{4}$ $5\frac{e}{4}$	0.150	75	75
$5\frac{e}{4}$ $6\frac{e}{4}$	0.100	50	49
$6\frac{e}{4}$ $7\frac{e}{4}$	0.054	27	30
$7\frac{e}{4}$ $8\frac{e}{4}$	0.028	14	17
$8\frac{e}{4}$ $9\frac{e}{4}$	0.014	7	9

Fig. 14.2. Experimental measurements of $|V(\delta t)|$.

Brownian motion of 500 particles. The table shows a comparison between the results expected for a two dimensional normal distribution and those observed. (Source, *Atoms*, English Edition, 1922.) The graphs in Figure 14.3 show what is happening.

Now suppose we follow the path of the same particle from $\mathbf{Y}(t_0)$ to $\mathbf{Y}(t_0 + 10^{-3}\delta t)$

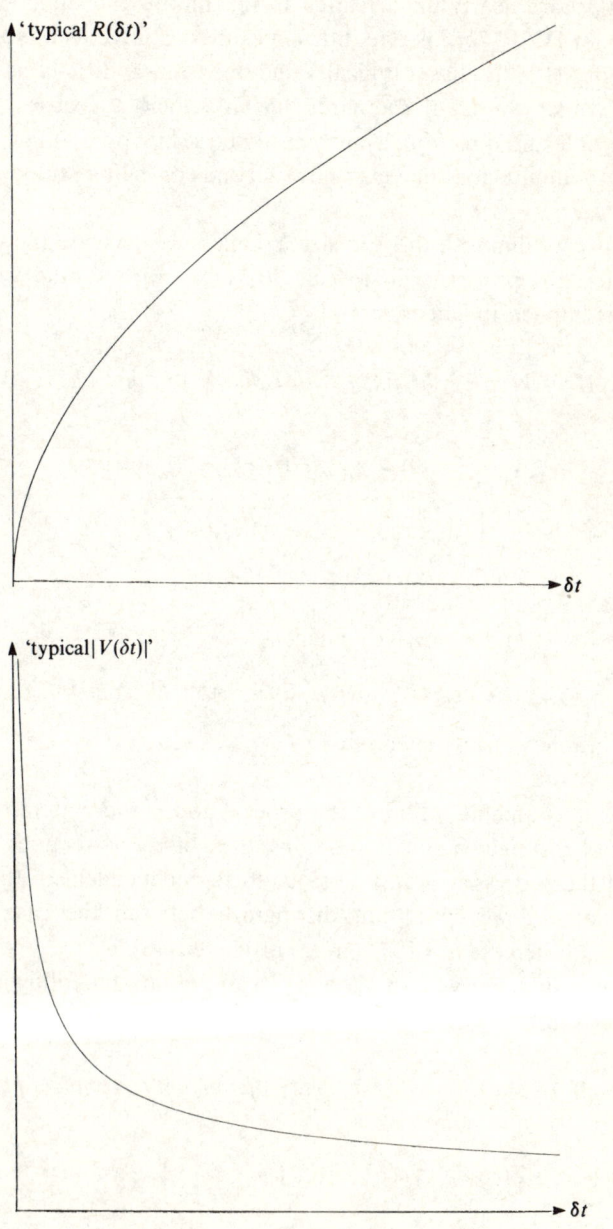

Fig. 14.3. Typical values of $R(\delta t)$ and $|V(\delta t)|$.

to $Y(t_0 + \delta t)$. It is clear that the effect of the first $10^{-3}\delta t \tau^{-1}$ steps from $Y(t_0)$ to $Y(t_0 + 10^{-3}\delta t)$ will be swamped by the remaining $999 \times 10^{-3}\delta t \tau^{-1}$ steps from $Y(t_0 + 10^{-3}\delta t)$ to $Y(t_0 + \delta t)$. Thus $Y(t_0 + 10^{-3}\delta t) - Y(t_0)$ and $Y(t_0 + \delta t) - Y(t_0)$ and so $V(10^{-3}\delta t)$ and $V(\delta t)$ are essentially uncorrelated both in magnitude and direction.

In the same way if $10^{-6}\delta t$ is large compared with τ we see that $V(10^{-6}\delta t)$, $V(10^{-3}\delta t)$, $V(\delta t)$ are nearly uncorrelated in magnitude and direction but now a typical value for $|V(10^{-6}\delta t)|$ is one thousand times as large as a typical value for $|V(\delta t)|$ (though $R(10^{-6}\delta t)$ has a typical value one thousand times as small as that for $R(\delta t)$). In other words, 'the apparent mean velocity.... varies in the *wildest way* in magnitude and direction, as may easily be seen... by noting the positions... from minute to minute and then every five seconds or, better still, every twentieth of a second'.

Now suppose we diminish the grid size h. Unless we decrease the length τ takes to make a step, the process will slow up. By how much should we diminish τ? The answer is implicit in Theorem 14.1.

Lemma 14.2. *If we keep $\tau^{-1}h^2$ fixed with value A say and let $h \to 0$ then, for any fixed δt,*

$$Pr\{(Y_{(\tau,h)}(t + \delta t) - Y_{(\tau,h)}(t))/(A\delta t)^{\frac{1}{2}} \in [a, b] \times [c, d]\}$$

$$\to \frac{1}{2\pi} \int_a^b \int_c^d \exp(-(x^2 + y^2)/2) \, dx \, dy$$

uniformly in t.

Proof. Set $n = t\tau^{-1}$ and $m = \delta t \tau^{-1}$. Then

$$(Y_{(\tau,h)}(t + \delta t) - Y_{(\tau,h)}(t))/(A\delta t)^{\frac{1}{2}} = (Y_{m+n} - Y_n)/(m^{\frac{1}{2}}h)$$

and the result follows from Theorem 14.1. ∎

Thus if we are watching a film of the process and h and τ are already so small that we cannot distinguish successive steps, then (if $\tau^{-1}h^2$ remains constant) the behaviour of the process will appear to us to be independent of the choice of h. (Thus, for example, we could not distinguish between the case $h = 10^{-3}$ cm, $\tau = 10^{-3}$ sec and the case $h = 10^{-4}$ cm, $\tau = 10^{-5}$ sec.)

It is thus plausible that we can allow $h \to 0$ and obtain the following result. (We take $A = 1$ for simplicity.)

Plausible Result 14.3. (i) *There is a family of random continuous paths $Z : \mathbb{R} \to \mathbb{R}^2$ such that*

$$Pr\{(Z(t + \delta t) - Z(t))/\delta t^{\frac{1}{2}} \in [a, b] \times [c, d]\} = \frac{1}{2\pi} \int_a^b \int_c^d \exp(-(x^2 + y^2)/2) \, dx \, dy,$$

independent of previous history.

Moreover, remembering the preceding discussion of $\mathbf{V}(t) = (\mathbf{Y}_{(\tau,h)}(t + \delta t) - \mathbf{Y}(t))/\delta t$, it seems not unreasonable that we have the following result.

Plausible Result 14.3. (ii) *For the family just described*

$$Pr\{\mathbf{Z}: \mathbb{R} \to \mathbb{R}^2 \text{ is nowhere differentiable}\} = 1.$$

Finally, remembering the ideas of Lemma 13.2 and Lemma 13.3 we would expect the following to be true.

Plausible Result 14.3. (iii) *Let* $\phi(\mathbf{x})$ *[$\mathbf{x} \in \Omega$] be the expectation of $G(\mathbf{Z}(t))$ given that* $\mathbf{Z}(0) = \mathbf{x}$ *and t is the first $s > 0$ such that $\mathbf{Z}(s) \in \partial\Omega$. Then $\nabla^2\phi = 0$ and $\phi(\mathbf{y}) = G(\mathbf{y})$* *for all* $\mathbf{y} \in \partial\Omega$.

Obviously a great deal of work must be done to make these plausible results into respectable theorems. However, I hope I have convinced the reader that a mathematical treatment of Brownian motion is possible, that the subjects of such a treatment must include nowhere differentiable curves, and that the end product of such investigations may involve very smooth functions such as ϕ in 14.3(iii).

Remark. Here and later we are mainly interested in two dimensional Brownian motion. However the one dimensional case is also instructive so I have included a discussion in Appendix G.

15

POINTWISE CONVERGENCE

In this chapter we shall obtain criteria for the convergence of Fourier series which apply to a much larger class of functions than those considered in Chapter 9.

We start by introducing a new sum intermediate between the Fourier sum $S_n(f)$ and the Féjer sum $\sigma_n(f)$. This sum introduced by de la Vallée Poussin is defined by

$$\sigma_{n,m}(f) = \frac{1}{m-n}((m+1)\sigma_m(f) - (n+1)\sigma_n(f))$$

(where $m > n \geqslant 0$ and f is Riemann integrable). Thus

$$\sigma_{n,m}(f) = \sum_{r=-m}^{-n-1} \frac{m+1-|r|}{m-n} \hat{f}(r)\exp irt + S_n(f,t) + \sum_{r=n+1}^{m} \frac{m+1-|r|}{m-n} \hat{f}(r)\exp irt.$$

(Figure 15.1 may help the reader to see what is going on. For the purpose of explaining the sketches when dealing with a sum $\Sigma \alpha_r \hat{f}(r)\exp irt$ we call the α_r 'multiplying factors'.)

The De La Vallée Poussin sum $\sigma_{kn,(k+1)n}$ inherits good convergence properties from the Fejér sum.

Lemma 15.1. (i) *If $f:\mathbb{T} \to \mathbb{C}$ is Riemann integrable and f is continuous at t then for each fixed integer k*

$$\sigma_{kn,(k+1)n}(f,t) \to f(t) \quad \text{as } n \to \infty.$$

(ii) *Moreover, if f is continuous on $\mathbb{T}$ then*

$$\sigma_{kn,(k+1)n}(f, \ ) \to f \text{ uniformly as } n \to \infty.$$

Proof. (i)

$$\sigma_{kn,(k+1)n}(f,t) = \left(k+1+\frac{1}{n}\right)\sigma_{(k+1)n}(f,t) - \left(k+\frac{1}{n}\right)\sigma_{kn}(f,t)$$

$$\to (k+1)f(t) - kf(t) = f(t)$$

as $n \to \infty$, using Theorem 2.3 (i).

(ii) Similar. ∎

56

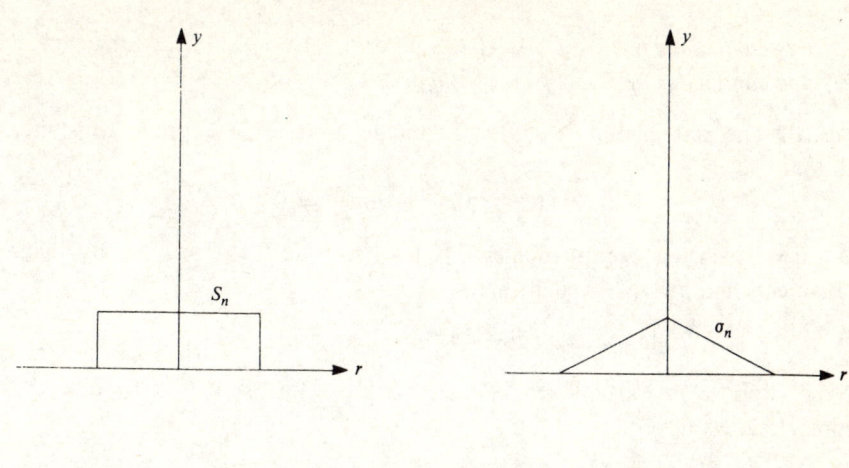

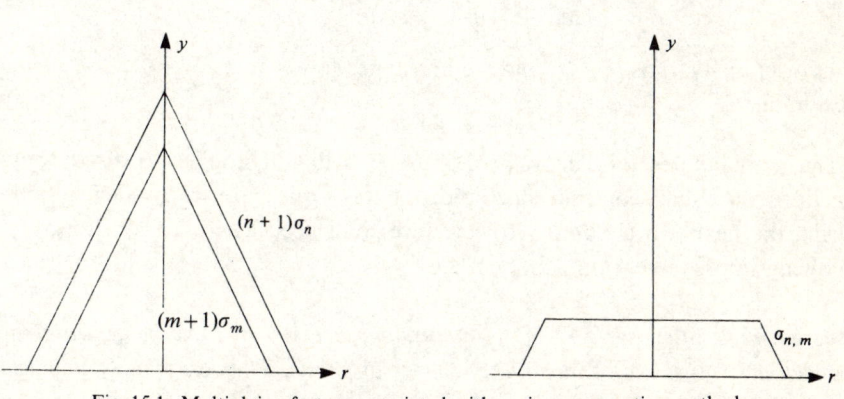

Fig. 15.1. Multiplying factors associated with various summation methods.

On the other hand if $|\hat{f}(r)|$ is not too large for $kn \leqslant |r| < (k+1)n$ then $\sigma_{kn,(k+1)n}(f)$ does not differ very greatly from $S_m(f)$ when $kn \leqslant m < (k+1)n$.

Lemma 15.2. *If* $f : \mathbb{T} \to \mathbb{C}$ *is Riemann integrable and* $|\hat{f}(r)| \leqslant A|r|^{-1}$ *for* $r \neq 0$ *then*

$$|\sigma_{kn,(k+1)n}(f,t) - S_m(f,t)| \leqslant 2Ak^{-1}$$

for all $t \in \mathbb{T}$ *and* $kn \leqslant m < (k+1)n$ [k, m, n, *positive integers*].
Proof. This is trivial. We just observe that

$$|\sigma_{kn,(k+1)n}(f,t) - S_m(f,t)| \leqslant \sum_{kn<|r|\leqslant(k+1)n} |\hat{f}(r)| \leqslant 2 \sum_{r=kn+1}^{(k+1)n} Ar^{-1} \leqslant 2An(kn)^{-1} = 2Ak^{-1}.$$

∎

Combining Lemmas 15.1 and 15.2 we get our new convergence criterion.

Theorem 15.3. *Suppose* $f : \mathbb{T} \to \mathbb{C}$ *is Riemann integrable and* $\hat{f}(r) = \mathrm{O}(1/|r|)$ *as* $r \to \infty$. *Then*

(i) *If f is continuous at t, $S_n(f,t) \to f(t)$ as $n \to \infty$.*

(ii) *If f is continuous on $\mathbb{T}, S_n(f,) \to f$ uniformly as $n \to \infty$.*

Proof. (i) The statement $\hat{f}(r) = O(|r|^{-1})$ means that there exists a constant A such that

$$|\hat{f}(r)| \leqslant A|r|^{-1} \quad \text{for} \quad r \neq 0.$$

If $\varepsilon > 0$ is given, choose an integer $k \geqslant 1 + 4A\varepsilon^{-1}$ (so $2Ak^{-1} < \varepsilon/2$). By Lemma 15.1 we can find an $n_0 \geqslant k$ such that

$$|\sigma_{kn,(k+1)n}(f,t) - f(t)| < \varepsilon/2 \quad \text{for all} \quad n \geqslant n_0.$$

Now suppose $m \geqslant kn_0$. Then $(k+1)n > m \geqslant kn$ for some $n \geqslant n_0$ and, using Lemma 15.2, we have

$$|S_m(f,t) - f(t)| \leqslant |\sigma_{kn,(k+1)n}(f,t) - f(t)| + |S_m(f,t) - \sigma_{kn,(k+1)n}(f,t)|$$
$$< \varepsilon/2 + 2Ak^{-1} < \varepsilon.$$

Thus $S_m(f,t) \to f(t)$ as $m \to \infty$ and we are done.

(ii) Similar   ∎

The main interest of Theorem 15.3 to us is that it applies to discontinuous functions. We shall continue our study of the Fourier series of such functions during the next two chapters. However we shall need Theorem 15.3 to give the following improvement on Theorem 9.6.

Theorem 15.4. *Suppose $f : \mathbb{T} \to \mathbb{C}$ is continuous everywhere and has a continuous bounded derivative except at a finite number of points. Then $S_n(f,) \to f$ uniformly.*
Proof. By Lemma 9.5, $\hat{f}(r) = O(|r|^{-1})$ so the result follows from Theorem 15.3. ∎

The reader, warned by Example 10.1 (where, after all, $\hat{h}(r) = O(|r|^{-2})$), will not expect rapid convergence when all that can be guaranteed is $\hat{h}(r) = O(|r|^{-1})$.

16

BEHAVIOUR AT
POINTS OF DISCONTINUITY I

Not all the functions which occur in classical physics are continuous. A typical discontinuous function with which we may have to deal is the 'sweep' or 'sawtooth' function h shown in Figure 16.1 given by

$$h(x) = x \quad -\pi < x < \pi,$$
$$h(\pi) = 0.$$

For this function we have $\hat{h}(0) = 0$ and

$$\hat{h}(r) = \frac{1}{2\pi}\int_{-\pi}^{\pi} x\exp(-irx)dx = \frac{1}{2\pi}\left[\frac{x\exp(-irx)}{-ir}\right]_{-\pi}^{\pi} + \frac{1}{2\pi}\int_{-\pi}^{\pi}\frac{\exp(-irx)}{ir}dx$$
$$= \frac{1}{2\pi}\frac{\pi\exp(-ir\pi) + \pi\exp(ir\pi)}{-ir} + 0 = \frac{(-1)^{r+1}}{ir} \quad \text{for } r \neq 0.$$

Thus $\hat{h}(r) = O(|r|^{-1})$ and so, by Theorem 15.3(i), $S_m(h, t) \to h(t)$ at all points t where h is continuous, i.e. for all points $t \neq \pi$. Further

$$S_m(h, t) = \sum_{r=1}^{m}\frac{(-1)^{r+1}}{ir}(\exp irt - \exp(-irt)) = \sum_{r=1}^{m}(-1)^{r+1}\frac{2}{r}\sin rt \quad \text{for all } t,$$

and so, in particular, $S_m(h, \pi) = 0 \to 0$ as $m \to \infty$. (This result could also have been obtained by a symmetry argument.)

Summing up, we have the following Lemma.

Lemma 16.1. *If h is the sawtooth function defined above, then $\hat{h}(r) = O(|r|^{-1})$ as $|r| \to \infty$ and $S_m(h, t) \to h(t)$ for $t \neq \pi$, $S_m(h, \pi) \to 0$ as $m \to \infty$.*

We now show that this kind of behaviour is typical of quite a large class of functions.

Lemma 16.2. *Suppose $f: \mathbb{T} \to \mathbb{C}$ is Riemann integrable with $\hat{f}(r) = O(|r|^{-1})$ as $|r| \to \infty$. Suppose further that $\lim_{t \to \pi-} f(t)$ and $\lim_{t \to \pi+} f(t)$ exist with values $f(\pi-)$*

59

Fourier series

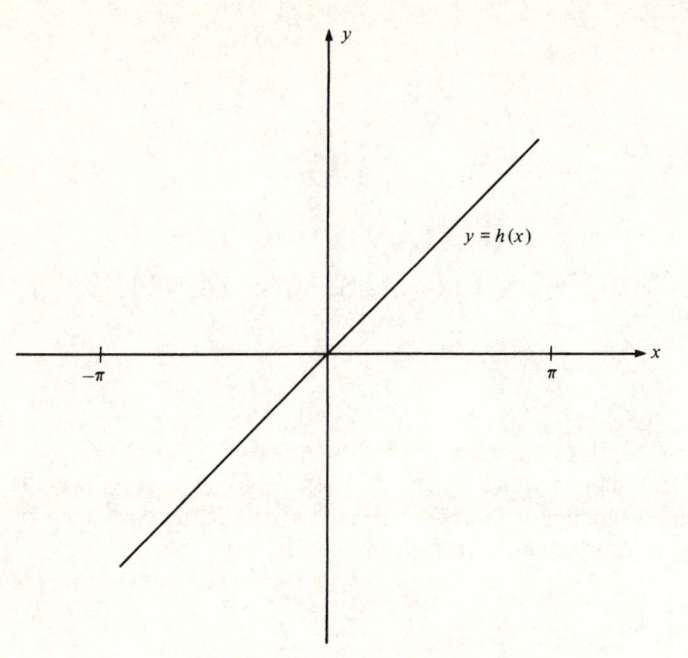

Fig. 16.1. The sawtooth function.

and $f(\pi +)$ say.

Then $$S_m(f, \pi) \to (f(\pi +) + f(\pi -))/2.$$

Proof. Define $g: \mathbb{T} \to \mathbb{C}$ by

$$g(t) = f(t) + (2\pi)^{-1}(f(\pi +) - f(\pi -))h(t) \quad \text{for } t \neq \pi,$$
$$g(\pi) = (f(\pi +) + f(\pi -))/2.$$

Then $\quad g(t) \to f(\pi -) + (2\pi)^{-1}(f(\pi +) - f(\pi -))\pi = g(\pi) \quad$ as $\quad t \to \pi -$

and, similarly, $$g(t) \to g(\pi) \quad \text{as} \quad t \to \pi +.$$

Thus g is continuous at π. But

$$\hat{g}(r) = \hat{f}(r) - (2\pi)^{-1}(f(\pi +) + f(\pi -))\hat{h}(r)$$

and so, since $\hat{f}(r) = \mathrm{O}(|r|^{-1})$ and $\hat{h}(r) = \mathrm{O}(|r|^{-1})$, we have $\hat{g}(r) = \mathrm{O}(|r|^{-1})$. Thus, by Theorem 15.3, we have $S_m(g, \pi) \to g(\pi)$. Hence,

$$S_m(f, \pi) - (2\pi)^{-1}(f(\pi +) - f(\pi -))S_m(h, \pi)$$
$$= S_m(g, \pi) \to g(\pi) = (f(\pi +) + f(\pi -))/2.$$

But $S_m(h, \pi) \to 0$ (indeed $S_m(h, \pi) = 0$) and so

$$S_m(f, \pi) \to (f(\pi +) + f(\pi -))/2. \qquad \blacksquare$$

The idea of the proof may be expressed as follows. 'Given a function f with a nasty singularity, express it as the sum $g + h$ of a nice function g and a known function h with the same kind of singularity.' It is an idea well worth remembering. Obviously π can be replaced by any other point of $\mathbb{T}$ to give the following theorem.

Theorem 16.3. *Suppose* $f : \mathbb{T} \to \mathbb{C}$ *is Riemann integrable with* $\hat{f}(r) = \mathrm{O}(|r|^{-1})$ *as* $|r| \to \infty$. *Then whenever* $\lim_{t \to x-} f(t)$ *and* $\lim_{t \to x+} f(t)$ *exist* $[x \in \mathbb{T}]$ *we have*

$$S_m(f, x) \to (\lim_{t \to x+} f(t) + \lim_{t \to x-} f(t))/2 \quad as \quad m \to \infty.$$

Proof. Observe that the convergence of a Fourier series to a function is unaffected by rotation. (If we write

$$f_y(t) = f(y + t) \ [t \in \mathbb{T}] \quad \text{then} \quad \hat{f}_y(r) = \exp iry \, \hat{f}(r), \text{ and}$$

$$S_m(f_y, t) = \sum_{r=-m}^{m} \exp iry \, \hat{f}(r) \exp irt = \sum_{r=-m}^{m} \hat{f}(r) \exp ir(y+t) = S_m(f, y+t).)$$

∎

Combining Theorem 16.3 and Lemma 9.5, we obtain the following result which fulfils our promise to prove Theorem 1.1 of the introduction.

Theorem 16.4. *Suppose* $f : \mathbb{T} \to \mathbb{C}$ *is continuous with a continuous bounded derivative except at a finite number of points. Then* $\lim_{t \to x-} f(t)$ *and* $\lim_{t \to x+} f(t)$ *exist for all* $x \in \mathbb{T}$ *and*

$$S_m(f, x) \to (\lim_{t \to x+} f(t) + \lim_{t \to x-} f(t))/2.$$

In particular if f is continuous at x we have

$$S_m(f, x) \to f(x) \quad as \quad m \to \infty.$$

Proof. The existence of $\lim_{t \to x-} f(t)$ and $\lim_{t \to x+} f(t)$ follows from the mean value theorem. By Lemma 9.5 $\hat{f}(r) = \mathrm{O}(|r|^{-1})$ so the convergence follows from Theorem 16.3. The final remark is obvious when we observe that if f is continuous at x then $\lim_{t \to x+} f(t)$ and $\lim_{t \to x-} f(t)$ both exist and have the value $f(x)$. ∎

17

BEHAVIOUR AT POINTS
OF DISCONTINUITY II

We have seen in Chapter 8 how Kelvin invented machines which could compute periodic functions from their Fourier series and conversely obtain the Fourier series of a given periodic function. One such machine was constructed by Michelson to work to a higher accuracy and to involve many more terms than previous models. (Michelson's ability to build and operate equipment to new standards of accuracy was legendary. Of his interferometer which he invented and used in the Michelson Morley experiments it was said that it was a remarkable instrument – provided you had Michelson to operate it. His experiments to measure the diameter of the nearest stars using an interferometer were not reproduced for 30 years.)

Michelson tested his machine by feeding in the first 80 Fourier coefficients of the sawtooth function h defined in Chapter 16. To his surprise the machine did not produce an exact sawtooth but instead added two little blips on either side of the discontinuity as shown in Figure 17.1. Even after making every effort to remove any mechanical defects which could account for them, the blips still remained. Finally hand calculation confirmed the existence of blips in $S_n(h, \ )$ close to the discontinuity. The effect of increasing n was to move the blips closer and closer to the discontinuity but they remained and their height (in absolute value) remained 17 or 18% above the correct absolute value. How could this be reconciled with Theorem 16.4 (or indeed Lemma 16.1)?

Gibbs in two letters to *Nature* (the second a correction of the first) clarified and resolved the issue. The difficulty is due to a confusion between '*the limit of the graphs* and ... *the graph of the limit* of the sum. A misunderstanding on this point is a natural consequence of the usage which allows us to omit the word *limit* in certain connections as when we speak of the sum of an infinite series.'

In other words $S_n(h, t) \to h(t)$ pointwise (after all, the blips move towards the discontinuity), but pointwise convergence of f_n to f does not imply that the graph of f_n starts to look like f for large n. The reader has already met more extreme examples of this when the notion of uniform convergence of a function g_n to g (which does imply that the graph of g_n starts to look like g) was introduced. For

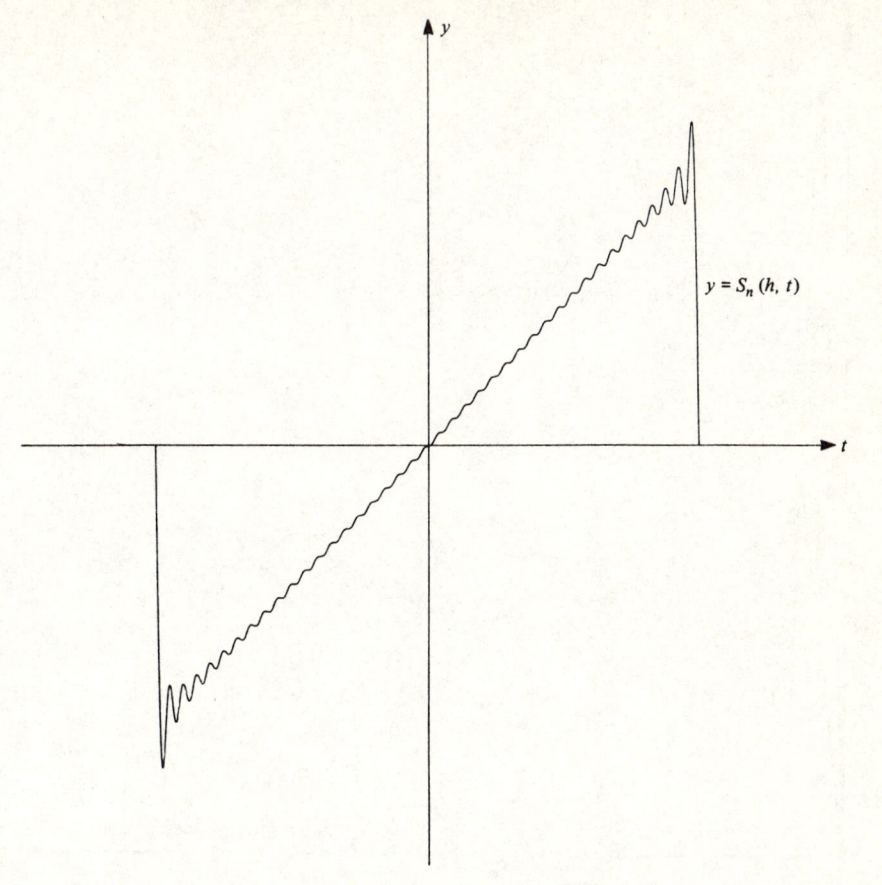

Fig. 17.1. A partial Fourier sum for the sawtooth function.

example, consider the 'witch's hats'

$$f_n(x) = n^2(1 - n|x - n^{-1}|) \quad \text{for } 0 \leqslant x \leqslant 2n^{-1},$$
$$f_n(x) = 0 \quad\quad\quad\quad\quad \text{otherwise.}$$

Then $f_n \to 0$ pointwise but the graph of f_n does not resemble 0. (see Figure 17.2).

In conclusion we may repeat once more that delicate and sophisticated results like Theorem 16.4 require much more care in use and interpretation than crude and unsubtle results like Theorem 9.1, and that the reader must always be careful to understand the limitations of any particular mode of convergence under discussion.

A full investigation of the 'Gibbs phenomenon' is not very difficult but neither is it very interesting. We shall therefore limit ourselves to demonstrating its reality.

Theorem 17.1. *If h is the sawtooth function defined by $h(x) = x[x \neq \pi]$, $h(\pi) = 0$ then*

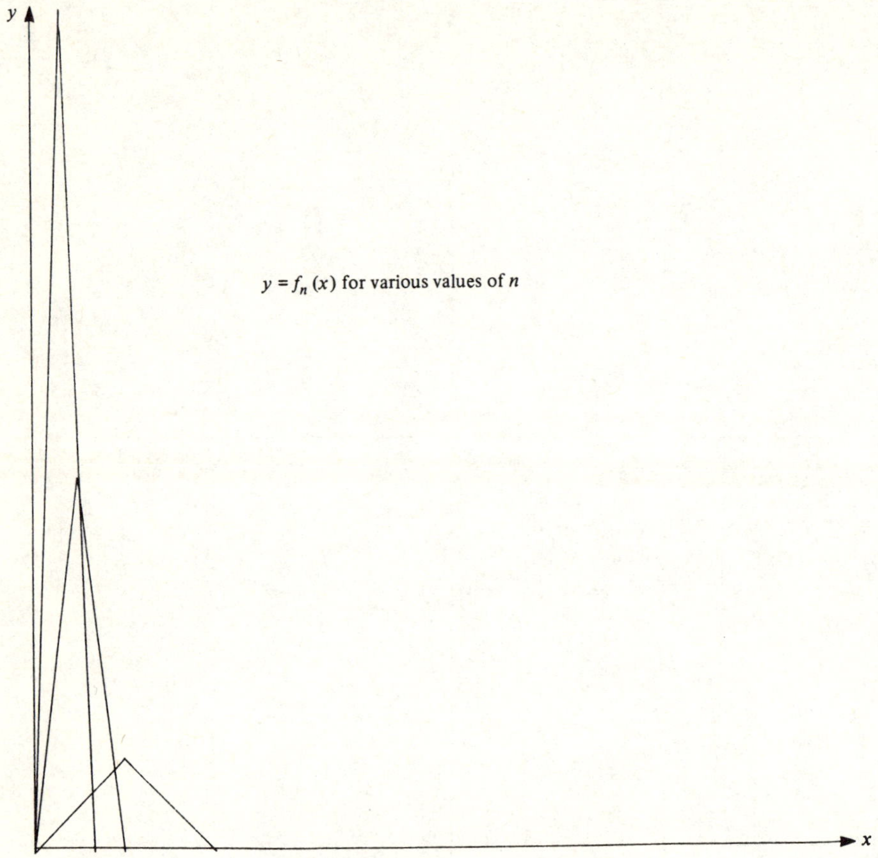

y = f_n (x) for various values of n

Fig. 17.2. The 'witch's hat' counter example.

$$S_n(h, \pi - \pi/n) \to A\pi,$$
$$S_n(h, -\pi + \pi/n) \to -A\pi \quad as \ n \to \infty,$$

where $A = 2/\pi \int_0^\pi (\sin x/x) \, dx > 1.17.$

Proof. At the beginning of Chapter 16 we saw that

$$S_n(h, x) = \sum_{r=1}^n (-1)^{r+1} \frac{2}{r} \sin rx$$

Thus $$S_n(h, \pi - \pi/n) = \sum_{r=1}^n \frac{2}{r} \sin \frac{r\pi}{n} = 2 \sum_{r=1}^n \frac{\pi}{n} \left(\frac{n}{r\pi} \sin \frac{r\pi}{n} \right) \to 2 \int_0^\pi \frac{\sin x}{x} dx$$

(using the standard results on the approximation of integrals by sums together with the observation that $\sin x/x$ is defined, continuous and bounded on $[0, \pi]$). Similarly,

$$S_n(h, -\pi + \pi/n) \to -2 \int_0^\pi (\sin x/x) dx,$$

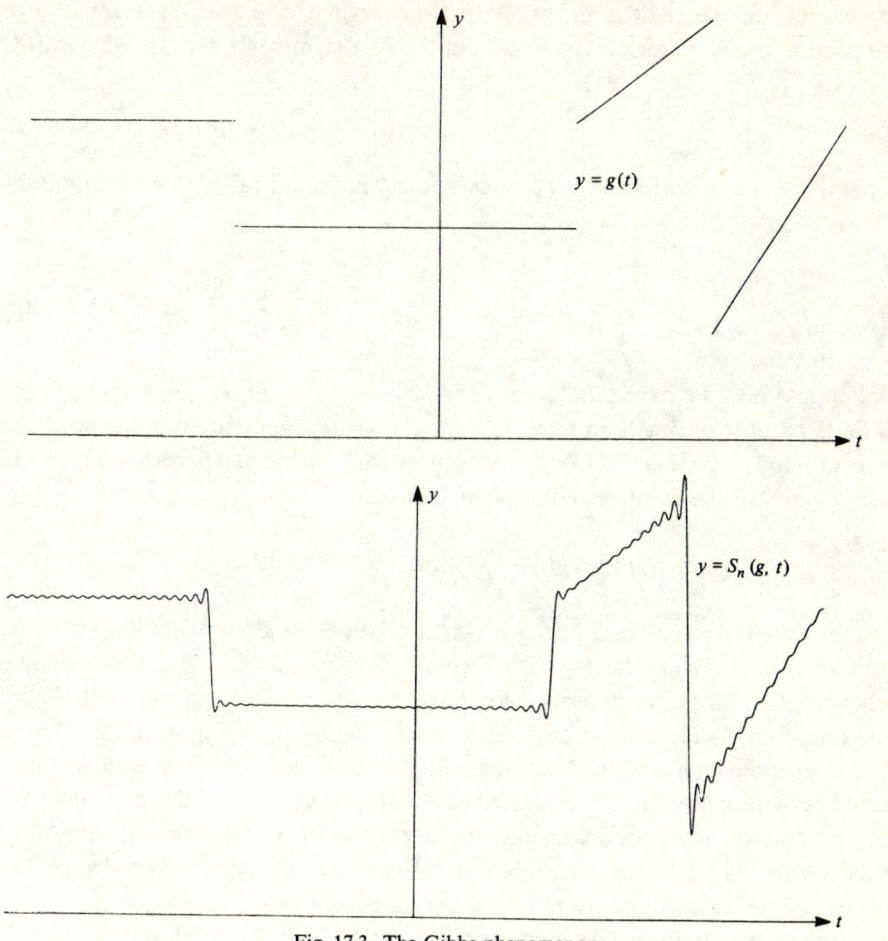

Fig. 17.3. The Gibbs phenomenon.

so all that remains to be done is to show that

$$\frac{2}{\pi}\int_0^\pi \frac{\sin x}{x}dx > 1.17.$$

This we do by direct numerical calculation. Since

$$\frac{\sin x}{x} = \sum_{r=0}^\infty \frac{(-1)^r x^{2r}}{(2r+1)!}$$

with the power series having infinite radius of convergence, we may integrate term by term to get

$$\int_0^\pi \frac{\sin x}{x}dx = \pi\left(1 - \frac{\pi^2}{3!3} + \frac{\pi^4}{5!5} - \frac{\pi^6}{7!7} + \cdots\right).$$

The series on the right is an oscillating decreasing series, so the error due to truncation is less, in absolute value, than the first term neglected. In other words

$$\left| \frac{2}{\pi} \int_0^\pi \frac{\sin x}{x} dx - 2 \sum_{r=0}^n \frac{\pi^{2r}(-1)^r}{(2r+1)^2(2r)!} \right| \leqslant \frac{2\pi^{2n+2}}{(2n+3)^2(2n+2)!}.$$

Taking $n = 4$ and performing the calculations on a hand calculator we obtain

$$\frac{2}{\pi} \int_0^\pi \frac{\sin x}{x} dx > 1.17,$$

as required.   ∎

Suppose now we have a function $f : \mathbb{T} \to \mathbb{C}$ with $\hat{f}(r) = O(|r|^{-1})$ which has only a finite number of discontinuities, at $x_1, x_2, \ldots x_N$ say, and suppose further that f is continuous on the left and on the right at each of these. Then the reader will easily verify (if she is interested) that we can write

$$f(t) = g(t) + \sum_{j=1}^N \lambda_j h(t - x_j) \quad [t \in \mathbb{T}],$$

where $\lambda_j \in \mathbb{C}[1 \leqslant j \leqslant N]$ and $g : \mathbb{T} \to \mathbb{C}$ is a continuous function with $\hat{g}(r) = O(|r|^{-1})$ and so, by Theorem 15.3 (i), with $S_n(g,) \to g$ uniformly on $\mathbb{T}$. Thus the same phenomenon which we described for h (of a blip overshooting by $8\frac{1}{2}$ to 9% of the total jump) will occur at each of the discontinuities (see Figure 17.3).

The phenomenon described in this chapter is called the 'Gibbs phenomenon' but could perhaps more fittingly be described as the 'Gibbs–Wilbraham phenomenon' since it had already been discovered and explained by an English mathematician called Wilbraham 60 years before. However this first discovery must have appeared as an isolated curiosity of no practical relevance and was soon forgotten.

During the early British development of radar it was decided to use the sawtooth function h to give the x coordinate on the oscilloscopes. The engineers produced h in the obvious way as a Fourier sum and the Gibbs–Wilbraham phenomenon was rediscovered yet again.

18

A FOURIER SERIES
DIVERGENT AT A POINT

Dirichlet's proof of Theorem 16.4 (or, more accurately, a result similar to it) left open the question as to whether the Fourier series of every Riemann integrable, or at least every continuous, function converged. At the end of his paper Dirichlet made it clear that he thought that the answer was yes (and that he would soon be able to prove it). During the next 40 years Riemann, Weierstrass and Dedekind also expressed their belief that the answer was positive.

It therefore came as a considerable surprise when Du Bois-Reymond produced a counter example.

Theorem 18.1 (Du Bois-Reymond). *There exists a continuous function* $f : \mathbb{T} \to \mathbb{C}$ *such that* $\limsup\limits_{n \to \infty} |S_n(f, 0)| = \infty$.

How could one set about getting such a counter example? One way is as follows.

We cannot immediately find a continuous function f with $\limsup\limits_{n \to \infty} |S_n(f, 0)| = \infty$ so let us set ourselves a more modest goal. Let us try and find a continuous function g with $\sup\limits_{t \in \mathbb{T}} |g(t)|$ small and yet $\sup\limits_{n} |S_n(g, 0)|$ very large and try to modify that to make a counter example. And if we cannot do that, let us content ourselves, for the moment with just a reasonably well behaved function h (continuous or otherwise) with $\sup\limits_{t \in \mathbb{T}} |h(t)|$ small and yet $\sup\limits_{n} |S_n(h, 0)|$ very large and then try to modify that.

How can we find a suitable h? Observe that (paralleling our treatment of Fejér sums in Chapter 2)

$$S_n(h, t) = \sum_{r=-n}^{n} \hat{h}(r) \exp irt = \sum_{r=-n}^{n} \frac{1}{2\pi} \int_{\mathbb{T}} h(x) \exp(-irx) \, dx \exp irt$$

$$= \frac{1}{2\pi} \int_{\mathbb{T}} h(x) \sum_{r=-n}^{n} \exp(ir(t-x)) \, dx = \frac{1}{2\pi} \int_{\mathbb{T}} h(x) D_n(t-x) \, dx,$$

67

where $D_n(s) = \sum_{r=-n}^{n} \exp irs$. In particular therefore

$$S_n(h,0) = \frac{1}{2\pi}\int_{\mathbb{T}} h(x)D_n(-x)\,dx.$$

Lemma 18.2.

$$D_n(s) = \frac{\sin(n+\frac{1}{2})s}{\sin\frac{1}{2}s} \quad (s \neq 0),$$

$$D_n(0) = 2n+1.$$

Proof. If $s \neq 0$, then

$$\sum_{r=-n}^{n} \exp irs = \exp(-ins)\sum_{r=0}^{2n} \exp irs = \exp(-ins)\frac{1-\exp(i(2n+1)s)}{1-\exp is}$$

$$= \frac{\exp(-i(n+\frac{1}{2})s) - \exp(i(n+\frac{1}{2})s)}{\exp(-i\frac{1}{2}s) - \exp(i\frac{1}{2}s)} = \frac{\sin(n+\frac{1}{2})s}{\sin\frac{1}{2}s}.$$

If $s = 0$, then $D_n(0) = 2n+1$ by direct computation. ∎

The function D_n which we sketch in Figure 18.1 is very badly behaved compared

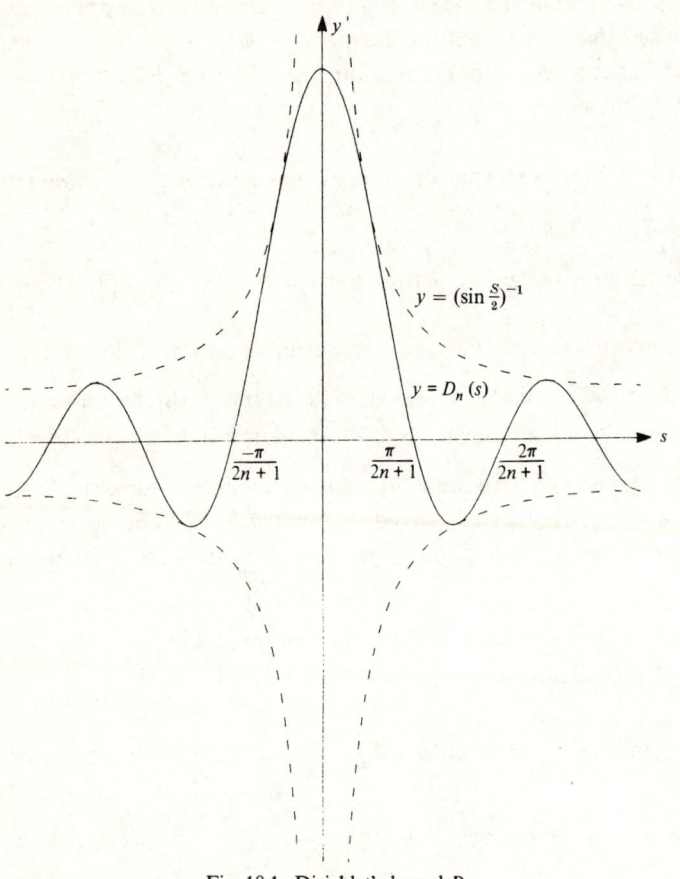

Fig. 18.1. Dirichlet's kernel D_n.

with K_n. Whilst it remains true that $(2\pi)^{-1}\int_{\mathbb{T}}D_n(S)\,ds = 1$, it is not true that D_n converges to zero at any point, it is not true that D_n is a positive function and, most important from our point of view, it is not true that $(2\pi)^{-1}\int_{\mathbb{T}}|D_n(s)|\,ds$ is bounded.

Indeed looking at the diagram it would appear that

$$\int_{\mathbb{T}}|D_n(s)|\,ds \geqslant \int_0^{\pi}|D_n(s)|\,ds \geqslant \int_{\pi/(2n+1)}^{\pi}|D_n(s)|\,ds \approx \int_{\pi/(2n+1)}^{\pi}\frac{1}{\sin s/2}\,ds$$

$$\approx \int_{\pi/(2n+1)}^{\pi}\frac{1}{s/2}\,ds \approx \int_{\pi/(2n+1)}^{\pi}\frac{1}{s}\,ds \approx \log n$$

(where $\approx$ means 'is roughly the same order of magnitude') and this turns out to be true.

Lemma 18.3.

$$\frac{1}{2\pi}\int_{\mathbb{T}}|D_n(s)|\,ds \geqslant \frac{4}{\pi^2}\log 2(n+1).$$

Proof. Recall that $y \geqslant \sin y \geqslant 2y/\pi$ for all $\pi/2 \geqslant y \geqslant 0$. Thus

$$\frac{1}{2\pi}\int_{\mathbb{T}}|D_n(s)|\,ds = \frac{1}{2\pi}\int_{-\pi}^{\pi}|D_n(s)|\,ds = \frac{1}{\pi}\int_0^{\pi}|D_n(s)|\,ds \geqslant \frac{2}{\pi}\int_0^{\pi}\left|\frac{\sin(n+\frac{1}{2})s}{s}\right|\,ds$$

$$= \frac{2}{\pi}\sum_{r=0}^{2n}\int_{r\pi/(2n+1)}^{(r+1)\pi/(2n+1)}\frac{|\sin(n+\frac{1}{2})s|}{s}\,ds$$

$$\geqslant \frac{2}{\pi}\sum_{r=0}^{2n}\frac{2n+1}{(r+1)\pi}\int_{r\pi/(2n+1)}^{(r+1)\pi/(2n+1)}|\sin(n+\frac{1}{2})s|\,ds$$

$$= \frac{2}{\pi}\sum_{r=0}^{2n}\frac{2n+1}{(r+1)\pi}\int_0^{\pi/(2n+1)}\sin(n+\frac{1}{2})s\,ds = \frac{4}{\pi^2}\sum_{r=0}^{2n}\frac{1}{r+1},$$

and since, as the reader probably knows,

$$\sum_{r=0}^{N-1}1/(r+1) \geqslant \sum_{r=1}^{N}\int_r^{r+1}1/x\,dx = \int_1^{N+1}1/x\,dx = \log(N+1),$$

we are done. ∎

We can now achieve our first goal on the road to a counter example.

Lemma 18.4. *Let* $h_n(s) = \operatorname{sgn} D_n(-s)$. *Then*

(i) h_n *is constant on the intervals* $(r\pi/(2n+1), (r+1)\pi/(2n+1))$,
(ii) $|h_n(s)| \leqslant 1$ *for all* s,
(iii) $S_n(h_n, 0) \geqslant 4/\pi^2 \log 2(n+1)$.

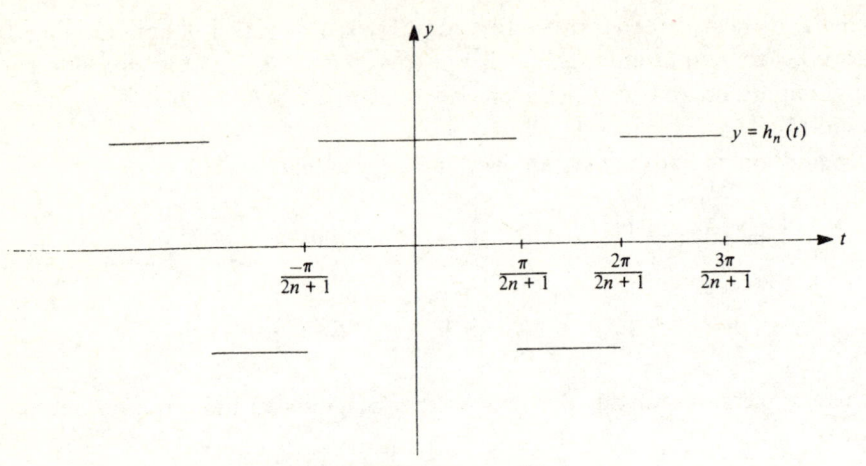

Fig. 18.2. Sketch of a possible h_n.

Proof. (We graph h_n in Figure 18.2.) By definition $h_n(s) = 1$ when $D_n(-s) > 0$, $h_n(s) = -1$ when $D_n(-s) < 0$ and $h_n(s) = 0$ when $D_n(-s) = 0$. Thus conditions (i) and (ii) (which say that h_n is reasonably well behaved) are trivial. Condition (iii) is also trivial since

$$S_n(h_n, 0) = \frac{1}{2\pi} \int_{\mathbb{T}} h_n(x) D_n(-x) dx = \frac{1}{2\pi} \int_{\mathbb{T}} |D_n(-x)| dx$$

$$= \frac{1}{2\pi} \int_{\mathbb{T}} |D_n(x)| dx \geqslant \frac{4}{\pi^2} \log 2(n+1). \qquad \blacksquare$$

The second goal is a very short distance from the first one if we use the following trivial observation.

Lemma 18.5. *If $g, h : \mathbb{T} \to \mathbb{C}$ are Riemann integrable then*

(i) $|\hat{g}(r) - \hat{h}(r)| \leqslant (2\pi)^{-1} \int_{\mathbb{T}} |g(t) - h(t)| dt$ *for all r,*
(ii) $|S_n(g, x) - S_n(h, x)| \leqslant (2n+1)/(2\pi) \int_{\mathbb{T}} |g(t) - h(t)| dt$ *for all $n \geqslant 0$.*

Proof. (i) Just observe that

$$|\hat{g}(r) - \hat{h}(r)| = \left| \frac{1}{2\pi} \int_{\mathbb{T}} (g(t) - h(t)) \exp(-irt) dt \right| \leqslant \frac{1}{2\pi} \int_{\mathbb{T}} |(g(t) - h(t)) \exp(-irt)| dt$$

$$= \frac{1}{2\pi} \int_{\mathbb{T}} |g(t) - h(t)| dt.$$

(ii) Just observe that, using (i),

$$|S_n(g, x) - S_n(h, x)| = \left| \sum_{r=-n}^{n} (\hat{g}(r) - \hat{h}(r)) \exp irx \right| \leqslant \sum_{r=-n}^{n} |(\hat{g}(r) - \hat{h}(r)) \exp irx|$$

$$= \sum_{r=-n}^{n} |\hat{g}(r) - \hat{h}(r)| \leqslant \frac{2n+1}{2\pi} \int_{\mathbb{T}} |g(t) - h(t)| \, dt. \qquad \blacksquare$$

Lemma 18.6. *For each $n \geqslant 0$ there exists a continuous function $g_n : \mathbb{T} \to \mathbb{R}$ such that*

(i) $|g_n(s)| \leqslant 1$ *for all s,*

(ii) $|S_n(g_n, 0)| \geqslant 4/\pi^2 \log(n+1)$.

Proof. Just construct a continuous function g_n such that $|g_n(s)| \leqslant 1$ for all s and

$$\frac{1}{2\pi} \int |g_n(s) - h_n(s)| \, ds \leqslant \frac{4^{-1}}{2n+1}.$$

(We sketch such a g_n in Figure 18.3.) Using Lemmas 18.4 and 18.5 (ii), we then have

$$|S_n(g_n, 0)| \geqslant |S_n(h_n, 0)| - |S_n(g_n, 0) - S_n(h_n, 0)| \geqslant \frac{4}{\pi^2} \log 2(n+1) - \tfrac{1}{4} \geqslant \frac{4}{\pi^2} \log(n+1). \qquad \blacksquare$$

It turns out to be more convenient to replace our well behaved continuous function g_n with $S_n(g_n, 0)$ large by a well behaved trigonometric polynomial G_n with $S_n(G_n, 0)$ large.

Lemma 18.7. *For each $n \geqslant 0$ there exists a trigonometric polynomial $G_n : \mathbb{T} \to \mathbb{C}$ such that*

(i) $|G_n(s)| \leqslant 2$ *for all $s \in \mathbb{T}$,*

(ii) $|S_n(G_n, 0)| \geqslant 4/\pi^2 \log(n+1) - 1$.

Proof. By Theorem 2.5 (which says that the trigonometric polynomials are uniformly dense in the continuous functions) we can find a trigonometric polynomial G_n

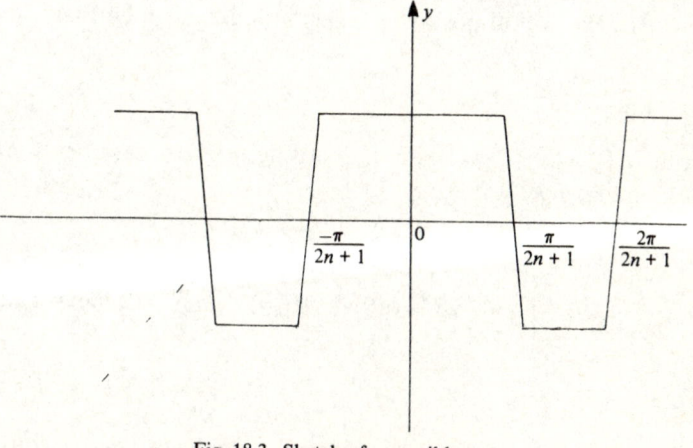

Fig. 18.3. Sketch of a possible g_n.

with

$$|G_n(s) - g_n(s)| \leqslant \frac{1}{2n+1} \quad \text{for all } s \in \mathbb{T}.$$

Trivially,

$$|G_n(s)| \leqslant |G_n(s) - g_n(s)| + |g_n(s)| \leqslant 2 \quad \text{for all } s \in \mathbb{T},$$

so (i) is true. Also

$$\frac{1}{2\pi} \int_{\mathbb{T}} |G_n(s) - g_n(s)| \, ds \leqslant \frac{1}{2n+1},$$

so, exactly as in Lemma 18.6,

$$|S_n(G_n, 0)| \geqslant |S_n(g_n, 0)| - |S_n(G_n, 0) - S_n(g_n, 0)| \geqslant \frac{4}{\pi^2} \log(n+1) - 1, \cdots$$

so (ii) is true.   ∎

Lemma 18.7 contains slightly more information than we need, so we simplify it in Lemma 18.7'.

Lemma 18.7'. *Given $A > 0$ we can find a trigonometric polynomial $G: \mathbb{T} \to \mathbb{C}$ and an integer $N \geqslant 0$ such that*

(i) $|G(s)| \leqslant 1$ *for all $s \in \mathbb{T}$,*
(ii) $|S_N(G, 0)| \geqslant A$.

Proof. Obvious from Lemma 18.7.   ∎

We can now prove Theorem 18.1 on the existence of a continuous f with $S_n(f, 0)$ divergent.
Proof of Theorem 18.1. By Lemma 18.7' we can find a sequence of trigonometric polynomials $H_1, H_2, \ldots$ and positive integers $n(1), n(2), \ldots$ such that

(i) $|H_k(s)| \leqslant 1$ for all $s \in \mathbb{T}$,
(ii) $|S_{n(k)}(H_k, 0)| \geqslant 2^{2k}$.

Let us write $H_k(t) = \sum_{r=-q(k)}^{q(k)} a_{kr} \exp irt$, choosing $q(k)$ so that $q(k) \geqslant q(k-1)$ and $q(k) > n(k)$, and set $p(k) = \sum_{j=1}^{k} (2q(j)+1)$.
Let $f_m(t) = \sum_{k=1}^{m} 2^{-k} \exp(ip(k)t) H_k(t)$. Since (if $m \geqslant n+1$)

$$|f_m(t) - f_n(t)| = \left| \sum_{k=n+1}^{m} 2^{-k} \exp(ip(k)t) H_k(t) \right| \leqslant \sum_{k=n+1}^{m} |2^{-k} \exp(ip(k)t) H_k(t)|$$

$$= \sum_{k=n+1}^{m} |2^{-k} H_k(t)| \leqslant \sum_{k=n-1}^{m} 2^{-k} \leqslant 2^{-n} \to 0 \quad \text{as } n \to \infty,$$

it follows by the general principle of uniform convergence that f_m converges uniformly to some function f. Since f_m is continuous, f is. Moreover $\exp(-irt) f_m(t) \to$

$\exp(-irt)f(t)$ uniformly and so

$$\hat{f}_m(r) = \frac{1}{2\pi}\int_{\mathbb{T}} \exp(-irt)\hat{f}_m(t)\,dt \to \frac{1}{2\pi}\int \exp(-irt)f(t)\,dt = \hat{f}(r).$$

Now observe that (because we have cunningly used the $\exp(ip(k)t)$ to 'spread out' the H_k), if $m \geq k$ and $|u| \leq q(k)$ is an integer, then

$$\hat{f}_m(p(k)+u) = 2^{-k}\hat{H}_k(u).$$

Thus

$$\hat{f}(p(k)+u) = 2^{-k}\hat{H}_k(u) \quad \text{for all } |u| \leq q(k).$$

A similar argument shows that

$$\hat{f}(r) = 0 \quad \text{for all } r < 0.$$

It follows that

$$|S_{p(k)+n(k)}(f,0) - S_{p(k)-n(k)}(f,0)| = \left| \sum_{u=-n(k)}^{n(k)} 2^{-k}\hat{H}_k(u)\exp(i(p(k)+u)t) \right|$$

$$= \left| \sum_{u=-n(k)}^{n(k)} 2^{-k}\hat{H}_k(u)\exp(iut) \right|$$

$$= 2^{-k}|S_{n(k)}(H_k,0)|$$

$$\geq 2^k \to \infty \quad \text{as } k \to \infty.$$

Thus

$$\lim_{k \to \infty}(\max(|S_{p(k)+n(k)}(f,0)|, |S_{p(k)-n(k)}(f,0)|)) = \infty$$

and so $\lim\sup_{n \to \infty}|S_n(f,0)| = \infty$. ∎

The general idea of (carefully) piling increasingly bad functions on top of one another to obtain a really nasty one is called 'the sliding hump method' or, in a slightly different context, 'the method of condensation of singularities'. It is a vague but powerful idea. (The most successful attempt to pin it down is the Banach–Steinhaus uniform boundedness theorem but even this, although a powerful and easily used result, does not quite capture the full force of the method.)

Finally, in case the reader wishes to see a version of the Du Bois-Reymond theorem with f real, we give the following easy corollary.

Lemma 18.8. *There exists a real valued continuous function*

$$F: \mathbb{T} \to \mathbb{R} \text{ with } \lim\sup_{n \to \infty}|S_n(F,0)| = \infty.$$

Proof. Take f as in Theorem 18.1. Write $f = f_1 + if_2$ where f_1 and f_2 are real valued continuous functions. Observe that $S_n(f,0) = S_n(f_1 + if_2, 0) = S_n(f_1,0) + iS_n(f_2,0)$ so at least one of $\lim\sup_{n \to \infty}|S_n(f_1,0)|$ and $\lim\sup_{n \to \infty}|S_n(f_2,0)|$ must be infinite.

19

POINTWISE CONVERGENCE,
THE ANSWER

After the publication of Du Bois-Reymond's counter example in 1876 mathematical opinion began to swing round to the belief that the Fourier series of a continuous function need not converge anywhere. This belief was strengthened by the publication in 1926 by Kolmogorov of a result which, translated into the language of this book, runs as follows.

Theorem 19.1 (Kolmogorov). *Given $A > 0$ we can find a trigonometric polynomial $G: \mathbb{T} \to \mathbb{R}$ such that*

(i) $G(s) \geqslant 0$ *for all* $s \in \mathbb{T}$, *and* $1 \geqslant (2\pi)^{-1} \int_{\mathbb{T}} G(s)\,ds$,
yet
(ii) $\sup_{n} |S_n(G, s)| \geqslant A$ *for all* $s \in \mathbb{T}$.

Proof. The proof is too long to give here.
(For details of this and other interesting results see Chapter VIII of Zygmund's *Trigonometric Series*.) ∎

In view of the resemblance between the statement of Lemma 18.7′ and Theorem 19.1 we would expect to obtain some analogue of Theorem 18.1. This is the case if we are prepared to extend our theory to cover not only Riemann integrable functions but the more general Lebesgue integrable functions.

Theorem 19.2 (Kolmogorov). *There is a Lebesgue integrable function $f: \mathbb{T} \to \mathbb{R}$ such that* $\lim_{n \to \infty} \sup |S_n(f, t)| = \infty$ *for all* $t \in \mathbb{T}$.

Proof. See Zygmund. ∎

Although the function constructed by Kolmogorov is not even Riemann integrable (indeed it is unbounded on every interval), it was now felt to be only a matter of time before someone constructed a continuous function whose Fourier series diverged everywhere.

Then in 1964 Carleson settled the matter completely and in an unexpected direction. To understand his result we need one definition.

Definition 19.3. *A set $E \subset \mathbb{T}$ is said to have measure zero if given any $\varepsilon > 0$ we can find intervals $I_1, I_2, \ldots$ of length $|I_1|, |I_2|, \ldots,$ say, such that*

(1) $\bigcup_{j=1}^{\infty} I_j \supset E,$
(2) $\sum_{j=1}^{\infty} |I_j| < \varepsilon.$

In other words a set has measure zero if given any $\varepsilon > 0$ we can cover it with a countable collection of intervals of total length less than ε.

Any countable set $\{x_1, x_2, x_3 \ldots\}$ has measure zero (consider $I_j = [x_j - 2^{-j-2}\varepsilon, x_j + 2^{-j-2}\varepsilon]$), so, for example, $\{2\pi x : x \in \mathbb{Q}\}$ is of measure zero. On the other hand no interval $[a, b]$ with $a \neq b$ can have measure zero.

Theorem 19.4 (Carleson). *If $f : \mathbb{T} \to \mathbb{C}$ is continuous (or even Riemann integrable) then $S_n(f, t) \to f(t)$ as $n \to \infty$ for all $t \notin E$ where E is some set of measure zero.*
Proof. This is still far too hard for an undergraduate text. See *Acta Mathematica*, Vol. 116, pages 135–57. ∎

About the same time Kahane and Katznelson proved a pretty and ingenious theorem which complements Carleson's result.

Theorem 19.5 (Kahane and Katznelson). *If E is a set of measure zero then there exist a continuous function $f : \mathbb{T} \to \mathbb{C}$ such that $\lim_{n \to \infty} \sup |S_n(f, t)| = \infty$ for all $t \in E$.*
Proof. The proof, which is by no means too hard for an undergraduate, will be found in Katznelson's *An Introduction to Harmonic Analysis*, Chapter 3. ∎

The problem of pointwise convergence is thus settled. There are few questions which have managed to occupy even a small part of humanity for 150 years. And of those questions, very few indeed have been answered with as complete and satisfactory an answer as Carleson has given to this one.

PART II
SOME DIFFERENTIAL EQUATIONS

20

THE UNDISTURBED DAMPED
OSCILLATOR DOES NOT EXPLODE

Many electrical circuits, complicated mechanical contrivances and black boxes in general are governed by a second order differential equation

$$x'' + p(x, x')x' + q(x) = f(t).$$

We often call $f(t)$ the *input* and $x(t)$ the *response*.

It is usually important that $x(t)$ remains bounded for all $t \geqslant 0$. To this end we usually require $q(x) \geqslant 0$ for $x \geqslant 0$, $q(x) < 0$ for $x < 0$ and $|q(x)| > \delta$ for all x large (mechanically speaking we want q to be a *restoring* force). Further we require $p(x, x') > 0$ (i.e. we want damping). If $f(t) = 0$ for $t > 0$ (i.e. if the system is *undisturbed*) these conditions by themselves are sufficient to prevent $|x(t)|$ and $|x'(t)|$ from becoming unbounded as $t \to \infty$.

The method of proof we adopt is due to Liapounov and makes good use of our physical feeling for the role of $p(x, x')x'$ and $q(x)$.

Theorem 20.1. *Suppose*

(i) $p: \mathbb{R}^2 \to \mathbb{R}$ *is a continuous function with* $p(u, v) \geqslant 0$ *for all* $u, v \in \mathbb{R}$,
(ii) $q: \mathbb{R} \to \mathbb{R}$ *is a continuous function with* $uq(u) \geqslant 0$ *for all* $u \in \mathbb{R}$ *and* $\int_0^y q(u) du \to \infty$
as $|y| \to \infty$.

Then, if $x: \mathbb{R} \to \mathbb{R}$ *is twice differentiable and satisfies*

$$x'' + p(x, x')x' + q(x) = 0,$$

we can find a $K > 0$ *(depending only on* $x(0)$ *and* $x'(0)$*) such that* $|x(t)|, |x'(t)| \leqslant K$ *for all* $t \geqslant 0$.

Proof. We take a hint from the law of conservation of energy. Define

$$V(t) = \frac{x'(t)^2}{2} + \int_0^{x(t)} q(u) du.$$

Differentiating we have

$$\frac{dV}{dt} = x'x'' + x'q(x) = x'(x'' + q(x)) = -p(x, x')x'^2 \leqslant 0.$$

79

Thus $V(t) \leqslant V(0)$ for all $t \geqslant 0$. But $x'(t)^2 \geqslant 0$ and $\int_0^{x(t)} q(u) \, du \geqslant 0$. It follows that, for all $t > 0$,

(a) $\int_0^{x(t)} q(u) \, du \leqslant V(0)$,

(b) $x'(t)^2/2 \leqslant V(0)$.

Since we assumed that $\int_0^y q(u) \, du \to \infty$ as $|y| \to \infty$, it follows from (a) that $|x(t)| \leqslant K_1$ for all $t \geqslant 0$ and some K_1 whilst (b) automatically gives $|x'(t)| \leqslant \sqrt{(2V(0))}$. ∎

Let us review the proof with the aid of diagrams. Any solution of $x'' + p(x, x')x' + q(x) = f(t)$ gives rise to a curve in the (u, v) plane given parametrically by $(u, v) = (x(t), x'(t))$ as t runs from 0 to ∞ (see Figure 20.1).

If we set

$$V_0(u, v) = \frac{v^2}{2} + \int_0^u q(s) \, ds,$$

then the closed curve $V_0(u, v) = K_0$ a constant divides the (u, v) plane into a

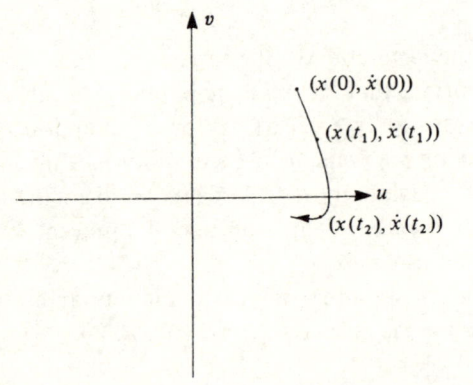

Fig. 20.1. Solution path for a damped oscillator.

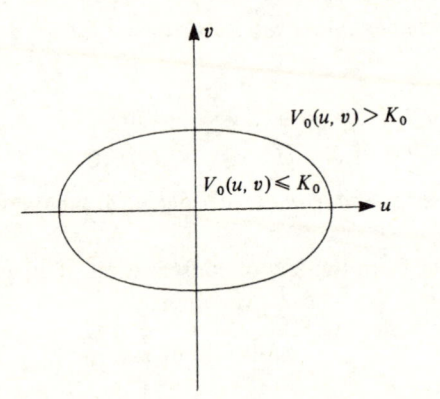

Fig. 20.2. Liapounov region for a damped oscillator.

bounded region $V_0(u, v) \leqslant K_0$ and an unbounded region $V_0(u, v) > K_0$ (see Figure 20.2). Since

$$\frac{dV_0}{dt}(x(t), x'(t)) = \frac{dV}{dt} \leqslant 0,$$

we know that V_0 is decreasing along the path $(u, v) = (x(t), x'(t))$ as t increases. Thus no path $(u, v) = (x(t), x'(t))$ can cross from $\{(u, v): V_0(u, v) \leqslant K_0\}$ to $\{(u, v): V_0(u, v) > K_0\}$ in the direction t increasing and any solution with $(x(0), x'(0)) \in \{(u, v): V_0(u, v) \leqslant K_0\}$ must have

$$(x(t), x'(t)) \in \{(u, v): V_0 \leqslant K_0\} \quad \text{for all } t \geqslant 0$$

(see Figure 20.3).

A special case of the argument above gives the following result.

Lemma 20.2. *Under the conditions of Theorem* 20.1, *if* $x(0) = x'(0) = 0$ *then* $x(t) = 0$ *for all* $t \geqslant 0$.

Proof. With the notation above

$$0 \leqslant (x'(t))^2 \leqslant 2V(0) = (x'(0))^2 + 2\int_0^{x(0)} q(s)\, ds = 0,$$

so that $x'(t) = 0$ for all $t \geqslant 0$ and so $x(t) = x(0) = 0$ for all $t \geqslant 0$. ∎

As a corollary we obtain a uniqueness theorem for the linear damped harmonic oscillator.

Lemma 20.3. *Let* a *and* b *be real numbers with* $a, b > 0$ *and let* $f: \mathbb{R} \to \mathbb{R}$ *be continuous. Then, if* $x: \mathbb{R} \to \mathbb{R}$ *and* $y: \mathbb{R} \to \mathbb{R}$ *are twice differentiable functions with*

$$x'' + ax' + bx = f(t), \quad y'' + ay' + by = f(t),$$

and $x(0) = y(0)$, $x'(0) = y'(0)$, *it follows that* $x(t) = y(t)$ *for all* $t \geqslant 0$.

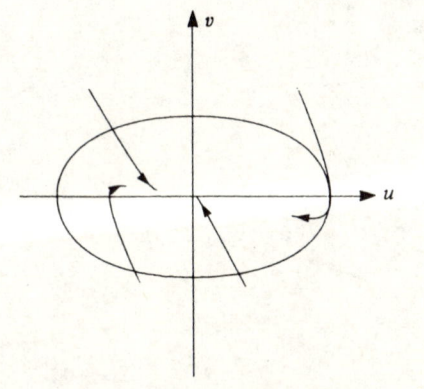

Fig. 20.3. Solution paths enter but cannot leave a Liapounov region.

Proof. Let $z(t) = x(t) - y(t)$. Then

$$z'' + az' + bz = 0, \ z(0) = z'(0) = 0,$$

and so, using Lemma 20.2, $z(t) = 0$ for all $t \geqslant 0$ as required.   ∎

   The reader may notice the absence of any treatment of the existence of solutions. The standard Picard (or contraction mapping) proof found in most textbooks on differential equations shows that all the equations discussed here and in the next few chapters do indeed have a solution.

21

THE DISTURBED DAMPED LINEAR OSCILLATOR DOES NOT EXPLODE

Although it is already helpful to know that solutions of

$$x'' + p(x, x')x' + q(x) = f(t)$$

remain bounded as $t \to \infty$ when $f(t) = 0$ for all $t \geq 0$, we know that in practice it is impossible to keep $f(t)$ accurately zero. What we really want to be sure of is that solutions remain bounded provided that $|f(t)|$ is always small. Better still, we would like to know that if we keep $|f(t)|$ (and $|x(0)|$, $|x'(0)|$) sufficiently small then $|x(t)|$ will not ever become very large.

At first sight it might appear obvious, that a small input must produce a small output. But for a child's swing a carefully timed succession of small pushes can produce a fairly spectacular output. Thus the truth of the conjecture is not evident.

However Levinson has shown that under mechanically plausible conditions on the damping $p(x, x')x'$ and the restoring force $q(x)$ such a result is true. The proof which is given in advanced texts on differential equations is certainly no harder than the hardest proofs in this book, but we shall not give it here. Instead we prove the result for the *linear* oscillator where $p(x, x')x' = ax'$, $q(x) = bx$ by a method which will also apply in the nearly linear case where $p(x, x')$ remains close to a and $q(x)/x$ close to b.

Theorem 21.1. *Let a and b be fixed real numbers with $a, b > 0$. Suppose $x : \mathbb{R} \to \mathbb{R}$ is twice differentiable and satisfies*

$$x'' + ax' + bx = f(t),$$

where f is continuous, $|f(t)| \leq \lambda$ for all $t \geq 0$ and $|x(0)|, |x'(0)| \leq \lambda [\lambda > 0]$. Then we can find an $R(\lambda)$ such that $|x(t)|, |x'(t)| \leq R(\lambda)$ for all $t \geq 0$. Moreover, we can choose $R(\lambda)$ in such a way that $R(\lambda) \to 0$ as $\lambda \to 0$.

The first part of the theorem tells us that bounded input produces bounded output. The last sentence tells us that sufficiently small input (and values of $|x(0)|$, $|x'(0)|$) produces satisfactorily small output. Repeated blowing on a child's swing, however cunningly timed, will not produce much effect.

83

Before beginning the proof proper, we make some observations. Suppose we try to imitate the proof of Theorem 20.1 by defining

$$V_0(u, v) = \frac{v^2}{2} + \int_0^u q(s)\,ds = \frac{v^2}{2} + \frac{bu^2}{2}$$

and

$$V(t) = V_0(x(t), x'(t)) = \frac{bx^2}{2} + \frac{x'^2}{2}.$$

Then, differentiating, we obtain

$$\frac{dV}{dt} = x'(x'' + q(x)) = x'(f(t) - p(x, x')x') = x'(f(t) - ax').$$

Thus contrary to what happened in Theorem 20.1 it is possible to have $dV/dt > 0$ when x' is small. In picturesque language we could say that at certain stages in the process more energy could be fed in by the input than is being removed by the damping.

However, although our initial choice of V_0 was based on physical intuition, there is no reason why we must restrict our choice to such a V_0. Looking more carefully at the proof of Theorem 20.1 we see that the properties of V_0 that we actually used were

(1) $V_0(u, v) \to \infty$ as $u^2 + v^2 \to \infty$.
(2) If $x(t)$ satisfies the appropriate equation then

$$\frac{d}{dt} V_0(x(t), x'(t)) < 0$$

for $x(t)^2 + x'(t)^2$ sufficiently large.

We therefore cast around for a function $U_0(u, v)$ which will have properties (1) and (2) in this case. After a certain amount of trial and error I came up with

$$U_0(u, v) = \frac{v^2}{2} + \frac{bu^2}{2} + \varepsilon uv,$$

i.e. $U_0(u, v) = V_0(u, v) + \varepsilon uv$ where $\varepsilon > 0$ is small compared with quantities like a and ba^{-1}.

As we said in the last chapter the ideas behind the method are due to Liapounov. It is these ideas, rather than the particular and unimportant detail of our proof, on which the reader should concentrate her attention.

Proof of Theorem 21.1. Choose some $\varepsilon > 0$ in such a way that $\varepsilon < \min(1, a/2, b/(200a), a/b)$ and take

$$U_0(u, v) = \frac{v^2}{2} + \frac{bu^2}{2} + \varepsilon uv.$$

We notice that

$$U_0(u, v) = ((v + \varepsilon u)^2/2) + ((b - \varepsilon^2)u^2/2),$$

so the regions $\{(u, v) : U_0(u, v) < k\}$ are the interiors of ellipses.

Writing $U(t) = U_0(x(t), x'(t))$ we see that, if $x(t)$ is the solution of the given differential equation, then

$$\frac{dU}{dt} = \frac{d}{dt}\left(\frac{x'^2}{2} + \frac{bx^2}{2} + \varepsilon xx'\right)$$

$$= x'(x'' + bx) + \varepsilon x'^2 + \varepsilon xx''$$

$$= x'(-ax' + f(t)) + \varepsilon x'^2 + \varepsilon xx''$$

$$= -(a - \varepsilon)x'^2 + x'f(t) + \varepsilon xx''$$

$$\leqslant -(a/2)x'^2 + \lambda|x'| + \varepsilon xx'',$$

since $|f(t)| \leqslant \lambda$ and $\varepsilon < a/2$.

We show, by estimating xx'', that, provided $(ax'(t))^2 + (bx(t))^2 \geqslant (10\lambda\varepsilon^{-2})^2$, we have $dU/dt < 0$. There are two cases to consider according as whether x' is small compared with x or not. (We introduced the extra factor $\varepsilon xx'$ to cope with problems when x' was small, but we must check that we have not thereby created other problems when x' is large.)

Case 1. Suppose $|bx| > 3|ax'|$. We note at once that

$$(bx)^2 > \tfrac{1}{2}((ax')^2 + (bx)^2) > \tfrac{1}{2}(10\lambda\varepsilon^{-2})^2$$

and so $|bx| \geqslant 5 \cdot 2^{1/2}\lambda\varepsilon^{-2} \geqslant 5 \cdot 2^{1/2}\lambda$. Thus examining the formula

$$x'' = -(ax' + bx - f(t))$$

we see that the bx term dominates both the ax' and the $f(t)$ term. In particular x'' and x have opposite signs and $|x''| > |bx|/2$. It follows that

$$|xx''| > bx^2/2 \geqslant 5\lambda\varepsilon^{-2}|x|/2^{\frac{1}{2}} \geqslant (5/2)^{\frac{1}{2}}\lambda\varepsilon^{-2}(3a|x'|/b) \geqslant 2\varepsilon^{-1}\lambda|x'|$$

(since $\varepsilon < a/b$). Thus, since $xx'' < 0$,

$$\frac{dU}{dt} \leqslant \frac{-a}{2}x'^2 + \lambda|x'| + \varepsilon xx'' < -\lambda|x'| \leqslant 0,$$

as required.

Case 2. Suppose $3|ax'| \geqslant |bx|$. Then

$$(ax')^2 \geqslant \tfrac{1}{10}((ax')^2 + (bx)^2) > \tfrac{1}{10}(10\lambda\varepsilon^{-2})^2,$$

and so

$$|ax'| > 3\lambda\varepsilon^{-2} \geqslant 3\lambda.$$

Thus, examining the formula

$$x'' = -(ax' + bx - f(t)),$$

we see that all the right hand terms are of the same order as ax' or smaller. In

particular $|x''| \leqslant 5|ax'|$ and so

$$|xx''| \leqslant 5|ax'||x| \leqslant 15a^2b^{-1}|x'|^2.$$

It follows that

$$\frac{dU}{dt} \leqslant -|x'|((a/2)|x'| - \lambda - 15a^2b^{-1}\varepsilon|x'|)$$

$$\leqslant -|x'|((a/2)|x'| - \lambda - (a/12)|x'|)$$

$$= -|x'|((5/12)|ax'| - \lambda)$$

$$< 0$$

(using the fact that $\varepsilon < b/(200a)$) and so we have completed the proof that $dU/dt < 0$ whenever $(ax')^2 + (bx)^2 \geqslant (10\lambda\varepsilon^{-2})^2$.

The remaining steps are easy. Choose $k(\lambda)$ so that the (interior of the) ellipse $\Sigma = \{(u,v):U_0(u,v) < k(\lambda)\}$ includes both the (interior of the) ellipse $\Sigma' = \{(u,v):(av)^2 + (bu)^2 < (10\lambda\varepsilon^{-2})^2\}$ and the square $S' = \{(u,v):|u|, |v| < \lambda\}$. Finally

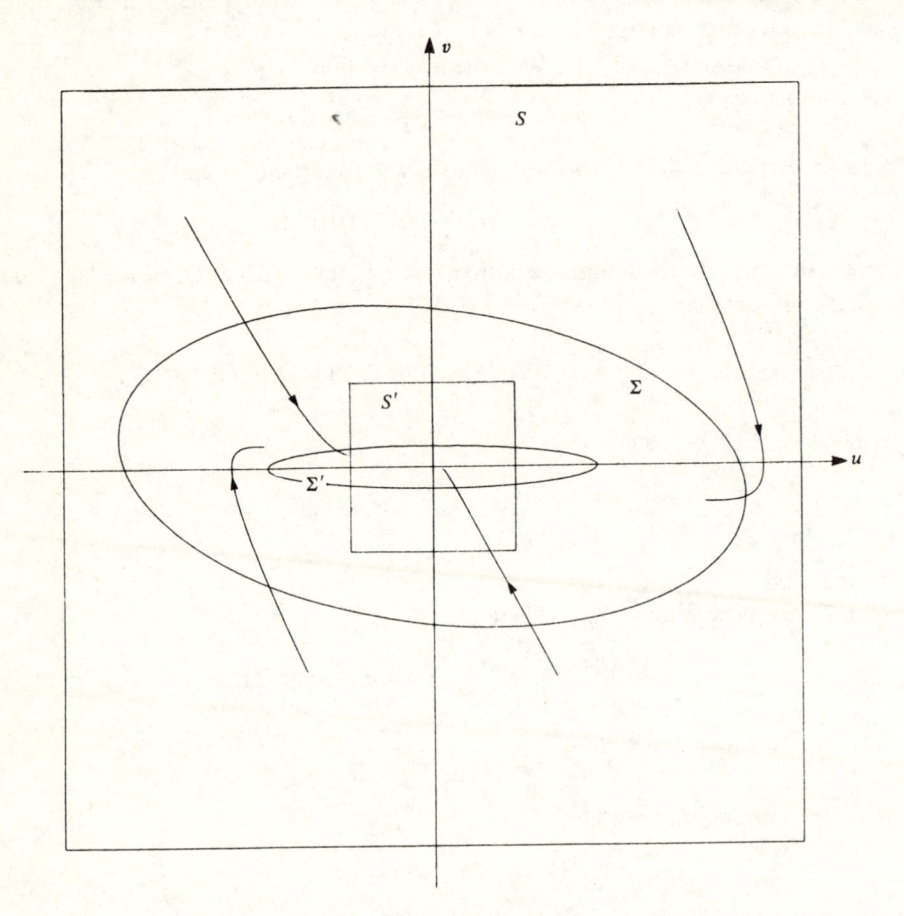

Fig. 21.1. Regions used in the proof of Theorem 21.1.

choose $R(\lambda)$ so that the square $S = \{(u, v):|u|, |v| < R(\lambda)\}$ contains Σ (see Figure 21.1).

Suppose $(u, v) = (x(t), x'(t))$ represents a path crossing the boundary

$$\partial\Sigma = \{(u, v): U_0(u, v) = k(\lambda)\} \text{ of } \Sigma \text{ at } t_0.$$

Since $(x(t_0), x'(t_0))\in\partial\Sigma$ we know that $(x(t_0), x'(t_0))\notin\Sigma'$, i.e. that $(ax'(t_0))^2 + (bx(t_0))^2 \geqslant (10\lambda\varepsilon^{-2})^2$ and so $dU/dt|_{t_0} < 0$. Thus the path must enter Σ. It follows that all paths $(u, v) = (x(t), x'(t))$ which start in Σ (i.e. which have $(x(0), x'(0))\in\Sigma$) must stay in Σ (i.e. have $(x(t), x'(t))\in\Sigma$ for all $t\geqslant 0$). In particular, if $(x(0), x'(0))\in S'\subset\Sigma$, it follows that $(x(t), x'(t))\in\Sigma\subset S$ for all $t\geqslant 0$, i.e. that, if $|x(0)|$, $|x'(0)| < \lambda$, then $|x(t)|, |x'(t)| < R(\lambda)$ which is what we set out to prove.

It is easy to see from the construction that as $\lambda\to 0$ we can allow $k(\lambda)\to 0$ and $R(\lambda)\to 0$. ∎

It should be fairly obvious that we made no use of the linearity of our equations and that the same argument (used with care) will give a rather more general theorem.

Theorem 21.2. *Let a_1, a_2, b_1 and b_2 be fixed real numbers with $a_2 > a_1 > 0$ and $b_2 > b_1 > 0$. Let $p:\mathbb{R}^2\to\mathbb{R}$ and $q:\mathbb{R}\to\mathbb{R}$ be continuous functions with*

(1) $a_2 \geqslant p(u, v) \geqslant a_1$ *for all $u, v\in\mathbb{R}$,*
(2) $b_2 u \geqslant q(u) \geqslant b_1 u$ *for all $u\geqslant 0$, and*
(3) $b_1 u \geqslant q(u) \geqslant b_2 u$ *for all $u\leqslant 0$.*

Suppose $x:\mathbb{R}\to\mathbb{R}$ is twice differentiable and satisfies

$$x'' + p(x, x')x' + q(x) = f(t),$$

where f is continuous, $|f(t)| \leqslant \lambda$ for all $t\geqslant 0$ and $|x(0)|, |x'(0)| \leqslant \lambda$ $[\lambda > 0]$. Then we can find a $R(\lambda)$ such that $|x(t)|, |x'(t)| \leqslant R(\lambda)$ for all $t\geqslant 0$. Moreover, we can choose $R(\lambda)$ in such a way that $R(\lambda)\to 0$ as $\lambda\to 0$.

However since we shall only use Theorem 21.1, we leave the proof to any reader who wants to check her understanding of our earlier proof.

22

TRANSIENTS

Consider the differential equation $x'(t) - x(t) = 0$ with solution $x(t) = x(0)e^t$. If $x(0) > 0$ then, even if $x(0)$ is very small, $x(t) \to \infty$ as $t \to \infty$. If $x(0) < 0$ then $x(t) \to -\infty$. Thus the ultimate behaviour of the solution depends crucially on the initial value $x(0)$. On the other hand, if we consider the diffential equation $x'(t) + x(t) = 0$ then $x(t) = x(0)e^{-t} \to 0$ as $t \to \infty$ whatever the value of $x(0)$ and so the ultimate behaviour of the solution is unaffected by the initial conditions.

Systems in which the effects of the initial conditions become negligible with the passage of time are obviously much easier to deal with. Here is Fourier talking about his work on the heat equation:

> The study of the theory of heat offers sufficient examples of the simple and constant effects of the general laws of nature; and if the order which establishes itself in these phenomena could be grasped by our senses it would produce a similar impression to that caused by a musical sound.
>
> The shapes of bodies are infinitely variable; the initial distribution of heat may be arbitrary and confused; but all these irregularities decrease rapidly and disappear with the passage of time. The progress of the phenomenon becomes more regular and more simple, at last remaining subject to a fixed law which is the same for all cases and no longer shows any measurable sign of the initial state.
>
> All observation confirms these conclusions. The argument from which they are derived clearly separates and expresses: 1) the general conditions, that is to say those which come from the natural properties of heat, 2) the accidental, but permanent, effect of the form or state of the surface, 3) the transient effect of the initial distribution.

In this section we consider the effect of the initial conditions on the damped linear oscillator $x'' + 2kx' + bx = f(t)$ $[k, b > 0]$. (We replace a by $2k$ to simplify the algebra.) We start with the case when $f(t) = 0$.

Lemma 22.1. *If $x : \mathbb{R} \to \mathbb{R}$ is twice differentiable with*
$$x'' + 2kx' + bx = 0 \quad [k, b \geqslant 0],$$

then, writing λ for the positive square root of $|k^2 - b|$, and choosing suitable real constants A, B,

(i) *If $k^2 > b$*
$$x(t) = A\exp(-(k-\lambda)t) + B\exp(-(k+\lambda)t).$$
(ii) *If $k^2 < b$*
$$x(t) = \exp(-kt)(A\cos\lambda t + B\sin\lambda t).$$
(iii) *If $k^2 = b$*
$$x(t) = \exp(-kt)(A + Bt).$$

Proof. In the expectation that the reader has seen this many times before we simply sketch the proof of (i). The proofs of (ii) and (iii) are similar. Let $g(t) = g_{AB}(t) = A\exp(-(k-\lambda)t) + B\exp(-(k+\lambda)t)$. Then

$$g''(t) + 2kg'(t) + bg(t) = A((k-\lambda)^2 - 2k(k-\lambda) + b)\exp(-(k-\lambda)t)$$
$$+ B((k+\lambda)^2 - 2k(k+\lambda) + b)\exp(-(k+\lambda))t = 0 \quad \text{for all} \quad t.$$

Moreover the equations $g(0) = x(0)$, $g'(0) = x'(0)$ become

$$A + B = x(0), \quad -(k-\lambda)A - (k+\lambda)B = x'(0),$$

so, if we put

$$A = (x'(0) + (k+\lambda)x(0))/2\lambda,$$
$$B = (-x'(0) - (k-\lambda)x(0))/2\lambda,$$

we know that

$$g''(t) + 2kg'(t) + bg(t) = 0, \quad g'(0) = x'(0), \quad g(0) = x(0),$$

and, so, by Lemma 20.3, $x(t) = g(t)$ $[t \geqslant 0]$. ∎

Lemma 22.2. *If $x: \mathbb{R} \to \mathbb{R}$ is twice differentiable with*
$$x'' + 2kx' + bx = 0 \quad [k, b > 0],$$
then $x(t) \to 0$ as $t \to \infty$.

Proof. Just look at the solutions in Lemma 22.1. ∎

We now make essential use of linearity to pass from Lemma 22.2 to Theorem 22.3.

Theorem 22.3. *Consider the differential equation*
$$x'' + 2kx' + bx = f(t)$$
with $k, b > 0$ and f continuous. If y_1 and y_2 are any two solutions then $y_1(t) - y_2(t) \to 0$ as $t \to \infty$.

Proof. Observe that writing $z = y_1 - y_2$ we have

$$z'' + 2kz' + bz = (y_1'' + 2ky_1' + by_1) - (y_2'' + 2ky_2' + by_2) = 0,$$

so, by Lemma 22.2, $z(t) \to 0$ as $t \to \infty$, as stated. ∎

As a typical example let us consider a meter with input $f(t)$ and output a meter reading $x(t)$ governed by $x'' + 2kx' + bx = f(t)$. Suppose that after remaining at 0 for a long time the input suddenly rises to 1 at time $t = 0$ and stays there. We have, in effect

$$x'' + 2kx' + bx = 1[t \geqslant 0], \quad x(0) = 0, \quad x'(0) = 0,$$

or, in a more convenient form, $x(t) = b^{-1} + e(t)$ where

$$e'' + 2ke' + be = 0[t \geqslant 0], \quad e(0) = -b^{-1}, e'(0) = 0.$$

It seems reasonable to refer to $z(t) = b^{-1}$ as the 'steady state solution', and to $e(t)$ as the 'transient solution'.

Using Lemma 22.1 we see that x takes one of the two forms shown in Figure 22.1. In both cases $x(t) \to b^{-1}$ as $t \to \infty$. Notice, however, that the transients do not die away instantaneously. If b is fixed and k is small then $x(t) \approx b^{-1}(1 - \exp(-kt)\cos b^{\frac{1}{2}}t)$ and $x(t) \to b^{-1}$ very slowly, whilst if k is large $x(t) \approx b^{-1}(1 - \exp(-bt/2k))$ and again $x(t) \to b^{-1}$ very slowly.

Lemma 22.4. *Consider the solution $x_k(t)$ of the differential equation*

$$x_k'' + 2kx_k' + bx_k = 1[t \geqslant 0], \quad x_k(0) = 0, \quad x_k'(0) = 0,$$

where $b > 0$ is fixed. Then given any $T > 0$ we can find an $\varepsilon(T) > 0$ such that $\sup_{t \geqslant T}|x_k(t) - b^{-1}| \geqslant \varepsilon(T)$ for all $k > 0$.

Proof. (If this is obvious to the reader she can omit the proof.) Since a direct algebraic proof is a bit messy, we use a more indirect approach. We note first that from the discussion in the previous paragraph we can clearly find $K_1 > K_2 > 0$ such that if $k \geqslant K_1$ or $K_2 \geqslant k > 0$ then $\sup_{t \geqslant T}|x_k(t) - b^{-1}| > b^{-1}/2$.

We now argue as follows. Suppose the lemma is false. Then for each $j \geqslant 1$ we can find a $k(j)$ with $|x_{k(j)}(t) - b^{-1}| < b^{-1}/2^{j+1}$ for all $t \geqslant T$. Since $k(j)$ must be in the closed bounded interval $[K_1, K_2]$, there must exists a subsequence $k(j(r))$

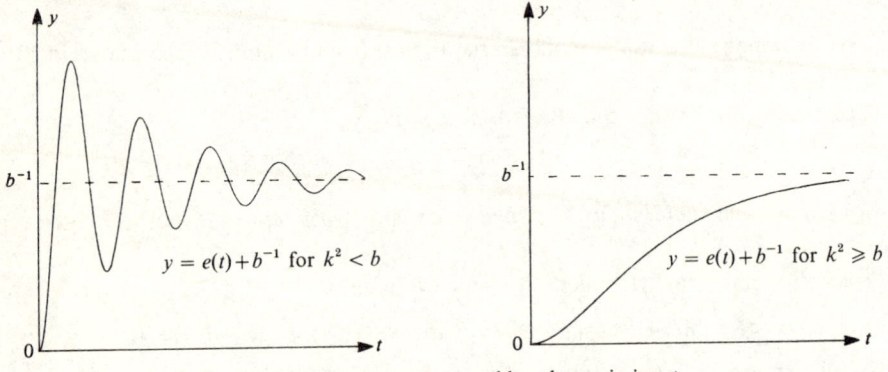

Fig. 22.1. Meter response to a sudden change in input.

converging to a limit $k(0)$ say. It is now easy to check that $x_{k(j(r))}(t) \to x_{k(0)}(t)$ as $r \to \infty$ for each t. Thus $x_{k(0)}(t) = b^{-1}$ for all $t \geq T$ which is impossible by inspection. ∎

Thus no choice of k will give 'ideal' performance. In practice a value of k close to $b^{\frac{1}{2}}$ seems to be most satisfactory.

Although we have used special features of the linear damped oscillator repeatedly in our proofs, the phenomenon of transience is, fortunately, rather common in 'damped' systems. That is to say that there are many systems for which, given two solutions $y_1(t)$ and $y_2(t)$ starting from different initial conditions, we know that $y_1(t) - y_2(t) \to 0$ as $t \to \infty$.

However, such behaviour is not universal even for damped systems as the following, rather unsubtle, counter example shows.

Example 22.5. *Consider the differential equation*

$$x'' + 2x' + q(x) = 0$$

where $\qquad q(x) = x - 1 \quad \text{for } x > 0, \quad q(x) = x + 1 \quad \text{for } x < 0,$

and $q(0) = 0$. Then, whatever the initial conditions, $x(t) \to 1$, or $x(t) \to -1$ or $x(t) \to 0$ as $t \to \infty$. However, if $x(0) = 0$, then

$$x(t) \to 1 \text{ if } x'(0) > 0; \quad x(t) \to -1 \text{ if } x'(0) < 0; \quad \text{and } x(t) \to 0 \text{ if } x'(0) = 0.$$

Proof. (If this is obvious to the reader she can omit the proof.) Consider first the case $x(0) = 0$, $x'(0) = v > 0$. Then for so long as $x(t)$ remains positive we have $x'' + 2x' + x = 1$ and so $x(t) = (A + Bt)e^{-t} + 1$ for some constants A and B. Since $x(0) = 0$ we have $A + 1 = 0$ and since $x'(0) = v$ we have $-A + B = v$ so $A = -1$, $B = v - 1$ and $x(t) = 1 - (1 - (v-1)t)e^{-t}$ for so long as $x(t)$ remains positive. But $1 - (1 - (v-1)t)e^{-t} > 1 - (1 + t)e^{-t} > 1 - 1 = 0$ for all $t > 0$ (since $e^t > 1 + t$ for $t > 0$) so $x(t) = 1 - (1 - (v-1)t)e^{-t}$ for all $t > 0$ and, of course, $x(t) \to 1$ as $t \to \infty$.

In the same way if $x(T) = 0$, $x'(T) > 0$ for some $T \geq 0$ then $x(t) = 1 - (1 - (v-1)$ $(t - T))e^{-(t-T)}$ for all $t \geq T$ and $x(t) \to 1$. On the other hand if $x(T) = 0$, $x'(T) < 0$ for some $T \geq 0$ then by the same reasoning $x(t) \to -1$. If $x(T) = 0$, $x'(T) = 0$ for some $T \geq 0$ then $x(t) = 0$ for all $t \geq T$ and $x(t) \to 0$ as $t \to \infty$. If $x(T) \neq 0$ for all $T \geq 0$ then either $x(t) > 0$ for all $t \geq 0$ so $x(t) = (A' + B't)e^{-t} + 1$ for some A', B' so $x(t) \to 1$, or $x(t) < 0$ for all $t \geq 0$ and $x(t) \to -1$ as $t \to \infty$ by a similar argument. ∎

It must be remembered not only that transient effects need not die away very fast (Lemma 22.4) but that they can also be extremely violent. Consider for example $x'' + 2kx' + bx = 0$ $[b, k > 0, b > k^2]$ with initial conditions $x(0) = 0$, $x'(0) = v$. Then $x(t) = \lambda^{-1} v \exp(-kt) \sin \lambda t$ where $\lambda = \sqrt{(b - k^2)}$. If k is very small compared with b, then although $x(t) \to 0$, $x(t)$ will swing for some time between maxima and minima close to $\pm b^{-\frac{1}{2}} v$.

If the solution of a mathematical model of some system contains transient effects

of sufficient violence, the model may cease to be accurate and the behaviour of the system becomes unpredictable. Thus the theoretical effect of a lightning strike on part of an electrical distribution grid is described as 'transient', but the actual effect may be to plunge New York or Paris into darkness.

In switching on a light bulb we cause a system to move from one steady state to another; it is the transient effects which mean that a bulb is most likely to fail when switched on. The same is true of electronic valves. During the Second World War a British team led by Max Newman obtained some success at breaking the highest German diplomatic codes using mechanical means to perform the large number of trial decodings involved. They proposed to speed up the work by using electronic circuitry instead. The new machine would contain 1500 valves, 10 or 20 times as many as any previous piece of electrical equipment. But even if the probability of any one valve failing when switched on was quite low, the probability that some valve out of 1500 would fail on switching on the machine would be very high.

The problem was solved by keeping the machine switched on all the time and Colossus went into highly successful service at the end of 1943. The precedent was followed with Colossus II (1944, conditional branching and 2500 valves) and the later peace time computers (ENIAC, EDSAC with over 8000 valves). But 8000 valves permanently switched on generate a great deal of heat and an essential component of at least one early computer installation was a graduate student with a bucket of sand.

23

THE LINEAR DAMPED OSCILLATOR
WITH PERIODIC INPUT

In this chapter we study the equation

$$x'' + 2kx' + bx = f(t) \quad [k, b > 0],$$

where $f: \mathbb{R} \to \mathbb{R}$ is a continuous periodic function with period $2\pi\omega^{-1}$ (i.e. with $f(t + 2\pi\omega^{-1}) = f(t)$ for all $t \in \mathbb{R}$). We shall in fact deal with the more realistic equation

$$x'' + 2kx' + bx = f(t) + \eta(t),$$

where all we know about the disturbing input η is that it is small and continuous.

As the reader presumably knows, one of the easiest ways to get started is to prove the following Lemma.

Lemma 23.1. *The equation*

$$x'' + 2kx' + bx = \exp i\omega t$$

has a solution $x_\omega : \mathbb{R} \to \mathbb{C}$ *given by*

$$x_\omega(t) = R_\omega \exp(i(\omega t + \theta_\omega)),$$

where

$$R_\omega = \frac{1}{\sqrt{((b - \omega^2)^2 + 4k^2\omega^2)}}$$

and

$$\cos\theta_\omega = \frac{b - \omega^2}{\sqrt{((b - \omega^2)^2 + 4k^2\omega^2)}}, \quad \sin\theta_\omega = \frac{-2k\omega}{\sqrt{((b - \omega^2)^2 + 4k^2\omega^2)}}$$

(taking positive square roots throughout).
Proof. Physical intuition, luck or good management lead us to try and find a solution of the form $x_\omega(t) = Z_\omega \exp(i\omega t)$. Substitution gives

$$Z_\omega((i\omega)^2 + 2k(i\omega) + b)\exp(i\omega t) = \exp(i\omega t),$$

so that

$$Z_\omega = \frac{1}{-\omega^2 + 2ki\omega + b} = R_\omega \exp i\theta_\omega.$$

93

It is a trivial matter to check that $x_\omega(t) = R_\omega \exp(i(\omega t + \theta_\omega))$ is indeed a solution.

■

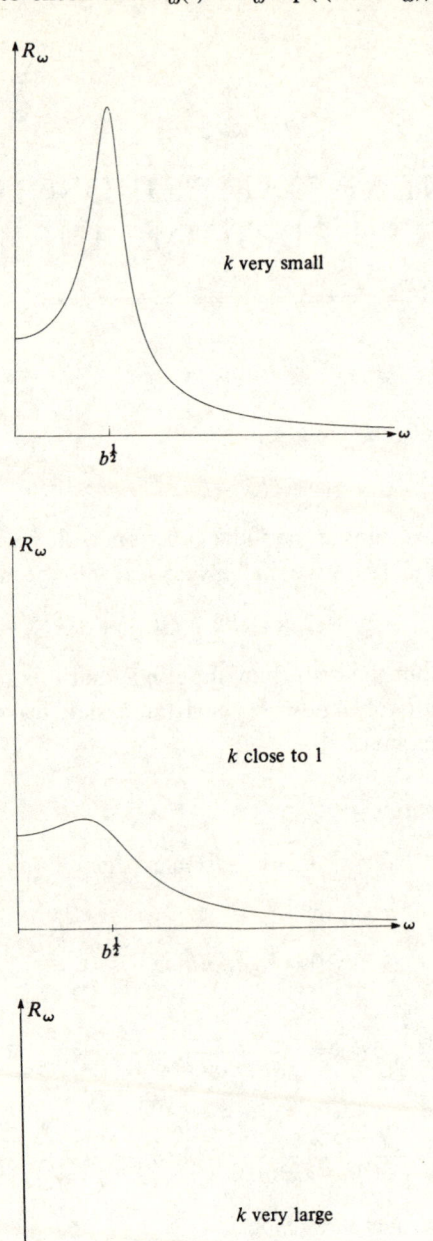

Fig. 23.1. Plot of the amplitude R_ω against frequency ω for various k.

We plot R_ω, θ_ω and Z_ω in Figures 23.1, 23.2 and 23.3. Notice that if the damping is small (i.e. k is much smaller than 1), then R_ω has a sharp *resonance peak* near the *natural frequency* $b^{\frac{1}{2}}$. If we are not too close to $b^{\frac{1}{2}}$, then (if k is small) R_ω behaves like $|b-\omega^2|^{-1}$. However, we remark that, even for low values of k, R_ω flattens out near $b^{\frac{1}{2}}$ and behaves like $(2kb^{\frac{1}{2}})^{-1}$ (i.e. is approximately constant). As k increases, the resonance peak becomes less and less pronounced and eventually vanishes for large k. We also note the existence (for all k and b) of a *phase change* θ_ω which runs from 0 to $-\pi/2$, when $\omega = b^{\frac{1}{2}}$, and then on to $-\pi$ as $\omega \to \infty$ (see Figure 23.2).

We now bring together all the information that we possess concerning the damped linear oscillator to prove the following theorem.

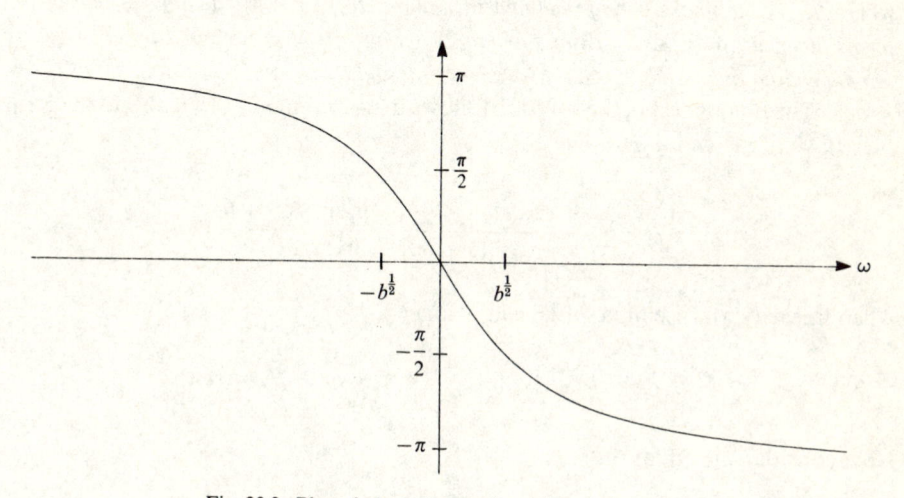

Fig. 23.2. Plot of phase change θ_ω against frequency ω.

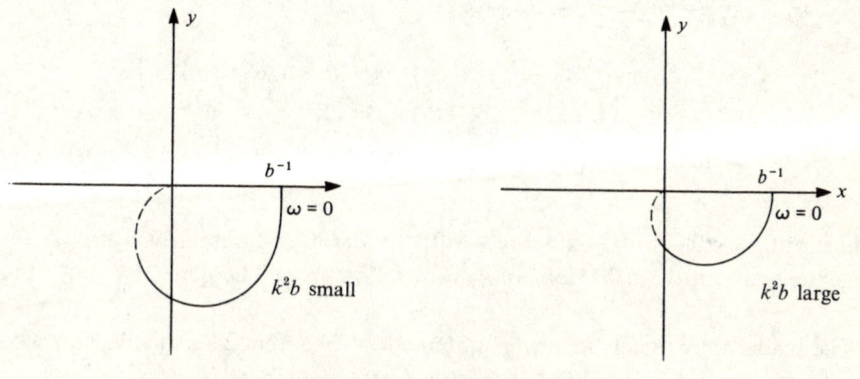

Fig. 23.3. Plot of Z_ω in the complex plane.

Theorem 23.2. *Suppose that*

$$f(t) = \tfrac{1}{2}A_0 + \sum_{r=1}^{n} (A_r \cos r\omega t + B_r \sin r\omega t) + \eta_1(t),$$

where $\eta_1 : \mathbb{R} \to \mathbb{R}$ is a continuous function with $\varepsilon = \sup_{t \in \mathbb{R}} |(\eta_1(t))|$ small. Then any solution of the equations

$$x'' + 2kx' + bx = f(t), \quad x(0) = u, \quad x'(0) = v$$

(where $k, b > 0$) is unique and has the form

$$x(t) = \frac{A_0}{2b} + \sum_{r=1}^{n} \frac{A_r \cos(r\omega t + \theta_{r\omega}) + B_r \sin(r\omega t + \theta_{r\omega})}{\sqrt{((b - r\omega)^2 + 4k^2 r^2 \omega^2)}} + e(t) + \eta_2(t),$$

where $\theta_{r\omega}$ is the phase change defined in Lemma 23.1, e is a transient of one of the forms given in (i), (ii) and (iii) of Lemma 22.1 and $|\eta_2(t)|$ is bounded for all $t \in \mathbb{R}$ by a δ depending on ε, k and b only. Moreover, if k and b are kept fixed $\delta(\varepsilon) \to 0$ as $\varepsilon \to 0$.
Proof. The uniqueness of the solution follows from Lemma 20.3. To obtain the form of $x(t)$ we argue as follows. Set

$$y(t) = \frac{A_0}{2b} + \sum_{r=1}^{n} \frac{A_r \cos(r\omega t + \theta_{r\omega}) + B_r \sin(r\omega t + \theta_{r\omega})}{\sqrt{((b - r\omega)^2 + 4k^2 r^2 \omega^2)}}.$$

Then linearity and Lemma 23.1 yield at once

$$y'' + 2ky' + by = \tfrac{1}{2}A_0 + \sum_{r=1}^{n} (A_r \cos r\omega t + B_r \sin r\omega t).$$

Now consider the equations

$$e'' + 2ke' + be = 0, \quad e(0) = u - y(0), \quad e'(0) = v - y'(0).$$

By Lemma 22.1 these equations have a unique solution, so we may consider $\eta_2 = x - y - e$. By linearity

$$\eta_2'' + 2k\eta_2' + b\eta_2 = (x'' + 2kx' + bx) - (y'' + 2ky' + by) - (e'' + 2ke' + be)$$

$$= \left(\tfrac{1}{2}A_0 + \sum_{r=0}^{n} (A_r \cos r\omega t + B_r \sin r\omega t) + \eta_1(t) \right)$$

$$- \left(\tfrac{1}{2}A_0 + \sum_{r=0}^{n} (A_r \cos r\omega t + B_r \sin r\omega t) - 0 \right) = \eta_1(t),$$

whilst $\eta_2(0) = x(0) - y(0) - e(0) = (u - y(0)) - e(0) = 0$ and, similarly, $\eta_2'(0) = 0$. Our statements about the behaviour of η_2 now follow from Theorem 21.1. ∎

The reader may prefer the neater statement of Theorem 23.2 which arises when we choose $\psi_r \in \mathbb{R}$, $C_r \in \mathbb{R}$ so that $A_r + iB_r = C_r e^{i\psi_r}$ to obtain

Theorem 23.3. *With the above notation, if*

$$f(t) = \tfrac{1}{2}C_0 + \sum_{r=1}^{n} C_r \cos(r\omega t - \psi_r) + \eta_1(t),$$

then

$$x(t) = \frac{C_0}{2b} + \sum_{r=1}^{n} \frac{C_r \cos(r\omega t + \theta_{r\omega} - \psi_r)}{\sqrt{((b - r\omega)^2 + 4k^2 r^2 \omega^2)}} + e(t) + \eta_2(t).$$

We have now shown how to deal with the problem set out in the opening paragraph of the chapter. Suppose we are dealing with a system

$$x'' + 2kx' + bx = f(t) + \eta(t),$$

where f is continuous with period $2\pi\omega^{-1}$ and η is small and continuous (but otherwise unknown). Then Theorem 2.5 (or, more precisely, Lemma 2.6 (iii)) tells us that we can approximate f uniformly to any desired degree of accuracy by a real trigonometric polynomial P (of period $2\pi\omega^{-1}$). There can be no advantage in approximating f to a much greater degree of accuracy than we know η so we choose a P such that

$$\sup_{t \in \mathbb{R}} |P(t) - f(t)| = \sup_{0 \leqslant t \leqslant 2\pi\omega^{-1}} |P(t) - f(t)|$$

is the same order as $\sup_{t \in \mathbb{R}} |\eta(t)|$. Setting $\eta_1 = \eta + (f - P)$ we have

$$x'' + 2kx' + bx = P(t) + \eta_1(t),$$

which is the case discussed in Theorem 23.2.

We notice that, if k is small and ω is close to $b^{\frac{1}{2}}$ (and to a lesser extent if $n\omega$ is close to $b^{\frac{1}{2}}$ for some small integer n), the output $x(t)$ may be much larger than would otherwise occur for a general input of the same magnitude. This phenomenon, as the reader no doubt is aware, is called resonance and may be desirable or undesirable.

Suspension bridges are 'lightly damped' structures and when subject to a periodic disturbance close to their natural frequency may oscillate so violently as to collapse. The Stephensons' first railway bridge across the Tyne was a suspension bridge which collapsed for this reason. The reader may have seen a film of the collapse under wind induced oscillation of the Tacoma Narrows suspension bridge 100 years later. (The film is shown regularly to engineering students but, to judge by the statements of the nuclear power community, any reduction in technological hubris is only temporary.)

In order to reduce the effect of resonance we may be able to increase k (which has resulted in some singularly ugly bridges) or change b so that for the kind of periodic input expected b is much smaller than ω^2. This was the approach adopted by Kelvin in designing his compass so that its natural frequency $b^{\frac{1}{2}}$ was less than the normal rolling frequency of the ship.

On the other hand, when we 'tune' a circuit to receive radio waves, we use

resonance to pick out a particular frequency from a general input. In the same way Helmholtz explained our ability to analyse sound by hypothesising a series of resonators inside the ear each tuned to a certain frequency.

In the treatment of the linear damped oscillator above, I have sought to do two connected things. First, I have tried to show that the Fourier method in this kind of application has nothing whatsoever to do with the convergence or divergence of certain infinite series. (Remember that, in accordance with Perrin's dictum, the input function f will not in general be 'smooth'.) Second, I have tried to show that the damped linear oscillator is a mathematically 'rugged' model in the sense that, for example, small changes in input produce small changes in output. It is thus likely that, even when it does not provide an exact model for a physical process, it will still provide useful information.

The use of mathematically attractive but 'unstable' models is extremely foolhardy. When a meteorologist announces the next ice age within 50 years, he is telling us nothing about the weather but a great deal about the stability of his model.

Postscript

Whilst preparing the proofs I wondered whether to add the Star Wars proposal for an infallible anti-missile defense as a further example of technological hubris. However in this case I suspect that those who promise such a system are not such fools as those who believe them.

24

A NON-LINEAR OSCILLATOR I

It is sometimes said that the great discovery of the nineteenth century was that the equations of nature were linear, and the great discovery of the twentieth century is that they are not. The object of the next three chapters is to show, by studying just one simple non-linear differential equation, some of the new phenomena and problems which can arise. Very little will be proved and, though I have not knowingly led the reader astray, the kind of reasoning employed can easily give false or misleading conclusions.

What sort of equation should we study? In order to make use of the insights obtained in the previous chapters it should be something of the form

$$x'' + p(x, x')x' + q(x) = f(t),$$

with appropriate p and q. Moreover it should be close to the linear form, so we shall take only one of the terms to be non linear. So long as $p(x, x')$ is always positive (so that we have damping and not excitation), it seems plausible to suppose that small non-linear variations in $p(x, x')x'$ will give rise to less interesting behaviour than will small non-linear variations in $q(x)$. Thus we shall study.

$$x'' + 2kx' + q(x) = f(t).$$

Expanding about the origin we may suppose

$$q(x) = a_0 + a_1 x + a_2 x^2 + a_3 x^3 + 0(x^4).$$

But for simplicity we want $xq(x) \geq 0$ so $a_0 = a_2 = 0$ and we have $q(x) = a_1 x + a_3 x^3 + 0(x^4)$. The simplest reasonable form of $q(x)$ to study is thus $q(x) = b(x + \eta x^3)$ with $b > 0$ and η small in absolute value (but not necessarily positive). Finally we must choose a suitable f. Since the case when $f(t) = F$, a constant, is easily dealt with by the methods of Chapter 20, we choose $f(t) = F \cos \omega t$ as a simple but typical input.

Our chain of reasonable choices has lead us to Duffing's equation

$$x'' + 2kx' + b(x + \eta x^3) = F \cos \omega t \quad [k, b > 0]. \tag{1}$$

We shall now seek solutions of this equation when k is small. When $\eta = 0$, Duffing's equation reduces to the equation

$$x'' + 2kx' + bx = F \cos \omega t, \tag{2}$$

which, as we saw in the previous chapter, has a periodic solution $Y(t) = A \cos(\omega t + \theta)$. Moreover, we know that for any solution x, the effect of damping is such that $x(t) - Y(t) \to 0$ as $t \to \infty$. On the strength of the analogy between (1) and (2) we are led to try and find a periodic solution $y(t)$ of (1) with period $2\pi\omega^{-1}$. (However the reader must be told that, although the suggestion is fruitful, the analogy obscures a remarkable phenomenon that we shall discuss in Chapter 26.)

Since y has period $2\pi\omega^{-1}$, it has the formal Fourier expansion

$$y(t) \sim \sum_{-\infty}^{\infty} a_r \exp(ir\omega t),$$

where
$$a_r = \frac{\omega}{2\pi} \int_0^{2\pi\omega^{-1}} y(t) \exp(-ir\omega t)\, dt.$$

But we can say rather more about y and thus about its Fourier expansion. Set

$$z(t) = -y(t + \pi\omega^{-1}).$$

Then
$$z''(t) + 2kz'(t) + b(z(t) + \eta(z(t))^3)$$
$$= -y''(t + \pi\omega^{-1}) - 2ky'(t + \pi\omega^{-1}) - b(y(t + \pi\omega^{-1}) + \eta(y(t + \pi\omega^{-1}))^3)$$
$$= -F\cos(w(t + \pi\omega^{-1})) = F\cos \omega t.$$

Thus y and z are both $2\pi\omega^{-1}$ periodic solutions of the same equation. Hence on the assumption that equation (1) has only one steady state solution (certainly true when $\eta = 0$ and very plausible in general on the grounds that it is maintained by 'external energy'), we have $y = z$ and so $y(t + \pi\omega^{-1}) = -y(t)$. Thus

$$a_r = \frac{\omega}{2\pi} \int_0^{\pi\omega^{-1}} y(t)\exp(-ir\omega t) + y(t + \pi\omega^{-1})\exp(-ir\omega(t + \pi\omega^{-1}))\, dt$$

$$= \frac{\omega}{2\pi} \int_0^{\pi\omega^{-1}} (1 + (-1)^{r+1})y(t)\exp(-ir\omega t)\, dt = 0$$

whenever r is even.

Further, since y is real, $a_r = a^*_{-r}$, and so we have

$$y(t) \sim \sum_{n=0}^{\infty} (c_{2n+1}\cos(2n+1)\omega t + d_{2n+1}\sin(2n+1)\omega t)$$

with c_{2n+1} and d_{2n+1} real. This we rewrite in a more suitable form for our application as

$$y(t) \sim \sum_{n=0}^{\infty} A_{2n+1}\cos((2n+1)\omega t + \theta_{2n+1}),$$

i.e.
$$y(t) \sim A_1 \cos(\omega t + \theta_1) + A_3 \cos(3\omega t + \theta_3) + \cdots,$$

where A_{2n+1} and θ_{2n+1} are real.

At this point we make the further assumptions that to a first approximation

$$y(t) = A_1 \cos(\omega t + \theta_1)$$

(i.e. the solution resembles the case when $\eta = 0$) and to a second approximation

$$y(t) = A_1 \cos(\omega t + \theta_1) + A_3 \cos(3\omega t + \theta_3).$$

If our calculations now give A_3/A_1 quite large, we will know that these assumptions must be false. (Of course, even if our calculations give A_3/A_1 small, it does not follow that we are near a true solution.)

Our next task is thus to seek an approximate solution of the form

$$y(t) = A_1 \cos(\omega t + \theta_1) + A_3 \cos(3\omega t + \theta_3)$$

with $|A_3 A_1^{-1}|$ small, to the equation

$$y'' + 2ky' + b(y + \eta y^3) = F \cos \omega t, \tag{3}$$

with η small. We have

$$y'' = -A_1 \omega^2 \cos(\omega t + \theta_1) - 9A_3 \omega^2 \cos(3\omega t + \theta_3)$$
$$2ky' = -2k\omega A_1 \sin(\omega t + \theta_1) - 6k\omega A_3 \sin(3\omega t + \theta_3),$$
$$by = bA_1 \cos(\omega t + \theta_1) + A_3 b \cos(3\omega t + \theta_3),$$

whilst neglecting terms in (A_3/A_1), $(A_3/A_1)^2$ and $(A_3/A_1)^3$ we have

$$\eta b y^3 = \eta b A_1^3 (\cos(\omega t + \theta_1) + (A_3/A_1)\cos(3\omega t + \theta_3))^3 \approx \eta b A_1^3 (\cos(\omega t + \theta_1))^3$$
$$= \eta b A_1^3 (\tfrac{3}{4}\cos(\omega t + \theta_1) + \tfrac{1}{4}\cos(3\omega t + 3\theta_1)).$$

Thus, neglecting terms as above, equation (3) yields

$$P(t) + Q(t) = 0, \tag{3'}$$

where
$$P(t) = -A_1(\omega^2 \cos(\omega t + \theta_1) + 2k\omega \sin(\omega t + \theta_1)$$
$$- b(1 + 3\eta A_1^2/4)\cos(\omega t + \theta_1)) - F \cos \omega t,$$

and
$$Q(t) = -A_3(9\omega^2 \cos(3\omega t + \theta_3) + 6k\omega \sin(3\omega t + \theta_3)$$
$$- b \cos(3\omega t + \theta_3)) + (A_1^3 b\eta/4)\cos(3\omega t + 3\theta_1).$$

We observe that $P(t) = a_1 \cos \omega t + b_1 \sin \omega t$, and $Q(t) = a_3 \cos 3\omega t + b_3 \sin 3\omega t$ for some $a_1, b_1, a_3, b_3 \in \mathbb{R}$. The condition $P(t) + Q(t) = 0$ thus implies

$$a_1 \cos \omega t + b_1 \sin \omega t + a_3 \cos 3\omega t + b_3 \sin 3\omega t = 0 \quad \text{for all} \quad t,$$

so that $a_1 = b_1 = a_3 = b_3 = 0$. Condition (3)′ thus implies

$$P(t) = 0, \tag{$3)_1$}$$
$$Q(t) = 0. \tag{$3)_2$}$$

But equations $(3)_1$ and $(3)_2$ are precisely those that arise when we use direct substitution to find solutions of the form $x_1(t) = A_1 \cos(\omega t + \theta_1)$, $x_3(t) = A_3 \cos(3\omega t + \theta_3)$ to the equations

$$x_1'' + 2kx_1' + b(1 + 3\eta A_1^2/4)x_1 = F \cos \omega t, \qquad (4)_1$$

$$x_3'' + 2kx_3' + bx_3 = (-A_1^3 b\eta/4) \cos(3\omega t + 3\theta_1). \qquad (4)_2$$

Equations of the type $(4)_1$ and $(4)_2$ were discussed in detail in the previous chapter where we saw that, under the conditions considered,

$$A_1 = \frac{F}{\sqrt{((b(1 + 3\eta A_1^2/4) - \omega^2)^2 + 4k^2\omega^2)}}.$$

Thus A_1^2 is a root of a cubic equation

$$A_1^2((b(1 + 3\eta A_1^2/4) - \omega^2)^2 + 4k^2\omega^2) = F^2,$$

$$\cos \theta_1 = \frac{b(1 + 3\eta A_1^2/4) - \omega^2}{\sqrt{((b(1 + 3\eta A_1^2/4) - \omega^2)^2 + 4k^2\omega^2)}},$$

$$\sin \theta_1 = \frac{-2k\omega}{\sqrt{((b(1 + 3\eta A_1^2/4) - \omega^2)^2 + 4k^2\omega^2)}},$$

$$A_3 = \frac{-A_1^3 b\eta/4}{\sqrt{((b - 9\omega^2)^2 + 36k^2\omega^2)}},$$

and

$$\theta_3 = 3\theta_1 + \phi_3,$$

where

$$\cos \phi_3 = \frac{b - 9\omega^2}{\sqrt{((b - 9\omega^2)^2 + 36k^2\omega^2)}},$$

$$\sin \phi_3 = \frac{-6k\omega}{\sqrt{((b - 9\omega^2)^2 + 36k^2\omega^2)}}.$$

We make the following remarks.

Remark 1. Had we gone ahead boldly and set $k = 0$, not only would the algebra have been simpler but the qualitative picture

$$x(t) = A_1 \cos \omega t + A_3 \cos 3\omega t,$$

with

$$A_1 = \frac{F}{b(1 + 3\eta A_1^2/4) - \omega^2}, \qquad A_3 = \frac{-A_1^3 b\eta/4}{b - 9\omega^2},$$

would have been the same.

Remark 2. We observe that

$$\frac{A_3}{A_1} = \frac{-A_1^2 b\eta/4}{\sqrt{((b - 9\omega^2)^2 + 36k^2\omega^2)}}$$

$$= \frac{-F^2 b\eta}{4((b(1 + 3\eta A_1^2/4) - \omega^2)^2 + 4k^2\omega^2)\sqrt{((b - 9\omega^2)^2 + 36k^2\omega^2)}}.$$

so that the condition $|A_3/A_1|$ small will be satisfied for small η and reasonable F. We will then have a 'solution' $y(t) = A_1 \cos(\omega t + \theta_1) + A_3 \cos(3\omega t + \theta_3)$ which satisfies Duffing's equation (equation (1)) to within an error of order $\eta b A_1^3 (A_3/A_1) = \eta b A_1^2 A_3$. This is evidence that a true solution exists which is close to $A_1 \cos(\omega t + \theta_1) + A_3 \cos(3\omega t + \theta_3)$. (But it is not conclusive evidence – consider the equation $x'' + \eta^2 x = 0$ with η small. The general solution is $x(t) = z(t)$ where $z(t) = A \cos(\eta t + \theta)$. On the other hand setting $z_\eta(t) = \eta^2 t \cos \eta t$, we have $|z_\eta''(t) + \eta^2 z_\eta(t)| \leqslant 2\eta^3$ although the ultimate behaviour of $z_\eta(t)$ for t large is very different from that of $z(t)$.)

25

A NON-LINEAR OSCILLATOR II

In the previous chapter we gave reasons for believing that Duffing's equation

$$x'' + 2kx' + b(x + \eta x^3) = F \cos \omega t$$

had solutions close to

$$x(t) = A \cos(\omega t + \theta),$$

where A^2 was a root of the cubic equation

$$A^2((b(1 + 3\eta A^2/4) - \omega^2)^2 + 4k^2\omega^2) = F^2.$$

In this chapter we investigate some of the consequences of this.

In order to simplify the algebra a bit, we put $b = 1$ (which will obviously not change the character of the solutions) and set $\varepsilon = 3\eta/4$, $\kappa = 2k$. With these modifications, we see that $u = A^2$ satisfies the equation $f(u, v) = F^2$ where $v = \omega^2$ and

$$f(u, v) = u(((1 + \varepsilon u) - v)^2 + \kappa^2 v) = \varepsilon^2 u^3 + 2\varepsilon(1 - v)u^2 + (\kappa^2 v + (1 - v)^2)u.$$

We are interested in the lightly damped case, so we shall assume that κ has the same order as ε. For the moment we will take $\varepsilon > 0$.

Let us graph $f(u, v)$ against u for various fixed values of $v > 0$. Since

$$\frac{\partial f}{\partial u} = 3\varepsilon^2 u^2 + 4\varepsilon(1 - v)u + (\kappa^2 v + (1 - v)^2),$$

we see that for v less than and not too close to 1 we get a set of increasing curves which move to the left as v increases (i.e. $f(u, v') < f(u, v)$ for $v' > v$) as shown in Figure 25.1.

If v is substantially greater than 1 then the equation $\partial f/\partial u = 0$ has two positive roots $v_1(u) < v_2(u)$ say. Since $f(u, v)$ is a cubic in u with positive leading coefficient this means that $f(u, v)$ has a maximum at $v_1(u)$ and a minimum at $v_2(u)$ say. Looking at the expression $f(u, v) = u((1 + \varepsilon u) - v)^2 + \kappa^2 v)$ we can convince ourselves that the maximum $f(u, v_1(u))$ and minimum $f(u, v_2(u))$ increase with u while the points where

104

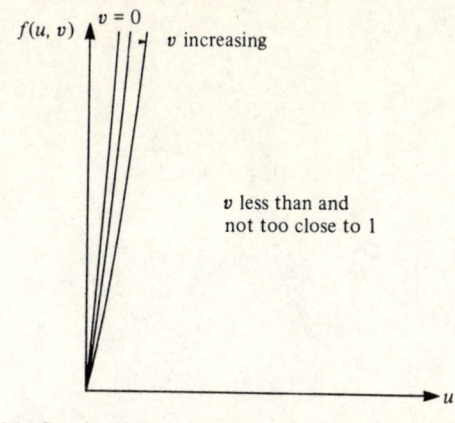

Fig. 25.1. Graphs of $f(\ ,v)$ for v less than and not too close to 1.

they are attained move to the left (i.e. $f(u',v_1(u')) > f(u,v_1(u))$, $f(u',v_2(u')) > f(u,v_2(u))$, $v_1(u') > v_1(u)$, $v_2(u') > v_2(u)$ for $u' > u$).

We are also interested in the relation between $f(u,v')$ and $f(u,v)$ as u varies whilst v' and v are kept fixed with $v' > v$. Since

$$f(u,v') - f(u,v) = 2\varepsilon(1-v')u^2 + (\kappa^2 v' + (1-v')^2)u - 2\varepsilon(1-v)u^2 + (\kappa^2 v + (1-v)^2)u$$
$$= u(v'-v)((\kappa^2 + v' + v - 2) - 2\varepsilon u),$$

we see that (provided $v \geqslant 1 + \kappa^2/2$)

$$f(u,v') > f(u,v) \quad \text{for} \quad 0 < u < (\kappa^2 v' + v - 2)/(2\varepsilon),$$
$$f(u,v') < f(u,v) \quad \text{for} \quad (\kappa^2 + v' + v - 2)/(2\varepsilon) < u.$$

We can now sketch the form of $y = f(u,v)$ for various fixed values of v substantially larger than 1 (Figure 25.2).

What about the form of $f(u,v)$ for v near 1. Investigation of the fine detail would

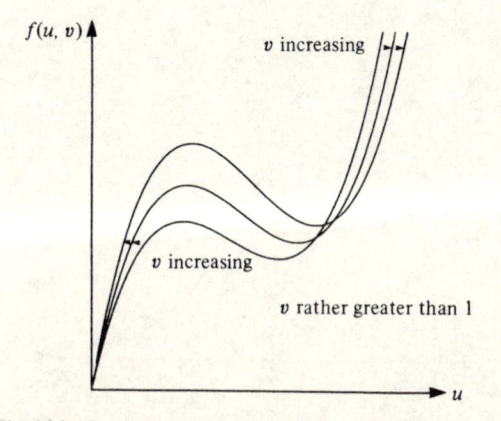

Fig. 25.2. Graphs of $f(\ ,v)$ for v substantially larger than 1.

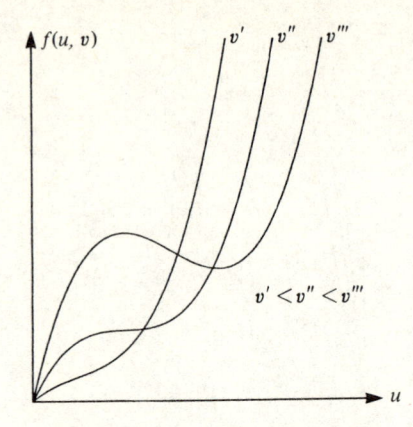

Fig. 25.3. Graphs of $f(\ ,v)$ for v near 1.

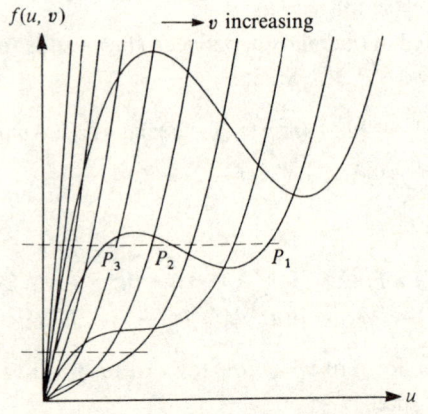

Fig. 25.4. Graphs of $f(\ ,v)$ for a wide range of v.

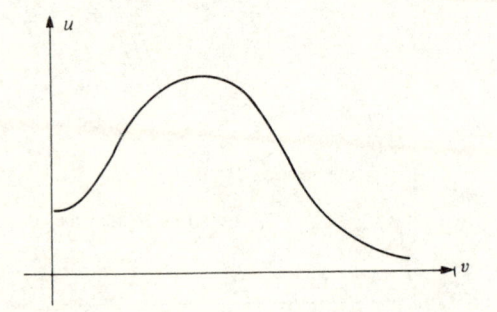

Fig. 25.5. Graph of u against v for $f(u,v) = F^2$ with F larger.

seem to include some rather unpleasant calculation but it is easy to see the rough way in which the behaviour of Figure 25.1 changes to that of Figure 25.2 (see Figure 25.3).

Since we want to know how the behaviour of u(i.e. A^2) depends on v(i.e. on ω^2) when $F^2 = f(u, v)$ is kept fixed, I have superimposed the various cases on one figure (Figure 25.4).

If F^2 is small then, looking at Figure 25.4, we see that the line $y = F^2$ crosses the

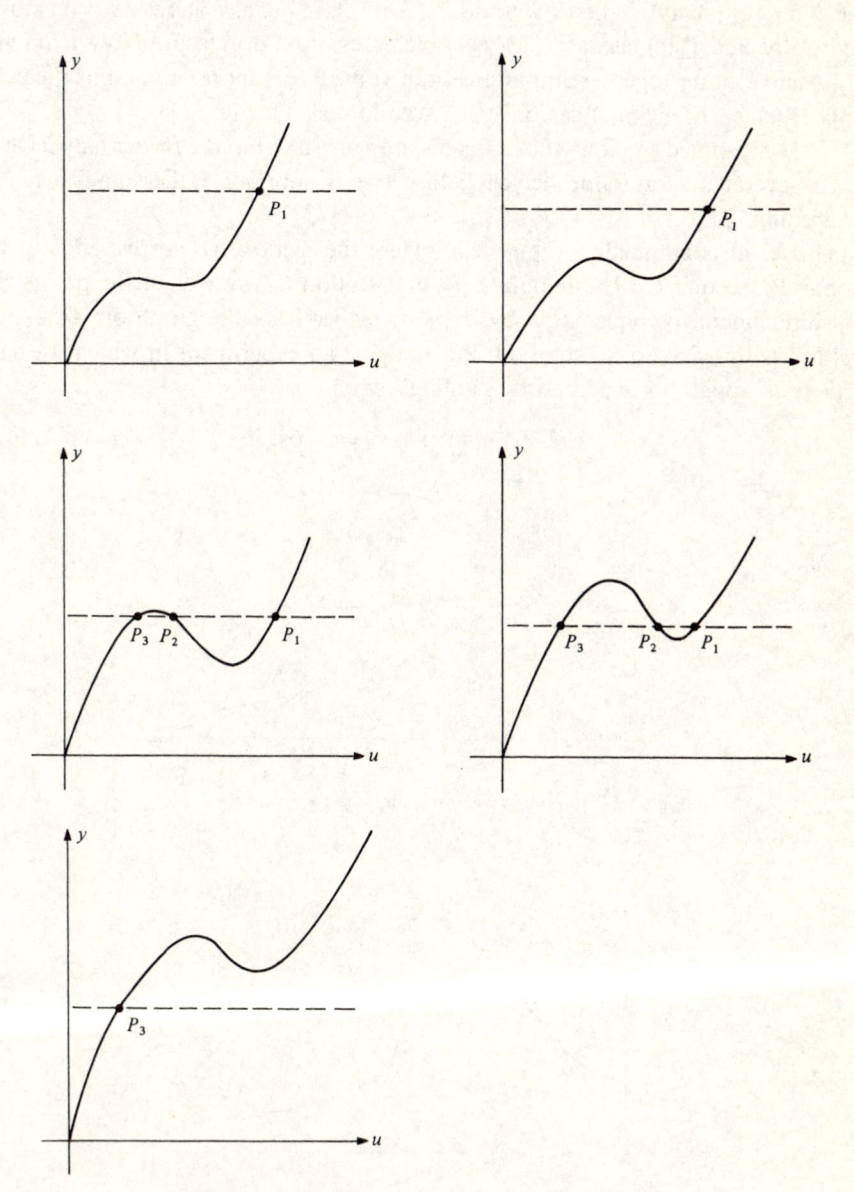

Fig. 25.6. Roots of $f(u, v) = F^2$ for increasing values of v.

curve $y = f(u, v)$ at a unique point $P_1(v) = (u, u(v))$. As v increases P_1 moves first to the right (i.e. $u(v)$ increases) and then back to the left with $u(v)$ decreasing to 0 as $v \to \infty$ (see Figure 25.5).

When F^2 is large things are much more interesting. When v is small the line $y = F^2$ still crosses the curve $y = f(u, v)$ at a unique point $P_1(v) = (u, u_1(v))$ which moves first right and then left but when v reaches a certain value $v = v_1$ there are two new points of intersection $P_2(v) = (u, u_2(v))$ and $P_3(v) = (u, u_3(v))$ (with $P_2(v)$ to the right of $P_3(v)$ (i.e. $u_2(v) > u_3(v)$ say). At first $P_2(v)$ and $P_3(v)$ are close but as v increases $P_2(v)$ moves rightward and $P_3(v)$ leftward (i.e. $u_2(v)$ increases, $u_3(v)$ decreases). Now $P_1(v)$ and $P_2(v)$ move closer together and at a certain value $v = v_2$ merge and vanish leaving only $P_3(v)$ (so $F^2 = f(u, v)$ has only one solution $u_3(v)$ for large v).

I have attempted to show what happens in Figure 25.6 but the reader may have to draw several diagrams for herself before she is sure she understands what is happening.

The result is graphed in Figure 25.7. Here the section AB corresponds to the points P_1, section CB to the points P_2, and section C onwards to the points P_3.

More concretely, replacing u by A^2, v by ω^2 we have the graph of $|A|$ against $|\omega|$ in Figure 25.8. So far so good. But imagine an experiment in which we have a piece of apparatus governed by Duffing's equation

$$x'' + 2kx' + b(x + \eta x^3) = F \cos \omega t,$$

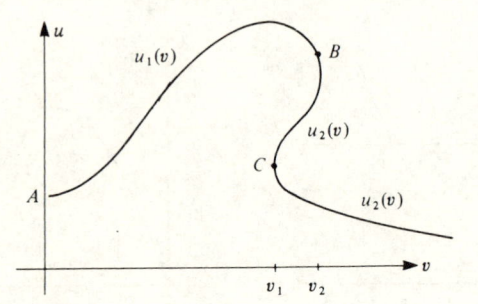

Fig. 25.7. Graph of u against v for $f(u, v) = F^2$ with F large.

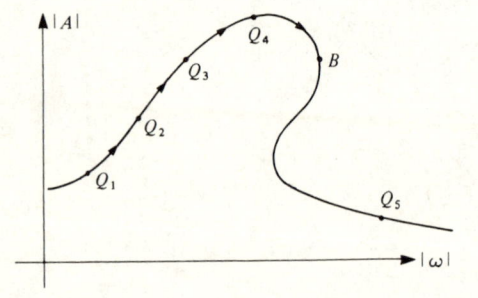

Fig. 25.8. 'Possible' output amplitudes $|A|$ against input frequency $|\omega|$.

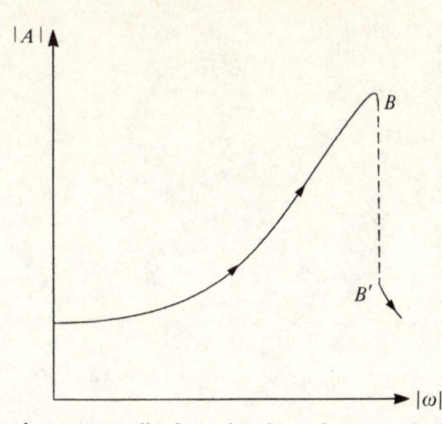

Fig. 25.9. Actual output amplitude against input frequency for $|\omega|$ increasing

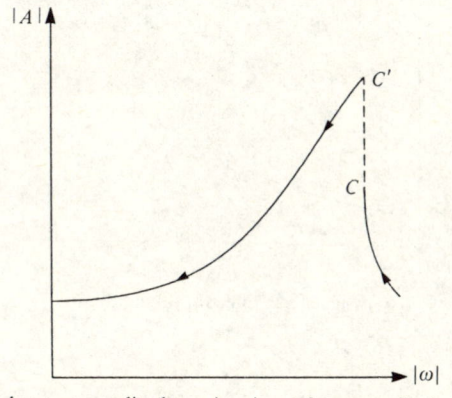

Fig. 25.10. Actual output amplitude against input frequency for $|\omega|$ decreasing.

in which we slowly change the frequency ω of the input. For small ω there is no problem, we are at some point like Q_1 on the graph. Then, as ω slowly increases we move steadily up the curve from Q_1 to Q_2, from Q_2 to Q_3, from Q_3 to Q_4 and then suddenly at the point marked B we run out of road – there is no continuous path to take as ω increases. Moreover, inspection shows that there is no continuous way in which $|A|$ can respond to a continuous increase in ω and no way in which we can travel continuously from Q to Q_5 with ω steadily increasing.

Now Figure 25.8 is the result of a long chain of heuristic reasoning, so we have no guarantee whatsoever of its correctness. The next step is to try the actual experiment (which is what the early investigators did) or at least a numerical simulation of it on a computer (which is what we would do now). What happens is sketched in Figure 25.9. The amplitude $|A|$ goes smoothly along the upper curve until we come to the point B, then jumps down to the lower curve at B' and then continues smoothly.

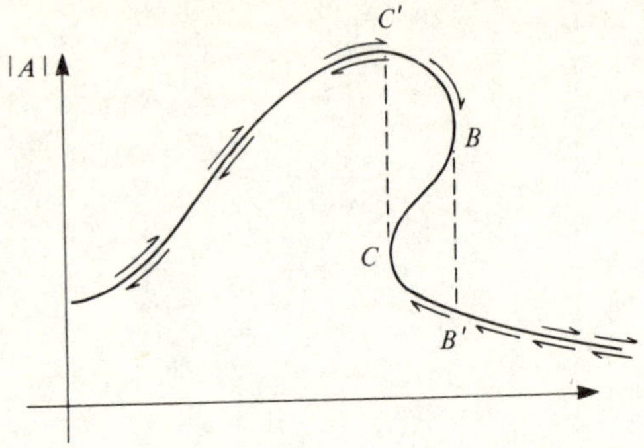

Some differential equations

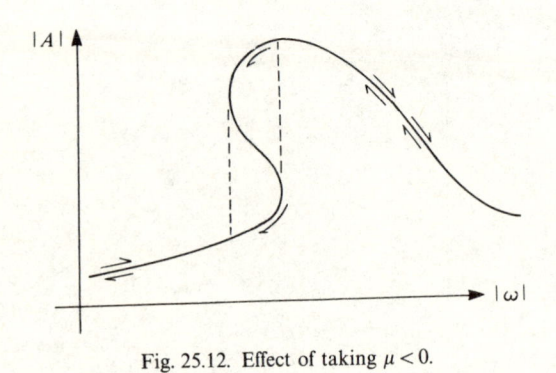

Fig. 25.11. Combination of the previous two figures.

Fig. 25.12. Effect of taking $\mu < 0$.

If we decrease ω from a large value towards 0 then, as we might now expect, the reverse occurs (with the jump from C to C') (see Figure 25.10). We combine Figures 25.9 and 25.10 in Figure 25.11. We note also that if we try to establish an oscillation corresponding to a point on the BC portion of the curve then, experimentally, the system will quickly jump to the corresponding $C'B$ or CB' point.

In some sense solutions corresponding to the upper part AB of the curve and the lower part C onwards are stable. If $|\omega|$ and F are kept steady and the system is slightly perturbed from such a state it will return. Solutions corresponding to the BC part are unstable, and if the system is close to but not exactly in such a state, it will move away from it.

The precise formulation and proof of the statements made above, although no harder than other tasks we have set ourselves in this book, would take us too far afield. We, therefore, confine ourselves to a few remarks. The first remark is simply that if η in Duffing's equation is negative then all that happens is that the curves 'bend the other way' (Figure 25.12).

The second remark is that, instead of keeping F fixed and varying ω we could vary both F and ω (slowly). We try to sketch the results in Figure 25.13. The

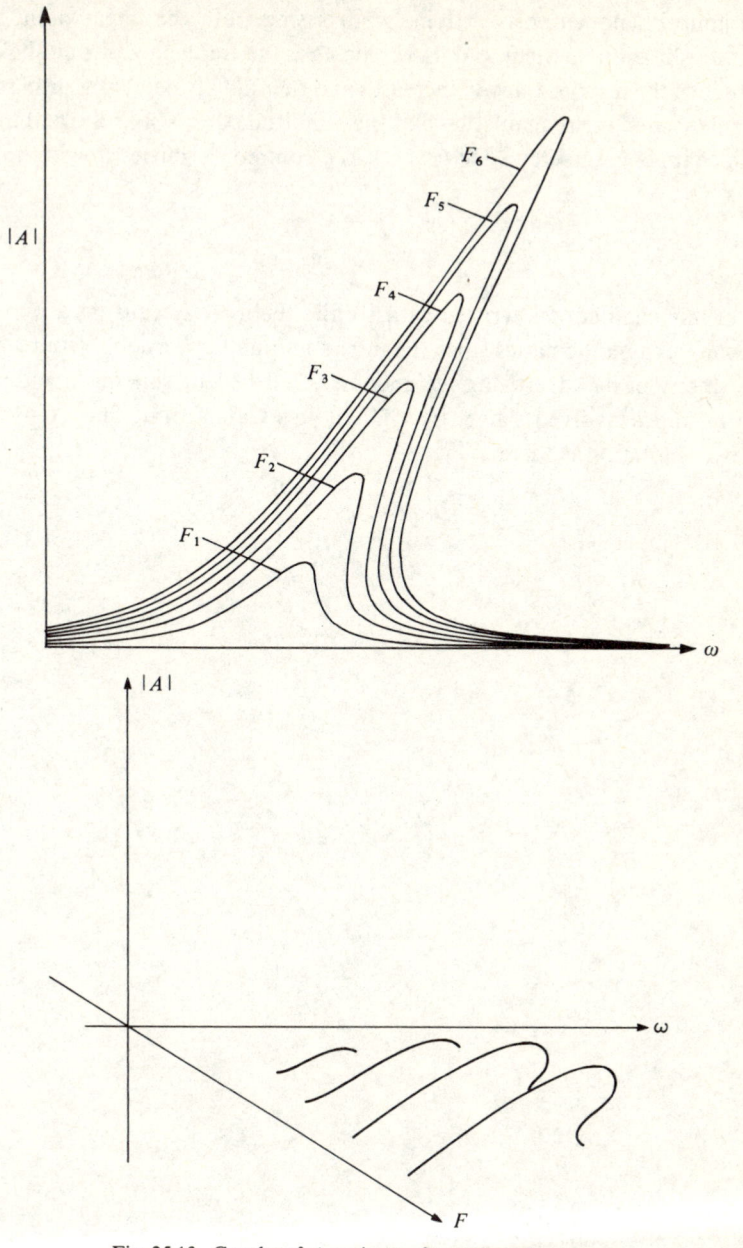

Fig. 25.13. Graphs of A against ω for various fixed F_1.

surface that emerges is the same as that which describes the bending of a long thin strut and several other apparently disparate physical and engineering phenomena. It can be shown that if the number of variables involved is small, only a finite number of 'typical' surfaces can exist and this (difficult) theorem is the basis of 'catastrophe theory'.

The jump phenomenon is extremely surprising from the linear point of view. We gently shake the system and as we increase the frequency of our shaking the amplitude of the response slowly increases and then slowly decreases until suddenly 'the whole system turns against us' and the amplitude drops sharply (and the phase lag also changes completely). However, it is a common occurrence with non-linear systems.

Postscript

When this chapter was written catastrophe theory was causing a popular stir unknown '... in mathematics since the introduction of cybernetics from which [*it*] derived many of its advertising techniques.' An excellent general introduction to the mathematics involved is given by V.I. Arnold's 'Catastrophe Theory' from which the above quotation is taken.

26

A NON-LINEAR
OSCILLATOR III

In the last two chapters we looked for the discussed solutions of Duffing's equation

$$x'' + 2kx' + b(x + \eta x^3) = F \cos \omega t,$$

which were close to a function of the form $A \cos(\omega t + \theta)$. The first sign that in so doing we may have overlooked possible solutions occurs when we consider the undamped $[k = 0]$ case.

Lemma 26.1. *Consider the equation*

$$x'' + b(x + \eta x^3) = F \cos \omega t \quad [b > 0],$$

with b, ω and η fixed, and $\eta > 0$. Provided that $(\omega/3)^2 > b$, there is a value F_0 of F depending on b, ω and η such that the equation

$$x'' + b(x + \eta x^3) = F_0 \cos \omega t$$

has a solution $x(t) = A_0 \cos \omega t/3$ with $A_0 \neq 0$.
Proof. Let $x(t) = A \cos \omega t/3$. Then

$$x'' + b(x + \eta x^3)$$

$$= A(-(\omega/3)^2 \cos(\omega t/3) + b \cos(\omega t/3) + b\eta A^2((\tfrac{3}{4})\cos(\omega t/3) + (\tfrac{1}{4})\cos \omega t))$$

$$= A((b - (\omega/3)^2) + (3b\eta/4)A^2)\cos \omega t/3 + (b\eta A^3/4)\cos \omega t = F \cos \omega t,$$

provided that
$$(b - (\omega/3)^2) + (3b\eta/4)A^2 = 0, \tag{1}$$

and
$$b\eta A^3/4 = F. \tag{2}$$

If $(\omega/3)^2 > b$, then equation (1) has a real non zero solution so we are done. ∎
Remark. Notice that under the conditions above $A \cos((\omega t + 2\pi)/3)$ and $A \cos((\omega t + 4\pi)/3)$ are also solutions.

Is Lemma 26.1 a trick of the arithmetic giving an effect which vanishes in the presence of damping (compare the solution $x(t) = (-t/2)\sin \omega t$ of the equation

113

$x'' + \omega^2 x = \cos \omega t$) or does a similar phenomenon occur when there is light damping?

(Notice that we cannot have an exact solution $x(t) = A \cos (\omega t/3 + \theta)$ (with $A \neq 0$) to the equation $x'' + 2kx' + b(x + \eta x^3) = F \cos \omega t$ when $k > 0$ [and $b, \eta > 0$], for then, matching terms of frequency $\omega/3$, we obtain

$$A(-(\omega/3)^2 \cos (\omega t/3 + \theta) - 2k(\omega/3) \sin (\omega t/3 + \theta) + (b + \eta A^2 b/4) \cos (\omega t/3 + \theta)) = 0,$$

and, if this equation held with $A \neq 0$, we would have a solution $z(t) = A \cos (\omega t/3 + \theta)$ of the undisturbed linear damped oscillator $z'' + 2kz' + (b + \eta A^2 b/4)z = 0$ with $z(t) \nrightarrow 0$ as $t \to \infty$, contrary to Lemma 22.2.)

If the reader tries the calculations, she will find, I think, that it is rather hard to reach a convincing conclusion on this point, even by the relaxed standards of argument adopted in the last two chapters. Once again the simplest approach is to try the actual numerical or practical experiment. It turns out that a solution close to $A_{1/3} \cos (\omega t/3 + \theta_{1/3}) + A_1 \cos (\omega t + \theta_1)$ exists if $\omega > \omega_0$ for some critical frequency ω_0 (compare the condition $(\omega/3)^2 > b$ in Lemma 26.1) and if k, b and F take a certain (surprisingly wide) range of values.

The solution is stable and the question, whether the ultimate state of the system is given by a solution of this type or one of the type discussed earlier, depends on the initial state of the system. Thus the effects of the initial conditions are not transient and we are in a situation closer to that described in Example 22.5 than to that associated with a linear damped oscillator. (An even closer analogy is with the system $x'' + 2x' + q(x) = \varepsilon \cos \omega t$ where ε is small but not zero and q is the function of Example 22.5.)

From the linear point of view the existence of a stable solution containing a *subharmonic* $A_{\frac{1}{3}} \cos (\omega t/3 + \theta_{\frac{1}{3}})$ is just as unexpected as the jump phenomenon discussed in the last chapter. In some way the system is using energy supplied by an input with period $2\pi\omega^{-1}$ to produce an output with period $6\pi\omega^{-1}$. (Notice how fishy this makes the argument in Chapter 24 by which we 'established' that

$$y(t) \sim \sum_{n=0}^{\infty} A_{2n+1} \cos ((2n + 1)\omega t + \theta_{2n+1}).)$$

Of course, the fact that subharmonics are unexpected does not always mean that they are unwelcome. In a quartz watch the designer is faced with the problem of converting the high frequency electrical oscillation produced by the quartz crystal to usable low frequency oscillations which are in some fixed ratio to the initial frequency. Circuits which produce subharmonics provide a means for doing this.

I hope that the last three chapters have shown that passing from the linear to the non-linear case raises an entirely new set of problems and difficulties. (A paper by Littlewood based on joint work with Cartwright on one type of non-linear differential equation runs to over 100 pages in *Acta Mathematica* (Vol. 98) and is

preceded by a 40 page paper of explanation.) Although a combination of semi-rigorous argument, general topological considerations and a great deal of numerical experimentation on a computer will eventually give a more or less reliable picture of the behaviour of any particular system, very few general insights have emerged and much remains to be done.

27

POISSON SUMMATION

In most of this book we treat the circle $\mathbb{T}$ and the Fourier series on it in isolation. But the circle can also be identified with the boundary $\partial D = \{e^{i\theta} : \theta \in \mathbb{T}\} = \{z \in \mathbb{C} : |z| = 1\}$ of the disc $D = \{z \in \mathbb{C} : |z| < 1\}$. In the next five chapters we shall look at Fourier series in this context.

Our basic result is the following which we shall interpret in several different ways.

Theorem 27.1. *Suppose $f : \partial D \to \mathbb{C}$ is continuous. Write $g(\theta) = f(\exp i\theta)$ and $a_n = \hat{g}(n)$. Then*

(i) $\sum_{n=-\infty}^{\infty} a_n r^{|n|} \exp in\theta$ *converges uniformly in θ for each $0 \leqslant r < 1$,*

(ii) $\sum_{n=-\infty}^{\infty} a_n r^{|n|} \exp in\theta \to f(\exp i\theta)$ *uniformly in θ as $r \to 1-$.*

That is to say $f(\exp i\theta)$ is the uniform limit as $r \to 1-$ of a function $\sum_{n=-\infty}^{\infty} a_n r^{|n|} \exp in\theta$ defined on D.

To prove Theorem 27.1 we need only prove the slightly simpler result.

Theorem 27.1′. *If $g : \mathbb{T} \to \mathbb{C}$ is continuous then*

(i) $\sum_{n=-\infty}^{\infty} \hat{g}(n) r^{|n|} \exp int$ *converges uniformly in t for each $0 \leqslant r < 1$,*

(ii) $\sum_{n=-\infty}^{\infty} \hat{g}(n) r^{|n|} \exp int \to g(t)$ *uniformly in t as $r \to 1-$.*

Proof of Theorem 27.1 from Theorem 27.1′. Just observe that if f and g are given as in the statement of Theorem 27.1 then g satisfies the conditions of Theorem 27.1′. ∎

Preliminary remarks on the proof of Theorem 27.1′. Since g is continuous on $\mathbb{T}$, g is bounded with $|g(t)| \leqslant M$ say. Thus

$$|\hat{g}(n)| \leqslant \frac{1}{2\pi} \int_{\mathbb{T}} |g(t) \exp(-int)| \, dt \leqslant M,$$

and so

$$|\hat{g}(n) r^{|n|} \exp int| \leqslant M r^{|n|}.$$

116

It is thus obvious by comparison that $\sum_{n=-\infty}^{\infty} \hat{g}(n)r^{|n|}\exp int$ converges uniformly to $P_r(g,t)$, say. We would now like to show that $P_r(g,t) \to g(t)$ as $r \to 1-$ in much the same way as we showed that

$$\sigma_N(g,t) = \sum_{n=-N}^{N} \hat{g}(n)((N+1-|n|)/(N+1))\exp int \to g(t)$$

in Theorem 2.3. Our first step is to write

$$P_r(g,t) = \sum_{n=-\infty}^{\infty} \frac{1}{2\pi}\int_{\mathbb{T}} g(x)\exp(-inx)\,dx\, r^{|n|}\exp int$$

$$= \frac{1}{2\pi}\int_{\mathbb{T}}\sum_{n=-\infty}^{\infty} g(x)r^{|n|}\exp(in(x-t))\,dx,$$

but this is term by term integration of an infinite sum (rather than the finite sum of Chapter 2) and requires justification.

First part of the proof of Theorem 27.1'. Since g is continuous on $\mathbb{T}$, g is bounded with $|g(x)| \leq M$ say for all $x \in \mathbb{T}$. Thus if $N(2) \geq N(1) \geq 0$, $P(2) \geq P(1) \geq 0$,

$$\left| \sum_{n=-P(2)}^{N(2)} g(x)r^{|n|}\exp(-in(x-t)) - \sum_{n=-P(1)}^{N(1)} g(x)r^{|n|}\exp(-in(x-t)) \right|$$

$$= \left| \sum_{n=-P(2)}^{-P(1)-1} g(x)r^{|n|}\exp(-in(x-t)) + \sum_{n=N(1)+1}^{N(2)} g(x)r^{|n|}\exp(-in(x-t)) \right|$$

$$\leq \sum_{n=-P(2)}^{-P(1)-1} |g(x)r^{|n|}\exp(-in(x-t))| + \sum_{n=N(1)+1}^{N(2)} |g(x)r^{|n|}\exp(-in(x-t))|$$

$$\leq \sum_{n=-P(2)}^{-P(1)-1} Mr^{|n|} + \sum_{n=N(1)+1}^{N(2)} Mr^{|n|}$$

$$\leq Mr(1-r)^{-1}(r^{P(1)} + r^{N(1)}) \to 0, \quad \text{as } N(1), P(1) \to \infty.$$

Thus by the general principle of uniform convergence $\sum_{n=-\infty}^{\infty} g(x)r^{|n|}\exp(-in(x-t))$ converges uniformly in x.

Thus we may integrate term by term with respect to x to obtain

$$\sum_{n=-\infty}^{\infty} \hat{g}(n)r^{|n|}\exp int$$

$$= \sum_{n=-\infty}^{\infty} \left(\frac{1}{2\pi}\int_{\mathbb{T}} g(x)\exp(-inx)\,dx\right)r^{|n|}\exp int$$

$$= \sum_{n=-\infty}^{\infty} \frac{1}{2\pi}\int_{\mathbb{T}} g(x)r^{|n|}\exp(in(t-x))\,dx$$

$$= \frac{1}{2\pi}\int_{\mathbb{T}}\sum_{n=-\infty}^{\infty} g(x)r^{|n|}\exp(in(t-x))\,dx = \frac{1}{2\pi}\int_{\mathbb{T}} g(x)\sum_{n=-\infty}^{\infty} r^{|n|}\exp(in(t-x))\,dx$$

$$= \frac{1}{2\pi}\int_{\mathbb{T}} g(x)P_r(t-x)\,dx = \frac{1}{2\pi}\int_{\mathbb{T}} g(t-y)P_r(y)\,dy$$

(making the substitution $y = t - x$) where $P_r(\theta) = \sum_{n=-\infty}^{\infty} r^{|n|} \exp in\theta$, and the convergence is uniform in t.

Thus, just as in Chapter 2, we are lead to interrupt the flow of the proof and examine the structure of P_r in some detail.

Lemma 27.2.

$$P_r(\theta) = \frac{1 - r^2}{1 - 2r\cos\theta + r^2} \quad [0 \leqslant r < 1].$$

Proof. $P_r(\theta) = \sum_{n=1}^{\infty} r^n \exp(-in\theta) + 1 + \sum_{n=1}^{\infty} r^n \exp(in\theta)$

$$= \frac{r \exp(-i\theta)}{1 - r\exp(-i\theta)} + 1 + \frac{r \exp i\theta}{1 - r\exp i\theta}$$

$$= \frac{(r\exp(-i\theta) - r^2) + (1 - r\exp(-i\theta) - r\exp i\theta + r^2) + (r\exp i\theta - r^2)}{1 - r\exp(-i\theta) - r\exp i\theta + r^2}$$

$$= \frac{1 - r^2}{1 - 2r\cos\theta + r^2}. \qquad \blacksquare$$

The function P_r has three vital properties.

Lemma 27.3. (i) $P_r(\theta) \geqslant 0$ *for all* $\theta \in \mathbb{T}$, $0 \leqslant r < 1$,

(ii) $P_r(\theta) \to 0$ *as* $r \to 1 -$ *uniformly outside* $[-\delta, \delta]$,
(iii) $(2\pi)^{-1} \int_{\mathbb{T}} P_r(\theta)\, d\theta = 1$ $[0 \leqslant r < 1]$.
Proof. (i) This is obvious from Lemma 27.2.
(ii) $1 - 2r\cos\theta + r^2 = (1 - r\cos\theta)^2 + r^2\sin^2\theta \geqslant (1 - r\cos\theta)^2 \geqslant (1 - |\cos\theta|)^2$ for all $0 \leqslant r < 1$ and thus, using Lemma 27.2,

$$0 \leqslant P_1(\theta) \leqslant \frac{1 - r^2}{(1 - |\cos\theta|)^2} \leqslant \frac{1 - r^2}{(1 - |\cos\delta|)^2} \to 0 \quad \text{as } r \to 1 -$$

uniformly outside $[-\delta, \delta]$.
(iii) Setting $g = 1$ in the formula obtained in the first part of the proof of Theorem 27.1 above we get

$$1 = \sum_{n=-\infty}^{\infty} \hat{g}(n) r^{|n|} \exp int$$

$$= \frac{1}{2\pi} \int_{\mathbb{T}} g(t - y) P_r(y)\, dy$$

$$= \frac{1}{2\pi} \int_{\mathbb{T}} P_r(y) \, dy,$$

as stated. ∎

Conclusion of proof of Theorem 27.1′. This follows exactly the pattern for Fejér's theorem (Theorem 2.3) and we leave it to the reader. ∎

Remark. We have shown that

$$\lim_{r \to 1-} \lim_{N \to \infty} \sum_{n=-N}^{N} \hat{g}(n) r^{|n|} \exp int$$

exists with value $g(t)$. However, the example of Theorem 18.1 shows that we cannot conclude that

$$\lim_{N \to \infty} \lim_{r \to 1-} \sum_{n=-N}^{N} \hat{g}(n) r^{|n|} \exp int \text{ exists}$$

(let alone that it has value $g(t)$), since

$$\lim_{r \to 1-} \sum_{n=-N}^{N} \hat{g}(n) r^{|n|} \exp int = S_N(g,t),$$

and the example given in Theorem 18.1 has $\lim_{N \to \infty} \sup |S_N(g,0)| = \infty$. An understandable failure to appreciate this subtle point led Poisson to believe that he had proved the convergence of all Fourier series.

The following lemma is based on Theorem 27.1 and adds very little to it but will be useful in the next chapter.

Lemma 27.4. *Suppose $f : \partial D \to \mathbb{C}$ is continuous. Write $g(\theta) = f(\exp i\theta)$ and $a_n = \hat{g}(n)$. Then* •

(i) $$f_1(z) = a_0/2 + \sum_{n=1}^{\infty} a_n z^n \quad \text{and} \quad f_2(z) = a_0/2 + \sum_{n=1}^{\infty} a_{-n} z^n$$
are well defined analytic functions in $D = \{z : |z| < 1\}$.
(ii) *If we take $F(z) = f_1(z) + f_2(z^*)$ for $z \in D$ and $F(z) = f(z)$ for $z \in \partial D$, then F is a well defined continuous function on the closed disc $\bar{D} = \{z : |z| \leqslant 1\}$.*
(iii) *If f is real valued so is F.*

Proof. (i) We either use the fact that $\sum_{n=-\infty}^{\infty} a_n r^{|n|} \exp in\theta$ converges or the fact that $|a_n| = |\hat{g}(n)| \leqslant M = \sup_{t \in \mathbb{T}} |g(t)|$ to show that $a_0/2 + \sum_{n=1}^{\infty} a_n z^n$ and $a_0/2 + \sum_{n=1}^{\infty} a_{-n} z^n$ have radius of convergence at least 1. Then we note that a power series is analytic within the radius of convergence.
(ii) F is clearly continuous at all z_0 with $|z_0| < 1$. We must show that it is continuous at z_0 with $|z_0| = 1$. Observe that

$$F(r \exp i\theta) = a_0/2 + \sum_{n=1}^{\infty} a_n r^{|n|} \exp in\theta + a_0/2 + \sum_{n=1}^{\infty} a_{-n} r^{|n|} \exp - in\theta$$

$$= \sum_{n=-\infty}^{\infty} a_n r^{|n|} \exp in\theta,$$

so, by Theorem 27.1, $F(r \exp i\theta) \to f(\exp i\theta) = F(\exp i\theta)$ uniformly as $r \to 1-$. Thus given $\varepsilon > 0$ we can find $\delta_1(\varepsilon) > 0$ such that $|\theta - \theta_0| < \delta_1$ implies $|f(\exp i\theta) - f(\exp i\theta_0)| < \varepsilon/2$ and $\delta_2(\varepsilon) > 0$ such that $1 - \delta_2 < r \leqslant 1$ implies $|F(r \exp i\theta) - F(\exp i\theta)| < \varepsilon/2$. It follows that if $|\theta - \theta_0| < \delta_1$ and $1 - \delta_2 < r \leqslant 1$ we have (since $F(\exp i\theta) = f(\exp i\theta)$)

$$|F(r \exp i\theta) - F(\exp i\theta_0)| < \varepsilon.$$

Thus, if $z_0 = \exp i\theta_0$, $|z_0 - w| < (\min(\delta_1(\varepsilon), \delta_2(\varepsilon), \frac{1}{2}))/2$ and $|w| \leqslant 1$, we have $|F(w) - F(z_0)| < \varepsilon$ so F is continuous in $\bar{D} = \{z : |z| \leqslant 1\}$.

(iii) Either use the fact that

$$\hat{g}(-n) = \frac{1}{2\pi} \int g(t) \exp int \, dt = \left(\frac{1}{2\pi} \int g(t) \exp - int \, dt \right)^* = \hat{g}(n)^*,$$

or the formula

$$\sum_{n=-\infty}^{\infty} \hat{g}(n) r^{|n|} \exp int = \frac{1}{2\pi} \int_{\mathbb{T}} g(t-y) P_r(y) \, dy,$$

obtained in the course of proving Theorem 27.1'.   ∎

Remark. The reader may wonder why, if mathematicians already knew about the Poisson kernel P_r, they did not use it to solve the problems considered in Chapter 1. It is certainly true that almost all of the results which we have proved using Fejér's theorem can be obtained from Theorem 27.1, but before Fejér nobody looked for this kind of unifying scheme so it is not strange that nobody found one. We may compare this with a group of people wandering round a darkened room bumping into furniture and each other. Suddenly someone finds a light switch and puts on the lights. Now everybody can see both the room and all the other light switches which could have been used.

28

DIRICHLET'S PROBLEM
FOR THE DISC

A, if not the, fundamental problem of potential theory is to solve Laplace's equation $\nabla^2 \phi = 0$ in some (open) region Ω subject to the condition $\phi = G$ on the boundary $\partial\Omega$. (This is called Dirichlet's problem.) In this chapter we shall see how the results on Poisson summation of Fourier series enable us to solve the problem in 2 dimensions when Ω is a disc. More specifically we shall solve the following problem.

Problem 28.1 (Dirichlet's problem for the disc). *Let*

$$D = \{(x,y):x^2 + y^2 < 1\}, \quad \bar{D} = \{(x,y):x^2 + y^2 \leqslant 1\}$$

and
$$\partial D = \{(x,y):x^2 + y^2 = 1\}.$$

Suppose $G:\partial D \to \mathbb{R}$ is continuous. Can we find a $\phi:\bar{D} \to \mathbb{R}$ such that

(i) *ϕ_{11} and ϕ_{22} exist with $\phi_{11} + \phi_{22} = 0$ at all points of D,*
(ii) *ϕ is continuous at all points of $\bar{D}$,*
(iii) *$\phi = G$ on ∂D?*

We establish the connection between the problem of this chapter and the results of the last by introducing a connection between analytic (i.e. $\mathbb{C} \to \mathbb{C}$ differentiable) functions and solutions of Laplace's equation (in 2 dimensions) with which the reader is probably familiar. If E is a set in $\mathbb{R}^2$, we write $E_\mathbb{C} = \{x + iy:(x,y)\in E\}$.

Lemma 28.2 (Cauchy Riemann equations). *Let Ω be an open region in $\mathbb{R}^2$ and $f:\Omega_\mathbb{C} \to \mathbb{C}$ a function. Write*

$$u(x,y) = \operatorname{Re} f(x+iy), v(x,y) = \operatorname{Im} f(x+iy) \text{ so } f(x+iy) = u(x,y) + iv(x,y)$$

for all $(x,y)\in\Omega$.
Then if f is analytic in $\Omega_\mathbb{C}$,

(i) *u, v are differentiable in Ω,*
(ii) *$u_1 = v_2, u_2 = -v_1$ in Ω,*
(iii) *$f'(x+iy) = u_1(x,y) - iu_2(x,y)$ in Ω.*

121

Proof. If $z \in \Omega_C$ write $f'(z) = A + Bi$ $[A, B \in \mathbb{R}]$. By definition

$$f(z + \eta) = f(z) + f'(z)\eta + \gamma(\eta)|\eta| \quad [z + \eta \in \Omega_C]$$

where $\gamma(\eta) \to 0$ as $\eta \to 0$. Thus writing $\eta = h + ik$ $[h, k \in \mathbb{R}]$ and $\alpha(h, k) = \operatorname{Re} \gamma(h + ik)$, $\beta(h, k) = \operatorname{Im} \gamma(h + ik)$ we have

$$
\begin{aligned}
u(x + h, y &+ k) + iv(x + h, y + k) \\
&= u(x, y) + iv(x, y) + (A + Bi)(h + ik) \\
&\quad + (\alpha(h, k) + i\beta(h, k))\sqrt{(h^2 + k^2)} \\
&= (u(x, y) + Ah - Bk + \alpha(h, k)\sqrt{(h^2 + k^2)}) \\
&\quad + i(v(x, y) + Ak + Bh + \beta(h, k)\sqrt{(h^2 + k^2)}).
\end{aligned}
$$

Thus, taking real parts,

$$u(x + h, y + k) = u(x, y) + Ah - Bk + \alpha(h, k)\sqrt{(h^2 + k^2)},$$

where $\alpha(h, k) \to 0$ as $(h, k) \to (0, 0)$. In other words, u is differentiable with $u_1 = A$, $u_2 = -B$. Similarly, v is differentiable with $v_1 = B$, $v_2 = A$. Conclusions (i), (ii) and (iii) follow at once. ∎

Using the (non-trivial) result that the derivative of an analytic function is itself analytic we obtain the following corollary.

Lemma 28.3. *Under the conditions of Lemma 28.2, u is infinitely differentiable and $\nabla^2 u = 0$ at all points of Ω.*
Proof. By conclusion (iii) of Lemma 28.2

$$f'(x + iy) = u_1(x, y) - iu_2(x, y).$$

But f' is analytic so that conclusions (i), (ii) and (iii) applied to f' show that u_1 and u_2 are differentiable in Ω with

$$f''(x + iy) = u_{11}(x, y) - iu_{12}(x, y) = -u_{22}(x, y) - iu_{21}(x, y).$$

Induction shows that u has derivatives of all orders and so is infinitely differentiable. Since $u_{11}(x, y) = -u_{22}(x, y)$ we have $u_{11} + u_{22} = 0$, i.e. $\nabla^2 u = 0$ in Ω. ∎

Lemma 28.4. *Let u and f be as in Lemma 28.2. Write $\tilde{u}(x, y) = u(x, -y)$ (so that $\tilde{u}(x, y) = \operatorname{Re} f((x + iy)^*))$. Then $\tilde{u}$ is infinitely differentiable and $\nabla^2 \tilde{u} = 0$ at all points of Ω.*
Proof. Trivially $\tilde{u}_1(x, y) = \tilde{u}_1(x, -y)$, $\tilde{u}_{11}(x, y) = u_{11}(x, -y)$, $\tilde{u}_2(x, y) = -u_2(x, -y)$ and $\tilde{u}_{22}(x, -y) = u_{22}(x, -y)$ so $\tilde{u}_{11} + \tilde{u}_{22} = u_{11} + u_{22} = 0$. The fact that $\tilde{u}$ is infinitely differentiable is also immediate. ∎

The relevance of the results of the last chapter to the problem in hand is now clear.

Theorem 28.5. *Suppose* $G : \partial D \to \mathbb{R}$ *is continuous. Set* $g(\theta) = G(\cos \theta, \sin \theta) \, [\theta \in \mathbb{T}]$ *and write* $\phi(r \cos \theta, r \sin \theta) = \sum_{n=-\infty}^{\infty} \hat{g}(n) r^{|n|} \exp in\theta$ *for* $0 \leqslant r < 1$, $\phi(\cos \theta, \sin \theta) = g(\theta)$. *Then* ϕ *is a real valued function on* $\bar{D}$ *such that*

(i) ϕ *is infinitely differentiable and* $\nabla^2 \phi = 0$ *at all points of* D,

(ii) ϕ *is continuous at all points of* $\bar{D}$,

(iii) $\phi = G$ *on* ∂D.

Proof. Let

$$f_{(1)}(z) = \hat{g}(0)/2 + \sum_{n=1}^{\infty} \hat{g}(n) z^n$$

and

$$f_{(2)}(z) = \hat{g}(0)/2 + \sum_{n=1}^{\infty} \hat{g}(-n) z^n.$$

Then, writing $F(z) = f_{(1)}(z) + f_{(2)}(z^*)$ for $|z| < 1$ and $F(\exp i\theta) = g(\theta)$ for $\theta \in \mathbb{T}$, we know from Lemma 27.4 that F is a well defined real continuous function on $\bar{D}_{\mathbb{C}} = \{z : |z| \leqslant 1\}$. But

$$\phi(r \cos \theta, r \sin \theta) = f_{(1)}(r \exp i\theta) + f_{(2)}((r \exp i\theta)^*) = F(r \exp i\theta) \quad \text{for } 0 \leqslant r < 1,$$

and $\phi(\cos \theta, \sin \theta) = g(\theta) = F(\exp i\theta)$ so ϕ is a well defined real continuous function on $\bar{D} = \{(x, y) : x^2 + y^2 \leqslant 1\}$.

Thus the only thing that remains to prove is condition (i). To do this we observe that, since ϕ is real,

$$\phi(x, y) = \operatorname{Re} \phi(x, y) = \operatorname{Re} F(x + iy) = \operatorname{Re} f_{(1)}(x + iy) + \operatorname{Re} f_{(2)}(x - iy).$$

Lemmas 28.3 and 28.4 now tell us that $\operatorname{Re} f_{(1)}(x + iy)$ and $\operatorname{Re} f_{(2)}(x - iy)$ satisfy Laplace's equation and so ϕ does. ∎

29

POTENTIAL THEORY
WITH SMOOTHNESS ASSUMPTIONS

The reader may have observed that we have left two questions open.

Problem 29.1. *We have seen that the Dirichlet problem for the disc has a solution. Is the solution unique?*

Problem 29.2. *We have seen that the real part of an analytic function satisfies Laplace's equation. If Φ is a solution of Laplace's equation in an open set Ω, can it be expressed as the real part of an analytic function?*

The answer to the first question is yes. The answer to the second is yes locally but no globally. Although we shall not need these facts in the remainder of the book it seems unfair not to give the reader some idea of why they are true. In this chapter we give an account of the matter assuming that the regions, functions, paths and so on are sufficiently well behaved for 'mathematical methods' proofs to work. In Chapter 31 we shall offer a more general treatment which requires no smoothness assumptions whatsoever. Ideally, the reader should read, or at least glance through, both accounts.

We start our simpler treatment with a uniqueness proof which resolves Problem 29.1. Throughout $\bar{\Omega}$ will denote the closure of a set Ω and $\partial\Omega$ its boundary (so $\bar{\Omega} = \Omega \cup \partial\Omega$).

Theorem 29.3. *Let Ω be a bounded open region in $\mathbb{R}^2$ with smooth boundary $\partial\Omega$ and let $G:\partial\Omega \to \mathbb{R}$ be continuous. Then if $\phi, \psi:\bar{\Omega} \to \mathbb{R}$ are functions with continuous second derivatives in $\bar{\Omega}$ satisfying*

(i) $\nabla^2\phi = 0$, $\nabla^2\psi = 0$ *on* Ω,

(ii) $\phi = G$, $\psi = G$ *on* $\partial\Omega$,

it follows that $\phi = \psi$.

Proof. (We use the notation $\psi_1 = \partial\psi/\partial x$, $\psi_2 = \partial\psi/\partial y$ etc.) Write $\omega = \phi - \psi$. Then $\nabla^2\omega = \nabla^2\phi - \nabla^2\psi = 0$ on Ω and $\omega = \phi - \psi = 0$ on $\partial\Omega$. Since ω and $\partial\Omega$ are sufficiently smooth, we can apply the divergence theorem to obtain

$$\iint_\Omega \Sigma \omega_i^2 \, dA = \iint_\Omega \Sigma((\omega\omega_i)_i - \omega\omega_{ii}) \, dA = \iint_\Omega \Sigma(\omega\omega_i)_i - \omega\nabla^2\omega \, dA$$

$$= \iint_\Omega (\Sigma\omega\omega_i)_i \, dA = \int_\Omega \omega\frac{\partial\omega}{\partial n} \, ds = 0.$$

But $\Sigma\omega_i^2 \geqslant 0$ so $\Sigma\omega_i^2 = 0$ in Ω, so $\omega_1 = \omega_2 = 0$ in Ω and ω is constant in Ω. Thus (by continuity) $\omega = 0$ and $\phi = \psi$. ∎

The proof given above is really an argument about the minimum of an energy type function (compare the proofs in Chapter 20) as the following extension makes clear.

Lemma 29.4. *If Ω, G, ϕ and ψ are as in Theorem 29.3 except that we no longer assume that $\nabla^2\psi = 0$ then*

$$\iint_\Omega \Sigma\psi_i^2 \, dA \geqslant \iint_\Omega \Sigma\phi_i^2 \, dA,$$

with equality if and only if $\psi = \phi$.

Proof. Write $\omega = \phi - \psi$. Then $\omega = \phi - \psi = 0$ on $\partial\Omega$ and since ω, ϕ and $\partial\Omega$ are sufficiently smooth we can apply the divergence theorem to obtain

$$\iint_\Omega \Sigma(\psi_i^2 - \phi_i^2 - \omega_i^2) \, dA$$

$$= \iint_\Omega \Sigma((\phi_i - \omega_i)^2 - \phi_i^2 - \omega_i^2) \, dA = 2\iint_\Omega \Sigma\omega_i\phi_i \, dA$$

$$= 2\iint_\Omega \Sigma((\omega\phi_i)_i - \omega\phi_{ii}) \, dA = 2\iint_\Omega \Sigma(\omega\phi_i)_i \, dA = 2\iint_\Omega \omega\frac{\partial\phi}{\partial n} \, ds = 0.$$

Thus $\iint_\Omega \Sigma\psi_i^2 \, dA = \iint_\Omega \Sigma\phi_i^2 \, dA + \iint_\Omega \Sigma\omega_i^2 \, dA$ and $\iint_\Omega \Sigma\psi_i^2 \, dA \geqslant \iint_\Omega \Sigma\phi_i^2 \, dA$ with equality if and only if $\iint_\Omega \Sigma\omega_i^2 \, dA = 0$ and so (as in the proof of Theorem 29.3) if and only if $\omega = 0$ and $\phi = \psi$. ∎

Conversely we can show that any (sufficiently smooth) ϕ with $\phi = G$ on $\partial\Omega$ which minimises $\iint_\Omega \Sigma\phi_i^2 \, dA$ satisfies Laplace's equation in Ω.

Lemma 29.5. *Let Ω be a bounded open region with smooth boundary $\partial\Omega$ and let $G:\partial\Omega \to \mathbb{R}$ be continuous. Let us write $\mathscr{G}$ for the set of continuous functions $\psi:\bar{\Omega} \to \mathbb{R}$ with continuous second derivatives on Ω satisfying $\psi = G$ on $\partial\Omega$. Then, if $\phi \in \mathscr{G}$ is such that*

$$\iint_\Omega \Sigma\psi_i^2 \, dA \geqslant \iint_\Omega \Sigma\phi_i^2 \, dA \quad \text{for all } \psi \in \mathscr{G},$$

it follows that $\nabla^2\phi = 0$ at all points of Ω.

Proof. Let $\mathscr{F}$ be the set of functions $\omega:\bar{\Omega}\to\mathbb{R}$ with continuous second derivatives and satisfying $\omega = 0$ on $\partial\Omega$. Then, if $h\in\mathbb{R}$ and $\omega\in\mathscr{F}$, it follows that $\phi + h\omega\in\mathscr{G}$ and so

$$0 \leqslant \iint_\Omega \Sigma(\phi + h\omega)_i^2\,dA - \iint_\Omega \Sigma\phi_i^2\,dA = 2h\iint_\Omega \Sigma\omega_i\phi_i\,dA + h^2\iint_\Omega \Sigma\omega_i^2\,dA.$$

But if $ah^2 + 2bh \geqslant 0$ for all $h\in\mathbb{R}$, it follows that $b = 0$. (For, if $b\neq 0$ and $a\leqslant 0$, then setting $h = -\operatorname{sgn} b$ we have $ah^2 + 2bh \leqslant -2|b| < 0$, whilst if $b\neq 0$ and $a > 0$, then, setting $h = -b/a$, we have $ah^2 + 2bh = ah(h + 2b/a) = -b^2/a < 0$.) Thus, using Green's theorem

$$0 = \iint_\Omega \Sigma\omega_i\phi_i\,dA = \iint_\Omega \Sigma(\omega\phi_i)_i\,dA - \iint_\Omega \omega\Sigma\phi_{ii}\,dA$$

$$= \int_{\partial\Omega} \omega\frac{\partial\phi}{\partial n}\,ds - \iint_\Omega \omega\nabla^2\phi\,dA = -\iint_\Omega \omega\nabla^2\phi\,dA.$$

Thus $\iint_\Omega \omega\nabla^2\phi = 0$ for all $\omega\in\mathscr{F}$ and so $\nabla^2\phi = 0$ at all points of Ω. ∎

Gauss, Green, Kelvin, Dirichlet and Riemann all considered it obvious that there must exist such a minimising ϕ and that, therefore, the problem of finding a ϕ with $\nabla^2\phi = 0$ on Ω, $\phi = G$ on $\partial\Omega$ is always soluble. However, the argument is fallacious since, as Weierstrass pointed out, it is not necessarily true that a positive integral actually attains its infimum (we shall given an example of this in Chapter 35). That Green and Kelvin should have missed this subtle point is in no way surprising but both Gauss and Dirichlet had criticised others for making similar assumptions. (Dirichlet's criticism arose in connection with the work of Steiner described in Chapter 35.)

All (mathematical) mistakes are obvious once they have been found. Thus, rather than leave the reader contemplating the obvious objection just given, let me point out that there is a second obvious objection to the argument above and invite her to find it. This second objection will be discussed in the next chapter.

We turn now to the resolution of Problem 29.2 (under suitable smoothness assumptions). The idea behind our attack is as follows. Suppose we know that $u:\mathbb{R}^2\to\mathbb{R}$ is the real part of an analytic function. More exactly suppose we can find $v:\mathbb{R}^2\to\mathbb{R}$ such that, writing $x = \operatorname{Re} z$, $y = \operatorname{Im} z$ throughout, if $f(z) = u(x, y) + iv(x, y)$ then $f:\mathbb{C}\to\mathbb{C}$ is analytic. Under these circumstances the Cauchy–Riemann equations (Lemma 28.2) tell us that $f'(z) = u_1(x, y) - iu_2(x, y)$ and so if $l(w)$ is a path from w to z_0 in $\mathbb{C}$

$$f(w) - f(z_0) = \int_{l(w)} f'(z)\,dz = \int_{l(w)} (u_1(x, y) - iu_2(x, y))\,dz.$$

This shows that if we are given a u which we hope is the real part of an analytic

function f then the only possible candidate for f is given by

$$f(w) - f(z_0) = \int_{l(w)} (u_1(x, y) - iu_2(x, y)) \, dz.$$

We can now give an example which shows that the answer to Problem 29.2 is no for a general Ω and ϕ. (Throughout we take positive square roots.)

Example 29.6. Let $R_1 > R_2 \geqslant 0$ and let Ω be the annulus $\{(x, y) : R_1 > (x^2 + y^2)^{\frac{1}{2}} > R_2\}$. Then, if $\phi(x, y) = \log(x^2 + y^2)^{\frac{1}{2}}$,

(i) $\nabla^2 \phi = 0$ in Ω.
(ii) *There does not exist a* $\psi : \Omega \to \mathbb{R}$ *such that* $f(x + iy) = \phi(x, y) + i\psi(x, y) [x, y \in \mathbb{R}]$ *defines an analytic function on* $\Omega_c = \{z : R_1 > |z| > R_2\}$.

Proof. We observe that if $(x, y) \neq (0, 0)$ then

$$\phi_1(x, y) = x(x^2 + y^2)^{-\frac{1}{2}}(x^2 + y^2)^{-\frac{1}{2}} = x(x^2 + y^2)^{-1},$$
$$\phi_{11}(x, y) = (x^2 + y^2)^{-1} - 2x^2(x^2 + y^2)^{-2} = (y^2 - x^2)(x^2 + y^2)^{-2},$$

and, similarly,

$$\phi_2(x, y) = y(x^2 + y^2)^{-1}, \phi_{22}(x, y) = (x^2 - y^2)(x^2 + y^2)^{-2}.$$

Thus $\phi_{11} + \phi_{22} = 0$ at all points of Ω.

On the other hand if ϕ is the real part of an analytic function f on Ω_c we know from the preceding discussion that

$$f(w) - f(z_0) = \int_l \phi_1(x, y) - i\phi_2(x, y) \, dz = \int_l \frac{x - iy}{x^2 + y^2} \, dz = \int_l \frac{1}{x + iy} \, dz = \int_l \frac{1}{z} \, dz,$$

where l is any path from z_0 to w in Ω_c. In particular taking $w = z_0 = (R_1 + R_2)/2$ and l to be the path $z = z_0 \exp i\theta$ as θ runs from 0 to 2π we obtain

$$0 = f(z_0) - f(z_0) = \int_0^{2\pi} \frac{1}{z_0 \exp i\theta} iz_0 \exp i\theta \, d\theta = \int_0^{2\pi} i \, d\theta = 2\pi i,$$

and the contradiction shows that no suitable f and so no suitable ψ can exist. ∎

Remark. The reader may find it helpful to observe that if $\Omega = \mathbb{R}^2 \setminus \{(x, 0) : x \leqslant 0\}$ the argument above gives $f(w) - f(1) = \int_l 1/z \, dz = \log z$, where $\log$ is a well defined analytic function on $\Omega_c = \mathbb{C} \setminus \{x : x \leqslant 0\}$ with $\log(r \exp i\theta) = \log r + i\theta$ $[-\pi < \theta < \pi, r > 0]$.

Next we note that the answer to Problem 28.2 is yes if Ω is a disc.

Lemma 29.7. Let Ω be a disc $\{(x, y) : (x - x_0)^2 + (y - y_0)^2 < R^2\}$ and $z_0 = x_0 + iy_0$. Then if $\phi : \Omega \to \mathbb{R}$ has continuous second derivatives and $\nabla^2 \phi = 0$, we can find

a $\psi : \Omega \to \mathbb{R}$ such that $f(x + iy) = \phi(x, y) + i\psi(x, y)$ defines an *analytic* function on $\Omega_c = \{z \in \mathbb{C} : |z - z_0| < R\}$.

Proof. Let C be a (well behaved) closed curve in Ω enclosing a region S. Then using the divergence theorem and the fact that $\phi_{12} = \phi_{21}$ we have

$$\int_C (\phi_1 - i\phi_2) \, dz = \int_C (\phi_1 - i\phi_2)(dx + i \, dy)$$

$$= \int_C (\phi_1 \, dx + \phi_2 \, dy) + i \int_C (-\phi_2 \, dx + \phi_1 \, dy)$$

$$= \iint_S (\phi_{11} + \phi_{22}) \, dA + i \iint_S (-\phi_{21} + \phi_{12}) \, dA$$

$$= \iint_S \nabla^2 \phi \, dA + i \iint_S 0 \, dA = \iint_S 0 \, dA + i \iint_S 0 \, dA = 0.$$

(Notice that the argument depends on the fact that $S \subset \Omega$ and so fails in Example 29.5.)

Suppose $w \in \Omega$ and $l(1)$ and $l(2)$ are well behaved curves from z_0 to w lying within Ω_c. Let C_c be the closed curve formed by following $l(1)$ from z_0 to w and then $l(2)$ reversed from w to z_0. Then by the result of the previous paragraph

$$\int_{l(1)} (\phi_1 - i\phi_2) \, dz - \int_{l(2)} (\phi_1 - i\phi_2) \, dz = \int_{C_c} \phi_1 - i\phi_2 \, dz = 0,$$

and so the integral $\int_l (\phi_1 - i\phi_2) \, dz$ takes the same value for all well behaved paths l from z_0 to w. We write $Q(w)$ for this value and take $f(w) = Q(w) + \phi(x_0, y_0)$.

But if $p : \Omega_c \to \mathbb{C}$ is continuous and $P(z) = \int_l p(w) \, dw$ is well defined (i.e. independent of the path l from z_0 to z) then, taking $l(\eta)$ to be the straight line path from z to $z + \eta$, we have $P(z + \eta) - P(z) = \int_{l(\eta)} p(w) \, dw$. Thus

$$\left| \frac{P(z + \eta) - P(z)}{\eta} - p(z) \right| = \left| \frac{1}{\eta} \int_{l(\eta)} (p(w) - p(z)) \, dw \right| \leqslant \sup_{w \in l(\eta)} |p(w) - p(z)| \to 0$$

$$\text{as } |\eta| \to 0,$$

and so P is analytic with $P'(z) = p(z)$.

Thus f is analytic with $f'(z) = Q'(z) = \phi_1(x, y) - i\phi_2(x, y)$. $[z = x + iy, \ x, \ y \in \mathbb{R}]$. Finally we note that

$$\text{Re } f(w) = \text{Re } Q(w) - \phi(x_0, y_0) = \text{Re} \int_l (\phi_1 - i\phi_2)(dx + i \, dy) - \phi(x_0, y_0)$$

$$= \int_l \phi_1 \, dx + \phi_2 \, dy - \phi(x_0, y_0) = \phi(x, y),$$

and we are done.   ∎

We note that f is only unique up to the addition of a purely imaginary constant.

A minor restatement of Lemma 29.7 shows why the answer to Problem 29.2 is locally affirmative.

Theorem 29.8. *Let Ω be an open set in $\mathbb{R}^2$ and let $\phi:\Omega\to\mathbb{R}$ have continuous second derivatives with $\nabla^2\phi=0$ on Ω. Then if $D=\{(x,y):(x-x_0)^2+(y-y_0^2)<\delta^2\}\subset\Omega$ we can find a $\psi:D\to\mathbb{R}$ such that $f(x+iy)=\phi(x,y)+i\psi(x,y)$ defines an analytic function in $D_{\mathbb{C}}=\{z:|z-z_0|<\delta\}$.*
Proof. Apply Lemma 29.7 to D and $\phi|D$ the restriction of ϕ to D. ∎

As a direct consequence we get the following important formula.

Theorem 29.9 (Gauss' mean value theorem). *Let Ω be an open set in $\mathbb{R}^2$ and let $\phi:\Omega\to\mathbb{R}$ have continuous second derivatives with $\nabla^2\phi=0$ on Ω. Then, if $\delta>\rho>0$ and*

$$D=\{(x,y):(x-x_0)^2+(y-y_0)^2<\delta^2\}\subset\Omega,$$

we have
$$\phi(x_0,y_0)=\frac{1}{2\pi\rho}\int_C\phi(x,y)\,ds,$$

where C is the circular path described by $(x,y)=(x_0+\rho\cos\theta,y_0+\rho\sin\theta)$ as θ runs from 0 to 2π.

(Thus the value of ϕ at the center (x_0,y_0) of the circle C is the mean of its values on C.)

Proof. Let f be as in Theorem 29.8. Then, writing $z=x+iy$, we have, by Cauchy's formula,

$$f(z_0)=\frac{1}{2\pi i}\int_C\frac{f(z)}{z-z_0}\,dz=\frac{1}{2\pi i}\int_0^{2\pi}\frac{f(z_0+\rho\exp i\theta)}{\rho\exp i\theta}i\rho\exp i\theta\,d\theta$$

$$=\frac{1}{2\pi\rho}\int_0^{2\pi}f(z_0+\rho\exp i\theta)\rho\,d\theta=\frac{1}{2\pi\rho}\int_C f(x+iy)\,ds.$$

Thus
$$\phi(x_0,y_0)+i\psi(x_0,y_0)=\frac{1}{2\pi\rho}\int_C\phi(x,y)+i\psi(x,y)\,ds$$

and taking real parts we obtain

$$\phi(x_0,y_0)=\frac{1}{2\pi\rho}\int_C\phi(x,y)\,ds,$$

as stated. ∎

As an immediate corollary we obtain an important maximum principle.

Theorem 29.10. *Let Ω be an open set in $\mathbb{R}^2$ and let $\phi:\Omega\to\mathbb{R}$ have continuous second derivatives with $\nabla^2\phi=0$ in Ω. Then ϕ cannot have a strict maximum in Ω.*
Proof. Let $(x_0,y_0)\in\Omega$. Since Ω is open we can find a $\delta>0$ with $\{(x-x_0)^2+$

$(y - y_0)^2 < \delta^2\} \subset \Omega.$ We shall show that given any $0 < \rho < \delta$ we can find an (x_1, y_1) with $(x_1 - x_0)^2 + (y_1 - y_0)^2 = \rho^2$ and $\phi(x_1, y_1) \geqslant \phi(x_0, y_0)$. For the function $g:[0, 2\pi] \to \mathbb{R}$ given by $g(t) = \phi(x_0 + \rho \cos t, \ y_0 + \rho \sin t)$ is continuous and so bounded and attains its bounds. In particular we can find a $t_1 \in [0, 2\pi]$, such that $g(t_1) \geqslant g(t)$ for all $t \in [0, 2\pi]$.

Thus, taking $(x_1, y_1) = (x_0 + \rho \cos t_1, \ y_0 + \rho \sin t_1)$, we have $(x_1 - x_0)^2 + (y_1 - y_0)^2 = \rho^2$ and, defining C as in Theorem 29.9,

$$\phi(x_0, y_0) = \frac{1}{2\pi\rho} \int_C \phi(x, y)\, ds = \frac{1}{2\pi\rho} \int_0^{2\pi} g(t)\rho\, dt$$

$$= \frac{1}{2\pi} \int_0^{2\pi} g(t)\, dt \leqslant \frac{1}{2\pi} \int_0^{2\pi} g(t_1)\, dt = g(t_1) = \phi(x_1, y_1).$$

Thus ϕ cannot have a strict maximum at (x_0, y_0) and we are done. ∎

30

AN EXAMPLE OF HADAMARD

In Chapter 29 we developed the theory of solutions ϕ of the equations $\nabla^2 \phi = 0$ on Ω, $\phi = G$ on $\partial\Omega$ on the assumption that ϕ, Ω and $\partial\Omega$ were sufficiently smooth. In particular we assumed that ϕ had continuous second derivatives on the closure $\bar{\Omega} = \Omega \cup \partial\Omega$ of Ω. This enabled us to use the divergence theorem to argue (at least for Ω bounded) as in the following sequence of formulae,

$$\iint_\Omega \Sigma \phi_i^2 \, dA = \iint_\Omega \Sigma((\phi\phi_i)_i - \phi\phi_{ii}) dA = \iint_\Omega (\Sigma(\phi\phi_i)_i - \phi\nabla^2\phi) dA$$

$$= \iint_\Omega \Sigma(\phi\phi_i)_i \, dA = \int_{\partial\Omega} \phi \frac{\partial\phi}{\partial n} \, ds.$$

This argument is entirely correct but as Hadamard pointed out in 1906 obscures a very important aspect of the problem. We revert to the simplest case described in Theorem 28.5 when $\Omega = D = \{(x,y): x^2 + y^2 < 1\}$.

Lemma 30.1. *Suppose* $G: \partial D \to \mathbb{R}$ *is continuous. Set* $g(\theta) = G(\cos\theta, \sin\theta)$ $[\theta \in \mathbb{T}]$ *and write*

$$\phi(r\cos\theta, r\sin\theta) = \sum_{n=-\infty}^{\infty} \hat{g}(n) r^{|n|} \exp in\theta \text{ for } 0 \leqslant r < 1,$$

$$\phi(\cos\theta, \sin\theta) = g(\theta).$$

Then ϕ *is a real valued function on* $\bar{D}$

(i) ϕ *is infinitely differentiable and* $\nabla^2 \phi = 0$ *at all points of D,*

(ii) ϕ *is continuous at all points of* $\bar{D}$*,*

(iii) $\phi = G$ *on* ∂D*.*

Further, writing $D(R) = \{(x,y): x^2 + y^2 \leqslant R^2\}$*, we have*

$$\iint_{D(R)} \Sigma\phi_j^2 \, dA = 2\pi \sum_{n=-\infty}^{\infty} |n| R^{2|n|} |\hat{g}(n)|^2 \quad \text{for all } 0 \leqslant R < 1.$$

131

Proof. Everything except the last sentence is a repetition of the statement of Theorem 28.5. To compute

$$\iint_{D(R)} \left(\frac{\partial\phi}{\partial x}\right)^2 + \left(\frac{\partial\phi}{\partial y}\right)^2 dA$$

we proceed as follows. First we convert from Cartesian coordinates (x, y) to polar coordinates (r, θ) observing that

$$\left(\frac{\partial\phi}{\partial x}\right)^2 + \left(\frac{\partial\phi}{\partial y}\right)^2 = \left(\frac{\partial\phi}{\partial r}\right)^2 + \left(\frac{1}{r}\frac{\partial\phi}{\partial\theta}\right)^2$$

and so

$$\iint_{D(R)} \left(\frac{\partial\phi}{\partial x}\right)^2 + \left(\frac{\partial\phi}{\partial y}\right)^2 dA = \iint_{D(R)} \left(\frac{\partial\phi}{\partial r}\right)^2 + \left(\frac{1}{r}\frac{\partial\phi}{\partial\theta}\right)^2 dA$$

$$= \int_0^R \int_0^{2\pi} \left(\frac{\partial\phi}{\partial r}\right)^2 + \left(\frac{1}{r}\frac{\partial\phi}{\partial\theta}\right)^2 r\, d\theta\, dr.$$

Since $|\hat{g}(n)| \leqslant \sup_{t\in\mathbb{T}} |g(t)|$, it follows that

$$\sum_{n=-\infty}^{\infty} |\hat{g}(n)r^{|n|} \exp in\theta| \quad \text{and} \quad \sum_{n=-\infty}^{\infty} |n\hat{g}(n)r^{|n|-1} \exp in\theta|$$

converge uniformly for $0 \leqslant r \leqslant R$, $\theta\in\mathbb{T}$. Thus we may differentiate term by term to get

$$\frac{\partial\phi}{\partial r} = \sum_{n\neq 0} |n|\hat{g}(n)r^{|n|-1} \exp in\theta,$$

$$\frac{1}{r}\frac{\partial\phi}{\partial\theta} = \sum_{n\neq 0} in\hat{g}(n)r^{|n|-1} \exp in\theta,$$

then multiply term by term to get

$$\left(\frac{\partial\phi}{\partial r}\right)^2 = \sum_{n,m\neq 0} |n||m|\hat{g}(n)\hat{g}(m)r^{|n|+|m|-2} \exp i(n+m)\theta$$

$$\left(\frac{1}{r}\frac{\partial\phi}{\partial\theta}\right)^2 = -\sum_{n,m\neq 0} nm\hat{g}(n)\hat{g}(m)r^{|n|+|m|-2} \exp i(n+m)\theta$$

(uniformly for $0 \leqslant r \leqslant R$, $\theta\in\mathbb{T}$) and finally integrate term by term twice to get

$$\int_0^{2\pi} \left(\frac{\partial\phi}{\partial r}\right)^2 r\, d\theta = \sum_{n,m\neq 0} \int_0^{2\pi} |n||m|\hat{g}(n)\hat{g}(m)r^{|n|+|m|-1} \exp i(n+m)\theta\, d\theta$$

$$= \sum_{n,m\neq 0} |n||m|\hat{g}(n)\hat{g}(m)r^{|n|+|m|-1} \int_0^{2\pi} \exp i(n+m)\theta\, d\theta$$

$$= \sum_{n\neq 0} |n||-n|\hat{g}(n)\hat{g}(-n)r^{|n|+|-n|-1} 2\pi = 2\pi \sum_{n\neq 0} n^2 \hat{g}(n)\hat{g}(-n)r^{2|n|-1},$$

$$\int_0^R \int_0^{2\pi} \left(\frac{\partial \phi}{\partial r}\right)^2 r d\theta \, dr = 2\pi \sum_{n \neq 0} n^2 \hat{g}(n)\hat{g}(-n) \int_0^R r^{2|n|-1} \, dr = \pi \sum_{n \neq 0} |n|\hat{g}(n)\hat{g}(-n)R^{2|n|},$$

and similarly

$$\int_0^R \int_0^{2\pi} \left(\frac{1}{r}\frac{\partial \phi}{\partial \theta}\right) r d\theta dr = -\pi \sum_{n \neq 0} (-n)(n)(1/|n|)\hat{g}(n)\hat{g}(-n)R^{2|n|}$$

$$= \pi \sum_{n \neq 0} |n|\hat{g}(n)\hat{g}(-n)R^{2|n|}.$$

Thus bearing in mind that g is real and so

$$\hat{g}(-n) = \frac{1}{2\pi}\int_0^{2\pi} \hat{g}(t)\exp int \, dt = \hat{g}(n)^*$$

we have

$$\iint_{D(R)} \sum \phi_j^2 \, dA = 2\pi \sum_{n=-\infty}^{\infty} |n|R^{2|n|}|\hat{g}(n)|^2. \qquad \blacksquare$$

Example 30.2 (Hadamard). *There exists a continuous function $G: \partial D \to \mathbb{R}$ and a real valued function ϕ on $\bar{D}$ such that*

(i) *ϕ is infinitely differentiable and $\nabla^2 \phi = 0$ at all points of D,*
(ii) *ϕ is continuous at all points of $\bar{D}$,*
(iii) *$\phi = G$ on ∂D,*

yet $\iint_D \sum \phi_j^2 \, dA$ diverges.

Proof. Let $c_n = 2^{-m}$ when $|n| = 2^{2m}$ $[m \geqslant 0]$, $c_n = 0$ otherwise. Clearly $\sum_{n=-\infty}^{\infty} |c_n| = \sum_{m=-\infty}^{\infty} 2^{-|m|}$ converges and so, by Theorem 9.2,

$$\sum_{n=-\infty}^{\infty} c_n \exp in\theta = \sum_{n=1}^{\infty} 2c_n \cos n\theta$$

converges uniformly to a real continuous function $g(\theta)$ with $\hat{g}(n) = c_n$. Set $G(\exp it) = g(t)$. Then taking ϕ as in Lemma 30.1 we have

$$\iint_{D(R)} \sum \phi_j^2 \, dA = 2\pi \sum_{n=-\infty}^{\infty} |n|R^{2|n|}|c_n|^2 = 4\pi \sum_{m=0}^{\infty} 2^{2m}R^{2^{2m+1}}2^{-2m}$$

$$= 4\pi \sum_{m=0}^{\infty} R^{2^{2m+1}} \to \infty \quad \text{as} \quad R \to 1-.$$

Thus $\iint_D \sum \phi_j^2 \, dA$ diverges $\qquad \blacksquare$

Hadamard's example does not invalidate the reasoning of the previous chapter. What happens is that although ϕ may be extended continuously to the whole of $\bar{D}$, the behaviour of the derivatives of ϕ becomes wilder and wilder as we approach the boundary. However Hadamard's example does justify us seeking answers to Problems 29.1 and 29.2 which do not make additional smoothness assumptions on ϕ.

31

POTENTIAL THEORY
WITHOUT SMOOTHNESS ASSUMPTIONS

Our object in this chapter is to recover the results of Chapter 29 with greater precision and no smoothness assumptions. However, our work will be informed by the results of the previous chapter and in particular by the knowledge that solutions of Laplace's equation $\nabla^2 \phi = 0$ can have no strict maxima at interior points (Theorem 29.10).

Lemma 31.1. *Let Ω be an open set in $\mathbb{R}^2$ with closure $\bar{\Omega}$ and boundary $\partial\Omega$. Suppose that $\bar{\Omega}$ is bounded and that $\phi: \bar{\Omega} \to \mathbb{R}$ satisfies the following conditions*

(i) *ϕ_{11} and ϕ_{22} exist with $\phi_{11} + \phi_{22} > 0$ at all points of Ω,*
(ii) *ϕ is continuous on $\bar{\Omega}$.*

Then there exists an $(x_0, y_0) \in \partial\Omega$ with $\phi(x_0, y_0) \geq \phi(x, y)$ for all $(x, y) \in \bar{\Omega}$.
Proof. Since ϕ is continuous on the closed bounded set $\bar{\Omega}$, it is bounded and attains its bounds (see, but only if really necessary, Theorem B.3 of Appendix B). Thus we can find an $(x_0, y_0) \in \bar{\Omega}$ with $\phi(x_0, y_0) \geq \phi(x, y)$. If $(x_0, y_0) \in \Omega$ then we can find a $\delta > 0$ such that $(x, y) \in \Omega$ whenever $|x - x_0|, |y - y_0| < \delta$. In particular $\phi(x_0, y)$ considered as a function of y is a twice differentiable function on $(y_0 - \delta, y_0 + \delta)$ with a maximum at y_0. Thus $(\partial^2\phi/\partial y^2)(x_0, y_0) \leq 0$ and similarly $(\partial^2\phi/\partial x^2)(x_0, y_0) \leq 0$ so $\phi_{11}(x_0, y_0) + \phi_{22}(x_0, y_0) \leq 0$, which contradicts hypothesis (i). By *reductio ad absurdum* $(x_0, y_0) \in \partial\Omega$ and we are done. ∎

Theorem 31.2 (A maximum principle for harmonic functions). *Let Ω be an open set in $\mathbb{R}^2$ with closure $\bar{\Omega}$ and boundary $\partial\Omega$. Suppose that $\bar{\Omega}$ is bounded and that $\phi: \bar{\Omega} \to \mathbb{R}$ satisfies the following conditions*

(i) *ϕ_{11} and ϕ_{22} exist with $\phi_{11} + \phi_{22} = 0$ at all points of Ω,*
(ii) *ϕ is continuous on $\bar{\Omega}$.*

Then there exists an $(x_0, y_0) \in \partial\Omega$ with $\phi(x_0, y_0) \geq \phi(x, y)$ for all $(x, y) \in \bar{\Omega}$.

Proof. To fix ideas we choose an R such that $(x, y) \in \bar{\Omega}$ implies $|x|, |y| \leqslant R$ and set $g(x, y) = x^2$. (However the argument below will work for any $g : \mathbb{R}^2 \to \mathbb{R}$ which is twice differentiable and has $g_{11} + g_{22} > 0$ everywhere.)

Since $\partial\Omega$ is a closed bounded set and ϕ is continuous on $\partial\Omega$ we can find (see Theorem B.3 if you must) $(x_0, y_0) \in \partial\Omega$ with $\phi(x_0, y_0) \geqslant \phi(x, y)$ for all $(x, y) \in \partial\Omega$. Pick $\varepsilon > 0$ and set $\psi(x, y) = \phi(x, y) + \varepsilon g(x, y)$. Then ψ satisfies the conditions of Lemma 31.1 (in particular,

$$\psi_{11} + \psi_{22} = \phi_{11} + \phi_{22} + \varepsilon g_{11} + \varepsilon g_{22} = 0 + 2\varepsilon > 0)$$

and so we can find an $(x_1, y_1) \in \partial\Omega$ with $\psi(x_1, y_1) \geqslant \psi(x, y)$ for all $(x, y) \in \bar{\Omega}$.

But $\phi(x_0, y_0) \geqslant \phi(x_1, y_1)$, by the definition of the first sentence of the last paragraph, and $g(x_1, y_1) \leqslant \varepsilon R^2$ since $(x_1, y_1) \in \bar{\Omega}$ and so $|x_1| \leqslant R$. Thus, since $g(x, y) \geqslant 0$ for all $(x, y) \in \bar{\Omega}$,

$$\phi(x_0, y_0) + \varepsilon R^2 \geqslant \psi(x_1, y_1) \geqslant \psi(x, y) \geqslant \phi(x, y),$$

and so $\phi(x_0, y_0) \geqslant \phi(x, y) - \varepsilon R^2$ for all $(x, y) \in \bar{\Omega}$ and all $\varepsilon > 0$. Letting $\varepsilon \to 0$ we see that $\phi(x_0, y_0) \geqslant \phi(x, y)$ for all $(x, y) \in \bar{\Omega}$. ∎

Our maximum principle gives us an immediate uniqueness theorem.

Theorem 31.3 (Uniqueness of solutions of Dirichlet's problem). *Let Ω be an open set in $\mathbb{R}^2$ with closure $\bar{\Omega}$ and boundary $\partial\Omega$. Suppose that $\bar{\Omega}$ is bounded and that $G : \partial\Omega \to \mathbb{R}$ is continuous. Then there exists at most one function $\phi : \bar{\Omega} \to \mathbb{R}$ satisfying the following conditions*

(i) *ϕ_{11} and ϕ_{22} exist with $\phi_{11} + \phi_{22} = 0$ at all points of Ω,*
(ii) *ϕ is continuous on $\bar{\Omega}$,*
(iii) *$\phi = G$ on $\partial\Omega$.*

Proof. Suppose $\phi, \psi : \bar{\Omega} \to \mathbb{R}$ satisfy conditions (i), (ii) and (iii). Set $\omega = \phi - \psi$. Then

(i)′ ω_{11} and ω_{22} exist with $\omega_{11} + \omega_{22} = 0$ at all points of Ω,
(ii)′ ω is continuous on $\bar{\Omega}$,
(iii)′ $\omega = 0$ on $\partial\Omega$.

Thus ω satisfies the conditions of Theorem 31.2 and so we can find $(x_0, y_0) \in \partial\Omega$ with $\omega(x_0, y_0) \geqslant \omega(x, y)$ for all $(x, y) \in \bar{\Omega}$. But, by condition (iii)′, $\omega(x_0, y_0) = 0$ so $0 \geqslant \omega(x, y)$ and $\psi(x, y) \geqslant \phi(x, y)$ for all $(x, y) \in \bar{\Omega}$. But by symmetry we must also have $\phi(x, y) \geqslant \psi(x, y)$ for all $(x, y) \in \bar{\Omega}$ and so $\phi(x, y) = \psi(x, y)$ for all $(x, y) \in \bar{\Omega}$ and our theorem holds. ∎

The reader will have noted our repeated use of the linearity of the expression $\phi_{11} + \phi_{22}$ (that is to say the fact that

$$(\lambda\phi + \mu\psi)_{11} + (\lambda\phi + \mu\psi)_{22} = \lambda(\phi_{11} + \phi_{22}) + \mu(\psi_{11} + \psi_{22})).$$

The following example shows that the hypothesis $\bar{\Omega}$ bounded cannot be dropped from the hypotheses of Theorem 31.3 (and so cannot be dropped from Theorem 31.2 or Lemma 31.1).

Example 31.4. *Let* $\Omega = \{(x,y):x>0\}$ *be the right half plane (and so* $\partial\Omega = \{(0,y):y\in\mathbb{R}\}$, $\bar{\Omega} = \{(x,y):x\geqslant 0\}$). *Then writing* $\phi(x,y) = x$ *and taking* $\lambda\in\mathbb{R}$ *we see that* $\lambda\phi$ *is real valued and*

(i) $\lambda\phi$ *is infinitely differentiable with* $\nabla^2\lambda\phi = 0$ *at all points of* $\mathbb{R}^2$ *and so at all points of* $\bar{\Omega}$,
(ii) $\lambda\phi = 0$ *on* $\partial\Omega$,
yet $\lambda\phi = \mu\phi$ *on* Ω *if and only if* $\lambda = \mu$.
Proof. Obvious. ∎

Remark. Notice that if we demand solutions u with $u(x,y)$ bounded as $(x,y)\to\infty$ then there is only one possible solution of the form $u(x,y) = \lambda x$, viz. $u = 0$.

The uniqueness theorem gives us a very strong hold over harmonic functions (that is to say, functions ϕ with $\phi_{11}+\phi_{22} = 0$). In particular, we can provide a new proof of Lemma 29.7 which avoids having to reason with 'well behaved' closed curves.

Lemma 31.6. *Let* Ω *be the disc* $\{(x,y):(x-x_0)^2+(y-y_0)^2<R^2\}$ *and* $\phi:\Omega\to\mathbb{R}$ *be a function such that* ϕ_{11} *and* ϕ_{22} *exist with* $\phi_{11}+\phi_{22} = 0$. *Then, if* $R>\rho>0$ *and* Ω' *is the disc* $\{(x,y):(x-x_0)^2+(y-y_0)^2<\rho^2\}$, *we can find a unique function* $\psi:\Omega'\to\mathbb{R}$ *such that* $\psi(x_0,y_0) = 0$ *and* $f(x+iy) = \phi(x,y)+i\psi(x,y)$ *defines an analytic function in the disc* $\Omega'_C = \{z\in\mathbb{C}:|z-z_0|<\rho\}$.
Proof (uniqueness). Suppose $\tau:\Omega'\to\mathbb{R}$ is such that $\tau(x_0,y_0) = 0$ and $g(x+iy) = \phi(x,y)+i\tau(x,y)$ is an analytic function in Ω'_C. Then $f-g$ is an analytic function in Ω'_C and applying the Cauchy–Riemann equation (Lemma 28.2) $(\psi-\tau)_1 = (\phi-\phi)_2 = 0_2 = 0$ and similarly $(\psi-\tau)_2 = 0$. Thus by the mean value theorem $\psi-\tau$ is constant, so

$$\psi(x,y)-\tau(x,y) = \psi(x_0,y_0)-\tau(x_0,y_0) = 0.$$

Proof (existence). Let us define a function u on the closure $\bar{D} = \{(x,y):x^2+y^2\leqslant 1\}$ of the unit disc $\{(x,y):x^2+y^2<1\}$ by $u(x,y) = \phi(x_0+\rho x,y_0+\rho y)$ and a function G on the boundary $\partial D = \{(x,y):x^2+y^2 = 1\}$ by $G = u|\partial D$. Set $g(\theta) = G(\cos\theta,\sin\theta)$. We observe that $G:\partial D\to\mathbb{R}$ is continuous and that

(i) u_{11} and u_{22} exist with $u_{11}+u_{22} = 0$ at all points of D,
(ii) u is continuous on $\bar{D}$,
(iii) $u = G$ on ∂D.

On the other hand, we know from Theorem 28.5 that if we write

$$w(r\cos\theta, r\sin\theta) = \sum_{n=-\infty}^{\infty} \hat{g}(n)r^{|n|}\exp in\theta,$$

$$w(\cos\theta, \sin\theta) = g(\theta),$$

then w is a real valued function such that

(i) w is infinitely differentiable with $\nabla^2 w = 0$ at all points of D,
(ii) w is continuous on $\bar{D}$,
(iii) $w = G$ on ∂D.

But by Theorem 31.3 there can be only one function satisfying these conditions and so $u = w$ on D and, in particular,

$$u(r\cos\theta, r\sin\theta) = \sum_{n=-\infty}^{\infty} \hat{g}(n)r^{|n|}\exp in\theta \quad \text{for all} \quad |r| < 1, \theta \in \mathbb{T}.$$

Since g is real $\hat{g}(-n) = (2\pi)^{-1}\int_g g(t)\exp int\,dt = \hat{g}(n)^*$ and (as we noted repeatedly in Chapter 27), we know that $|\hat{g}(n)| \leqslant \sup_{t \in \mathbb{T}}|g(t)|$ so that, writing $a_n = \hat{g}(n)$, we have $a_0/2 + \sum_{r=1}^{\infty} a_r z^r$ convergent to an analytic function $h_{(1)}(z)$, say, and $a_0/2 + \sum_{r=1}^{\infty} a_r^* z^r$ convergent to an analytic function $h_{(2)}(z)$ say for all $|z| < 1$. But u is real and

$$u(r\cos\theta, r\sin\theta) = h_{(1)}(re^{i\theta}) + h_{(2)}((re^{i\theta})^*)$$

so

$$u(r\cos\theta, r\sin\theta) = \text{Re}(h_{(1)}(re^{i\theta}) + h_{(2)}((re^{i\theta})^*)) = \text{Re}\,h_{(1)}(re^{i\theta}) + \text{Re}\,h_{(2)}((re^{i\theta})^*)$$

$$= 2\,\text{Re}\,h_{(1)}(re^{i\theta}) \quad \text{for all} \quad 0 \leqslant r < 1, \theta \in \mathbb{T}.$$

In particular setting $h(z) = 2h_{(1)}(z) - \text{Im}\,a_0$ we see that h is analytic in $\{z:|z|<1\}$, that $\text{Re}\,h(x + iy) = u(x, y)$ and that $\text{Im}\,h(0) = \text{Im}\,a_0 - \text{Im}\,a_0 = 0$.

Setting $f(z) = h(\rho^{-1}(z - z_0))$ we see that f is a well defined analytic function on $\Omega'_{\mathbb{C}}$ with

$$\text{Re}\,f(x + iy) = \text{Re}\,h(\rho^{-1}(x - x_0) + i\rho^{-1}(y - y_0))$$

$$= u(\rho^{-1}(x - x_0), \rho^{-1}(y - y_0)) = \phi(x_0 + \rho\rho^{-1}(x - x_0), y_0 + \rho\rho^{-1}(y - y_0))$$

$$= \phi(x, y) \quad \text{for all} \quad (x, y) \in \Omega' \text{ and } \text{Im}\,f(x_0 + iy_0) = \text{Im}\,h(0) = 0 \qquad \blacksquare$$

We add for the sake of completeness rather than utility the following minor extension.

Lemma 31.7. *Let Ω be the disc $\{(x, y):(x - x_0)^2 + (y - y_0)^2 < R^2\}$ and $\phi:\Omega \to \mathbb{R}$ be a function such ϕ_{11} and ϕ_{22} exist with $\phi_{11} + \phi_{22} = 0$ in Ω. Then we can find a unique function $\psi:\Omega \to \mathbb{R}$ such that $\psi(x_0, y_0) = 0$ and $f(x + iy) = \phi(x, y) + i\psi(x, y)$ defines an analytic function on Ω.*

Proof. Let Ω_ρ be the disc $\{(x, y):(x - x_0)^2 + (y - y_0)^2 < \rho^2\}$. By Lemma 31.6 we know that for each $R > \rho > 0$ we can find a unique $\psi_\rho:\Omega_\rho \to \mathbb{R}$ such that $\psi_\rho(x_0, y_0) = 0$

and $f_\rho(x + iy) = \phi(x, y) + i\psi_\rho(x, y)$ defines an analytic function on $\Omega_{\rho C} = \{z : |z - z_0|$ $< \rho\}$. But if $R > \rho(1) \geqslant \rho(2) > 0$, we know that $\psi_{\rho(1)}|\Omega_{\rho(2)}$ (the restriction of $\psi_{\rho(1)}$ to $\Omega_{\rho(2)}$) takes the value 0 at (x_0, y_0) and is the imaginary part of an analytic function $f_{\rho(1)}|\Omega_{\rho(2)C}$ whose real part is $\phi|\Omega_{\rho(2)}$. Thus by the uniqueness condition $\psi_{\rho(2)} = \psi_{\rho(1)}|\Omega_{\rho(2)}$ and (emerging from a forest of notation)

$$f_{\rho(2)}(z) = f_{\rho(1)}(z) \text{ whenever } |z - z_0| < \rho(2).$$

Thus we may define a function f on Ω_C by the formula

$$f(z) = f_\rho(z) \text{ whenever } |z - z_0| < \rho.$$

If $z_1 \in \Omega_C$ then $|z_1 - z_0| < R$ so we may find a ρ with $|z_1 - z_0| < \rho < R$. Since $f(z) = f_\rho(z)$ for all $|z - z_0| < \rho$ and f_ρ is analytic at z_1, it follows that f is analytic at z_1. Also taking $z_1 = x_1 + iy_1$ (with $x_1, y_1 \in \mathbb{R}$ as usual) we have $\operatorname{Re} f(z_1) = \operatorname{Re} f_\rho(z_1) = \phi(x_1, y_1)$ as required. Since $\operatorname{Im} f(z_0) = \operatorname{Im} f_{R/2}(z_0) = \psi_{R/2}(x_0, y_0) = 0$ it follows that $\psi = \operatorname{Im} f$ is a function of the type desired. The uniqueness of ψ may be proved just as in Lemma 31.6. ∎

Thus, exactly as in Chapter 29, we can deduce that any function ϕ satisfying $\phi_{11} + \phi_{22} = 0$ is locally the real part of an analytic function.

Theorem 31.8. *Let Ω be an open set in $\mathbb{R}^2$ and let $\phi : \Omega \to \mathbb{R}$ be a function such that ϕ_{11} and ϕ_{22} exist with $\phi_{11} + \phi_{22} = 0$ in Ω. Then if $D = \{(x, y) : (x - x_0)^2 + (y - y_0)^2 < R^2\} \subset \Omega$ we can find a $\psi : D \to \mathbb{R}$ such that $f(x + iy) = \phi(x, y) + i\psi(x, y)$ defines an analytic function in $D_C = \{z : |z - z_0| < R\}$.*
Proof. Just apply Lemma 31.7. ∎

As an immediate corollary we obtain the following remarkable result.

Theorem 31.9. *Let Ω be an open set in $\mathbb{R}^2$ and let $\phi : \Omega \to \mathbb{R}$ be a function such that ϕ_{11} and ϕ_{22} exist with $\phi_{11} + \phi_{22} = 0$ in Ω. Then ϕ is infinitely differentiable in Ω.*
Proof. Let $(x_0, y_0) \in \Omega$. Pick an $R > 0$ such that $D = \{(x, y) : (x - x_0)^2 + (y - y_0)^2 < R^2\} \subset \Omega$. By Theorem 31.8 we can find an analytic function f on $D_C = \{z : |z - z_0| < R\}$ with $\operatorname{Re} f(x + iy) = \phi(x, y)$ on D. By Lemma 28.3, ϕ is infinitely differentiable on D and in particular at (x_0, y_0). Since (x_0, y_0) was an arbitrary point of Ω, we are done. ∎

For the sake of completeness we recall Theorem 29.9 although no new proof is required.

Theorem 31.10 (Gauss' mean value theorem). *Let Ω and ϕ be as in Theorem 31.9. Then if $R > \rho > 0$ and*

$$D = \{(x, y) : (x - x_0)^2 + (y - y_0)^2 < R^2\} \subset \Omega$$

we have

$$\phi(x_0, y_0) = \frac{1}{2\pi\rho} \int_C \phi(x, y) \, ds,$$

where C is a circular path described by $(x, y) = (x_0 + \rho \sin \theta, y_0 + \rho \cos \theta)$ as θ runs from 0 to 2π.

Proof. As for Theorem 29.9. ∎

In Chapter 28 we solved Dirichlet's problem for the disc (Problem 28.1). The (local) characterisation of solutions of Laplace's equation as the real part of analytic functions contained in Theorem 31.8 enables us to solve the problem for a wide range of open sets.

Theorem 31.11. *Let $D_C = \{z : |z| < 1\}$ and let $\bar{D}_C = \{z : |z| \leqslant 1\}$ be its closure, $\partial D_C = \{z : |z| = 1\}$ its boundary. Let Ω_C be an open set in $\mathbb{C}$ with closure $\bar{\Omega}_C$ and boundary $\partial \Omega_C$. Suppose that there exists a bijective map $T : \bar{\Omega}_C \to \bar{D}_C$ such that*

(a) *T is continuous on $\bar{\Omega}_C$ and T^{-1} is continuous on $\bar{D}_C$,*
(b) *$T(\partial \Omega_C) = \partial D_C$,*
(c) *T is analytic on Ω_C and so T^{-1} is analytic on D_C.*

Then, writing $\Omega = \{(x, y) \in \mathbb{R}^2 : x + iy \in \Omega_C\}$ and so on, if $G : \partial \Omega \to \mathbb{R}$ is continuous, we can find a $\phi : \bar{\Omega} \to \mathbb{R}$ such that

(i) *ϕ is infinitely differentiable with $\nabla^2 \phi = 0$ at all points of Ω,*
(ii) *ϕ is continuous at all points of $\bar{\Omega}$,*
(iii) *$\phi = G$ on $\partial \Omega$.*

Proof. The proof will involve a great deal of switching between $\mathbb{C}$ and $\mathbb{R}^2$ so we introduce a function $S : \mathbb{C} \to \mathbb{R}^2$ given by $S(x + iy) = (x, y)$ $[x, y \in \mathbb{R}]$. Define $H : \partial D \to \mathbb{R}$ by $H = GST^{-1}S^{-1}$. (Thus $H(x, y) = G(\text{Re}(T^{-1}(x + iy), \text{Im}(T^{-1}(x + iy))).$) Then H is continuous since G, S, T^{-1} and S^{-1} are. It follows from Theorem 28.5 that we can find a $u : \bar{D} \to \mathbb{R}$ such that

(i)′ *u is infinitely differentiable with $\nabla^2 u = 0$ at all points of D,*
(ii)′ *u is continuous at all points of $\bar{D}$,*
(iii)′ *$u = H$ on ∂D.*

Define $\phi : \bar{\Omega} \to \mathbb{R}$ by $\phi = uSTS^{-1}$. Then since u, S, T and S^{-1} are continuous we have

(ii) *u is continuous at all points of $\bar{\Omega}$ whilst, automatically,*
(iii) *$\phi = uSTS^{-1} = HSTS^{-1} = G$ on ∂D.*

All that remains is to prove (i). To this end let (x_0, y_0) be an arbitrary point of Ω. Then

$$(x_1, y_1) = STS^{-1}(x_0, y_0) \in D,$$

and so we can find a $\delta > 0$ such $D' = \{(x, y) \in \mathbb{R}: (x - x_1)^2 + (y - y_1)^2 < \delta^2\} \subset D$.
Condition (i)' allows us to apply Lemma 31.7 to show the existence of an analytic
$f:D'_\mathbb{c} \to \mathbb{C}$ with $\operatorname{Re} fS^{-1} = u$ on D' and so with

$$\operatorname{Re} fTS^{-1} = \operatorname{Re} fS^{-1}STS^{-1} = uSTS^{-1} = \phi$$

on $ST^{-1}D'_\mathbb{c}$. But $(x_1, y_1) \in D$ so, by the continuity of S and T^{-1}, we can find a
$\delta_1 > 0$ such that $|x - x_0|^2 + |y - y_0|^2 < \delta_1^2$ implies $(x, y) \in ST^{-1}D'_\mathbb{c}$. Thus $\phi(x, y) =$
$\operatorname{Re} fT(x + iy)$ for all (x, y) with $(x - x_0)^2 + (y - y_0)^2 < \delta^2$. But since f and T are
analytic so is fT and we have thus shown that ϕ is, locally, the real part of an
analytic function. Thus using Lemma 28.3 we see that

(i) ϕ is infinitely differentiable with $\nabla^2 \phi = 0$ at all points of Ω.   ∎

However Dirichlet's problem is not soluble for all bounded open domains. The
first example to show this was published by Zaremba in 1911 and is shockingly
simple.

Example 31.12. *Let $\Omega = \{(x, y): 0 < x^2 + y^2 < 1\}$ be the punctured disc. Then Ω is an
open bounded set with boundary*

$$\partial\Omega = \{(x, y): x^2 + y^2 = 1\} \cup \{(0, 0)\}$$

and closure

$$\bar\Omega = \{(x, y): x^2 + y^2 \leqslant 1\}.$$

*If we define $G:\partial\Omega \to \mathbb{R}$ by $G(x, y) = 0$ for $x^2 + y^2 = 1$ and $G(0, 0) = 1$ then G is
continuous on $\partial\Omega$. However we cannot find a $\phi:\bar\Omega \to \mathbb{R}$ such that*

 (i) *ϕ is twice differentiable with $\nabla^2 \phi = 0$ at all points of Ω,*
 (ii) *ϕ is continuous at all points of $\bar\Omega$,*
 (iii) *$\phi = G$ on $\partial\Omega$.*

Proof. In order to exploit the symmetries of the problem we use polar coordinates
with origin the center of $\bar\Omega$. Observe that if ϕ has properties (i), (ii), (iii) then so does
ϕ_ω where $\phi_\omega(r, \theta) = \phi(r, \theta + \omega)$. By the uniqueness property (Theorem 31.3) we see
that $\phi_\omega = \phi$ and so $\phi(r, \theta) = \phi(r, \theta + \omega)$ for all θ and ω. It follows that $\phi(r, \theta) = f(r)$
where conditions (i), (ii) and (iii) imply

 (i)' f is twice differentiable with $\dfrac{1}{r}\dfrac{d}{dr}\left(r\dfrac{df}{dr}\right) = 0$ for all $0 < r < 1$.

 (ii)' f is continuous on $[0, 1]$,
 (iii)' $f(0) = 1$, $f(1) = 0$.

Condition (i)' gives $f(r) = A + B\log r$ for some constants A and B. Since this is
incompatible with conditions (ii)' and (iii)' we are done.   ∎

Thus the minimum energy argument for the existence of a solution outlined

after Lemma 29.5 is not merely incomplete but can, under certain circumstances, lead to a false conclusion. We shall return briefly to this point in Chapter 35 (Example 35.5).

We began Chapter 29 with two problems. We have answered these completely but in doing so we have suggested a whole new range of questions. For example

(1) To what extent can our results be generalised to higher dimensions?
(2) To what extent can the objections of Weierstrass, Zaremba and Hadamard be bypassed and the elegant energy arguments of Dirichlet and his predecessors be saved?
(3) In Chapters 13 and 14 we exhibited a strong connection between Brownian motion and Dirichlet's problem. How should we interpret results like Theorem 31.11 and Example 31.12 in terms of Brownian motion?

The above questions also have been completely solved (in fact the reader should be able to make considerable progress with (1) and (3)) but the questions that they have given rise to in turn are still subjects for research.

PART III
ORTHOGONAL SERIES

32

MEAN SQUARE APPROXIMATION I

Let $C(\mathbb{T})$ be the set of continuous functions $f:\mathbb{T}\to\mathbb{C}$. Write $(f,g)=(2\pi)^{-1}\int_{\mathbb{T}}f(t)g(t)^*\,dt$ where z^* denotes the complex conjugate of z.

Lemma 32.1. *If $f,g,h\in C(\mathbb{T})$ and $\lambda,\mu\in\mathbb{C}$ then*

(i) $(f,g)=(g,f)^*$,

(ii) $(\lambda f+\mu g,h)=\lambda(f,h)+\mu(g,h)$,

(iii) (f,f) *is real and* $(f,f)\geqslant 0$,

(iv) *If* $(f,f)=0$ *then* $f=0$.

Proof.

(i)
$$(g,f)^*=\frac{1}{2\pi}\left(\int_{\mathbb{T}}g(t)f(t)^*\,dt\right)^*=\frac{1}{2\pi}\int_{\mathbb{T}}(g(t)f(t)^*)^*\,dt$$

$$=\frac{1}{2\pi}\int_{\mathbb{T}}f(t)^{**}g(t)^*\,dt=\frac{1}{2\pi}\int_{\mathbb{T}}f(t)g(t)^*\,dt=(f,g).$$

(ii)
$$(\lambda f+\mu g,h)=\frac{1}{2\pi}\int_{\mathbb{T}}(\lambda f(t)+\mu g(t))h^*(t)\,dt$$

$$=\frac{1}{2\pi}\int_{\mathbb{T}}\lambda f(t)h^*(t)+\mu g(t)h^*(t)\,dt=\lambda(f,h)+\mu(g,h).$$

(iii)
$$(f,f)=\frac{1}{2\pi}\int_{\mathbb{T}}f(t)f(t)^*\,dt=\frac{1}{2\pi}\int_{\mathbb{T}}|f(t)|^2\,dt\geqslant 0.$$

(iv) Recall that, if g is a continuous positive function on an interval $[a,b]$ with $a<b$, then $\int_a^b g(x)\,dx=0$ implies $g(x)=0$ for all $x\in[a,b]$. Setting $g(x)=|f(x)|^2$ we see that if $(f,f)=0$ then $|f(x)|^2=0$ for all $x\in\mathbb{T}$, so $f(x)=0$ for all $x\in\mathbb{T}$ and we are done. ∎

In the language of abstract algebra $C(\mathbb{T})$ is an (infinite dimensional) vector space

145

with $(\ ,\ )$ as an inner product. All inner products satisfy a very important inequality, special cases of which were discovered by Cauchy, Schwarz, Buniakowski and, probably, many others.

Lemma 32.2 (The Cauchy, Schwarz, Buniakowski inequality). *If f, $g \in C(\mathbb{T})$ then $|(f,g)|^2 \leqslant (f,f)(g,g)$ with equality if and only if $\lambda f + \mu g = 0$ for some $\lambda, \mu \in \mathbb{C}$ not both zero.*

Proof. If $f = g = 0$, then there is nothing to prove, so we may suppose without loss of generality that $f \neq 0$ and so, by Lemma 32.1 (iv), $(f,f) \neq 0$. Thus using the various results of Lemma 32.1, we have

$$0 \leqslant (\lambda f + \mu g, \lambda f + \mu g) = \lambda(f, \lambda f + \mu g) + \mu(g, \lambda f + \mu g)$$

$$= \lambda(\lambda f + \mu g, f)^* + \mu(\lambda f + \mu g, g)^* = \lambda(\lambda(f,f) + \mu(g,f))^* + \mu(\lambda(f,g) + \mu(g,g))^*$$

$$= \lambda\lambda^*(f,f) + \lambda\mu^*(g,f)^* + \mu\lambda^*(f,g)^* + \mu\mu^*(g,g)$$

$$= \left(\lambda(f,f)^{\frac{1}{2}} + \mu \frac{(g,f)}{(f,f)^{\frac{1}{2}}}\right)\left(\lambda(f,f)^{\frac{1}{2}} + \mu \frac{(g,f)}{(f,f)^{\frac{1}{2}}}\right)^* + \mu\mu^*\left((g,g) - \frac{|(g,f)|^2}{(f,f)}\right)$$

$$= \left|\lambda(f,f)^{\frac{1}{2}} + \mu \frac{(g,f)}{(f,f)^{\frac{1}{2}}}\right|^2 + |\mu|^2\left((g,g) - \frac{|(g,f)|^2}{(f,f)}\right) \quad \text{for all } \lambda, \mu \in \mathbb{C}.$$

In particular, taking $\mu = 1$, $\lambda = -(g,f)/(f,f)$, we see that

$$0 \leqslant \left((g,g) - \frac{|(g,f)|^2}{(f,f)}\right),$$

and so $|(g,f)|^2 \leqslant (f,f)(g,g)$, i.e. $|(f,g)|^2 \leqslant (f,f)(g,g)$ (since $(f,g) = (g,f)^*$) with equality only if

$$0 = (\lambda f + \mu g, \lambda f + \mu g),$$

i.e. (using Lemma 32.1 (iv)) only if

$$\lambda f + \mu g = 0,$$

where μ and λ have the values chosen at the beginning of the paragraph.

Conversely, if $\lambda f + \mu g = 0$ for some λ and μ with, say, $\lambda \neq 0$ then $f = -\lambda^{-1}\mu g$ and

$$|(f,g)|^2 = |\lambda^{-1}\mu|^2(g,g)^2 = ((\lambda^{-1}\mu)(\lambda^{-1}\mu)^*(g,g))(g,g) = (f,f)(g,g),$$

so we are done. ∎

Every inner product has an associated norm $\|\ \ \|_2$ given by $\|f\|_2 = (f,f)^{\frac{1}{2}}$ (where the positive square root is taken).

Lemma 32.3. *If $f, g \in C(\mathbb{T})$ and $\lambda \in \mathbb{C}$ then*

(i) $\|\lambda f\|_2 = |\lambda|\, \|f\|_2$,

(ii) $\|f\|_2 \geqslant 0$ with equality if and only if $f = 0$,

(iii) *(triangle inequality)* $\|f\|_2 + \|g\|_2 \geqslant \|f + g\|_2$.

Proof. (i) $\|\lambda f\|_2^2 = (\lambda f, \lambda f) = \lambda(f, \lambda f) = \lambda(\lambda f, f)^* = \lambda(\lambda(f, f))^* = \lambda\lambda^*(f, f)^* = |\lambda|^2(f, f) = |\lambda|^2 \|f\|_2^2.$

(ii) Immediate from the definition and Lemma 32.1 (iv).

(iii) Using Lemma 32.1 and Lemma 32.2 (the Cauchy, Schwartz, Buniakowski inequality) we have

$$\begin{aligned}
\|f + g\|_2^2 &= (f + g, f + g) = (f, f) + (g, f) + (f, g) + (g, g) \\
&= (f, f) + (f, g)^* + (f, g) + (g, g) = \|f\|_2^2 + 2\operatorname{Re}(f, g) + \|g\|_2^2 \\
&\leqslant \|f\|_2^2 + 2|(f, g)| + \|g\|_2^2 \leqslant \|f\|_2^2 + 2\|f\|_2\|g\|_2 + \|g\|_2^2 \\
&= (\|f\|_2 + \|g\|_2)^2,
\end{aligned}$$

and the result follows. $\blacksquare$

In the context of this part of the book the exponentials $e_n(t) = \exp int[t \in \mathbb{T}]$ have one outstanding characteristic: they are orthonormal.

Lemma 32.4. (i) $(e_n, e_n) = 1,$

(ii) $(e_n, e_m) = 0$ for $n \neq m.$

Proof. Trivial. $\blacksquare$

The great advantage of inner product spaces is that they allow the use of geometric analogy. For example, consider

Question A. *What values of $\lambda_{-n}, \lambda_{-n+1}, \ldots, \lambda_n \in \mathbb{C}$ (if any) minimise $\|f - \sum_{j=-n}^n \lambda_j e_j\|_2$?*

In geometrical language this may be restated as

Question B. *Which points g (if any) in the subspace $E = \{\sum_{j=-n}^n \lambda_j e_j : \lambda_j \in \mathbb{C}\}$ minimise $\|f - g\|_2$?*

Our geometrical intuition suggests the following answer.

Hypothesis C. *There is a unique point $g_0 \in E$ with $\|f - g\|_2 > \|f - g_0\|_2$ for all $g \in E$, $g \neq g_0$. This g_0 is given by the condition that $f - g_0$ be perpendicular to E.*

What does it mean to say that $f - g$ is perpendicular to E? Geometrically it means that $f - g_0$ is perpendicular to each $h \in E$, or, following the analogy with the physicist's inner product on $\mathbb{R}^3$, that $(f - g_0, h) = 0$ for each $h \in E$. In particular since $e_j \in E$ it follows that $(f - g_0, e_j) = 0$ and so $(f, e_j) = (g_0, e_j)$ for each $-n \leqslant j \leqslant n$. But if $g_0 \in E$ then $g_0 = \sum_{k=-n}^n \mu_k e_k$ for some $\mu_k \in \mathbb{C}$ $[-n \leqslant k \leqslant n]$, where we have $(g_0, e_j) = \sum_{j=-n}^n \mu_k(e_k, e_j) = \mu_j$, so that $\mu_j = (f, e_j)[-n \leqslant j \leqslant n]$ and we have

$g_0 = \sum_{j=-n}^{n} (f, e_j) e_j$. Hypothesis C can thus be rewritten algebraically to give the following theorem.

Theorem 32.5. *If* $g_0 = \sum_{j=-n}^{n} (f, e_j) e_j$ *and* $g = \sum_{j=-n}^{n} \lambda_j e_j$ *then*

$$\|f\|_2^2 \geqslant \sum_{j=-n}^{n} |(f, e_j)|^2$$

and

$$\|f - g\|_2 \geqslant \|f - g_0\|_2 = \sqrt{\left(\|f\|_2^2 - \sum_{j=-n}^{n} |(f, e_j)|^2 \right)}$$

with equality if and only if $\lambda_j = (f, e_j) [-n \leqslant j \leqslant n]$.
Proof.

$$\|f - g\|_2^2 = \left(f - \sum_{j=-n}^{n} \lambda_j e_j, f - \sum_{j=-n}^{n} \lambda_j e_j \right)$$

$$= (f, f) - \sum_{j=-n}^{n} \lambda_j (e_j, f) - \sum_{j=-n}^{n} \lambda_j^* (f, e_j) + \sum_{j=-n}^{n} \sum_{k=-n}^{n} \lambda_j \lambda_k^* (e_j, e_k)$$

$$= (f, f) - \sum_{j=-n}^{n} (\lambda_j (f, e_j)^* + \lambda_j^* (f, e_j)) + \sum_{j=-n}^{n} \lambda_j \lambda_j^*$$

$$= (f, f) - \sum_{j=-n}^{n} (f, e_j)(f, e_j)^* + \sum_{j=-n}^{n} (\lambda_j - (f, e_j))(\lambda_j - (f, e_j))^*$$

$$= (f, f) - \sum_{j=-n}^{n} |(f, e_j)|^2 + \sum_{j=-n}^{n} |\lambda_j - (f, e_j)|^2$$

$$\geqslant (f, f) - \sum_{j=-n}^{n} |(f, e_j)|^2 = \|f\|_2^2 - \sum_{j=-n}^{n} |(f, e_j)|^2$$

with equality if and only if $\lambda_j = (f, e_j) [-n \leqslant j \leqslant n]$. Setting $\lambda_j = (f, e_j)$, we thus obtain

$$\|f - g_0\|_2^2 = \|f\|_2^2 - \sum_{j=-n}^{n} |(f, e_j)|^2,$$

so

$$\|f\|_2^2 - \sum_{j=-n}^{n} |(f, e_j)|^2 \geqslant 0$$

and the full result now follows. ∎

Let us call

$$\|f - g\|_2 = \sqrt{\left(\frac{1}{2\pi} \int |f(t) - g(t)|^2 \, dt \right)}$$

the mean square distance between f and g. Theorem 32.5 can now be rewritten as follows.

Theorem 32.6. *Let* $f: \mathbb{T} \to \mathbb{C}$ *be continuous. Then, among all functions of the form*

$\sum_{j=-n}^{n} \lambda_j e_j$, *the best mean square approximation to f is given by* $S_n(f,)$. *Further,*

$$\frac{1}{2\pi}\int_{\mathbb{T}}\left|f(t) - \sum_{j=-n}^{n} \lambda_j \exp ijt\right|^2 dt \geq \frac{1}{2\pi}\int_{\mathbb{T}} |f(t)|^2 dt - \sum_{j=-n}^{n} |\hat{f}(j)|^2, \quad \text{for all } \lambda_j \in \mathbb{C}.$$

Proof. Observe that $\hat{f}(j) = (f, e_j)$ and apply Theorem 32.5. ∎

Thus, although the Fourier sum $S_n(f,)$ may not be a good *uniform* approximation (see Chapter 11 and elsewhere) it is always the best *mean square* approximation to f.

33

MEAN SQUARE APPROXIMATION II

Returning, briefly, to more general arguments, we see that Theorem 32.5 implies an inequality which is valid for any orthonormal set in an inner product space.

Theorem 33.1 (Bessel's inequality). *If* $f \in C(\mathbb{T})$ *then*

$$\|f\|_2^2 \geqslant \sum_{n=-\infty}^{\infty} |(f,e_n)|^2.$$

Proof. By Theorem 32.5

$$\|f\|_2^2 \geqslant \sum_{n=-m}^{m} |(f,e_n)|^2$$

so, since $|(f,e_n)|^2 \geqslant 0$, it follows by basic theorems of analysis that $\sum_{n=-\infty}^{\infty} |(f,e_n)|^2$ converges and $\|f\|_2^2 \geqslant \sum_{n=-\infty}^{\infty} |(f,e_n)|^2$. ∎

Translated back into the language of Fourier series Theorem 33.1 takes the following form.

Lemma 33.2. *If* $f:\mathbb{C} \to \mathbb{T}$ *is continuous then*

$$\frac{1}{2\pi}\int |f(t)|^2 \, dt \geqslant \sum_{n=-\infty}^{\infty} |\hat{f}(n)|^2.$$

Proof. Just rewrite Theorem 33.1. ∎

In the next chapter we shall see that if we make use of special (and non trivial) properties of Fourier series we can replace the inequality of Lemma 33.2 by equality. However, in the remainder of this chapter we shall show that even restricting ourselves to the much simpler Bessel's inequality we can draw interesting conclusions not only about mean square but also about uniform approximation by trigonometric polynomials.

Lemma 33.3. (i) *If* $f:\mathbb{T}\to\mathbb{C}$ *is continuous and* $\lambda_j\in\mathbb{C}$ *then*

(i) *If* $f:\mathbb{T}\to\mathbb{C}$ *is continuous and* $\lambda_j\in\mathbb{C}$ *then*

$$\frac{1}{2\pi}\int_{\mathbb{T}}\left|f(t)-\sum_{j=-n}^{n}\lambda_j\exp ijt\right|^2 dt \geqslant \sum_{|j|>n}|\hat{f}(j)|^2.$$

(ii) *If* $f:\mathbb{T}\to\mathbb{C}$ *is continuous and* $\lambda_j\in\mathbb{C}$ *then*

$$\sup_{t\in\mathbb{T}}\left|f(t)-\sum_{j=-n}^{n}\lambda_j\exp ijt\right| \geqslant \sqrt{\left(\sum_{|j|>n}|\hat{f}(j)|^2\right)}.$$

Proof. (i) Set
$$g(t)=f(t)-\sum_{j=-n}^{n}\lambda_j\exp ijt.$$

Then $\hat{g}(j)=\hat{f}(j)$ for all $|j|>n$ and so, using Lemma 33.2,

$$\frac{1}{2\pi}\int_{\mathbb{T}}\left|f(t)-\sum_{j=-n}^{n}\lambda_j\exp ijt\right|^2 dt = \frac{1}{2\pi}\int_{\mathbb{T}}|g(t)|^2 dt$$

$$\geqslant \sum_{j=-\infty}^{\infty}|\hat{g}(j)|^2 \geqslant \sum_{|j|>n}|\hat{g}(j)|^2 = \sum_{|j|>n}|\hat{f}(j)|^2.$$

(Alternatively we could have obtained this result by combining Theorem 32.6 and Lemma 33.2.)

(ii) Observe that

$$\left(\sup_{t\in\mathbb{T}}\left|f(t)-\sum_{j=-n}^{n}\lambda_j\exp ijt\right|\right)^2 = \sup_{t\in\mathbb{T}}\left|f(t)-\sum_{j=-n}^{n}\lambda_j\exp ijt\right|^2$$

$$= \frac{1}{2\pi}\int_{\mathbb{T}}\sup_{t\in\mathbb{T}}\left|f(t)-\sum_{j=-n}^{n}\lambda_j\exp ijt\right|^2 ds$$

$$\geqslant \frac{1}{2\pi}\int_{\mathbb{T}}\left|f(s)-\sum_{j=-n}^{n}\lambda_j\exp ijs\right|^2 ds \geqslant \sum_{|j|>n}|\hat{f}(j)|^2$$

using (i). ∎

Applying this to the 'plucked string' function h of Chapter 10 we get the following result.

Example 33.4. *Consider the function* $h:\mathbb{T}\to\mathbb{R}$ *given by* $h(t)=\pi/2-|t|[0\leqslant|t|\leqslant\pi]$. *Then for any choice of* $\lambda_j\in\mathbb{C}$ *we have*

$$\sup_{t\in\mathbb{T}}\left|h(t)-\sum_{j=-n}^{n}\lambda_j\exp ijt\right| \geqslant \frac{2}{\pi3^{\frac{1}{2}}}\frac{1}{(n+3)^{\frac{3}{2}}}.$$

Proof. From Chapter 10 (or by redoing the calculation) we know that

$$\hat{h}(r)=\begin{cases}0 & \text{for } r \text{ even,}\\ 2/\pi r^2 & \text{for } r \text{ odd.}\end{cases}$$

Thus, using Lemma 33.3,

$$\left(\sup_{t\in\mathbb{T}}\left|h(t)-\sum_{j=-n}^{n}\lambda_j\exp ijt\right|\right)^2\geqslant\sum_{|r|>n}|\hat{h}(r)|^2\geqslant 2(2/\pi)^2\sum_{r\geqslant n+1,\text{rodd}}r^{-4}.$$

But

$$2\sum_{r\geqslant n+1,\text{rodd}}r^{-4}\geqslant\sum_{r\geqslant n+2}r^{-4}\geqslant\sum_{r\geqslant n+2}(r(r+1)(r+2)(r+3))^{-1}$$

$$=3^{-1}\sum_{r=n+2}^{\infty}((r(r+1)(r+2))^{-1}-((r+1)(r+2)(r+3))^{-1})$$

$$=3^{-1}((n+2)(n+3)(n+4))^{-1}\geqslant 3^{-1}(n+3)^{-3},$$

and the result follows. ∎

Thus any good uniform approximation to h by a trigonometric polynomial must have very high degree.

The final result of this chapter requires several, more or less well known, facts for its proof. For the convenience of the reader these are presented in a couple of preliminary lemmas.

Lemma 33.5. (i) *if x_j, $y_j\in\mathbb{C}$ then*

$$\left|\sum_{j=1}^{n}x_jy_j^*\right|\leqslant\left(\sum_{j=1}^{n}|x_j|^2\right)^{\frac{1}{2}}\left(\sum_{j=1}^{n}|y_j|^2\right)^{\frac{1}{2}}.$$

(ii) *(Cauchy's inequality) If a_j, $b_j\in\mathbb{C}$ then*

$$\sum_{j=1}^{n}|a_j||b_j|\leqslant\left(\sum_{j=1}^{n}|a_j|^2\right)^{\frac{1}{2}}\left(\sum_{j=1}^{n}|b_j|^2\right)^{\frac{1}{2}}.$$

Proof. (i) If $\mathbf{x},\mathbf{y}\in\mathbb{C}^n$, set $(\mathbf{x},\mathbf{y})=\sum_{j=1}^{n}x_jy_j^*$. It is easy to check that

    (a) $(\mathbf{x},\mathbf{y})=(\mathbf{y},\mathbf{x})^*$,
    (b) $(\lambda\mathbf{x}+\mu\mathbf{y},\mathbf{w})=\lambda(\mathbf{x},\mathbf{w})+\mu(\mathbf{y},\mathbf{w})$,
    (c) $(\mathbf{x},\mathbf{x})$ is real and $(\mathbf{x},\mathbf{x})\geqslant 0$,
    (d) If $(\mathbf{x},\mathbf{x})=0$ then $\mathbf{x}=\mathbf{0}$,

whenever $\mathbf{x},\mathbf{y},\mathbf{w}\in\mathbb{C}^n$ and $\lambda,\mu\in\mathbb{C}$. Thus exactly the same proof that gave Lemma 32.2 from Lemma 32.1 shows that

$$|(\mathbf{x},\mathbf{y})|^2\leqslant(\mathbf{x},\mathbf{x})(\mathbf{y},\mathbf{y}),$$

i.e. that

$$\left|\sum_{j=1}^{n}x_jy_j^*\right|^2\leqslant\sum_{j=1}^{n}|x_j|^2\sum_{j=1}^{n}|y_j|^2,$$

which is the result required.
(ii) Set $x_j=|a_j|$, $y_j=|b_j|$ in (i). ∎

Lemma 33.6.

(i) $\sum_{n=1}^{\infty} 1/n^2$ *converges*
(ii) $\sum_{n=0}^{\infty} 1/(2n+1)^2$ *converges*
(iii) $\sum_{n=0}^{\infty} 1/(2n+1)^2 = \pi^2/8$
(iv) $\sum_{n=1}^{\infty} 1/n^2 = \pi^2/6$

Proof. (i) Note, for example, that

$$\sum_{n=1}^{N} (n(n+1))^{-1} = \sum_{n=1}^{N} (n^{-1} - (n+1)^{-1}) = 1 - (N+1)^{-1}$$

converges. But $0 \leqslant n^{-2} \leqslant 2(n(n+1))^{-1} [n \geqslant 1]$ so $\sum_{n=1}^{\infty} n^{-2}$ converges by the comparison test.

(ii) Similarly $(2n+1)^{-2} \leqslant (n+1)^{-2}$ for $n \geqslant 0$.

(iii) Let $h(t) = \pi/2 - |t| [0 \leqslant t \leqslant \pi]$. Then as we noted in Chapter 10

$$\hat{h}(r) = \begin{cases} 0 & \text{for } r \text{ even,} \\ 2/\pi r^2 & \text{for } r \text{ odd.} \end{cases}$$

Using (ii) we see that $\Sigma|\hat{h}(r)|$ is absolutely convergent and thus, by Theorem 9.2,

$$S_n(h, t) \to h(t) \quad \text{for all } t \in \mathbb{T}.$$

In particular, $S_n(h, 0) \to h(0)$, i.e. $\sum_{r=-\infty}^{\infty} 2/\pi(2r+1)^2 = \pi/2$, and so $\sum_{r=0}^{\infty} 1/(2r+1)^2 = \pi^2/8$.

(iv) $$\pi^2/8 = \sum_{r=0}^{\infty} 1/(2r+1)^2 = \sum_{n=1}^{\infty} 1/n^2 - \sum_{n=1}^{\infty} 1/(2n)^2 = \left(\frac{3}{4}\right) \sum_{n=1}^{\infty} 1/n^2$$

whence the result. ∎

We can now obtain an improvement on some of the results of Chapter 9.

Theorem 33.7. *Suppose* $f : \mathbb{T} \to \mathbb{C}$ *has a continuous derivative. Then*

$$\sum_{r=-\infty}^{\infty} |\hat{f}(r)| \leqslant |\hat{f}(0)| + \left(\frac{\pi}{6} \int_{-}^{} |f'(t)|^2 \, dt \right)^{\frac{1}{2}}.$$

(*Thus in particular* $\sum_{-\infty}^{\infty} \hat{f}(r) \exp irt$ *is uniformly absolutely convergent.*)

Proof. Recall that

$$\hat{f'}(r) = \frac{1}{2\pi} \int_{-\pi}^{\pi} f'(t) \exp(-irt) \, dt$$

$$= \left[\frac{1}{2\pi} f(t) \exp(-irt) \right]_{-\pi}^{\pi} + \frac{ir}{2\pi} \int_{-\pi}^{\pi} f(t) \exp(-irt) \, dt = ir\hat{f}(r).$$

Thus, using Lemma 33.5 (ii) (Cauchy's inequality), Lemma 33.2 (Bessel's inequality),

and Lemma 33.6 (iv) we have

$$\sum_{r=-n}^{n} |\hat{f}(r)| = |\hat{f}(0)| + \sum_{0 \neq |r| \leqslant n} r^{-1} |r\hat{f}(r)|$$

$$\leqslant |\hat{f}(0)| + \left(\sum_{0 \neq |r| \leqslant n} r^{-2}\right)^{\frac{1}{2}} \left(\sum_{0 \neq |r| \leqslant n} |r\hat{f}(r)|^2\right)^{\frac{1}{2}}$$

$$= |\hat{f}(0)| + \left(2\sum_{r=1}^{n} r^{-2}\right)^{\frac{1}{2}} \left(\sum_{0 \neq |r| \leqslant n} |\hat{f}'(r)|^2\right)^{\frac{1}{2}}$$

$$\leqslant |\hat{f}(0)| + \left(2\sum_{r=1}^{n} r^{-2}\right)^{\frac{1}{2}} \left(\frac{1}{2\pi}\int_{\mathbb{T}} |f'(t)|^2 \, dt\right)^{\frac{1}{2}}$$

$$\leqslant |\hat{f}(0)| + \left(\frac{\pi^2}{3}\right)^{\frac{1}{2}} \left(\frac{1}{2\pi}\int_{\mathbb{T}} |f'(t)|^2 \, dt\right)^{\frac{1}{2}}$$

$$= |\hat{f}(0)| + \left(\frac{\pi}{6}\right)^{\frac{1}{2}} \left(\int_{\mathbb{T}} |f'(t)|^2 \, dt\right)^{\frac{1}{2}}$$

for all $n \geqslant 1$. Thus

$$\sum_{r=-\infty}^{\infty} |\hat{f}(r)| \leqslant |\hat{f}(0)| + \left(\frac{\pi}{6}\right)^{\frac{1}{2}} \left(\int_{\mathbb{T}} |f'(t)|^2 \, dt\right)^{\frac{1}{2}},$$

as claimed.   ∎

34

MEAN SQUARE CONVERGENCE

In Chapter 32 we saw, using arguments which applied to any orthonormal sequence, that the mean square distance $\| f - \sum_{r=-n}^{n} \lambda_r e_r \|_2$ between $f \in C(\mathbb{T})$ and a trigonometric polynomial $\sum_{r=-n}^{n} \lambda_r e_r$ is minimised when $\sum_{r=-n}^{n} \lambda_r e_r = S_n(f,)$. However, it does not follow from the fact that something is a best approximation that it is a good approximation. In the next theorem we show that $\| f - S_n(f,) \|_2 \to 0$ as $n \to \infty$, but to do this we make use of special and fairly deep properties of Fourier series. (In what follows we write $S_n f = S_n(f,)$ to simplify the notation.)

Theorem 34.1. *Let $f: \mathbb{T} \to \mathbb{C}$ be continuous. Then*

$$\| f - S_n f \|_2 \to 0 \quad as \ n \to \infty.$$

Proof. The basic result on Fourier series that we use is Theorem 2.5 which says that the trigonometric polynomials are uniformly dense in $C(\mathbb{T})$. Thus given $\varepsilon > 0$ we can find a trigonometric polynomial $P = \sum_{j=-m}^{m} a_j e_j$, say, with $|P(t) - f(t)| < \varepsilon$ for all $t \in \mathbb{T}$. In particular,

$$\| P - f \|_2 = \left(\frac{1}{2\pi} \int |P(t) - f(t)|^2 \, dt \right)^{\frac{1}{2}} \leqslant \varepsilon.$$

Now, if $n \geqslant m$, then $S_n(P) = P$ and so

$$\| f - S_n f \|_2 \leqslant \| f - P \|_2 + \| P - S_n P \|_2 + \| S_n P - S_n f \|_2$$

$$= \| f - P \|_2 + \| S_n P - S_n f \|_2 = \| f - P \|_2 + \| S_n(P - f) \|_2 \leqslant 2 \| f - P \|_2 \leqslant 2\varepsilon$$

(the inequality $\| S_n g \|_2 \leqslant \| g \|_2$ coming from Theorem 32.5). We have thus shown that $\| f - S_n f \|_2 \to 0$ as $n \to \infty$. ∎

We say that $S_n f$ *converges in mean square* to f. Notice that the example of Chapter 18 shows that mean square convergence need not imply uniform or even pointwise convergence. (As a simpler example, set $h_n(x) = \sqrt{(n(1 - n^2 |x|))}$ for $|x| \leqslant n^{-2}$, $h_n(x) = 0$ otherwise. Then $\| h_n - 0 \|_2 = \| h_n \|_2 = (2\pi n)^{-\frac{1}{2}} \to 0$ as $n \to \infty$ but $h_n(0) \to \infty$ as $n \to \infty$.)

155

As a direct consequence of the theorem above we can replace the inequality in Bessel's inequality by equality.

Theorem 34.2 (Parseval's formula). *If* $f:\mathbb{T}\to\mathbb{C}$ *is continuous then*

$$\|f\|_2^2 = \sum_{r=-\infty}^{\infty} |\hat{f}(r)|^2.$$

(In other words, $(2\pi)^{-1}\int|f(t)|^2\,dt = \sum_{r=-\infty}^{\infty}|\hat{f}(r)|^2$.)

Proof. Much as in Theorem 32.5 (with $\lambda_j = \hat{f}(j)$) we have

$$\|f - S_n f\|_2^2$$
$$= (f - S_n f, f - S_n f) = (f,f) - (S_n f, f) - (S_n f, f)^* + (S_n f, S_n f)$$
$$= (f,f) - \left(\sum_{j=-n}^{n} \hat{f}(j)e_j, f\right) - \left(\sum_{j=-n}^{n} \hat{f}(j)e_j, f\right)^* + \left(\sum_{j=-n}^{n} \hat{f}(j)e_j, \sum_{k=-n}^{n} \hat{f}(k)e_k\right)$$
$$= (f,f) - \sum_{j=-n}^{n} \hat{f}(j)\hat{f}(j)^* - \sum_{j=-n}^{n} \hat{f}(j)\hat{f}(j)^* + \sum_{j=-n}^{n} \hat{f}(j)\hat{f}(j)^*$$
$$= \|f\|_2^2 - \sum_{j=-n}^{n} |\hat{f}(j)|^2.$$

Since $\|f - S_n f\|_2^2 \to 0$, it follows that $\sum_{j=-n}^{n}|\hat{f}(j)|^2$ converges to $\|f\|_2^2$. ∎

We remark that the reasoning by which we deduced Theorem 34.2 from Theorem 34.1 could be reversed to deduce Theorem 34.1 from Theorem 34.2. We also note the strong family resemblance between Theorem 34.2 and the theorem of Pythagoras which states that the square on the hypotenuse of a right angled triangle is equal to the sum of the squares on the other two sides.

If f and g are trigonometric polynomials of degree less than N, say, then simple manipulation shows that

$$(f,g) = \frac{1}{2\pi}\int_{\mathbb{T}} f(t)g(t)^*\,dt = \frac{1}{2\pi}\int_{\mathbb{T}} \left(\sum_{r=-N}^{N} \hat{f}(r)\exp irt\right)\left(\sum_{s=-N}^{N} \hat{g}(s)\exp ist\right)^*\,dt$$

$$= \sum_{r=-N}^{N}\sum_{s=-N}^{N} \hat{f}(r)\hat{g}(s)^* \frac{1}{2\pi}\int_{\mathbb{T}} \exp i(r-s)t\,dt = \sum_{r=-N}^{N} \hat{f}(r)\hat{g}(r)^* = \sum_{r=-\infty}^{\infty} \hat{f}(r)\hat{g}(r)^*.$$

In the next theorem we shall show that in fact the formula $(f,g) = \sum_{r=-\infty}^{\infty} \hat{f}(r)\hat{g}(r)^*$ holds for all continuous f and g. The key to its proof lies in the observation that the special case when $f = g$ has already been proved in Theorem 34.2 and that the general case may be obtained from it by simple manipulation along the lines laid out in the next lemma.

Lemma 34.3. (i) *If* $z_1, z_2 \in \mathbb{C}$
$$z_1 z_2^* = \tfrac{1}{4}(|z_1 + z_2|^2 - |z_1 - z_2|^2 + i|z_1 + iz_2|^2 - i|z_1 - iz_2|^2).$$

(ii) *If f, g: $\mathbb{T} \to \mathbb{C}$ are continuous*

$$(f, g) = \tfrac{1}{4}(\| f + g \|_2^2 - \| f - g \|_2^2 + i \| f + ig \|_2^2 - i \| f - ig \|_2^2).$$

Proof. (i) Observe first that

$$|z_1 + z_2|^2 = (z_1 + z_2)(z_1 + z_2)^* = |z_1|^2 + z_1 z_2^* + z_1^* z_2 + |z_2|^2,$$

so that

$$|z_1 + z_2|^2 - |z_1 - z_2|^2 = 2(z_1 z_2^* + z_1^* z_2),$$

from which we deduce that

$$(|z_1 + z_2|^2 - |z_1 - z_2|^2) + i(|z_1 + iz_2|^2 - |z_1 - iz_2|^2)$$
$$= 2(z_1 z_2^* + z_1^* z_2) + 2i(- iz_1 z_2^* + iz_1^* z_2) = 4z_1 z_2^*.$$

(ii) Similar. ∎

Theorem 34.4. *If f, g: $\mathbb{T} \to \mathbb{C}$ are continuous then*

$$(f, g) = \sum_{r = -\infty}^{\infty} \hat{f}(r) \hat{g}(r)^*.$$

In other words,

$$\frac{1}{2\pi} \int_{\mathbb{T}} f(t) g(t)^* \, dt = \sum_{r = -\infty}^{\infty} \hat{f}(r) \hat{g}(r)^*.$$

Proof. Using Lemma 34.3 and Theorem 34.2,

$$\sum_{r = -n}^{n} \hat{f}(r) \hat{g}(r)^* = \frac{1}{4} \left(\sum_{r = -n}^{n} |\hat{f}(r) + \hat{g}(r)|^2 - \sum_{r = -n}^{n} |\hat{f}(r) - \hat{g}(r)|^2 \right.$$

$$\left. + i \sum_{r = -n}^{n} |\hat{f}(r) + i\hat{g}(r)|^2 - i \sum_{r = -n}^{n} |\hat{f}(r) - i\hat{g}(r)|^2 \right)$$

$$= \frac{1}{4} \left(\sum_{r = -n}^{n} |(f + g)\hat{\,}(r)|^2 - \sum_{r = -n}^{n} |(f - g)\hat{\,}(r)|^2 \right.$$

$$\left. + i \sum_{r = -n}^{n} |(f + ig)\hat{\,}(r)|^2 - i \sum_{r = -n}^{n} |(f - ig)\hat{\,}(r)|^2 \right)$$

$$\to \frac{1}{4} (\| f + g \|_2^2 - \| f - g \|_2^2 + i \| f + ig \|_2^2 - i \| f - ig \|_2^2)$$

$$= (f, g) \quad \text{as } n \to \infty,$$

which is the required result. ∎

Before leaving this chapter the reader should convince herself that any two of Theorems 34.1, 34.2 and 34.4 can readily be deduced from the third. We remind the reader, once again, that (with the exception of Lemma 34.3 which holds for

any inner product space) the results of this chapter have only been proved for Fourier series and need not hold for general orthonormal systems.

(As a simple example let $s_n = (e_n - e_{-n})/2^{1/2}i$. Then the $s_1, s_2, \ldots$ form an orthonormal sequence in $C(\mathbb{T})$ but $(1, s_j) = 0$ for all $j \geq 1$ and so $\|1 - \sum_{j=1}^{n}(1, s_j)s_j\|_2 = 1 \nrightarrow 0$ as $n \to \infty$, whilst $\sum_{j=1}^{\infty}|(1, s_j)|^2 = 0 \neq \|1\|_2^2$. Thus the analogues of Theorems 34.1, 34.2 and 34.4 all fail.)

35

THE ISOPERIMETRIC PROBLEM I

In ancient times the extent of a city or an armed camp was often given in terms of its perimeter (so that a town would be described as requiring so many thousand paces to walk round). In the same way, according to Proclus, certain socialistic communities used to divide land so that each family received a plot of equal perimeter and it may have been in this context that it was first realised that a square contains a much greater area than a long thin rectangle of the same perimeter.

Once it was understood that figures with the same perimeter may contain different areas it was natural to ask whether there exists a figure of maximum area. It is not hard to guess that the answer is a circle but a guess is not a proof. The isoperimetric problem thus asks for a proof that among all figures of equal perimeter the circle has greatest area.

This question formed the subject of one of the last substantial investigations of the golden age of Greek geometry. In it Zenodorus proved that the circle has greater area than any polygon of the same perimeter.

We might expect that a purely geometrical approach could not go much further in the absence of precise notions of area and length. However, in 1841 Steiner showed how simple geometric considerations could be used to prove the following theorem.

Theorem 35.1. *If there exists a figure Γ_0 whose area is never less than that of any other figure of the same perimeter, then Γ_0 is a circle.*

Steiner gave five proofs of this result. We sketch a proof based on his first argument. (Since this is a detour I have not hesitated to argue informally but it should be clear that the argument can be made as rigorous as the reader requires.)

We start with a preliminary simplification.

Lemma 35.2. *With the notation above, Γ_0 is convex.*

Sketch of proof. Suppose Γ_0 is not convex. Then we can find a straight line intersecting the boundary of Γ_0 at A, B, C, D (and possibly elsewhere) in such a way

159

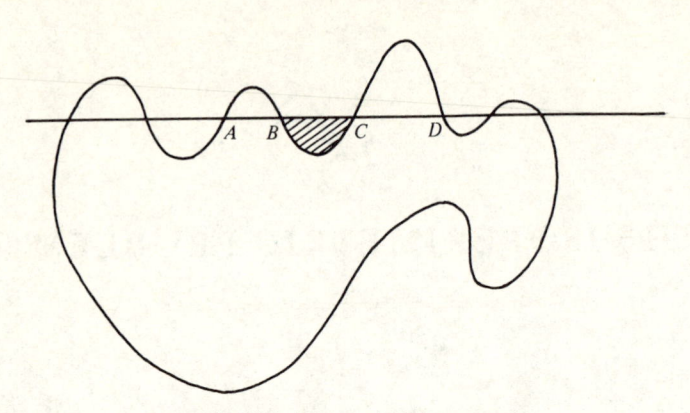

Fig. 35.1. Why must Γ_0 be convex?

that the segments AB and CD lie within Γ_0 and the segment BC outside (see Figure 35.1). If we replace the arc BC of the boundary by the segment BC we obtain a curve of smaller perimeter (since a straight line is the shortest distance between two points) and larger area (since the area enclosed by the straight line segment BC and the arc CB is now included). This contradicts the definition of Γ_0.   ∎

Next, instead of investigating Γ_0 directly, we construct new more symmetric figures as follows. Let A be a point on the perimeter of Γ_0 and B that point on the perimeter of Γ_0 whose distance from A along the boundary is exactly half the length of the perimeter of Γ_0. We see that the boundary of Γ_0 is made up of two arcs γ_1 and γ_2 from A to B each of length half the perimeter of Γ_0. Since Γ_0 is convex γ_1 and γ_2 only intersect the straight line segment γ_0 from A to B at A and B.

If O is the midpoint of AB we take γ_j' to be the reflection of γ_j in O and Γ_j to be the figure bounded by the arcs γ_j and $\gamma_j'[j = 1, 2]$. (The construction is illustrated in Figure 35.2.)

Lemma 35.3. Γ_1 and Γ_2 have the same perimeter and area as Γ_0. In particular the areas of Γ_1 and Γ_2 are never less than that of any other figure with the same perimeter.

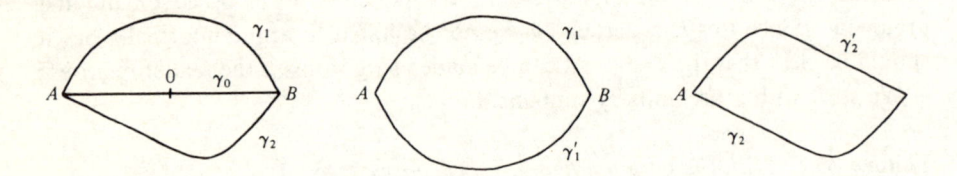

Fig. 35.2. Construction of Γ_1 and Γ_2 from Γ.

Proof. Observe that (in a faintly abusive notation)

$$\text{perimeter } \Gamma_j = \text{length } \gamma_j + \text{length } \gamma'_j = 2 \text{ length } \gamma_j = \text{perimeter } \Gamma_0 \quad [j=1,2]$$

as required. It follows by the maximal property of Γ_0 that

$$\text{area } \Gamma_0 \geqslant \text{area } \Gamma_1, \text{ area } \Gamma_2.$$

But by symmetry

$$\text{area } \Gamma_j = 2 \text{ area bounded by } \gamma_j \text{ and } \gamma_0 \quad [j=1,2],$$

so that

$$\text{area } \Gamma_1 + \text{area } \Gamma_2$$

$$= 2(\text{area bounded by } \gamma_1 \text{ and } \gamma_0 + \text{area bounded by } \gamma_2 \text{ and } \gamma_0) = 2 \text{ area } \Gamma_0,$$

so, in fact, area $\Gamma_1 = $ area $\Gamma_2 = $ area Γ_0. ∎

Lemma 35.4. Γ_1 *and* Γ_2 *are circles.*

Sketch of proof. Recall that if Σ is a closed curve and A and B are fixed points such that the angle $A\hat{C}B$ is a right angle for all $C \in \Sigma$ then Σ is a circle with diameter AB. (If you do not know this result it can be obtained by simple trigonometry. It suffices to show that, if O is the mid point of AB, then OC has constant length.) We shall thus have proved the lemma if we show that if C is any point on the boundary of $\Gamma_j [j=1,2]$ and if A and B are the points used in the construction of Γ_j above, then $A\hat{C}B$ is a right angle.

To prove this we consider C_* the reflection of C in O (notice that the symmetric construction of Γ_j ensures that C_* is also on the boundary of Γ_j). We can thus consider Γ_j to be made up of distinct pieces T_1, T_2, T_3, T_4, T_5 with T_5 the parallelogram $ACBC_*$, T_1 the part of Γ sitting on AC (i.e. the part of Γ bounded by the straight line AC and the shorter arc τ_1 along the boundary of Γ connecting A and C), T_2 that sitting on CB, T_3 that on BC_* and T_4 that on C_*A (see Figure 35.3).

Suppose that $A\hat{C}B$ is not a right angle. Consider a new diagram Γ'_j consisting of a rectangle $T'_5 = A'C'B'C'_*$ (with the length of the side $A'C'$ equal to the length AC, the length $C'B'$ equal to the length CB and so on) together with figures T'_1, T'_2, T'_3, T'_4, congruent to T_1, T_2, T_3, T_4 and sitting on the sides $A'C', C'B', B'C'_*, C'_*A$

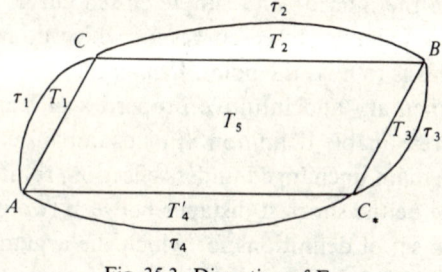

Fig. 35.3. Dissection of Γ_j.

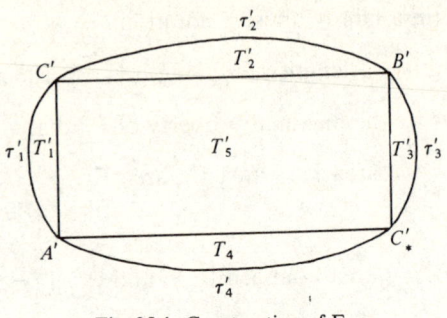

Fig. 35.4. Construction of Γ_j.

of the rectangle, as shown in Figure 35.4. (The reader might find it helpful to think
of a flexible framework composed of rods AC, CB, BC_*, C_*A' to which are rigidly
attached pieces of cardboard T_1, T_2, T_3, T_4. By moving the rods we try to get a
figure of maximum area.)

Comparing the figures Γ_j and Γ'_j we see that the curved arcs T_1 from A to C,
T'_1 from A' to C' have the same lengths and that, similarly T_2 and T'_2, T_3 and
T'_3, T_4 and T'_4 have the same lengths. Thus Γ_j and Γ'_j have the same perimeter.
On the other hand, since the area of a parallelogram is the product of the base
and the height,

$$\text{area } T'_5 > \text{area } T_5$$

and so, since $T_k = \text{area } T'_k$ for $1 \leqslant k \leqslant 4$,

$$\text{area } \Gamma'_j = \sum_{k=1}^{5} \text{area } T'_k = \sum_{k=1}^{4} \text{area } T_k + \text{area } T'_5 > \sum_{k=1}^{5} \text{area } T_k = \text{area } \Gamma_j.$$

This contradicts the maximum property of Γ_j so, by *reductio ad absurdum*, $A\hat{C}B$
is a right angle and we are done.   ■

Proof of Theorem 35.1. Since Γ_1 and Γ_2 are circles, γ_1 and γ'_1 must have been
semicircular arcs and so Γ is a circle.   ■

The reader may still feel that the arguments above are too vague and imprecise.
Do we mean by figure the interior of a simple closed curve, or do we allow self
intersection? What about simple closed curves for which no reasonable definition
of length exists? But this is to miss the point. I claim that our arguments are based
on only the most elementary and intuitive properties of length and area which
should work for any reasonable definition. (For example our arguments require
that area and length remain unchanged under reflection, rotation and translation,
and that a straight line be the shortest distance between two points on the plane.)
Thus if we are given a set of definitions for which the arguments of this chapter
fail, we may well reject the definitions rather than the argument.

Notice that Theorem 35.1 simply states that if a figure Γ_0 exists with maximum area for a fixed perimeter it must be a circle. But, as Dirichlet pointed out to Steiner, it does not show that such a Γ_0 exists. (Consider the following 'proof' that 1 is the largest integer. Let n_0 be the largest integer. If $n_0 > 1$ then $n_0^2 > n_0$ contradicting the definition of n_0. Thus $n_0 = 1$.)

If the reader has looked at Chapter 29 she will recall that Dirichlet's own work on potential theory continued a similar flaw. The next example should be read in conjunction with Lemma 29.4 and Example 31.12 which it illuminates.

Example 35.5. *Let $\Omega = \{(x,y):0 < x^2 + y^2 < 1\}$ be the punctured disc. Then Ω is an open bounded set with boundary $\partial\Omega = \{(x,y):x^2 + y^2 = 1\} \cup \{(0,0)\}$ and closure $\bar{\Omega} = \{(x,y):x^2 + y^2 \leqslant 1\}$. If we define $G:\partial\Omega \to \mathbb{R}$ by $G(x,y) = 0$ for $x^2 + y^2 = 1$ and $G(0,0) = 1$ then G is continuous on $\partial\Omega$. Let us write $\mathcal{G}$ for the set of continuous functions $\psi:\bar{\Omega} \to \mathbb{R}$ with continuous second derivatives on Ω satisfying $\psi = G$ on $\partial\Omega$. Then there does not exist a $\phi \in \mathcal{G}$ such that*

$$\iint_{\Omega} \Sigma \psi_i^2 dA \geqslant \iint_{\Omega} \Sigma \phi_i^2 dA \quad \text{for all} \quad \psi \in \mathcal{G}.$$

Proof. The main step is to show $\inf_{\psi \in \mathcal{G}} \iint_{\Omega} \Sigma \phi_i^2 dA = 0$. Examination of the proof for Example 31.12 suggests that this could be done by using $\psi \in \mathcal{G}$ which are close, in some way, to a scalar multiple of $\log(x^2 + y^2)^{\frac{1}{2}}$. A little experimentation may suggest proceeding as follows.

Choose a twice continuously differentiable function $f:\mathbb{R} \to \mathbb{R}$ such that $f(x) = 1$ $[1 \geqslant x], 2 \geqslant f(x) \geqslant 1 [2 \geqslant x \geqslant 1]$ and $f(x) = x [x \geqslant 2]$. Let $K = \sup_{2 \geqslant x \geqslant 1} |f'(x)|$. Now set $f_\varepsilon(x) = \varepsilon f(\varepsilon^{-1}x)[\frac{1}{2} > \varepsilon > 0]$. Observe that $f_\varepsilon(x) = \varepsilon[\varepsilon \geqslant x], 2\varepsilon \geqslant f_\varepsilon(x) \geqslant \varepsilon[2\varepsilon \geqslant x \geqslant \varepsilon]$, $f_\varepsilon(x) = x[x \geqslant 2\varepsilon]$ and $\sup_{2\varepsilon \geqslant x \geqslant \varepsilon} |f'_\varepsilon(x)| = K$. It follows that if $g_\varepsilon(x) = \log f_\varepsilon(x)/\log\varepsilon$ then $g_\varepsilon(x) = 1 [\varepsilon \geqslant x]$, $\log 2\varepsilon/\log\varepsilon \leqslant g_\varepsilon(x) \leqslant 1 [2\varepsilon \geqslant x \geqslant \varepsilon]$, $g_\varepsilon(x) = \log x/\log\varepsilon [x \geqslant 2\varepsilon]$. We note that g_ε is twice continuously differentiable and that

$$|g'_\varepsilon(x)| = |f'_\varepsilon(x)|/(|f_\varepsilon(x)| |\log\varepsilon|) \leqslant K/\varepsilon|\log\varepsilon| \quad \text{for} \quad 2\varepsilon \geqslant x \geqslant \varepsilon.$$

Now let $\frac{1}{2} > \varepsilon > 0$ and set $\psi_\varepsilon(x,y) = g_\varepsilon((x^2 + y^2)^{\frac{1}{2}})$ (Note that ψ_ε looks like $(\log\varepsilon)^{-1}\log(x^2 + y^2)^{\frac{1}{2}}$ except when $x^2 + y^2$ is very small.) Since $g_\varepsilon(1) = 0$ and $g_\varepsilon(0) = 1$ it follows that $\psi_\varepsilon \in \mathcal{G}$. Further,

$$\iint_{\Omega} \Sigma \psi_{\varepsilon i}^2 dA = \int_0^{2\pi} \int_0^1 g'_\varepsilon(r)^2 r dr d\theta = 2\pi \int_0^1 g'_\varepsilon(r)^2 r dr$$

$$= 2\pi \left(\int_0^\varepsilon g'_\varepsilon(r)^2 r dr + \int_\varepsilon^{2\varepsilon} g'_\varepsilon(r)^2 r dr + \int_{2\varepsilon}^1 g'_\varepsilon(r)^2 r dr \right)$$

$$= 2\pi\left(\int_0^\varepsilon 0\,dr + \int_\varepsilon^{2\varepsilon} g'_\varepsilon(r)^2 r\,dr + \int_{2\varepsilon}^1 \frac{1}{r(\log \varepsilon)^2}\,dr \right)$$

$$\leqslant 2\pi\left(\int_\varepsilon^{2\varepsilon} \frac{K^2}{\varepsilon^2 (\log \varepsilon)^2} r\,dr + \int_{2\varepsilon}^1 \frac{1}{r(\log \varepsilon)^2}\,dr \right)$$

$$= 2\pi\left(\frac{3K^2}{4(\log \varepsilon)^2} - \frac{\log 2\varepsilon}{(\log \varepsilon)^2} \right) \to 0 \quad \text{as} \quad \varepsilon \to 0+.$$

Thus $\inf_{\psi \in \mathscr{G}} \iint_\Omega \Sigma \psi_i^2\, dA = 0.$

In particular, if $\phi \in \mathscr{G}$ and $\iint_\Omega \Sigma \psi_i^2\, dA \geqslant \iint_\Omega \Sigma \phi_i^2\, dA$ for all $\psi \in \mathscr{G}$ it follows that $\iint_\Omega \Sigma \phi_i^2\, dA = 0$. Thus, since $\Sigma \phi_i^2$ is continuous on Ω, $\Sigma \phi_i^2 = 0$ on Ω so $\phi_1 = \phi_2 = 0$ on Ω and ϕ is constant on Ω. This is incompatible with the boundary conditions so we are done. ∎

Remark. The reader may check that the 3-dimensional analogue of this example (dealing with $\iiint_\Omega \Sigma_{i=1}^3 \psi_i^2\, dV$) is even easier to prove and that the one-dimensional analogue is false. (The expression $\int_{-1}^0 \psi'(x)^2\, dx + \int_0^1 \psi'(x)^2\, dx$ subject to $\psi(0) = 1$, $\psi(1) = \psi(-1) = 0$ is minimised by taking $\psi(x) = 1 - |x|$.)

The next example is much more trivial but closer to the subject of the chapter.

Example 35.6. *Let T be the 'pencil shaped' cylinder, illustrated in Figure 35.5, formed by rotating the curve*

$$\begin{aligned} y &= x \qquad [0 \leqslant x \leqslant 1], \\ y &= 1 \qquad [1 \leqslant x], \end{aligned}$$

about the positive real axis. Then curves of perimeter 2π can bound arbitrarily large areas on T.

Proof. The curve Γ formed by rotating the point $(x, y) = (K + 1, 1)$ about the real axis lies on T, has perimeter 2π and encloses an area of at least $K\pi[K > 0]$. ∎

The reader may care to consider the surfaces enclosed by rotating curves of the type shown in Figure 35.6.

It is worth noting that the elementary methods of the calculus of variation only

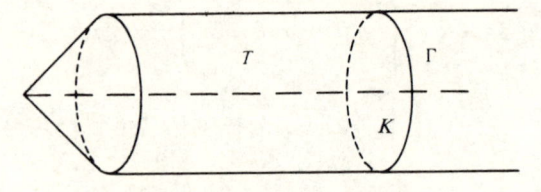

Fig. 35.5. The pencil-shaped cyclinder.

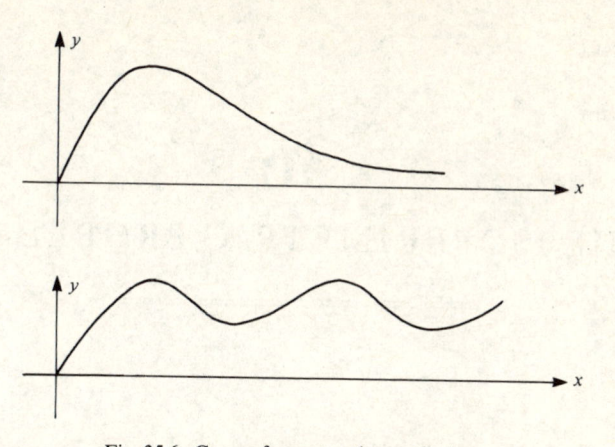

Fig. 35.6. Curves for generating cyclinders.

find local stationary points rather than global maxima or minima and are thus irrelevant to the question of *proving* results of the type discussed. It was only with the work of Weierstrass and his successors that there emerged a clear picture both of the problems involved in proving the existence of extrema and their solution. The modern notion of compactness may be traced back to one of the main methods used. We shall give an easy example of the kind of technique involved in Chapter 44.

36

THE ISOPERIMETRIC PROBLEM II

In the last chapter we showed how geometric methods could be used to attack the problem of proving that amongst all figures of given perimeter the circle has greatest area. In this chapter we give a completely different approach due to Hurwitz which uses Fourier analysis.

The nature of this proof requires greater precision than we used in the last chapter, so we start with a preliminary discussion. We remark that there are many ways of presenting a closed curve in the plane. Consider, for example, the maps $(x_1, y_1): \mathbb{T} \to \mathbb{R}^2$ and $(x_2, y_2): \mathbb{T} \to \mathbb{R}^2$ given by

$$(x_1(t), y_1(t)) = (\cos \pi^{-2} t^3, \sin \pi^{-2} t^3) \quad [-\pi \leqslant t < \pi]$$

and

$$(x_2(t), y_2(t)) = (\sin 8t, \cos 8t) \quad [0 \leqslant t \leqslant \pi/8],$$
$$(x_2(t), y_2(t)) = (0, -1) \quad [\pi/8 \leqslant t \leqslant 7\pi/8],$$
$$(x_2(t), y_2(t)) = (\sin 8t, \cos 8t) \quad [7\pi/8 \leqslant t \leqslant 2\pi].$$

Both represent the unit circle but not in a very natural manner.

A more natural procedure is to parametrise our curves by arc length s, so that $(x(s), y(s))$ is the point at arc length $Ls/2\pi$ from $(x(0), y(0))$ along the curve in one direction. (If arc length is not defined, then the curve is not relevant to the isoperimetric problem.) We shall deal only with non-intersecting closed curves, i.e. those given by maps $(x, y): \mathbb{T} \to \mathbb{R}$ such that $(x(s_1), y(s_1)) = (x(s_2), y(s_2))$ implies $s_1 = s_2$.

We now show how to prove the isoperimetric theorem when we restrict ourselves to curves whose arc length parametrisations are smooth.

Theorem 36.1. (i) *Let* $(x, y): \mathbb{T} \to \mathbb{R}^2$ *be an anticlockwise arc length parametrisation of a nonintersecting closed curve* Γ *of length* L. *Suppose that* x *and* y *are continuously differentiable. Then if* A *is the area enclosed by the curve* Γ, *we have*

$$L^2 - 4\pi A \geqslant 2\pi^2 \sum_{n \neq 0} (|n\hat{x}(n) - i\hat{y}(n)|^2 + |n\hat{y}(n) + i\hat{x}(n)|^2 + (n^2 - 1)(|\hat{x}(n)|^2 + |\hat{y}(n)|^2)).$$

(ii) *In particular* $L^2 \geqslant 4\pi A$ *with equality if and only if* Γ *is a circle.*

166

Proof. Because we have chosen an arc length parametrisation

$$(\dot{x}(s))^2 + (\dot{y}(s))^2 = \left(\frac{dx}{ds}\right)^2 + \left(\frac{dy}{ds}\right)^2 = \left(\frac{L}{2\pi}\frac{ds}{ds}\right)^2 = \left(\frac{L}{2\pi}\right)^2$$

and so

$$\frac{1}{2\pi}\int_T (\dot{x}(s))^2 + (\dot{y}(s))^2 \, ds = \frac{1}{2\pi}\int_T \left(\frac{L}{2\pi}\right)^2 ds.$$

Now, by Parseval's formula (Theorem 34.2),

$$\frac{1}{2\pi}\int (\dot{x}(s))^2 \, ds = \frac{1}{2\pi}\int_T |\dot{x}(s)|^2 \, ds = \sum_{n=-\infty}^{\infty} |\hat{\dot{x}}(n)|^2,$$

so, since

$$\hat{\dot{x}}(n) = \frac{1}{2\pi}\int_0^{2\pi'} \dot{x}(s)\exp(-ins)\, ds$$

$$= \frac{1}{2\pi}[x(s)\exp(-ins)]_0^{2\pi} + \frac{in}{2\pi}\int_T x(s)\exp(-ins)\, ds = in\hat{x}(n),$$

we have

$$\frac{1}{2\pi}\int_T (\dot{x}(s))^2 \, ds = \sum_{n=-\infty}^{\infty} |n\hat{x}(n)|^2$$

and similarly

$$\frac{1}{2\pi}\int_T (\dot{y}(s))^2 \, ds = \sum_{n=-\infty}^{\infty} |n\hat{y}(n)|^2.$$

Thus

$$\left(\frac{L}{2\pi}\right)^2 = \sum_{n=-\infty}^{\infty} n^2(|\hat{x}(n)|^2 + |\hat{y}(n)|^2).$$

On the other hand, we have (for an anticlockwise parametrisation)

$$A = -\int_T y\frac{dx}{ds}\, ds = -\int_T y(s)\dot{x}(s)ds,$$

so (since x is real and so $x(s)^* = x(s)$) the extended Parseval's formula of Theorem 34.4 yields

$$\frac{A}{2\pi} = -\frac{1}{2\pi}\int_T y(s)\dot{x}(s)ds = -\sum_{n=-\infty}^{\infty} \hat{y}(n)\hat{\dot{x}}(n)^* = \sum_{n=-\infty}^{\infty} in\hat{y}(n)\hat{x}(n)^*.$$

As it stands this expression is unsymmetrical, but taking complex conjugates (or observing that $A = \int_T x(s)\dot{y}(s)ds$) we get

$$\frac{A}{2\pi} = -\sum_{n=-\infty}^{\infty} in\hat{x}(n)\hat{y}(n)^*$$

and
$$\frac{A}{\pi} = \sum_{n=-\infty}^{\infty} in(\hat{x}(n)^*\hat{y}(n) - \hat{x}(n)\hat{y}(n)^*).$$

Thus

$$\frac{L^2 - 4\pi A}{2\pi^2} = \sum_{n=-\infty}^{\infty} (2n^2(|\hat{x}(n)|^2 + |\hat{y}(n)|^2) + 2in(\hat{x}(n)\hat{y}(n)^* - \hat{x}(n)^*\hat{y}(n))$$

$$= \sum_{n\neq 0} (2n^2(\hat{x}(n)\hat{x}(n)^* + \hat{y}(n)\hat{y}(n)^*) + 2in(\hat{x}(n)\hat{y}(n)^* - \hat{x}(n)^*\hat{y}(n))$$

$$= \sum_{n\neq 0} ((n\hat{x}(n) - i\hat{y}(n))(n\hat{x}(n) - i\hat{y}(n))^*$$

$$+ (n\hat{y}(n) + i\hat{x}(n))(n\hat{y}(n) + i\hat{x}(n))^* + (n^2 - 1)(\hat{x}(n)\hat{x}(n)^* + \hat{y}(n)\hat{y}(n)^*))$$

$$= \sum_{n\neq 0} (|n\hat{x}(n) - i\hat{y}(n)|^2 + |n\hat{y}(n) + i\hat{x}(n)|^2$$

$$+ (n^2 - 1)(|\hat{x}(n)|^2 + |\hat{y}(n)|^2)),$$

as stated.

(ii) We thus have $L^2 \geqslant 4\pi A$ with equality if and only if

(1)$_n$ $n\hat{x}(n) - i\hat{y}(n) = 0$,
(2)$_n$ $n\hat{y}(n) + i\hat{x}(n) = 0$,
(3)$_n$ $(n^2 - 1)\hat{x}(n)^2 = 0$,
(4)$_n$ $(n^2 - 1)\hat{y}(n)^2 = 0$,   for all n.

Conditions (3)$_n$ and (4)$_n$ show that $\hat{x}(n) = \hat{y}(n) = 0$ for all $n \neq 0, 1, -1$. Condition (2)$_1$ shows that $\hat{y}(1) = -i\hat{x}(1)$. Since x and y are real, $\hat{x}(0)$ and $\hat{y}(0)$ are real whilst $\hat{x}(-1) = \hat{x}(1)^*$ and $\hat{y}(-1) = \hat{y}(1)^* = i\hat{x}(1)^*$. Setting $x_0 = \hat{x}(0)$, $y_0 = \hat{y}(0)$ and choosing $R \geqslant 0$ and θ real such that $R \exp i\theta = 2\hat{x}(1)$, we obtain

$$(x(s), y(s)) = (x_0 + R\cos(s + \theta), y_0 + R\sin(s + \theta))   \text{ for all } s \in \mathbb{T},$$

so Γ is indeed a circle.   ∎

In the absence of a fairly lengthy discussion of the general notions of length and area, we cannot expect to give a complete argument for the extension to general curves. We can however see how such an argument would run.

Plausible Result 36.2. *Let* $(x, y): \mathbb{T} \to \mathbb{R}$ *be an arc length parametrisation of non-intersecting curve* Γ *of length L enclosing an area A. Then we can find a sequence* $\Gamma_1, \Gamma_2, \Gamma_3, \ldots$ *of non-intersecting closed curves with arc length parametrisations* $(x_m, y_m): \mathbb{T} \to \mathbb{R}$ *such that* x_m, y_m *are continuously differentiable and*

(i) $x_m \to x$, $y_m \to y$ *uniformly on* $\mathbb{T}$
(ii) *If A_m is the area enclosed by* Γ_m *then* $A_m \to A$
(iii) *If L_m is the length of* Γ_m *then* $L_m \to L$.

Remark. The reader may wonder (correctly) whether, for example, (ii) might not be a consequence of (i). However, all that is important here is to agree that we should be unhappy with any definition of length and area for which 36.2 fails.

Provided the reader is willing to accept the above plausible result (at least on a temporary basis) we can now see how to generalise Theorem 36.1.

Theorem 36.3 (i) *Let* $(x, y): \mathbb{T} \to \mathbb{R}^2$ *be an arc length parametrisation of a non-intersecting closed curve* Γ *of length L. Then if A is the area enclosed by* Γ *we have*

$$L^2 - 4\pi A \geqslant 2\pi^2 \sum_{n \neq 0} |n\hat{x}(n) - i\hat{y}(n)|^2 + |n\hat{y}(n) - i\hat{x}(n)|^2 + (n^2 - 1)(|\hat{x}(n)|^2 + |\hat{y}(n)|^2).$$

(ii) *In particular* $L^2 \geqslant 4\pi A$ *with equality if and only if* Γ *is a circle.*

Proof. (i) Pick Γ_m and (x_m, y_m) as in 36.2 and fix some integer $N \geqslant 1$ for the time being. By Theorem 36.1 we have

$$L_m^2 - 4\pi A_m \geqslant 2\pi^2 \Bigg(\sum_{n \neq 0} |n\hat{x}_m(n) - i\hat{y}_m(n)|^2 + |n\hat{y}_m(n) - i\hat{x}_m(n)|^2 \\ + (n^2 - 1)(|\hat{x}_m(n)|^2 + |\hat{y}_m(n)|^2) \Bigg),$$

so that, trivially,

$$L_m^2 - 4\pi A_m \geqslant 2\pi \Bigg(\sum_{0 \neq |n| \leqslant N} |n\hat{x}_m(n) - i\hat{y}_m(n)|^2 + |n\hat{y}_m(n) - i\hat{x}_m(n)|^2 \\ + (n^2 - 1)(|\hat{x}_m(n)|^2 + |\hat{y}_m(n)|^2) \Bigg).$$

But for each fixed n we know that, as $m \to \infty$,

$$|\hat{x}(n) - \hat{x}_m(n)| = \left| \frac{1}{2\pi} \int (x(s) - x_m(s)) \exp(-ins) \, ds \right| \leqslant \sup_{s \in \mathbb{T}} |x(s) - x_m(s)| \to 0,$$

so, allowing $m \to \infty$ and keeping N fixed, we have

$$L^2 - 4\pi A \geqslant 2\pi \Bigg(\sum_{0 \neq |n| \leqslant N} |n\hat{x}(n) - i\hat{y}(n)|^2 + |n\hat{y}(n) - i\hat{x}(n)|^2 \\ + (n^2 - 1)(|\hat{x}(n)|^2 + |\hat{y}(n)|^2) \Bigg).$$

We now observe that the last inequality is true for all N and so

$$L^2 - 4\pi A \geqslant 2\pi \Bigg(\sum_{n \neq 0} |n\hat{x}(n) - i\hat{y}(n)|^2 + |n\hat{y}(n) - i\hat{x}(n)|^2 + (n^2 - 1)(|\hat{x}(n)|^2 + |\hat{y}(n)|^2) \Bigg),$$

as desired.

(ii) This part of the proof is exactly the same as that of Theorem 36.1 (ii). ∎

Theorem 36.3 (ii) says that among all closed non-intersecting curves of given length the circle is the (unique) curve enclosing the greatest area and this is precisely the result desired.

37

THE STURM–LIOUVILLE EQUATION I

In his biography of Fourier, Herivel reproduces various copies of letters found among Fourier's papers. One of the most interesting (Letter XXI in *Joseph Fourier*, Herivel, Oxford 1975) unfortunately lacks the name of the addressee, but was probably intended for Lagrange.

In it Fourier describes how after obtaining Fourier series in solutions of special problems

I recognised that the development would also apply to an arbitrary function and I arrived by a different method at the same equation

$$\phi(x) = \sin x \int \phi(x) \sin x \, dx + \sin 2x \int \phi(x) \sin 2x \, dx + \cdots$$

which I had obtained previously. I transmitted this part of my work two years ago to M. Biot and M. Poisson who then knew the use I was making of it to express the integrals [i.e. solutions] of partial differential equations in trigonometrical or exponential series: they did not point out to me that d'Alembert or Euler had employed these integrations to develop a trigonometric solution. I was ignorant of this fact myself or I had entirely forgotten it; it was in attempting to verify a third theorem that I employed the procedure which consists in multiplying by $\cos x \, dx$ the two sides of the equation

$$\phi(x) = a_0 + a_1 \cos x + a_2 \cos 2x + \cdots .$$

and integrating between $x = 0$ and $x = \pi$. I am sorry not to have known the name of the mathematician who first made use of this method because I would have cited him. Regarding the researches of d'Alembert and Euler could one not add that if they knew this expansion they made but a very imperfect use of it. They were both persuaded that an arbitrary... function could never be resolved in a series of this kind, and it does not seem that any one had developed a constant in cosines of multiple arcs [i.e. found $a_1, a_2, \ldots$, with $1 = a_1 \cos x + a_2 \cos 2x \ldots$ for $-\pi/2 < x < \pi/2$] the first problem which I had to solve in the theory of heat. It was also necessary to know the limits between which this development took place.

For example it has to be realised that the equation

$$x/2 = \sin x - \tfrac{1}{2}\sin 2x + \tfrac{1}{3}\sin 3x - \cdots$$

is no longer true when the value of x is between π and 2π. However the second side of the equation is still a convergent series but the sum is not equal to $x/2$....

Fourier then points out that although Euler gave the trigonometric expansion for $x/2$ shown above he failed to specify the range over which it was true.

Finally, this development of a function in sines and cosines of multiple arcs is only a particular case among those which I have had to treat, and these latter offered analytical difficulties of a very different order. It was necessary, for example, for determining the movement of heat in a cylindrical body to develop an arbitrary function in a series whose terms depended on a transcendental function given by a differential equation of the second order. I beg you, sir, to be good enough to examine this part of my work which is really the only part worthy of your attention.

Other people before Fourier had used expansions of the form $f(x) \sim \sum_{r=-\infty}^{\infty} a_r \exp irt$ but Fourier's work extended this idea in two totally new ways. One was the Fourier integral which we shall discuss later and the other (to which he draws Lagrange's attention) marked the birth of Sturm–Liouville theory.

One of the problems that Fourier set himself was the study of the cooling of a sphere. Let us suppose that the temperature $\theta(\mathbf{x}, t)$ of a homogeneous sphere centre O radius R depends only on $r = |\mathbf{x}|$ and t so that the heat equation $K\nabla^2\theta = \partial\theta/\partial t$ becomes

$$\frac{\partial^2\theta}{\partial r^2} + \frac{2}{r}\frac{\partial\theta}{\partial r} = K^{-1}\frac{\partial\theta}{\partial t}. \tag{1}$$

Fourier considered the case in which the heat loss at the boundary is proportional to the excess temperature, i.e.

$$\left.\frac{\partial\theta}{\partial r}\right|_{r=R} = -h\theta(R, t). \tag{2}$$

We now apply the technique of separation of variables by seeking solutions of the form $\theta(r, t) = \rho(r)\tau(t)$. Equation (1) becomes

$$\rho''(r)\tau(t) + \frac{2}{r}\rho'(r)\tau(t) = K^{-1}\rho(r)\tau'(t),$$

so that (ignoring the possibility that $\tau(t)$ might be zero)

$$\frac{1}{\rho(r)}\left(\rho''(r) + \frac{2}{r}\rho'(r)\right) = K^{-1}\frac{\tau'(t)}{\tau(t)} \quad [0 < r \leqslant R].$$

Since $K(\tau'(t)/\tau(t))$ is independent of r, it follows that $(1/\rho(r))(\rho''(r) + (2/r)\rho'(r))$ is a

constant, $-\lambda^2$, say. Thus

$$\rho''(r) + \frac{2}{r}\rho'(r) + \lambda^2\rho(r) = 0 \tag{3}$$

and

$$K^{-1}\tau'(t) + \lambda^2\tau(t) = 0. \tag{4}$$

In addition equation (3) must be supplemented by two extra conditions. Firstly, condition (2) gives

$$\rho'(R) = -h\rho(R), \tag{3a}$$

whilst physical reality demands that θ be bounded for all $R \geqslant r \geqslant 0$ and so there exists an M such that

$$|\rho(r)| \leqslant M \quad \text{for all} \quad 0 < r \leqslant R. \tag{3b}$$

Equation (4) is easily solved to give $\tau(t) = A_1 \exp(-\lambda^2 Kt)$ (we shall take $A_1, A_2, \ldots$ as arbitrary constants) but (3) requires more thought. The key lies in the substitution $\gamma(r) = r\rho(r)$. Then $\gamma'(r) = r\rho'(r) + \rho(r)$ and $\gamma''(r) = r\rho''(r) + 2\rho'(r)$ so that

$$\gamma''(r) + \lambda^2\gamma(r) = 0, \tag{5}$$

whilst conditions (3a) and (3b) give

$$R\gamma'(R) = (1 - Rh)\gamma(R), \tag{5a}$$

$$\gamma(r) \to 0 \quad \text{as } r \to 0 + . \tag{5b}$$

Equation (5) has the general solution $\gamma(r) = A_2 e^{i\lambda r} + A_3 e^{-i\lambda r}$ but condition (5b) shows that $A_2 + A_3 = 0$ so that $\gamma(r) = A_4 \sin \lambda r$. Finally, condition (5a) yields

$$R\lambda \cos \lambda R = (1 - Rh)\sin \lambda R, \tag{6}$$

so that λ must be chosen to be a root of

$$\frac{R\lambda}{\tan R\lambda} = 1 - Rh. \tag{7}$$

Direct substitution shows that, indeed,

$$\theta(r, t) = \frac{1}{r}\exp(-\lambda^2 Kt)\sin \lambda r$$

is a solution whenever λ is a root of (7). Fourier now notes that (7) has an infinity of real roots $\lambda_1, \lambda_2, \ldots$, say, and proposes that all solutions of the problem have the form

$$\theta(r, t) = \sum_{n=1}^{\infty} \frac{a_n}{r}\exp(-\lambda_n^2 Kt)\sin \lambda_n r.$$

Thus if the initial heat distribution is given by

$$\theta(r, 0) = f(r),$$

we must be able to find $a_1, a_2, \ldots$ with

$$rf(r) = \sum_{n=1}^{\infty} a_n \sin \lambda_n r. \tag{8}$$

To find $a_1, a_2, \ldots$ Fourier remarks that if λ_m and λ_n are roots of (7), then

$$\frac{R\lambda_n}{\tan R\lambda_n} = 1 - Rh = \frac{R\lambda_m}{\tan R\lambda_m},$$

and so

$$\lambda_n \sin \lambda_m R \cos \lambda_n R - \lambda_m \sin \lambda_n R \cos \lambda_m R = 0,$$

whence, if $m \neq n$,

$$\int_0^R \sin \lambda_n r \sin \lambda_m r \, dr$$

$$= \frac{1}{2} \int_0^R \cos(\lambda_n - \lambda_m)r - \cos(\lambda_n + \lambda_m)r \, dr = \frac{1}{2}\left[\frac{\sin(\lambda_n - \lambda_m)r}{\lambda_n - \lambda_m} - \frac{\sin(\lambda_n + \lambda_m)r}{\lambda_n + \lambda_m} \right]_0^R$$

$$= \frac{1}{2}\left[\frac{\sin \lambda_n R \cos \lambda_m R - \sin \lambda_m R \cos \lambda_n R}{\lambda_n - \lambda_m} - \frac{\sin \lambda_m R \cos \lambda_n R + \sin \lambda_n R \cos \lambda_m R}{\lambda_n + \lambda_m} \right]$$

$$= (\lambda_n \sin \lambda_m R \cos \lambda_n R - \lambda_m \sin \lambda_n R \cos \lambda_m R)/(\lambda_m^2 - \lambda_n^2) = 0,$$

whilst

$$\int_0^R (\sin \lambda_n r)^2 \, dr = \frac{1}{2}\int_0^R (1 - \cos 2\lambda_n r) \, dr = \left(\frac{R}{2} - \frac{\sin 2\lambda_n R}{4\lambda_n} \right).$$

Thus multiplying both sides of (8) by $\sin \lambda_m r$ and integrating term by term we get

$$\int_0^R rf(r) \sin \lambda_m r \, dr = \sum_{n=1}^{\infty} a_n \int_0^R \sin \lambda_n r \sin \lambda_m r \, dr = a_m \left(\frac{R}{2} - \frac{\sin 2\lambda_m R}{4\lambda_m} \right),$$

so that (if our arguments are valid)

$$a_m = 2 \int_0^R rf(r) \sin \lambda_m r \, dr \bigg/ \left(R - \frac{\sin 2\lambda_m R}{2\lambda_m} \right).$$

Not surprisingly this way of proceeding came in for serious criticism. In particular it was objected that Fourier had not shown that all the roots of (7) are real. The force of this objection becomes apparent when we notice that if λ is complex then

$$\frac{1}{r}\exp(-\lambda^2 Kt)\sin \lambda r$$

will be oscillatory as t increases giving a physically implausible solution.

The specific problem referred to in Fourier's letter is that of the axially symmetric flow of heat in a homogeneous cylinder of radius R. Here the heat equation becomes

$$\frac{\partial^2 \theta}{\partial r^2} + \frac{1}{r}\frac{\partial \theta}{\partial r} = K^{-1}\frac{\partial \theta}{\partial t}$$

and we have the boundary conditions

$$\frac{\partial \theta}{\partial r}\bigg|_{r=R} = -h\theta(R,t).$$

Trying separation of variables as before, we take $\theta(r,t) = \rho(r)\tau(t)$ and obtain

$$r\rho''(r) + \rho'(r) + \lambda^2 r\rho(r) = 0, \qquad (9)$$

$$K^{-1}\tau'(t) + \lambda^2\tau(t) = 0, \qquad (10)$$

$$\rho'(R) = -h\rho(R) \qquad (9a)$$

$$|\rho'(r)|, |\rho(r)| \leqslant M \quad [0 < r \leqslant R]. \qquad (9b)$$

In this case Fourier was able to show that there exists a sequence of real numbers $\lambda_0 < \lambda_1 < \cdots$ such that, if $\lambda = \lambda_j$, then (9), (9a) and (9b) can be simultaneously satisfied by a function u_j. He showed further that

$$\int_0^R r u_m(r) u_n(r)\, dr = 0 \quad \text{for } m \neq n,$$

suggesting the expansion

$$f(r) = \sum_{n=0}^{\infty} a_n u_n(r)$$

with
$$\int_0^R r f(r) u_m(r)\, dr = a_m \int_0^R r u_m(r)^2\, dr.$$

(The knowledgeable reader will perhaps recognise the u_m as Bessel functions.)

38

LIOUVILLE

In Chapters 39 and 40 we shall see how Sturm and Liouville extended the ideas of Fourier discussed in the last chapter. Liouville worked in many other mathematical fields as well and his results here were often so simple and basic that they have been completely absorbed into the general body of mathematics. It thus seems appropriate to recall some of his achievements.

At the age of 27 he founded the *Journal des mathématiques pures et appliquées*, which became, under his editorship, one of the major journals of the nineteenth century. He edited and published Galois' manuscripts in his journal and, just as importantly, gave a series of lectures organising and interpreting Galois theory for the general mathematical public.

In complex variable theory he used simple general arguments to bring order to the subject of multiply periodic functions. The result, called by his name which states that a bounded analytic function is constant, was known to Cauchy, but Liouville was the first to demonstrate its power. Some of the flavour of his work in this field is conveyed in the following theorem.

Recall that if $g(z) = \exp z$ and $w = 2\pi i$ then g is an analytic function with $g(z) = g(z + w)$ for all $z \in \mathbb{C}$. We call g *singly periodic*. It is natural to ask whether we can find a, non trivial, *doubly periodic* analytic function f, that is to say an analytic function f and $w_1, w_2 \in \mathbb{C}$ such that w_1 and w_2 are not real multiples of each other and such that $f(z) = f(z + w_1) = f(z + w_2)$ for all $z \in \mathbb{C}$.

Theorem 38.1. *Let w_1, w_2 be non zero complex numbers with $w_1/w_2 \notin \mathbb{R}$. Suppose $f : \mathbb{C} \to \mathbb{C}$ is an analytic function with $f(z) = f(z + w_1) = f(z + w_2)$ for all $z \in \mathbb{C}$. Then f is constant.*

Proof. (This is not Liouville's original proof which used Fourier analysis rather than complex variable theory.) Since w_1/w_2 is not real it follows that we can write any $z \in \mathbb{C}$ in the form $z = \mu_1 w_1 + \mu_2 w_2$ with $\mu_1, \mu_2 \in \mathbb{R}$. But we can write $\mu_1 = \lambda_1 + n_1$, $\mu_2 = \lambda_2 + n_2$ with $n_1, n_2 \in \mathbb{Z}$ and $0 \leqslant \lambda_1, \lambda_2 < 1$ and, by the periodicity of f, we will have

$$f(z) = f((\lambda_1 + n_1)w_1 + (\lambda_2 + n_2)w_2) = f(\lambda_1 w_1 + \lambda_2 w_2).$$

175

The behaviour of f on $\mathbb{C}$ is thus entirely given by its behaviour on the parallelogram $\Gamma_1 = \{\lambda_1 w_1 + \lambda_2 w_2 : 0 \leqslant \lambda_1, \lambda_2 < 1\}$.

Now we observe that f is analytic and so $|f|$ is continuous on the closed bounded parallelogram $\Gamma_2 = \{\lambda_1 w_1 + \lambda_2 w_2 : 0 \leqslant \lambda_1, \lambda_2 \leqslant 1\}$. Thus (consult Appendix B if you must) $|f|$ is bounded on Γ_2 with $|f(z)| \leqslant M$ say for all $z \in \Gamma_2$. Since $\Gamma_1 \subseteq \Gamma_2$, it follows that $|f(z)| \leqslant M$ for all $z \in \Gamma_1$ and so by the first paragraph $|f(z)| \leqslant M$ for all $z \in \mathbb{C}$. Liouville's theorem now tells us that f is constant and we are done.   ∎

Remark. This does not close the subject of doubly periodic functions. If we consider meromorphic functions (i.e. if we allow f to have poles), then we can find many doubly periodic functions f. Their study formed one of the major themes of nineteenth-century analysis.

Liouville did important work on differential geometry introducing, for example, the notion of geodesic curvature. He was one of the founders of analytic number theory. He worked in theoretical physics and mechanics (there is a Liouville's theorem at the beginning of statistical mechanics) and in integral and partial differential equations. He produced a theory of fractional differentiation enabling a meaning to be assigned to d^α/dx^α when α is not a positive integer. Answering a question which must have occurred to the reader, he showed that integrals like $\int_0^t e^{-x^2/2}\, dx$ cannot be expressed by means of 'elementary' functions like exp, cos, log and polynomials. He published over 400 papers in all (although some of his most important contributions appeared only in the works of others). His pupils included Briot and Hermite.

In 1844 Liouville discovered the first proof of the following remarkable result.

Theorem 38.2. *Transcendental numbers exist.*

Recall that a real number is called *algebraic* if it is the root of a polynomial equation
$$a_n x^n + a_{n-1} x^{n-1} + \cdots + a_0 = 0$$
where $n \geqslant 1$, $a_n \neq 0$ and $a_j \in \mathbb{Z}\,[0 \leqslant j \leqslant n]$. Any number which is not algebraic is called *transcendental*. By 1840 many people suspected that numbers like π and e were not algebraic but nobody could show that even one transcendental number actually existed.

Liouville's proof is simple and unexpected. The key lies in the following theorem.

Theorem 38.3. *Let y be a real number such that*
$$a_n y^n + a_{n-1} y^{n-1} + \cdots + a_0 = 0,$$
where $n \geqslant 1$, $a_n \neq 0$ and $a_j \in \mathbb{Z}\,[0 \leqslant j \leqslant n]$. Then there exists a $K > 0$ such that, if $p, q \in \mathbb{Z}$ and $q \geqslant 1$, then either $y = p/q$ or
$$\left| \frac{p}{q} - y \right| \geqslant \frac{K}{q^n}.$$

In other words algebraic numbers cannot be approximated very well by rational numbers.

Proof. Let $f(x) = \sum_{j=0}^{n} a_j x^j$. Choose an interval $I = [y - \delta, y + \delta]$ surrounding y so that y is the only root of f in I (i.e. $f(x) \neq 0$ for all x with $y - \delta \leqslant x < y$ or $y < x \leqslant y + \delta$) and choose an M so that $|f'(x)| \leqslant M$ for all $x \in I$.

Suppose now that $p/q \in I$ and $p/q \neq y$. Then $f(p/q) \neq 0$. In other words,

$$a_n \left(\frac{p}{q} \right)^n + a_{n-1} \left(\frac{p}{q} \right)^{n-1} + \cdots + a_0 \neq 0,$$

and so

$$a_n p^n + a_{n-1} p^{n-1} q + \cdots + a_0 q^n \neq 0.$$

But $a_n p^n + a_{n-1} p^{n-1} q + \cdots + a_0 q^n$ is an integer so

$$|a_n p^n + a_{n-1} p^{n-1} q + \cdots + a_0 q^n| \geqslant 1.$$

It follows that

$$|f(p/q)| = \left| \frac{a_n p^n + a_{n-1} p^{n-1} q + \cdots + a_0 q^n}{q^n} \right|$$

$$= |a_n p^n + a_{n-1} p^{n-1} q + \cdots + a_0 q^n| / |q|^n$$

$$\geqslant 1/|q|^n.$$

But by the mean value theorem we know that there is a ξ between p/q and y such that

$$|f(p/q) - f(y)| = |f'(\xi)||p/q - y|.$$

Since $f(y) = 0$, $|f(p/q)| \geqslant 1/q^n$ and $|f'(\xi)| \leqslant M$ this gives

$$|y - p/q| \geqslant 1/(Mq^n).$$

Thus if $p/q \neq y$ then either $p/q \notin I$ and so $|y - p/q| \geqslant \delta \geqslant \delta/q^n$, trivially, or $p/q \in I$ and $|y - p/q| \geqslant M^{-1}/q^n$. Setting $K = \min(M^{-1}, \delta)$ we have the stated result. ∎

We can now actually write down a transcendental number (and so prove Theorem 38.2).

Theorem 38.4. *The number*

$$y = 10^{-1!} + 10^{-2!} + 10^{-3!} + \cdots$$

is transcendental.

Proof. Suppose not. Then y satisfies an equation

$$a_n y^n + a_{n-1} y^{n-1} + \cdots + a_0 = 0,$$

with $n \geqslant 1$, $a_n \neq 0$, $a_j \in \mathbb{Z}\,[0 \leqslant j \leqslant n]$. Thus by Theorem 37.2 there exists a $K > 0$ such

that

$$\left| \frac{p}{q} - y \right| \geqslant \frac{K}{q^n}$$

whenever $p, q \in \mathbb{Z}$, $q \geqslant 1$ and $y \neq p/q$.

In particular let us take $q = 10^{m!}$ and $p = 10^{m!}(10^{-1!} + 10^{-2!} + \cdots + 10^{-m!})$. We see that $p/q - y = \sum_{r=m+1}^{\infty} 10^{-r!}$ so $y \neq p/q$. Thus

$$\left| \frac{p}{q} - y \right| \geqslant \frac{K}{q^n},$$

and so

$$K \leqslant q^n \left| \frac{p}{q} - y \right| = 10^{n \cdot m!} \sum_{r=m+1}^{\infty} 10^{-r!} \leqslant 10^{n \cdot m!} 10^{-(m+1)!} \sum_{s=0}^{\infty} 10^{-s}$$

$$\leqslant 2 \cdot 10^{n \cdot m!} \cdot 10^{-(m+1)!} = 2 \cdot 10^{(n-m-1) \cdot m!} \to 0 \quad \text{as } m \to \infty.$$

Thus $K = 0$ and we have a contradiction.   ∎

What we have shown, in effect, is that $y = \sum_{r=1}^{\infty} 10^{-r!}$ is extremely well approximable by rational numbers and so must be transcendental!

> Once when lecturing to a class... [Kelvin] used the word 'mathematician', and then interrupting himself asked his class: 'Do you know what a mathematician is?' Stepping to the blackboard he wrote upon it:
>
> $$\int_{-\infty}^{\infty} e^{-x^2} dx = \sqrt{\pi}.$$
>
> Then putting his finger on what he had written he turned to his class and said: 'A mathematician is one to whom *that* is as obvious as that twice two makes four is to you. Liouville was a mathematician.'
> (From Sylvanus Thompson's *Life of Lord Kelvin*, p. 1139.)

39

THE STURM–LIOUVILLE
EQUATION II

Following Fourier it was found that many differential equations of physics, and in particular many of the equations resulting from the separation of variables in partial differential equations and from the solution of variational problems took the form

$$\frac{d}{dx}\left(p(x)\frac{du}{dx}\right) - (\lambda r(x) + q(x))u = 0,$$

subject to certain end conditions. (Thus in Chapter 37, equation (5) we have $p(x) = r(x) = 1$, $q(x) = 0$ and in equation (9) we have $p(x) = x$, $r(x) = x$, $q(x) = 0$. We note further that if $P(x) \neq 0$ the apparently more general equation

$$P(x)\frac{d^2y}{dx} + W(x)\frac{dy}{dx} - (\lambda R(x) + Q(x))y = 0$$

may be rewritten as

$$p(x)\frac{d^2y}{dx^2} + p'(x)\frac{dy}{dx} - \frac{(\lambda R(x) + Q(x))}{P(x)}p(x)y = 0,$$

with $p(x) = \exp \int_a^x W(t)/P(t)\, dt$ and so as

$$\frac{d}{dx}\left(p(x)\frac{dy}{dx}\right) - (\lambda r(x) + q(x))y = 0,$$

where $\qquad r(x) = R(x)p(x)/P(x)$, $q(x) = Q(x)p(x)/P(x)$.)

Moreover it was found that (if $p(x)$ and $r(x)$ had constant sign) it was often possible to use solutions of this equation in expansions analogous to Fourier series. In a series of papers Sturm and Liouville produced a theory which accounted for and organised this analogy. We shall repeat part (but only part) of their work using the concept of an inner product space introduced in Chapters 32 and 33.

In what follows we shall consider only equations of the type

$$\frac{d}{dx}\left(p(x)\frac{du}{dx}\right) - (\lambda r(x) + q(x))u = 0 \qquad (1)$$

on $[a, b]$ subject to the boundary conditions

$$\alpha_a u(a) + \beta_a u'(a) = 0, \tag{2}$$

$$\alpha_b u(b) + \beta_b u'(b) = 0. \tag{3}$$

We suppose that p is real and once continuously differentiable, that r and q are real and continuous, and that $\alpha_a, \beta_a, \alpha_b, \beta_b$ are real and $(\alpha_a, \beta_a) \neq (0, 0)$, $(\alpha_b, \beta_b) \neq (0, 0)$. Further we demand that $p(x)$, $r(x) > 0$ for all $x \in [a, b]$.

(The attentive reader may observe that in spite of their apparent generality these conditions do not cover equations (9), (9a) and (9b) of Chapter 37. However she will find it an interesting exercise to check that although our *hypotheses* do not cover this case, our *method* can be readily extended to cover this case as well.)

We start with the following modification of the opening of Chapter 32.

Let $C([a, b])$ be the set of continuous functions $f : [a, b] \to \mathbb{C}$. Write

$$(f, g) = \int_a^b r(t) f(t) g(t)^* dt,$$

where z^* denotes the complex conjugate of z.

Lemma 39.1. (*cf. Lemma* 32.1.) *If $f, g, h \in C([a, b])$ and $\lambda, \mu \in \mathbb{C}$, then*

(i) $(f, g) = (g, f)^*$
(ii) $(\lambda f + \mu g, h) = \lambda(f, h) + \mu(g, h)$
(iii) (f, f) *is real and* $(f, f) \geqslant 0$
(iv) *If* $(f, f) = 0$, *then* $f = 0$.

Proof. As for Lemma 32.1. We use the hypothesis $r(t) > 0$ for all $t \in [a, b]$ in part (iv). Here we have $\int_a^b r(t) |f(t)|^2 dt = (f, f) = 0$, so, since $r(t) |f(t)|^2$ is a continuous positive function of t, $r(t) |f(t)|^2 = 0$ for all $t \in [a, b]$ and so $|f(t)|^2 = 0$ and $f(t) = 0$ for all $t \in [a, b]$. ∎

Lemma 39.2 (The Cauchy, Schwarz, Buniakowski inequality). *If f, $g \in C([a, b])$, then $|(f, g)|^2 \leqslant (f, f)(g, g)$ with equality if and only if $\lambda f + \mu g = 0$ for some $\lambda, \mu \in \mathbb{C}$ not both zero.*

In other words

$$\left| \int_a^b r(t) f(t) g(t)^* dt \right|^2 \leqslant \int_a^b r(t) |f(t)|^2 dt \int_a^b r(t) |g(t)|^2 dt.$$

Proof. Exactly as in Lemma 32.2. ∎

Associated with our inner product we have a norm given by $\|f\|_2 = (f, f)^{\frac{1}{2}}$ (where the positive square root is taken).

Lemma 39.3. (*cf. Lemma* 32.3) *If $f, g \in C(\mathbb{T})$ and $\lambda \in \mathbb{C}$, then*

(i) $\|\lambda f\|_2 = |\lambda| \|f\|_2$

(ii) $\|f\|_2 \geqslant 0$ *with equality if and only if* $f = 0$

(iii) *(triangle inequality)* $\|f\|_2 + \|g\|_2 \geqslant \|f + g\|_2$.

Thus, for example, Lemma 39.3 (iii) gives

$$\left(\int_a^b r(t)|f(t)|^2 dt\right)^{\frac{1}{2}} + \left(\int_a^b r(t)|g(t)|^2 dt\right)^{\frac{1}{2}} \geqslant \left(\int_a^b r(t)|f(t) + g(t)|^2 dt\right)^{\frac{1}{2}}.$$

Proof. Exactly as in Lemma 32.3. ■

So far our results have dealt only with the description of a new inner product. We now turn to matters more closely connected with the Sturm–Liouville equation itself. If u is a twice continuously differentiable function on $[a, b]$ let us write

$$Lu = \frac{1}{r(x)}\left[\frac{d}{dx}\left(p(x)\frac{du}{dx}\right) - q(x)u\right].$$

We note that L is linear in the sense that $L(\lambda u + \mu v) = \lambda Lu + \mu Lv$ for all u, v twice continuously differentiable and $\lambda, \mu \in \mathbb{C}$.

For the sake of conciseness let us also write E for the set of continuously twice differentiable functions u such that

$$\alpha_a u(a) + \beta_a u'(a) = 0,$$
$$\alpha_b u(b) + \beta_b u'(b) = 0.$$

If $u \in E$ is a non-zero solution of the Sturm–Liouville equation we have $Lu = \lambda u$ and we call u an *eigen function*, λ an *eigen value* of L. The key to our treatment of L lies in the following formula.

Lemma 39.4. *If* $u, v \in E$ *then* $(Lu, v) = (u, Lv)$.

Proof. Integrating by parts twice we have

$$(Lu, v) = \int_a^b ((pu')' - qu)v^* dx = \int_a^b (pu')'v^* dx - \int_a^b quv^* dx$$

$$= [pu'v^*]_a^b - \int_a^b pu'v^{*\prime} dx - \int_a^b quv^* dx$$

$$= [pu'v^*]_a^b - \int_a^b (pv^{*\prime})u' dx - \int_a^b quv^* dx$$

$$= [pu'v^*]_a^b - [pv^{*\prime}u]_a^b + \int_a^b (pv^{*\prime})'u dx - \int_a^b quv^* dx$$

$$= [pu'v^* - pv^{*\prime}u]_a^b + \int_a^b (pv^{*\prime})'u - quv^* dx$$

$$= [p(u'v^* - v^{*\prime}u)]_a^b + \int_a^b u((pv')' - qv)^* dx = [p(u'v^* - v'^*u)]_a^b + (u, Lv).$$

Now

$$\alpha_a(u'(a)v(a)^* - v'(a)^*u(a)) = \alpha_a u'(a)v(a)^* - \alpha_a v'(a)^*u(a)$$
$$= -\beta_a u'(a)v(a)^* + \beta_a v'(a)^*u'(a) = 0,$$

and, similarly, $\beta_a(u'(a)v(a)^* - v'(a)^*u(a)) = 0$ so since α_a and β_α are not both zero, $u'(a)v(a)^* - v'(a)^*u(a) = 0$. Similarly, $u'(b)v(b)^* - v'(b)^*u(b) = 0$, so that

$$[p(u'v^* - v'^*u)]_a^b = p(b)(u'(b)v(b)^* - v'(b)^*u(b)) - p(a)(u'(a)v(a)^* - v'(a)^*u(a)) = 0,$$

and so $(Lu, v) = (u, Lv)$. ∎

In the last chapter but one we noted that Fourier was unable to prove that if $u'' + \lambda^2 u = 0$, $Ru'(R) = (1 - Rh)u(R)$, $u(0) = 0$ then λ^2 must be real (and so, by inspection of equation (7), positive). (Indeed the problem was only resolved by Poisson 20 years later.) Lemma 39.4 enables us to prove this result in a general context.

Theorem 39.5. *The eigen values of L are real.*
Proof. Suppose $u \in E$, $\lambda \in \mathbb{C}$, u is not identically 0 and $Lu = \lambda u$. Then $\|u\|_2^2 \neq 0$ and

$$\lambda \|u\|_2^2 = \lambda(u, u) = (\lambda u, u) = (Lu, u) = (u, Lu) = (u, \lambda u) = \lambda^*(u, u) = \lambda^* \|u\|_2^2,$$

so $\lambda = \lambda^*$. ∎

Remark. We note that $u \in E$ for some suitable choice of $\alpha_a, \alpha_b, \beta_a, \beta_b$, if and only if $u(a)$ and $u'(a)$ are in a real ratio and $u(b)$ and $u'(b)$ are in a real ratio.

Next we prove a result which essentially includes both Lemma 32.4 and the orthogonality results of Chapter 37. Let us call u and v *orthogonal* if $(u, v) = 0$.

Theorem 39.6. *Eigen functions associated with distinct eigen values are orthogonal.*
Proof. Suppose $u, v \in E$, λ, $\mu \in \mathbb{C}$, $\lambda \neq \mu$, u and v are not identically zero and $Lu = \lambda u$, $Lv = \mu v$. Then (recalling from Lemma 34.5 that λ and μ are real) we have $\lambda(u, v) = (\lambda u, v) = (Lu, v) = (u, Lv) = (u, \mu v) = \mu(u, v)$ so, since $\lambda \neq \mu$, $(u, v) = 0$. ∎

The reader may have noticed a close resemblance to the treatment of self adjoint (Hermitian) matrices. This is not accidental.

We remark that if v is an eigen function associated with an eigen value λ then so is $u = v/\|v\|_2$ and in addition $\|u\|_2 = 1$. We call a sequence $u_1, u_2, \ldots$ orthonormal if

$$(u_n, u_n) = 1, (u_n, u_m) = 0 \quad \text{for } n \neq m.$$

Given a sequence $\lambda_1, \lambda_2, \ldots$ of distinct eigen values, the result above shows that we can find an associated sequence of orthonormal eigen functions $u_1, u_2, \ldots$.

It is now easy to grasp the relevance of the analogues of Theorem 32.5 and Theorem 33.1.

Theorem 39.7. *Let $u_1, u_2, \ldots$ be an orthonormal sequence. If $g_0 = \sum_{j=1}^n (f, u_j)u_j$ and $g = \sum_{j=1}^n \lambda_j u_j$, then $\|f\|_2^2 \geqslant \sum_{j=1}^n |(f, u_j)|^2$ and*

$$\|f - g\|_2 \geqslant \|f - g_0\|_2 = \sqrt{\left(\|f\|_2^2 - \sum_{j=1}^n |(f, u_j)|^2 \right)}.$$

Proof. Exactly as for Theorem 32.5. ∎

Thus the Fourier expansion $\sum_{j=1}^n \mu_j u_j$ with $\mu_j = \int_a^b r(x) f(x) u_j(x)^* dx$ gives the minimum distance of $\sum_{j=1}^n \lambda_j u_j$ from f in a certain norm.

Theorem 39.8 (Bessel's inequality). *Let $u_1, u_2, \ldots$ be an orthonormal sequence. If $f \in C([a, b])$ then*

$$\|f\|_2^2 \geqslant \sum_{n=1}^\infty |(f, u_n)|^2.$$

Proof. Exactly as for Theorem 33.1. ∎

Thus if $v_1, v_2 \ldots$ are eigen functions with distinct eigen values $\lambda_1, \lambda_2, \ldots$ we have

$$\int_a^b r(x) |f(x)|^2 dx \geqslant \sum_{n=1}^\infty \left| \int_a^b r(x) f(x) v_n^*(x) dx \right|^2 \Big/ \int_a^b r(x) |v_n(x)|^2 dx.$$

The discussion above shows the way towards a uniform 'Fourier mean square' development of the theory of the Sturm–Liouville equations but leaves two major questions unanswered. The first question was answered by Sturm.

Theorem A. *The eigen values form a sequence $\lambda_1, \lambda_2, \ldots$ with $|\lambda_j| \to \infty$. To each eigen value λ_j there corresponds (up to multiplication by a scalar of modulus 1) exactly one eigen function u_j with $\|u_j\|_2 = 1$.*

The second question asks whether a result corresponding to Theorem 34.1 holds in general. That is to say, do we have the following theorem?

Theorem B. *Let $f \in C([a, b])$. Then*

$$\left\| f - \sum_{j=1}^n (f, u_j)u_j \right\|_2 \to 0 \quad \text{as } n \to \infty.$$

Notice that Theorem B is immediately complemented by the following results corresponding to Theorems 34.2 and 34.3.

Theorem 39.9 (Parseval's formula). *Suppose $u_1, u_2, \ldots$ is an orthonormal sequence such that $\|f - \sum_{j=1}^n (f, u_j)\|_2 \to 0$ as $n \to \infty$ for all $f \in C([a, b])$. Then*

(i) *If $f \in C([a, b])$,*

$$\|f\|_2^2 = \sum_{j=1}^\infty |(f, u_j)|^2$$

Orthogonal series

(ii) *If* $f, g \in C([a, b])$

$$(f, g) = \sum_{j=1}^{\infty} (f, u_j)(g, u_j)^*.$$

Proof. As for Theorem 34.2 and 34.3. ∎

Although Sturm and Liouville were able to make some progress in the direction of Theorem B and although it is possible to give ad hoc proofs for particular cases (as we shall do in the next chapter), a rigorous proof for the general result was not given until the beginning of the twentieth century. Indeed the search for such a proof constituted one of those major problems which give direction to mathematics.

40
ORTHOGONAL POLYNOMIALS

In Chapter 39 we discussed the inner product space $C([a,b])$ with $(f,g) = \int_a^b r(t)f(t)g(t)^* dt$ under the condition that r was a strictly positive continuous function on $[a,b]$. In fact the conditions on r can be relaxed somewhat. Suppose simply that r is defined and continuous on (a,b), that $r(x) > 0$ for $x \in (a,b)$ and that

$$\int_a^b r(t)dt = \lim_{\eta(1),\eta(2) \to 0+} \int_{a+\eta(1)}^{b-\eta(2)} r(t)dt$$

converges.

Lemma 40.1. (*cf. Lemma 39.1.*) *If* $f, g \in C([a,b])$ *then*

$$(f,g) = \int_a^b r(t)f(t)g(t)^* dt = \lim_{\eta(1),\eta(2) \to 0+} \int_{a+\eta(1)}^{b-\eta(2)} r(t)f(t)g(t)^* dt$$

is well defined. If $f, g, h \in C([a,b])$ *and* $\lambda, \mu \in \mathbb{C}$ *then*

(i) $(f,g) = (g,f)^*$
(ii) $(\lambda f + \mu g, h) = \lambda(f,h) + \mu(g,h)$
(iii) (f,f) *is real and* $(f,f) \geqslant 0$
(iv) *If* $(f,f) = 0$ *then* $f = 0$.

Proof. Since f and g are continuous on the closed bounded interval $[a,b]$, they are bounded with $|f(t)|, |g(t)| \leqslant M$ for all $t \in [a,b]$, say. Thus $|r(t)f(t)g(t)^*| \leqslant M^2 r(t)$ and the comparison theorem for integrals shows that $\int_a^b r(t)f(t)g(t)^* dt$ converges. Parts (i), (ii) and (iii) are easy to verify. To check (iv), observe that, since $r(t)|f(t)|^2 \geqslant 0$, it follows that if $(f,f) = 0$ then $\int_a^b r(t)|f(t)|^2 dt = 0$ and so $f(t) = 0$ for all $t \in [c,d]$ whenever $a < c < d < b$. Since f is continuous it follows that $f(t) = 0$ for all $t \in [a,b]$. ∎

Thus all the general results on inner product spaces proved in the previous chapter apply in this case as well. (We sometimes call r the *weight function*.) We noted there that although it was easy to show that for any orthonormal series

185

$u_0, u_1, \ldots$ say $\| f - \sum_{j=0}^n \lambda_j u_j \|_2$ is minimised by the choice $\lambda_j = (f, u_j)$, it is often hard to decide whether $\| f - \sum_{j=0}^n (f, u_j) u_j \|_2 \to 0$ as $n \to \infty$. There is, however, one simple result which is often useful.

Theorem 40.2. *Suppose u_j is a polynomial of degree exactly j. Then if $u_0, u_1, \ldots$ are orthonormal*

$$\left\| f - \sum_{j=0}^n (f, u_j) u_j \right\|_2 \to 0 \quad \text{for all } f \in C([a, b]).$$

Our proof of Theorem 40.2 is based on a preliminary lemma.

Lemma 40.3. *Suppose u_j is a polynomial of degree exactly j. Then, if $u_0, u_1, \ldots$ are orthonormal and P is a polynomial of degree m, or less,*

$$P = \sum_{j=0}^m (P, u_j) u_j.$$

Proof. We first prove by induction that $P = \sum_{j=0}^m \lambda_j u_j$ for some $\lambda_j \in \mathbb{C}$ and then show that $\lambda_j = (P, u_j)$.

Step 1. Suppose p_m is a polynomial of degree m, or less, with $p_m(t) = \sum_{r=0}^m a_r t^r$. If $u_m(t) = \sum_{r=0}^m b_r t^r$ then, since u_m has degree exactly m, $b_m \neq 0$ and so $p_m = b_m^{-1} a_m u_m + p_{m-1}$ where $p_{m-1} = p_m - b_m^{-1} a_m u_m$ is a polynomial of degree $m - 1$, or less. It follows by induction that $p_m = \sum_{j=0}^m \lambda_j u_j$ for some $\lambda_j \in \mathbb{C}$.

Step 2. By Step 1, $P = \sum_{j=0}^m \lambda_j u_j$. Thus

$$(P, u_k) = \left(\sum_{j=0}^m \lambda_j u_j, u_k \right) = \sum_{j=0}^m \lambda_j (u_j, u_k) = \lambda_k \quad [0 \leqslant k \leqslant m],$$

and $P = \sum_{j=0}^m (P, u_j) u_j$, as stated. ∎

We can now turn to the proof of Theorem 40.2. This is an adaptation of the proof of Theorem 34.1 but instead of using Theorem 2.5 on uniform approximation by trigonometric polynomials we use Theorem 4.3 (Weierstrass) on uniform approximation by polynomials.

Proof of Theorem 40.2. By the theorem of Weierstrass (Theorem 4.3) we know that given any $\varepsilon > 0$ we can find a polynomial P of degree m say such that $|P(t) - f(t)| < \varepsilon$ for all $t \in [a, b]$. In particular

$$\| P - f \|_2 = \left(\int_a^b r(t) |P(t) - f(t)|^2 dt \right)^{\frac{1}{2}} \leqslant K\varepsilon$$

where $K = (\int_a^b r(t) dt)^{\frac{1}{2}}$.

Let us write $R_n(g) = \sum_{j=0}^n (g, u_j) u_j$. Then, if $n \geqslant m$, Lemma 40.3 tells us that $R_n(P) = P$ and so

$$\|f - \sum_{j=0}^{n} (f,u_j)u_j\|_2 = \|f - R_n f\|_2 \leqslant \|f - P\|_2 + \|P - R_n P\|_2 + \|R_n P - R_n f\|_2$$

$$= \|f - P\|_2 + \|R_n P - R_n f\|_2 = \|f - P\|_2 + \|R_n(P - f)\|_2$$

$$= 2\|f - P\|_2 \leqslant 2K\varepsilon$$

(the inequality $\|R_n g\|_2 \leqslant \|g\|_2$ coming from Theorem 39.7). We have thus shown that $\|f - R_n f\|_2 \to 0$ as $n \to \infty$. ∎

Exactly as in Chapter 34 we obtain the results corresponding to Theorems 34.2 and 34.3.

Theorem 40.4 (Parseval's formula). *Suppose u_j is a polynomial of degree exactly j. Then, if $u_0, u_1, \ldots$ are orthonormal,*

(i) *If $f \in C([a,b])$*

$$\|f\|_2^2 = \sum_{j=0}^{\infty} |(f, u_j)|^2.$$

(ii) *If $f, g \in C([a,b])$*

$$(f, g) = \sum_{j=0}^{\infty} (f, u_j)(g, u_j)^*.$$

Proof. As for Theorems 34.2 and 34.3. ∎

Orthogonal polynomials play an important role in classical analysis and give rise to a variety of beautiful formulae. We have already met one such system of polynomials in Chapter 5 where we defined the Tchebychev polynomials T_n.

Lemma 40.5. (i) *There exists a polynomial T_n (of degree exactly n) such that*

$$\cos n\theta = T_n(\cos \theta).$$

(ii) *The polynomials $2^{-\frac{1}{2}} T_0, T_1, T_2, \ldots$ are orthonormal on $[-1,1]$ with respect to the weight function $r(x) = 2\pi^{-1}(1 - x^2)^{-\frac{1}{2}}$ (the positive root being taken).*

Proof. (i) See Lemma 5.3
(ii) Making the substitution $\cos \theta = x$ with θ running from π to 0 as x runs from -1 to 1 we obtain $dx = -\sin \theta d\theta = -(1 - x^2)^{\frac{1}{2}} d\theta$ and

$$\int_{-1}^{1} T_n(x) T_m(x) r(x) dx = \frac{-2}{\pi} \int_{\pi}^{0} T_n(\cos \theta) T_m(\cos \theta) d\theta$$

$$= \frac{2}{\pi} \int_{0}^{\pi} \cos(n\theta) \cos(m\theta) d\theta = \begin{cases} 0 & \text{if } m \neq n \\ 1 & \text{if } m = n \end{cases} \quad [m, n \neq 0].$$

The cases $m = 0$ and/or $n = 0$ are also easy to check. ∎

Developing this theme we could consider the study of Fourier series as a special case of the study of orthogonal polynomials.

The next theorem shows that any weight function r has as associated sequence of orthogonal polynomials. (It was first obtained by Gram who generalised an idea of Schmidt.)

Theorem 40.6. *Suppose that $r:(a,b) \to \mathbb{R}$ is continuous, that $r(x) > 0$ for $x \in (a,b)$ and $\int_a^b r(t)\,dt$ converges. Then*

(i) *There exists a sequence of polynomials $u_0, u_1, \ldots$ orthonormal with respect to r and with u_j of degree exactly j.*

(ii) *If $u_0, u_1, \ldots$ and $v_0, v_1, \ldots$ are two sequences of polynomials orthonormal with respect to r and with u_j and v_j of degree exactly $j [j \geqslant 0]$, then we can find $|\lambda_j| = 1$ with $v_j = \lambda_j u_j [j \geqslant 0]$.*

Proof (i). We proceed by induction. Write $h_j(x) = x^j$. Suppose that $u_0, u_1, \ldots, u_{j-1}$ have been constructed with $(u_k, u_l) = 0$ for $l \neq k$, $(u_l, u_l) = 1$ and u_l of degree exactly $l [0 \leqslant l, k \leqslant j-1]$. We remark that $\sum_{k=0}^{j-1}(h_j, u_k)u_k$ is a polynomial of degree at most $j-1$ and so $w_j = h_j - \sum_{k=0}^{j-1}(h_j, u_k)u_k$ is a polynomial of degree exactly j. We note that

$$(w_j, u_l) = (h_j, u_l) - \sum_{k=0}^{j-1}(h_j, u_k)(u_k, u_l) = (h_j, u_l) - (h_j, u_l) = 0 \text{ for all } 0 \leqslant l \leqslant j-1.$$

Since w_j has degree exactly j, $w_j \neq 0$ and so $\|w_j\|_2 \neq 0$. Setting $u_j = \|w_j\|_2^{-1}w_j$ we see that u_j is a polynomial of degree exactly j with $(u_j, u_l) = 0$ for $0 \leqslant l \leqslant j-1$ and $(u_j, u_j) = 1$ so we can continue with the induction.

(ii) Suppose $v_j = \lambda_j u_j$ with $|\lambda_j| = 1$ for all $0 \leqslant j \leqslant m-1$. By Lemma 40.3,

$$v_m = \sum_{j=0}^{m}(v_m, u_j)u_j = (v_m, u_m)u_m + \sum_{j=0}^{m-1}(v_m, \lambda_j^{-1}v_j)u_j$$

$$= (v_m, u_m)u_m + \sum_{j=0}^{m-1}(\lambda_j^{-1})^*(v_m, v_j)u_j = (v_m, u_m)u_m.$$

Thus $v_m = \lambda_m u_m$ for some $\lambda_m \in \mathbb{C}$. Since $1 = \|v_m\|_2 = |\lambda_m| \|u_m\|_2 = |\lambda_m|$ we have $|\lambda_m| = 1$, as required. The result follows by induction. ∎

In the next chapter we shall consider the system of orthonormal polynomials which arise when we choose the simplest possible weight function $r(x) = 1$ on the interval $[-1, 1]$. The following, entirely computational, lemma shows that our system consists of scalar multiples of the *Legendre polynomials*

$$P_n(x) = \frac{1}{2^n n!}\frac{d^n}{dx^n}(x^2 - 1)^n.$$

Lemma 40.7. (i) *P_n is a real polynomial of degree exactly n.*

(ii) *If* $n \geqslant r \geqslant 0$ *then* $(d^r/dx^r)(x^2-1)^n = (x^2-1)^{n-r}Q_{n,r}(x)$ *where* $Q_{n,r}(x)$ *is a polynomial.*

(iii) *If* $n > r \geqslant 0$ *then* $(d^r/dx^r)(x^2-1)^n$ *takes the value* 0 *when* $x=1$ *or* $x=-1$.

(iv) *If* $n \geqslant r \geqslant 0$ *and* $m \geqslant 0$

$$\int_{-1}^{1} \frac{d^n}{dx^n}(x^2-1)^n \frac{d^m}{dx^m}(x^2-1)^m dx = (-1)^r \int_{-1}^{1} \frac{d^{n-r}}{dx^{n-r}}(x^2-1)^n \frac{d^{m+r}}{dx^{m+r}}(x^2-1)^m dx.$$

(v) $\int_{-1}^{1}(x^2-1)^n dx = (-1)^n 2^{n+1} n! / \prod_{r=0}^{n}(2r+1)$.

(vi) $\int_{-1}^{1} P_n(x)P_m(x)dx = 0$ *if* $n \neq m$.

(vii) $\int_{-1}^{1} P_n(x)^2 dx = 2/(2n+1)$.

Proof. (i) Observe that $(x^2-1)^n$ is a polynomial of degree exactly $2n$.

(ii) If $n > r \geqslant 0$ and $d^r/dx^r(x^2-1)^n = (x^2-1)^{n-r}Q_{n,r}(x)$ then

$$\frac{d^{r+1}}{dx^{r+1}}(x^2-1)^n = (x^2-1)^{n-r-1}(2x(n-r)Q_{n,r}(x) + (x^2-1)Q'_{n,r}(x)),$$

so the result follows by induction.

(iii) Follows from (ii).

(iv) If $n > r \geqslant 0$ and $m \geqslant 0$ then integrating by parts and using (iii) we have

$$\int_{-1}^{1} \frac{d^{n-r}}{dx^{n-r}}(x^2-1)^n \frac{d^{m+r}}{dx^{m+r}}(x^2-1)^m dx = \left[\frac{d^{n-r-1}}{dx^{n-r-1}}(x^2-1)^n \frac{d^{m+r}}{dx^{m+r}}(x^2-1)^m \right]_{-1}^{1}$$

$$- \int_{-1}^{1} \frac{d^{n-r-1}}{dx^{n-r-1}}(x^2-1)^n \frac{d^{m+r+1}}{dx^{m+r+1}}(x^2-1)^m dx$$

$$= - \int_{-1}^{1} \frac{d^{n-r-1}}{dx^{n-r-1}}(x^2-1)^n \frac{d^{m+r+1}}{dx^{m+r+1}}(x^2-1)^m dx.$$

Now use induction.

(v) Set $I_n = \int_{-1}^{1}(x^2-1)^n dx$. Integrating by parts we have

$$I_n = [x(x^2-1)^n]_{-1}^{1} - \int_{-1}^{1} 2nx^2(x^2-1)^{n-1} dx$$

$$= - \int_{-1}^{1} 2nx^2(x^2-1)^{n-1} dx = -2n(I_n + I_{n-1}) \text{ so } I_n = (2n/(2n+1))I_{n-1}.$$

Since $I_0 = 2$, this gives the result by induction.

(vi) Without loss of generality, suppose $n > m$. Setting $r = m+1$ in (iv), we get

$$\int_{-1}^{1} \frac{d^n}{dx^n}(x^2-1)^n \frac{d^m}{dx^m}(x^2-1)^m \, dx$$

$$= (-1)^{m+1} \int_{-1}^{1} \frac{d^{n-m-1}}{dx^{n-m-1}}(x^2-1)^n \frac{d^{2m+1}}{dx^{2m+1}}(x^2-1)^m \, dx$$

$$= \int_{-1}^{1} 0 \, dx = 0.$$

(vii) Setting $r = m = n$ in (iv) we get

$$\int_{-1}^{1} \left(\frac{d^n}{dx^n}(x^2-1)^n \right)^2 dx = (-1)^n \int_{-1}^{1} (x^2-1)^n \frac{d^{2n}}{dx^{2n}}(x^2-1)^n dx$$

$$= (-1)^n(2n)! \int_{-1}^{1} (x^2-1)^n dx = 2^{n+1}n!(2n)! \bigg/ \prod_{r=0}^{n}(2r+1)$$

$$= 2(2^n n!)/(2n+1),$$

by (v). Thus $\|P_n\|_2^2 = 2/(2n+1)$. ∎

Although it is reassuring to have an explicit formula the only fact that we shall actually need about the Legendre polynomials is that all their roots are real, simple and lie in $[-1,1]$. And this is most easily proved as a general result on real orthogonal polynomials.

Theorem 40.8. *Suppose* $u_0, u_1, u_2 \ldots$ *are orthonormal polynomials with respect to a weight function* r *on an interval* $[a,b]$ *and that* u_j *is a real polynomial of degree exactly* $j [\, j \geqslant 0]$. *Then* u_n *has* n *simple roots all lying in* (a,b).
Proof. Suppose not. Then u_n changes sign less than n times on (a,b) at $x_1, x_2, \ldots, x_k$ say, with $a < x_1 < x_2 < \cdots < x_k < b$. Set $Q(x) = \prod_{j=0}^{k}(x-x_j)$. Then Qu_n is either strictly positive on all of (a,b) except at $x_1, x_2, \ldots, x_k$ or strictly negative except at $x_1, x_2, \ldots, x_k$. Thus

$$(Q, u_n) = \int_{a}^{b} Q(t)u_n(t)r(t) \, dt \neq 0.$$

But Q is of degree $k < n$ and so, using Lemma 40.3,

$$(Q, u_n) = \left(\sum_{j=0}^{k} (Q, u_j)u_j, u_n \right) = \sum_{j=0}^{k} (Q, u_j)(u_j, u_n) = 0.$$

The theorem follows by *reductio ad absurdum.* ∎

41

GAUSSIAN QUADRATURE

Consider the problem of the numerical evaluation of $\int_{-1}^{1} f(x)dx$ with f continuous. (Since the change of variable

$$x = \frac{2}{b-a}\left(t - \frac{(b+a)}{2}\right)$$

converts the integral $\int_{a}^{b} g(t)dt$ to one of the form $\int_{-1}^{1} f(x)dx$ there is no advantage in considering a different range of integration.)

One obvious method is to use the approximation

$$T_{2n}f = \frac{1}{n}\sum_{r=-n}^{n-1} f\left(\frac{2r+1}{2n}\right).$$

This has the advantage of simplicity and the certainty that $T_{2n}f \to \int_{-1}^{1} f(x)dx$ as $n \to \infty$. A disadvantage becomes clear when we consider the particular case $g(x) = x^2$. We then have

$$T_{2n}g = n^{-1}\sum_{r=-n}^{n-1} ((2r+1)/2n)^2 = 2^{-1}n^{-3}\sum_{r=0}^{n-1} (2r+1)^2$$

$$= 2^{-1}n^{-3}\sum_{r=0}^{n-1}(4r^2 + 4r + 1) = 2^{-1}n^{-3}\sum_{r=0}^{n-1}(4r(r+1) + 1)$$

$$= 2^{-1}n^{-3}\sum_{r=0}^{n-1}(\tfrac{4}{3}(r(r+1)(r+2) - (r-1)r(r+1)) + 1)$$

$$= 2^{-1}n^{-3}(\tfrac{4}{3}(n-1)n(n+1) + n) = \tfrac{2}{3} - (1/6)n^{-2}.$$

Thus

$$\left| T_{2n}g - \int_{-1}^{1} g(x)dx \right| = \frac{1}{6n^2}$$

and we would have to take $n \geqslant 130$ (involving over 250 evaluations of g) to get an error of less than 10^{-5}.

We would like to have a method of numerical integration which converges faster

191

when the function f is sufficiently well behaved. Bearing in mind that one criterion of good behaviour is good approximation by polynomials, an obvious strategy is to seek methods which are exact for polynomials of low degree.

Lemma 41.1. *Let* $-1 \leqslant x_1 < x_2 < \cdots < x_n \leqslant 1$.   *Then   there   exist   unique* $A_1, A_2, \ldots, A_n$ *such that*

$$\int_{-1}^{1} P(x)dx = \sum_{j=1}^{n} A_j P(x_j)$$

whenever P is a polynomial of degree $n-1$ or less.
Proof. Write $Q_j(x) = \prod_{k \neq j}(x - x_k)/\prod_{k \neq j}(x_j - x_k)$. Then $Q_j(x_k) = 0$ for $k \neq j$ and $Q_j(x_j) = 1$. It follows that writing

$$R(x) = P(x) - \sum_{j=1}^{n} P(x_j)Q_j(x),$$

we have $R(x_k) = 0$ for all $1 \leqslant k \leqslant n$. Thus, if P is a polynomial of degree $n-1$ or less, we see that R is a polynomial of degree at most $n-1$ which vanishes at n points. Thus R is identically zero and $P(x) = \sum_{j=1}^{n} P(x_j)Q_j(x)$ for all x. It follows that

$$\int_{-1}^{1} P(x)dx = \sum_{j=1}^{n} \int_{-1}^{1} Q_j(x)dx \, P(x_j)$$

and, setting $A_j = \int_{-1}^{1} Q_j(x)dx$, we have the required formula.
    Conversely, if $\int_{-1}^{1} P(x)dx = \sum_{j=1}^{n} B_j P(x_j)$ for all polynomials of degree $n-1$ or less then, taking $P(x) = Q_j(x)$, we obtain $\int_{-1}^{1} Q_j(x)dx = B_j$. We have thus shown uniqueness. ∎

We now have a method of numerical integration by the sum $\sum_{j=1}^{n} A_j f(x_j)$ which involves n evaluations and is exact for polynomials of degree $n-1$ or less.
    However, Gauss discovered that we can do much better by a careful choice of the x_j.

Theorem 41.2 (Gauss). *Let $x_1, x_2, \ldots, x_n$ be the roots of the nth Legendre polynomial P_n. Then we can find $A_1, A_2, \ldots, A_n$ such that*

$$\int_{-1}^{1} P(x)dx = \sum_{j=1}^{n} A_j P(x_j)$$

whenever P is a polynomial of degree $2n-1$ or less.
    Before giving the proof, we recall the facts we shall need from the previous chapter. We defined the Legendre polynomial P_n by $P_n(x) = (1/2^n n!)(d^n/dx^n)(x^2 - 1)^n$. The following results were proved.

Lemma 41.3. *Write $p_n = ((2n + 1)/2)^{\frac{1}{2}} P_n$.*

(i) $p_0, p_1, \ldots$ *form an orthonormal sequence (with respect to the weight function $r(x) = 1$ on $[-1, 1]$).*

(ii) p_j *is a real polynomial of degree exactly* j.

(iii) *The zeros of* p_n *and so of* P_n *are real and simple and lie in* $(-1, 1)$.

(iv) $\int_{-1}^{1} Q(x)P_n(x)dx = 0$ *whenever* Q *is a polynomial of degree* $n - 1$ *or less*.

Proof. (i) This is Lemma 40.7 (vi) and (vii).

(ii) This is Lemma 40.7 (i).

(iii) Use Theorem 40.8.

(iv) By Lemma 40.3, $Q = \sum_{j=0}^{n-1}(Q, p_j)p_j$ so that

$$(Q, p_n) = \sum_{j=0}^{n-1} (Q, p_j)(p_j, p_n) = 0$$

and so

$$(Q, P_n) = 0. \qquad \blacksquare$$

Proof of Theorem 41.2. Let $A_1, A_2, \ldots, A_n$ be chosen as in Lemma 41.1. If P is a polynomial of degree at most $2n - 1$, then simple division gives $P = QP_n + R$ where Q and R are polynomials of degree at most $n - 1$.

Applying first Lemma 41.3 (iv) and then Lemma 41.1 we have

$$\int_{-1}^{1} P(x)dx = \int_{-1}^{1} Q(x)P_n(x)dx + \int_{-1}^{1} R(x)dx = \int_{-1}^{1} R(x)dx = \sum_{j=1}^{n} A_j R(x_j).$$

But $P_n(x_j) = 0$ by the choice of x_j so,

$$P(x_j) = Q(x_j)P_n(x_j) + R(x_j) = R(x_j).$$

Thus

$$\int_{-1}^{1} P(x)dx = \sum_{j=1}^{n} A_j P(x_j),$$

as required. $\qquad \blacksquare$

Not surprisingly, the choice of $x_1, x_2, \ldots, x_n$ is unique.

Lemma 41.4. *Suppose* $-1 \leqslant x_1 < x_2 < \cdots < x_n \leqslant 1$ *and* $A_1, A_2, \ldots, A_n$ *are given so that*

$$\int_{-1}^{1} P(x)dx = \sum_{j=1}^{n} A_j P(x_j)$$

for every polynomial of degree $2n - 1$ *or less. Then* $x_1, x_2, \ldots, x_n$ *are the roots of the* nth *Legendre polynomial.*

Proof. Let $T(x) = \prod_{j=1}^{n}(x - x_j)$. Then Tp_r is a polynomial of degree $n + r$ and so

$$\int_{-1}^{1} T(x)p_r(x)dx = \sum_{j=1}^{n} A_j T(x_j)p_r(x_j) = \sum_{j=1}^{n} A_j 0 p_r(x_j) = 0 \quad \text{for all } 0 \leqslant r \leqslant n - 1.$$

Since T is a polynomial of degree n, Lemma 40.3 yields $T = \sum_{r=0}^{n}(T, p_r)p_r = (T, p_n)p_n$ so T is a scalar multiple of p_n. Hence $x_1, x_2, \ldots, x_n$ are the zeros of P_n and so of P_n. $\qquad \blacksquare$

However, the fact that a method of numerical integration is exact for polynomials of fairly high degree does not by itself imply that the method is a good one. Suppose, for example, that we take $x_j = -1 + 2(j-1)/(n-1)$ $[1 \leqslant j \leqslant n]$ and fix the A_j as in Lemma 41.1, so that

$$\int_{-1}^{1} P(x)\,dx = \sum_{j=0}^{n} A_j P(x_j)$$

for all polynomials of degree $n-1$ or less. The resulting formula for numerical integration

$$T_n(f) = \sum_{j=0}^{n} A_j f(x_j)$$

is called the Newton–Cotes quadrature formula of order n.

For $n=2$ we get the trapezoidal rule $\int_{-1}^{1} f(x)\,dx = f(-1) + f(1)$, and for $n=3$ we get Simpson's rule $\int_{-1}^{1} f(x)\,dx = \frac{1}{3}(f(-1) + 4f(0) + f(1))$. However when n is large the situation becomes more complicated. For example, if $n=21$,

$$T_n(f) = \sum_{j=0}^{21} A_j f(x_j) = B_0 f(0) + \sum_{r=1}^{10} B_r(f(r/10) + f(-r/10)),$$

where $B_r = H_r/D$,

$H_{10} =$	1947 01402 41329
$H_9 =$	18792 60903 80000
$H_8 =$	$-$ 38935 81914 77500
$H_7 =$	1 98596 91593 40000
$H_6 =$	$-$ 6 20894 88358 89375
$H_5 =$	17 01938 77765 17504
$H_4 =$	$-$ 37 38973 46712 90000
$H_3 =$	68 86928 75743 20000
$H_2 =$	$-$ 105 49901 48137 01250
$H_1 =$	136 32452 17984 40000
$H_0 =$	$-$ 148 19252 66072 80936

and $D = 8232\ 27205\ 40240$ (these figures are taken from Appendix 4 of Kopal, *Numerical Analysis*).

Evaluation of $\int_{-1}^{1} f(x)\,dx$ using $T_{21}(f)$ will thus usually involve the computation of a small quantity by cancellation of two large quantities, an undesirable feature in any method. But things are much worse than this. If we take $f(x) = 1/(1 + 16x^2)$ then f is a fairly well behaved function with $\int_{-1}^{1} f(x)\,dx = [\frac{1}{4}\tan^{-1} 4x]_{-1}^{1} = \frac{1}{2}\tan^{-1} 4 = 0.66291\ldots$ (correct to the number of figures stated). But (again to the number of figures stated) $T_{21}(f) = -1.9379\ldots$ and we have a negative approximation to a positive integral.

Thus we need more information before we can judge the quality of Gaussian quadrature. The necessary results were found by Stieltjes.

Lemma 41.5. *Let* $x_1, x_2, \ldots, x_n$ *and* $A_1, A_2, \ldots, A_n$ *be chosen as in Theorem* 41.2. *Then*

(i) $A_1, A_2, \ldots, A_n > 0$.

(ii) $\sum_{j=1}^{n} A_j = 2$.

Proof. (i) Taking $P(x) = \prod_{k \neq j}(x - x_k)^2$, we see that P is a polynomial of degree less than $2n - 1$ so that

$$\int_{-1}^{1} P(x)dx = \sum_{l=1}^{n} A_l P(x_l) = A_j P(x_j).$$

But $P(x) > 0$ for $x \notin \{x_k : k \neq j\}$ so that $P(x_j) > 0$ and (since P is continuous) $\int_{-1}^{1} P(x)dx > 0$. Thus $A_j > 0$ for each $1 \leqslant j \leqslant n$.
(ii) Setting $P(x) = 1$ we have

$$2 = \int_{-1}^{1} P(x)dx = \sum_{j=1}^{n} A_j P(x_j) = \sum_{j=1}^{n} A_j,$$

as required. ∎

(The reader may care to recall, from Chapter 2, the importance we attached to the positivity of the Fejér kernel K_n in the formula

$$\sigma_n(f, t) = \frac{1}{2\pi} \int_{\mathbb{T}} f(t - y) K_n(y) dy.)$$

We can now prove Stieltjes' result on the convergence of Gaussian quadrature. We write $G_n(f) = \sum_{j=1}^{n} A_j f(x_j)$ where $x_1, x_2, \ldots, x_n$ and $A_1, A_2, \ldots, A_n$ are chosen as in Theorem 41.2.

Theorem 41.6 (Stieltjes). (i) *If* $\mathscr{P}_{2n-1}$ *denotes the set of polynomials of degree at most* $2n - 1$ *then*

$$\left| G_n(f) - \int_{-1}^{1} f(x)dx \right| \leqslant 4 \inf \left\{ \sup_{-1 \leqslant t \leqslant 1} |f(t) - P(t)| : P \in \mathscr{P}_{2n-1} \right\}.$$

(ii) $G_n(f) \to \int_{-1}^{1} f(x)dx$ *as* $n \to \infty$.
Proof. (i) Let $P \in \mathscr{P}_{2n-1}$. Then $G_n(P) = \int_{-1}^{1} P(x)dx$ so

$$\left| G_n(f) - \int_{-1}^{1} f(x)dx \right| = \left| (G_n(f) - G_n(P)) - \left(\int_{-1}^{1} f(x)dx - \int_{-1}^{1} P(x)dx \right) \right|$$

$$\leqslant |G_n(f) - G_n(P)| + \left| \int_{-1}^{1} f(x)dx - \int_{-1}^{1} P(x)dx \right|$$

$$= \left| \sum_{j=1}^{n} A_j(f(x_j) - P(x_j)) \right| + \left| \int_{-1}^{1} (f(x) - P(x))dx \right|$$

$$\leqslant \sum_{j=1}^{n} A_j |f(x_j) - P(x_j)| + \int_{-1}^{1} |f(x) - P(x)| dx$$

$$\leqslant \sum_{j=1}^{n} A_j \sup_{-1 \leqslant t \leqslant 1} |f(t) - P(t)| + 2 \sup_{-1 \leqslant t \leqslant 1} |f(t) - P(t)|$$

$$= 4 \sup_{-1 \leqslant t \leqslant 1} |f(t) - P(t)|,$$

and the result follows.

(ii) The theorem of Weierstrass (Theorem 4.3) shows

$$\inf \left\{ \sup_{-1 \leqslant t \leqslant 1} |f(t) - P(t)| : P \in \mathscr{P}_{2n-1} \right\} \to 0 \quad \text{as} \quad n \to \infty. \quad \blacksquare$$

Theorem 41.6 states that Gaussian quadrature will give good results whenever the continuous function f is well approximable by polynomials, but that in any case it always converges to the correct answer as $n \to \infty$.

The reader may wonder how it behaves in the case of the function $f(x) = 1/(1 + 16x^2)$ mentioned earlier. Using the values of A_j and x_j provided in the *Handbook of Mathematical Functions* by Abramowitz and Stegun we get (to five-figure accuracy)

$$G_{10}(f) = 0.65431 \ldots$$

(which is in error by less than 1.5% by comparison with the true answer $2^{-1} \tan^{-1} 4 = 0.66291 \ldots$) and

$$G_{20}(f) = 0.66284 \ldots$$

(which is in error by less than 0.01%).

In the days before electronic computers all calculations had to be made using printed tables and, at best, a glorified adding machine. Under these circumstances the fact that the x_j are not simple rational numbers reduced Gaussian quadrature to the status of a curiosity. However, such considerations are irrelevant when dealing with an electronic computer and now Gaussian quadrature is one of the chief methods employed for numerical integration.

42

LINKAGES

During the last third of the eighteenth century Watt embarked on a total redesign of the Newcomen steam engine turning it from a special purpose machine to an economical general purpose power source. This involved the solution not just of one but many technical problems each one requiring either a new idea (like the separate condenser) or the astute adaptation of an old one (like the centrifugal governor which we shall briefly return to in Chapter 72).

One of the problems that had to be overcome was that of turning the rectilinear motion of a piston into the circular motion of a wheel and vice versa. Spurred on by his partner Boulton, Watt patented five different solutions. The one we shall discuss is called Watt's parallelogram. Watt is reported to have been prouder of this invention than of any other.

In its simplest form it consists of three bars AB, BC and CD linked as shown in Figure 42.1 with A and D fixed but B and C free to move. The 'tracer point' Q is on BC. A little experiment will show the reader that as C rotates round D, Q performs quite a complicated motion. However, for many purposes it is sufficient that the path of Q is approximately linear for rotations of CD through a small angle $\theta(|\theta| < 20°$ say).

Watt discovered (presumably by repeated trial calculations) that such approximate linear motion would result if he took Q to be the mid point of BC, both AB and CD of the same length and chose the length of BC in such a way that when AB and CD were parallel then BC was perpendicular both to AB and CD (see Figure 42.2).

The importance of this idea lay in the fact that the mechanism did not require precision engineering whereas alternative ideas involving metal rods sliding through metal guides would stretch the resources of an eighteenth-century craftsman to the limit.

Many people tried to find linkages which give a closer approximation to linearity or indeed gave exact straight lines. Tchebychev was fascinated by the problem and was led by it to results of an entirely new kind which we shall consider in the next chapter.

197

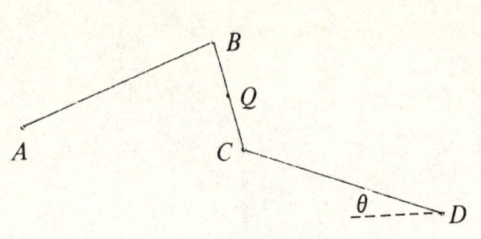

Fig. 42.1. Watt's linkage.

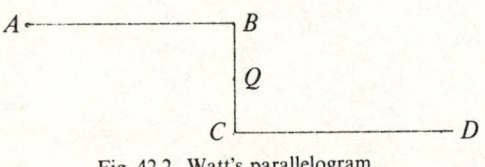

Fig. 42.2. Watt's parallelogram.

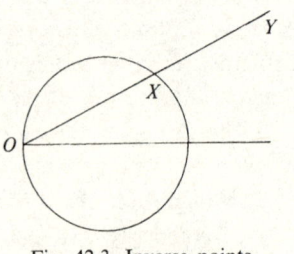

Fig. 42.3. Inverse points.

However, although this discussion is mainly intended to introduce Tchebychev's idea, I cannot resist continuing with this topic for a little. For many years it was considered that the problem of a linkage to convert circular into straight line motion was insoluble. Tchebychev himself seems to have been of this opinion. Yet in 1864 a lieutenant in the French army called Peaucellier obtained a simple solution which we shall now describe.

We need as a preliminary lemma a result which in 1860 (or indeed 1960) would have been common knowledge.

Lemma 42.1. *Let O be a fixed point in the plane and let X, Y be points varying, as shown in Figure 42.3, in such a way that*

(i) *OXY is a straight line*
(ii) *$|OX||OY| = k$ where $|OX|$ is the length of OX, $|OY|$ of OY and k is a strictly positive constant.*

Then, as X describes a circle through O, Y describes a straight line.

We give two proofs.

Proof A. Let us work in the complex plane. Recall that the map T_1 given by

$T_1 z = 1/z$ takes circles through the origin to straight lines, that the map T_2 given by $T_2(re^{i\theta}) = re^{-i\theta}$ (reflection in the real axis) takes straight lines to straight lines and that the map T_3 given by $T_3 z = kz$ (dilatation) takes straight lines to straight lines. Thus $T_3 T_2 T_1$ takes circles through the origin to straight lines. But if X corresponds to z and Y to w then $w = T_3 T_2 T_1 z$, so the result follows. ∎

Proof B. Let O be the origin of a system of polar coordinates such that the circle traced by X (coordinates (r, θ) say) is given by $r = 2R \cos \theta$. Then the coordinates (s, θ) of Y are given by $sr = k$, and so by $s = (k/2R) \sec \theta$. Thus Y traces out a straight line. ∎

We now describe Peaucellier's mechanism which is illustrated in Figure 42.4. We take six bars OA, OC, AB, BC, CD, DA linked as shown. The point O is fixed but A, B, C and D may move freely. The lengths of the four rods AB, BC, CD and DA are equal as are those of the rods OA and OC.

Lemma 42.2. *In the arrangement described above we have*

(i) *OBD is a straight line*
(ii) $|OB||OD| = k$ *for some constant* $k > 0$.

Proof. That OBD is a straight line is clear from symmetry. To prove (ii) we consider the point E at the intersection of OD and AC as shown in Figure 42.5.

Note first that, by symmetry, all the angles $A\hat{E}B, B\hat{E}C, C\hat{E}D, D\hat{E}A$ are equal and so are right angles. Applying Pythagoras's theorem to the right angled triangles BEA and OEA we obtain

$$|BA|^2 = |BE|^2 + |EA|^2,$$

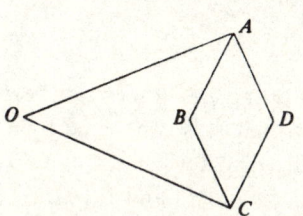

Fig. 42.4. Peaucellier's mechanism.

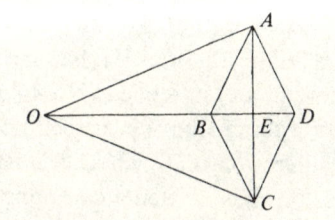

Fig. 42.5. Diagram for Peaucellier's proof.

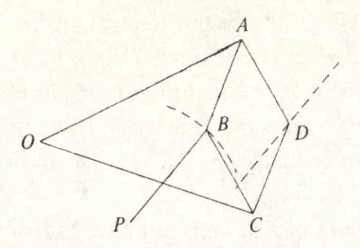

Fig. 42.6. Peaucellier's mechanism in action.

and
$$|OA|^2 = |OE|^2 + |EA|^2.$$

Thus, eliminating $|EA|^2$ between the equations, we get

$$|OA|^2 - |BA|^2 = |OE|^2 - |BE|^2 = (|OE| - |BE|)(|OE| + |BE|).$$

But, by symmetry,

$$|BE| = |ED| \quad \text{so} \quad |OE| + |BE| = |OE| + |ED| = |OD|$$

whilst
$$|OE| - |BE| = |OB|$$

and the equation above becomes

$$|OA|^2 - |BA|^2 = |OB||OD|.$$

Since $|OA|$ and $|BA|$ are fixed this proves (ii).   ∎

Combining Lemmas 42.1 and 42.2 we obtain the desired result.

Theorem 42.3 (Peaucellier). *If in the arrangement above B is constrained to move in (part of) a circle through O, then D will describe (part of) a straight line.*
Proof. Immediate.   ∎

An arrangement for illustrating the result can be obtained by modifying the mechanism of Figure 42.4 so that B is connected by a rod PB to a fixed point P whose distance from O is equal to the length of PB, as in Figure 42.6. Then B will describe (part of) a circle round P, and D (part of) a straight line.

Because of the advances in workshop technique Peaucellier's discovery was of less practical importance than had been hoped, but the elegance and simplicity of this solution to a problem which had baffled so many aroused great enthusiasm.

Sylvester (*Collected Works*, Vol. 3 Paper 2) recalls that when he showed a working model to Kelvin: '[he] nursed it as if it had been his own child, and when a motion was made to relieve him of it, replied "No! I have not had nearly enough of it – it is the most beautiful thing I have ever seen in my life"'.

Remark. The correct transliteration of Tchebychev's Russian name is a matter of some controversy. Philip Davis has written a charming book in which this forms the central theme. Without accepting Davis's preferred spelling I do agree with him that only admirers of Čaykovskiy's music are entitled to write Čebysev.

43

TCHEBYCHEV AND
UNIFORM APPROXIMATION I

In Chapter 32 we found that the best mean square approximation P to a continuous function f by trigonometric polynomials of order n was $P(t) = \sum_{r=-n}^{n} \hat{f}(r) \exp irt$. The form of the result is simple and the proof easy because of the particular nature of mean square approximation. For other types of approximation the problem of best (or even of good) approximation may be much harder.

The great Russian mathematician Tchebychev loved to study and construct mechanisms of all sorts. As we noted in the previous chapter, he was deeply interested in possible improvements to Watt's parallelogram and wrote a series of papers on the subject.

In Figure 43.1 we recall the form of Watt's parallelogram (turning it through $90°$ with respect to Figure 42.1). The accompanying very free sketch shows the distance $y = g(\theta)$ of Q from some fixed straight UV when CD makes an angle θ to the perpendicular to UV. It is clear that after we have fixed A, D and the straight line UV together with the general arrangement of the three links AB, BC, CD (with Q lying on CB) we can still change the lengths $\lambda_1 = AB$, $\lambda_2 = BQ$, $\lambda_3 = QC$ and $\lambda_4 = CD$.

Thus the distance y of Q from UV is given by $y = g(\theta, \lambda_1, \lambda_2, \lambda_3, \lambda_4)$ where we are free to choose λ_1, λ_2, λ_3 and λ_4. Our object is to minimise the deviation y of Q from the straight line path UV as θ runs from ϕ to ψ, say. In other words we wish to minimise

$$\sup_{\phi \leqslant \theta \leqslant \psi} |g(\theta, \lambda_1, \lambda_2, \lambda_3, \lambda_4)|$$

by a suitable choice of λ_1, λ_2, λ_3, λ_4.

More generally we may investigate the problem of minimising

$$\sup_{a-h \leqslant t \leqslant a+h} |g(t, \lambda_1, \lambda_2, \ldots, n) - f(t)|$$

by a suitable choice of $\lambda_1, \lambda_2, \ldots, \lambda_n$ when g and f are well behaved. If we demand an exact solution this programme looks overambitious, but, remembering the

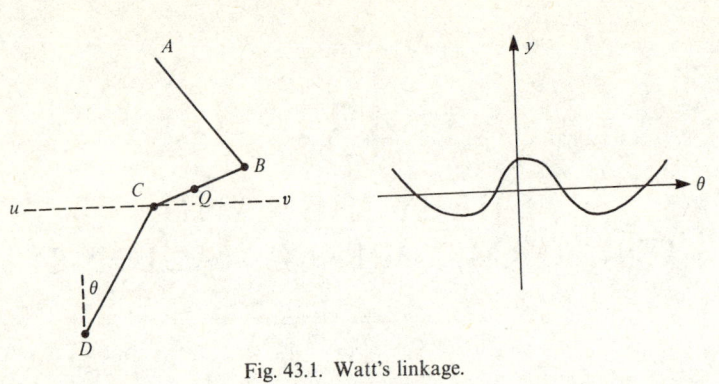

Fig. 43.1. Watt's linkage.

original problem of Watt's parallelogram, an approximate solution may be easier
to find and still be interesting.

If we are prepared to accept an approximate solution, then an obvious first step
is to expand g in a Taylor series about a to obtain

$$g(t) \approx k_0 + k_1(t-a) + k_2(t-a)^2 + \cdots + k_{n-1}(t-a)^{n-1},$$

where we ignore nth and higher order terms. Speaking rather vaguely, we might
expect our free choice of $\lambda_1, \lambda_2, \ldots, \lambda_n$ to translate as a free choice of $k_0, k_1, \ldots, k_{n-1}$
and we might hope to get an (approximate) solution to our original problem by
showing how to choose $k_0, k_1, \ldots, k_{n-1}$ in such a way as to minimise

$$\sup_{a-h \leqslant t \leqslant a+h} \left| \sum_{r=0}^{n-1} k_r(t-a)^r - f(t) \right|.$$

We have thus arrived at Tchebychev's problem.

Problem (Tchebychev). *Let $f:[a,b] \to \mathbb{R}$ be continuous. Characterise the polynomials
P of degree $n-1$ which minimise* $\sup_{a \leqslant t \leqslant b} |f(t) - P(t)|$.

On the face of it this problem, even though simpler, than the previous ones still
looks too hard. However, Tchebychev was able to find a simple usable criterion,
although he could only prove it in certain cases. Let us write $\mathcal{P}_{n-1}$ for the set of
polynomials of degree $n-1$ or less.

Definition 43.1. *Let $f:[a,b] \to \mathbb{R}$ be continuous. We say that $P \in \mathcal{P}_{n-1}$ satisfies the
Tchebychev equal ripple criterion (with respect to f and n) if we can find $a \leqslant y_0 <
y_1 < \cdots < y_n \leqslant b$ such that*

(i) $|f(y_j) - P(y_j)| = \sup_{t \in [a,b]} |P(t) - f(t)| \qquad [0 \leqslant j \leqslant n]$

(ii) $f(y_j) - P(y_j) = -(f(y_{j-1}) - P(y_{j-1})) \quad [1 \leqslant j \leqslant n]$.

(Thus $|f - P|$ has a global maximum at each of the points $y_0, y_1, \ldots, y_n$ but $f - P$
is alternately a maximum and a minimum, see Figure 43.2.)

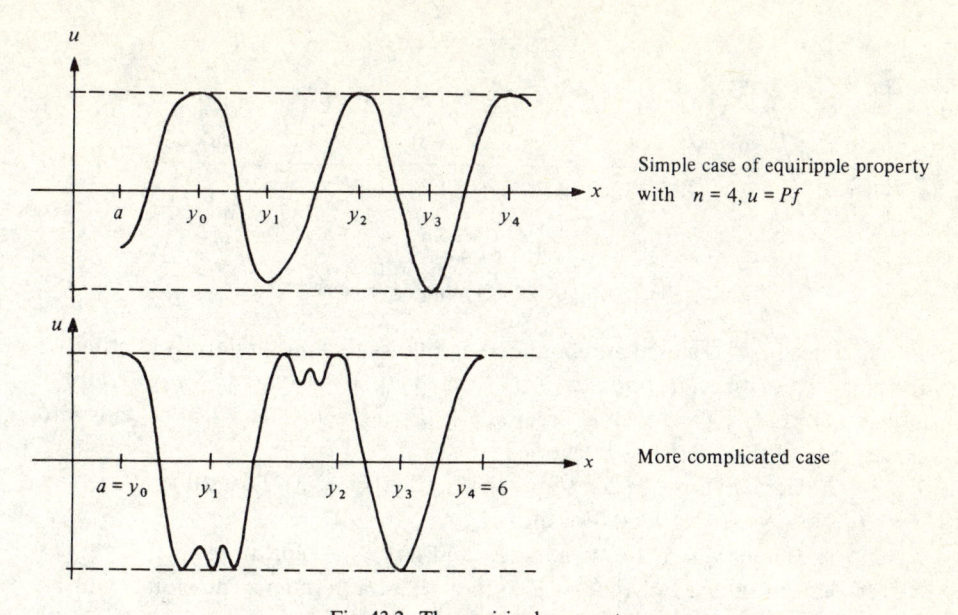

Simple case of equiripple property with $n = 4, u = Pf$

More complicated case

Fig. 43.2. The equiripple property.

Theorem 43.2. *Let $f:[a,b] \to \mathbb{R}$ be continuous. Then if $P \in \mathscr{P}_{n-1}$ is such that*

$$\sup_{t \in [a,b]} |P(t) - f(t)| \leqslant \sup_{t \in [a,b]} |Q(t) - f(t)| \quad \text{for all } Q \in \mathscr{P}_{n-1}$$

it follows that P satisfies the equal ripple criterion.

(More briefly, any solution of Tchebychev's problem satisfies Tchebychev's criterion.)

The key to the proof of Theorem 43.2 lies in the following lemma whose proof is unfortunately somewhat tedious. I suggest that the reader ignores this proof for the time being and simply convinces herself that the statement is plausible. She can return to it (or, even better, provide her own proof) later if she wants.

Lemma 43.3. *Let $g:[a,b] \to \mathbb{R}$ be continuous with $\sup_{t \in [a,b]} |g(t)| = M > 0$. Then we can find an integer $k \geqslant 1$ and points*

$$a = x(0) \leqslant y(0) < x(1) < y(1) < x(2) < \cdots < x(k) < y(k) \leqslant x(k+1) = b$$

and γ taking the value 0 or 1 such that

(i) $g(x(j)) = 0$ *for* $1 \leqslant j \leqslant k$,
(ii) $(-1)^{j+\gamma} g(y(j)) = M$ *for* $0 \leqslant j \leqslant k$,
(iii) $(-1)^{j+\gamma} g(t) > -M$ *for all* $x(j-1) \leqslant t \leqslant x(j)$ $[1 \leqslant j \leqslant k+1]$.

(Thus the $x(j)$ divide $[a,b]$ into intervals on which, alternately, g takes the value M at least once but never the value $-M$, or g takes the value $-M$ at least once but never the value M. We show the kind of thing that can happen in Figure 43.3.)

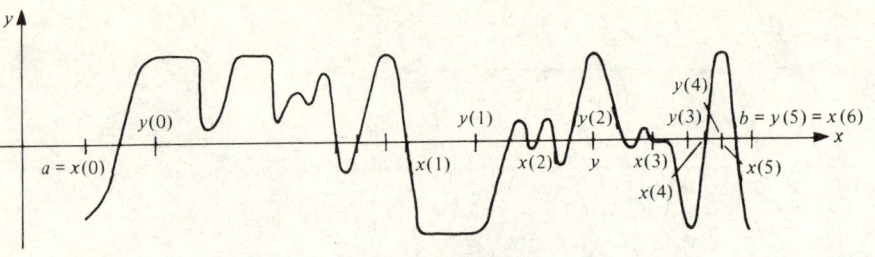

Fig. 43.3. Graph illustrating Lemma 43.3.

Proof. Since $[a, b]$ is closed and bounded, it follows that g is uniformly continuous In particular, we can find an $N \geqslant 1$ such that $|u - v| \leqslant (b - a)/N$ implies $|g(u) - g(v)| \leqslant M/2$. Thus each interval $I_n = [n(b - a)/N, (n + 1)(b - a)/N]$ falls into exactly one of the following categories.

Type $- 1$. Here $g(y) = - M$ for some $y \in I_n$ and $g(t) \leqslant - M/2$ for all $t \in I_n$.

Type 0. Here $- M < g(t) < M$ for all $t \in I_n$.

Type $+ 1$. Here $g(y) = M$ for some $y \in I_n$, and $g(t) \geqslant M/2$ for all $t \in I_n$.

Notice that an interval of type $- 1$ cannot share a common end point with an interval of type $+ 1$.

Since $[a, b]$ is closed and bounded, $|g|$ attains its supremum and so not all intervals can be of type 0. We can therefore find integers

$$0 = s(0) \leqslant r(1) \leqslant s(1) < r(2) \leqslant s(2) < r(3) \leqslant s(3) < r(4) \leqslant \cdots$$
$$\cdots < r(k - 1) \leqslant s(k - 1) < r(k) \leqslant s(k) \leqslant r(k + 1) = N - 1,$$

and γ taking the value 0 or 1 such that

(a) If $0 \leqslant n < r(1)$, $s(k) < n \leqslant N - 1$ or $s(j - 1) < n < r(j)$ then I_n is of type 0.

(b) If $r(j) \leqslant n \leqslant s(j)$ then I_n is of type 0 or $(- 1)^{j + \gamma}$.

(c) If $n = r(j)$ or $n = s(j)$ and $1 \leqslant j \leqslant k$ then I_n is of type $(- 1)^{j + \gamma}$.

Since $(- 1)^{j + \gamma} g(s(j)(b - a)/N) \geqslant M/2$ and $(- 1)^{j + \gamma} g(r(j + 1)(b - a)/N) \leqslant - M/2$, it follows by the intermediate value theorem that we can find an $x(j)$ with $s(j)(b - a)/N < x(j) < r(j + 1)(b - a)/N$ and $g(x(j)) = 0$ $[1 \leqslant j \leqslant k - 1]$. Set $x(0) = a$, $x(k + 1) = b$. Using conditions (a) and (b) we see that

$$(- 1)^{j + \gamma} g(t) > - M \quad \text{for all } x(j - 1) \leqslant t \leqslant x(j) \quad [1 \leqslant j \leqslant k + 1],$$

whilst by condition (c) there exists a $y(j)$ (belonging for example to $I_{s(j)}$) such that

$$(- 1)^{j + \gamma} g(y(j)) = M \quad [0 \leqslant j \leqslant k]. \quad \blacksquare$$

Let me repeat that although it is necessary to understand the statement of Lemma 43.3 to understand what follows, the proof itself adds little but rigour.

Proof of Theorem 43.2 (the necessity of Tchebychev's criterion). Suppose P does not satisfy Tchebychev's criterion. We shall show how to find a $Q \in \mathscr{P}_{n - 1}$ with

$$\sup_{t \in [a, b]} |Q(t) - f(t)| < \sup_{t \in [a, b]} |P(t) - f(t)|.$$

Observe first that $P \neq f$ and so $\sup_{t \in [a,b]} |P(t) - f(t)| = M > 0$. By Lemma 43.3 we can find an integer $k \geq 1$ and points

$$a = x(0) \leq y(0) < x(1) < y(1) < \cdots < x(k) < y(k) \leq x(k+1) = b$$

and γ taking the value 0 or 1 such that

(i) $P(x(j)) - f(x(j)) = 0$ for $1 \leq j \leq k$
(ii) $(-1)^{j+\gamma} P(y(j)) = M$ for $0 \leq j \leq k$
(iii) $(-1)^{j+\gamma}(P(t) - f(t)) > -M$ for all $x(j-1) \leq t \leq x(j)[1 \leq j \leq k+1]$.

Since $P - f$ is continuous on the closed interval $[x(j-1), x(j)]$ it attains its bounds and so condition (iii) gives

(iii)' $(-1)^{j+\gamma}(P(t) - f(t)) \geq -M + \varepsilon(j)$ for all $x(j-1) \leq t \leq x(j)$ and some $\varepsilon(j) > 0$ $[1 \leq j \leq k+1]$.

Setting $\varepsilon = \min_{1 \leq j \leq k+1} \varepsilon(j)$ we obtain

(iii)'' $(-1)^{j+\gamma}(P(t) - f(t)) \geq -M + \varepsilon$ for all $x(j-1) \leq t \leq x(j)$ $[1 \leq j \leq k+1]$.

So far we have not used our hypothesis that P does not satisfy the equal ripple criterion. But now comparing Definition 43.1 with condition (ii) we obtain from it the crucial fact that $k \leq n-1$. The way ahead is now clear. Consider the polynomial $h(x) = (-1)^{\gamma+1} \eta \prod_{j=1}^{k}(x - x(j))$ with $\eta = \varepsilon/2(b-a)^k$. We observe that

(iv) h is a polynomial of degree $k \leq n-1$,
(v) $|h(t)| \leq \eta \prod_{j=1}^{k} |t - x(j)| \leq \eta(b-a)^k \leq \varepsilon/2$ for all $t \in [a,b]$,
(vii) $h(x(j)) = 0$ for $1 \leq j \leq k$,
(vii) $(-1)^{j+\gamma} h(t) > 0$ for $x(j-1) < t < x(j)$.

Thus, defining $Q = P - h$, we see that $Q \in \mathcal{P}_{n-1}$ and, using (vi) and (ii), that

(viii) $Q(x(j)) - f(x(j)) = P(x(j)) - f(x(j)) - h(x(j)) = 0$ for $1 \leq j \leq k$,

whilst, using (v) and (iii)'',

(ix) $(-1)^{j+\gamma}(Q(t) - f(t)) = (-1)^{j+\gamma}(P(t) - f(t) - h(t)) \geq -M + \varepsilon - \varepsilon/2 = -M + \varepsilon/2$ for all $x(j-1) < t < x(j)$,

and, finally, using (vii) and the definition of M,

(x) $(-1)^{j+\gamma}(Q(t) - f(t)) = (-1)^{j+\gamma}(P(t) - f(t) - h(t)) \leq M - |h(t)| < M$,

again for all $x(j-1) < t < x(j)$.

Putting (viii), (ix) and (x) together we see that $|Q(t) - f(t)| < M$ for all $t \in [a,b]$ and so, since every continuous function on $[a,b]$ attains its bounds,

$$\sup_{t \in [a,b]} |Q(t) - f(t)| < M = \sup_{t \in [a,b]} |P(t) - f(t)|. \qquad \blacksquare$$

We can also prove the sufficiency of Tchebychev's criterion.

Theorem 43.4. *Let $f:[a,b] \to \mathbb{R}$ be continuous and suppose that $P \in \mathscr{P}_{n-1}$ satisfies the equal ripple criterion. Then*

$$\sup_{t \in [a,b]} |f(t) - P(t)| \leqslant \sup_{t \in [a,b]} |f(t) - Q(t)| \quad \text{for all } Q \in \mathscr{P}_{n-1}.$$

Proof. By Definition 43.1 we can find $a \leqslant y_0 < y_1 < \cdots < y_n \leqslant b$ such that, writing $M = \sup_{t \in [a,b]} |f(t) - P(t)|$,

$$f(y_j) - P(y_j) = (-1)^{j+\gamma} M \quad \text{for some } \gamma \in \{0,1\} \quad \text{and all } 0 \leqslant j \leqslant n.$$

Suppose, if possible, that there exists a $Q \in \mathscr{P}_{n-1}$ with $\sup_{t \in [a,b]} |f(t) - Q(t)| = m < M$. Then $Q - P \in \mathscr{P}_{n-1}$ also, and

$$(-1)^{j+\gamma}(Q(y_j) - P(y_j)) = (-1)^{j+\gamma}((f(y_j) - P(y_j)) - (f(y_j) - Q(y_j)))$$
$$= M + (-1)^{j+\gamma}(f(y_j) - Q(y_j)) \geqslant M - m > 0.$$

Thus $(Q-P)(y_{j-1})$ and $(Q-P)(y_j)$ have opposite sign and by the intermediate value theorem there exists an x_j with $y_{j-1} < x_j < y_j$ and $(Q-P)(x_j) = 0$ $[1 \leqslant j \leqslant n]$. We have thus shown that $Q - P$ is a polynomial of degree at most $n-1$ with n distinct zeros $x_1, x_2, \ldots, x_n$. Such a polynomial must be identically zero so $Q = P$ and we have a contradiction. ∎

Remark. By tightening the arguments above we can extract still more information. Suppose we had assumed only that $\sup_{t \in [a,b]} |f(t) - Q(t)| \leqslant M$. We would then only obtain

$$(-1)^{j+\gamma}(Q(v_j) - P(y_j)) \geqslant 0.$$

But if $Q(y_j) - P(y_j) = 0$ then y_j is a maximum of $(-1)^{j+\gamma}(Q-P)$. A careful counting of roots, which the reader is invited to perform, establishes that, if multiple roots are counted multiply, then in this case also $Q - P$ has at least n zeros and so $Q = P$. Thus, using Theorem 43.2, the minimising polynomial P, if it exists, is unique.

Theorem 43.2 and 43.4 combine to give the following single theorem.

Theorem 43.5. *Let $f:[a,b] \to \mathbb{R}$ be continuous. Then $P \in \mathscr{P}_{n-1}$ satisfies*

$$\sup_{t \in [a,b]} |f(t) - P(t)| \leqslant \sup_{t \in [a,b]} |f(t) - Q(t)| \quad \text{for all } Q \in \mathscr{P}_{n-1}$$

if and only if P satisfies the Tchebychev equal ripple criterion.
Proof. Immediate. ∎

The Tchebychev criterion is certainly simple. In Chapter 45 (which does not depend on Chapter 44 and so can be read at once) we shall show that it is useful.

44

THE EXISTENCE OF
THE BEST APPROXIMATION

The careful reader may have noticed that, although in the last chapter we gave a necessary and sufficient condition for a real polynomial P of degree n or less to minimise $\sup_{t\in[a,b]} |f(t) - P(t)|$ we did not show that such a polynomial existed.

There are, apparently similar, problems for which no solution exists.

Example 44.1. *Let $\mathcal{P}$ be the set of all real polynomials. Then there does not exist a $P\in\mathcal{P}$ with*

$$\sup_{t\in[-1,1]} |P(t) - \exp t| \leqslant \sup_{t\in[-1,1]} |Q(t) - \exp t| \quad \textit{for all } Q\in\mathcal{P}.$$

Proof. Suppose P is a polynomial of degree n. Then

$$(P - \exp)^{(n+1)}(0) = P^{(n+1)}(0) - \exp^{(n+1)}(0) = 0 - \exp 0 = -1,$$

and so $P(t) - \exp t$ is not identically 0 on $[-1,1]$. It follows that $\sup|P(t) - \exp t| = \varepsilon > 0$. Choosing an integer $N \geqslant \varepsilon^{-1} + 2$ and setting $Q(t) = \sum_{r=0}^{N} t^r/r!$ we have

$$|\exp t - Q(t)| = \left| \sum_{r=N+1}^{\infty} t^r/r! \right| \leqslant \sum_{r=N+1}^{\infty} |t|^r/r!$$

$$\leqslant \sum_{r=N+1}^{\infty} 1/r! \leqslant ((N+1)!)^{-1} \sum_{k=0}^{\infty} (N+1)^{-k}$$

$$\leqslant ((N+1)!)^{-1} \sum_{k=0}^{\infty} 3^{-k} \leqslant 2/(N+1)! < \varepsilon \quad \text{for all } t\in[-1,1].$$

Thus $Q\in\mathcal{P}$, yet

$$\sup_{t\in[-1,1]} |Q(t) - \exp t| < \sup_{t\in[-1,1]} |P(t) - \exp t|. \qquad \blacksquare$$

(For a deeper example see Example 35.5.)

Even when a minimum existed it was often beyond the mathematical resources

of the first three-quarters of the nineteenth century to prove the fact. However, the work of Weierstrass and his successors has changed the status of many such problems from the intractable to the routine. Because of this I have felt free to avoid giving existence proofs for minima in connection with Dirichlet's problem in Chapter 31 (where the proof is hard but interesting) and in connection with the isoperimetric problem in Chapter 35 (where the proof is routine).

None the less it seems reasonable to include one example of this type for the reader's inspection. This chapter is therefore devoted to a proof of the following theorem. (As in the previous chapter we shall work in $\mathbb{R}$ rather than $\mathbb{C}$ and write $\mathscr{P}_n$ for the set of real polynomials of degree n or less.)

Theorem 44.2. *If $f:[a,b] \to \mathbb{R}$ is continuous then there exists a $P \in \mathscr{P}_n$ with*

$$\sup_{t \in [a,b]} |P(t) - f(t)| \leqslant \sup_{t \in [a,b]} |Q(t) - f(t)| \quad \text{for all } Q \in \mathscr{P}_n.$$

The reader should note that no other part of the book depends on this chapter and that this chapter does not depend on any other part of the book. In particular this chapter in independent both of Chapter 43 and Chapter 45. Our arguments depend on the following crucial result whose proof it is assumed the reader will be able to supply.

Theorem 44.3. *Let $I = [-R, R]$. Then if $g:I^m \to \mathbb{R}$ is continuous g is bounded and attains its bounds. That is to say, we can find $y_1, y_2 \in I^m$ with*

$$g(y_1) \leqslant g(x) \leqslant g(y_2) \quad \text{for all } x \in I^m.$$

Proof. Even if she does not already know this result the reader should be able to obtain it by generalising the proof used in the case $m = 1$. If she has to consult Appendix B (Theorem B3) then this chapter is not for her. ∎

In order to make use of Theorem 44.3 we need the following lemma.

Lemma 44.4. *Suppose $f:[a,b] \to \mathbb{R}$ is continuous. Write*

$$g(a_0, a_1, \ldots, a_n) = \sup_{t \in [a,b]} \left| f(t) - \sum_{r=0}^{n} a_r t^r \right|.$$

Then $g:\mathbb{R}^{n+1} \to \mathbb{R}$ is a well defined continuous function.

Proof. Since $f(t) - \sum_{r=0}^{n} a_r t^r$ is a continuous function on the closed bounded interval $[a,b]$, $\sup_{t \in [a,b]} |f(t) - \sum_{r=0}^{n} a_r t^r|$ exists and g is well defined. Choosing $c > \max(|a|, |b|) + 1$ we observe that

$$\left| \left| f(t) - \sum_{r=0}^{n} a_r t^r \right| - \left| f(t) - \sum_{r=0}^{n} b_r t^r \right| \right| \leqslant \left| \left(f(t) - \sum_{r=0}^{n} a_r t^r \right) - \left(f(t) - \sum_{r=0}^{n} b_r t^r \right) \right|$$

$$= \left| \sum_{r=0}^{n} (a_r - b_r) t^r \right| \leqslant \sum_{r=0}^{n} |a_r - b_r| |t|^r = c^n \sum_{r=0}^{n} |a_r - b_r| \leqslant \varepsilon$$

for all $t \in [a, b]$ whenever $|a_r - b_r| \leqslant \varepsilon/((n+1)c^n)$ for all $0 \leqslant r \leqslant n$. Thus

$$|g(a_0, a_1, \ldots, a_n) - g(b_0, b_1, \ldots, b_n)| \leqslant \varepsilon$$

whenever $|a_r - b_r| < \varepsilon((n+1)c^n)$ for all $0 \leqslant r \leqslant n$ and we have shown that g is continuous. ∎

Our proof of Theorem 44.2 splits into two parts both of which utilise Lemma 44.4 and Theorem 44.3. The first part of the proof is contained in the following lemma.

Lemma 44.5. *Suppose* $f:[a, b] \rightarrow \mathbb{R}$ *is continuous. Then we can find an* $R > 0$ *such that*

$$\sup_{t \in [a,b]} \left| f(t) - \sum_{r=0}^{n} a_r t^r \right| \geqslant \sup_{t \in [a,b]} |f(t)| + 1$$

whenever $\max_{0 \leqslant r \leqslant n} |a_r| \geqslant R$.

The proof of Lemma 44.5 itself divides into three steps the first two of which form the two parts of the next lemma.

Lemma 44.6. (i) *For each* $0 \leqslant k \leqslant n$ *there exists an* $\varepsilon_k > 0$ *such that*

$$\sup_{t \in [a,b]} \left| \sum_{r=0}^{n} a_r t^r \right| \geqslant \varepsilon_k$$

whenever $a_k = 1$ *and* $|a_r| \leqslant 1$ *for all* $0 \leqslant r \leqslant n$.
(ii) *There exists an* $\varepsilon > 0$ *such that*

$$\sup_{t \in [a,b]} \left| \sum_{r=0}^{n} a_r t^r \right| \geqslant \varepsilon R$$

whenever $\max_{0 \leqslant r \leqslant n} |a_r| \geqslant R$.

Proof. (i) Let

$$h(a_0, a_1, \ldots, a_n) = \sup_{t \in [a,b]} \left| \sum_{r=0}^{n} a_r t^r \right|.$$

Applying Lemma 44.4 with $f = 0$, we see that $h:\mathbb{R}^{n+1} \rightarrow \mathbb{R}$ is continuous. In particular, setting

$$h_k(a_0, a_1, \ldots, a_{k-1}, a_{k+1}, \ldots, a_n) = h(a_0, a_1, \ldots, a_{k-1}, 1, a_{k+1}, \ldots, a_k),$$

we see that $h_k:\mathbb{R}^n \rightarrow \mathbb{R}$ is continuous. Applying Theorem 44.3 with $g = h_k$, $m = n$ and $R = 1$, we see that there exist $b_r \in [-1, 1]$ $[0 \leqslant r \leqslant n, r \neq k]$ with

$$h_k(b_0, b_1, \ldots, b_{k-1}, b_{k+1}, \ldots, b_n) \leqslant h_k(a_0, a_1, \ldots, a_{k-1}, a_{k+1}, \ldots, a_n)$$

whenever $a_r \in [-1, 1]$.
But if we set $b_k = 1$, we know that $\sum_{r=0}^{n} b_r t^r$ is a non zero polynomial and so

$$\varepsilon_k = h_k(b_0, b_1, \ldots, b_{k-1}, b_{k+1}, \ldots, b_n)$$

$$= \sup_{t \in [a,b]} \left| \sum_{r=0}^{n} b_r t^r \right| > 0$$

and we are done.

(ii) Let $\varepsilon = \min(\varepsilon_0, \varepsilon_1, \ldots, \varepsilon_n)$. Choose $1 \le k \le n$ such that $|a_k| \ge |a_r|$ for all $0 \le r \le n$. If $a_k = 0$ there is nothing to prove. If $a_k \ne 0$ then setting $b_r = a_r a_k^{-1}$ we see that $b_k = 1, |b_r| \le 1 \; [0 \le r \le n]$ and so, by the first part,

$$\sup_{t \in [a,b]} \left| \sum_{r=0}^{n} a_r t^r \right| = \sup_{t \in [a,b]} \left| a_k \sum_{r=0}^{n} b_r t^r \right| = |a_k| \sup_{t \in [a,b]} \left| \sum_{r=0}^{n} b_r t^r \right| \ge |a_k| \varepsilon_k \ge |a_k| \varepsilon.$$

The result follows. ∎

Proof of Lemma 44.5. Take ε as in Lemma 44.6 (ii) and set $R = (1 + 2 \sup_{t \in [a,b]} |f(t)|) \varepsilon^{-1}$.

Then

$$\sup_{t \in [a,b]} \left| f(t) - \sum_{r=0}^{n} a_r t^r \right| \ge \sup_{t \in [a,b]} \left| \sum_{r=0}^{n} a_r t^r \right| - \sup_{t \in [a,b]} |f(t)|$$

$$\ge \varepsilon R - \sup_{t \in [a,b]} |f(t)|$$

$$= \left(1 + 2 \sup_{t \in [a,b]} |f(t)| \right) - \sup_{t \in [a,b]} |f(t)| = a + \sup_{t \in [a,b]} |f(t)|$$

whenever $\max_{0 \le r \le n} |a_r| \ge R$, as stated. ∎

Lemma 44.5 allows us to restrict our search for a minimum to the closed bounded cube $\{\mathbf{a} : |a_r| \le R [0 \le r \le n]\}$ and so to apply Theorem 44.3 directly.

Proof of Theorem 44.2. Take R as in Lemma 44.5. By Lemma 44.4 and Theorem 44.3 we can find $b_0, b_1, \ldots, b_n \in [-R, R]$ such that

$$\sup_{t \in [a,b]} \left| f(t) - \sum_{r=0}^{n} b_r t^r \right| \le \sup_{t \in [a,b]} \left| f(t) - \sum_{r=0}^{n} a_r t^r \right|$$

whenever $a_0, a_1, \ldots, a_n \in [-R, R]$, i.e. whenever $\max_{0 \le r \le n} |a_r| \le R$.

In particular, setting $a_0 = a_1 = \cdots = a_n = 0$, we see that

$$\sup_{t \in [a,b]} \left| f(t) - \sum_{r=0}^{n} b_r t^r \right| \le \sup_{t \in [a,b]} |f(t)|,$$

and so, by Lemma 44.5,

$$\sup_{t \in [a,b]} \left| f(t) - \sum_{r=0}^{n} b_r t^r \right| < \sup_{t \in [a,b]} \left| f(t) - \sum_{r=0}^{n} a_r t^r \right|$$

whenever $\max_{0 \leqslant r \leqslant n} |a_r| \geqslant R$. Thus

$$\sup_{t \in [a,b]} \left| f(t) - \sum_{r=0}^{n} b_r t^r \right| \leqslant \sup_{t \in [a,b]} \left| f(t) - \sum_{r=0}^{n} a_r t^r \right|$$

for any $a_0, a_1, \ldots, a_n$ whatsoever. Theorem 44.2 follows on setting $P(t) = \sum_{r=0}^{n} b_r t^r$.

∎

Remark. To understand this proof fully the reader must be clear in her own mind as to why we needed Lemma 44.5 and why it was possible to prove that lemma using Theorem 44.3. If she is familiar with the proof of the fundamental theorem of algebra (every polynomial has a root in $\mathbb{C}$) which starts by showing that the modulus $|P(z)|$ of a polynomial has a minimum this may be helpful in understanding the plan of the proof above.

45

TCHEBYCHEV AND UNIFORM
APPROXIMATION II

In Chapter 43 we showed that the best uniform approximation to a (real, continuous) function f by (real) polynomials P of degree $n-1$ is characterised by the equal ripple property (i.e. the appearance of $n+1$ points at which $f-P$ takes alternately maximum and minimum values $\pm M$).

We shall follow Tchebychev ('Sur les questions de minima qui se rattachent a la répresentation approximative des fonctions' included in his *Collected Works*) by using this result to find P explicitly in the important case when $f(x) = x^n$ and $[a, b] = [-1, 1]$.

Problem 45.1. *Find the real polynomial P of degree $n-1$ which minimises*
$$\sup_{t \in [-1, 1]} |t^n - P(t)|.$$

Resolution of Problem. Set $F(t) = t^n - P(t)$ observing that F is a polynomial of degree n. By Theorem 43.5 (the equal ripple criterion) we can find $-1 \leqslant y_0 < y_1 < \cdots < y_n \leqslant 1$ and an L such that

$$|F(t)| \leqslant L \qquad \text{for all } t \in [-1, 1],$$
$$F(y_j) = (-1)^{j+\gamma} L \quad [0 \leqslant j \leqslant n]$$

(where $\gamma = 0$ or $\gamma = 1$).

We now examine the behaviour of F^2 and F' at each of the y_j.

Case 1 $(1 \leqslant j \leqslant n-1)$. If $1 \leqslant j \leqslant n-1$, then y_j is an interior point of $[-1, 1]$ and so since F^2 has a maximum of value L^2 there, it follows that $F^2 - L^2$ has a zero of multiplicity m_j at least two. Since F has a maximum or minimum, F' has a zero (and so a zero of multiplicity $k_j \geqslant 1$).

Case 2 $(j = 0)$. If $y_0 > -1$, then y_0 is an interior point and as before $F^2 - L^2$ has a zero of multiplicity $m_0 \geqslant 2$, and F' has a zero of multiplicity $k_0 \geqslant 1$. If $y_0 = -1$ then $F^2 - L^2$ still has a zero at y_0 but it may be simple and all we can say is that its multiplicity is $m_0 \geqslant 1$. It is now possible that F' has no zero at all at y_0 (and we will then set $k_0 = 0$).

212

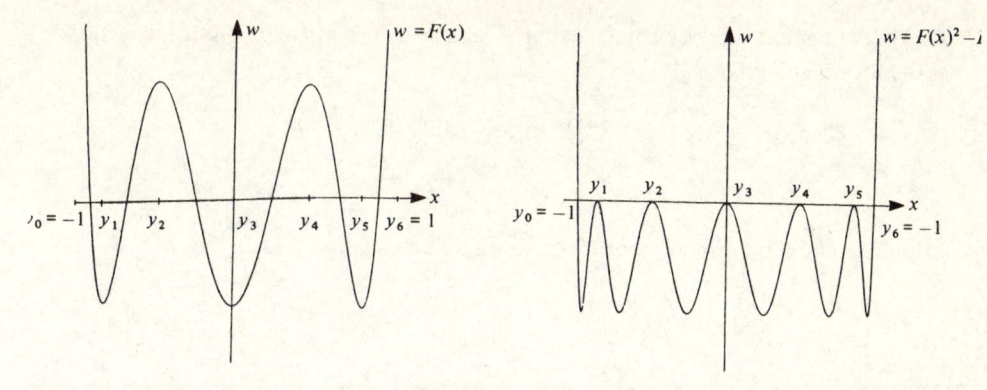

Fig. 45.1. Graphs of F and $F^2 = L^2$ when $n = 6$. Note that $m_0 = m_6 = 1$ and $k_0 = k_6 = 0$.

Case 3 ($j = n$). With the notation and the reasoning above, we see that if $y_n < 1$ then $m_n \geqslant 2$, $k_n \geqslant 1$ but that if $y_n = 1$ all we know is that $m_n \geqslant 1$, $k_n \geqslant 0$. (We illustrate the behaviour of F when $n = 6$ in Figure 45.1. Note that $m_0 = m_6 = 1$ and $k_0 = k_6 = 0$.)

Now $F^2 - L^2$ is a non zero polynomial of degree $2n$. (It is non-zero because, for example, its leading term is x^{2n}.) Thus $F^2 - L^2$ has at most $2n$ zeros, multiple zeros being counted multiply. Thus $2n \geqslant \sum_{j=0}^{n} m_j$ and since $m_j \geqslant 2$ for $n - 1 \geqslant j \geqslant 1$ and $m_j \geqslant 1$ for $j = 0$ and $j = n$ it follows that, in fact, $m_j = 2$ for $n - 1 \geqslant j \geqslant 1$ and $m_0 = m_n = 1$. (Thus $y_0 = -1$ and $y_n = 1$.) The zeros of $F^2 - L^2$ are thus

(i) double zero, at $y_1, y_2, \ldots, y_{n-1}$,
(ii) single zero at $y_0 = -1$, $y_n = 1$,

and so $F(x)^2 - L^2 = K_1(x^2 - 1)\prod_{j=1}^{n-1}(x - y_j)^2$ for some constant K_1. Since the leading term in $F(x)^2 - L^2$ is x^{2n} it follows that $K_1 = 1$ and

$$F(x)^2 - L^2 = (x^2 - 1) \prod_{j=1}^{n-1} (x - y_j)^2.$$

Arguing in a similar manner, we observe that F' is a non zero polynomial of degree $n - 1$ and so has at most $n - 1$ zeros. Thus $k_0 = k_n = 0$, $k_1 = k_2 = \cdots = k_{n-1} = 1$ and F' has single zeros at $y_1, y_2, \ldots, y_{n-1}$. (This could also have been deduced from the results for $F^2 - L^2$.) Thus $F'(x) = K_2\prod_{j=1}^{n-1}(x - y_j)$ for some constant K_2. Comparing the coefficients of x^{n-1}, we get $K_2 = n$ and

$$F'(x) = n \prod_{j=1}^{n-1} (x - y_j).$$

We thus obtain a differential equation

$$L^2 - F(x)^2 = n^{-2}(1 - x^2)(F'(x))^2,$$

which we shall try to solve. (From here our argument diverges from that of Tchebychev who uses methods involving continued fractions to find F.) Our

working is complicated by the fact that F' changes sign at each y_j but inspection of the formula $F'(x) = n\prod_{j=1}^{n}(x - y_j)$ shows that

$$(-1)^{n-j+1}F'(x) > 0 \quad \text{in } (y_{j-1}, y_j),$$

and so $n(L^2 - F(x)^2)^{\frac{1}{2}} = (-1)^{n-j}(1 - x^2)^{\frac{1}{2}}F'(x) \quad \text{for } y_{j-1} < x < y_j$

(taking the positive square root throughout). Thus since

$$L^2 - F(x)^2 \neq 0, 1 - x^2 \neq 0 \quad \text{for} \quad y_{j-1} < x < y_j,$$

it follows that $\dfrac{n\,dx}{(1-x^2)^{\frac{1}{2}}} = \dfrac{(-1)^{n-j}F'(x)\,dx}{(L^2 - F(x)^2)^{\frac{1}{2}}},$

and so

$$n\cos^{-1}x = (-1)^{n-j}\cos^{-1}(L^{-1}F(x)) + k_j$$

for $y_{j-1} < x < y_j$ and some constant k_j. The next two paragraphs are devoted to matching up and evaluating the k_j and are not terribly interesting.

By continuity we have

$$n\cos^{-1}y_j = (-1)^{n-j}\cos^{-1}(L^{-1}F(y_j)) + k_j, \tag{1$_j$}$$

$$n\cos^{-1}y_{j-1} = (-1)^{n-j}\cos^{-1}(L^{-1}F(y_{j-1})) + k_j. \tag{2$_j$}$$

Now $(2)_{j+1}$ gives

$$n\cos^{-1}y_j = (-1)^{n-j-1}\cos^{-1}(L^{-1}F(y_j)) + k_{j+1},$$

and so comparing this with $(1)_j$ we have

$$k_{j+1} = 2(-1)^{n-j}\cos^{-1}(L^{-1}F(y_j)) + k_j,$$

i.e. $\left.\begin{array}{ll} k_{j+1} = k_j & \text{if } F(y_j) = L, \\ k_{j+1} = k_j + 2(-1)^{n-j}\pi & \text{if } F(y_j) = -L. \end{array}\right\}$

However, we know (for example, by inspecting the behaviour of F') that F is a maximum when $n - j$ is even and a minimum when $n - j$ is odd. Thus

$$\left.\begin{array}{ll} k_{j+1} = k_j & \text{if } n-j \text{ is even,} \\ k_{j+1} = k_j - 2\pi & \text{if } n-j \text{ is odd.} \end{array}\right\}$$

Finally $(1)_n$ yields $n\cos^{-1}1 = \cos^{-1}1 + k_n$, i.e. $k_n = 0$ and so $k_j = [(n-j+1)/2]2\pi$ (where $[t]$ is the integral part of t).

We thus have

$$n\cos^{-1}x = (-1)^{n-j}\cos^{-1}(L^{-1}F(x)) + [(n-j+1)/2]2\pi$$

whenever $y_{j-1} \leqslant x \leqslant y_j[1 \leqslant j \leqslant n]$ and so, setting $x = \cos\theta$,

$$L^{-1}F(\cos\theta) = \cos(n\theta).$$

Thus $F(x) = LT_n(x)$ where T_n is the nth Tchebychev polynomial (see Lemma 5.3) and the required polynomial is $P(t) = t^n - LT_n(t)$ where L is such as to make P of degree $n - 1$ (i.e. L^{-1} is the coefficient of x^n in $T_n(x)$). ∎

It is, I hope, instructive to examine such a derivation, but of course once the result is known we can verify it directly much more rapidly. To fix ideas we repeat the statement of Lemma 5.3

Lemma 45.2. *There exists a unique, real polynomial T_n (of degree n and leading coefficient 1 if $n = 0$, 2^{n-1} if $n \geqslant 1$) such that*

$$\cos n\theta = T_n(\cos \theta) \quad \text{for all } \theta \in \mathbb{R}.$$

Proof. See Lemma 5.3. ∎

Theorem 45.3. *If $n \geqslant 1$ then the equation $F_{n-1}(t) = t^n - 2^{-n+1}T_n(t)$ defines a polynomial F_{n-1} of degree $n - 1$ with the property that*

$$2^{-n+1} = \sup_{t \in [-1, 1]} |t^n - F_{n-1}(t)| \leqslant \sup_{t \in [-1, 1]} |t^n - P(t)|$$

for all real polynomials of degree $n - 1$.

Proof. Lemma 45.2 tells us that F_{n-1} has degree $n - 1$. Setting $y_j = \cos((n-j)\pi/n)$ we have

$$-1 = y_0 < y_1 < \cdots < y_n = 1,$$
$$T_n(y_j) = \cos(n(n-j)\pi/n) = (-1)^{n-j},$$

whilst $\qquad\qquad |T_n(\cos \theta)| = |\cos(n \cos \theta)| \leqslant 1 \quad \text{for all } \theta,$

and so $\qquad\qquad\quad |T_n(t)| \leqslant 1 \quad \text{for all } -1 \leqslant t \leqslant 1.$

Since $\qquad\qquad\quad t^n - F_{n-1}(t) = 2^{-n+1}T_n(t),$

we have $\qquad\quad y_j^n - F_{n-1}(y_j) = (-1)^{n-j}2^{-n+1},$
$$|t^n - F_{n-1}(t)| \leqslant 2^{-n+1} \quad \text{for all } -1 \leqslant t \leqslant 1,$$

and the result follows by the Tchebychev equal ripple criterion (Theorem 43.5). ∎

One important application of Tchebychev's ideas occurs in connection with electronic computers and calculators. Here we may wish to have a function like e^x available 'at the touch of a button' and so a program to evaluate e^x has to be permanently stored. It is then important that the computation time is short and that the program takes up as little room as possible. (The relative importance of these two requirements will vary. In early scientific hand calculators the shortage of space in the memory meant that the second consideration dominated. In a larger machine the first may be more important.)

The natural way to generate values for a function $f(x)$ for x in a given range

$[a, b]$ is by means of a polynomial approximation

$$P(x) = a_0 + a_1 x + a_2 x^2 + \cdots + a_n x^n$$

or, in a more convenient form for the machine,

$$P(x) = a_0 + x(a_1 + x(a_2 + x(a_3 + \cdots + x(a_{n-1} + a_n x)\ldots))).$$

This requires the storage of $n + 1$ numbers $a_0, a_1, \ldots, a_n$ and the performance of about $2n$ arithmetic operations so we are obviously interested in finding a polynomial P of as low degree as possible which still gives a sufficiently good approximation to f on $[a, b]$.

As an illustration of what can be done, consider the following problem.

Problem 45.4. *Find a polynomial of low degree which approximates* $\exp x$ *on* $[-1, 1]$ *with an error of less than* 10^{-5}.

Discussion. (Notice that we only ask for a good solution, not for the best.) Since $\exp x = \sum_{r=0}^{\infty} x^r / r!$, an obvious first choice is $Q_n(x) = \sum_{r=0}^{n} x^r / r!$. We observe that

$$\exp 1 - Q_n(1) = \sum_{r=n+1}^{\infty} 1/r!$$

and so $$\exp 1 - Q_7(1) \geqslant 1/8! + 1/9! \geqslant 2.75 \times 10^{-5}$$

so that, if we wish to use Q_n, we must choose $n \geqslant 8$ and use a polynomial of degree at least 8.

On the other hand

$$|\exp(x) - Q_8(x)| = \left| \sum_{r=9}^{\infty} x^r / r! \right| \leqslant \sum_{r=9}^{\infty} |x|^r / r! \leqslant \sum_{r=9}^{\infty} 1/r! \leqslant (1/9!) \sum_{r=0}^{\infty} 10^{-r}$$

$$\leqslant (1/9!)(9/8) = 1/(8.8!) \leqslant 3.11 \times 10^{-6}.$$

But Theorem 45.3 tells us that $x^8/8!$ (the highest term in $Q_8(x)$) can be very well approximated by $(x^8 - 2^{-7} T_8(x))/8!$ and, indeed,

$$|x^8/8! - (x^8 - 2^{-7} T_8(x))/8!| = 2^{-7} |T_8(x)|/8!$$

$$\leqslant 2^{-7}/8! \leqslant 1.94 \times 10^{-7} \quad \text{for all} \quad x \in [-1, 1].$$

Thus writing $P_7(x) = Q_8(x) - 2^{-7} T_8(x)/8!$ we see that P_7 is a polynomial of degree 7 with

$$|\exp x - P_7(x)| \leqslant |\exp x - Q_8(x)| + |Q_8(x) - P_7(x)|$$

$$\leqslant 3.11 \times 10^{-6} + 1.94 \times 10^{-7} \leqslant 3.31 \times 10^{-6}.$$

But we can go further

$$T_8(x) = 128 x^8 - 256 x^6 + 160 x^4 - 32 x^2 + 1$$

(Tchebychev polynomials are tabulated, as Chebychev polynomials, in, e.g.,

Abramowitz and Stegun *Handbook of Mathematical Functions*), and so

$$P_7(x) = x^7/7! + \sum_{r=0}^{6} b_r x^r,$$

which suggests replacing $x^7/7!$ by $(x^7 - 2^{-6}T_7(x))/7!$ i.e. considering $P_6(x) = P_7(x) - 2^{-6}T_7(x)/7!$ Then

$$|P_7(x) - P_6(x)| = 2^{-6}|T_7(x)|/7! \leqslant 2^{-6}/7!$$
$$= 1/(8.8!) \leqslant 3.11 \times 10^{-6},$$

and so

$$|\exp x - P_6(x)| \leqslant |\exp x - P_7(x)| + |P_7(x) - P_6(x)|$$
$$\leqslant 3.31 \times 10^{-6} + 3.11 \times 10^{-6} = 6.42 \times 10^{-6}$$

so P_6 is a polynomial of degree 6 which approximates exp on $[-1, 1]$ to an accuracy of better than 10^{-5}.

(Notice however that although P_7 is the best uniform polynomial approximation of degree 7 to Q_8, and P_6 is the best approximation of degree 6 to P_7 it does not follow, and in fact it is not true, that P_6 is the best approximation to Q_8 of order 6.)

Thus with only a little work we have found a polynomial $P_6(x) = \sum_{r=0}^{6} a_r x^r$ which is of smaller degree than the Taylor polynomial $Q_7(x) = \sum_{r=0}^{7} x^r/r!$ and four times as accurate.

If the reader is interested we remark that

$$T_7(x) = 64x^7 - 112x^5 + 56x^3 - 7x,$$

and so to the order of accuracy shown

$a_6 = 0.00143849$	$a_5 = 0.00868056$
$a_4 = 0.04163566$	$a_3 = 0.16649306$
$a_2 = 0.50000620$	$a_1 = 1.00002170$
$a_0 = 0.99999981.$	

PART IV
FOURIER TRANSFORMS

46

INTRODUCTION

The idea that a periodic function can be decomposed into the sum of simpler periodic functions was already in the air when Fourier showed how to exploit it as a powerful tool for the understanding of nature. It is now built into the commonsense of our society. Overtones, harmonics, wireless frequencies, trade cycles, beats, Moog synthesisers, band filters; all these things are now part of our cultural background.

Most of the rest of the book will be devoted to a startling and interesting generalisation (which first appears in Fourier's work), that of the Fourier integral. Recall that if $f: \mathbb{T} \to \mathbb{C}$ we write

$$\hat{f}(r) = \frac{1}{2\pi} \int_{\mathbb{T}} f(t) \exp(-irt)dt \qquad [r \in \mathbb{Z}].$$

The idea of the Fourier transform is that if we have a function $f: \mathbb{R} \to \mathbb{C}$ we can extract information about f by looking at

$$\hat{f}(\zeta) = \int_{-\infty}^{\infty} f(t) \exp(-i\zeta t)dt \qquad [\zeta \in \mathbb{R}].$$

Does an analogy hold between the Fourier coefficients of a function on $\mathbb{T}$ and the Fourier transform $\hat{f}$ of a function f on $\mathbb{R}$? The answer is yes but the analogy is not perfect. It can be improved but only by introducing ideas beyond the scope of this book. We thus have a heuristic principle.

Slogan 46.1. *Try to argue about Fourier transforms in the same way as about Fourier series but do not expect your arguments to work every time.*

Let us see how the slogan works in practice. The first problem arises at once. How can we be sure that $\hat{f}(\zeta)$ exists? The following is a satisfactory, simple (though not quite complete) answer.

Lemma 46.2. *Suppose $f: \mathbb{R} \to \mathbb{C}$ is Riemann integrable on every interval $[a, b]$ and*

$\int_{-\infty}^{\infty} |f(t)| \, dt$ *converges. Then*

$$\hat{f}(\zeta) = \int_{-\infty}^{\infty} f(t) \exp(-it\zeta) \, dt \text{ is well defined for all } \zeta \in \mathbb{R}.$$

Proof. Observe that $|f(t)\exp(it\zeta)| = |f(t)|$ so, since any absolutely convergent (Riemann) infinite integral is convergent, it follows that

$$\int_{-\infty}^{\infty} f(t) \exp(-it\zeta) \, dt \text{ converges.} \qquad \blacksquare$$

In some ways $\hat{f}$ is better behaved than f.

Lemma 46.3. *If f satisfies the conditions of Lemma 41.2 then $\hat{f}$ is uniformly continuous.*

Proof. Let $\varepsilon > 0$ be given. Since $\int_{-\infty}^{\infty} |f(t)| \, dt$ converges we can find an $R(\varepsilon) > 0$ such that

$$\int_{|t| \geqslant R(\varepsilon)} |f(t)| \, dt \leqslant \varepsilon/4$$

On the other hand, since f is Riemann integrable on $[-R(\varepsilon), R(\varepsilon)]$, it is bounded on that interval with $\sup_{t \in [-R(\varepsilon), R(\varepsilon)]} |f(t)| = K(\varepsilon)$, say.

Thus

$$|\hat{f}(\zeta) - \hat{f}(\eta)| = \left| \int_{-\infty}^{\infty} f(t)(e^{-i\zeta t} - e^{-i\eta t}) \, dt \right|$$

$$\leqslant \left| \int_{|t| \geqslant R(\varepsilon)} f(t)(e^{-i\zeta t} - e^{-i\eta t}) \, dt \right| + \left| \int_{-R(\varepsilon)}^{R(\varepsilon)} f(t)(e^{-i\zeta t} - e^{-i\eta t}) \, dt \right|$$

$$\leqslant \int_{|t| \geqslant R(\varepsilon)} |f(t)(e^{-i\zeta t} - e^{-i\eta t})| \, dt + \int_{-R(\varepsilon)}^{R(\varepsilon)} |f(t)(e^{-i\zeta t} - e^{-i\eta t})| \, dt$$

$$\leqslant 2 \int_{|t| \geqslant R(\varepsilon)} |f(t)| \, dt + 2R(\varepsilon) \sup_{t \in [-R(\varepsilon), R(\varepsilon)]} |f(t)(e^{-i\zeta t} - e^{-i\eta t})|$$

$$\leqslant \varepsilon/2 + 2R(\varepsilon)K(\varepsilon) \sup_{t \in [-R(\varepsilon), R(\varepsilon)]} |e^{-i\zeta t} - e^{-i\eta t}| \leqslant \varepsilon/2 + 2R(\varepsilon)K(\varepsilon) \sup_{t \in [-R(\varepsilon), R(\varepsilon)]} |\zeta t - \eta t|$$

$$\leqslant \varepsilon/2 + 4R(\varepsilon)^2 K(\varepsilon) |\zeta - \eta| \leqslant \varepsilon.$$

whenever $|\zeta - \eta| \leqslant \varepsilon/(8R(\varepsilon)^2 K(\varepsilon) + 1)$, and we are done. $\qquad \blacksquare$

Unfortunately $\hat{f}$ is worse behaved than f in other ways.

Example 46.4. *Let $f(x) = 1$ for $x \in [-1, 1]$, $f(x) = 0$ otherwise. Then f satisfies the conditions of Lemma 41.6 (and in particular $\int_{-\infty}^{\infty} |f(t)| \, dt$ converges) but $\int_{-\infty}^{\infty} |\hat{f}(\zeta)| \, d\zeta$ diverges.*

Proof. It is obvious that f satisfies the conditions of Lemma 46.2. On the other hand, if $\zeta \neq 0$,

$$\hat{f}(\zeta) = \int_{-1}^{1} e^{-i\zeta t} dt = \left[\frac{e^{-i\zeta t}}{-i\zeta} \right]_{-1}^{1} = \frac{e^{i\zeta} - e^{-i\zeta}}{i\zeta}$$

$$= \frac{2 \sin \zeta}{\zeta},$$

and so

$$\int_{-(N+1)\pi}^{(N+1)\pi} |\hat{f}(\zeta)| d\zeta \geqslant \sum_{n=0}^{N} \int_{n\pi + \pi/4}^{n\pi + 3\pi/4} |\hat{f}(\zeta)| d\zeta \geqslant \sum_{n=0}^{N} \int_{n\pi + \pi/4}^{n\pi + 3\pi/4} \frac{1}{2^{\frac{1}{2}}\zeta} d\zeta$$

$$\geqslant \sum_{n=0}^{N} \frac{\pi/2}{2^{\frac{1}{2}}(n+1)\pi} = \frac{1}{2^{3/2}} \sum_{n=0}^{N} \frac{1}{n+1} \to \infty,$$

so $\int_{-\infty}^{\infty} |\hat{f}(\zeta)| d\zeta$ diverges. ∎

Starting from this example it is possible to go further and construct examples in which f is continuous and satisfies the conditions of Lemma 46.2, yet $\int_{-\infty}^{\infty} |\hat{f}(\zeta)| d\zeta$ still fails to converge. It is also possible to use the ideas of Chapter 18 and produce a continuous function f satisfying the conditions of Lemma 46.2 for which $\lim \sup_{R\to\infty} \int_{-R}^{R} \hat{f}(\zeta) \exp(-i\zeta t) dt = \infty$, and so, automatically, $\int_{-\infty}^{\infty} |\hat{f}(\zeta)| d\zeta$ diverges.

We shall evade these difficulties by using rather strong hypotheses in our theorems. Even with the tools at our disposal these hypotheses can be substantially relaxed. (To show the reader how this could be done I have taken a particular result, Theorem 51.7 and shown in Appendix C how to extend it.) However, my object throughout is to give the flavour of the subject rather than to provide a compendium of best results. Thus for simplicity we shall restrict f and, where convenient, $\hat{f}$ to lie in the set $L^1 \cap C$ defined below. (The notation $L^1 \cap C$ is chosen for consistency with more advanced work.)

Definition 46.5. *The set $L^1 \cap C$ consists of all continuous functions $f : \mathbb{R} \to \mathbb{C}$ such that $\int_{-\infty}^{\infty} |f(t)| dt$ converges.*

If we wish to develop the theory of Fourier transforms by analogy with our development of Fourier series, the first task is to find a result corresponding to Fejér's theorem (Theorem 2.3). (The reader may find it useful at this point to review Chapter 2.)

Theorem 46.6. (i) *If $f \in L^1 \cap C$ then*

$$\frac{1}{2\pi} \int_{-R}^{R} \left(1 - \frac{|\zeta|}{R} \right) \hat{f}(\zeta) \exp(i\zeta t) ds \to f(t) \text{ as } R \to \infty \text{ for each } t \in \mathbb{R}.$$

(ii) *Moreover, if f is uniformly continuous and bounded, the convergence is uniform.*

Notice that we need the continuity of f, demonstrated in Lemma 46.3, to show that the integral is well defined. The analogy between

$$\sum_{r=-n}^{n} \frac{n+1-|r|}{n+1} \hat{f}(r) \exp irt$$

and

$$\frac{1}{2\pi} \int_{-R}^{R} \left(1 - \frac{|\zeta|}{R}\right) \hat{f}(\zeta) \exp i\zeta t \, dt$$

should be obvious to the reader.

Let us try to immitate the proof of Theorem 2.3 as follows.

Attempted proof of Theorem 46.6. (The argument is, in fact, correct, but the step marked (*) where we change the order of integration requires further justification.)

$$\frac{1}{2\pi} \int_{-R}^{R} \left(1 - \frac{|\zeta|}{R}\right) \hat{f}(\zeta) \exp i\zeta t \, d\zeta$$

$$= \frac{1}{2\pi} \int_{-R}^{R} \left(1 - \frac{|\zeta|}{R}\right) \left(\int_{-\infty}^{\infty} f(x) \exp(-i\zeta x) dx\right) \exp i\zeta t \, d\zeta$$

$$= \frac{1}{2\pi} \int_{-R}^{R} \left(\int_{-\infty}^{\infty} \left(1 - \frac{|\zeta|}{R}\right) f(x) \exp i(t-x)\zeta dx\right) d\zeta$$

$$\overset{(*)}{=} \frac{1}{2\pi} \int_{-\infty}^{\infty} \left(\int_{-R}^{R} \left(\frac{1-|\zeta|}{R}\right) f(x) \exp i(t-x)\zeta d\zeta\right) dx$$

$$= \int_{-\infty}^{\infty} f(x) \left(\frac{1}{2\pi} \int_{-R}^{R} \left(\frac{1-|\zeta|}{R}\right) \exp i(t-x)\zeta d\zeta\right) dx$$

$$= \int_{-\infty}^{\infty} f(x) \tilde{K}_R(t-x) dx = \int_{-\infty}^{\infty} f(t-y) \tilde{K}_R(y) dy,$$

where

$$\tilde{K}_R(t) = \frac{1}{2\pi} \int_{-R}^{R} \left(\frac{1-|\zeta|}{R}\right) \exp i\zeta t \, d\zeta$$

and we make the linear change of variable $y = t - x$.

To evaluate $\tilde{K}_R$ we note that, making the linear change of variable $\eta = -\zeta$, we have

$$\int_{-R}^{0} \left(\frac{1-|\zeta|}{R}\right) \exp i\zeta t d\zeta = \int_{-R}^{0} \left(1 + \frac{\zeta}{R}\right) \exp i\zeta t d\zeta = -\int_{R}^{0} \left(1 - \frac{\eta}{R}\right) \exp(-i\eta t) d\eta$$

$$= \int_{0}^{R} \left(1 - \frac{\eta}{R}\right) \exp(-i\eta t) d\eta,$$

and so, if $t \neq 0$,

$$\tilde{K}_R(t) = \frac{1}{2\pi} \int_0^R \left(1 - \frac{|\zeta|}{R}\right) \exp i\zeta t d\zeta + \frac{1}{2\pi} \int_{-R}^0 \left(1 - \frac{|\zeta|}{R}\right) \exp i\zeta t d\zeta$$

$$= \frac{1}{2\pi} \int_0^R \left(1 - \frac{\zeta}{R}\right) \exp i\zeta t d\zeta + \frac{1}{2\pi} \int_0^R \left(1 - \frac{\zeta}{R}\right) \exp(-i\zeta t) d\zeta$$

$$= \frac{1}{\pi} \int_0^R \left(1 - \frac{\zeta}{R}\right) \cos \zeta t d\zeta = \frac{1}{\pi} \left[\left(1 - \frac{\zeta}{R}\right) \frac{\sin \zeta t}{t}\right]_0^R + \frac{1}{\pi R t} \int_0^R \sin \zeta t d\zeta$$

$$= 0 + \frac{1}{\pi R t^2} [-\cos \zeta t]_0^R = \frac{1 - \cos Rt}{\pi R t^2} = \frac{R}{2\pi} \left(\frac{\sin Rt/2}{Rt/2}\right)^2.$$

If $t = 0$ then

$$\tilde{K}_R(0) = \frac{1}{2\pi} \int_{-R}^R \left(1 - \frac{\zeta}{R}\right) d\zeta = \frac{R}{2\pi}.$$

This is very promising since it looks as though $\tilde{K}_R$ can be made to play the same role as K_n did in our proof of Theorem 2.3. It therefore becomes important to justify the change in the order of integration at step (*).

47

CHANGE IN THE
ORDER OF INTEGRATION I

The following example, which the reader may find slightly disturbing, shows why we hesitated over the step (*) in our attempted proof of Theorem 46.6.

Example 47.1

(i)
$$\int_1^\infty \frac{x^2 - y^2}{(x^2 + y^2)^2}\, dy = \left[\frac{y}{x^2 + y^2}\right]_1^\infty = \frac{1}{1 + x^2}\ [x \geqslant 1].$$

(ii)
$$\int_1^\infty \left(\int_1^\infty \frac{x^2 - y^2}{(x^2 + y^2)^2}\, dy\right) dx = \frac{\pi}{4}.$$

(iii)
$$\int_1^\infty \left(\int_1^\infty \frac{x^2 - y^2}{(x^2 + y^2)^2}\, dx\right) dy = -\frac{\pi}{4}.$$

Proof. The reader should do the simple calculations for herself so that she may convince herself that there is nothing up my sleeve. ∎

In fact, the example above is just another example of the 'paradox' '$\pi/4 = \infty - \infty = -(\infty - \infty) = -\pi/4$' and there are simple general conditions under which changing the order of integration presents no problem. Indeed if we confine ourselves to continuous functions on closed bounded rectangles no conditions whatsoever are needed.

Lemma 47.2. *If $f:[a,b] \times [c,d] \to \mathbb{C}$ is continuous then*

$$\int_a^b \left(\int_c^d f(x,y)dy\right) dx = \int_c^d \left(\int_a^b f(x,y)dx\right) dy.$$

Our proof of this depends on the following result.

Theorem 47.3. *If $f:[a,b] \times [c,d] \to \mathbb{C}$ is continuous it is uniformly continuous.*
(That is to say that given $\varepsilon > 0$ we can find a $\delta > 0$ such that $|f(x_1,y_1) - f(x_2,y_2)| < \varepsilon$

whenever $|x_1 - x_2| < \delta$, $|y_1 - y_2| < \delta$ and $(x_1, y_1), (x_2, y_2) \in [a, b] \times [c, d]$.)

Proof. Even if the reader does not already know this result she should be able to obtain it by generalising the proof used in the one dimensional case. If all else fails she will find it in Appendix B (Theorem B.5). ∎

Notice that, strictly speaking we need Theorem 47.3 (or something like it) to show that $\int_c^d f(x, y) dy$ is continuous since

$$\left| \int_c^d f(x_1, y) dy - \int_c^d f(x_2, y) dy \right| \leqslant (d - c) \sup_{y \in [c, d]} |f(x_1, y) - f(x_2, y)| \to 0 \quad \text{as } x_1 \to x_2$$

and so Riemann integrable.

Proof of Lemma 47.2. Our proof proceeds by stages.

Step 1. If I is an interval we define $\chi_I : \mathbb{R} \to \mathbb{R}$ by

$$\chi_I(t) = 1 \qquad \text{if } t \in I,$$
$$\chi_I(t) = 0 \qquad \text{otherwise.}$$

If I and J are intervals, then, clearly,

$$\int_a^b \left(\int_c^d \chi_I(x) \chi_J(y) dy \right) dx = \int_a^b \chi_I(x) \left(\int_c^d \chi_J(y) dy \right) dx = \int_a^b \chi_I(x) dx \int_c^d \chi_J(y) dy$$

$$= \int_c^d \left(\int_a^b \chi_I(x) \chi_J(y) dx \right) dy.$$

Step 2. Suppose $I(k), J(k)$ are intervals and $\lambda_k \in \mathbb{C}$ $[1 \leqslant k \leqslant n]$. Then, if $g(x, y) = \sum_{k=1}^n \lambda_k \chi_{I(k)}(x) \chi_{J(k)}(y)$, we have, using linearity and step 1,

$$\int_a^b \left(\int_c^d g(x, y) dy \right) dx = \int_a^b \sum_{k=1}^n \lambda_k \left(\int_c^d \chi_{I(k)}(x) \chi_{J(k)}(y) dy \right) dx$$

$$= \sum_{k=1}^n \lambda_k \int_a^b \left(\int_c^d \chi_{I(k)}(x) \chi_{J(k)}(y) dy \right) dx$$

$$= \sum_{k=1}^n \lambda_k \int_c^d \left(\int_a^b \chi_{I(k)}(x) \chi_{J(k)}(y) dx \right) dy$$

$$= \int_c^d \left(\int_a^b g(x, y) dx \right) dy.$$

Step 3. Suppose now that $f : [a, b] \times [c, d] \to \mathbb{C}$ is continuous. Then, by Theorem 47.3, f is uniformly continuous and given $\varepsilon > 0$ we can find an integer $N \geqslant 1$ such that

$$|f(x_1, y_1) - f(x_2, y_2)| \leqslant \varepsilon$$

whenever $\qquad |x_1 - x_2| \leqslant (b - a)/N, |y_1 - y_2| \leqslant (c - d)/N$

and $\qquad (x_1, y_1), (x_2, y_2) \in [a, b] \times [c, d].$

In particular, taking

$$I(k) = [a + (k-1)(b-a)/N, a + k(b-a)/N],$$
$$J(l) = [c + (l-1)(d-c)/N, c + l(d-c)/N],$$

for $1 \leqslant k, l < N-1$ and

$$I(N) = [a + (N-1)(b-a)/N, b],$$
$$J(N) = [c + (N-1)(d-c)/N, d],$$
$$\lambda_{kl} = f(a + (k-1)(b-a)/N, c + (l-1)(d-c)/N),$$

for $1 \leqslant k, l \leqslant N$ we know that

$$|\lambda_{kl} - f(x, y)| \leqslant \varepsilon.$$

whenever $(x, y) \in I(k) \times J(l)$ and $1 \leqslant k, l \leqslant N$.

It follows that, writing

$$g(x, y) = \sum_{k=1}^{N} \sum_{l=1}^{N} \lambda_{kl} \chi_{I(k)}(x) \chi_{J(l)}(y),$$

we have $\qquad |g(x, y) - f(x, y)| = |\lambda_{kl} - f(x, y)| \leqslant \varepsilon$

whenever $\qquad (x, y) \in I(k) \times J(l)$ and $1 \leqslant k, l \leqslant N$.

Thus $\qquad |g(x, y) - f(x, y)| \leqslant \varepsilon$ for all $x, y \in [a, b] \times [c, d]$,

so that

$$\left| \int_c^d g(x, y) dy - \int_c^d f(x, y) dy \right| = \left| \int_c^d (g(x, y) - f(x, y)) dy \right| \leqslant \varepsilon(d-c) \text{ for all } x \in [a, b],$$

and so

$$\left| \int_a^b \left(\int_c^d g(x, y) dy \right) dx - \int_a^b \left(\int_c^d f(x, y) dy \right) dx \right|$$
$$= \left| \int_a^b \left(\int_c^d g(x, y) dy - \int_c^d f(x, y) dy \right) dx \right| \leqslant \varepsilon(d-c)(b-a).$$

Similarly,

$$\left| \int_c^d \left(\int_a^b g(x, y) dx \right) dy - \int_c^d \left(\int_a^b f(x, y) dx \right) dy \right| \leqslant \varepsilon(d-c)(b-a),$$

and so, since by Step 2

$$\int_a^b \left(\int_c^d g(x, y) dy \right) dx = \int_c^d \left(\int_a^b g(x, y) dx \right) dy,$$

it follows that

$$\left| \int_a^b \left(\int_c^d f(x,y)dy \right) dx - \int_c^d \left(\int_a^b f(x,y)dx \right) dy \right| \leqslant 2\varepsilon(d-c)(b-a).$$

But ε was arbitrary so

$$\int_a^b \left(\int_c^d f(x,y)dy \right) dx = \int_c^d \left(\int_a^b f(x,y)dx \right) dy,$$

as claimed. ∎

The pattern of proof is a familiar one (compare, for example, the proof of Weyl's equidistribution theorem in Chapter 3) in which we prove the result for a simple function $\chi_I(x)\chi_J(y)$, then extend it to sums of simple functions $\sum_{k=1}^n \lambda_k \chi_{I(k)}(x)\chi_{J(k)}(y)$, and finally use the fact that any continuous function f may be approximated by sums g_n of simple functions to obtain the full result.

Why then can the result fail when we replace the bounded rectangle $[a,b] \times [d,c]$ by an unbounded set? Roughly speaking, we can say that, if the convergence of expressions like $\int_{-\infty}^{\infty} (\int_{-\infty}^{\infty} f(x,y)dx)dy$ and $\int_{-\infty}^{\infty} (\int_{-\infty}^{\infty} f(x,y)dy)dx$ is due to 'accidental cancellation at infinity', then there is no reason why the two expressions obtained by interchanging the order of integration should be equal. However, if the convergence is not 'accidental', then we can change the order of integration. The situation may be summed up in a slogan and a theorem.

Slogan 47.4. *If*
$$\int_{-\infty}^{\infty} \left(\int_{-\infty}^{\infty} |f(x,y)|dy \right) dx < \infty$$

then
$$\int_{-\infty}^{\infty} \left(\int_{-\infty}^{\infty} f(x,y)dx \right) dy = \int_{-\infty}^{\infty} \left(\int_{-\infty}^{\infty} f(x,y)dy \right) dx.$$

Theorem 47.5. *Suppose $f:\mathbb{R}^2 \to \mathbb{C}$ is continuous and*

(A1) $\int_{-\infty}^{\infty} |f(x,y)|dy$ *is a continuous function of x.*
(A2) $\int_{-\infty}^{\infty} |f(x,y)|dx$ *is a continuous function of y.*

Then if

(B) $\int_{-\infty}^{\infty}(\int_{-\infty}^{\infty} |f(x,y)|dx)dy$ *converges*

it follows that $\int_{-\infty}^{\infty}(\int_{-\infty}^{\infty} f(x,y)dx)dy$ and $\int_{-\infty}^{\infty}(\int_{-\infty}^{\infty} f(x,y)dy)dx$ converge with

$$\int_{-\infty}^{\infty} \left(\int_{-\infty}^{\infty} f(x,y)dy \right) dx = \int_{-\infty}^{\infty} \left(\int_{-\infty}^{\infty} f(x,y)dx \right) dy.$$

In the next chapter I shall try to explain why this result is true and prove it (as Theorem 48.8). However, I suggest that the reader should simply read enough to understand what is going on and then pass on to Chapter 49.

48

CHANGE IN THE
ORDER OF INTEGRATION II

In the previous chapter we saw that changing the order of integration presents no difficulty when the range of integration is bounded (Lemma 47.2) but may present problems when the range is unbounded (Example 47.1).

In order to understand the unbounded case the reader may find it helpful to consider the analogous but simpler problem of interchanging infinite sums.

Example 48.1. *Let*

$$a_{00} = 1, a_{r0} = 0 \ [r \geqslant 1], a_{rs} = -2^{-r} \ [0 \leqslant s < 2^{r-1}, r \geqslant 1],$$
$$a_{rs} = 2^{-r} \ [2^{r-1} \leqslant s < 2^r, r \geqslant 1], a_{rs} = 0 \ otherwise.$$

Then $\sum_{r=0}^{\infty} a_{rs}$ converges for all s, $\sum_{s=0}^{\infty} a_{rs}$ converges for all r, $\sum_{s=0}^{\infty} \sum_{r=0}^{\infty} a_{rs}$ converges and $\sum_{r=0}^{\infty} \sum_{s=0}^{\infty} a_{rs}$ converges, yet

$$\sum_{s=0}^{\infty} \sum_{r=0}^{\infty} a_{rs} \neq \sum_{r=0}^{\infty} \sum_{s=0}^{\infty} a_{rs}.$$

Table 48.1

1	$-1/2$	$-1/4$	$-1/8$	$-1/16$	$-1/32$	$\cdots$
0	$1/2$	$-1/4$	$-1/8$	$-1/16$	$-1/32$	$\cdots$
0	0	$1/4$	$-1/8$	$-1/16$	$-1/32$	$\cdots$
0	0	$1/4$	$-1/8$	$-1/16$	$-1/32$	$\cdots$
0	0	0	$1/8$	$-1/16$	$-1/32$	$\cdots$
0	0	0	$1/8$	$-1/16$	$-1/32$	$\cdots$
$\vdots$	$\vdots$	$\vdots$	$\vdots$	$\vdots$	$\vdots$	

Proof. We set out the values of a_{rs} in Table 48.1. Looking at this table, it is clear that $\sum_{s=0}^{\infty} a_{rs}$ converges to 0 for all r and $\sum_{r=0}^{\infty} a_{rs}$ is a finite sum with value 0 for $s \geqslant 1$ and value 1 for $s = 0$. Thus $\sum_{r=0}^{\infty} \sum_{s=0}^{\infty} a_{rs} = \sum_{r=0}^{\infty} 0$ converges to 0 and $\sum_{s=0}^{\infty} \sum_{r=0}^{\infty} a_{rs} = 1 + \sum_{s=1}^{\infty} 0$ converges to 1. ∎

Remark 1. I have picked this example more or less at random. A simpler example

(in which, however, $a_{rs} \nrightarrow 0$ as $r, s \to \infty$) is given by $a_{rr} = 1$, $a_{rr+1} = -1$, $a_{rs} = 0$ otherwise.

Remark 2. If we take

$$g(x,y) = (\tfrac{1}{2} - |x|)(\tfrac{1}{2} - |y|) \quad \text{for } |x|, |y| \leqslant \tfrac{1}{2}, g(x,y) = 0 \text{ otherwise,}$$

and put

$$f(x,y) = \sum_{s=0}^{\infty} \sum_{r=0}^{\infty} a_{rs} g(x - r - 1, y - s - 1),$$

then as the reader may easily check we get a well defined continuous function $f : \mathbb{R}^2 \to \mathbb{R}$ with

$$\int_0^\infty \left(\int_0^\infty f(x,y) dx \right) dy \neq \int_0^\infty \left(\int_0^\infty f(x,y) dy \right) dx.$$

The reader probably knows the following condition for the interchange of summation.

Lemma 48.2. *Let $a_{rs} \in \mathbb{C}$. If $\sum_{r=-\infty}^{\infty} |a_{rs}|$ converges for each s and $\sum_{s=-\infty}^{\infty} (\sum_{r=-\infty}^{\infty} |a_{rs}|)$ converges, then $\sum_{r=-\infty}^{\infty} a_{rs}$ converges for each s, $\sum_{s=-\infty}^{\infty} a_{rs}$ converges for each r, $\sum_{s=-\infty}^{\infty} (\sum_{r=-\infty}^{\infty} a_{rs})$ converges, $\sum_{r=-\infty}^{\infty} (\sum_{s=-\infty}^{\infty} a_{rs})$ converges and*

$$\sum_{r=-\infty}^{\infty} \left(\sum_{s=-\infty}^{\infty} a_{rs} \right) = \sum_{s=-\infty}^{\infty} \left(\sum_{r=-\infty}^{\infty} a_{rs} \right).$$

One way of proving Lemma 48.2 is to remark that it suffices to prove a special case.

Lemma 48.3. *Let $a_{rs} \geqslant 0$. If $\sum_{r=-\infty}^{\infty} a_{rs}$ converges for each s and $\sum_{s=-\infty}^{\infty} (\sum_{r=-\infty}^{\infty} a_{rs})$ converges then $\sum_{s=-\infty}^{\infty} a_{rs}$ converges for each r and $\sum_{r=-\infty}^{\infty} (\sum_{s=-\infty}^{\infty} a_{rs})$ converges with*

$$\sum_{r=-\infty}^{\infty} \left(\sum_{s=-\infty}^{\infty} a_{rs} \right) = \sum_{s=-\infty}^{\infty} \left(\sum_{r=-\infty}^{\infty} a_{rs} \right).$$

Proof of Lemma 48.2 from Lemma 48.3. Suppose a_{rs} satisfies the conditions of Lemma 48.2. Set $a_{rs} = u_{rs} + iv_{rs}$ with $u_{rs}, v_{rs} \in \mathbb{R}$ and define $a_{rs}^{(1)}, a_{rs}^{(2)}, a_{rs}^{(3)}, a_{rs}^{(4)}$ as follows

$$
\begin{aligned}
a_{rs}^{(1)} &= u_{rs} && \text{if } u_{rs} \geqslant 0, & a_{rs}^{(1)} &= 0 && \text{otherwise,} \\
a_{rs}^{(2)} &= -u_{rs} && \text{if } u_{rs} \leqslant 0, & a_{rs}^{(2)} &= 0 && \text{otherwise,} \\
a_{rs}^{(3)} &= v_{rs} && \text{if } v_{rs} \geqslant 0, & a_{rs}^{(3)} &= 0 && \text{otherwise,} \\
a_{rs}^{(4)} &= -v_{rs} && \text{if } v_{rs} \leqslant 0, & a_{rs}^{(4)} &= 0 && \text{otherwise.}
\end{aligned}
$$

Then $0 \leqslant a_{rs}^{(j)} \leqslant |a_{rs}|$ and so by comparison $\sum_{r=-\infty}^{\infty} a_{rs}^{(j)}$ converges with

$0 \leqslant \sum_{r=-\infty}^{\infty} a_{rs}^{(j)} \leqslant \sum_{r=-\infty}^{\infty} |a_{rs}|$. Thus, by comparison again, $\sum_{s=-\infty}^{\infty} \sum_{r=-\infty}^{\infty} a_{rs}^{(j)}$ converges.

It follows that $a_{rs}^{(j)}$ satisfies the hypotheses of Lemma 43.3 and so $\sum_{s=-\infty}^{\infty} a_{rs}^{(j)}$ converges for each r, $\sum_{r=-\infty}^{\infty} \sum_{s=-\infty}^{\infty} a_{rs}^{(j)}$ converges and

$$\sum_{r=-\infty}^{\infty} \left(\sum_{s=-\infty}^{\infty} a_{rs}^{(j)} \right) = \sum_{s=-\infty}^{\infty} \left(\sum_{r=-\infty}^{\infty} a_{rs}^{(j)} \right) \qquad \text{for } j = 1, 2, 3, 4.$$

Now
$$a_{rs} = a_{rs}^{(1)} - a_{rs}^{(2)} + i a_{rs}^{(3)} - i a_{rs}^{(4)}$$

so, for example,

$$\sum_{s=-M}^{N} a_{rs} = \sum_{s=-M}^{N} a_{rs}^{(1)} - \sum_{s=-M}^{N} a_{rs}^{(2)} + i \sum_{s=-M}^{N} a_{rs}^{(3)} - i \sum_{s=-M}^{N} a_{rs}^{(4)}$$

$$\to \sum_{s=-\infty}^{\infty} a_{rs}^{(1)} - \sum_{s=-\infty}^{\infty} a_{rs}^{(2)} + i \sum_{s=-\infty}^{\infty} a_{rs}^{(3)} - i \sum_{s=-\infty}^{\infty} a_{rs}^{(4)},$$

as $N, M \to \infty$, i.e. $\sum_{s=-\infty}^{\infty} a_{rs}$ converges with

$$\sum_{s=-\infty}^{\infty} a_{rs} = \sum_{s=-\infty}^{\infty} a_{rs}^{(1)} - i \sum_{s=-\infty}^{\infty} a_{rs}^{(2)} + \sum_{s=-\infty}^{\infty} a_{rs}^{(3)} - i \sum_{s=-\infty}^{\infty} a_{rs}^{(4)}.$$

Similarly, writing $\sigma(1) = 1$, $\sigma(2) = -1$, $\sigma(3) = i$, $\sigma(4) = -i$ we see that $\sum_{r=-\infty}^{\infty} \sum_{s=-\infty}^{\infty} a_{rs}$ converges to $\sum_{k=1}^{4} \sigma(k) \sum_{r=-\infty}^{\infty} \sum_{s=-\infty}^{\infty} a_{rs}^{(k)}$, $\sum_{r=-\infty}^{\infty} a_{rs}$ converges to $\sum_{k=1}^{4} \sigma(k) \sum_{r=-\infty}^{\infty} a_{rs}^{(k)}$, and $\sum_{s=-\infty}^{\infty} \sum_{r=-\infty}^{\infty} a_{rs}$ converges to $\sum_{k=1}^{4} \sigma(k) \sum_{s=-\infty}^{\infty} \sum_{r=-\infty}^{\infty} a_{rs}^{(k)}$. Thus, applying Lemma 48.2 to the $a_{rs}^{(k)}$,

$$\sum_{r=-\infty}^{\infty} \sum_{s=-\infty}^{\infty} a_{rs} = \sum_{k=1}^{4} \sigma(k) \sum_{r=-\infty}^{\infty} \sum_{s=-\infty}^{\infty} a_{rs}^{(k)} = \sum_{k=1}^{4} \sigma(k) \sum_{s=-\infty}^{\infty} \sum_{r=-\infty}^{\infty} a_{rs}^{(k)}$$

$$= \sum_{s=-\infty}^{\infty} \sum_{r=-\infty}^{\infty} a_{rs}. \qquad \blacksquare$$

Lemma 48.3 follows in turn from the following result.

Lemma 48.4. *Let* $a_{rs} \geqslant 0$. *Then*

(i) *If* $\sum_{r=-\infty}^{\infty} a_{rs}$ *converges for each* s *and* $\sum_{s=-\infty}^{\infty} \sum_{r=-\infty}^{\infty} a_{rs}$ *converges it follows that* $\sum_{s=-N}^{N} \sum_{r=-N}^{N} a_{rs}$ *tends to a limit as* $N \to \infty$.

(ii) *If* $\sum_{s=-N}^{N} \sum_{r=-N}^{N} a_{rs} \to K$ *as* $N \to \infty$ *it follows that* $\sum_{r=-\infty}^{\infty} a_{rs}$ *converges for each* s *and that* $\sum_{s=-\infty}^{\infty} \sum_{r=-\infty}^{\infty} a_{rs}$ *converges to* K.

Proof of Lemma 48.3 from Lemma 48.4. Suppose a_{rs} obeys the hypotheses of Lemma 48.3. Then, using Lemma 48.4(i), we see that $\sum_{s=-N}^{N} \sum_{r=-N}^{N} a_{rs} \to K$ as $N \to \infty$ for some K. Further, using Lemma 48.4(ii), we know that $K = \sum_{s=-\infty}^{\infty} \sum_{r=-\infty}^{\infty} a_{rs}$. But $\sum_{s=-N}^{N} \sum_{r=-N}^{N} a_{rs} = \sum_{r=-N}^{N} \sum_{s=-N}^{N} a_{rs}$ (since we can certainly change the order of summation for finite sums) so $\sum_{r=-N}^{N} \sum_{s=-N}^{N} a_{rs} \to K$ as $N \to \infty$. Thus, using Lemma 48.4(ii), with the roles of r and s interchanged, we

see that $\sum_{s=-\infty}^{\infty} a_{rs}$ converges for all r and $\sum_{r=-\infty}^{\infty} \sum_{s=-\infty}^{\infty} a_{rs}$ converges to K, giving $\sum_{r=-\infty}^{\infty} \sum_{s=-\infty}^{\infty} a_{rs} = \sum_{s=-\infty}^{\infty} \sum_{r=-\infty}^{\infty} a_{rs}$ as required. ∎

Thus we have reduced our problem to that of proving Lemma 48.4.
Proof of Lemma 48.4. (i) Observe that

$$\sum_{s=-N}^{N} \sum_{r=-N}^{N} a_{rs} \leqslant \sum_{s=-N}^{N} \sum_{r=-\infty}^{\infty} a_{rs} \leqslant \sum_{s=-\infty}^{\infty} \sum_{r=-\infty}^{\infty} a_{rs},$$

so that $\sum_{s=-N}^{N} \sum_{r=-N}^{N} a_{rs}$ is an increasing function of N bounded above and thus tends to a limit as $N \to \infty$.

(ii) If $\sum_{r=-\infty}^{\infty} a_{rs(0)}$ diverges for some $s(0)$ then (since the $a_{rs(0)}$ are positive) we can find an $N \geqslant |s(0)|$ such that $\sum_{r=-N}^{N} a_{rs(0)} \geqslant K+1$ and so $\sum_{r=-N}^{N} \sum_{s=-N}^{N} a_{rs} \geqslant \sum_{r=-N}^{N} a_{rs(0)} \geqslant K+1$ giving a contradiction. Thus $\sum_{r=-\infty}^{\infty} a_{rs}$ converges for all s.

Since $\sum_{r=-\infty}^{\infty} a_{rs} \geqslant 0$, it follows that either $\sum_{s=-\infty}^{\infty} \sum_{r=-\infty}^{\infty} a_{rs}$ diverges or $\sum_{s=-\infty}^{\infty} \sum_{r=-\infty}^{\infty} a_{rs}$ converges to a limit L. We show first that $\sum_{s=-\infty}^{\infty} \sum_{r=-\infty}^{\infty} a_{rs}$ converges to a limit $L \leqslant K$. For, if not, we can find an $\eta > 0$ and an integer $M \geqslant 1$ with $\sum_{s=-M}^{M} \sum_{r=-\infty}^{\infty} a_{rs} \geqslant K + \eta$. But for each $-M \leqslant s \leqslant M$ we can find an $N(s)$ such that

$$\sum_{r=N(s)}^{N(s)} a_{rs} \geqslant \sum_{r=-\infty}^{\infty} a_{rs} - \eta/(4M+2),$$

and so, setting $N = \max(M, N(-M), N(-M+1), \ldots, N(M))$, we have

$$K \geqslant \sum_{s=-N}^{N} \sum_{r=-N}^{N} a_{rs} \geqslant \sum_{s=-M}^{M} \sum_{r=-N}^{N} a_{rs}$$

$$\geqslant \sum_{s=-M}^{M} \left(\sum_{r=-\infty}^{\infty} a_{rs} - \eta/(4M+2) \right)$$

$$\geqslant (K+\eta) - \eta/2 = K + \eta/2 > K,$$

so, by contradiction, $\sum_{s=-\infty}^{\infty} \sum_{r=-\infty}^{\infty} a_{rs}$ converges to a limit $L \leqslant K$. But

$$L = \sum_{s=-\infty}^{\infty} \sum_{r=-\infty}^{\infty} a_{rs} \geqslant \sum_{s=-N}^{N} \sum_{r=-\infty}^{\infty} a_{rs} \geqslant \sum_{s=-N}^{N} \sum_{r=-N}^{N} a_{rs} \to K \quad \text{as } N \to \infty, \text{ so } L \geqslant K,$$

whence $L = K$ and we are done. ∎

We use the proofs for infinite sums as the model for our proofs for infinite integrals. First we prove a result corresponding to Lemma 48.4.

Lemma 48.5. *Let $f : \mathbb{R}^2 \to \mathbb{R}$ be a continuous function with $f(x, y) \geqslant 0$ for all $(x, y) \in \mathbb{R}^2$. Suppose that*

(A1) $\int_{-\infty}^{\infty} f(x, y) dy$ *is continuous function of x,*
(A2) $\int_{-\infty}^{\infty} f(x, y) dx$ *is a continuous function of y.*

Then
(i) If $\int_{-\infty}^{\infty}(\int_{-\infty}^{\infty}f(x,y)dx)dy$ converges, it follows that $\int_{-R}^{R}(\int_{-R}^{R}f(x,y)dx)dy$ tends to a limit as $R \to \infty$.
(ii) If $\int_{-R}^{R}(\int_{-R}^{R}f(x,y)dx)dy \to K$ as $R \to \infty$, it follows that $\int_{-\infty}^{\infty}(\int_{-\infty}^{\infty}f(x,y)dx)dy$ converges to K.

Remark. The reader will notice that although this lemma follows the same pattern as Lemma 48.4, the hypotheses are stronger. To some extent this is unavoidable as the following example shows. Take

$$f(x,y) = (1 - (1 + y^2)|x|) \text{ for } |x| \leqslant (1 + y^2)^{-1}.$$

Then $\int_{-\infty}^{\infty} f(x,y)dx = (1 + y^2)^{-1}$ is continuous and

$$\int_{-\infty}^{\infty}\int_{-\infty}^{\infty}f(x,y)dxdy = \int_{-\infty}^{\infty}\frac{1}{1+y^2}dy = [\tan^{-1}y]_{-\infty}^{\infty} = \pi$$

yet

$$\int_{-\infty}^{\infty} f(0,y)dy = \int_{-\infty}^{\infty} 1dy$$

diverges.
Our proof will make use of the following simple result.

Lemma 48.6. *If* $f:\mathbb{R} \to \mathbb{C}$ *is continuous then, for each fixed R,* $\int_{-R}^{R}f(x,y)dx$ *is a continuous function of y.*

Proof. Let $y_0 \in \mathbb{R}$. By Theorem 47.3, f is uniformly continuous on $[-R,R] \times [y_0 - 1, y_0 + 1]$ and so

$$\left| \int_{-R}^{R} f(x,y_0)dx - \int_{-R}^{R} f(x,y)dx \right|$$

$$\leqslant 2R \sup_{x \in [-R,R]} |f(x,y_0) - f(x,y)|$$

$$\to 0 \quad \text{as } y \to y_0. \qquad \blacksquare$$

Proof of Lemma 48.5 (i) Observe that

$$\int_{-R}^{R}\left(\int_{-R}^{R} f(x,y)dx\right)dy \leqslant \int_{-R}^{R}\left(\int_{-\infty}^{\infty} f(x,y)dx\right)dy$$

$$\leqslant \int_{-\infty}^{\infty}\left(\int_{-\infty}^{\infty} f(x,y)dx\right)dy,$$

so that $\int_{-R}^{R}(\int_{-R}^{R}f(x,y)dx)dy$ is an increasing function of R bounded above and thus tends to a limit as $R \to \infty$.
(ii) (Because our hypotheses are stronger there is nothing corresponding to the first paragraph of the proof of Lemma 48.4(ii).)
Since $\int_{-\infty}^{\infty} f(x,y)dx \geqslant 0$, it follows that either $\int_{-\infty}^{\infty}(\int_{-\infty}^{\infty}f(x,y)dx)dy$ diverges or

$\int_{-\infty}^{\infty}(\int_{-\infty}^{\infty}f(x,y)dx)dy$ converges to a limit L. We show first that $\int_{-\infty}^{\infty}(\int_{-\infty}^{\infty}f(x,y)dx)dy$ converges to a limit $L \leq K$. For, if not, we can find an $\eta > 0$ and a $T > 0$ with $\int_{-T}^{T}(\int_{-\infty}^{\infty}f(x,y)dx)dy \geq K + \eta$.

The argument now becomes more complicated than in Lemma 48.4. We start by noting that for each $v \in [-T, T]$ we can find an $R(v) > 0$ such that

$$\int_{-\infty}^{\infty}f(x,v)dx - \int_{-R(v)}^{R(v)}f(x,v)dx \leq \frac{\eta}{4T+2}.$$

But $\int_{-\infty}^{\infty}f(x,y)dx$ is continuous by hypothesis (A2) and $\int_{-R(v)}^{R(v)}f(x,y)dx$ is continuous by Lemma 43.6, so $\int_{-\infty}^{\infty}f(x,y)dx - \int_{-R(v)}^{R(v)}f(x,y)dx$ is a continuous function of y and we can find a $\delta(v) > 0$ with

$$\int_{-\infty}^{\infty}f(x,y)dx - \int_{-R(v)}^{R(v)}f(x,y)dx \leq \frac{\eta}{2T+2} \quad \text{whenever } |v - y| < \delta(v).$$

The open intervals $(v - \delta(v), v + \delta(v))$ form a cover of the closed bounded interval $[-T, T]$ and so by the theorem of Heine Borel Lebesgue we can find

$$v(1), v(2), \ldots, v(M) \in [-T, T]$$

with

$$\bigcup_{j=1}^{M}(v(j) - \delta(v(j)), v(j) + \delta(v(j))) \supseteq [T, T].$$

In particular, if $R = \max(T, R(v(1)), R(v(2)), \ldots, R(v(M)))$, we have

$$\int_{-\infty}^{\infty}f(x,y)dx - \int_{-R}^{R}f(x,y)dy$$

$$\leq \int_{-\infty}^{\infty}f(x,y)dx - \int_{-R(v(j))}^{R(v(j))}f(x,y)dx \leq \frac{\eta}{2T+2} \quad \text{for all } y \in [-T, T].$$

Thus, giving rather more detail than the reader probably requires,

$$K \geq \int_{R}^{R}\left(\int_{-R}^{R}f(x,y)dx\right)dy \geq \int_{-T}^{T}\left(\int_{-R}^{R}f(x,y)dx\right)dy$$

$$\geq \int_{-T}^{T}\left(\int_{-\infty}^{\infty}f(x,y)dx - \frac{\eta}{2T+2}\right)dy = \int_{-T}^{T}\left(\int_{-\infty}^{\infty}f(x,y)dx\right)dy - \frac{2T\eta}{2T+2}$$

$$> K + \eta - \eta = K.$$

and, by contradiction, $\int_{-\infty}^{\infty}(\int_{-\infty}^{\infty}f(x,y)dx)dy$ converges to a limit $L \leq K$. But

$$L = \int_{-\infty}^{\infty}\left(\int_{-\infty}^{\infty}f(x,y)dx\right)dy \geq \int_{-R}^{R}\left(\int_{-\infty}^{\infty}f(x,y)dx\right)dy$$

$$\geq \int_{-R}^{R}\left(\int_{-R}^{R}f(x,y)dx\right)dy \to K \text{ as } R \to \infty,$$

so $L \geq K$ whence $L = K$ and we are done. ∎

We can now prove a result corresponding to Lemma 48.3.

Lemma 48.7. *Let* $f:\mathbb{R}^2 \to \mathbb{R}$ *be a continuous function with* $f(x,y) \geqslant 0$ *for all* $(x,y) \in \mathbb{R}^2$. *Suppose that*

(A1) $\int_{-\infty}^{\infty} f(x,y)dy$ *is a continuous function of* x,

(A2) $\int_{-\infty}^{\infty} f(x,y)dx$ *is a continuous function of* y.

Then, if

(B) $\int_{-\infty}^{\infty} (\int_{-\infty}^{\infty} f(x,y)dx)dy$ *converges,*

it follows that $\int_{-\infty}^{\infty} (\int_{-\infty}^{\infty} f(x,y)dy)dx$ *converges with* $\int_{-\infty}^{\infty} (\int_{-\infty}^{\infty} f(x,y)dy)dx = \int_{-\infty}^{\infty} (\int_{-\infty}^{\infty} f(x,y)dx)dy$.

Proof. This follows the proof of Lemma 48.3 and is left to the reader. Instead of the relation $\sum_{r=-N}^{N} \sum_{s=-N}^{N} a_{rs} = \sum_{s=-N}^{N} \sum_{r=-N}^{N} a_{rs}$ we use Theorem 47.2 to give

$$\int_{-R}^{R} \left(\int_{-R}^{R} f(x,y)dx \right) dy = \int_{-R}^{R} \left(\int_{-R}^{R} f(x,y)dy \right) dx. \qquad \blacksquare$$

We can now state and prove a theorem corresponding to Lemma 48.2. (This is the result promised at the end of Chapter 47 as Theorem 47.5.)

Theorem 48.8. *Suppose* $f:\mathbb{R}^2 \to \mathbb{C}$ *is continuous and*

(A1) $\int_{-\infty}^{\infty} |f(x,y)|dy$ *is a continuous function of* x.

(A2) $\int_{-\infty}^{\infty} |f(x,y)|dx$ *is a continuous function of* y.

Then

(A1)' $\int_{-\infty}^{\infty} f(x,y)dy$ *is a continuous function of* x.

(A2)' $\int_{-\infty}^{\infty} f(x,y)dx$ *is a continuous function of* y.

If, further,

(B) $\int_{-\infty}^{\infty} (\int_{-\infty}^{\infty} |f(x,y)|dx)dy$ *converges,*

then

$$\int_{-\infty}^{\infty} \left(\int_{-\infty}^{\infty} f(x,y)dx \right) dy \quad and \quad \int_{-\infty}^{\infty} \left(\int_{-\infty}^{\infty} f(x,y)dy \right) dx$$

converge with

$$\int_{-\infty}^{\infty} \left(\int_{-\infty}^{\infty} f(x,y)dx \right) dy = \int_{-\infty}^{\infty} \left(\int_{-\infty}^{\infty} f(x,y)dy \right) dx.$$

Proof. We follow the pattern of proof for Lemma 48.2. Write $f(x,y) = u(x,y) + iv(x,y)$ with $u(x,y), v(x,y) \in \mathbb{R}$ and define $f_j:\mathbb{R}^2 \to \mathbb{R}$ for $j = 1,2,3,4$ as follows:

$$f_1(x, y) = u(x, y) \quad \text{if } u(x, y) \geq 0, \quad f_1(x, y) = 0 \quad \text{otherwise,}$$

$$f_2(x, y) = -u(x, y) \quad \text{if } u(x, y) \leq 0, \quad f_2(x, y) = 0 \quad \text{otherwise,}$$

$$f_3(x, y) = v(x, y) \quad \text{if } v(x, y) \geq 0, \quad f_3(x, y) = 0 \quad \text{otherwise,}$$

$$f_4(x, y) = -v(x, y) \quad \text{if } v(x, y) \leq 0, \quad f_4(x, y) = 0 \quad \text{otherwise.}$$

We now wish to show that f_j satisfies the conditions of Lemma 48.7. Clearly, $f_j(x, y) \geq 0$ for all $(x, y) \in \mathbb{R}^2$ and, since

$$|f_j(x_1, y_1) - f_j(x_2, y_2)| \leq |f(x_1, y_1) - f(x_2, y_2)|,$$

we see that f_j is continuous. Again, since $0 \leq f_j(x, y) \leq |f(x, y)|$, it follows by the comparison test that $\int_{-\infty}^{\infty} f_j(x, y)dx$ and $\int_{-\infty}^{\infty} f_j(x, y)dy$ converge. The only difficulty that arises lies in proving that f_j satisfies conditions (A1) and (A2) of Lemma 48.7

To see this we argue as follows. Let $y_0 \in \mathbb{R}$ and $\varepsilon > 0$ be given. Then we can find an $R > 0$ with

$$0 \leq \int_{-\infty}^{\infty} |f(x, y_0)|dx - \int_{-R}^{R} |f(x, y_0)|dx < \varepsilon/6.$$

Arguing as in the proof of Lemma 48.5, we observe that $\int_{-\infty}^{\infty} |f(x, y_0)|dx$ is continuous in x by hypothesis and $\int_{-R}^{R} |f(x, y_0)|dx$ is continuous by Lemma 48.6, so we can find a $\delta_1 > 0$ such that

$$0 \leq \int_{-\infty}^{\infty} |f(x, y)|dx - \int_{-R}^{R} |f(x, y)|dx < \varepsilon/3 \quad \text{for } |y - y_0| < \delta_1,$$

and so (since $|f(x, y)| \geq f_j(x, y) \geq 0$),

$$0 \leq \int_{-\infty}^{\infty} f_j(x, y)dx - \int_{-R}^{R} f_j(x, y)dx < \varepsilon/3 \quad \text{for all } |y - y_0| < \delta_1.$$

But, by Lemma 48.5, $\int_{-R}^{R} f(x, y)dx$ is continuous, so we can find a $0 < \delta_2 < \delta_1$ with

$$\left| \int_{-R}^{R} f_j(x, y)dx - \int_{-R}^{R} f_j(x, y_0)dx \right| < \varepsilon/3 \quad \text{for all} \quad |y - y_0| < \delta_2.$$

Thus

$$\left| \int_{-\infty}^{\infty} f_j(x, y)dx - \int_{-\infty}^{\infty} f_j(x, y_0)dx \right|$$

$$\leq \left| \int_{-\infty}^{\infty} f_j(x, y)dx - \int_{-R}^{R} f_j(x, y)dx \right| + \left| \int_{-R}^{R} f_j(x, y)dx - \int_{-R}^{R} f_j(x, y_0)dx \right|$$

$$+ \left| \int_{-R}^{R} f_j(x, y_0)dx - \int_{-\infty}^{\infty} f_j(x, y_0)dx \right| \leq \frac{\varepsilon}{3} + \frac{\varepsilon}{3} + \frac{\varepsilon}{3} = \varepsilon \quad \text{for all} \quad |y - y_0| < \delta.$$

Thus

(A1)$_j$ $\int_{-\infty}^{\infty} f_j(x,y)dx$ is continuous.

Similarly,

(A2)$_j$ $\int_{-\infty}^{\infty} f_j(x,y)dy$ is continuous.

Finally, since $0 \leqslant \int_{-\infty}^{\infty} f_j(x,y)dx \leqslant \int_{-\infty}^{\infty} |f(x,y)|dx$, it follows by the comparison test that, if $\int_{-\infty}^{\infty}(\int_{-\infty}^{\infty} |f(x,y)|dx)dy$ converges, so does $\int_{-\infty}^{\infty}(\int_{-\infty}^{\infty} f_j(x,y)dx)dy$.

The result required can now be read off just as in Lemma 48.2 by observing that $f = f_1 - f_2 + if_3 - if_4$ and applying Lemma 48.7 to each of the f_j. ∎

This concludes the main business of the chapter but there is a further chain of ideas which the proofs above call to mind and which the reader may find amusing if she has not seen them before.

Let us write

$$\chi_{S(R)}(x,y) = 1 \text{ if } |x|,|y| \leqslant R, \; \chi_{S(R)}(x,y) = 0 \qquad \text{otherwise}$$
$$\chi_{D(R)}(x,y) = 1 \text{ if } x^2 + y^2 \leqslant R^2, \chi_{D(R)}(x,y) = 0 \quad \text{otherwise.}$$

Lemma 48.9. *Let $f: \mathbb{R}^2 \to \mathbb{R}$ be a continuous function with $f(x,y) \geqslant 0$ for all $(x,y) \in \mathbb{R}^2$. Under the conditions (A1), (A2), (B) of Lemma 48.7 we have*

(i) $\int_{-\infty}^{\infty}(\int_{-\infty}^{\infty} \chi_{S(R)}(x,y)f(x,y)dx)dy \to \int_{-\infty}^{\infty}(\int_{-\infty}^{\infty} f(x,y)dx)dy,$

(ii) $\int_{-\infty}^{\infty}(\int_{-\infty}^{\infty} \chi_{D(R)}(x,y)f(x,y)dx)dy \to \int_{-\infty}^{\infty}(\int_{-\infty}^{\infty} f(x,y)dx)dy.$

Proof. (i) Observe that

$$\int_{-\infty}^{\infty}\left(\int_{-\infty}^{\infty} \chi_{S(R)}(x,y)f(x,y)dx\right)dy = \int_{-R}^{R}\left(\int_{-R}^{R} f(x,y)dx\right)dy.$$

The result now follows by applying Lemma 48.5 (i) and (ii).
(ii) We observe that

$$\chi_{S(R)}(x,y) \geqslant \chi_{D(R)}(x,y) \geqslant \chi_{S(R/2)}(x,y),$$

so that

$$\chi_{S(R)}(x,y)f(x,y) \geqslant \chi_{D(R)}(x,y)f(x,y) \geqslant \chi_{S(R/2)}(x,y)f(x,y)$$

and

$$\int_{-\infty}^{\infty}\left(\int_{-\infty}^{\infty} \chi_{S(R)}(x,y)f(x,y)dx\right)dy$$
$$\geqslant \int_{-\infty}^{\infty}\left(\int_{-\infty}^{\infty} \chi_{D(R)}(x,y)f(x,y)dx\right)dy$$
$$\geqslant \int_{-\infty}^{\infty}\left(\int_{-\infty}^{\infty} \chi_{S(R/2)}(x,y)f(x,y)dx\right)dy.$$

Allowing $R \to \infty$ and using (i) we have the result. ∎

Remark. The condition $f(x, y) \geqslant 0$ is only used to simplify the proof. The reader is invited to state and prove the approximate result without this condition.

As a corollary we get the following formula (which we saw near the end of Chapter 38).

Lemma 48.10.

$$\int_{-\infty}^{\infty} e^{-x^2/2}\, dx = (2\pi)^{\frac{1}{2}}.$$

Proof. Since

$$0 \leqslant e^{-x^2/2} \leqslant e^{-|x|}, \text{ for } |x| \geqslant 2$$

it follows by comparison that $\int_{-\infty}^{\infty} e^{-x^2/2}\, dx$ converges to L say. Setting

$$f(x, y) = e^{-x^2/2} e^{-y^2/2}$$

we have

$$\int_{-\infty}^{\infty} f(x, y)\, dx = L e^{-y^2/2}, \quad \int_{-\infty}^{\infty} f(x, y)\, dy = L e^{-x^2/2}$$

and

$$\int_{-\infty}^{\infty} \left(\int_{-\infty}^{\infty} f(x, y)\, dx \right) dy = \int_{-\infty}^{\infty} L e^{-y^2/2}\, dy = L^2,$$

so the conditions of Lemma 48.7, and so of Lemma 48.9, are satisfied. Lemma 48.9 (ii) thus yields

$$\int_{-\infty}^{\infty} \left(\int_{-\infty}^{\infty} \chi_{D(R)}(x, y) f(x, y)\, dx \right) dy \to L^2 \quad \text{as } R \to \infty.$$

But changing from Cartesian to polar coordinates

$$\int_{-\infty}^{\infty} \left(\int_{-\infty}^{\infty} \chi_{D(R)}(x, y) f(x, y)\, dx \right) dy = \iint_{x^2 + y^2 < R^2} \exp -(x^2 + y^2)/2\, dx\, dy$$

$$= \int_0^R \int_0^{2\pi} \exp(-r^2/2) r\, dr\, d\theta = 2\pi \int_0^R r \exp(-r^2/2)\, dr$$

$$= 2\pi [\exp(-r^2/2)]_0^R = 2\pi(1 - \exp(-R^2/2)) \to 2\pi \quad \text{as } R \to \infty,$$

so $L^2 = 2\pi$ and the result follows. ∎

Remark. The acute reader may find it inconsistent to prove Lemma 47.2 and quote a formula for change from Cartesian to polar coordinates whose proof must be of the same order of difficulty. She is correct and is invited to remove this blemish by a proof along the lines of Lemma 47.2. (She may find the calculations easier if she replaces $\chi_I(x)\chi_J(y)$ in Step 1 by a $\chi(r, \theta)$ with $\chi(r, \theta) = 1$ for $r(2) \geqslant r \geqslant r(1)$, $\theta(2) \geqslant \theta \geqslant \theta(1)$. $\chi(r, \theta) = 0$ otherwise.)

49

FEJÉR'S THEOREM FOR FOURIER TRANSFORMS

Let us set

$$\Delta_R(\zeta) = 1 - \frac{|\zeta|}{R} \quad \text{for} \quad |\zeta| \leqslant R,$$

$$\Delta_R(\zeta) = 0 \quad \text{otherwise,}$$

and $F(x, \zeta) = (2\pi)^{-1}\Delta_R(\zeta) \exp i\zeta(t - x)f(x)[\zeta, x \in \mathbb{R}]$. The problem of justifying step
(*) in our attempted proof of Fejér's theorem (Theorem 46.6) at the end of Chapter 46
is then reduced to showing that

$$\int_{-\infty}^{\infty} \left(\int_{-\infty}^{\infty} F(x, \zeta)dx \right) d\zeta = \int_{-\infty}^{\infty} \left(\int_{-\infty}^{\infty} F(x, \zeta)dx \right) d\zeta.$$

But

(A1) $$\int_{-\infty}^{\infty} |F(x, \zeta)|dx = \int_{-\infty}^{\infty} (2\pi)^{-1}\Delta_R(\zeta)|f(x)|dx = \frac{1}{2\pi}\int_{-\infty}^{\infty}|f(x)|dx\Delta_R(\zeta),$$

which is a continuous function of ζ, and

(A2) $$\int_{-\infty}^{\infty} |F(x, \zeta)|d\zeta = |f(x)|(2\pi)^{-1}\int_{-\infty}^{\infty} \Delta_R(\zeta)d\zeta = \frac{R}{2\pi}|f(x)|,$$

which is a continuous function of x, whilst

(B) $$\int_{-\infty}^{\infty} \left(\int_{-\infty}^{\infty} |F(x, \zeta)|d\zeta \right)dx = \frac{R}{2\pi}\int_{-\infty}^{\infty}|f(x)|dx$$

converges, so, by Theorem 48.8, we may interchange the order of integration in (*).
Thus the conclusions we obtained at the end of Chapter 46 are valid.

Lemma 49.1. *If $f \in L^1 \cap C$ and $t \in \mathbb{R}$ then*

$$\frac{1}{2\pi}\int_{-R}^{R} \left(1 - \frac{|\zeta|}{R} \right)\hat{f}(\zeta)\exp(i\zeta t)d\zeta = \int_{-\infty}^{\infty} f(t - x)\tilde{K}_R(x)dx,$$

240

where
$$\tilde{K}_R(t) = \frac{R}{2\pi}\left(\frac{\sin Rt/2}{Rt/2}\right)^2$$

if $t \neq 0$ and $\tilde{K}_R(0) = (R/2\pi)$.

Proof. See the attempted proof of Theorem 46.6. ∎

If the reader now runs through the proof of Fejér's theorem for Fourier series (Chapter 2) in her mind, she will be led to try and prove that $\tilde{K}_R$ has three vital properties. (Notice that, since $\mathbb{R}$ is unbounded, condition (ii) has to be expanded by comparison with Lemma 2.2(ii).)

Lemma 49.2. (i) $\tilde{K}_R(s) \geqslant 0$ *for all $s \in \mathbb{R}$*

(ii) $\tilde{K}_R(s) \to 0$ *uniformly outside $[-\delta, \delta]$*

and
$$\int_{|s|\geqslant\delta} \tilde{K}_R(s)ds \to 0 \qquad \text{as } R \to \infty \quad \text{for all } \delta > 0.$$

(iii) $\int_{-\infty}^{\infty} \tilde{K}_R(s)ds = 1$.

Proof. (i) This is trivial.

(ii) $|\tilde{K}_R(s)| \leqslant (R/2\pi)((1/(Rs/2))^2 = 2/\pi Rs^2 \leqslant 2/\pi R\delta^2 \to 0$

as $R \to \infty$ for all $|s| > \delta$, whilst
$$\int_{|s|>\delta} \tilde{K}_R(s)ds \leqslant \frac{2}{\pi R}\int_{|s|>\delta}\frac{1}{s^2}ds = \frac{4}{\pi R}\int_{\delta}^{\infty}\frac{1}{s^2}ds = \frac{4}{\pi R\delta} \to 0 \quad \text{as } R \to \infty \text{ also.}$$

(iii) By the linear change of variable $t = Rs$ we see that
$$\int_{-\infty}^{\infty}\tilde{K}_R(s)ds = \int_{-\infty}^{\infty}\frac{R}{2\pi}\left(\frac{\sin(Rs/2)}{Rs/2}\right)^2 ds$$
$$= \int_{-\infty}^{\infty}\frac{1}{2\pi}\left(\frac{\sin t/2}{t/2}\right)^2 dt = \int_{-\infty}^{\infty}\tilde{K}_1(t)dt. \qquad ∎$$

Remark. (The main thing that interests us is the fact that $\int_{-\infty}^{\infty}\tilde{K}_R(s)ds$ is a constant, independent of R. The actual value of this constant appears only as a scalar factor in our equations. However, as the next paragraph shows, it is not very hard to compute $\int_{-\infty}^{\infty}\tilde{K}_1(s)ds$ directly and in the discussion following we shall give an easy indirect method for obtaining it.)

We compute $\int_{-\infty}^{\infty}\tilde{K}_1(s)ds$ by comparing $\tilde{K}_n$ and the function K_{n+1} of Chapter 2. Observe first that keeping $\pi > \delta > 0$ fixed we have
$$\int_{-\infty}^{\infty}\tilde{K}_1(t)dt - \int_{-\delta}^{\delta}\tilde{K}_{n+1}(t)dt = \int_{-\infty}^{\infty}\tilde{K}_{n+1}(t)dt - \int_{-\delta}^{\delta}\tilde{K}_{n+1}(t)dt$$
$$= \int_{|t|>\delta}\tilde{K}_{n+1}(t)dt \to 0 \qquad \text{as } n \to \infty,$$

by Lemma 44.2 (ii), so that

$$\int_{-\delta}^{\delta} \tilde{K}_{n+1}(t)dt \to \int_{-\infty}^{\infty} \tilde{K}_1(t)dt,$$

whilst by Lemma 2.2 (ii),

$$\frac{1}{2\pi}\int_{-\pi}^{\pi} K_n(t)dt - \frac{1}{2\pi}\int_{-\delta}^{\delta} K_n(t)dt \to 0,$$

so, since $(1/2\pi)\int_{-\pi}^{\pi}K_n(t)dt = 1$ (by Lemma 2.2 (iii)), it follows that

$$\frac{1}{2\pi}\int_{-\delta}^{\delta} K_n(t)\,dt \to 1 \quad \text{as } n \to \infty.$$

On the other hand,

$$K_n(t) = \frac{1}{n+1}\left(\frac{\sin(n+1)t/2}{\sin t/2}\right)^2$$

and

$$\tilde{K}_{n+1}(t) = \frac{1}{2\pi(n+1)}\left(\frac{\sin(n+1)t/2}{t/2}\right)^2,$$

so that

$$\tilde{K}_{n+1}(t) = \frac{1}{2\pi}K_n(t)\left(\frac{\sin t/2}{t/2}\right)^2 \quad \text{for all } 0 < |t| < \pi,$$

and so

$$\left|\frac{1}{2\pi}\int_{-\delta}^{\delta} K_n(t)dt - \int_{-\delta}^{\delta}\tilde{K}_{n+1}(t)dt\right| = \left|\frac{1}{2\pi}\int_{-\delta}^{\delta}K_n(t)\left(1 - \left(\frac{\sin t/2}{t/2}\right)^2\right)dt\right|$$

$$\leqslant \sup_{0<|t|<\delta}\left|1 - \left(\frac{\sin t/2}{t/2}\right)^2\right|\frac{1}{2\pi}\int_{-\delta}^{\delta}K_n(t)dt$$

$$\leqslant \sup_{0<|t|<\delta}\left|1 - \left(\frac{\sin t/2}{t/2}\right)^2\right|.$$

Thus letting $n \to \infty$, and using the result of the previous paragraph,

$$\left|1 - \int_{-\infty}^{\infty} \tilde{K}_1(t)dt\right| \leqslant \sup_{0<|t|<\delta}\left|1 - \left(\frac{\sin t/2}{t/2}\right)^2\right|,$$

so, allowing $\delta \to 0$, we have $1 - \int_{-\infty}^{\infty}\tilde{K}_1(t)dt| \leqslant 0$, and $\int_{-\infty}^{\infty}\tilde{K}_1(t)dt = 1$, as stated. ∎

Fejér's theorem for Fourier transforms now follows in much the same way as his theorem for Fourier series.

Theorem 49.3. (i) *If $f \in L^1 \cap C$, then*

$$\frac{1}{2\pi}\int_{-R}^{R}\left(1 - \frac{|\zeta|}{R}\right)\hat{f}(\zeta)\exp(i\zeta t)d\zeta \to f(t)$$

as $R \to \infty$ for each $t \in \mathbb{R}$.

(ii) *Moreover, if f is uniformly continuous and bounded, the convergence is uniform.*

Proof. (i) By Lemmas 49.1 and 49.2

$$\left| \frac{1}{2\pi} \int_{-R}^{R} \left(1 - \frac{|\zeta|}{R} \right) \hat{f}(\zeta) \exp(i\zeta t) d\zeta - f(t) \right| = \left| \int_{-\infty}^{\infty} f(t-x) \tilde{K}_R(x) dx - f(t) \right|$$

$$= \left| \int_{-\infty}^{\infty} (f(t-x) - f(t)) \tilde{K}_R(x) dx \right| \leqslant \left| \int_{|x|>\delta} (f(t-x) - f(t)) \tilde{K}_R(x) dx \right|$$

$$+ \left| \int_{|x|\leqslant\delta} (f(t-x) - f(t)) \tilde{K}_R(x) dx \right|$$

$$\leqslant \left| \int_{|x|>\delta} f(t-x) \tilde{K}_R(x) dx \right| + \left| \int_{|x|>\delta} f(t) \tilde{K}_R(x) dx \right|$$

$$+ \left| \int_{|x|\leqslant\delta} (f(t-x) - f(t)) \tilde{K}_R(x) dx \right|$$

$$\leqslant \sup_{|x|>\delta} \tilde{K}_R(x) \int_{|x|>\delta} |f(t-x)| dx + |f(t)| \int_{|x|>\delta} \tilde{K}_R(x) dx$$

$$+ \sup_{|x|\leqslant\delta} |f(t-x) - f(t)| \int_{|x|\leqslant\delta} \tilde{K}_R(x) dx$$

$$\leqslant \sup_{|x|>\delta} \tilde{K}_R(x) \int_{-\infty}^{\infty} |f(s)| ds + |f(t)| \int_{|x|>\delta} \tilde{K}_R(x) dx$$

$$+ \sup_{|x|\leqslant\delta} |f(t-x) - f(t)|.$$

We see that, given $\varepsilon > 0$, we can use the continuity of f at t to find a $\delta(\varepsilon) > 0$ with $\sup_{|x|\leqslant\delta(\varepsilon)} |f(t-x) - f(t)| < \varepsilon/3$ and then, using Lemma 49.2(ii), we can find an R_0 depending on $\delta(\varepsilon)$ such that

$$\sup_{|x|>\delta(\varepsilon)} \tilde{K}_R(x) < \varepsilon \left/ \left(3 \int_{-\infty}^{\infty} |f(s)| ds + 1 \right) \right.,$$

and

$$\int_{|x|>\delta(\varepsilon)} \tilde{K}_R(x) dx < \varepsilon/(3|f(t)| + 1),$$

and so

$$\left| \frac{1}{2\pi} \int_{-R}^{R} \left(1 - \frac{|\zeta|}{R} \right) \hat{f}(\zeta) \exp(i\zeta t) d\zeta - f(t) \right| < \varepsilon/3 + \varepsilon/3 + \varepsilon/3 = \varepsilon$$

for all $R \geqslant R_0$.

(ii) Simply observe that under the hypotheses we can choose $\delta(\varepsilon)$ and R_0 in (i) independent of t. ∎

More important than the result is the following simple consequence.

Theorem 49.4 (*a uniqueness theorem for Fourier transforms*). *If* $f, g \in L^1 \cap C$ *and* $\hat{f} = \hat{g}$ *then* $f = g$.

Proof. If $\hat{f}(\zeta) = \hat{g}(\zeta)$ for all $\zeta \in \mathbb{R}$ then

$$0 = \frac{1}{2\pi} \int_{-R}^{R} \left(1 - \frac{|\zeta|}{R}\right) \hat{f}(\zeta) \exp(i\zeta t) dt$$

$$- \frac{1}{2\pi} \int_{-R}^{R} \left(1 - \frac{|\zeta|}{R}\right) \hat{g}(\zeta) \exp(i\zeta t) dt \to f(t) - g(t) \quad \text{as } R \to \infty,$$

and so $f(t) = g(t)$ for all $t \in \mathbb{R}$.   ∎

In the next chapter we illustrate the use of Fourier transforms in studying sums of independent random variables. The reader who is not interested in this may nonetheless wish to study Lemma 50.2 (which does not depend on the rest of the chapter) where we compute two important Fourier transforms.

50

SUMS OF INDEPENDENT
RANDOM VARIABLES

Suppose that X is a random variable taking values on $\mathbb{R}$ and having continuous probability density f_X. Since $f_X(t) \geqslant 0$ for all $t \in \mathbb{R}$ and $\int_{-\infty}^{\infty} f_X(t)dt = 1$, we see that $f_X \in L^1 \cap C$. By definition the expectation $\mathbb{E}(e^{-iX\zeta})$ of $e^{-iX\zeta}$ is given by

$$\mathbb{E}(e^{-iX\zeta}) = \int_{-\infty}^{\infty} f_X(t) \exp(-i\zeta t)dt = \hat{f}_X(\zeta) \quad [\zeta \in \mathbb{R}].$$

By a natural but slightly unfortunate historical accident probabilists tend to work with the *characteristic function* $\mathbb{E}(e^{iX\zeta}) = \hat{f}_X(-\zeta)$. However, this creates no problems (except for the substantial body of students who learn about Fourier transforms in applied courses and characteristic functions in probability courses without ever seeing the connection).

Now suppose $X(1)$ and $X(2)$ are independent random variables taking values on $\mathbb{R}$ and having bounded continuous probability densities $f_{X(1)}$ and $f_{X(2)}$. Then $e^{-iX(1)\zeta}$ and $e^{-iX(2)\zeta}$ are independent random variables and so

$$\mathbb{E}(e^{-i(X(1)+X(2))\zeta}) = \mathbb{E}(e^{-iX(1)\zeta}e^{-iX(2)\zeta}) = \mathbb{E}(e^{-iX(1)\zeta})\mathbb{E}(e^{-iX(2)\zeta}) = \hat{f}_{X(1)}(\zeta)\hat{f}_{X(2)}(\zeta).$$

Now it is a – very plausible – result of probability theory that $X(1) + X(2)$ also has bounded continuous probability density $f_{X(1)+X(2)}$, say, so

$$\hat{f}_{X(1)+X(2)}(\zeta) = \hat{f}_{X(1)}(\zeta)\hat{f}_{X(2)}(\zeta).$$

In general, if $X(1), X(2), \ldots, X(n)$ are all independent random variables on $\mathbb{R}$ with bounded continuous probability densities, we have

$$\hat{f}_{X(1)+X(2)+\cdots+X(n)}(\zeta) = \prod_{r=1}^{n} \hat{f}_{X(r)}(\zeta),$$

and, in the special case when $X(1), X(2), \ldots, X(n)$ are all identically distributed (like the random variable X, say),

$$\hat{f}_{X(1)+X(2)+\cdots+X(n)}(\zeta) = (\hat{f}_X(\zeta))^n.$$

But if Y is a random variable taking values in $\mathbb{R}$ with continuous probability density

245

and $\lambda \in \mathbb{R}$, $\lambda \neq 0$

$$\hat{f}_{\lambda Y}(\zeta) = \mathbb{E}(e^{-i\lambda Y\zeta}) = \mathbb{E}(e^{-i(\lambda\zeta)Y}) = \hat{f}(\lambda\zeta)$$

so

$$\hat{f}_{\lambda(X(1)+X(2)+\cdots+X(n))}(\zeta) = (\hat{f}_X(\lambda\zeta))^n.$$

(Strictly speaking we must check that, if Y has a continuous probability density, f_Y say, then so does λY. To see this observe that, if $\lambda > 0$, then making the linear substitution $s = \lambda t$, we have

$$Pr(a \leqslant \lambda Y < b) = Pr(\lambda^{-1}a \leqslant Y < \lambda^{-1}b) = \int_{\lambda^{-1}a}^{\lambda^{-1}b} f_Y(t)dt = \int_a^b f_Y(\lambda^{-1}s)\lambda^{-1}ds,$$

so λY has continuous probability density $\lambda^{-1}f_Y(\lambda^{-1}s)$. If $\lambda < 0$,

$$Pr(a \leqslant \lambda Y < b) = Pr(\lambda^{-1}b < Y \leqslant \lambda^{-1}a) = \int_{\lambda^{-1}b}^{\lambda^{-1}a} f_Y(t)dt$$

$$= \int_b^a f_Y(\lambda^{-1}s)\lambda^{-1}ds = \int_a^b (-\lambda^{-1}f_Y(\lambda^{-1}s))ds,$$

and so λY has continuous probability density $-\lambda^{-1}f_Y(\lambda^{-1}s)$.)

We summarise our conclusions as follows.

Lemma 50.1. *Let X be a random variable on $\mathbb{R}$ with bounded continuous probability density f_X. Suppose $X(1), X(2), \ldots, X(n)$ are independent random variables on $\mathbb{R}$ with the same probability distribution as X. Write $S(n) = X_1 + X_2 + \cdots + X_n$. Then $\lambda S(n)$ has bounded continuous probability density $f_{\lambda S(n)}$ with*

$$\hat{f}_{\lambda S(n)}(\zeta) = (\hat{f}_X(\lambda\zeta))^n.$$

Proof. The argument above is correct but requires a firmer foundation in probability theory than we can give it here. An equivalent result which contains no reference to probability theory will be found together with a rigorous proof in Theorem 51.7(iv) of the next chapter. ∎

Let us compute $\hat{f}_X$ in two interesting cases.

Lemma 50.2 (i) (*Normal distribution*). *If $f_X(t) = (1/\sqrt{(2\pi)})\exp(-t^2/2)$ then $\hat{f}_X(\zeta) = \exp(-\zeta^2/2)$.*
(ii) (*Cauchy distribution*). *If $f_X(t) = (1/\pi)(1+t^2)^{-1}$ then $\hat{f}_X(\zeta) = \exp(-|\zeta|)$.*
Proof.

$$\hat{f}_X(\zeta) = (2\pi)^{-\frac{1}{2}}\int_{-\infty}^{\infty} \exp(-i\zeta t)\exp(-t^2/2)dt$$

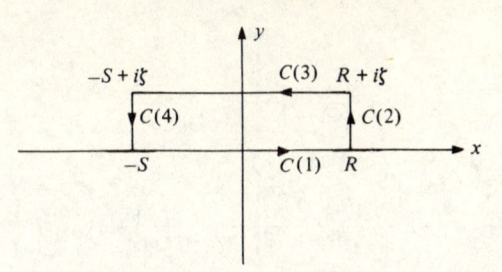

Fig. 50.1. Contour for the evaluation of $\int_{-\infty}^{\infty} \exp(-(t+i\zeta)^2/2)dt$.

$$= (2\pi)^{-\frac{1}{2}} \int_{-\infty}^{\infty} \exp(-((t+i\zeta)^2/2) - \zeta^2/2)dt$$

$$= (2\pi)^{-\frac{1}{2}} \exp(-\zeta^2/2) \int_{-\infty}^{\infty} \exp(-(t+i\zeta)^2/2)dt.$$

Our task reduces to evaluating $\int_{-\infty}^{\infty} \exp(-(t+i\zeta)^2/2)dt$ and this we do by complex variable methods.

Consider the function $g(z) = \exp(-z^2/2)$ $[z \in \mathbb{C}]$ and the contour C in $\mathbb{C}$ shown in Figure 50.1 consisting of the straight lines $C(1)$ from $-S$ to R, $C(2)$ from R to $R + i\zeta$, $C(3)$ from $R + i\zeta$ to $-S + i\zeta$ and $C(4)$ from $-S + i\zeta$ to $-S$. Since g is analytic in $\mathbb{C}$, Cauchy's theorem gives

$$\sum_{j=1}^{4} \int_{C(j)} g(z)\,dz = \int_C g(z)\,dz = 0.$$

But $\int_{-\infty}^{\infty} \exp(-x^2/2)dx = (2\pi)^{\frac{1}{2}}$ (if the reader does not know this result she can find it proved as Lemma 48.10) so

$$\int_{C(1)} g(z)dz = \int_{-S}^{R} \exp(-x^2/2)dx \to (2\pi)^{\frac{1}{2}} \quad \text{as } R, S \to \infty.$$

Further,

$$\left| \int_{C(2)} g(z)dz \right| \leqslant \text{length } C(2) \sup_{z \in C(2)} |g(z)|$$

$$= \text{length } C(2) \sup_{z \in C(2)} |\exp((-R^2 + y^2)/2 + iRy)|$$

$$= |\zeta| \exp((-R^2 + \zeta^2)/2) \to 0 \quad \text{as } R \to \infty,$$

so $\int_{C(2)} g(z)dz \to 0$ and, similarly, $\int_{C(4)} g(z)dz \to 0$ as $R, S \to \infty$. Thus

$$\int_{-S}^{R} \exp(-(t+i\zeta)^2/2)\,dt = -\int_{C(3)} g(z)\,dz$$

$$= \sum_{j=1,2,4} \int_{C(j)} g(z)\,dz \to (2\pi)^{\frac{1}{2}} \quad \text{as } R, S \to \infty,$$

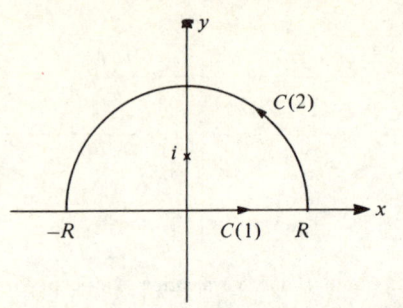

Fig. 50.2. Contour for the evaluation of $\int\limits_{-\infty}^{\infty} \dfrac{\exp(-i\zeta t)}{1+t^2}\,dt$.

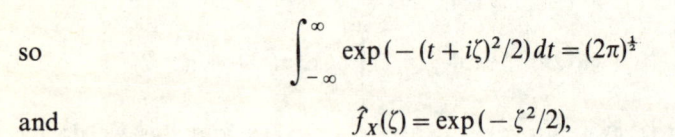

so
$$\int_{-\infty}^{\infty} \exp(-(t+i\zeta)^2/2)\,dt = (2\pi)^{\frac{1}{2}}$$

and
$$\hat{f}_X(\zeta) = \exp(-\zeta^2/2),$$

as claimed.

(ii) We shall also use complex variable methods to evaluate

$$\hat{f}_X(\zeta) = \frac{1}{\pi}\int_{-\infty}^{\infty} \frac{\exp(-i\zeta t)}{1+t^2}\,dt.$$

We consider the cases $\zeta \leqslant 0$ and $\zeta > 0$ separately. Suppose first that $\zeta \leqslant 0$. Consider the function

$$g(z) = \frac{\exp(-i\zeta z)}{1+z^2}$$

and the contour C shown in Figure 50.2 consisting of the straight line $C(1)$ from $-R$ to R and the semicircle $C(2)$ described by $Re^{i\theta}$ as θ runs from 0 to π. (We take $R > 1$ throughout.)

Since g is analytic within and in a neighbourhood of C except for a pole at i it follows from Cauchy's formula that

$$\int_{C(1)} g(z)\,dz + \int_{C(2)} g(z)\,dz = \int_{C} g(z)\,dz = 2\pi i \times \text{residue of } g \text{ at } i.$$

Since
$$g(z) = \frac{h(z)}{z-i},$$

where
$$h(z) = \frac{\exp(-i\zeta z)}{z+i}$$

is analytic in a neighbourhood of i, the residue of g at i is $h(i) = \exp\zeta/2i$ and

$$\int_{C(1)} g(z)\,dz + \int_{C(2)} g(z)\,dz = \pi\exp\zeta.$$

Next we observe that if $z \in C(2)$, then

$$\left| \frac{1}{z^2 + 1} \right| \leqslant \frac{1}{|z|^2 - 1} \leqslant \frac{1}{R^2 - 1},$$

whilst writing $z = x + iy$ $[x, y \in \mathbb{R}]$ we have $y \geqslant 0$ and so, since $\zeta \leqslant 0$,

$$|\exp - i\zeta z| = |\exp(-i\zeta x + \zeta y)| = \exp \zeta y \leqslant 1.$$

Thus, if $z \in C(2)$, $|g(z)| \leqslant (R^2 - 1)^{-1}$. It follows that

$$\left| \int_{C(2)} g(z) \, dz \right| \leqslant \text{length } C(2) \times \sup_{z \in C(2)} |g(z)| \leqslant \frac{\pi R}{R^2 - 1} \to 0 \quad \text{as } R \to \infty.$$

(We have written out the argument in greater detail than usual in order to draw the reader's attention to the fact that it fails if $\zeta > 0$.)

Using the results above we see that

$$\int_{-R}^{R} \frac{\exp(-i\zeta t)}{1 + t^2} \, dt = \int_{C(1)} g(z) \, dz = \pi \exp \zeta - \int_{C(2)} g(z) \, dz \to \pi \exp \zeta,$$

as $R \to \infty$, so that

$$\int_{-\infty}^{\infty} \frac{\exp(-i\zeta t)}{1 + t^2} \, dt = \pi \exp \zeta \quad \text{for } \zeta \leqslant 0.$$

To compute

$$\int_{-\infty}^{\infty} \frac{\exp(-i\zeta t)}{1 + t^2} \, dt \quad \text{when } \zeta > 0$$

we can either repeat the argument with the semicircle $C(2)$ in the upper half plane replaced by one in the lower plane or we can argue as follows.

$$\int_{-\infty}^{\infty} \frac{\cos \zeta t}{1 + t^2} \, dt = \int_{-\infty}^{\infty} \text{Re} \left(\frac{\exp(i\zeta t)}{1 + t^2} \right) dt = \text{Re} \int_{-\infty}^{\infty} \frac{\exp - i\zeta t}{1 + t^2} \, dt$$

$$= \text{Re}(\pi \exp \zeta) = \pi \exp \zeta \quad \text{for } \zeta \leqslant 0$$

and

$$\int_{-\infty}^{\infty} \frac{\sin \zeta t}{1 + t^2} \, dt = -\int_{-\infty}^{\infty} \text{Im} \left(\frac{\exp(-i\zeta t)}{1 + t^2} \right) dt = -\text{Im} \int_{-\infty}^{\infty} \frac{\exp(-i\zeta t)}{1 + t^2} \, dt$$

$$= -\text{Im}(\pi \exp \zeta) = 0 \quad \text{for } \zeta \leqslant 0.$$

Thus, if $\zeta \geqslant 0$ (and so $-\zeta \leqslant 0$), we have

$$\exp(-i\zeta t) = \cos \zeta t - i \sin \zeta t = \cos(-\zeta)t + i \sin(-\zeta)t,$$

and so

$$\int_{-\infty}^{\infty} \frac{\exp - i\zeta t}{1+t^2} dt = \int_{-\infty}^{\infty} \frac{\cos(-\zeta)t}{1+t^2} dt + i \int_{-\infty}^{\infty} \frac{\sin(-\zeta)t}{1+t^2} dt = \pi \exp(-\zeta).$$

Combining the results for $\zeta \leqslant 0$ and $\zeta \geqslant 0$ we have $\hat{f}(\zeta) = \exp(-|\zeta|)$, as stated. ∎

Remark. The reader might be worried by the fact that we compute

$$\int_{-\infty}^{\infty} \frac{\exp(-i\zeta t)}{1+t^2} dt$$

as the limit of

$$\int_{-R}^{R} \frac{\exp(-i\zeta t)}{1+t^2} dt \quad \text{as } R \to \infty$$

rather than

$$\int_{-R}^{S} \frac{\exp(-i\zeta t)}{1+t^2} dt \quad \text{as } R, S \to \infty.$$

But we already know (for example, by comparison with the convergent integral

$$\int_{-\infty}^{\infty} \frac{1}{1+t^2} dt = [\tan^{-1} t]_{-\infty}^{\infty} = \pi)$$

that

$$\int_{-\infty}^{\infty} \frac{\exp(-i\zeta t)}{1+t^2} dt$$

converges, so we are only concerned with finding its value.

Let us draw the consequences of Lemmas 50.1 and 50.2.

Lemma 50.3. (i) *If* $X_1, X_2, \ldots, X_n$ *are independent random variables with identical* $N(0,1)$ *normal probability densities* $f(t) = (2\pi)^{-\frac{1}{2}} \exp(-t^2/2)$, *then* $(X_1 + X_2 + \cdots + X_n)/n^{\frac{1}{2}}$ *has the same probability density.*

(ii) *If* $X_1, X_2, \ldots, X_n$ *are independent random variables with identical Cauchy probability densities* $f(t) = \pi^{-1}(1+t^2)^{-1}$, *then* $(X_1 + X_2 + \cdots + X_n)/n$ *has the same probability density.*

Proof. (i) Let us write $S(n) = X_1 + X_2 + \cdots + X_n$. By Lemmas 50.1 and 50.2 (i)

$$\hat{f}_{S(n)/\sqrt{n}}(\zeta) = (\hat{f}(n^{-\frac{1}{2}}\zeta))^n = (\exp(-(n^{-\frac{1}{2}}\zeta)^2/2))^n = \exp(-\zeta^2/2) = \hat{f}(\zeta),$$

so, by Theorem 49.4 (the uniqueness theorem for Fourier transforms), $f_{S(n)/\sqrt{n}} = f$, as stated.

(ii) By Lemmas 50.1 and 50.2 (ii)

$$\hat{f}_{S(n)/n}(\zeta) = (\hat{f}(n^{-1}\zeta))^n = (\exp(-n^{-1}|\zeta|))^n = \exp(-|\zeta|) = \hat{f}(\zeta),$$

so, by Theorem 49.4, $f_{S(n)/n} = f$, as stated. ∎

Lemma 50.3(ii) gives rise to an important counter example. Let us recall the following simple sequence of observations due to Tchebychev.

Theorem 50.4. (i) (*Tchebychev's inequality*). *Suppose* Y *is a real valued random variable such that* $\mathbb{E}Y^2$ *exists. Then*

$$Pr\{|Y| > a\} \leqslant \mathbb{E}Y^2/a^2 \qquad [a > 0].$$

(ii) *If* $X_1, X_2, \ldots, X_n$ *are independent random variables all with the same distribution as* X, *say, and* $\mathbb{E}X$, *and* var X (*the variance of* X) *exist then, writing* $S(n) = X_1 + X_2 + \cdots + X_n$, *we have*

$$Pr\{|(S(n)/n) - \mathbb{E}X| < \varepsilon\} \to 1 \quad \text{as } n \to \infty \text{ for all } \varepsilon > 0.$$

Proof. (i) Define a new random variable by the conditions $Z = a^2$ if $|Y| > a$ $Z = 0$ otherwise. Then $0 \leqslant Z \leqslant Y^2$ and so $\mathbb{E}Z \leqslant \mathbb{E}Y^2$. But

$$\mathbb{E}Z = a^2 Pr\{Z = a^2\} + 0 \, Pr\{Z = 0\} = a^2 Pr\{|Y| > a\},$$

so $a^2 Pr\{|Y| > a\} = \mathbb{E}Z \leqslant \mathbb{E}Y^2$, which gives the desired result.

(ii) Set $Y_n = \sum_{j=1}^n X_j - n\mathbb{E}X$. Then $\mathbb{E}Y = \sum_{j=1}^n \mathbb{E}X_j - n\mathbb{E}X = 0$ and, since the variance of the sum of independent random variable is the sum of their variance, $\mathbb{E}Y_n^2 = \text{var } Y_n = \sum_{j=1}^n \text{var } X_j = n \text{ var } X$. Thus since $Y_n = n(S(n)/n - \mathbb{E}X)$ part (i) yields

$$Pr\{|S(n)/n - \mathbb{E}X| \geqslant \varepsilon\} = Pr\{|Y_n| \geqslant n\varepsilon\}$$

$$\leqslant \mathbb{E}Y_n^2/(n\varepsilon)^2 = n(\text{var } X)/(n\varepsilon)^2 = \varepsilon^{-2}(\text{var } X)/n \to 0 \quad \text{as } n \to \infty,$$

and so
$$Pr\{|S(n)/n - \mathbb{E}X| < \varepsilon\} \to 1 \quad \text{as } n \to \infty. \qquad \blacksquare$$

We contrast this with the situation for the Cauchy distribution.

Lemma 50.5. *Let* $X_1, X_2, \ldots, X_n$ *be independent random variables with identical Cauchy probability densities* $f(t) = \pi^{-1}(1 + t^2)^{-1}$. *Then*

$$Pr\left\{n^{-1} \sum_{j=1}^n X_j \in (a, b)\right\}$$

is independent of n *and, in particular, although the probability density* f *is symmetric about* 0,

$$Pr\left\{\left|n^{-1} \sum_{j=1}^n X_j\right| < \varepsilon\right\} \nrightarrow 1 \quad \text{as } n \to \infty \text{ for any } \varepsilon > 0.$$

Proof. By Lemma 50.3(ii), $n^{-1} \sum_{j=1}^n X_j$ has exactly the same distribution as X_1 and so

$$Pr\left\{n^{-1} \sum_{j=1}^n X_j \in (a, b)\right\} = Pr\{X_1 \in (a, b)\} \quad \text{for all } n. \qquad \blacksquare$$

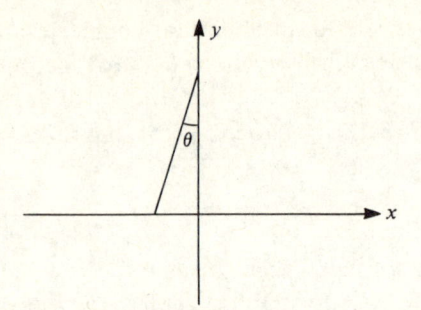

Fig. 50.3. Mechanism for producing a Cauchy distribution.

Observe that, since $t^2/(1+t^2) \to 1$ as $|t| \to \infty$, $(1/\pi)\int_{-\infty}^{\infty} t^2/(1+t^2)dt$, cannot converge so that, if X is a Cauchy distribution, $\mathbb{E}X^2$ fails to converge. Indeed, since

$$\frac{1}{\pi}\int_{-s}^{R} \frac{t}{1+t^2}dt = \frac{1}{2\pi}[\log(1+t^2)]_S^R = \frac{1}{2\pi}\log\left(\frac{1+R^2}{1+S^2}\right)$$

fails to converge, we cannot even define $\mathbb{E}X$ (at least as $(1/\pi)\int_{-\infty}^{\infty} t/(1+t^2)dt$).

The reader may perhaps appreciate the content of Lemma 50.5 better if she imagines herself as an experimenter recording the independent outcomes $X_1, X_2, \ldots$ of an experiment governed by a Cauchy distribution and attempting to make sense of the data by considering the average $\bar{X}_n = n^{-1}\sum_{j=1}^{n} X_j$ of the first n trials. However large she takes n, $\bar{X}_n$ will show no tendency to settle down.

The reader may be inclined to dismiss the example above by claiming that the Cauchy distribution cannot occur in practice. The following example, artificial as it is, may shake her confidence.

Example 50.6. *We work in* $\mathbb{R}^2$. *A point source at* $(0,1)$ *emits a particle at a random angle* θ *to the line joining the source to the origin which hits the x axis at a point* $(X,0)$ *(see Figure* 50.3*). If* θ *is uniformly distributed on* $(-\pi/2, \pi/2)$, *then* X *has the Cauchy probability density* $f(t) = \pi^{-1}(1+t^2)^{-1}$.
Proof. Observe that

$$Pr\{X < t\} = Pr\{\tan\theta < t\} = Pr\{\theta < \tan^{-1}t\} = \frac{1}{\pi}\tan^{-1}t + \tfrac{1}{2},$$

and so X has probability density

$$f(t) = \frac{d}{dt}\left(\frac{1}{\pi}\tan^{-1}t + \frac{1}{2}\right) = \frac{1}{\pi(1+t^2)}. \qquad \blacksquare$$

Bearing in mind the ease with which we obtained the results given in Lemma 50.3 it will be clear why the Fourier transform is sometimes made to occupy the position in probability theory which in other branches of mathematics is taken by thought.

51

CONVOLUTION

In my desire to show off the methods and results of Chapter 50 to their best advantage, I deliberately ignored the fairly heavy analytic theory which is needed to make the probability theory work. In particular, contrary to the impression the reader may have received, it is not trivial to show that the sum of 2 random variables may itself be considered as a random variable. We need, in fact, a theorem along the following lines.

Theorem 51.1. *Let X and Y be independent random variables, taking values in $\mathbb{R}$, with bounded continuous probability densities f and g. Then $X + Y$ can be defined as a random variable, taking values in $\mathbb{R}$, with bounded continuous probability density $f*g$ where*

$$f*g(t) = \int_{-\infty}^{\infty} f(t-x)g(x)\,dx.$$

We shall not pause to pull this particular probabilistic chestnut out of the fire. Instead we shall define a more general notion of convolution $f*g(t) = \int_{-\infty}^{\infty} f(t-x) g(x)\,dx$ and re-prove the key result $(f*g)\hat{}\,(\zeta) = \hat{f}(\zeta)\hat{g}(\zeta)$ of Chapter 50 in this context.

Definition 51.2. *If $f, g:\mathbb{R} \to \mathbb{C}$ are Riemann integrable on each closed interval $[a, b]$ and $\int_{-\infty}^{\infty} |f(t-x)g(x)|\,dx$ converges for each t, we write*

$$f*g(t) = \int_{-\infty}^{\infty} f(t-x)g(x)\,dx.$$

Remark. Since f is Riemann integrable on $[t-R, t+S]$ and g is Riemann integrable on $[-S, R]$, it follows that $h(x) = f(t-x)g(x)$ defines a Riemann integrable function on $[-S, R]$. Since $\int_{-\infty}^{\infty} |h(x)|\,dx$ converges, so does $\int_{-\infty}^{\infty} h(x)\,dx$.

It will probably take the reader some time to appreciate the importance of convolution. The following routine remarks are unlikely to help and may be ignored.

Lemma 51.3. *Let us write* $(f,g) \in \mathscr{C}$ *whenever the conditions of Definition 51.2 hold.*

(i) *If* $(f,g) \in \mathscr{C}$ *then* $(g,f) \in \mathscr{C}$ *and* $f*g = g*f$.

(ii) *If* $(f,h) \in \mathscr{C}, (g,h) \in \mathscr{C}$ *and* $\lambda, \mu \in \mathscr{C}$ *then* $(\lambda f + \mu g, h) \in \mathscr{C}$ *and* $(\lambda f + \mu g)*h = \lambda(f*h) + \mu(g*h)$.

(iii) *If* $f,g: \mathbb{R} \to \mathbb{C}$ *are Riemann integrable on each* $[a,b]$, f *is bounded and* $\int_{-\infty}^{\infty} |g(x)|dx$ *converges, then* $(f,g) \in \mathscr{C}$ *and* $f*g$ *is bounded.*

Proof. (We just sketch these proofs)

(i) Observe that, making the linear change of variable $y = t - x$, we have

$$\int_{-S}^{R} g(t-x)f(x)dx = -\int_{t+S}^{t-R} g(y)f(t-y)dy = \int_{t-R}^{t+S} f(t-y)g(y)dy,$$

and that, if t is kept fixed, $t + S \to \infty$, $t - R \to -\infty$ as $R, S \to \infty$.

(ii) Trivial

(iii) Let $|f(x)| \leq M$ for all $x \in \mathbb{R}$. Then $|f(t-x)g(x)| \leq M|g(x)|$ and, by the comparison theorem for integrals, $\int_{-\infty}^{\infty} |f(t-x)g(x)|dx$ converges and

$$\left| \int_{-\infty}^{\infty} f(t-x)g(x)dx \right| \leq M \int_{-\infty}^{\infty} |g(x)|dx \quad \text{for all} \quad t \in \mathbb{R}. \qquad \blacksquare$$

More to the point we can place the results of Chapter 49 in a more general context.

Theorem 51.4. *Suppose* $K_R: \mathbb{R} \to \mathbb{R}$ *is a Riemann integrable function with the following properties:*

(i) $K_R(s) \geq 0$ *for all* $s \in \mathbb{R}$.

(ii) $K_R(s) \to 0$ *uniformly outside* $[-\delta, \delta]$ *and* $\int_{|s| \geq \delta} K_R(s)ds \to 0$ *as* $R \to \infty$ *for all* $\delta > 0$.

(iii) $\int_{-\infty}^{\infty} K_R(s)ds = 1$.

Then if $f: \mathbb{R} \to \mathbb{C}$ *is Riemann integrable on each* $[a,b]$ *and bounded on* $\mathbb{R}$ *it follows that*

$$f*K_R(t) \to f(t) \quad \text{as} \quad R \to \infty$$

whenever f *is continuous at* t. *Moreover, if* f *is uniformly continuous, the convergence is uniform.*

Proof. The reader should by now be able to provide her own. If not, she can consult the proof of Theorem 49.3. $\qquad \blacksquare$

Theorem 51.5. *Suppose* $K: \mathbb{R} \to \mathbb{R}$ *is a Riemann integrable function with the following properties*

(i) $K(s) \geq 0$ *for all* $s \in \mathbb{R}$.

(ii) $sK(s) \to 0$ *as* $|s| \to \infty$.

(iii) $\int_{-\infty}^{\infty} K(s)\,ds = 1$.

Let us write

$$K_R(s) = RK(Rs) \quad [s \in \mathbb{R}].$$

Then if $f : \mathbb{R} \to \mathbb{C}$ *is Riemann integrable on each* $[a,b]$ *and bounded on* $\mathbb{R}$ *it follows that*

$$f * K_R(t) \to f(t) \quad \text{as} \quad R \to \infty$$

whenever f *is continuous at* t. *Moreover if* f *is uniformly continuous the convergence is uniform.*

Proof. Observe that

(i) $K_R(s) = RK(Rs) > 0$.

(ii) Given any $\varepsilon > 0$ we can find a $T(\varepsilon) > 0$ such that $|tK(t)| \leqslant \varepsilon$ for any $|t| > T(\varepsilon)$. Thus, if $R > T(\varepsilon)\delta^{-1}$, it follows that $|Rs| > T(\varepsilon)$, and

$$|K_R(s)| = |RK(Rs)| \leqslant \delta^{-1}|Rs\,K(Rs)| \leqslant \delta^{-1}\varepsilon \quad \text{for all} \quad |s| > \delta.$$

Also, making the linear change of variable $t = Rs$, we have

$$\int_{|s| \geqslant \delta} K_R(s)\,ds = \int_{|s| \geqslant \delta} RK(Rs)\,ds = \int_{|t| \geqslant \delta/R} K(t)\,dt \to \int_{-\infty}^{\infty} K(t)\,dt = 1 \quad \text{as } R \to \infty.$$

(iii) The same argument gives $\int_{-\infty}^{\infty} K_R(s)\,ds = 1$.
Thus Theorem 51.4 applies. ∎

The reader should draw some diagrams like those of Figure 2.1 for suitable K. She may also like to try and show that Theorem 51.4 remains true if we replace condition (i) by

(i)' $\int_{-\infty}^{\infty}|K(s)|\,ds$ converges.

However (and this is a longer though not a harder exercise), a construction along the lines of Chapter 18 shows that if

(i)'' $\int_{-\infty}^{\infty}|K(s)|\,ds$ diverges,

then we can find a bounded continuous f with $\limsup_{R \to \infty}|K_R * f(0)| = \infty$.

We now return to the ideas of Chapter 50 but we require a preliminary lemma.

Lemma 51.6. (i) *Suppose* $f : \mathbb{R} \to \mathbb{C}$ *is continuous and bounded and that* $g : \mathbb{R} \to \mathbb{C}$ *is Riemann integrable on each* $[a,b]$ *with* $\int_{-\infty}^{\infty}|g(x)|\,dx$ *convergent. Then* $f * g$ *is continuous and bounded.*
(ii) *If, further,* f *is uniformly continuous, so is* $f * g$.
Proof. (i) $|f(t)| \leqslant M$ for all $t \in \mathbb{R}$. We observe that $f * g$ exists and is bounded by Lemma 51.2 (iii) or direct recalculation. To show that $f * g$ is continuous at x we argue as follows.

$$|f*g(x) - f*g(y)| = \left| \int_{-\infty}^{\infty} f(x-t)g(t) - f(y-t)g(t) \, dt \right|$$

$$\leqslant \int_{-\infty}^{\infty} |f(x-t) - f(y-t)| |g(t)| \, dt = \int_{|t| \leqslant R} |f(x-t) - f(y-t)| |g(t)| \, dt$$

$$+ \int_{R < |t|} |f(x-t) - f(y-t)| |g(t)| \, dt \leqslant \sup_{|t| \leqslant R} |f(x-t) - f(y-t)| \int_{|t| \leqslant R} |g(t)| \, dt$$

$$+ \int_{R < |t|} (|f(x-t)| + |f(y-t)|) |g(t)| \, dt$$

$$\leqslant \sup_{|t| \leqslant R} |f(x-t) - f(y-t)| \int_{-\infty}^{\infty} |g(t)| \, dt + 2M \int_{R < |t|} |g(t)| \, dt.$$

Now $\int_{-\infty}^{\infty} |g(t)| \, dt$ converges so, given any $\varepsilon > 0$, we can find an $R > 0$ with

$$\int_{R < |t|} |g(t)| \, dt \leqslant \frac{\varepsilon}{4M + 1}.$$

Since f is continuous on the closed bounded interval $[x - 1 - R, x + 1 + R]$, f is uniformly continuous on $[x - 1 - R, x + 1 + R]$ and so we can find a $1 > \delta > 0$ such that

$$|f(u) - f(v)| \leqslant \varepsilon \left/ \left(2 \int_{-\infty}^{\infty} |g(t)| \, dt + 1 \right) \right.$$

whenever $|u - v| < \delta$ and $u, v \in [x - 1 - R, x + 1 + R]$. In particular, then,

$$|f(x-s) - f(y-s)| \leqslant \varepsilon \left/ \left(2 \int_{-\infty}^{\infty} |g(t)| \, dt + 1 \right) \right.$$

whenever $|x - y| < \delta$ and $s \in [-R, R]$.
   Thus if $|x - y| < \delta$

$$|f*g(x) - f*g(y)| \leqslant \sup_{|s| \leqslant R} |f(x-s) - f(y-s)| \int_{-\infty}^{\infty} |g(t)| \, dt$$

$$+ 2M \int_{R < |t|} |g(t)| \, dt \leqslant \varepsilon/2 + \varepsilon/2 = \varepsilon$$

and we have shown that $f*g$ is continuous.
(ii) Trivial by direct computation.   ∎

   We can now express some of the ideas of Chapter 50 in terms of convolution. Recall that $L^1 \cap C$ is the set of continuous function $f : \mathbb{R} \to \mathbb{C}$ with $\int_{-\infty}^{\infty} |f(x)| \, dx$ convergent.

Theorem 51.7. *Suppose f and g are bounded functions with $f, g \in L^1 \cap C$. Then*

(i) $f*g \in L^1 \cap C$ *and $f*g$ is bounded.*

(ii) $\int_{-\infty}^{\infty} f*g(y)dy = \int_{-\infty}^{\infty} f(y)dy \int_{-\infty}^{\infty} g(y)dy.$

(iii) $\int_{-\infty}^{\infty} |f*g(y)|dy \leqslant \int_{-\infty}^{\infty} |f(y)|dy \int_{-\infty}^{\infty} |g(y)|dy.$

(iv) $(f*g)\hat{\ }\,(\zeta) = \hat{f}(\zeta)\hat{g}(\zeta)$ *for all* $\zeta \in \mathbb{R}.$

Proof. (i) and (ii) We know from Lemma 51.6 that $f*g$ is a well defined continuous bounded function. Further, applying Lemma 51.6 to $|f|*|g|$, we have that

(A1)
$$\int_{-\infty}^{\infty} |f(x-t)g(t)|dt = |f|*|g|(x)$$

is a continuous function of x.

But, trivially,

(A2)
$$\int_{-\infty}^{\infty} |f(x-t)g(t)|dx = \int_{-\infty}^{\infty} |f(x-t)|dx|g(t)| = \int_{-\infty}^{\infty} |f(s)|ds|g(t)|$$

is a continuous function of t, and

(B)
$$\int_{-\infty}^{\infty} \left(\int_{-\infty}^{\infty} |f(x-t)g(t)|dx \right)dt = \int_{-\infty}^{\infty} \int_{-\infty}^{\infty} |f(s)|ds|g(t)|dt$$

$$= \int_{-\infty}^{\infty} |f(s)|ds \int_{-\infty}^{\infty} |g(t)|dt$$

converges.

Thus, by Theorem 48.8,

$$\int_{-\infty}^{\infty} f*g(x)dx = \int_{-\infty}^{\infty} \left(\int_{-\infty}^{\infty} f(x-t)g(t)dt \right)dx$$

converges and

$$\int_{-\infty}^{\infty} f*g(x)dx = \int_{-\infty}^{\infty} \left(\int_{-\infty}^{\infty} f(x-t)g(t)dt \right)dt = \int_{-\infty}^{\infty} f(s)ds \int_{-\infty}^{\infty} g(t)dt.$$

(iii) Observe that

$$|f*g(x)| = \left| \int_{-\infty}^{\infty} f(x-t)g(t)dt \right| \leqslant \int_{-\infty}^{\infty} |f(x-t)g(t)|dt$$

$$= |f|*|g|(x), \quad \text{for all } x \in \mathbb{R}, \text{ so}$$

$$\int_{-\infty}^{\infty} |f*g(x)|dx \leqslant \int_{-\infty}^{\infty} |f|*|g|(x)dx = \int_{-\infty}^{\infty} |f(x)|dx \int_{-\infty}^{\infty} |g(x)|dx,$$

using (ii).

(iv) Write $e_\zeta(s) = \exp - i\zeta s$ and $e_\zeta f(s) = e_\zeta(s)f(s)$. Trivially $e_\zeta f$ and $e_\zeta g$ are bounded functions with $e_\zeta f, e_\zeta g \in L^1 \cap C$. Thus, by part (ii),

$$(f*g)\hat{\ }\,(\zeta) = \int_{-\infty}^{\infty} \left(\int_{-\infty}^{\infty} f(x-t)g(t)dt \right)e_\zeta(x)dx$$

$$= \int_{-\infty}^{\infty} \left(\int_{-\infty}^{\infty} e_{\zeta}(x) f(x-t) g(t) dt \right) dx$$

$$= \int_{-\infty}^{\infty} \left(\int_{-\infty}^{\infty} (e_{\zeta}(x-t) f(x-t))(e_{\zeta}(t) g(t)) dt \right) dx$$

$$= \int_{-\infty}^{\infty} e_{\zeta}(y) f(y) dy \int_{-\infty}^{\infty} e_{\zeta}(y) g(y) dy = \hat{f}(\zeta) \hat{g}(\zeta). \qquad \blacksquare$$

Remark 1. In connection with the probabilistic interpretation of this result in Chapter 50 we note that, if $f(t), g(t) \geq 0$ for all $t \in \mathbb{R}$, then $f * g(t) \geq 0$ and that by Theorem 51.7 (ii), if $\int_{-\infty}^{\infty} f(t) dt = \int_{-\infty}^{\infty} g(t) dt = 1$, then $\int_{-\infty}^{\infty} f * g(t) dt = 1$. Hence, if f and g are continuous bounded probability density functions, so is $f * g$.

Remark 2. With a little extra work it is possible to weaken the hypotheses of Theorem 51.7 (and result like it) quite considerably. If the reader is interested, we refer her to Appendix C. However, the results proved above are entirely sufficient for the needs of the rest of the book and give just as good an idea of the flavour of the subject as does the appendix.

52

CONVOLUTION ON $\mathbb{T}$

As we have seen there is often a close analogy between analysis on $\mathbb{T}$ and analysis on $\mathbb{R}$ but the results for $\mathbb{T}$ may be easier to state and prove than the corresponding results for $\mathbb{R}$. We would thus expect that the study of convolution on $\mathbb{T}$ should present few problems and this is indeed the case.

Definition 52.1. *If $f, g : \mathbb{T} \to \mathbb{C}$ are Riemann integrable functions then we write*

$$f * g(t) = \frac{1}{2\pi} \int_{\mathbb{T}} f(t - x) g(x) \, dx.$$

We shall confine our discussion to the case when f and g are both continuous but if the reader is interested she can use the methods of Appendix C to extend our results to any Riemann integrable f and g. (Thus, for example, she could adapt our next result to prove that, if f and g are Riemann integrable, $f * g$ is continuous.)

Theorem 52.2. *Suppose $f, g : \mathbb{T} \to \mathbb{C}$ are continuous functions. Then*

(i) $f * g$ *is continuous.*
(ii) $(2\pi)^{-1} \int_{\mathbb{T}} f * g(y) \, dy = (2\pi)^{-1} \int_{\mathbb{T}} f(y) \, dy \, (2\pi)^{-1} \int g(y) \, dy.$
(iii) $(2\pi)^{-1} \int_{\mathbb{T}} |f * g(y)| \, dy \leqslant (2\pi)^{-1} \int_{\mathbb{T}} |f(y)| \, dy (2\pi)^{-1} \int_{\mathbb{T}} |g(y)| \, dy.$
(iv) $(f * g)^{\hat{}}(n) = \hat{f}(n) \hat{g}(n).$

Proof. The proofs are simpler versions of those required for Lemma 51.6(i) and Theorem 51.7 and so I leave them as near compulsory exercises for the reader. ∎

Theorem 52.2 (iv) gives us a simple way of proving our next results.

Lemma 52.3. *Suppose that $f, g, h : \mathbb{T} \to \mathbb{C}$ are continuous and $\lambda \in \mathbb{C}$. Then*

(i) $f * g = g * f,$
(ii) $f * (g * h) = (f * g) * h,$

259

(iii) $f*(g+h)=f*g+f*h$,

(iv) $(\lambda f)*g=\lambda(f*g)$.

Proof. These are all proved in the same way. We do (ii) and leave the others to the reader.

To prove (ii) observe that, by Theorem 52.2(i) all of $g*h$, $f*(g*h)$, $f*g$ and $(f*g)*h$ are continuous. Repeated use of Theorem 52.2(iv) combined with the associative law of multiplication for $\mathbb{C}$ yield,

$$(f*(g*h))\hat{}(n)=\hat{f}(n)(g*h)\hat{}(n)=\hat{f}(n)(\hat{g}(n)\hat{h}(n))=(\hat{f}(n)\hat{g}(n))\hat{h}(n)=((f*g)*h)\hat{}(n).$$

Thus using the uniqueness of the Fourier coefficients of continuous functions proved as Theorem 2.4 we have $f*(g*h)=(f*g)*h$ as required. ∎

Conclusions (i), (ii) and (iii) of Lemma 52.3 (together with the trivial fact that the continuous functions $C(\mathbb{T})$ form an Abelian group under addition) show that $(C(\mathbb{T}),+,*)$ is a commutative ring. It is natural to ask whether this ring has an identity.

Question. *Does there exist a continuous function* $k:\mathbb{T}\to\mathbb{C}$ *such that* $k*f=f$ *for every continuous function* $f:\mathbb{T}\to\mathbb{C}$?

The answer is no and we shall prove this by using a simple but important result called the Riemann–Lebesgue lemma.

Theorem 52.4 (the Riemann–Lebesgue lemma). *Let* $f:\mathbb{T}\to\mathbb{C}$ *be continuous. Then* $\hat{f}(n)\to 0$ *as* $|n|\to\infty$.

We shall give two proofs, both along familiar lines.

First Proof. Given any $\varepsilon>0$, we know by Theorem 2.5 (which states that the trigonometric polynomials are uniformly dense in the continuous functions on $\mathbb{T}$) that there exists a trigonometric polynomial $P(t)=\sum_{r=-m}^{m}a_r\exp irt$, say, with $\sup_{t\in\mathbb{T}}|P(t)-f(t)|\leq\varepsilon$. Thus, if $|n|>m$, $\hat{P}(n)=0$ and

$$|\hat{f}(n)|=|\hat{f}(n)-\hat{P}(n)|=|(f-P)\hat{}(n)|=\frac{1}{2\pi}\left|\int_{\mathbb{T}}(f-P)(t)\exp int\,dt\right|$$

$$\leq\frac{1}{2\pi}\int_{\mathbb{T}}|(f-P)(t)|\,dt\leq\varepsilon. \qquad\blacksquare$$

Second Proof. Let $I=[a,b)$ be an interval with $0\leq b-a<2\pi$. Then writing $\chi_I(t)=1$ if $t\in I$, $\chi_I(t)=0$ otherwise, we have

$$|\hat{\chi}_I(n)|=\frac{1}{2\pi}\left|\int_a^b\exp-int\,dt\right|=\frac{1}{2\pi}\left|\left[\frac{\exp-int}{in}\right]_a^b\right|$$

$$=\frac{1}{2\pi|n|}|\exp(-ina)-\exp(inb)|\leq\frac{1}{\pi|n|}\to 0\quad\text{as}\quad|n|\to\infty.$$

It follows that if $I(1), I(2), \ldots, I(k)$ are intervals in $\mathbb{T}$ and $\lambda_1, \lambda_2, \ldots, \lambda_k \in \mathbb{C}$, then

$$\left| \left(\sum_{j=1}^{k} \lambda_j \chi_{I(j)} \right)^{\wedge}(n) \right| = \left| \sum_{j=1}^{k} \lambda_j \hat{\chi}_{I(j)}(n) \right| \leqslant \sum_{j=1}^{k} |\lambda_j| |\hat{\chi}_{I(j)}(n)| \to 0 \quad \text{as } |n| \to \infty.$$

Now f is continuous on $\mathbb{T}$ and so uniformly continuous. Thus given $\varepsilon > 0$, we can find an integer $N \geqslant 1$ such that $|x - y| \leqslant 2\pi/N$ implies $|f(x) - f(y)| \leqslant \varepsilon$. It follows that, writing $\lambda_j = f(2\pi j/N)$ and $I(j) = [2\pi(j-1)/N, 2\pi j/N] [1 \leqslant j \leqslant N]$ and $F = \sum_{j=1}^{N} \lambda_j \chi_{I(j)}$, we have $|F(t) - f(t)| \leqslant \varepsilon$ for all $t \in \mathbb{T}$. But by the first paragraph $\hat{F}(n) \to 0$ as $|n| \to \infty$ so we can find an $n_0(\varepsilon)$ such that $|\hat{F}(n)| < \varepsilon$ and

$$|\hat{f}(n)| \leqslant |\hat{f}(n) - \hat{F}(n)| + |\hat{F}(n)| \leqslant \frac{1}{2\pi} \int_{\mathbb{T}} |f(t) - F(t)| dt + \varepsilon \leqslant 2\varepsilon$$

whenever $n \geqslant n_0(\varepsilon)$. ∎

We can now show that convolution has no unit.

Theorem 52.5. *There does not exist a continuous function* $k : \mathbb{T} \to \mathbb{C}$ *such that* $k * f = f$ *for every continuous function* $f : \mathbb{T} \to \mathbb{C}$.

Proof. Suppose that k had the stated properties. Since $k * f = f$, Theorem 52.2 (iv) yields $\hat{k}(n)\hat{f}(n) = \hat{f}(n)$ for all f and all n. In particular, setting $f(t) = \exp imt$, we have $\hat{k}(m) = 1$ for all m. But the Riemann–Lebesgue lemma (Theorem 52.4) shows that $\hat{k}(m) \to 0$ as $|m| \to \infty$ and we have a contradiction. ∎

In the absence of an identity, the existence of 'approximate identities' becomes important.

Theorem 52.6. *Suppose* $K_R : \mathbb{T} \to \mathbb{R}$ *is a continuous function with the following properties*

(i) $K_R(s) \geqslant 0$ *for all* $s \in \mathbb{T}$,
(ii) $K_R(s) \to 0$ *uniformly outside* $[-\delta, \delta]$ *for all* $\delta > 0$ *as* $R \to \infty$,
(iii) $(2\pi)^{-1} \int_{\mathbb{T}} K_R(t) dt = 1$.

Then, if $f : \mathbb{T} \to \mathbb{C}$ *is continuous,*

$$f * K_R \to f \text{ uniformly as } R \to \infty.$$

Proof. By now, the reader should be able to do this for herself. If not she can consult the proof of Theorem 2.3. ∎

We have seen two important examples of 'approximate identities' in Fejér's kernel K_n of Chapter 2 and Poisson's kernel P_r of Chapter 27. An illuminating example of failure to satisfy the conditions of Theorem 52.6 is given by Dirichlet's kernel D_n of Chapter 18.

There is another aspect of convolution which we have not yet discussed, but

which accounts for much of the concept's utility. This is the 'mixing property'.

Slogan 52.7. *The function $f*g$ has the good properties both of f and g.*

As a simple example we have the following direct consequence of Theorem 52.2 (iv).

Lemma 52.8. *Suppose $f:\mathbb{T}\to\mathbb{C}$ is continuous and P is a trigonometric polynomial. Then $f*P$ is a trigonometric polynomial.*

Proof. If P is a trigonometric polynomial then we can find an N with $\hat{P}(n)=0$ and so with $(f*P)\hat{\ }(n)=\hat{f}(n)\hat{P}(n)=0$ for all $n>|N|$. By Theorem 9.1 (or a simpler argument) $f*P(t)=\sum_{r=-N}^{N}\hat{f}(r)\hat{P}(r)\exp irt$ and $f*P$ is a trigonometric polynomial. ∎

The next result is a very important example of our slogan in action.

Theorem 52.9. (i) *If $f:\mathbb{T}\to\mathbb{C}$ is differentiable with continuous derivative and $g:\mathbb{T}\to\mathbb{C}$ is continuous, then $f*g$ is differentiable with continuous derivative $(f*g)'=f'*g$.*
(ii) *If $f:\mathbb{T}\to\mathbb{C}$ is r times continuously differentiable and $g:\mathbb{T}\to\mathbb{C}$ is s times continuously differentiable, then $f*g$ is $r+s$ times continuously differentiable with $(f*g)^{(r+s)}=f^{(r)}*g^{(s)}$.*

Proof. (i) (This is a fairly delicate argument and the reader may wish to check her understanding by rewriting it for herself). Since f' is continuous on $\mathbb{T}$ it follows that f' is uniformly continuous and so, given any $\varepsilon>0$, we can find a $\delta(\varepsilon)>0$ such that $|f'(u)-f'(v)|<\varepsilon$ whenever $|u-v|<\delta(\varepsilon)$. But the mean value theorem gives $f(y)-f(x)=f'(\zeta)(y-x)$ and so $|f(y)-f(x)-f'(x)(y-x)|=|f'(\zeta)-f'(x)||y-x|$ for some ζ between x and y. Thus, if $|y-x|<\delta(\varepsilon)$, we have $|\zeta-x|<\delta(\varepsilon)$ and $|f(y)-f(x)-f'(x)(y-x)|<\varepsilon|y-x|$.

It follows that, if $|\eta|<\delta(\varepsilon)$,

$$|f*g(u+\eta)-f*g(u)-\eta f'*g(u)|$$

$$=\left|\frac{1}{2\pi}\int_{\mathbb{T}}(f(u+\eta-t)-f(u-t)-\eta f'(u-t))g(t)dt\right|$$

$$\leqslant \sup_{u,t\in\mathbb{T}}|f(u+\eta-t)-f(u-t)-\eta f'(u-t)|\sup_{t\in\mathbb{T}}|g(t)|$$

$$=\sup_{x\in\mathbb{T}}|f(x+\eta)-f(x)-\eta f'(x)|\sup_{t\in\mathbb{T}}|g(t)|$$

$$\leqslant (\varepsilon\sup_{t\in\mathbb{T}}|g(t)|)\eta.$$

and $|((f*g)(u+\eta)-(f*g)(u))/\eta-f'*g(u)|\leqslant\varepsilon\sup_{t\in\mathbb{T}}|g(t)|$. Since ε was arbitrary it follows that

$(f*g(u+\eta)-f*g(u))/\eta\to f'*g(u)$ as $\eta\to0$ and we have the desired result.

(ii) Use induction ∎

All the new results of this chapter transfer back to bounded $L^1\cap C$ functions on the real line $\mathbb{R}$. To extend Theorem 52.9 requires some work which will be done in the next chapter. The extensions of Lemma 52.3, Theorem 52.4 and Theorem 52.5 are easy and I have left much of the proofs to the reader.

Lemma 52.10. *Suppose that f, g and h are bounded functions with $f, g, h \in L^1\cap C$ and that $\lambda\in\mathbb{C}$. Then*

(i) $f*g=g*f$,

(ii) $f*(g*h)=(f*g)*h$,

(iii) $f*(g+h)=f*g+f*h$,

(iv) $(\lambda f)*g=\lambda(f*g)$.

Proof. Left to the reader (look at the proof of Lemma 52.3 if necessary). ∎

Theorem 52.11 (Riemann–Lebesgue lemma). *If $f\in L^1\cap C$, then $\hat{f}(\zeta)\to0$ as $|\zeta|\to\infty$.*
Sketch of proof. Let $\varepsilon>0$ be given. Since $\int_{-\infty}^{\infty}|f(x)|dx$ converges, we can find an $R>0$ such that

$$\int_{|x|>R}|f(x)|dx\leqslant\varepsilon/4.$$

Since $[-R,R]$ is compact and f is continuous, it follows that f is uniformly continuous on $[-R,R]$. In particular, choosing N large enough and writing

$$\lambda_j=f(-R+2jR/N),\ I(j)=[-R+2(j-1)R/N,\ -R+2jR/N)$$

and

$$F=\sum_{j=1}^{N}\lambda_j\chi_{I(j)},$$

we have (just as in the second proof of Theorem 52.4)

$$|F(t)-f(t)|\leqslant\varepsilon/4R\quad\text{for all}\quad t\in[-R,R).$$

It follows that

$$\int_{-R}^{R}|F(t)-f(t)|dt\leqslant\varepsilon/2\quad\text{and}\quad\int_{-\infty}^{\infty}|F(t)-f(t)|dt\leqslant\varepsilon,$$

so

$$|\hat{F}(\zeta)-\hat{f}(\zeta)|\leqslant\varepsilon.$$

The argument now follows that of the second proof of Theorem 52.4 ∎

(There is also an analogue of our first proof of Theorem 52.4 but we leave it to the reader as an interesting exercise to find it.)

Theorem 52.12. *There does not exist a bounded $k \in L^1 \cap C$ such that $k * f = f$ for every bounded $f \in L^1 \cap C$.*

Proof. Pick any bounded $F \in L^1 \cap C$ with $\int_{-\infty}^{\infty} F(t) dt \neq 0$. Then $\hat{F}(0) = \int_{-\infty}^{\infty} F(t) dt \neq 0$ and, writing $F_\zeta(t) = F(t) \exp i\zeta t$, we see that F_ζ is a bounded function, that $F_\zeta \in L^1 \cap C$ and $\hat{F}_\zeta(\zeta) = \hat{F}(0) \neq 0$. Since $k * F_\zeta = F_\zeta$, it follows that $\hat{k}(\zeta) \hat{F}_\zeta(\zeta) = (k * F_\zeta)^{\wedge}(\zeta) = \hat{F}_\zeta(\zeta)$ so, since $\hat{F}_\zeta(\zeta) \neq 0$, $\hat{k}(\zeta) = 1$ for all $\zeta \in \mathbb{R}$. This contradicts the Riemann–Lebesgue lemma (Theorem 52.11), so we are done. ∎

53

DIFFERENTIATION UNDER THE INTEGRAL

Consider the formal manipulations

$$(f*g)'(x) = \frac{d}{dx}\int_{-\infty}^{\infty} f(x-t)g(t)\,dt = \int_{-\infty}^{\infty} \frac{\partial}{\partial x}(f(x-t)g(t))\,dt$$

$$= \int_{-\infty}^{\infty} f'(x-t)g(t)\,dt = f'*g(t)$$

and

$$\frac{d\hat{f}}{d\zeta}(\zeta) = \frac{d}{d\zeta}\int_{-\infty}^{\infty} f(t)\exp(-i\zeta t)\,dt = \int_{-\infty}^{\infty} \frac{\partial}{\partial \zeta}(f(t)\exp(-i\zeta t))\,dt$$

$$= \int_{-\infty}^{\infty} (-itf(t))\exp(-i\zeta t)\,dt = -i(Jf)\hat{}(\zeta),$$

where $(Jf)(t) = tf(t)$.

In order to justify such manipulations we need to be able to differentiate under the integral sign, that is to say we need to justify the interchange of two limiting operations. But we already know (from, e.g., Example 47.1 and the remark just before Lemma 27.4) that such a procedure may lead us badly astray. In this chapter we shall look for simple conditions under which the interchanges needed in the first paragraph can be made without any problems.

The key to understanding both the conditions and their proofs lies in the observation that, whereas integration is a 'smoothing, well behaved' operation for which we expect things to go well, differentiation is the reverse. Thus we use conditions and proofs which replace a problem involving both differentiation and integration by a problem involving integrations only. The sketch proof which follows may make this remark clear.

Theorem 53.1. *If $f:\mathbb{R}\to\mathbb{C}$ is once differentiable and f, f' and g are bounded functions with $f, f', g\in L^1\cap C$, then $f*g$ is differentiable with $(f*g)' = f'*g$.*
Sketch of proof. (We shall give a complete second proof later in the chapter.)

265

Step I. Observe that

(A1) $$\int_0^y |f'(x-t)g(t)|\,dx = |g(t)| \int_0^y |f'(x-t)|\,dx = |g(t)| \int_{-t}^{y-t} |f'(u)|\,du$$

$$= |g(t)| \left(\int_0^{y-t} |f'(u)|\,du + \int_{-t}^0 |f'(u)|\,du \right)$$

is a continuous function of t, and that, using Theorem 51.7(i),

(A2) $$\int_{-\infty}^\infty |f'(x-t)g(t)|\,dt = |f'|*|g|(x)$$

is a continuous function of x. Further, using Theorem 51.7 (ii),

(B) $$\int_0^y \left(\int_{-\infty}^\infty |f'(x-t)g(t)|\,dt \right) dx = \int_0^y |f'|*|g|(x)\,dx \leqslant \int_{-\infty}^\infty |f'|*|g|(x)\,dx$$

$$= \int_{-\infty}^\infty |f'(x)|\,dx \int_{-\infty}^\infty |g(x)|\,dx < \infty.$$

It follows by a simple modification of Theorem 48.8 that

$$\int_0^y f'*g(x)\,dx = \int_0^y \left(\int_{-\infty}^\infty f'(x-t)g(t)\,dt \right) dx = \int_{-\infty}^\infty \left(\int_0^y f'(x-t)g(t)\,dx \right) dt$$

$$= \int_{-\infty}^\infty g(t) \left(\int_0^y f'(x-t)\,dx \right) dt$$

$$= \int_{-\infty}^\infty g(t)(f(y-t)-f(-t))\,dt = f*g(y)-f*g(0).$$

Step 2. Since $f'*g$ is continuous, $\int_0^y f'*g(x)\,dx$ is a differentiable function of y with derivative $f'*g(y)$. Thus the expression

$$\int_0^y f'*g(x)\,dx = f*g(y)-f*g(0),$$

obtained above, shows that $f*g$ is differentiable with $(f*g)'(y)=f'*g(y)$.   ∎

We see that the main part of the proof was concerned with obtaining the expression involving integrals at the end of Step 1. Once this was done we just differentiated to obtain the desired result.

The same point is evident in the usual proof of the following result (which the reader probably knows).

Lemma 53.2. *Let $f_1, f_2, \ldots : \mathbb{R} \to \mathbb{C}$ be once differentiable functions with continuous derivatives $f'_1, f'_2, \ldots$. Suppose that $f_n \to f$ pointwise and $f'_n \to g$ uniformly on each interval $[a, b]$. Then f is once differentiable with continuous derivative g.*

We obtain Lemma 53.2 as a consequence of a result on integration which the reader ought to know already.

Theorem 53.3. *Let $g_1, g_2, \ldots : [a, b] \to \mathbb{C}$ be continuous functions with $g_n \to g$ uniformly on $[a, b]$. Then g is continuous and*

$$\int_a^b g_n(t)\,dt \to \int_a^b g(t)\,dt \quad \text{as } n \to \infty.$$

Proof. The proof is in all the standard texts but if the reader cannot supply it for herself this would suggest serious gaps in her undetstanding of uniform convergence whose mending should take priority over the contents of this chapter. ∎

Proof of Lemma 53.2. Write $g_n = f'_n$ and take $[a, b]$ to be the interval $[0, y]$ (if $y \geqslant 0$) or $[y, 0]$ if $(y < 0)$. Then the fundamental theorem of the calculus and Theorem 53.3 give

$$f_n(y) = \int_0^y f'_n(t)\,dt - f_n(0) \to \int_0^y g(t)\,dt - f(0).$$

But $f_n(y) \to f(y)$ as $n \to \infty$, so

$$f(y) = \int_0^y g(t)\,dt - f(0).$$

Applying Theorem 53.3 again with $[a, b] = [-|y| - 1, |y| + 1]$, we see that g is continuous at each $y \in \mathbb{R}$ and so $\int_0^y g(t)\,dt$ is differentiable with derivative $g(y)$. It follows that f is differentiable with $f'(y) = g(y)$ for all $y \in \mathbb{R}$, as stated. ∎

We now prove our main results on differentiating under the integral sign. In what follows we write $g_2(x, t) = \partial g / \partial t(x, t)$ if it exists.

Theorem 53.4. *Let $g : \mathbb{R} \times \mathbb{R} \to \mathbb{C}$ be a continuous function such that g_2 exists and is continuous. Then $\int_a^b g(x, t)\,dx$ is differentiable with*

$$\frac{d}{dt} \int_a^b g(x, t)\,dx = \int_a^b \frac{\partial g}{\partial t}(x, t)\,dt.$$

Proof. Since $g_2 : \mathbb{R} \times \mathbb{R} \to \mathbb{C}$ is continuous, Lemma 47.2 tells us that we may interchange the order of integration to get

$$\int_0^y \left(\int_a^b g_2(x, t)\,dx \right) dt = \int_a^b \left(\int_0^y g_2(x, t)\,dt \right) dx$$

$$= \int_a^b (g(x, y) - g(x, 0))\,dx = \int_a^b g(x, y)\,dx - \int_a^b g(x, 0)\,dx.$$

We have thus shown that

$$\int_a^b g(x,y)dx = \int_0^y G(t)dt + \text{constant},$$

where $G(t) = \int_a^b g_2(x,t)dx$, but, before we can conclude that $\int_a^b g(x,y)dx$ is differentiable with derivative $G(y)$, we must show that G is continuous at each $y \in \mathbb{R}$.

This is not hard to do. Choose $R \geqslant \max(|a|,|b|,|y|) + 1$. Then (see Appendix B, Theorem B.5, if you must, but it would be better to supply your own proof) g_2 is uniformly continuous on $[-R,R] \times [-R,R]$ and given any $\varepsilon > 0$ we can find a $\delta(\varepsilon)$ with $1 > \delta(\varepsilon) > 0$ such that $(x,t_1),(x_2,t_2) \in [-R,R] \times [-R,R]$ and $|x_1 - x_2|$, $|t_1 - t_2| < \delta(\varepsilon)$ imply $|(g(x_1,t_1) - g(x_2,t_2)| < \varepsilon$. In particular, if $|t - y| < \delta(\varepsilon)$, we have

$$|G(t) - G(y)| = \left| \int_a^b g_2(x,t) - g_2(x,y)\,dx \right| \leqslant \int_a^b |g_2(x,t) - g_2(x,y)|\,dx \leqslant (b-a)\varepsilon.$$

Thus G is continuous and $\int_a^b g(x,y)dx$ is differentiable with

$$\frac{d}{dy}\int_a^b g(x,y)dy = G(y) = \int_a^b g_2(x,y)dy. \qquad \blacksquare$$

Remark 1. The reader is now in a position to give a second proof of Theorem 52.9 and it would be a good exercise for her to do so.

Remark 2. (This is rather sophisticated and can be omitted without loss.) In connection with Remark 1 the reader may notice that our original proof of Theorem 52.9 used the mean value theorem explicitly but the new proof does not. However, the statement $\int_0^y g_2(x,t)dt = g(x,y) - g(x,0)$ is proved by using the fundamental theorem of the calculus and the fundamental theorem of the calculus requires the mean value theorem for its proof.

Theorem 53.5. *Let $g : \mathbb{R} \times \mathbb{R} \to \mathbb{C}$ be a continuous function such that g_2 exists and is continuous. Suppose $\int_{-\infty}^{\infty} |g(x,t)|dx$ and $\int_{-\infty}^{\infty} |g_2(x,t)|dx$ exist for each t and that $\int_{|x|>R} |g_2(x,t)|dx \to 0$ as $R \to \infty$ uniformly in t on each $[a,b]$. Then $\int_{-\infty}^{\infty} g(x,t)dx$ is differentiable with*

$$\frac{d}{dt}\int_{-\infty}^{\infty} g(x,t)dx = \int_{-\infty}^{\infty} \frac{\partial g}{\partial t}(x,t)dx.$$

Proof. Set

$$f_n(t) = \int_{-n}^{n} g(x,t)dx, \; f(t) = \int_{-\infty}^{\infty} g(x,t)dt.$$

Then

$$|f_n(t) - f(t)| = \left| \int_{|x|>n} g(x,t)dx \right| \leqslant \int_{|x|>n} |g(x,t)|dx \to 0,$$

so $f_n \to f$ pointwise. On the other hand, by Theorem 53.4, f_n is differentiable with

$$f'_n(t) = \frac{d}{dt} \int_{-n}^{n} g(x,t)dx = \int_{-n}^{n} g_2(x,t)dx$$

(so f'_n is continuous by the second part of the proof of Theorem 53.4). Further

$$\left| f'_n(t) - \int_{-\infty}^{\infty} g_2(x,t)dx \right| = \left| \int_{|x|>n} g_2(x,t)dx \right| \to 0$$

uniformly on each $[a,b]$ by hypothesis. Thus, by Lemma 53.2, $f(t) = \int_{-\infty}^{\infty} g(x,t)dx$ is differentiable with

$$\frac{d}{dt} \int_{-\infty}^{\infty} g(x,t)dx = \int_{-\infty}^{\infty} g_2(x,t)dt. \qquad \blacksquare$$

We can now justify the formal manipulations of the first paragraph under quite wide conditions.

Second proof of Theorem 53.1. Let $G(x,t) = f(x-t)g(t)$. Then $G:\mathbb{R} \times \mathbb{R} \to \mathbb{C}$ is continuous and $G_1 = \partial G/\partial x$ exists and is continuous with value $f'(x-t)g(t)$. Since f' is bounded (with $|f'(s)| \leq M$ for all s say),

$$\int_{|t|>R} |G_1(x,t)|dt = \int_{|t|>R} |f'(x-t)g(t)|dt \leq M \int_{|t|>R} |g(t)|dt \to 0$$

uniformly on $\mathbb{R}$. Similarly $\int_{-\infty}^{\infty} |G(x,t)|dt \to 0$ is bounded on $\mathbb{R}$. Thus, by Theorem 53.5, $f*g(x) = \int_{-\infty}^{\infty} G(x,t)dt$ is differentiable with

$$(f*g)'(x) = \frac{d}{dx} \int_{-\infty}^{\infty} G(x,t)dt = \int_{-\infty}^{\infty} \frac{\partial G}{\partial x}(x,t)dt = \int_{-\infty}^{\infty} f'(x-t)g(t)dt = f'*g(x). \qquad \blacksquare$$

Lemma 53.6. *Suppose $f:\mathbb{R} \to \mathbb{C}$ is given. Let us write $(Jf)(t) = tf(t)$. Then if $f, Jf \in L^1 \cap C$ it follows that $\hat{f}$ is differentiable with $d\hat{f}/d\zeta(\zeta) = -i(Jf)\hat{}(\zeta)$.*

(Compare the statement and proof of Theorem 9.4.)

Proof. Let $G(x,\zeta) = f(x)\exp(-ix\zeta)$. Then $G:\mathbb{R} \times \mathbb{R} \to \mathbb{C}$ is continuous and $G_2(x,\zeta)$ exists with value $-i(Jf)(x)\exp(-ix\zeta)$. Now

$$\int_{|x|>R} |G_2(x,\zeta)|dx = \int_{|x|>R} |Jf(x)|dx \to 0$$

uniformly on $\mathbb{R}$ and $\int_{-\infty}^{\infty} |G(x,\zeta)|dx = \int_{-\infty}^{\infty} |f(x)|dx$ is bounded on $\mathbb{R}$. Thus, by Theorem 53.5, $\hat{f}(\zeta) = \int_{-\infty}^{\infty} G(x,\zeta)dx$ is differentiable and

$$\frac{d\hat{f}}{d\zeta}(\zeta) = \frac{d}{d\zeta} \int_{-\infty}^{\infty} G(x,\zeta)dx = \int_{-\infty}^{\infty} \frac{\partial}{\partial \zeta} G(x,\zeta)dx$$

$$= \int_{-\infty}^{\infty} -i(Jf)(x)\exp(-i\zeta x)dx = -i(Jf)\hat{}(\zeta). \qquad \blacksquare$$

54

LORD KELVIN

Fourier believed that the main aim of mathematics should be the understanding of nature and that the purpose of understanding nature should be the benefit of mankind. Yet, interesting though Fourier's own work was, he could point to no direct practical benefit from it. The credit for showing how powerful were the tools that Fourier had forged belongs, above all, to William Thomson, Lord Kelvin.

The life of William Thomson was linked with the University of Glasgow from 1832, when his father became professor there. Thomson was not a late developer. At the age of ten he and his 12-year-old brother enrolled in the University of Glasgow. Prizes in Greek, logic (what we would now call philosophy), mathematics, astronomy and physics marked his progress. 'A boy', he said later, 'should have learnt by the age of twelve to write his own language with accuracy and some elegance; he should have a reading knowledge of French, should be able to translate Latin and easy Greek authors and should have some acquaintance with German.'

Towards the end of his time as a student in Glasgow he fell under the influence of an inspiring physics teacher called Nichol.

...after I had attended in 1839 Nichol's... Class, I had become filled with the utmost admiration for the splendour and poetry of Fourier. Nichol was not a mathematician and did not profess to have really read Fourier, but he was capable of perceiving his greatness and of understanding what he was driving at, and of making us appreciate it. I asked Nichol if he thought I could read Fourier. He replied 'perhaps'. He thought the book a work of most transcendent merit. So on the Ist of May the very day when the prizes were given, I took Fourier out of the University Library, and in a fortnight I had mastered it – gone right through it. (This and the next quotation are from Silvanus P. Thompson's *Life of Lord Kelvin*.)

At the age of seventeen Thomson entered Cambridge University. (It was not uncommon for good students to first study at a Scottish University and then transfer to Cambridge. A cynic might say that people like Maxwell and Thomson were educated in Scotland and examined in Cambridge.) The examinations in Cambridge were fiercely competitive exercises in problem solving against the clock. The best

candidates were trained as for an athletics contest. Thomson's coach might declare that his candidate had mathematical abilities which would outshine any man in England but the coach of Thomson's rival boasted, more pertinently, that he had a candidate who could beat any man in Europe in an examination. Thomson (like Maxwell later) came second. During his time in Cambridge he had however won the Colquhoun sculls as a rower, published sixteen papers and, most important of all, obtained from his coach, a day before he left, two copies of Green's *Essay on the Application of Mathematical Analysis to the Theories of Electricity and Magnetism.*

Thomson's father had by now conceived the ambition of seeing his son elected to a chair in Glasgow. To improve his chances he was sent to France to learn experimental work in the laboratory of Regnault, to meet as many influential people as possible and, since the electors would place great emphasis on teaching ability, to learn the art of giving popular lectures. Thomson's abilities won him a ready welcome and his copies of Green's *Essay* created a great stir.

Eager steps were heard without and with a hasty rap upon the door a panting visitor rushed in. It was Sturm, in a state of high excitement. 'Vouz avez le Mémoire de Green', he exclaimed, 'M. Liouville me l'a dit!'. The *Essay* was produced, and Sturm eagerly scanned its contents, turning over page after page. 'Ah voilà mon affaire', he cried, jumping from his seat as he caught sight of the formula in which Green had anticipated his theorem of the equivalent distribution.

The methods of Green and Fourier formed the foundation of Thomson's mathematical methods. The foundations for his deepest physical insights were laid in Paris. Liouville encouraged him to investigate the relation between Faraday's 'lines of force' view of electrical phenomena and the classical continental 'action at a distance', and, by reading Clapeyron, he became aware of Sadi Carnot's view of heat.

Meanwhile his father's campaign had borne fruit and at the age of twenty two, swept in on a wave of testimonials from, among others, De Morgan, Cayley, Hamilton, Boole, Sylvester, Stokes, Regnault and Liouville, Thomson became Professor of Natural Sciences at the University of Glasgow. Nor were his sponsors mistaken. For the next 20 years Thomson was to dominate physics.

There is a curious anonymity about physical ideas which makes it difficult to appreciate fully the originality of their discoverers. A novel always retains the signature of its author, a proof in pure mathematics may still delight us by its ingenuity or power after a hundred (consider Liouville's proof given as Theorem 38.2 of the existence of transcendental numbers) or a thousand years (consider Euclid's proof that there exist an infinity of primes), but the physics breakthrough of today is the common sense of tomorrow.

In electricity Thomson provided the link between Faraday and Maxwell. He was able to mathematise Faraday's laws and to show the formal analogy between problems in heat flow and electricity. Thus the work of Fourier on heat immediately gave rise to theorems on electricity and the work of Green on potential theory

immediately gave rise to theorems on heat flow. Similarly methods used to deal with linear and rotational displacements in elastic solids could be applied to give results on electricity and magnetism.

Maxwell, who from the end of his undergraduate days conducted a constant correspondence with Thomson first as pupil to master and then as equal to equal, felt sure that Thomson must have a complete theory of electrodynamics 'lying in loose papers and neglected only till you have worked out Heat or got a little spare time'. (The correspondence was published as *Origins of Clerk Maxwell's Electric Ideas*, Larmor, Cambridge University Press, 1937. It includes a useful note from Maxwell on the habits of peacocks.) However, Thomson did not have such a theory and could never fully accept Maxwell's ideas.

Thomson's other major contribution to fundamental physics was his combination of the almost forgotten work of Carnot with the work of Joule on the conservation of energy to lay the foundations of Thermodynamics. Thomson himself was particularly proud of an argument developed with his brother which predicted the fall in the freezing point of water under pressure.

Other discoveries and inventions of Thomson are dealt with elsewhere in this book. We note in passing the discovery of what is now called Stokes' theorem and the first mathematical discription of the oscillation of an electric circuit. After his work on the Atlantic cables (to be described in Chapters 65 and 66) he turned increasingly towards the practical applications of physics. He acted as consultant engineer to undersea telegraph and overland electric power transmission schemes (for example he worked out the optimum diameter of a transmission line). He was also unsurpassed as a designer of the rugged but accurate instruments required by the new electrical industries and took an active part in running the firm of Kelvin and Wright which manufactured them.

As a teacher he was inspiring rather than methodical. By a tradition which he supported, his audience included non specialists such as future doctors and theologians. Since he refused to talk down to his audience, some of them may not have understood very much but they all appreciated that they were being taught by a great man. Particularly enjoyed were lectures on acoustics (illustrated by a performance on the French horn), impulse (in which a large rifle was fired at a ballistic pendulum) and an experiment in which a large rubber sheet was slowly filled with water until it burst.

His teaching laboratory was the first in Britain. (The proposal to introduce experimental work into the Cambridge undergraduate courses aroused strong opposition. 'If [the student] does not believe the statements of his tutor – probably a clergyman of mature knowledge, recognised ability and blameless character – his suspicion is irrational and manifests a want of the power of appreciating evidence, a want fatal to his success in that branch of science which he is supposed to be cultivating.' (Todhunter, First essay in '*The Conflict of Studies*', Macmillan 1873; since Todhunter is no fool the essay is worth reading.))

In conjunction with his friend Tait he wrote the famous *Thomson and Tait* (or *T*

and T' as it was known by everybody from Maxwell downwards) *Treatise on Natural Philosophy* . Originally intended as three moderately sized volumes which would cover all of mathematical and experimental physics, it became, as these things tend to become, two large volumes on mechanics.

As Maxwell said in his review of the second edition, the great merit of this work lay in rescuing a large chunk of physics from the mathematicians. Thanks to Thomson and Tait

> dynamical theorems have been dragged out of the sanctuary of profound mathematics in which they lay so long enshrined and have been set to do all kinds of work easy as well as difficult, throughout the whole range of physical science, ...
>
> The two northern wizards were the first who, without compunction or dread, uttered in their mother tongue the true and proper names of those dynamical concepts which the magicians of old were wont to invoke only by the aid of muttered symbols and inarticulate equations. And now the feeblest among us can repeat the words of power and take part in dynamical discussions which but a few years ago we should have left for our betters.
>
> (*Maxwell's Collected Works, p. 782, Vol.* I.)

As in mechanics so elsewhere Thomson's career marked the break away of mathematical physics and mathematical engineering from mathematics into separate disciplines with their own methods and ideas. Thomson's contemporaries considered him almost a second Newton, but he would, I think, have been happy with the modern estimate of him as Fourier's successor and (with Faraday) Maxwell's predecessor.

55

THE HEAT EQUATION

In this chapter we shall try to solve the heat equation

$$\frac{\partial \phi}{\partial t}(x,t) = K \frac{\partial^2 \phi}{\partial x^2}(x,t)$$

for $x \in \mathbb{R}$, $t > 0$ subject to the given initial condition

$$\phi(x,t) \to G(x) \quad \text{as} \quad t \to 0+.$$

It is probable that the reader has seen this problem solved before but if not, she should spend a little time attacking it for herself. She will then appreciate the achievement of Laplace (building on earlier work of Poisson) in finding a simple general solution.

Laplace's result seems to have been a major source of inspiration in leading Fourier to the idea of the Fourier transform and the first use that Fourier made of his transform method was to recover Laplace's solution. We too will use Fourier transforms though we shall not follow Fourier's original argument.

We shall need two easy results on Fourier transforms.

Lemma 55.1. (i) *If* $f \in L^1 \cap C$ *and* $R > 0$ *then, writing* $f_R(x) = Rf(Rx)$ $[x \in \mathbb{R}]$, *we have* $f_R \in L^1 \cap C$ *and* $\hat{f}_R(\zeta) = \hat{f}(\zeta/R)$.
(ii) *If* $f : \mathbb{R} \to \mathbb{C}$ *is once differentiable with* $f, f' \in L^1 \cap C$ *and* $f(x) \to 0$ *as* $|x| \to \infty$, *then* $(f')^\wedge(\zeta) = i\zeta \hat{f}(\zeta)$.

Proof. (i) It is easy to check that $f_R \in L^1 \cap C$. By making the linear substitution $t = Rx$ we obtain

$$\hat{f}_R(\zeta) = \int_{-\infty}^{\infty} f(Rx) \exp(-i\zeta x) R\,dx = \int_{-\infty}^{\infty} f(t) \exp(-i\zeta R^{-1} t)\,dt = \hat{f}(R^{-1}\zeta).$$

(The reader may recall a similar computation directly preceding the statement of Lemma 50.1.)
(ii) Integrating by parts we obtain

274

$$(f')^{\wedge}(\zeta) = \int_{-\infty}^{\infty} f'(x) \exp{-i\zeta x}\,dx$$

$$= [f(x)(-i\zeta \exp{-i\zeta x})]_{-\infty}^{\infty} + i\zeta \int_{-\infty}^{\infty} f(x) \exp{-i\zeta x}\,dx = i\zeta \hat{f}(\zeta),$$

as stated. ∎

Equations like the heat equation

$$\frac{\partial \phi}{\partial t}(x,t) = K \frac{\partial^2 \phi}{\partial x^2}(x,t)$$

involve functions of two variables and a consequent risk of confusion between operations with respect to the first and operations with respect to the second variable. We introduce some new notation to try to prevent this.

Definition 55.2. *If* $g : \mathbb{R} \times \mathbb{R} \to \mathbb{C}$ *is such that* $g(\ ,t) \in L^1 \cap C$ *for some given* $t \in \mathbb{R}$, *we write*

$$(\mathcal{F}_1 g)(\zeta, t) = \int_{-\infty}^{\infty} g(x,t) \exp{-i\zeta x}\,dx.$$

(Thus $\mathcal{F}_1 g(\ ,t)$ is the Fourier transform of $g(\ ,t)$ for each fixed t.)
We also write

$$(D_1 \phi)(x,t) = \frac{\partial \phi}{\partial x}(x,t)$$

and

$$(D_2 \phi)(x,t) = \frac{\partial \phi}{\partial t}(x,t).$$

In what follows we shall try to find solutions to the heat equation $D_2 \phi = K D_1 D_1 \phi$. Although our arguments can be justified (provided we impose certain restrictions on ϕ) our primary concern is to guess a solution whose correctness can then be verified by direct substitution.

We start by taking the $\mathcal{F}_1$ transform of each side of the heat equation to get

$$(\mathcal{F}_1(D_2 \phi))(\zeta, t) = K(\mathcal{F}_1(D_1 D_1 \phi))(\zeta, t).$$

But by Lemma 55.1(ii) we have (at least if ϕ is sufficiently well behaved)

$$\mathcal{F}_1(D_1 D_1 \phi)(\zeta, t) = (i\zeta)^2 (\mathcal{F}_1 \phi)(\zeta, t),$$

whilst (assuming that we can differentiate under the integral sign)

$$(\mathcal{F}_1(D_2 \phi))(\zeta, t) = \int_{-\infty}^{\infty} \frac{\partial \phi}{\partial t}(x,t) \exp(-i\zeta x)\,dx = \int_{-\infty}^{\infty} \frac{\partial}{\partial t}(\phi(x,t) \exp(-i\zeta x))\,dx$$

$$= \frac{\partial}{\partial t} \int_{-\infty}^{\infty} \phi(x,t) \exp(-i\zeta x)\,dx = (D_2(\mathcal{F}_1 \phi))(\zeta, t).$$

Thus (if ϕ is sufficiently well behaved)

$$\frac{\partial}{\partial t}\mathcal{F}_1\phi(\zeta,t) = -K\zeta^2\mathcal{F}_1\phi(\zeta,t).$$

Now the equation

$$u'(t) = -K\zeta^2 u(t)$$

has the general solution $u(t) = A(\zeta)\exp(-K\zeta^2 t)$ where $A(\zeta)$ is an arbitrary constant so

$$(\mathcal{F}_1\phi)(\zeta,t) = A(\zeta)\exp(-K\zeta^2 t)$$

for some function $A:\mathbb{R}\to\mathbb{C}$.

If ϕ is sufficiently well behaved, for example if

$$\int_{-\infty}^{\infty}|\phi(x,t)-\phi(x,0)|dx\to 0 \quad \text{as} \quad t\to 0+,$$

then

$$(\mathcal{F}_1\phi)(\zeta,t)\to\mathcal{F}_1\phi(\zeta,0) \quad \text{as} \quad t\to 0+.$$

Thus, since

$$\exp(-K\zeta^2 t)\to 1 \quad \text{as} \quad t\to 0+,$$

we have

$$(\mathcal{F}_1\phi)(\zeta,0) = A(\zeta),$$

and so

$$(\mathcal{F}_1\phi)(\zeta,t) = (\mathcal{F}_1\phi)(\zeta,0)\exp(-K\zeta^2 t).$$

Let us now fix t and write $F(x) = \phi(x,t)$, $G(x) = \phi(x,0)$ so that our equation becomes

$$\hat{F}(\zeta) = \hat{G}(\zeta)\exp(-K\zeta^2 t),$$

i.e.

$$\hat{F}(\zeta) = (2\pi)^{\frac{1}{2}}\hat{G}(\zeta)E((2Kt)^{\frac{1}{2}}\zeta),$$

where $E(x) = (2\pi)^{-\frac{1}{2}}\exp(-x^2/2)$. We know from Lemma 50.2(i) that $\hat{E}(\zeta) = (2\pi)^{\frac{1}{2}}E(\zeta)$ and from Lemma 55.1(i) that $\hat{E}((2Kt)^{\frac{1}{2}}\zeta) = \hat{E}_{1/\sqrt{(2Kt)}}(\zeta)$, and so

$$\hat{F}(\zeta) = \hat{G}(\zeta)\hat{E}_{1/\sqrt{(2Kt)}}(\zeta).$$

Thus, using the convolution formula $(f*g)\hat{\ }(\zeta) = \hat{f}(\zeta)\hat{g}(\zeta)$ obtained in Theorem 51.7(iv), we have

$$\hat{F}(\zeta) = (G*E_{1/\sqrt{(2Kt)}})\hat{\ }(\zeta),$$

and, by the uniqueness of the Fourier transform (Theorem 49.4), it follows that

$$F(x) = (G*E_{1/\sqrt{(2Kt)}})(x),$$

i.e.

$$\phi(x,t) = (G*E_{1/\sqrt{(2Kt)}})(x).$$

If we go back through the argument above dotting the *i*s and crossing the *t*s we obtain a theorem along the following lines. (We write $\mathbb{R}^+ = \{t\in\mathbb{R}:t>0\}$.)

Theorem 55.3. *Let $G \in L^1 \cap C$. Suppose $\phi : \mathbb{R} \times \mathbb{R}^+ \to \mathbb{C}$ is such that $D_1\phi$, $D_1D_1\phi$ and $D_2\phi$ exist with*

(i) $\phi(\ ,t), D_1\phi(\ ,t), D_1D_1\phi(\ ,t), D_2\phi(\ ,t) \in L^1 \cap C$ and
(ii) $\phi(x,t), D_1\phi(x,t) \to 0$ as $|x| \to \infty$ for each $t \in \mathbb{R}^+$.

Then if

(iii) $\partial\phi/\partial t(x,t) = K(\partial^2\phi/\partial t^2)(x,t)$ for each $(x,t) \in \mathbb{R} \times \mathbb{R}^+$ and
(iv) $\int_{-\infty}^{\infty} |\phi(x,t) - G(x)| dx \to 0$ as $t \to 0+$

it follows that

$$\phi(x,t) = (G * E_{1/\sqrt{(2Kt)}})(x)$$

$$= \frac{1}{2\sqrt{(\pi Kt)}} \int_{-\infty}^{\infty} G(x-y)\exp(-y^2/2Kt)\,dy \text{ for all } (x,t) \in \mathbb{R} \times \mathbb{R}^+.$$

Proof. This is left to the reader and will enable her to review the argument above. ∎

Theorem 55.3 is a uniqueness theorem of a sort but too much must not be read into it. We shall return to the question of uniqueness in Chapter 67. For the moment we shall concentrate on verifying that $\phi(x,t) = (G * E_{1/\sqrt{(2Kt)}})(x)$ is indeed a solution for the heat equation on $\mathbb{R} \times \mathbb{R}^+$ subject to the initial condition $\phi(x,t) \to G(x)$ as $t \to 0$.

The methods used in obtaining Theorem 55.3 might lead us to suppose that G has to be restricted to lie in $L^1 \cap C$ but the verification below shows that a much wider choice of G is possible.

Theorem 55.4. *Suppose that $G : \mathbb{R} \to \mathbb{C}$ is continuous and bounded. Then, if*

$$\phi(x,t) = G * E_{1/\sqrt{(2Kt)}}(x) \quad [t > 0],$$

it follows that ϕ is an infinitely differentiable function on $\mathbb{R} \times \mathbb{R}^+$ with

(i) $\partial\phi/\partial t(x,t) = K(\partial^2\phi/\partial x^2)(x,t)$,
(ii) $\phi(x,t) \to G(x)$ as $t \to 0+$.

Proof. Set

$$f(x,y,t) = G(y)2^{-1}(\pi Kt)^{-\frac{1}{2}}\exp(-(x-y)^2/4Kt),$$

so that $\phi(x,t) = \int_{-\infty}^{\infty} f(x,y,t)\,dy$. Our proof splits into three parts.
Step 1. We first use Theorem 53.5 to show that ϕ is infinitely differentiable. To do this we prove that

$$\frac{\partial^n}{\partial t^n}\frac{\partial^m}{\partial x^m} f(x,\ ,t) \in L^1 \cap C$$

for each fixed $x \in \mathbb{R}$, $t \in \mathbb{R}^+$. Observe first that a simple induction shows that

$$\frac{\partial^n}{\partial t^n}\frac{\partial^m}{\partial x^m} f(x,y,t) = G(y)P_{nm}(t,y)\exp(-(x-y)^2/4Kt),$$

where P_{nm} is a polynomial in y whose coefficients are continuous functions of t (i.e. $P_{nm}(t, y) = \sum_{j=0}^{N(n,m)} a_{nmj}(t) y^j$).

It follows that, if t is in a fixed interval $[T_1, T_2]$ with $T_2 > T_1 > 0$,

$$\exp(-|y|)P_{nm}(t, y) \to 0 \text{ uniformly as } |y| \to \infty.$$

But $2|y| - ((x-y)^2/4Kt) \to -\infty$ uniformly as $|y| \to \infty$ if x and t are kept within fixed intervals, $[T_1, T_2]$ and $[X_1, X_2]$, so

$$\exp(2|y|)\exp(-(x-y)^2/4Kt) \to 0$$

and

$$\exp|y| \frac{\partial^n}{\partial t^n} \frac{\partial^m}{\partial x^m} f(x, y, t) \to 0 \quad \text{as} \quad |y| \to \infty$$

for fixed $(x, t) \in \mathbb{R} \times \mathbb{R}^+$. We deduce that

$$\left| \frac{\partial^n}{\partial t^n} \frac{\partial^m}{\partial x^m} f(x, y, t) \right| \leqslant \exp - |y|$$

for large $|y|$ and so by the comparison theorem for integrals

$$\frac{\partial^n}{\partial t^n} \frac{\partial^m}{\partial x^m} f(x,\ , t) \in L^1 \cap C$$

and

$$\int_{|y| > R} \frac{\partial^n}{\partial t^n} \frac{\partial^m}{\partial x^m} f(x, y, t) \, dy \to 0$$

as $R \to \infty$ uniformly for $t \in [T_1, T_2]$, $x \in [X_1, X_2]$.

Repeated use of Theorem 53.5 now shows that ϕ is infinitely differentiable with

$$\frac{\partial^n}{\partial t^n} \frac{\partial^m}{\partial x^m} \phi(x, t) = \int_{-\infty}^{\infty} \frac{\partial^n}{\partial t^n} \frac{\partial^m}{\partial x^m} f(x, y, t) \, dy.$$

Step 2. Having shown that we can differentiate under the integral sign, it is now only a matter of direct calculation to check that ϕ satisfies the heat equation. We have

$$\frac{\partial \phi}{\partial t}(x, t) - K \frac{\partial^2 \phi}{\partial x^2}(x, t) = \int_{-\infty}^{\infty} \left(\frac{\partial f}{\partial t}(x, y, t) - K \frac{\partial^2 f}{\partial x^2}(x, y, t) \right) dy,$$

$$\frac{\partial f}{\partial t}(x, y, t) = \left(-\frac{1}{2t^{3/2}} + \frac{(x-y)^2}{4Kt^{5/2}} \right) \left(\frac{G(y)}{2(\pi K)^{1/2}} \exp\left(-\frac{(x-y)^2}{4Kt} \right) \right)$$

$$\frac{\partial f}{\partial x}(x, y, t) = \left(-\frac{(x-y)}{2Kt^{3/2}} \right) \left(\frac{G(y)}{2(\pi K)^{1/2}} \exp\left(-\frac{(x-y)^2}{4Kt} \right) \right)$$

$$\frac{\partial^2 f}{\partial x^2}(x, y, t) = \left(-\frac{1}{2Kt^{3/2}} + \frac{(x-y)^2}{4K^2t^{5/2}} \right) \left(\frac{G(y)}{2(\pi K)^{1/2}} \exp\left(-\frac{(x-y)^2}{4Kt} \right) \right),$$

so that

$$\frac{\partial f}{\partial t} - K \frac{\partial^2 f}{\partial x^2} = 0 \quad \text{and} \quad \frac{\partial \phi}{\partial t} - K \frac{\partial^2 \phi}{\partial x^2} = 0 \text{ in } \mathbb{R} \times \mathbb{R}^+$$

as required.

Step 3. All that remains is to prove that $\phi(x, t) \to G(x)$ as $t \to 0 +$. But here we are on familiar ground since E has the following properties

(i) $E(s) \geqslant 0$ for all $s \in \mathbb{R}$,
(ii) $sE(s) \to 0$ as $|s| \to \infty$,
(iii) $\int_{-\infty}^{\infty} E(s)ds = 1$ (by Lemma 48.10),

so by a chain of reasoning summarised in Theorem 51.5

$$G * E_{1/\sqrt{(2Kt)}}(x) \to G(x) \quad \text{as } t \to 0 +. \qquad \blacksquare$$

We have just seen another confirmation of Slogan 52.8. The function $E_{1/\sqrt{(2Kt)}}(x)$ is a solution of the heat equation and the convolution $G * E_{1/\sqrt{(2Kt)}}$ inherits this good property. It is easy to generalise our particular example.

Lemma 55.5. *Let* $\psi : \mathbb{R}^n \to C$ *be twice differentiable with*

$$\psi(\ , x_2, x_3, \ldots, x_n), D_i\psi(\ , x_2, x_3, \ldots, x_n),$$
$$D_i D_j \psi(\ , x_2, x_3, \ldots, x_n) \in L^1 \cap C \quad [1 \leqslant i, j \leqslant n]$$

and let $G : \mathbb{R} \to C$ *be a bounded continuous function. Then if*

$$\sum_{j=1}^{n} \sum_{i=1}^{n} a_{ij} D_i D_j \psi = 0$$

and we write

$$\phi(x_1, x_2, \ldots, x_n) = \int_{-\infty}^{\infty} \psi(x_1 - y, x_2, x_3, \ldots, x_n) G(y) dy,$$

it follows that ϕ *also satisfies the equation*

$$\sum_{j=1}^{n} \sum_{i=1}^{n} a_{ij} D_i D_j \phi = 0.$$

Proof. Left as a recommended exercise for the reader. It need hardly be added that this lemma is not given as an end in itself but as an example of the kind of thing that can be done, to be modified and improved as the occasion demands. $\blacksquare$

In the same vein we leave it to the reader to check the steps in the following simple extension of Theorem 55.4 to the case when G has a finite number of simple discontinuities.

Lemma 55.6. (i) *Let* $g: \mathbb{R} \times \mathbb{R}^+ \to \mathbb{C}$ *be a continuous function such that* g_2 *exists and is continuous. Suppose* $\int_c^\infty |g(x,t)| dx$ *and* $\int_c^\infty |g_2(x,t)| dx$ *exist for each t and that* $\int_R^\infty |g_2(x,t)| dx \to 0$ *as* $R \to \infty$ *uniformly in t on each* $[a,b]$. *Then* $\int_c^\infty g(x,) dx$ *is differentiable with*

$$\frac{d}{dt} \int_c^\infty g(x,t) dx = \int_c^\infty g_2(x,t) dx.$$

(ii) *If* $H_c(x) = 1$ *for* $x > c$, $H_c(x) = 0$ *otherwise then, writing*

$$\psi_c(x,t) = H_c * E_{1/\sqrt{(2Kt)}}(x),$$

i.e. $$\psi_c(x,t) = \frac{1}{2(\pi Kt)^{\frac{1}{2}}} \int_c^\infty \exp\left(-\frac{(x-y)^2}{4Kt}\right) dy,$$

we know that ψ_c *is an infinitely differentiable function with*
  (a) $(\partial \psi_c/\partial t)(x,t) = K(\partial^2 \psi_c/\partial x^2)(x,t)$,
  (b) $\psi(x,t) \to 1 \quad$ *for* $x > c$,
      $\psi(c,t) \to \frac{1}{2}$,
and
      $\psi(x,t) \to 0 \quad$ *for* $x < c$,
  as $t \to 0+$.

(iii) *Let* $G: \mathbb{R} \to \mathbb{C}$ *be a bounded function, continuous except possibly at* $c(1)$, $c(2), \ldots, c(n)$ *and left and right continuous everywhere. Then if*

$$\phi(x,t) = G * E_{1/\sqrt{(2Kt)}}(x) \quad [t > 0],$$

it follows that ϕ *is an infinitely differentiable function on* $\mathbb{R} \times \mathbb{R}^+$ *with*

  (a) $(\partial \phi/\partial t)(x,t) = K(\partial^2 \phi/\partial x^2)(x,t)$,
  (b) $\phi(x,t) \to G(x)$ *for* $x \notin \{c(1), c(2), \ldots, c(n)\}$,

  $$\phi(c(j),t) \to \lim_{h \to 0+} (G(c(j) + h) + G(c(j) - h))/2 \quad [1 \leq j \leq n]$$

as $t \to 0+$.

Proof. (i) Use the proof of Theorem 53.5 as a model.

(ii) Use the proof of Theorem 55.4 as a model.
(iii) Let us write

$$\lambda(j) = \lim_{h \to 0+} (G(c(j) + h) - G(c(j) - h))/2 \quad [1 \leq j \leq n].$$

Then, taking

$$F(x) = G(x) - \sum_{j=1}^n \lambda(j) H_{c(j)}(x) \quad \text{for} \quad x \notin \{c(1), c(2), \ldots, c(n)\}$$

and $$F(c(j)) = \lim_{h \to 0+} F(c(j) - h),$$

we see that F is a continuous bounded function. We now observe that

$$G = F + \sum_{j=1}^{n} \lambda(j) H_{c(j)},$$

so that

$$G * E_{1/\sqrt{(2Kt)}} = F * E_{1/\sqrt{(2Kt)}} + \sum_{j=1}^{n} \lambda(j) \psi_{c(j)},$$

and so applying Theorem 55.4 to $F * E_{1/\sqrt{(2Kt)}}$ and part (ii) to the $\psi_{c(j)}$ we obtain the stated result. ∎

Later on we shall need an explicit formula for $\partial \psi_c / \partial t$.

Lemma 55.7. (i) *If $f : \mathbb{R} \to \mathbb{C}$ is twice differentiable with $f, f' \in L^1 \cap C$ and $f(x) \to 0$ as $x \to -\infty$ then $H_0 * f$ is differentiable with $(H_0 * f)'(x) = f(x)$.*
(ii) *If ψ_0 is as in Lemma 55.6 then*

$$\frac{\partial \psi_0}{\partial t}(x, t) = -\frac{x}{4(\pi K)^{\frac{1}{2}} t^{3/2}} \exp\left(-\frac{x^2}{4Kt}\right).$$

Proof. (i) By the methods of Lemma 55.6, $H_0 * f$ is differentiable with $(H_0 * f)' = H_0 * f'$. Thus

$$(H_0 * f)'(x) = \int_{-\infty}^{\infty} H_0(x - t) f'(t)\, dt = \int_{-\infty}^{x} f'(t)\, dt = f(x).$$

(ii) By Lemma 55.6 (i) and part (i)

$$\frac{\partial \psi_0}{\partial t}(x, t) = K \frac{\partial^2 \psi_0}{\partial x^2}(x, t) = K \frac{\partial^2}{\partial x^2}(H_0 * E_{1/\sqrt{(2Kt)}})(x) = K \frac{\partial}{\partial x} E_{1/\sqrt{(2Kt)}}(x)$$

$$= K \frac{\partial}{\partial x}\left(\frac{1}{2(\pi Kt)^{\frac{1}{2}}} \exp\left(-\frac{x^2}{4Kt}\right)\right) = \frac{-x}{4(\pi K)^{\frac{1}{2}} t^{3/2}} \exp\left(-\frac{x^2}{4Kt}\right). \quad ∎$$

56

THE AGE OF THE EARTH I

It is a very natural assumption that the earth and the species on it, including man, were all created at the same time. The age of the earth would then be approximately the length of man's recorded history, a few thousand years, and its geography and landscape essentially unchanged since the beginning.

However in the first decades of the nineteenth century geological evidence for great changes in the past began to build up. Large areas of land had once been under water, mountain ranges had been thrown up from lowlands and the evidence of fossils showed the past existence of species with no living counterparts.

At first it seemed that such vast changes must have been the outcome of tremendous and violent forces. According to this view the world that we live in has been shaped by a series of cataclysms (the last of which was often identified with Noah's flood) and the age of the earth must be reckoned in tens of thousands of years.

However Lyell in his *Principles of Geology* presented detailed and convincing arguments for an opposing view. He sought to 'explain the former changes by causes now in operation'. According to his theory processes such as slow erosion by wind and water, gradual deposition of sediment by rivers and the cumulative effect of earthquakes and volcanic action combined over very long periods of time to produce the vast changes recorded in the earth's surface.

Lyell's theory was called 'uniformitarian' because it rejected the idea that geological processes had been more violent in the past than they are today. But the rejection of vast forces was only made possible by the acceptance of vast times. The rate of action of very slow forces is, obviously, very hard to measure but uniformitarian theories demanded that the age of the earth be measured certainly in terms of millions of years, probably in hundreds of millions, perhaps in thousands of millions of years or more.

Lyell was unable to produce a uniformitarian biological theory to complement his uniformitarian geological theory. He was able to account for the disappearance of species in the geological record but not for the emergence of new species. A

282

solution to this problem was provided by Darwin with his theory of evolution by natural selection.

Darwin's theory too, required a geat age for the earth. 'He who can read Sir Charles Lyell's grand work on the Principles of Geology,... yet does not admit how incomprehensible vast has been the past period of time, may at once close this volume' (Darwin, *On the Origin of Species*, First Edition, *p.* 282). To allow time for natural selection to operate, the age of the earth must be measured in many hundreds of millions of years.

But such demands for endless time run counter to the laws of thermodynamics. Every day the sun radiates immense amounts of energy. By the law of conservation of energy there must be some source of this energy and when the source is exhausted the process must cease. Kelvin, as one of the founders of thermodynamics, was fascinated by this problem. Simple calculation shows that chemical processes (such as in the burning of coal) are totally insufficient as a source and he was forced to conclude that the only available mechanism was the conversion of gravitational potential energy into heat as the sun contracted.

On this assumption his calculations showed that '[it is] on the whole most probable that the sun has not illuminated the earth for 100 000 000 years and almost certain that [it] has not done so for 500 000 000 years'. Moreover, Kelvin believed that the sun was cooling so that the climate would become more extreme (and the associated geological processes more violent) as we go back in history.

However, Kelvin's most compelling argument, though still based on the principle of conservation of energy, concerned the earth rather than the sun. It is well known that the temperature of the earth increases with depth and

this implies a continual loss of heat from the interior, by conduction outwards through or into the upper crust. Hence, since the upper crust does not become hotter from year to year there must be a... loss of heat from the whole earth. It is possible that no cooling may result from this loss of heat but only an exhaustion of potential energy which in this case could scarcely be other than chemical.

But there is no reasonable mechanism to keep a chemical reaction going at a steady pace for millions of years so, Kelvin concludes, 'the... view... that the earth is merely a warm chemically inert body cooling, is clearly to be preferred in the present state of science'. (The quotations are from Appendices D and E of Thomson [Kelvin] and Tait '*Treatise on Natural Philosophy*'.)

However high the initial temperature of the earth, geological history could only begin when the outer crust had solidified. Did the earth solidify first from the outside leaving a liquid core or did convection currents ensure that the earth solidified at a uniform temperature throughout? Many early Victorian geologists pictured the earth as a molten mass on which floated a thin skin of solidified rock and this plausible picture survives in the popular mind today. However, Kelvin and others had already shown that such a picture was dynamically inconsistent

with known facts. (Indeed, Kelvin's computations revealed a body with the rigidity of a steel ball of similar size.) The rejection of this model together with the fact that molten rock contracts on solidifying led Kelvin to believe that the earth was a solid body and that it had solidified at a more or less uniform temperature. In the next chapter we shall see how, starting from this hypothesis, he was able to estimate the age of the earth.

57

THE AGE OF THE EARTH II

Let us examine the mathematical consequences of Kelvin's model of a cooling earth. We take the earth to be the sphere $\{\mathbf{x}:|\mathbf{x}| < R\}$ and write the temperature at a point $\mathbf{x}$ and time t as $\psi(\mathbf{x}, t)$. Assuming the composition of the earth to be, roughly, homogeneous the temperature distribution will be governed, roughly, by Fourier's heat equation

$$\frac{\partial \psi}{\partial t}(\mathbf{x}, t) = K\nabla^2 \psi(\mathbf{x}, t).$$

We have already decided to take the initial temperature (at time $t = 0$ say) to be constant,
   i.e.

$$\psi(\mathbf{x}, 0) = \theta_0 \quad \text{for all } |\mathbf{x}| < R,$$

but we still have to decide on our other boundary condition. Kelvin gives several convincing arguments for the choice

$$\psi(\mathbf{x}, t) = C \quad \text{for all } |\mathbf{x}| = R, t > 0,$$

where C is a constant and it is, of course, part of the uniformitarian position that the temperature of the earth's surface has not fluctuated very much during geological time. (This is even more important to the biological than to the geological theory. If the climate became too hot or too cold life would be wiped out.) Choosing a suitable temperature scale we may simplify the last equation by setting $C = 0$ to obtain

$$\psi(\mathbf{x}, t) = 0 \quad \text{for all } |\mathbf{x}| = R, t > 0.$$

The equations for a cooling sphere set out above are not difficult to solve and Kelvin had already solved them in a different context. But preliminary calculations show that the cooling effect for the earth can still be only skin deep. (That is to say that, at depths small compared with the radius of the earth, $\psi(\mathbf{x}, t)$ is, at the present time, very close to its initial value ψ_0.) Under these circumstances the

mathematics becomes easier to interpret if we neglect the curvature of the earth. Writing $\theta(y, t)$ for the temperature of the earth at depth y and time t our problem then reduces to the one dimensional problem

$$\frac{\partial \theta}{\partial t}(y, t) = K \frac{\partial^2 \theta}{\partial y^2}(y, t) \quad \text{for all } y > 0, t > 0,$$

$$\theta(y, 0) = \theta_a \qquad\qquad \text{for all } y > 0,$$

$$\theta(0, t) = 0 \qquad\qquad \text{for all } t > 0.$$

We shall solve these equations by using the results of Chapter 55 together with one of Kelvin's favourite mathematical tricks – the method of images. A vivid illustration of the method of images is given by the following example. Suppose we wish to find the behaviour of a plume of smoke from a tall factory chimney set in the middle of a large plane (see Figure 57.1). In addition to the general equations describing the behaviour of the smoke we must satisfy a boundary condition which states that smoke cannot pass through the ground.

To get round this problem we consider a reflected factory chimney producing a reflected plume of smoke as shown (Figure 57.2). There will then be no net flow

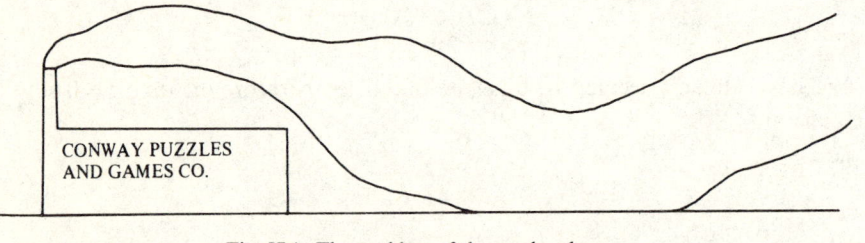

CONWAY PUZZLES
AND GAMES CO.

Fig. 57.1. The problem of the smoke plume.

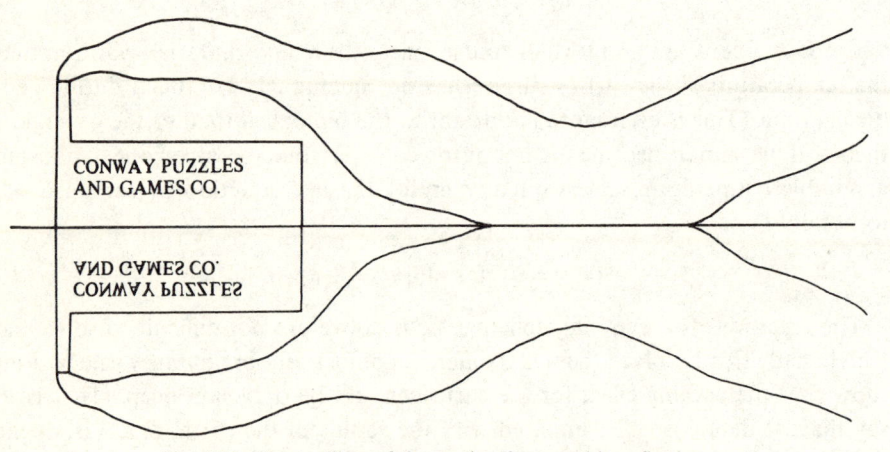

CONWAY PUZZLES
AND GAMES CO.

ᴠɴD ᴄᴠɯᴇꙅ ᴄᴏˑ
ᴄᴏɯɯᴠʎ ᗤɒꙅꙅꞀᴇꙅ

Fig. 57.2. The problem of the smoke plume and its reflection.

of smoke across the plane of reflection and the upper half of the solution to the new problem will correspond to the solution of our original problem.

Using a similar idea we obtain the following result.

Lemma 57.1. *Let $\theta_0 \in \mathbb{R}$ and*

$$\theta(y, t) = \frac{\theta_0}{2\sqrt{(\pi Kt)}} \int_0^\infty \left\{ \exp\left(-\frac{(y-w)^2}{4Kt} \right) - \exp\left(-\frac{(y+w)^2}{4Kt} \right) \right\} dw$$

for all $y \geq 0$, $t > 0$. Then $\theta : [0, \infty) \times (0, \infty) \to \mathbb{R}$ is an infinitely differentiable function with

(i) $(\partial\theta/\partial t)(y, t) = K(\partial^2\theta/\partial y^2)(y, t)$,
(ii) $\theta(y, t) \to \theta_0$ as $t \to 0+$ for all $y > 0$,
(iii) $\theta(0, t) = 0$ for all $t > 0$.

Proof. Let

$$G(x) = \theta_0 \quad \text{for } x > 0, G(0) = 0$$

and

$$G(x) = -\theta_0 \quad \text{for } x < 0.$$

Then, taking

$$\phi(x, t) = G * E_{1/\sqrt{(2Kt)}}(x) \quad \text{for } x \in \mathbb{R}, t > 0,$$

we know from Lemma 55.6 that ϕ is an infinitely differentiable function on $\mathbb{R} \times \mathbb{R}^+$ with

(i)′ $\partial\phi/\partial t = K(\partial^2\phi/\partial x^2)$,
(ii)′ $\phi(x, t) \to G(x)$ for all $x \in \mathbb{R}$ as $t \to 0+$.

By symmetry we have $\phi(x, t) = -\phi(-x, t)$ for all $x \in \mathbb{R}$ and so in particular

(iii)′ $\phi(0, t) = 0$ for all $t > 0$.

We observe that

$$\phi(x, t) = \frac{1}{2\sqrt{(\pi Kt)}} \int_{-\infty}^\infty G(w) \exp\left(-\frac{(x-w)^2}{4Kt} \right) dw$$

$$= \frac{\theta_0}{2\sqrt{(\pi Kt)}} \left(\int_0^\infty \exp\left(-\frac{(x-w)^2}{4Kt} \right) dw - \int_{-\infty}^0 \exp\left(-\frac{(x-w)^2}{4Kt} \right) dw \right)$$

$$= \frac{\theta_0}{2\sqrt{(\pi Kt)}} \left(\int_0^\infty \exp\left(-\frac{(x-w)^2}{4Kt} \right) - \exp\left(-\frac{(x+w)^2}{4Kt} \right) dw \right),$$

and so by inspection ϕ is real valued. Setting $\theta(y, t) = \phi(y, t)$ for all $y \geq 0$, $t > 0$ the stated result follows. ∎

Ignoring questions of uniqueness as physically irrelevant we have thus obtained the temperature of the earth at depth y and time t in the form

$$\theta(y, t) = \frac{\theta_0}{2\sqrt{(\pi Kt)}} \left(\int_0^\infty \exp\left(-\frac{(y-w)^2}{4Kt} \right) - \exp\left(-\frac{(y+w)^2}{4Kt} \right) dw \right).$$

We can make no direct measurements of the earth's temperature at great depths but we can measure the rate of increase of underground temperature as we mine deeper into the earth's crust. Our interest is therefore in

$$\frac{\partial \theta}{\partial y}(0, t) = \left[\frac{\theta_0}{2\sqrt{(\pi K t)}} \int_0^\infty -\frac{2(y-w)}{4Kt} \exp\left(-\frac{(y-w)^2}{4Kt} \right) \right.$$

$$\left. + \frac{2(y+w)}{4Kt} \exp\left(-\frac{(y+w)^2}{4Kt} \right) dw \right]_{y=0}$$

$$= \frac{\theta_0}{2\sqrt{(\pi K t)}} \int_0^\infty \frac{w}{Kt} \exp\left(-\frac{w^2}{4Kt} \right) dw$$

$$= \frac{\theta_0}{\sqrt{(\pi K t)}} \left[-\exp\left(-\frac{w^2}{4Kt} \right) \right]_0^\infty = \frac{\theta_0}{\sqrt{(\pi K t)}}$$

(so that, writing $v = (\partial \theta / \partial y)(0, t)$, we have

$$t = \theta_0^2 (\pi K)^{-1} v^{-2}).$$

(This result could also have been obtained using Lemma 55.7 (ii).)

If the earth is indeed more or less homogeneous then the conductivity K and melting point θ_0 can be determined by measurements on samples of surface rock whilst measurements in mines or special borings will give v. Taking the best available measurements Kelvin arrived at an estimate of 100 000 000 years as the age of the earth. Even allowing for uncertainties in his data, he felt confident that the age of the earth could not be more than 400 000 000 years old and his calculations of the sun's age referred to in the last chapter confirmed that an age of 100 000 000 years was more likely.

58

THE AGE OF THE EARTH III

The reception of Kelvin's calculations is described by J.D. Burchfield in his very interesting book *Lord Kelvin and The Age of the Earth* (Macmillan, 1975) on which this chapter is based. Before Kelvin few geologists had thought deeply about the problem of geological time partly from lack of data but mainly perhaps because the human mind finds it hard to contrast a vast time of ten million years with a vast time of a thousand million years.

As the force of Kelvin's argument sank in, geologists began to examine their uniformitarian theories more critically. Certainly if the rate of geological change had been the same throughout geological history as that observed today, 100 000 000 years would not suffice to shape the present state of the earth's surface. But as Kelvin himself pointed out, if the earth is losing energy then in earlier times geological forces like earthquakes and volcanoes were probably much more violent. Thus by modifying uniformitarian theory so as to allow past geological forces, though still the same in kind, to be different in degree from those observed today it was possible to speed up the rate of geological change in the past and compress the earth's geological history into Kelvin's 100 000 000 years.

The problems posed to Darwin's theory of evolution by natural selection were more serious. 'I am greatly troubled at the short duration of the world according to [Kelvin], for I require for my theoretical views a very long period before the Cambrian formation (Darwin quoted in Burchfield, p. 75).' In the fifth edition of *The Origin of Species*, Darwin attempted to adjust to the new time scale by allowing greater scope for evolution by processes other than natural selection. He also pointed out that more extreme geological conditions might increase the pressures of natural selection and so increase the rate of evolution.

But in the end he was forced to ask for a suspension of judgement. In the final chapter of *The Origin of Species* in which Darwin reviewed the argument of his book he now added the following words.

With respect to the lapse of time not having been sufficient since our planet was consolidated for the assumed amount of organic change, and this objection, as

289

argued by [Kelvin], is probably one of the gravest yet advanced, I can only say, firstly that we do not know at what rate species change as measured by years, and secondly, that many philosophers are not as yet willing to admit that we know enough of the constitution of the universe and of the interior of our globe to speculate with safety on its past duration.

(Darwin *The Origin of Species*, Sixth Edition, p. 409.)

Kelvin reinforced his case with a third argument drawn from the tidal retardation of the earth's rotation. But here he exposed the chief weakness of his arguments. Huxley was able to quote Laplace, Adams and earlier work of Kelvin himself as disagreeing both about the cause and the extent of this supposed phenomenon.

I do not presume (Huxley wrote) to throw the slightest doubt upon the accuracy of any of these calculations made by such distinguished mathematicians as those who have made the suggestions I have cited. On the contrary, it is necessary to my argument to assume that they are all correct. But I desire to point out that this seems to be one of the many cases in which the admitted accuracy of mathematical processes is allowed to throw a wholly inadmissible appearance of authority over the results obtained by them. Mathematics may be compared to a mill of exquisite workmanship, which grinds you stuff of any degree of fineness; but nevertheless, what you get out depends on what you put in; and as the grandest mill in the world will not extract wheat-flour from peascods, so pages of formulae will not get a definite result out of loose data.

(*Quarterly Journal of the Geological Society of London*, Vol. 25, 1869)

However Kelvin's estimates were the best available and for the next thirty years geology took its time from physics, and biology its time from geology. Tait only echoed the general opinion when he wrote: 'Let us then hear no more nonsense about the interference of mathematicians in matters with which they have no concern; rather let them be lauded for condescending from their proud preeminence to help out of the rut the too ponderous waggon of some scientific brother (quoted in Burchfield, p. 93.).'

But even as the geologists readjusted their time scales to fit Kelvin's first estimate for the age of the earth, he and his followers began to adjust it down until at the end of the nineteenth century the best physical estimates of the age of the earth and sun were about 20 million years whilst the minimum the geologists could allow was closer to Kelvin's original 100 million years.

Then in 1904 Rutherford announced that the radio-active decay of radium was accompanied by the release of immense amounts of energy and speculated that this could replace the heat lost from the surface of the earth. Kelvin's argument would then only give a minimum for the earth's age. 'The discovery of the radio-active elements... thus increases the possible limit of the duration of life on this planet, and allows the time claimed by the geologist and biologist for the process of evolution. (Rutherford quoted in Burchfield, p. 164.)'

Rutherford was fond of recounting the story of his lecture to the Royal Institution,

I came into the room which was half dark, and presently spotted Lord Kelvin in the audience and realised I was in for trouble at the last part of the speech dealing with the age of the earth, where my views conflicted with his. To my relief, Kelvin fell fast asleep, but as I came to the important point, I saw the old bird sit up, open an eye and take a baleful glance at me! Then a sudden inspiration came, and I said Lord Kelvin had limited the age of the earth, *provided no new source of heat was discovered*. That prophetic utterance refers to what we are now considering tonight, radium! Behold! the old boy beamed upon me.

(Eve *Rutherford*, Macmillan 1939, p. 107.)

A problem for the geologists was now replaced by a problem for the physicists. How could the principle of conservation of energy be reconciled with the immense release of energy which occurs in radioactive decay? The answer was provided by a theory which was just beginning to be gossiped about. 'One day... Rutherford began twitting Wien about relativity. Wien explained that Newton was wrong in the matter of relative motion, which was not the joint velocities $u + v$, but that expression, according to Einstein must be divided by $1 + uv/c^2$. Wien added, "But no Anglo Saxon can understand relativity!" "No!" laughed Rutherford, "they have too much good sense" (Eve, p. 193).' Einstein's theory of relativity extended the principle of conservation of energy by taking matter as a form of energy. It is the conversion of matter to heat which maintains the earth's internal temperature and supplies the energy radiated by the sun.

Radio-active dating now leads geologists to give the earth an age of about two or three thousand million years. Indeed they now have so much time that they believe that the present day must be a period of particularly intense geological activity. And in this way Darwin's theories have received a remarkable confirmation.

59

WEIERSTRASS'S PROOF OF
WEIERSTRASS'S THEOREM

Weierstrass was well acquainted with the solution of the heat equation obtained in Chapter 55.

Theorem 59.1. *Suppose that* $G: \mathbb{R} \to \mathbb{C}$ *is a uniformly continuous bounded function. Then if*

$$\phi(x, t) = G * E_{1/\sqrt{(2Kt)}}(x) \quad [t > 0],$$

it follows that ϕ *is an infinitely differentiable function on* $\mathbb{R} \times \mathbb{R}^+$ *with*

(i) $(\partial\phi/\partial t)(x, t) = K(\partial^2\phi/\partial x^2)(x, t)$,
(ii) $\phi(x, t) \to G(x)$ *uniformly in* x *as* $t \to 0+$.

Proof. As for Theorem 55.4, but making use of the fact that G is now uniformly continuous. ∎

Let us see how he made use of this kind of result to prove the theorem discussed in Chapter 4. (Recall that Weierstrass's discovery preceded Fejér's theorem by 20 years.)

Theorem 59.2 (Weierstrass). *If* $f: [a, b] \to \mathbb{C}$ *is continuous and* $\varepsilon > 0$, *we can find a polynomial P with*

$$\sup_{t \in [a,b]} |P(t) - f(t)| < \varepsilon.$$

Proof. We divide the proof into simple steps.
Step 1. Let $R = 1 + \max(|a|, |b|)$. We can certainly find $F: \mathbb{R} \to \mathbb{C}$ continuous with $F(t) = f(t)$ if $t \in [a, b]$ and $F(t) = 0$ if $|t| > R$. (For example, set $F(t) = f(t)$ if $t \in [a, b]$, $F(t) = f(b)(b + 1 - t)$ if $t \in [b, b + 1]$, $F(t) = f(a)(t - a + 1)$ if $t \in [a - 1, a]$ and $F(t) = 0$ otherwise.) Then F is automatically uniformly continuous and bounded (with $|F(t)| \leqslant M$ for all t, say) on the closed bounded interval $[-R, R]$ and so on $\mathbb{R}$. It follows that $F * E_{1/\eta} \to F$ uniformly on $\mathbb{R}$ as $\eta \to 0+$ (see Theorem 59.1 above) and so

292

we can find a $K > 0$, depending on ε, with $|F * E_K(t) - F(t)| < \varepsilon/2$ for all $t \in \mathbb{R}$.

Step 2. We know that $\sum_{r=0}^{n} x^r/r! \to \exp x$ uniformly on any interval $[-S, S]$. (Observe for example that

$$\left| \exp x - \sum_{r=0}^{n} x^r/r! \right| = \left| \sum_{r=n+1}^{\infty} x^r/r! \right| \leqslant \sum_{r=n+1}^{\infty} |x|^r/r! \leqslant (|x|^{n+1}/(n+1)!) \sum_{k=0}^{\infty} |x|^k/(n+1)^k$$

$$\leqslant |x|^{n+1}/((1 - |x|/(n+1))(n+1)!) \leqslant 2|S|^{n+1}/(n+1)!$$

whenever $|x| \leqslant S$ and $2S \leqslant n$.) It follows that writing

$$Q_{2n}(x) = K^{-1}(2\pi)^{-\frac{1}{2}} \sum_{r=0}^{n} (-K^2 x^2/2)^r/r!$$

we have

$$Q_{2n}(x) \to K^{-1}(2\pi)^{-\frac{1}{2}} \exp(-K^2 x^2/2) = E_K(x) \quad \text{uniformly on } [-2R, 2R].$$

Thus we can find an N such that

$$|Q_{2N}(x) - E_K(x)| \leqslant \varepsilon/((2M+1)(2R+1)) \quad \text{for all } |x| \leqslant 2R,$$

and so (using the fact that $F(x) = 0$ for all $|x| \geqslant R$) we have

$$|F * E_K(t) - F * Q_{2N}(t)| = |F * (E_K - Q_{2N})(t)|$$

$$= \left| \int_{-\infty}^{\infty} F(x)(E_K(t-x) - Q_{2N}(t-x)) dx \right| = \left| \int_{-R}^{R} F(x)(E_K(t-x) - Q_{2N}(t-x)) dx \right|$$

$$\leqslant 2R \sup_{x \in [-R,R]} |F(x)| \sup_{x \in [-R,R]} \sup_{s \in [-R,R]} |E_K(s-x) - Q_{2N}(s-x)|$$

$$\leqslant 2RM \sup_{y \in [-2R,2R]} |E_K(y) - Q_{2N}(y)| < \varepsilon/2 \quad \text{for all } t \in [-R, R].$$

Thus

$$|F * Q_{2N}(t) - f(t)| = |F * Q_{2N}(t) - F(t)|$$

$$\leqslant |F * Q_{2N}(t) - F * E_K(t)| + |F * E_K(t) - F(t)| < \varepsilon \quad \text{for all } t \in [a, b].$$

Step 3. Having shown that $F * Q_{2N}$ is a good uniform approximation to f on $[a, b]$, all that remains is to show that $F * Q_{2N}$ is a polynomial. By definition, Q_{2N} is a polynomial of degree $2N$ with $Q_{2N}(x) = \sum_{r=0}^{2N} a_r x^r$ say. Thus $Q_{2N}(t-x) = \sum_{r=0}^{2N} b_r(x) t^r$ where $b_r : \mathbb{R} \to \mathbb{C}$ is a polynomial of degree at most $2N$. It follows that

$$F * Q_{2N}(t) = \int_{-\infty}^{\infty} F(x) Q_{2N}(t-x) dx = \int_{-R}^{R} F(x) \sum_{r=0}^{2N} b_r(x) t^r dx$$

$$= \sum_{r=0}^{2N} \left(\int_{-R}^{R} F(x) b_r(x) dx \right) t^r,$$

as required. Setting $P = F * Q_{2N}$ we obtain the stated result. ∎

We notice in passing yet another confirmation of Slogan 52.7.

Lemma 59.3. *Suppose* $f: \mathbb{R} \to \mathbb{C}$ *is continuous and* $x^{n+1+\varepsilon} f(x) \to 0$ *as* $|x| \to \infty$ *for some* $\varepsilon > 0$. *Then, if* P_n *is a polynomial of degree at most* n, *so is* $P_n * f$.
Proof. Follow Step 3 above. The condition $x^{n+1+\varepsilon} f(x) \to 0$ as $|x| \to \infty$ ensures that $\int_{-\infty}^{\infty} |f(x) b_r(x)| \, dx$ converges. ∎

One of the readers of the manuscript has suggested the following faster alternative proof.
Sketch alternative proof. The function $g: \mathbb{R} \to \mathbb{C}$ is a polynomial of degree at most n if and only if $g^{(n+1)} = 0$. A slight strengthening of the arguments of Chapter 53 gives $(P_n * f)^{(n+1)} = P_n^{(n+1)} * f = 0 * f = 0$ so $P_n * f$ is a polynomial of degree at most n. ∎

Fejér discovered his theorem at the age of 19, Weierstrass published this theorem at the age of 70. With time the reader may come to appreciate why many mathematicians regard the second circumstance as even more romantic and heart warming than the first.

60

THE INVERSION FORMULA

In the first part of this book we saw that for sufficiently well behaved functions $f: \mathbb{T} \to \mathbb{C}$ we have

$$f(t) = \sum_{n=-\infty}^{\infty} \hat{f}(n) \exp int.$$

The analogous result for Fourier transforms was found by Fourier in the course of his researches into the heat equation and later, but independently, by Cauchy in the course of his researchers into water waves. Briefly, if $f: \mathbb{R} \to \mathbb{C}$ is sufficiently well behaved, then

$$f(t) = \frac{1}{2\pi} \int_{-\infty}^{\infty} \hat{f}(\zeta) \exp i\zeta t \, dt.$$

In other words

$$f(t) = (2\pi)^{-1} f^{\wedge \wedge}(-t)$$

or, writing, $\check{f}(t) = f(-t)$,

$$\check{f} = (2\pi)^{-1} f^{\wedge \wedge}.$$

To try and justify the formula one could argue roughly as follows. We know from results on Fourier series that if $g: (-\pi, \pi] \to \mathbb{C}$ is well behaved, then

$$g(t) = \sum_{n=-\infty}^{\infty} \frac{1}{2\pi} \int_{-\pi}^{\pi} g(x) \exp(-inx) \, dx \exp int$$

for all $t \in (-\pi, \pi]$. A simple change of variable argument shows that if $h: (-L\pi, L\pi] \to \mathbb{C}$ is well behaved, then

$$h(t) = \sum_{n=-\infty}^{\infty} \frac{1}{2\pi L} \int_{-L\pi}^{L\pi} h(x) \exp\left(-\frac{inx}{L}\right) dx \exp\frac{int}{L}$$

for all $t \in (-L\pi, L\pi)$ (set $g(s) = h(Ls)$ for $s \in (-\pi, \pi]$). It follows that if $f: \mathbb{R} \to \mathbb{C}$ is well behaved and $t \in (-\pi L, \pi L]$, then

$$f(t) = \sum_{n=-\infty}^{\infty} \frac{1}{2\pi L} \int_{-L\pi}^{L\pi} f(x) \exp\left(-\frac{inx}{L}\right) dx \exp\frac{int}{L}.$$

So far the argument is flawless but we must now head into deeper waters. If L is sufficiently large, we expect that

$$\int_{-L\pi}^{L\pi} f(x)\exp\left(-\frac{inx}{L}\right)dx \approx \int_{-\infty}^{\infty} f(x)\exp\left(-\frac{inx}{L}\right)dx = \hat{f}\left(\frac{n}{L}\right),$$

and so, with luck,

$$f(t) \approx (2\pi L)^{-1}\sum_{n=-\infty}^{\infty}\hat{f}(nL^{-1})\exp(int\,L^{-1}).$$

Taking $L = N$ a large integer we may rewrite our expression as

$$f(t) \approx (2\pi N)^{-1}\sum_{n=-\infty}^{\infty}\hat{f}(nN^{-1})\exp(int\,N^{-1})$$

$$= (2\pi N)^{-1}\sum_{r=-\infty}^{\infty}\sum_{k=0}^{N-1}\hat{f}((k+rN)N^{-1})\exp(i(k+rN)tN^{-1})$$

$$= (2\pi N)^{-1}\sum_{r=-\infty}^{\infty}\sum_{k=0}^{N-1}\hat{f}(kN^{-1}+r)\exp(i(kN^{-1}+r)t).$$

But for large N we know that

$$N^{-1}\sum_{k=0}^{N-1}\hat{f}(kN^{-1}+r)\exp(i(kN^{-1}+r)t) \approx \int_{r}^{r+1}\hat{f}(\zeta)\exp(i\zeta t)\,d\zeta,$$

and so, we hope

$$f(t) \approx (2\pi)^{-1}\sum_{r=-\infty}^{\infty}\int_{r}^{r+1}\hat{f}(\zeta)\exp(i\zeta t)\,d\zeta = \frac{1}{2\pi}\int_{-\infty}^{\infty}\hat{f}(\zeta)\exp(i\zeta t)\,dt.$$

If we take N larger and larger we may hope that the approximation will go on improving, yielding equality in the limit.

Suggestive though this argument is, it will only carry conviction to those who believe that the infinite sum of small things is always small (except when it is large). It is also rather hard to make the argument above rigorous. (I suggest that the reader spends half an hour trying this. Even if she does not succeed she will benefit from the experience. If she does succeed she is probably wasting her time reading this book.)

Fortunately we have another line of argument to hand. Pursuing the analogy with the methods of the first part of this book we use Fejér's theorem as the key to our inversion theorem.

Theorem 60.1 (The Inversion Theorem) *If $f \in L^1 \cap C$ and $\hat{f} \in L^1 \cap C$ then*

$$f(t) = \frac{1}{2\pi}\int_{-\infty}^{\infty}\hat{f}(\zeta)\exp i\zeta t\,d\zeta,$$

the integral being uniformly absolutely convergent.

Proof. Since

$$\left| \frac{1}{2\pi} \int_{-\infty}^{\infty} \hat{f}(\zeta) \exp i\zeta t \, d\zeta - \frac{1}{2\pi} \int_{-R}^{S} \hat{f}(\zeta) \exp i\zeta t \, d\zeta \right|$$

$$= \frac{1}{2\pi} \left| \int_{\zeta > S \text{ or } \zeta < -R} \hat{f}(\zeta) \exp i\zeta t \, d\zeta \right| \leqslant \frac{1}{2\pi} \int_{\zeta > S \text{ or } \zeta < -R} |\hat{f}(\zeta) \exp i\zeta t| \, d\zeta$$

$$= \frac{1}{2\pi} \int_{\zeta > S \text{ or } \zeta < -R} |\hat{f}(\zeta)| \, d\zeta \to 0 \quad \text{as } R, S \to \infty,$$

the uniform convergence of the integral is trivial and we can concentrate on showing that, indeed,

$$\frac{1}{2\pi} \int_{-\infty}^{\infty} \hat{f}(\zeta) \exp i\zeta t \, dt = f(t).$$

Write

$$\Delta_R(\zeta) = (1 - |\zeta|/R) \text{ for } |\zeta| \leqslant R, \Delta_R(\zeta) = 0 \text{ otherwise.}$$

We know from Theorem 49.3 (i) (Fejér's theorem for integrals) that

$$\frac{1}{2\pi} \int_{-\infty}^{\infty} \Delta_R(\zeta) \hat{f}(\zeta) \exp i\zeta t \, d\zeta \to f(t) \quad \text{as } R \to \infty.$$

If we can show also that

$$\frac{1}{2\pi} \int_{-\infty}^{\infty} \Delta_R(\zeta) \hat{f}(\zeta) \exp i\zeta t \, d\zeta \to \frac{1}{2\pi} \int_{-\infty}^{\infty} \hat{f}(\zeta) \exp i\zeta t \, d\zeta,$$

then the result will follow.

This is very easy and the reader might prefer to try and do it for herself. Let $\varepsilon > 0$ be given. Then since $\int_{-\infty}^{\infty} |\hat{f}(\zeta)| \, d\zeta$ converges we can find an $S > 0$ such that $\int_{\zeta > S} |\hat{f}(\zeta)| \, d\zeta < \varepsilon$. If now $R > \varepsilon^{-1} S$ we have

$$\left| \int_{-\infty}^{\infty} \hat{f}(\zeta) \exp i\zeta t \, d\zeta - \int_{-\infty}^{\infty} \Delta_R(\zeta) \hat{f}(\zeta) \exp i\zeta t \, d\zeta \right| = \left| \int_{-\infty}^{\infty} (1 - \Delta_R(\zeta)) \hat{f}(\zeta) \exp i\zeta t \, d\zeta \right|$$

$$\leqslant \int_{-\infty}^{\infty} |(1 - \Delta_R(\zeta)) \hat{f}(\zeta) \exp i\zeta t| \, d\zeta = \int_{-\infty}^{\infty} (1 - \Delta_R(\zeta)) |\hat{f}(\zeta)| \, d\zeta$$

$$= \int_{|\zeta| < S} (1 - \Delta_R(\zeta)) |\hat{f}(\zeta)| \, d\zeta + \int_{|\zeta| \geqslant S} (1 - \Delta_R(\zeta)) |\hat{f}(\zeta)| \, d\zeta$$

$$\leqslant \int_{|\zeta| < S} \frac{S}{R} |\hat{f}(\zeta)| \, d\zeta + \int_{|\zeta| \geqslant S} |\hat{f}(\zeta)| \, d\zeta \leqslant \varepsilon \int_{|\zeta| < S} |\hat{f}(\zeta)| \, d\zeta + \varepsilon$$

$$\leqslant \varepsilon \left(1 + \int_{-\infty}^{\infty} |\hat{f}(\zeta)| \, d\zeta \right).$$

Since ε was arbitrary we are done. ∎

The reader should give a proof of Theorem 9.1 by a similar method. The results and proofs of the next lemma run parallel to those of Lemma 9.5 and Theorem 9.6, whilst the theorem that follows runs parallel to Theorem 15.3 and Theorem 15.4.

Theorem 60.2. (i) *If $f:\mathbb{R}\to\mathbb{C}$ is n times differentiable with $f, f',...,f^{(n)}\in L^1\cap C$ and $f^{(r)}(x)\to 0$ as $|x|\to\infty$ for each $0\leqslant r\leqslant n-1$ then*

$$|\hat{f}(\zeta)|\leqslant A|\zeta|^{-n}\quad\text{for all }\zeta\neq 0,$$

where $A=\int_{-\infty}^{\infty}|f^{(n)}(x)|dx$.
(ii) *If $f:\mathbb{R}\to\mathbb{C}$ is twice differentiable with $f, f', f''\in L^1\cap C$ and $f(x), f'(x)\to 0$ as $|x|\to\infty$ then*

$$f(t)=\frac{1}{2\pi}\int_{-\infty}^{\infty}\hat{f}(\zeta)\exp i\zeta t\,d\zeta,$$

the integral being uniformly absolutely convergent.
Proof. (i) Repeated integration by parts.
(ii) By (i)$|\hat{f}(\zeta)|\leqslant A|\zeta|^{-2}$ for all $\zeta\neq 0$. Thus $\hat{f}\in L^1\cap C$ and Theorem 60.1 applies. ∎

Theorem 60.3. (i) *If $f\in L^1\cap C$ and $|\hat{f}(\zeta)|\leqslant A|\zeta|^{-1}$ for all $\zeta\neq 0$ and for some $A\geqslant 0$ then for each $t\in\mathbb{R}$*

$$\frac{1}{2\pi}\int_{-R}^{R}\hat{f}(\zeta)\exp i\zeta t\,d\zeta\to f(t)\quad\text{as }R\to\infty.$$

(ii) *If f is once differentiable with $f, f'\in L^1\cap C$ and $f'(x)\to 0$ as $x\to\infty$ then for each $t\in\mathbb{R}$*

$$\frac{1}{2\pi}\int_{-R}^{R}\hat{f}(\zeta)\exp i\zeta t\,d\zeta\to f(t)\quad\text{as }R\to\infty.$$

Remark. It should not be necessary to remind the reader that we cannot conclude the convergence of $\int_{-R}^{S}g(\zeta)d\zeta$ as $R, S\to\infty$ from the convergence of $\int_{-R}^{R}g(\zeta)d\zeta$ as $R\to\infty$. This limitation means that Theorem 60.3 must be used with extreme caution.
Proof. (i) The proof modelled on the proof of Theorem 15.3 is left as a recommended exercise to the reader.
(ii) By Lemma 60.2 (i), f satisfies the conditions of (i). ∎

We end this chapter with an observation which may be omitted by the impatient reader but which would form a nice revision exercise. In Lemma 49.2 (iii) we evaluated $A=\int_{-\infty}^{\infty}\tilde{K}_1(s)ds$ where

$$\tilde{K}_1(s)=\frac{1}{2\pi}\left(\frac{\sin s/2}{s/2}\right)^2$$

and in Lemma 48.10 we evaluated $B=\int_{-\infty}^{\infty}e^{-x^2/2}dx$. Even without knowing these values, the lines of reasoning which produced Theorem 49.3 (i) (Fejér's theorem) and Theorem 60.1 would give the following modified results.

Theorem 49.3′. *Under the conditions of Theorem 49.3*

$$\frac{1}{2\pi} \int_{-R}^{R} \left(1 - \frac{|\zeta|}{R}\right) \hat{f}(\zeta) \exp i\zeta t \, d\zeta \to Af(t).$$

Theorem 60.1′. *Under the conditions of Theorem 60.1*

$$\frac{1}{2\pi} \int_{-\infty}^{\infty} \hat{f}(\zeta) \exp i\zeta t \, d\zeta = Af(t).$$

On the other hand, the reasoning behind Lemma 50.2 gives the following.

Lemma 50.2′. *If* $E(t) = (2\pi)^{-\frac{1}{2}} \exp(-t^2/2)$ *then* $E(t) = E(-t)$ *and* $\hat{E}(\zeta) = BE(\zeta)$.

From Theorem 60.1′ we have $E^{\wedge\wedge}(-t) = 2\pi AE(t)$ whilst from Lemma 50.2′ we know that $E^{\wedge\wedge}(-t) = BE^{\wedge}(-t) = B^2 E(-t) = B^2 E(t)$ so (since $E(0) \neq 0$) $B^2 = 2\pi A$. Thus one of our computations was unnecessary and we can deduce Lemma 49.2 (iii) from Lemma 48.10 or vice versa.

61
SIMPLE DISCONTINUITIES

This chapter is included for the sake of completeness rather than for any intrinsic interest and the results contained in it will not be used in any essential way elsewhere. However, any reader who has just done a first course in complex variables may find some amusement in seeing how we use the methods established there.

Our object is to prove the following (weak) analogue of Theorem 16.3. (Recall that '$g(x) = O(G(x))$ as $|x| \to \infty$' is just an abbreviation for '$|g(x)| \leqslant A|G(x)|$' for some A and all sufficiently large $|x|$'.)

Theorem 61.1. *Suppose $f: \mathbb{R} \to \mathbb{C}$ is continuous except at a finite number of points $x(1), x(2), \ldots, x(n)$, say, where $\lim_{\eta \to 0+} f(x(j) + \eta)$ and $\lim_{\eta \to 0+} f(x(j) - \eta)$ exist. Then, if $\int_{-\infty}^{\infty} |f(t)| dt$ converges and $\hat{f}(\zeta) = O(|\zeta|^{-1})$ as $|\zeta| \to \infty$, it follows that*

$$\frac{1}{2\pi} \int_{-R}^{R} \hat{f}(\zeta) \exp i\zeta t \, d\zeta \to f(t) \text{ for } t \notin \{x(1), x(2), \ldots, x(n)\},$$

$$\frac{1}{2\pi} \int_{-R}^{R} \hat{f}(\zeta) \exp i\zeta x(j) d\zeta \to \frac{1}{2}(\lim_{\eta \to 0+} f(x(j) + \eta) + \lim_{\eta \to 0+} f(x(j) - \eta)).$$

Just as in Chapter 16 we first investigate a 'typical' discontinuous function in detail. Let

$$\begin{aligned} h(x) &= 1 - x && \text{for } 0 < x \leqslant 1, \\ h(x) &= -1 - x && \text{for } -1 \leqslant x < 0, \\ h(x) &= 0 && \text{otherwise,} \end{aligned}$$

as shown in Figure 61.1. For this function we have $\hat{h}(0) = 0$ and

$$\hat{h}(\zeta) = \int_{-\infty}^{\infty} h(x) \exp - i\zeta x \, dx$$

$$= \int_{0}^{1} (1 - x) \exp - i\zeta x \, dx + \int_{-1}^{0} (-1 - x) \exp - i\zeta x \, dx.$$

300

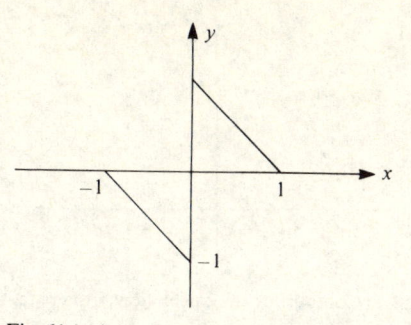

Fig. 61.1. A 'sawtooth' function on the line.

Making the linear substitution $y = -x$ gives

$$\hat{h}(\zeta) = \int_0^1 (1-x)\exp - i\zeta x\,dx - \int_1^0 (-1+y)\exp i\zeta y\,dy$$

$$= \int_0^1 (1-x)\exp(-i\zeta)x + (-1+x)\exp(i\zeta x)\,dx$$

$$= -2i\int_0^1 (1-x)\sin\zeta x\,dx$$

$$= -2i\left(\left[-(1-x)\frac{\cos\zeta x}{\zeta}\right]_0^1 - \int_0^1 \frac{\cos\zeta x}{\zeta}\,dx\right)$$

$$= -2i\left(\frac{1}{\zeta} - \frac{\sin\zeta}{\zeta^2}\right) \quad \text{for } \zeta \neq 0.$$

We must now try and find out what happens to $\int_{-R}^R \hat{h}(\zeta)\exp i\zeta t\,d\zeta$ as $R \to \infty$. The stages by which we do so are set out below but, rather than look at the proofs given, the reader may like to treat the matter as a long exercise in complex integration.

Lemma 61.2.

(i) $\displaystyle \int_{-R}^{-\varepsilon} + \int_{\varepsilon}^R \frac{e^{ia\zeta}}{\zeta}\,d\zeta \to \begin{cases} \pi i & \text{if } a > 0 \\ 0 & \text{if } a = 0 \\ -\pi i & \text{if } a < 0 \end{cases}$ as $R \to \infty,\ \varepsilon \to 0+$.

(ii) $\displaystyle \int_{-R}^{-\varepsilon} + \int_{\varepsilon}^R \frac{e^{ia\zeta}-1}{\zeta^2}\,d\zeta \to \begin{cases} -\pi a & \text{if } a \geqslant 0 \\ \pi a & \text{if } a \leqslant 0 \end{cases}$ as $R \to \infty,\ \varepsilon \to 0+$.

(iii) $\displaystyle \int_{-R}^{-\varepsilon} + \int_{\varepsilon}^R \hat{h}(\zeta)\exp i\zeta t\,d\zeta \to 2\pi h(t)$ as $R \to \infty,\ \varepsilon \to 0+$ for all $t \in \mathbb{R}$.

(iv) $\displaystyle \int_{-\varepsilon}^{\varepsilon} \hat{h}(\zeta)\exp i\zeta t\,d\zeta \to 0$ as $\varepsilon \to 0$ for all $t \in \mathbb{R}$.

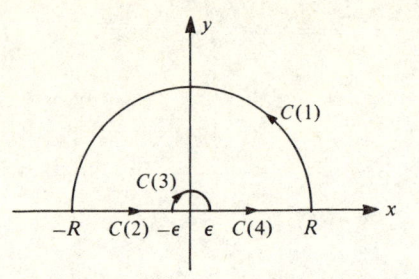

Fig. 61.2. Contour for the calculations of Lemma 61.2.

(v) $(2\pi)^{-1} \displaystyle\int_{-R}^{R} \hat{h}(\zeta) \exp i\zeta t \, d\zeta \to h(t)$ as $R \to \infty$ for all $t \in \mathbb{R}$.

Proof. (i) Suppose first that $a > 0$. Write $f(z) = e^{iaz}/z$ so that f is analytic in C except for a pole at $z = 0$. Let C be the contour of Figure 61.2 made up of the semicircle $C(1)$ described by $Re^{i\theta}$ as θ runs from 0 to π, the straight line $C(2)$ from $-R$ to $-\varepsilon$, the semicircle $C(3)$ described by $\varepsilon e^{i\theta}$ as θ runs from π to 0 and the straight line $C(4)$ from ε to R. (We take $R > \varepsilon > 0$ throughout.)

Since f is analytic within and in a neighbourhood of C, Cauchy's theorem yields

$$\int_{C(1)} + \int_{C(2)} + \int_{C(3)} + \int_{C(4)} f(z)\, dz = \int_C f(z)\, dz = 0,$$

so that

$$\int_{-R}^{-\varepsilon} + \int_{\varepsilon}^{R} \frac{e^{ia\zeta}}{\zeta} d\zeta = \int_{C(2)} + \int_{C(4)} f(z)\, dz = -\int_{C(1)} f(z)\, dz - \int_{C(3)} f(z)\, dz.$$

To examine the behaviour of

$$\int_{C(1)} f(z)\, dz = \int_0^\pi \frac{\exp(iaR \exp i\theta)}{R \exp i\theta} iR \exp i\theta \, d\theta$$

$$= i \int_0^\pi \exp(iaR \exp i\theta)\, d\theta \quad \text{as } R \to \infty,$$

we use a procedure which is probably familiar to the reader under the name of Jordan's lemma. Observe that, since $0 \le 2\theta/\pi \le \sin\theta$ for all $0 \le \theta \le \pi/2$, we have

$$|\exp(iaR \exp i\theta)| = |\exp(iaR(\cos\theta + i\sin\theta)|$$

$$= |\exp(iaR \cos\theta)||(\exp(-aR \sin\theta)|$$

$$= \exp(-aR \sin\theta) \le \exp(-2aR\theta/\pi) \quad \text{(for } 0 \le \theta \le \pi/2),$$

and so

$$\left| i \int_0^{\pi/2} \exp(iaR \exp i\theta) \, d\theta \right| \leqslant \int_0^{\pi/2} |\exp(iaR \exp i\theta)| \, d\theta \leqslant \int_0^{\pi/2} \exp(-2aR\theta/\pi) \, d\theta$$

$$= \left[-\frac{\pi}{2aR} \exp(-2aR\theta/\pi) \right]_0^{\pi/2} \leqslant \frac{\pi}{2aR}.$$

A similar computation using the fact that $0 \leqslant 2(\pi - \theta)/\pi \leqslant \sin \theta$ (or any one of a number of similar observations) show that

$$\left| i \int_{\pi/2}^{\pi} \exp(iaR \exp i\theta) \, d\theta \right| \leqslant \frac{\pi}{2aR},$$

whence

$$\left| \int_{C(1)} f(z) \, dz \right| \leqslant \frac{\pi}{aR} \to 0 \quad \text{as } R \to \infty.$$

On the other hand, we know that $e^{iaz} \to 1$ as $z \to 0$ so that

$$\int_{C(3)} f(z) \, dz = \int_{\pi}^{0} \frac{\exp(ia\varepsilon \exp i\theta)}{\varepsilon \exp i\theta} i\varepsilon \exp i\theta \, d\theta = i \int_{\pi}^{0} \exp(ia\varepsilon \exp i\theta) \, d\theta$$

$$\to i \int_{\pi}^{0} 1 \, d\theta = -\pi i \quad \text{as } \varepsilon \to 0+.$$

Thus if $a > 0$

$$\int_{-R}^{-\varepsilon} + \int_{\varepsilon}^{R} \frac{e^{ia\zeta}}{\zeta} \, d\zeta = -\int_{C(1)} f(z) \, dz - \int_{C(3)} f(z) \, dz \to \pi i \quad \text{as } \varepsilon \to 0+, R \to \infty.$$

If $a = 0$, then, since $1/\zeta$ is an odd function of ζ,

$$\int_{-R}^{-\varepsilon} + \int_{\varepsilon}^{R} \frac{e^{ia\zeta}}{\zeta} \, d\zeta = \int_{-R}^{-\varepsilon} \frac{1}{\zeta} \, d\zeta + \int_{\varepsilon}^{R} \frac{1}{\zeta} \, d\zeta = 0 \to 0$$

automatically. Finally if $a < 0$ then $-a > 0$ and

$$\int_{-R}^{-\varepsilon} + \int_{\varepsilon}^{R} \frac{e^{ia\zeta}}{\zeta} \, d\zeta = \int_{-R}^{-\varepsilon} + \int_{\varepsilon}^{R} \left(\frac{e^{-ia\zeta}}{\zeta} \right)^* \, d\zeta = \left(\int_{-R}^{-\varepsilon} + \int_{\varepsilon}^{R} \frac{e^{-ia\zeta}}{\zeta} \, d\zeta \right)^*$$

$$\to (-\pi i)^* = \pi i \quad \text{as } \varepsilon \to 0+, R \to \infty.$$

(Alternatively, we could redo the calculation with a contour in the lower rather than the upper half plane. The reader shoud check that she understands why this modification has to be made.)

(ii) This is very similar to (i). Writing $g(z) = (e^{iaz} - 1)/z^2 [a > 0]$ and proceeding as before we get

$$\left| \int_{C(1)} g(z) \, dz \right| \leqslant \text{length } C(1) \times \sup_{z \in C(1)} |g(z)| = \pi R 2 R^{-2} = 2\pi R^{-1} \to 0 \quad \text{as } R \to \infty,$$

whilst since $(e^{iaz} - 1)/z \to ia$ as $z \to 0$,

$$\int_{C(3)} g(z)\,dz = i \int_{\pi}^{0} \frac{\exp(ia\varepsilon \exp i\theta) - 1}{\varepsilon \exp i\theta}\,d\theta \to i \int_{\pi}^{0} ia\,d\theta = a\pi \quad \text{as } \varepsilon \to 0+.$$

We leave the remaining details to the reader.

(iii)   Since $\hat{h}(\zeta) \exp i\zeta t = -2i\left(\dfrac{1}{\zeta} - \dfrac{\sin \zeta}{\zeta^2}\right)\exp i\zeta t$

$$= -2i\frac{\exp i\zeta t}{\zeta} + \frac{\exp i\zeta(1 + t)}{\zeta^2} - \frac{\exp i\zeta(-1 + t)}{\zeta^2}$$

$$= -2i\frac{\exp i\zeta t}{\zeta} + \frac{\exp i\zeta(1 + t) - 1}{\zeta^2} - \frac{\exp i\zeta(-1 + t) - 1}{\zeta^2}$$

we have

$$\int_{-R}^{-\varepsilon} + \int_{\varepsilon}^{R} \hat{h}(\zeta) \exp i\zeta t\,d\zeta$$

$$= \int_{-R}^{\varepsilon} + \int_{\varepsilon}^{R} -\frac{2i\exp i\zeta t}{\zeta} + \frac{\exp i\zeta(1 + t) - 1}{\zeta^2} - \frac{\exp i\zeta(-1 + t) - 1}{\zeta^2}\,d\zeta$$

$$\to = \begin{cases} -2i(i\pi) & -\pi(1 + t) & +\pi(-1 + t) & \text{if } t \geqslant 1 \\ -2i(i\pi) & -\pi(1 + t) & -\pi(-1 + t) & \text{if } 1 > t > 0 \\ -2i0 & -\pi & +\pi & \text{if } t = 0 \\ -2i(-i\pi) & -\pi(1 + t) & -\pi(-1 + t) & \text{if } 0 > t > -1 \\ -2i(-i\pi) & +\pi(1 + t) & -\pi(-1 + t) & \text{if } -1 \geqslant t \end{cases}$$

$$= 2\pi h(t)$$

as required.

(iv) By direct inspection (or by using the general inequality $|\hat{f}(\zeta)| \leqslant \int_{-\infty}^{\infty} |f(t)|\,dt$) we know that $\hat{h}$ is bounded and so

$$\left| \int_{-\varepsilon}^{\varepsilon} \hat{h}(\zeta) \exp i\zeta t\,d\zeta \right| \leqslant 2\varepsilon \sup_{\zeta \in \mathbb{R}} |\hat{h}(\zeta)| \to 0 \quad \text{as } \varepsilon \to 0+.$$

(v) From (iii) and (iv)

$$\int_{-R}^{R} \hat{h}(\zeta) \exp i\zeta t\,d\zeta = \int_{-R}^{-\varepsilon} + \int_{\varepsilon}^{R} + \int_{-\varepsilon}^{\varepsilon} \hat{h}(\zeta) \exp i\zeta t\,d\zeta$$

$$\to 2\pi h(t) + 0 \quad \text{as } R \to \infty \text{ and } \varepsilon \to 0+. \qquad \blacksquare$$

Notice once again that we are talking about the 'principal value'

$$\lim_{R \to \infty} \frac{1}{2\pi} \int_{-R}^{R} \hat{h}(\zeta) \exp i\zeta t\,d\zeta.$$

As the following computation shows, the integral $(1/2\pi)\int_{-\infty}^{\infty} \hat{h}(\zeta) \exp i\zeta t\,d\zeta$ fails to converge when $t = 0$. (If the reader is prepared to accept this without proof there is no particular call for her to go into the details.)

Example 61.3. *If h is defined as above and $k, l > 0$ then*

$$\frac{1}{2\pi} \int_{-kR}^{lR} \hat{h}(\zeta) d\zeta \to (2\pi)^{-1} \log k/l$$

and $\quad \dfrac{1}{2\pi} \displaystyle\int_{-R}^{R \log R} \hat{h}(\zeta) d\zeta \to -\infty, \dfrac{1}{2\pi} \displaystyle\int_{-R \log R}^{R} \hat{h}(\zeta) d\zeta \to \infty \quad$ *as* $R \to \infty$.

Proof. We concentrate on showing that $\int_{-kR}^{lR} \hat{h}(\zeta) d\zeta \to \log k/l$ as $R \to \infty$. Observe first that $|\sin \zeta / \zeta^2| \leqslant \zeta^{-2}$, so, by comparison, $\int_{|\zeta|>1} (\sin \zeta / \zeta^2) d\zeta$ converges absolutely and so converges. In other words $\int_{-R}^{-1} + \int_{1}^{S} (\sin \zeta / \zeta^2) d\zeta$ converges to a limit $\int_{|\zeta|>1} (\sin \zeta / \zeta^2) d\zeta$ as $R, S \to \infty$. In particular, taking $R = S$ and observing that $\sin \zeta / \zeta^2$ is an odd function, we have

$$0 = \int_{-R}^{-1} + \int_{1}^{R} \frac{\sin \zeta}{\zeta^2} d\zeta \to \int_{|\zeta|>1} \frac{\sin \zeta}{\zeta^2} d\zeta \quad \text{so} \quad \int_{|\zeta|>1} \frac{\sin \zeta}{\zeta^2} d\zeta = 0$$

and $\quad \displaystyle\int_{-R}^{-1} + \int_{1}^{S} \frac{\sin \zeta}{\zeta^2} d\zeta \to 0$ as $R, S \to \infty$.

We note that since

$$\hat{h}(\zeta) = -\frac{1}{\zeta} + \frac{\sin \zeta}{\zeta^2}$$

is also an odd function,

$$\int_{-1}^{1} \hat{h}(\zeta) d\zeta = 0.$$

Thus

$$\int_{-kR}^{lR} \hat{h}(\zeta) d\zeta = \int_{-kR}^{-1} \hat{h}(\zeta) d\zeta + \int_{-1}^{1} \hat{h}(\zeta) d\zeta + \int_{1}^{lR} \hat{h}(\zeta) d\zeta = \int_{-kR}^{-1} + \int_{1}^{lR} \hat{h}(\zeta) d\zeta$$

$$= -\left(\int_{-kR}^{-1} + \int_{1}^{lR} \frac{1}{\zeta} d\zeta \right) + \left(\int_{-kR}^{-1} + \int_{1}^{lR} \frac{\sin \zeta}{\zeta^2} d\zeta \right)$$

$$= -\left([\log(-\zeta)]_{-kR}^{-1} + [\log \zeta]_{1}^{lR} \right) + \left(\int_{-kR}^{-1} + \int_{1}^{lR} \frac{\sin \zeta}{\zeta^2} d\zeta \right)$$

$$= -\log l/k + \left(\int_{-kR}^{-1} + \int_{1}^{lR} \frac{\sin \zeta}{\zeta^2} d\zeta \right) \to \log k/l \quad \text{as } R \to \infty.$$

The other two limits are obtained similarly and are left as a simple exercise. ∎

We now return to the proof of Theorem 61.1. We require a simple preliminary lemma.

Lemma 61.4. *Let* $h_x(t) = h(t - x)$. *Then if* $h \in L^1 \cap C$,

(i) $\hat{h}_x(\zeta) = \exp - i\zeta x\, \hat{h}(\zeta)$,

(ii) $(2\pi)^{-1} \int_{-R}^{R} \hat{h}_x(\zeta) \exp i\zeta t\, d\zeta \to h_x(t)$ *as* $R \to \infty$,

(iii) $|\hat{h}_x(\zeta)| = O(|\zeta|^{-1})$ *as* $|\zeta| \to \infty$.

Proof. (i) Making the linear substitution $s = t - x$ we have

$$\hat{h}_x(\zeta) = \int_{-\infty}^{\infty} h_x(t) \exp - i\zeta t\, dt = \int_{-\infty}^{\infty} h(t - x) \exp - i\zeta t\, dt$$

$$= \int_{-\infty}^{\infty} h(s) \exp(- i\zeta(s + x)) ds = \exp(- i\zeta x) \int_{-\infty}^{\infty} h(s) \exp - i\zeta s\, ds$$

$$= \exp - i\zeta x\, \hat{h}(\zeta),$$

as stated,

(ii) $(2\pi)^{-1} \int_{-R}^{R} \hat{h}_x(\zeta) \exp i\zeta t\, d\zeta = \int_{-R}^{R} \hat{h}(\zeta) \exp i(t - x)\zeta\, d\zeta \to h(t - x) = h_x(t)$ as $R \to \infty$
(using (i) and Lemma 61.2 (v)).

(iii) $|\hat{h}_x(\zeta)| = |\hat{h}(\zeta)| = |1/\zeta - \sin \zeta/\zeta^2| \leqslant 1/|\zeta| + |\sin \zeta/\zeta^2| \leqslant 1/|\zeta| + 1/|\zeta|^2$
$\leqslant 2/|\zeta|$ for all $|\zeta| \geqslant 1$. ∎

The proof of Theorem 61.1 now follows the pattern laid out in Chapter 16. The reader may prefer to write out her own proof rather than use the one given below.
Proof of Theorem 61.1 Set

$$\lambda_j = (\lim_{\eta \to 0+} f(x(j) + \eta) - \lim_{\eta \to 0+} f(x(j) - \eta))/2,$$

$$g(t) = f(t) - \sum_{j=1}^{n} \lambda_j h_{x(j)}(t) \text{ for } t \notin \{x(1), x(2), \dots, x(n)\},$$

$$g(x(j)) = (\lim_{\eta \to 0+} f(x(j) + \eta) + \lim_{\eta \to 0+} f(x(j) - \eta))/2.$$

Then it is easy to check that g is continuous and (since $\int_{-\infty}^{\infty} |f(t)| dt$ and $\int_{-\infty}^{\infty} |h_{x(j)}(t)| dt$ converge) $\int_{-\infty}^{\infty} |g(t)| dt$ converges. Further,

$$\hat{g}(\zeta) = \hat{f}(\zeta) - \sum_{j=1}^{n} \lambda_j \hat{h}_{x(j)}(\zeta) = O(|\zeta|^{-1}),$$

since $\hat{f}(\zeta) = O(|\zeta|^{-1})$ and, by Lemma 61.4 (iii), $\hat{h}(\zeta) = O(|\zeta|^{-1})$ as $|\zeta| \to \infty$. Thus, by Theorem 60.3 (i),

$$\frac{1}{2\pi} \int_{-R}^{R} \hat{g}(\zeta) \exp i\zeta t\, dt \to g(t) \text{ as } R \to \infty \text{ for all } t \in \mathbb{R}.$$

But
$$\hat{f}(\zeta) = \hat{g}(\zeta) + \sum_{j=1}^{n} \lambda_j \hat{h}_{x(j)}(\zeta),$$

so, using Lemma 61.4 (ii), we obtain

$$\frac{1}{2\pi}\int_{-R}^{R}\hat{f}(\zeta)\exp i\zeta t\,d\zeta = \frac{1}{2\pi}\int_{-R}^{R}\hat{g}(\zeta)\exp i\zeta t\,d\zeta + \sum_{j=1}^{n}\frac{\lambda_j}{2\pi}\int_{-R}^{R}\hat{h}_j(\zeta)\exp i\zeta t\,d\zeta$$

$$\to g(t) + \sum_{j=1}^{n}\lambda_j h_j(t)$$

$$= \begin{cases} f(t) & \text{if } t\notin\{x(1),x(2),\ldots,x(n)\} \\ (\lim_{\eta\to 0+} f(t+\eta) + \lim_{\eta\to 0+} f(t-\eta))/2 & \text{if } t\in\{x(1),x(2),\ldots,x(n)\}, \end{cases}$$

as required. ∎

Remark. The reader will, no doubt, have observed that, if f is continuous at t, then $f(t) = \lim_{\eta\to 0+} (f(t+\eta)+f(t-\eta))/2$ so the conclusion of the theorem could be written more briefly in the style of Chapter 16 as

$$\frac{1}{2\pi}\int_{-R}^{R}\hat{f}(\zeta)\exp i\zeta t\,d\zeta \to (\lim_{\eta\to 0+} f(t+\eta) + \lim_{\eta\to 0+} f(t-\eta))/2 \quad \text{for all } t\in\mathbb{R}.$$

As a corollary we obtain the following uniqueness theorem.

Theorem 61.5. *Suppose $f_k:\mathbb{R}\to\mathbb{C}$ is continuous except at a finite set of points $A(k)$ where $\lim_{\eta\to 0+} f_k(a+\eta)$ and $\lim_{\eta\to 0+} f_k(a-\eta)$ exist, and that $\int_{-\infty}^{\infty}|f_k(t)|\,dt$ converges $[k=1,2]$. Then if $\hat{f}_1(\zeta) = \hat{f}_2(\zeta)$ for all $\zeta\in\mathbb{R}$ it follows that $f_1(t) = f_2(t)$ for all $t\notin A(1)\cup A(2)$.*

Proof. Let $f = f_1 - f_2$. Then f is continuous except, possibly, at points of $A(1)\cup A(2)$ and $\lim_{\eta\to 0+} f(a+\eta)$, $\lim_{\eta\to 0+} f(a-\eta)$ exist for all $a\in A(1)\cup A(2)$. Further, $\int_{-\infty}^{\infty}|f(t)|\,dt$ converges and

$$\hat{f}(\zeta) = \hat{f}_1(\zeta) - \hat{f}_2(\zeta) = 0 \quad \text{for all } \zeta\in\mathbb{R}.$$

Since $\hat{f}(\zeta) = 0 = O(|\zeta|^{-1})$ we can apply Theorem 61.1 to obtain

$$f(t) = \lim_{R\to\infty}\int_{-R}^{R}\hat{f}(\zeta)\exp i\zeta t\,d\zeta = \lim_{R\to\infty}\int_{-R}^{R} 0\,d\zeta = 0 \quad \text{for all } t\notin A(1)\cup A(2),$$

as claimed. ∎

We could go further and discuss the Gibbs phenomenon for Fourier transforms, or the possibility of an infinite number of simple discontinuities, or the pointwise convergence when f is merely Riemann integrable. But we shall not.

62

HEAT FLOW IN A SEMI-INFINITE ROD

In Chapter 57 we solved the problem of heat flow in a semi-infinite rod initially at a uniform temperature θ_0 whose end is held at a fixed temperature 0. Less picturesquely but more precisely we proved the following lemma

Lemma 62.1. *Let* $\theta_0 \in \mathbb{R}$ *and*

$$\theta(y,t) = \frac{\theta_0}{2\sqrt{(\pi K t)}} \int_0^\infty \left\{ \exp\left(-\frac{(y-w)^2}{4Kt} \right) - \exp\left(-\frac{(y+w)^2}{4Kt} \right) \right\} dw$$

for all $y \geq 0$, $t > 0$. *Then* $\theta:[0,\infty) \times (0,\infty) \to \mathbb{R}$ *is an infinitely differentiable function with*

(i) $(\partial\theta/\partial t)(y,t) = K(\partial^2\theta/\partial y^2)(y,t)$,
(ii) $\theta(y,t) \to \theta_0$ *as* $t \to 0 +$ *for all* $y > 0$,
(iii) $\theta(0,t) = 0$ *for all* $t > 0$.

Proof. This is Lemma 57.1. ∎

By adding a constant we see that a solution to the problem in which the rod is initially at a uniform temperature 0 and the end is held at temperature θ_1 is

$$\theta(y,t) = \theta_1 - \frac{\theta_1}{2\sqrt{(\pi K t)}} \int_0^\infty \left\{ \exp\left(-\frac{(y-w)^2}{4Kt} \right) - \exp\left(-\frac{(y+w)^2}{4Kt} \right) \right\} dw.$$

In this chapter we shall attack the more general problem in which the rod is initially at a uniform temperature 0 and the end is held at a varying temperature $f(t)$. More explicitly we shall try to solve the equations

(i) $(\partial\theta/\partial t)(y,t) = K(\partial^2\theta/\partial y^2)(y,t)[y,t > 0]$,
(ii) $\theta(y,t) \to \theta_0$ *as* $t \to 0 +$ for $y > 0$,
(iii) $\theta(y,t) \to f(t)$ *as* $y \to 0 +$ for $t > 0$.

The discussion that follows is heuristic, though there is only one major gap in the reasoning. (I shall point this gap out when we come to it.)

Write $H_s(t) = 0$ for $t \leqslant s$, $H_s(t) = 1$ for $s < t$. We have solved the problem above for $f(t) = \theta_1 H_0(t)$ and this can be made the basis for a tentative solution when $f(t) = \lambda H_s(t) [s > 0]$. For, if we set

$$\theta_s(y, t) = 0 \quad \text{for } t \leqslant s,$$

$$\theta_s(y, t) = 1 - \frac{1}{2\sqrt{(\pi K(t - s))}} \int_0^\infty \left\{ \exp\left(-\frac{(y - w)^2}{4K(t - s)} \right) \right.$$

$$\left. - \exp\left(-\frac{(y + w)^2}{4K(t - s)} \right) \right\} dw \quad \text{for } t > s$$

then, certainly, $\lambda \theta_s$ satisfies conditions (i) and (iii) (with $f = \lambda H_s$) except possibly when $t = s$. (In fact it also satisfies them when $t = s$ but since the argument is heuristic, we shall not pause to verify this.)

The heat equation is linear and so, in particular, if

$$0 < s(1) < s(2) < \cdots < s(n) \quad \text{and} \quad \theta = \sum_{j=1}^n \lambda_j \theta_{s(j)},$$

(i) $(\partial\theta/\partial t)(y, t) = K(\partial^2\theta/\partial y^2)(y, t)$ for $y > 0$, $t > 0$, $t \neq s(1), \ldots, s(n)$,
(ii) $\theta(y, t) = 0$ for $0 \leqslant t < s(1)$,
(iii) $\theta(y, t) \rightarrow \sum_{j=1}^k \lambda_j$ as $y \rightarrow 0+$ for $s(k) < t < s(k + 1)$,
$\theta(y, t) \rightarrow \sum_{j=1}^n \lambda_j$ as $y \rightarrow 0+$ for $s(n) < t$.

Ignoring the (not very difficult) convergence problems involved, we may suspect that for reasonable choices of a sequence $0 < s(1) < s(2) < \cdots$ and $\lambda_j \in \mathbb{C}$ the infinite sum $\theta = \sum_{j=1}^\infty \lambda_j \theta_{s(j)}$ will also satisfy

(i) $(\partial\theta/\partial t)(y, t) = K(\partial^2\theta/\partial y^2)(y, t)$ for $y > 0, t > 0, t \neq s(j)$,
(ii) $\theta(y, t) = 0$ for $0 \leqslant t < s(1)$,
(iii) $\theta(y, t) \rightarrow \sum_{j=1}^k \lambda_j$ as $y \rightarrow 0+$ for $s(k) < t < s(k + 1)$.

If we now take a well behaved function f and some small $\delta > 0$ then, putting $s(j) = j\delta$, $\lambda_1 = f(\delta)$ and $\lambda_j = f((j + 1)\delta) - f(j\delta)$, we conjecture that if

$$(*) \quad \phi_\delta(y, t) = \sum_{j=1}^\infty (f((j + 1)\delta) - f(j\delta))\theta_{j\delta}(y, t),$$

then

(i) $(\partial\phi_\delta/\partial t)(y, t) = K(\partial^2\phi_\delta/\partial y^2)$ for $y > 0, t > 0, t \neq s(j)$,
(ii) $\phi_\delta(y, t) = 0$ for $0 < t < \delta$,
(iii) $\phi_\delta(y, t) \rightarrow f(j\delta)$ as $y \rightarrow 0+$ for $j\delta < t < (j + 1)\delta$.

We now let $\delta \rightarrow 0$, and it is here that there is the widest gap in our argument. With luck we expect that

$$\phi_\delta(y, t) = \sum_{j=1}^\infty \frac{f((j + 1)\delta) - f(j\delta)}{\delta} \theta_{j\delta}(y, t)\delta \rightarrow \int_0^\infty f'(s)\theta_s(y, t)ds = \phi(y, t), \text{ say}$$

and that

(i) $(\partial\phi/\partial t)(y,t) = K(\partial^2\phi/\partial y^2)(y,t)$ for $y>0, t>0$,

(ii) $\phi(y,t)\to 0$ as $t\to 0+$ for $y>0$,

(iii) $\phi(y,t)\to f(t)$ as $y\to 0+$ for $t>0$.

To get an expression for ϕ involving f rather that f' we integrate by parts to obtain

$$\phi(y,t) = \int_0^\infty f'(s)\theta_s(y,t)\,ds = [f(s)\theta_s(y,t)]_0^\infty - \int_0^\infty f(s)\left(\frac{\partial}{\partial s}\theta_s(y,t)\right)ds$$

$$= -\int_0^\infty f(s)\left(\frac{\partial}{\partial s}\theta_s(y,t)\right)ds = -\int_0^t f(s)\left(\frac{\partial}{\partial s}\theta_s(y,t)\right)ds,$$

since $\theta_s(y,t) = 0$ for $s>t$.

The reader might like to pause to consider what we have done so far, but a further simplification is possible. Writing

$$\psi_0(x,t) = \frac{1}{2(\pi Kt)^{\frac12}}\int_0^\infty \exp\left(-\frac{(x-w)^2}{4Kt}\right)dw,$$

we see that

$$\theta_s(y,t) = \theta_0(y,t-s) = 1 - \psi_0(y,t-s) + \psi_0(-y,t-s),$$

and so, by direct computation, or an appeal to Lemma 55.7 (ii), we have

$$\frac{\partial\theta_s}{\partial s}(y,t) = -D_2\theta_0(y,t-s) = D_2\psi_0(y,t-s) - D_2\psi_0(-y,t-s)$$

$$= \frac{-y}{4\pi^{\frac12}K^{\frac12}(t-s)^{3/2}}\exp\left(\frac{-y^2}{4K(t-s)}\right) - \frac{y}{4\pi^{\frac12}K^{\frac12}(t-s)^{3/2}}\exp\left(-\frac{(-y)^2}{4K(t-s)}\right)$$

$$= \frac{-y}{2\pi^{\frac12}K^{\frac12}(t-s)^{3/2}}\exp\left(\frac{-y^2}{4K(t-s)}\right) \text{ for } t>s.$$

In this way we have arrived at a putative solution

$$\phi(y,t) = \int_0^t f(s)\frac{y}{2\pi^{\frac12}K^{\frac12}(t-s)^{3/2}}\exp\left(-\frac{y^2}{4K(t-s)}\right)ds$$

to our problem. It remains to check our guess.

Theorem 62.2. *Suppose that $f:(0,\infty)\to\mathbb{C}$ is continuous and bounded. Then if we define $\phi:(0,\infty)\times(0,\infty)\to\mathbb{C}$ by*

$$\phi(x,t) = \int_0^t f(s)\frac{x}{2\pi^{\frac12}K^{\frac12}(t-s)^{3/2}}\exp\left(-\frac{x^2}{4K(t-s)}\right)ds,$$

if follows that ϕ is an infinitely differentiable function on $(0,\infty)\times(0,\infty)$ with

(i) $(\partial\phi/\partial t)(x,t) = K(\partial^2\phi/\partial x^2)(x,t)$ *for* $x, t > 0$,

(ii) $\phi(x,t) \to 0$ *as* $t \to 0+$ *for* $x > 0$,

(iii) $\phi(x,t) \to f(t)$ *as* $x \to 0+$ *for* $t > 0$.

Proof. The proof follows the same pattern as Theorem 55.4. It is not very interesting and I suggest that the reader just skims through it.

Step 1. (We show that we can differentiate under the integral sign.) Set

$$G(s,x,t) = f(s) \frac{x}{2\pi^{\frac{1}{2}} K(t-s)^{3/2}} \exp\left(-\frac{x^2}{4K(t-s)}\right)$$

if $t > s > 0$, $G(s,x,t) = 0$ otherwise. We claim that $G : \mathbb{R} \times \mathbb{R}^+ \times \mathbb{R}^+ \to \mathbb{C}$ is infinitely differentiable in x and t. This the reader will readily allow except possibly when $t = s$. To deal with the case $t = s$ we observe that the same proof as in Example 4.2 shows that, if

$$g(u) = u^{-3/2} \exp(-1/u) \quad \text{for } u > 0,$$
$$g(u) = 0 \quad \text{for } u \leqslant 0,$$

then g is infinitely differentiable everywhere including 0.

Since $G(s,x,t) = 0$ for $s \notin [0,t]$, it follows that $D_2^j D_3^k G(s,x,t) = 0$ for $s \notin [0,t]$ so that $D_2^j D_3^k G(\ ,x,t) \in L^1 \cap C$ and, if $t \in [T_1, T_2]$,

$$\int_{|s| \geqslant R} |D_2^j D_3^k G(s,x,t)| \, ds = 0 \quad \text{for} \quad R > \max(|T_1|, |T_2|).$$

Thus using the appropriate modification to Theorem 53.4 repeatedly we see that ϕ is infinitely differentiable with

$$\frac{\partial^j}{\partial x^j} \frac{\partial^k}{\partial t^k} \phi(x,t) = \int_{-\infty}^{\infty} D_2^j D_3^k G(s,x,t) \, ds.$$

Step 2. If we show that $\qquad D_2 G(x,s,t) = K D_3^2 G(x,s,t)$,

it will follow from Step 1 that ϕ satisfies the heat equation. We can verify this directly but it is simpler to observe that writing

$$E_{1/\sqrt{(2Kt)}}(x) = \frac{1}{2\pi^{\frac{1}{2}}(Kt)^{\frac{1}{2}}} \exp\left(-\frac{x^2}{4Kt}\right),$$

we have $\qquad G(x,s,t) = f(s) \dfrac{\partial}{\partial x} E_{1/\sqrt{(2K(t-s))}}(x) \quad$ for $\quad t > s > 0$.

But we know that $E_{1/\sqrt{(2Kt)}}(x)$ satisfies the heat equation and so

$$\frac{\partial}{\partial t} \frac{\partial}{\partial x} E_{1/\sqrt{(2Kt)}}(x) = \frac{\partial}{\partial x} \frac{\partial}{\partial t} E_{1/\sqrt{(2Kt)}}(x) = K \frac{\partial}{\partial x} \frac{\partial}{\partial x^2} E_{1/\sqrt{(2Kt)}}(x)$$

$$= K \frac{\partial}{\partial x^2} \frac{\partial}{\partial x} E_{1/\sqrt{(2Kt)}}(x)$$

and the required result now follows.

Step 3. Let

$$P(u) = \frac{1}{(2\pi)^{\frac{1}{2}} u^{3/2}} \exp\left(-\frac{1}{2u}\right) \qquad u > 0,$$

$$P(u) = 0 \qquad\qquad\qquad\qquad u \leqslant 0,$$

(i.e. let $P(u) = 2\pi^{-\frac{1}{2}} g(2u)$). Then $P: \mathbb{R} \to \mathbb{R}$ is continuous and has the following properties

(i) $P(u) \geqslant 0$ for all u,

(ii) $u(P(u)) \to 0$ as $|u| \to \infty$,

(iii) $\displaystyle\int_{-\infty}^{\infty} P(u)\,du = \frac{1}{(2\pi)^{\frac{1}{2}}} \int_{0}^{\infty} \frac{1}{u^{\frac{3}{2}}} \exp\left(-\frac{1}{2u}\right)du = \frac{1}{(2\pi)^{\frac{1}{2}}} \int_{\infty}^{0} -2\exp\left(-\frac{v^2}{2}\right)dv$

$$= \frac{2}{(2\pi)^{\frac{1}{2}}} \int_{0}^{\infty} \exp\left(-\frac{v^2}{2}\right)dv = 1,$$

making the substitution $v^2 = 1/u$ and remembering (for example from Lemma 48.10) that $\int_{-\infty}^{\infty} \exp(-t^2/2)\,dt = (2\pi)^{\frac{1}{2}}$.

Writing $\tilde{f}(t) = f(t)$ for $t \geqslant 0$, $\tilde{f}(t) = 0$ for $t < 0$ we see that with the notation of Theorem 51.5 we have

$$\phi(x,t) = \int_{0}^{t} \tilde{f}(s) P_{2K/x^2}(t-s)\,ds = \int_{-\infty}^{\infty} \tilde{f}(s) P_{2K/x^2}(t-s)\,ds = \tilde{f} * P_{2K/x^2}(t)$$

$$\to \tilde{f}(t) = f(t) \quad \text{as } x \to 0+ \text{ for all } t > 0.$$

All that remains is to verify that $\phi(x,t) \to 0$ as $t \to 0+$ for all $x > 0$. But

$$\phi(x,t) = \tilde{f} * P_{2K/x^2}(t) = \int_{-\infty}^{\infty} \tilde{f}(t-s) P_{2K/x^2}(s)\,ds = \int_{0}^{t} f(t-s) P_{2K/x^2}(s)\,ds,$$

so

$$|\phi(x,t)| \leqslant \sup_{u \in T} |f(u)| \int_{0}^{t} P_{2K/x^2}(x)\,ds \to 0 \quad \text{as } t \to 0+,$$

and we are done. $\qquad\qquad\qquad\qquad\qquad\qquad\qquad\qquad\qquad\qquad\qquad\blacksquare$

Suppose that we change the temperature at the end of the rod for a short time and then return it to its initial value. How will the resulting 'heat pulse' travel down the rod? Ignoring as usual the problem of uniqueness, we shall consider the question in the following form.

Problem. *Suppose $f:(0,\infty) \to \mathbb{R}$ is a continuous function with $f(t) = 0$ for $t \geqslant a$. What can we say about the behaviour of $\phi(x,t)$?*

We can make two remarks at once.

(1) If $f(u) > 0$ for all $0 < u < a$ then

$$\phi(x,t) = \int_0^t f(s) P_{2K/x^2}(t-s)\,ds > 0 \quad \text{for all } t > 0,\, x > 0,$$

i.e. there is an instantaneous effect at all points.

(2) However, if $Kt \ll x^2$ (i.e. if Kt is small compared with x^2), then

$$|\phi(x,t)| \leqslant \sup_{u\in\mathbb{R}} |f(u)| \int_0^t P_{2K/x^2}(s)\,ds = \sup_{u\in\mathbb{R}} |f(u)| \int_0^{2Kt/x^2} P(v)\,dv$$

$$\leqslant \sup_{u\in\mathbb{R}} |f(u)| \sup_{v\in\mathbb{R}} |P(v)| 2Kt/x^2 \ll 1,$$

so the effect is negligible at distances x which are large compared with $\sqrt{(Kt)}$.

If t/a is of the order of 1 (or less) then the behaviour of $\phi(x,t)$ depends on the precise behaviour of f. But if $t \gg a$ then

$$\phi(x,t) = \int_0^t f(s) P_{2K/x^2}(t-s)\,ds = \int_0^a f(s) P_{2K/x^2}(t-s)\,ds \approx \int_0^a f(s) P_{2K/x^2}(t)\,ds$$

(since
$$P_{2K/x^2}(t-s) = \frac{x}{2\pi^{\frac{1}{2}}(K(t-s))^{\frac{3}{2}}} \exp\left(-\frac{x^2}{K(t-s)}\right)$$

$$\approx \frac{x}{2\pi^{\frac{1}{2}}(Kt)^{\frac{3}{2}}} \exp\left(-\frac{x^2}{Kt}\right) \quad \text{for all } 0 \leqslant s \leqslant a).$$

Hence
$$\phi(x,t) \approx \int_0^a f(s)\,ds\, P_{2K/x^2}(t).$$

Thus we have the following remark.

(3) If $t \gg a$ the 'shape' of the pulse $\phi(x,t)$ is independent of the 'shape' of f and its magnitude depends only on $\int_0^a f(s)\,ds$ (and, of course, x and t).

Using Remark (2) we obtain the following version of (3).

(4) If $x^2 \gg Ka$ then the shape of the pulse $\phi(x,)$ depends only on x and its magnitude depends only on $\int_0^a f(s)\,ds$ and x.

Thus the three pulses in Figure 62.1 produce much the same pulse $\phi(x,) \approx AP_{2K/x^2}(\ )$ (with $A = \int_0^a f(s)\,ds$) for large x.

To describe the pulse shape in more detail we make some preliminary observations.

Lemma 62.3. (i) *Let*

$$P(u) = (2\pi)^{-\frac{1}{2}} u^{-\frac{3}{2}} \exp(-1/2u) \quad [u > 0].$$

Then P increases from 0 to $M = (2\pi)^{-\frac{1}{2}} 3^{\frac{3}{2}} \exp(-3/2)$ as u runs from 0 to $T_0 = 1/3$ and then decreases to 0 as u runs from T_0 to ∞. In particular if $0 < \delta < 1$ there exist unique $T_1(\delta)$ and $T_2(\delta)$ with $0 < T_1(\delta) < 1/3 < T_2(\delta)$ and $P(u) \geqslant \delta M$ for $u \in [T_1(\delta), T_2(\delta)]$, $P(u) < \delta M$ otherwise.

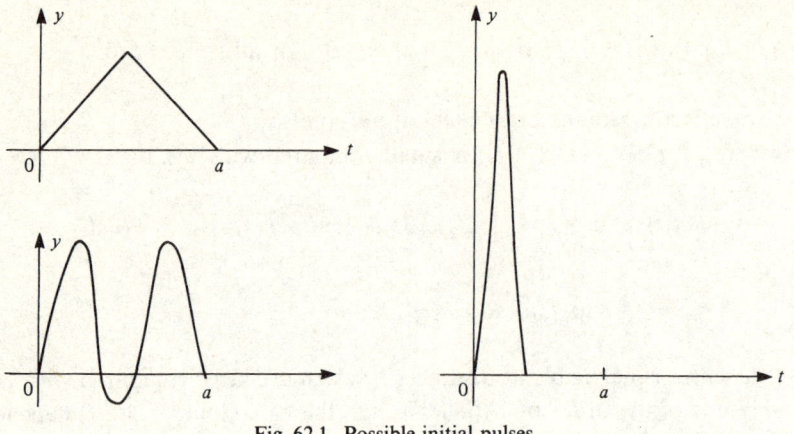

Fig. 62.1. Possible initial pulses.

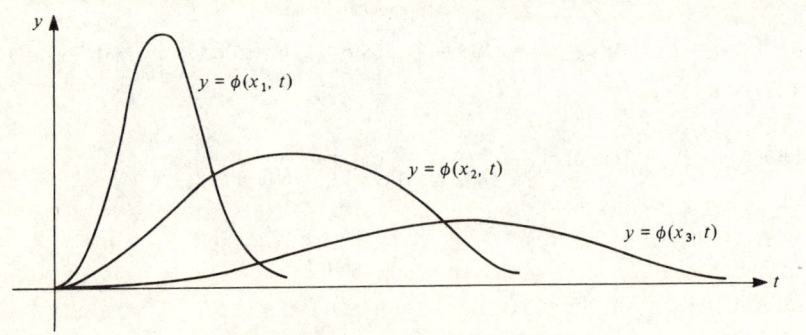

Fig. 62.2. Pulse shapes at various distances.

(ii) *Let* $Q(t) = AP_{2K/x^2}(t)[t > 0]$ *where* $A, K, x > 0$. *Then with the notation of* (i), *Q increases from 0 to* $(2K/x^2)MA$ *as u runs from 0 to* $(x^2/2K)T_0$ *and then decreases to 0 as u runs to* ∞. *If* $0 < \delta < 1$ *then*

$$Q(t) \geqslant 2K\delta MA/x^2 \text{ for } t \in [(x^2/2K)T_1(\delta), (x^2/2K)T_2(\delta)], Q(t) < \delta MA/x^2 \text{ otherwise.}$$

Proof. (i) Observe that

$$P'(u) = \frac{1}{(2\pi)^{\frac{1}{2}}}\left(-\frac{3}{2u^{\frac{5}{2}}} + \frac{1}{2u^{\frac{7}{2}}}\right)\exp\left(-\frac{1}{2u}\right) = \frac{1}{(2\pi)^{\frac{1}{2}}}\frac{1-3u}{2u^{\frac{7}{2}}}\exp\left(-\frac{1}{2u}\right).$$

(ii) Since $Q(t) = (A2K/x^2)P(2Kt/x^2)$ this follows directly from (i). ∎

Rewriting Lemma 62.3 in vaguer terms we obtain our concluding remark.

(5) If $x^2 \gg Ka$ then the pulse rises for a time proportional to x^2 (and inversely proportional to K) to a maximum which is inversely proportional to x^2 (and proportional to K) and then declines to 0 (Figure 62.2). The length of time during which the pulse is greater than a given fraction of its maximum is proportional to x^2 (and inversely proportional to K).

63

A SECOND APPROACH

In Chapter 62 we were faced with the equations

$$\frac{\partial \phi}{\partial t}(x, t) = K \frac{\partial^2 \phi}{\partial x^2}(x, t) \quad \text{for } x > 0,$$

$$\phi(x, t) \to f(t) \qquad \text{as } x \to 0+.$$

(It will be helpful to ignore for the time being the restriction $t > 0$ and the condition $\phi(x, t) \to 0$ as $t \to 0+$.) We found a solution to the general problem by building it up from solutions found in Chapter 57 to the simpler problem when $f(t) = H_s(t)$.

In this chapter we shall build up general solutions starting from the solutions found in Chapter 7 in the particular case when $f(t) = \exp i\zeta t$. The results obtained will be exactly the same as those of Chapter 62 but the reader may find it helps her understanding of these results to see both ways of obtaining them.

Lemma 63.1. *Write*

$$G(y) = \exp((-1 - i)|y|^{\frac{1}{2}}) \quad \text{for } y > 0,$$
$$G(0) = 1,$$
$$G(y) = \exp((-1 + i)|y|^{\frac{1}{2}}) \quad \text{for } y < 0$$

(taking positive square roots throughout). Then, if

$$\theta_\zeta(x, t) = G(\zeta x^2 / 2K) \exp i\zeta t,$$

it follows that

(i) $(\partial \theta_\zeta / \partial t)(x, t) = K(\partial^2 \theta_\zeta / \partial x^2)(x, t)$,
(ii) $\theta_\zeta(0, t) = \exp i\zeta t$,
(iii) $\theta_\zeta(x, t) \to 0$ *as* $x \to \infty$.

Proof. Direct verification. (A derivation when $\zeta > 0$ is given in Lemma 7.1. The derivation when $\zeta < 0$ is similar.) ∎

Now, since θ_ζ is a well behaved solution of the heat equation, so is $\sum_{j=1}^{n} \lambda_j \theta_{\zeta(j)}$,

315

and so, we might hope, would be

$$\phi(x, t) = \int_{-\infty}^{\infty} F(\zeta)\theta_{\zeta}(x, t)\,d\zeta = \int_{-\infty}^{\infty} F(\zeta)G(\zeta x^2/2K)\exp i\zeta t\,d\zeta, \tag{1}$$

at least if F is sufficiently well behaved. If we try to satisfy the boundary condition

$$\phi(0, t) = f(t),$$

our equation becomes

$$f(t) = \int_{-\infty}^{\infty} F(\zeta)G(0)\exp i\zeta t\,dt = \int_{-\infty}^{\infty} F(\zeta)\exp i\zeta t\,dt,$$

so $f(t) = \hat{F}(-t)$ and, applying the inversion formula (Theorem 60.1), we have $\hat{f}(\zeta) = f^{\vee\,\vee}(-\zeta) = 2\pi F(\zeta)$ so $\hat{f} = 2\pi F$ and

$$\phi(x, t) = \frac{1}{2\pi}\int_{-\infty}^{\infty} \hat{f}(\zeta)G(\zeta x^2/2K)\exp i\zeta t\,d\zeta. \tag{2}$$

The reader should pause at this point and ask herself if we could not have guessed this directly by inspection of the formula (1).

Returning to (2) we make the linear substitution $\omega = \zeta x^2/2K$ to obtain

$$\phi(x, t) = \frac{2K}{x^2}\frac{1}{2\pi}\int_{-\infty}^{\infty} \hat{f}(2\omega K/x^2)G(\omega)\exp(2i\omega tK/x^2)\,d\omega. \tag{3}$$

Using this formula alone, we can recover most of the information about heat pulses that we obtained in Chapter 62. Observe first that $|G(\omega)| = \exp(-|\omega|^{\frac{1}{2}}) \to 0$ rapidly as $\omega \to \infty$. Thus we may hope that

$$\phi(x, t) \approx \frac{2K}{x^2}\frac{1}{2\pi}\int_{-R}^{R} \hat{f}(2\omega K/x^2)G(\omega)\exp(2i\omega tK/x^2)\,d\omega$$

for some large R.

In other words, since, as we saw in Chapter 7, high frequency oscillations are damped out very rapidly with depth, the value of $\phi(x, t)$ is determined essentially by the low frequency components and so by the values of $\hat{f}(\zeta)$ with $|\zeta| \leqslant 2RK/x^2$. But if $f(t) = 0$ for $t \in [0, a]$ then $\hat{f}(\zeta) = \int_0^a f(t)\exp - i\zeta t\,d\zeta \to \int_0^a f(t)\,dt = A$, say, as $\zeta \to 0$ and so $\hat{f}(\zeta) \approx A$ for small ζ. In other words the values of the low frequency components of a short pulse are essentially independent of the shape of that pulse.

Combining the results of the last two paragraphs we see that if $x^2 \gg Ka$ then

$$\phi(x, t) \approx \frac{2K}{x^2}\frac{1}{2\pi}\int_{-R}^{R} \hat{f}(2\omega K/x^2)G(\omega)\exp(2i\omega tK/x^2)\,d\omega$$

$$\approx \frac{2K}{x^2}\frac{1}{2\pi}\int_{-R}^{R} AG(\omega)\exp(2i\omega tK/x^2)\,d\omega \approx \frac{AK}{\pi x^2}\hat{G}(-2tK/x^2).$$

Thus we see that (for large x)

(a) the 'shape' of the pulse at x is essentially independent of the 'shape' of the pulse at 0,
(b) the 'size' of the pulse at x is inversely proportional to x^2 (and proportional to K),
(c) the length of time before the pulse reaches a significant fraction of its maximum value at x and the length of time during which it remains significant are both proportional to x^2 (and inversely proportional to K).

To get (2) into a form resembling that of Theorem 62.2 we temporarily write $S(\zeta) = G(-\zeta x^2/2K)$ and proceed as follows. By the inversion formula

$$S^{\wedge\wedge}(\zeta) = 2\pi S(-\zeta),$$

whilst by the convolution formula (Theorem 51.7 (iv))

$$\hat{f}(\zeta)S(-\zeta) = (2\pi)^{-1}\hat{f}(\zeta)S^{\wedge\wedge}(\zeta) = (2\pi)^{-1}(f*\hat{S})^{\wedge}(\zeta).$$

Thus

$$\phi(x, t) = \frac{1}{2\pi}\int_{-\infty}^{\infty} \hat{f}(\zeta)S(-\zeta)\exp i\zeta t \, d\zeta$$

$$= \left(\frac{1}{2\pi}\right)^2 \int_{-\infty}^{\infty} (f*\hat{S})^{\wedge}(\zeta)\exp i\zeta t \, d\zeta = (2\pi)^{-1}(f*\hat{S})(t),$$

using the inversion formula again. But, making the linear substitution $\omega = -\zeta x^2/2K$, we have

$$\hat{S}(t) = \int_{-\infty}^{\infty} G(-\zeta x^2/2K)\exp(-i\zeta t)\,d\zeta$$

$$= \frac{2K}{x^2}\int_{-\infty}^{\infty} G(\omega)\exp(2i\omega tK/x^2)\,d\omega = \frac{2K}{x^2}\,\hat{G}\left(-\frac{2tK}{x^2}\right)$$

(or we could have used Lemma 55.1 (i)). Thus if $f(t) = 0$ for $t \leqslant 0$ we have

$$\phi(x, t) = \frac{1}{2\pi}\int_{-\infty}^{\infty} f(s)\hat{S}(t-s)\,ds = \frac{K}{\pi x^2}\int_0^{\infty} f(s)\hat{G}(-2K(t-s)/x^2)\,ds.$$

We will thus have obtained a solution identical with that of Theorem 62.2 if we can show that

$$\hat{G}(u) = \frac{(2\pi)^{\frac{1}{2}}}{(-u)^{\frac{3}{2}}}\exp\left(\frac{1}{2u}\right) \quad \text{for } u < 0,$$

$$\hat{G}(u) = 0 \qquad\qquad \text{for } u \geqslant 0$$

(i.e. in the notation of the lemma below, that $\hat{G} = 2\pi P$).

This concludes the conceptually interesting part of the chapter since what remains is a mere computation. However, I was unable to do this calculation unaided and it is possible that the reader in turn would like some help in this matter. (Though, obviously, she should try to do it by herself first and if she succeeds so much the better.)

Lemma 63.2. (i) *If a and b are real with $a > 0$, $b \geqslant 0$ then*

$$\int_0^\infty \exp(-a^2 u^2 + ib^2 u^{-2}) \, du = \frac{\pi^{\frac{1}{2}}}{2a} \exp(-2^{\frac{1}{2}}ab(1-i)),$$

$$\int_0^\infty \exp(-a^2 u^2 - ib^2 u^{-2}) \, du = \frac{\pi^{\frac{1}{2}}}{2a} \exp(-2^{\frac{1}{2}}ab(1+i)).$$

(ii) *If*

$$P(t) = \frac{1}{(2\pi)^{\frac{1}{2}}(-t)^{\frac{3}{2}}} \exp\left(-\frac{1}{2t}\right) \quad \text{for } t < 0,$$

$$P(t) = 0 \qquad\qquad\qquad\qquad \text{for } t \geqslant 0,$$

then $\hat{P}(\zeta) = G(-\zeta)$ for all $\zeta \in \mathbb{R}$, where

$$G(\zeta) = \exp(-(1+i)|\zeta|^{\frac{1}{2}}) \quad \text{for } \zeta \geqslant 0,$$
$$G(\zeta) = \exp(-(1-i)|\zeta|^{\frac{1}{2}}) \quad \text{for } \zeta < 0.$$

(iii) *With the notation of* (ii), *$\hat{G}(t) = 2\pi P(t)$ for all $t \in \mathbb{R}$.*

Proof. (i) Throughout the proof we shall treat a as a constant and b as a variable. If $S > \eta > 0$, let us write

$$I(a, b, \eta, S) = \int_\eta^S \exp(-a^2 u^2 + ib^2 u^{-2}) \, du.$$

Since $|\exp(-au^2 + ib^2 u^{-2})| = \exp(-a^2 u^2)$, a comparison test shows that $\int_0^\infty \exp(-a^2 u^2 + ib^2 u^{-2}) \, du$ exists, with value $I(a, b)$, say, and that $I(a, b, \eta, S) \to I(a, b)$, uniformly in b, as $\eta \to 0$, $S \to \infty$.

By Theorem 53.4 we may differentiate under the integral sign to get

$$\frac{\partial I}{\partial b}(a, b, \eta, S) = \int_\eta^S \frac{\partial}{\partial b} \exp(-a^2 u^2 + ib^2 u^{-2}) \, du$$

$$= 2ib \int_\eta^S \frac{1}{u^2} \exp(-a^2 u^2 + ib^2 u^{-2}) \, du.$$

If $b \neq 0$, the substitution, $v = ba^{-1} u^{-1}$ now gives

$$\frac{\partial I}{\partial b}(a, b, \eta, S) = -2ia \int_{ba^{-1}\eta^{-1}}^{ba^{-1}S^{-1}} \exp(-b^2 v^{-2} + ia^2 v^2) \, dv$$

$$= 2ia \int_{ba^{-1}S^{-1}}^{ba^{-1}\eta^{-1}} \exp(ia^2 v^2 - b^2 v^{-2}) \, dv.$$

The reader may be tempted at this point to try and make the further substitution $u = i^{\frac{1}{2}}v$ but this is not permissible since it changes the path along which the integral is evaluated and not merely (as a permissible substitution would) the speed with which the path is traversed. However we can achieve a similar effect by the legal methods of contour integration.

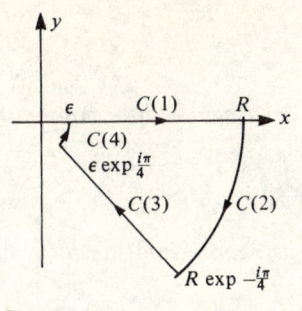

Fig. 63.1. Contour for the calculations of Lemma 63.2.

Let $f(z) = \exp(-a^2 z^2 + ib^2 z^{-2})$ so that f is analytic in the complex plane except at 0. Let $R > \varepsilon > 0$ and let C be the contour of Figure 63.1 composed of the straight line $C(1)$ from ε to R, the segment of a circle $C(2)$ traced out by $R \exp i\theta$ as θ runs from 0 to $-\pi/4$, the straight line $C(3)$ from $R \exp - i\pi/4$ to $\varepsilon \exp - i\pi/4$ and the segment of a circle $C(4)$ traced out by $\varepsilon \exp i\theta$ as θ runs from $-\pi/4$ to 0. By Cauchy's theorem

$$0 = \int_C f(z)\,dz = \int_{C(1)} + \int_{C(2)} + \int_{C(3)} + \int_{C(4)} f(z)\,dz$$

and we now proceed to consider the $\int_{C(j)} f(z)\,dz$ one at a time.

Since $\operatorname{Re}(-a^2 z^2 + ib^2 z^{-2}) \leqslant 0$ and $|f(z)| \leqslant 1$ for all $z \in C(4)$, it follows that

$$\left| \int_{C(4)} f(z)\,dz \right| \leqslant \text{length } C(4) \times \sup_{z \in C(4)} |f(z)| = \frac{\pi\varepsilon}{4} \to 0 \quad \text{as } \varepsilon \to 0.$$

The estimates for $C(2)$ are more delicate and follow the idea of Jordan's lemma (just as in Lemma 61.2). Observe that $\sin\psi \geqslant 2\psi/\pi$, and so

$$\cos\psi = \sin(\pi/2 - \psi) \geqslant 2(\pi/2 - \psi)/\pi = 1 - 2\psi/\pi \quad \text{for } 0 \leqslant \psi \leqslant \pi/2.$$

Thus

$$\operatorname{Re}(-a^2(R \exp i\theta)^2 + ib^2(R \exp i\theta)^{-2}) = \operatorname{Re}(-a^2 R^2 \exp 2i\theta + ib^2 R^{-2} \exp - 2i\theta)$$

$$= -a^2 R^2 \cos 2\theta + b^2 R^{-2} \sin 2\theta \leqslant -a^2 R^2 \cos(-2\theta)$$

$$\leqslant -a^2 R^2 (1 + 4\theta/\pi) \quad \text{for } -\frac{\pi}{4} \leqslant \theta \leqslant 0$$

and

$$\left| \int_{C(2)} f(z)\,dz \right| = \left| \int_0^{-\pi/4} f(R \exp i\theta) iR \exp i\theta\,d\theta \right|$$

$$\leqslant R \int_{-\pi/4}^0 |f(R \exp i\theta)|\,d\theta \leqslant \int_{-\pi/4}^0 \exp(-a^2 R^2 (1 + 4\theta/\pi))\,d\theta$$

$$= R \left[\frac{\pi}{4R^2 a^2} \exp(-a^2 R^2 (1 + 4\theta/\pi)) \right]_{-\pi/4}^0 \leqslant \frac{\pi}{4Ra^2} \to 0 \quad \text{as } R \to \infty.$$

On the other hand

$$\int_{C(1)} f(z)\,dz = \int_{\varepsilon}^{R} \exp(-a^2v^2 + ib^2v^{-2})\,dv \to I(a,b) \quad \text{as } R \to \infty, \varepsilon \to 0,$$

and

$$\int_{C(3)} f(z)\,dz = \int_{R}^{\varepsilon} \exp(-a^2(v\exp - i\pi/4)^2 + ib^2(v\exp - i\pi/4)^{-2})\exp(-i\pi/4)\,dv$$

$$= -\exp(-i\pi/4)\int_{\varepsilon}^{R} \exp(ia^2v^2 - b^2v^{-2})\,dv.$$

It follows that

$$\exp(-i\pi/4)\int_{\varepsilon}^{R} \exp(ia^2v^2 - b^2v^{-2})\,dv = -\int_{C(3)} f(z)\,dz$$

$$= \int_{C(1)} f(z)\,dz + \int_{C(2)} f(z)\,dz + \int_{C(4)} f(z)\,dz$$

$$\to I(a,b) + 0 + 0 = I(a,b) \quad \text{as } \varepsilon \to 0, R \to 0.$$

Looking back over our computations we see that the convergence is uniform in b. But, as we showed in the second paragraph of the proof,

$$\frac{\partial I}{\partial b}(a,b,\eta,S) = 2ia\int_{ba^{-1}s^{-1}}^{ba^{-1}\eta^{-1}} \exp(ia^2v^2 - b^2v^{-2})\,dv.$$

Thus, choosing c and d with $0 < c < d$, we have

$$\frac{\partial I}{\partial b}(a,b,n^{-1},n) = 2ia\int_{ba^{-1}n^{-1}}^{ba^{-1}n} \exp(ia^2v^2 - b^2v^{-2})\,dv \to 2ia\exp(i\pi/4)/I(a,b)$$

$$= 2a\exp(3i\pi/4)I(a,b) \quad \text{as } n \to \infty \text{ uniformly for } b \in [c,d].$$

Since we already know that $I(a,b,n^{-1},n) \to I(a,b)$ as $n \to \infty$, a standard result (Theorem 53.4) now tells us that

$$\frac{\partial I}{\partial b}(a,b) = 2a\exp(3i\pi/4)I(a,b)$$

whenever $b \in [c,d]$. Since c and d were arbitrary we have

$$\frac{\partial I}{\partial b}(a,b) = 2a\exp(3i\pi/4)I(a,b)$$

for all $a, b > 0$.

Obtaining the differential equation

$$\frac{\partial I}{\partial b}(a,b) = 2a\exp(3i\pi/4)I(a,b) \quad [a,b > 0]$$

in b, represents a major step forward in finding $I(a, b)$. We have, immediately, $I(a, b) = A(a) \exp((2a \exp 3i\pi/4)b) = A(a) \exp(2^{\frac{1}{2}}(-1 + i)ab)$ for all a, $b > 0$, where $A(a)$ is to be determined.

But $I(a, b) \to I(a, 0)$ as $b \to 0+$. (To see this, write

$$g(u) = \exp(-a^2u^2 + ib^2u^{-2}) - \exp(-a^2u^2)$$

and choose $R > \varepsilon > 0$. Then

$$|g(u)| = \exp(-a^2u^2)|1 - \exp(ib^2u^{-2})|,$$

so

$$|g(u)| \leqslant 2\exp(-a^2u^2) \quad \text{for all } u > 0,$$

and $g(u) \to 0$ uniformly on $[\varepsilon, R]$ as $b \to 0+$. Thus

$$|I(a, b) - I(a, 0)| \leqslant \int_0^\infty |g(u)| \, du \leqslant \int_0^\varepsilon + \int_R^\infty |g(u)| \, du + \int_\varepsilon^R |g(u)| \, du$$

$$\leqslant \int_0^\varepsilon + \int_R^\infty 2\exp(-a^2u^2) \, du + \int_\varepsilon^R |g(u)| \, du$$

$$\to \int_0^\varepsilon + \int_R^\infty 2\exp(-a^2u^2) \, du \quad \text{as } b \to 0+,$$

so

$$\limsup_{b \to 0+} |I(a, b) - I(a, 0)| \leqslant \left(\int_0^\varepsilon + \int_R^\infty 2\exp(-a^2u^2) \, du \right).$$

Allowing $\varepsilon \to 0+$ and $R \to \infty$ we obtain $\limsup_{b \to 0+} |I(a, b) - I(a, 0)| = 0$, which is the required result.)

Thus

$$A(a) \exp(2^{-\frac{1}{2}}(1 - i)ab) \to I(a, 0) \quad \text{as } b \to 0+,$$

and so

$$A(a) = I(a, 0) = \int_0^\infty \exp(-a^2v^2) \, dv$$

$$= \frac{1}{2^{\frac{1}{2}}a} \int_0^\infty \exp(-u^2/2) \, du = \frac{1}{2^{\frac{3}{2}}a} \int_{-\infty}^\infty \exp(-u^2/2) \, du = \frac{(2\pi)^{\frac{1}{2}}}{2^{\frac{3}{2}}a} = \frac{\pi^{\frac{1}{2}}}{2a}$$

(making the change of variable $u = 2^{\frac{1}{2}}av$ and using the value of $\int_{-\infty}^\infty \exp(-x^2/2) \, dx$ found, for example, in Lemma 48.10). It follows that

$$I(a, b) = \frac{\pi^{\frac{1}{2}}}{2a} \exp(2^{\frac{1}{2}}(1 - i)ab) \quad \text{for all } a \geqslant 0, b > 0,$$

and we are done.

To find $\int_0^\infty \exp(-a^2u^2 - ib^2u^{-2}) \, du$ we can either repeat the computation or observe that

$$\int_0^\infty \exp(-a^2u^2 - ib^2u^{-2})\,du = \int_0^\infty (\exp(-a^2u^2 + ib^2u^{-2}))^* \, du$$

$$= \left(\int_{-\infty}^\infty \exp(-a^2u^2 + ib^2u^{-2})\,du \right)^*.$$

(ii) Making the substitution $u = (-t)^{\frac{1}{2}}$ we have

$$\hat{P}(\zeta) = \frac{1}{(2\pi)^{\frac{1}{2}}} \int_{-\infty}^0 \frac{1}{t^{3/2}} \exp\left(\frac{1}{2t}\right) \exp(-i\zeta t)\,dt = \frac{2}{(2\pi)^{\frac{1}{2}}} \int_0^\infty \exp\left(-\frac{u^2}{2} + i\zeta u^{-2}\right) du,$$

so by part (i) with $a = 2^{-\frac{1}{2}}$, $b = |\zeta|^{\frac{1}{2}}$ we have

$$\hat{P}(\zeta) = \begin{cases} \exp(-2^{\frac{1}{2}}2^{-\frac{1}{2}}|\zeta|^{\frac{1}{2}}(1-i)) & \text{if } \zeta \geqslant 0 \\ \exp(-2^{\frac{1}{2}}2^{-\frac{1}{2}}|\zeta|^{\frac{1}{2}}(1+i)) & \text{if } \zeta < 0 \end{cases}$$

$$= G(-\zeta).$$

(iii) Let us write $f^\vee(t) = f(-t)$. The linear change of variable $u = -t$ shows that, if $f \in L^1 \cap C$

$$f^{\vee\wedge}(\zeta) = \int_{-\infty}^\infty f(-t)\exp - i\zeta t\,dt = \int_{-\infty}^\infty f(u)\exp i\zeta u\,du = f^{\wedge\vee}(\zeta).$$

Thus, since $G, P \in L^1 \cap C$, we can use part (ii), which shows $\hat{P} = \check{G}$, and the inversion formula, which shows $2\pi P = P^{\wedge\wedge\vee}$, to obtain

$$\check{G} = G^{\vee\wedge\vee} = P^{\wedge\wedge\vee} = 2\pi P,$$

completing the calculation.   ∎

Remark 1. Lemma 63.2 can be generalised to show that $\int_0^\infty \exp(-\alpha u^2 - \beta u^{-2})\,du$ converges to $(\pi/2\alpha^{\frac{1}{2}})\exp(2(\alpha\beta)^{\frac{1}{2}})$ whenever $\operatorname{Re}\alpha > 0$, $\operatorname{Re}\beta \geqslant 0$. We leave it as an instructive exercise to the reader to determine which square roots of $\alpha\beta$ and α have to be taken to make the formula correct.

Remark 2. Faced with a proof like the one above the novice mathematician tends to adopt an attitude of respectful bafflement. 'I could never do anything like that.' But if we look at the proof above carefully we see that it contains only one idea – that of examining $(\partial I/\partial b)(a, b)$. (The one idea is, however, rather cunning.) The rest of the proof consists of prefabricated mathematical blocks (a contour integration, an application of Theorem 53.4, and so on) which rigorise the formal argument set out below.

$$\frac{\partial I}{\partial b}(a, b) = \int_0^\infty \frac{\partial}{\partial b} \exp(-a^2u^2 + ib^2u^{-2})\,du = 2ib\int_0^\infty u^{-2}\exp(-a^2u^2 + ib^2u^{-2})\,du$$

$$= 2ia\int_\infty^0 \exp(ia^2v^2 - b^2v^{-1})\,dv \quad [v = ba^{-1}u^{-1}]$$

$$= -2ia \int_0^\infty \exp(ia^2v^2 - b^2v^{-2})\,dv$$

$$= 2i^{\frac{1}{2}}a \int_0^\infty \exp(-a^2w^2 + ib^2w^{-2})\,dw \quad [w = i^{\frac{1}{2}}v]$$

$$= 2i^{\frac{1}{2}}aI(a,b),$$

and so $I(a,b) = I(a,0)\exp(2i^{\frac{1}{2}}ab)$.

The need for a rigorous rather than a formal argument is best seen by looking at the behaviour of

$$u^{-2}\exp(-a^2u^2 + ib^2u^{-2}) \quad \text{as } u \to 0+.$$

The convergence of

$$\int_0^\infty u^{-2}\exp(-a^2u^2 + ib^2u^{-2})\,du$$

is due to fairly delicate cancellation and the arguments used reflect this.

Remark 3. If we were not interested in an independent derivation of the results of Chapter 62 but only in obtaining the results of Lemma 63.2, then the following chain of arguments might well have occurred to the reader. Suppose we knew that the problem solved in Theorem 62.2 had a unique solution under fairly wide conditions (we will return to this question in Chapter 67). Then by showing that the functions $f*P_{2K/x^2}$ and $(2\pi)^{-1}f*(\hat{G})_{2K/x^2}$ both solved this problem we could deduce that $f*P_{2K/x^2} = (2\pi)^{-1}f*(\hat{G})_{2K/x^2}$ for a large class of f. In particular, choosing the f to be members of a sequence of 'approximate identities' in a sense like that of Theorem 52.6, we could deduce without any trouble that $P_{2K/x^2} = (2\pi)^{-1}\hat{G}_{2K/x^2}$ and so $P = (2\pi)^{-1}\hat{G}$. The reader is invited to examine the not insuperable problems of making this argument rigorous.

64

THE WAVE EQUATION

Among partial differential equations in two variables three have historically proved of particular interest

(a) Laplace's potential equation $\partial^2\phi/\partial x^2 + \partial^2\phi/\partial y^2 = 0$,
(b) Fourier's heat equation $\partial^2\phi/\partial x^2 - \partial\phi/\partial y = 0$,
(c) the wave equation $\partial^2\phi/\partial x^2 - \partial^2\phi/\partial y^2 = 0$.

In this book we have studied the first two equations both for their own sake and to give examples of the power of the techniques of Fourier analysis (as in Chapters 7, 14, 15, 28 to 31 and Chapter 55 onwards). We shall look at the wave equation briefly in the present chapter but only to compare it to the heat equation. The reader who knows anything at all about the wave equation will find nothing new.

We first make the substitution $y = ct$ (with $c > 0$) to transform the wave equation to its usual form

$$\frac{\partial^2\phi}{\partial t^2}(x, t) = c^2 \frac{\partial^2\phi}{\partial x^2}(x, t).$$

The following problem is analogous to the one solved for the heat equation in Chapter 55.

Problem. *Solve the wave equation*

$$\frac{\partial^2\phi}{\partial t^2} = c^2 \frac{\partial^2\phi}{\partial x^2} \qquad \text{for } x \geqslant 0, \ t \geqslant 0$$

subject to the boundary conditions

$$\phi(x, 0) = u(x), \qquad \frac{\partial\phi}{\partial t}(x, 0) = v(x).$$

A beautiful solution of this problem was given by d'Alembert. Set $X(x, t) = x + ct$ and $Y(x, t) = x - ct$. Then

$$\frac{\partial\phi}{\partial t} = \frac{\partial X}{\partial t}\frac{\partial\phi}{\partial X} + \frac{\partial Y}{\partial t}\frac{\partial\phi}{\partial Y} = c\frac{\partial\phi}{\partial X} - c\frac{\partial\phi}{\partial Y},$$

and (assuming that ϕ is sufficiently smooth to ensure that $\partial^2\phi/\partial X\partial Y = \partial^2\phi/\partial Y\partial X$)

$$\frac{\partial^2\phi}{\partial t^2} = c\frac{\partial}{\partial x}\left(c\frac{\partial\phi}{\partial X} - c\frac{\partial\phi}{\partial Y}\right) - c\frac{\partial}{\partial Y}\left(c\frac{\partial\phi}{\partial X} - c\frac{\partial\phi}{\partial Y}\right)$$

$$= c^2\left(\frac{\partial^2\phi}{\partial X^2} - 2\frac{\partial^2\phi}{\partial Y\partial X} + \frac{\partial^2\phi}{\partial Y^2}\right).$$

Similarly

$$\frac{\partial^2\phi}{\partial x^2} = \frac{\partial^2\phi}{\partial X^2} + 2\frac{\partial^2\phi}{\partial Y\partial X} + \frac{\partial^2\phi}{\partial Y^2}.$$

Thus the wave equation may be written as

$$c^2\left(\frac{\partial^2\phi}{\partial X^2} - 2\frac{\partial^2\phi}{\partial Y\partial X} + \frac{\partial^2\phi}{\partial Y^2}\right) = c^2\left(\frac{\partial^2\phi}{\partial X^2} + 2\frac{\partial^2\phi}{\partial Y\partial X} + \frac{\partial^2\phi}{\partial Y^2}\right),$$

i.e.

$$\frac{\partial^2\phi}{\partial Y\partial X} = 0.$$

Integrating with respect to Y we obtain

$$\frac{\partial\phi}{\partial X} = f(X),$$

where f is some general (sufficiently smooth) function, and then integrating with respect to X we obtain

$$\phi = \int_0^X f(s)ds + g(Y),$$

where g is some (sufficiently smooth) function. Substituting for X and Y we obtain the solution

$$\phi(x, t) = \int_0^{x+ct} f(s)ds + g(x - ct).$$

The apparent asymmetry is due to the fact that we integrated first with respect to Y and then with respect to X. We restore the symmetry by rewriting the solution as

$$\phi(x, t) = F(x + ct) + G(x - ct),$$

with $F(y) = \int_0^y f(s)ds$ (and so, in particular, $F(0) = 0$), $G(y) = g(y)$.

Putting in the boundary conditions $\phi(x, 0) = u(x), (\partial\phi/\partial t)(x, 0) = v(x)$ we obtain

$$F(x) + G(x) = u(x)$$

(and so, in particular, $G(0) = u(0)$) whilst

$$cF'(x) - cG'(x) = v(x).$$

Thus differentiating the first of the two equations we get $F'(x) + G'(x) = u'(x)$ and so,

adding,

$$2cF'(x) = cu'(x) + v(x).$$

Integrating both sides of the equation we get

$$2cF(y) = c(u(y) - u(0)) + \int_0^y v(s)ds.$$

Similarly we have $2cG'(x) = cu'(x) - v(x)$ and

$$2c(G(y) - G(0)) = c(u(y) - u(0)) - \int_0^y v(s)ds$$

giving

$$2cG(y) = cu(y) - \int_0^y v(s)\,ds.$$

Thus

$$\phi(x, t) = F(x + ct) + G(x - ct)$$

$$= \frac{1}{2c}\left(c(u(x + ct) + u(x - ct)) + \int_0^{x+ct} v(s)\,ds - \int_0^{x-ct} v(s)\,ds \right)$$

$$= \tfrac{1}{2}(u(x + ct) + u(x - ct)) + \frac{1}{2c}\int_{x-ct}^{x+ct} v(s)\,ds.$$

We summarise (and check) these results in the following theorem.

Theorem 64.1. (i) *If $F, G : \mathbb{R} \to \mathbb{C}$ are twice differentiable then, writing*

$$\phi(x, t) = F(x + ct) + G(x - ct),$$

we have

$$\frac{\partial^2 \phi}{\partial t^2} = c^2 \frac{\partial^2 \phi}{\partial x^2}.$$

(ii) *If $u : \mathbb{R} \to \mathbb{C}$ is twice differentiable and $v : \mathbb{R} \to \mathbb{C}$ is once differentiable then, writing*

$$\phi(x, t) = \frac{u(x + ct) + u(x - ct)}{2} + \frac{1}{2c}\int_{x-ct}^{x+ct} v(s)\,ds,$$

we have

$$\frac{\partial^2 \phi}{\partial t^2} = c^2 \frac{\partial \phi}{\partial x^2}, \quad \phi(x, 0) = u(x)$$

and

$$\frac{\partial \phi}{\partial t}(x, 0) = v(x).$$

Proof. Direct verification which is left to the reader.   ∎

We could also have obtained the results above (though for a very restricted class of ϕ, u and v) by the methods of Chapter 55. In order to follow and compare the

argument, the reader should reread Chapter 55 as far as Theorem 55.3. (If she is not interested, she can skip to the remarks preceding Theorem 64.3.)

We recall a simple but useful fact about Fourier transforms.

Lemma 64.2. *Let* $h_x(t) = h(t - x)$. *Then, if* $h \in L^1 \cap C$, *we have*

$$\hat{h}_x(\zeta) = \exp(-i\zeta x)\hat{h}(\zeta).$$

Proof. Left to the reader (or see Lemma 61.4(i)). ∎

We try to solve the wave equation

$$(D_2 D_2 \phi)(x, t) = c^2 (D_1 D_1 \phi)(x, t),$$

in exactly the same way as we solved the heat equation in Chapter 55. (Indeed it would be a good idea for the reader to see how far she can get in this way before reading what follows.)

Taking the Fourier transform of each side we get

$$\mathscr{F}_1(D_2 D_2 \phi)(\zeta, t) = c^2 (\mathscr{F}_1(D_1 D_1 \phi)(\zeta, t)).$$

But by Lemma 55.1(ii) we have (at least if ϕ is sufficiently well behaved)

$$\mathscr{F}_1(D_2 D_2 \phi)(\zeta, t) = c^2 (i\zeta)^2 (\mathscr{F}_1 \phi)(\zeta, t),$$

and so (assuming that we can differentiate under the integral sign)

$$(D_2 D_2 \mathscr{F}_1 \phi)(\zeta, t) = c^2 (i\zeta)^2 (\mathscr{F}_1 \phi)(\zeta, t),$$

i.e.
$$\frac{\partial^2}{\partial t^2}(\mathscr{F}_1 \phi)(\zeta, t) = -(c^2 \zeta^2)(\mathscr{F}_1 \phi)(\zeta, t).$$

Solving this differential equation for $(\mathscr{F}_1 \phi)(\zeta, \ )$ we obtain

$$(\mathscr{F}_1 \phi)(\zeta, t) = A(\zeta) \cos c\zeta t + B(\zeta) \sin c\zeta t,$$

where $A: \mathbb{R} \to \mathbb{C}$ and $B: \mathbb{R} \to \mathbb{C}$ are to be determined. Using the boundary conditions we have

$$A(\zeta) = (\mathscr{F}_1 \phi)(\zeta, 0) = \hat{u}(\zeta),$$

whilst

$$c\zeta B(\zeta) = \frac{\partial}{\partial t}(\mathscr{F}_1 \phi)(\zeta, 0) = (D_2 \mathscr{F}_1 \phi)(\zeta, 0) = (\mathscr{F}_1 D_2 \phi)(\zeta, 0) = \hat{v}(\zeta),$$

giving

$$(\mathscr{F}_1 \phi)(\zeta, t) = \hat{u}(\zeta) \cos c\zeta t + \frac{\hat{v}(\zeta)}{c\zeta} \sin c\zeta t.$$

In general, even if $v \in L^1 \cap C$, it is extremely unlikely that its integral $V(y) =$

$\int_{-\infty}^{y} v(s)\,ds$ also lies in $L^1 \cap C$. However, since our argument is heuristic, let us confine ourselves to the case when $V \in L^1 \cap C$. Then, by Lemma 55.1(i), since $V' = v$, we have $\hat{v}(\zeta) = i\zeta \hat{V}(\zeta)$ and

$$
\begin{aligned}
(\mathscr{F}_1\phi)(\zeta, t) &= \hat{u}(\zeta)\cos\zeta ct + ic^{-1}\hat{V}(\zeta)\sin\zeta ct \\
&= \tfrac{1}{2}((\hat{u}(\zeta)e^{i\zeta ct} + \hat{u}(\zeta)e^{-i\zeta ct}) + c^{-1}(\hat{V}(\zeta)e^{i\zeta ct} - \hat{V}(\zeta)e^{-i\zeta ct})) \\
&= \tfrac{1}{2}((\hat{u}_{-ct}(\zeta) + \hat{u}_{ct}(\zeta)) + c^{-1}(\hat{V}_{-ct}(\zeta) - \hat{V}_{ct}(\zeta))) \\
&= (\tfrac{1}{2}((u_{ct} + u_{-ct}) - c^{-1}(V_{ct} - V_{-ct})))\,\hat{}(\zeta)
\end{aligned}
$$

(using Lemma 64.2). Thus by the uniqueness of the Fourier transform (Theorem 49.4) we have

$$
\begin{aligned}
\phi(x, t) &= (\tfrac{1}{2}((u_{ct} + u_{-ct}) - c^{-1}(V_{ct} - V_{-ct})))(x) \\
&= \tfrac{1}{2}(u(x + ct) - u(x - ct)) - \frac{1}{2c}\left(\int_{-\infty}^{x-ct} v(s)\,ds - \int_{-\infty}^{x+ct} v(s)\,ds\right) \\
&= \tfrac{1}{2}(u(x + ct) - u(x - ct)) + \frac{1}{2c}\left(\int_{x-ct}^{x+ct} v(s)\,ds\right).
\end{aligned}
$$

The reader may feel that the comparison of the two methods for solving the wave equation does not show the Fourier analytic method in a particularly favourable light. But she should observe that the Fourier transform gives a general method obviously applicable to a wide choice of linear partial differential equations and it is only to be expected that, in special cases, special methods may prove more attractive than general ones.

Comparison of Theorem 55.3 and Theorem 64.1 reveals two interesting and related differences between our solutions of the heat and wave equations.

(1) The solutions of the heat equation become smoother as we go forward in time. In particular, whatever the character of the boundary function $f(\ ) = \lim_{t \to 0+} \phi(\ , t)$, $\phi(\ , t)$ is infinitely differentiable for all $t > 0$. This is not true of the wave equation.

(2) The behaviour of our solutions of the wave equation can be traced backward as well as forward in time. In particular if u is twice differentiable and v once differentiable then

$$
\phi(x, t) = \tfrac{1}{2}(u(x + ct) + u(x - ct)) + \frac{1}{2c}\int_{x-ct}^{x+ct} v(s)\,ds
$$

satisfies the wave equation

$$
\frac{\partial^2 \phi}{\partial t^2}(x, t) = c^2\frac{\partial^2 \phi}{\partial x^2}(x, t)
$$

for all $t \in \mathbb{R}$ (and not only $t > 0$) as well as the boundary conditions

$$
\phi(x, 0) = u(x), \frac{\partial \phi}{\partial t}(x, 0) = v(x).
$$

On the other hand, the remark (1) above shows that this cannot be true for our solution of the heat equation. (For example, if ϕ is a solution of the heat equation in $t > -1$ of the type given in Theorem 55.3, then we cannot have

$$\phi(x,0) = f(x) \quad \text{with} \quad f(x) = (1-x^2)^3 \quad \text{for } |x| < 1, \ f(x) = 0 \text{ otherwise,}$$

since, although f is reasonably well behaved, it is not infinitely differentiable.)

Because the solution to the wave equation given in Theorem 64.1 is so simple, it is easy to solve these problems for the wave equation which are analogous to those for the heat equation in Chapters 55 and 62.

Theorem 64.3. *Suppose that* $u:[0,\infty)\to\mathbb{C}$ *is twice differentiable with* $u(0)=0$ *and (right second derivative)* $u''(0)=0$ *whilst* $v:[0,\infty)\to\mathbb{C}$ *is once differentiable with* $v(0)=0$. *Then, writing*

$$\tilde{v}(s) = u(s), \tilde{v}(s) \quad \text{for} \quad s \geqslant 0, \tilde{u}(s) = -u(-s), \tilde{v}(s) = -v(-s)$$

for $s < 0$ *and*

$$\phi(x,t) = \frac{\tilde{u}(x+ct)+\tilde{u}(x-ct)}{2} + \frac{1}{2c}\int_{x-ct}^{x+ct}\tilde{v}(s)ds \quad \text{for all } x \geqslant 0,$$

we have

$$\frac{\partial^2\phi}{\partial t^2}(x,t) = c^2\frac{\partial^2\phi}{\partial x^2}(x,t) \quad \text{for all } x > 0,$$

and

$$\phi(x,0) = u(x), \frac{\partial\phi}{\partial t}(x,0) = v(x) \quad \text{for all } x > 0,$$

whilst

$$\phi(0,t) = 0 \quad \text{for all } t.$$

Proof. This uses the method of images in the same way that we used it in Lemma 57.1. Observe that $\tilde{u}$ is certainly twice differentiable except at 0. But at 0

$$\text{left derivative } \tilde{u} = \lim_{h\to0+}\frac{\tilde{u}(-h)-\tilde{u}(0)}{-h} = \lim_{h\to0+}\frac{-u(h)-0}{-h}$$

$$= \lim_{h\to0+}\frac{u(h)-u(0)}{h} = \text{right derivative } u,$$

so $\tilde{u}$ is once differentiable at 0. Again at 0
left second derivative

$$\tilde{u} = \lim_{h\to0+}\frac{\tilde{u}'(-h)-\tilde{u}'(0)}{-h}$$

$$= \lim_{h\to0+}\frac{u'(h)-u'(0)}{-h} = -(\text{right second derivative } u) = 0,$$

so $\tilde{u}$ is twice differentiable at 0 with $\tilde{u}''(0)=0$. Thus $\tilde{u}$ is everywhere twice differentiable and similarly $\tilde{v}$ is once differentiable everywhere.

Thus writing

$$\tilde{\phi}(x,t) = \frac{\tilde{u}(x+ct)+\tilde{u}(x-ct)}{2} + \frac{1}{2c}\int_{x-ct}^{x+ct}\tilde{v}(s)ds \quad \text{for all } x,t \in \mathbb{R},$$

we know from Theorem 64.1 that $\tilde{\phi}$ satisfies the wave equation for $x,t \in \mathbb{R}$ and that

$$\tilde{\phi}(x,0) = \tilde{u}(x), \frac{\partial\tilde{\phi}}{\partial t}(x,0) = \tilde{v}(x).$$

Further,

$$\tilde{\phi}(0,t) = \frac{\tilde{u}(ct)+\tilde{u}(-ct)}{2} + \frac{1}{2c}\int_{-ct}^{ct}\tilde{v}(s)\,ds$$

$$= \frac{u(ct)-u(ct)}{2} + \frac{1}{2c}\left(\int_{0}^{ct}v(s)\,ds - \int_{0}^{ct}v(s)\,ds\right) = 0 \quad \text{for all } t.$$

The required result follows.   ∎

It is even easier to guess the result corresponding to Theorem 62.2.

Theorem 64.4. *Suppose that $f:[0,\infty\} \to \mathbb{C}$ is continuous and twice differentiable with $f(0)=f'(0)=f''(0)=0$. Then writing*

$$\phi(x,t) = f(t-c^{-1}x) = f(-c^{-1}(x-ct)) \quad \text{for } ct \geqslant x, x \geqslant 0 = 0 \quad \text{for } x \geqslant ct, x \geqslant 0$$

we have

(i) $(\partial^2\phi/\partial t^2)(x,t) = c^2(\partial^2\phi/\partial x^2)(x,t)$ *for all $x>0$,*

(ii) $\phi(x,0) = (\partial\phi/\partial t)(x,0)$ *for all $x \geqslant 0$,*

(iii) $\phi(0,t) = f(t)$ *for all $t \geqslant 0$.*

Proof. Trivial verification.   ∎

In spite of the fact that the form of Theorem 64.4 is immediately obvious by inspection, it is informative to see how the method of Chapter 63 applies in this case.

Let us try to find a solution of the wave equation

$$\frac{\partial^2\phi}{\partial t^2} = c^2\frac{\partial^2\phi}{\partial x^2},$$

subject to the boundary condition $\phi(0,t) = f(t)$ in the simple case when $f(t) = \exp i\zeta t$. Seeking a solution of the form $\phi(x,t) = g(x)\exp i\zeta t$, we obtain $(i\zeta)^2 g(x)\exp i\zeta t = c^2 g''(x)\exp i\zeta t$ whence $g''(x)+c^{-2}\zeta^2 g(x) = 0$ and $g(x) = A\exp ic^{-1}\zeta x + B\exp -ic^{-1}\zeta x$. Thus we find $\phi(x,t) = A\exp i\zeta c^{-1}(x+ct) + B\exp -i\zeta c^{-1}(x-ct)$. We now make the remark (which is to some extent inspired by the knowledge of the solutions found elsewhere in this chapter) that the solution $\phi(x,t) = A\exp i\zeta(x+ct)$ with $B=0$ corresponds to inward travelling waves coming from infinity whilst the

solution $\phi(x, t) = B \exp - i\zeta(x - ct)$ corresponds to outward travelling waves.

We therefore try to build up solutions to the problem with general f from the simple solutions $B \exp - i\zeta c^{-1}(x - ct)$ to the special problem when $f(t) = \exp i\zeta t$. Let us try

$$\phi(x, t) = \int_{-\infty}^{\infty} B(\zeta) \exp - i\zeta c^{-1}(x - ct) d\zeta.$$

Setting $x = 0$ we obtain

$$f(t) = \int_{-\infty}^{\infty} B(\zeta) \exp ic\zeta t \, d\zeta = \hat{B}(-ct),$$

and so (making the substitution $u = -ct$)

$$\hat{f}(\zeta) = \int_{-\infty}^{\infty} \hat{B}(-ct) \exp(-i\zeta t) dt = \frac{1}{c} \int_{-\infty}^{\infty} \hat{B}(u) \exp(i\zeta c^{-1} u) du = \frac{1}{c} \hat{\hat{B}}(-\zeta c^{-1}).$$

Thus applying the inversion formula (Theorem 60.1) we have

$$B(\zeta c^{-1}) = (2\pi)^{-1} B^{\wedge\wedge}(-\zeta c^{-1}) = (2\pi)c^{-1} \hat{f}(\zeta),$$

and so $B(\zeta) = (2\pi)c^{-1} \hat{f}(c\zeta)$. Substituting back (and setting $\omega = c\zeta$) we obtain

$$\phi(x, t) = \frac{c}{2\pi} \int_{-\infty}^{\infty} \hat{f}(c\zeta) \exp - i\zeta(x - ct) d\zeta = \frac{1}{2\pi} \int_{-\infty}^{\infty} \hat{f}(\omega) \exp(i\omega c^{-1}(x - ct)) d\omega$$

$$= \frac{1}{2\pi} f^{\wedge\vee}(c^{-1}(x - ct)) = f(-c^{-1}(x - ct)) = f(t - c^{-1}x).$$

Thus in complete contrast with the case of the heat equation there is no damping. A pulse $\phi(x, t) = f(t - c^{-1}x)$ travels at a well defined speed c with no change of shape or magnitude with time. (In particular if $f(t) = 0$ for $t \notin [a, b]$ then $\phi(x, t) = 0$ for $t \notin [a + c^{-1}x, b + c^{-1}x]$.)

65

THE TRANSATLANTIC CABLE I

At the beginning of the nineteenth century a new industrial Europe was taking shape. Its capitalists knew that time is money and its politicians that time is power. During the Napoleonic wars the French and then the English set up systems of semaphores for transmitting military information but with this exception messages could travel no faster than a horse or ship. The invention of the electric telegraph in the 1830s changed all this and within ten years a network of telegraph wires covered England, western Europe and the more settled part of the USA. (Telegraphs were particularly important for the efficient and safe control of the new railways.)

It was thus inevitable that attempts would be made to provide underwater links between the various separate systems. The first cable between Britain and France was laid in 1850. The operators found the greatest difficulty in transmitting even a few words. After 12 hours the enterprise was brought to an abrupt conclusion when a trawler accidentally caught and cut the cable. Undeterred the railway engineer Crampton put up money for a new attempt with a heavier armoured cable of his own design.

Crampton's cable was a complete success and a spurt of submarine cable laying began. On the whole, the short lines worked, but their operators found that signals could not be transmitted along submarine cables as fast as along land lines without becoming confused. (It was the attempt to transmit at normal speed which had led to the jumbled messages along the first Anglo French cable.) The practical men just noted this fact as an annoyance while the scientific men observed that Faraday had predicted some retardation on account of the increased capacitance of undersea cables.

The record of the longer lines was, at best, mixed. Inspite of this an American called Cyrus J. Fields proposed a telegraph line linking Europe and America. Oceanographic surveys showed that the bottom of the Atlantic was suitable for cable laying and by connecting circuits of the existing land telegraph lines people had produced a telegraph line of the length of the proposed cable through which signals had been passed extremely rapidly. The British government offered a subsidy and money was

rapidly raised. Contracts were placed and the business of spinning 2500 miles of cable put in hand.

The mechanical problems of laying such a cable were considerable. Airy, the Astronomer Royal stated that 'it was a mathematical impossibility to submerge the cable successfully at so great a depth'. (Airy had a bad track record – he failed to look for Neptune when it was predicted by Adams, he disbelieved Hamilton's prediction of double refraction, he predicted that the Crystal Palace would be destroyed by hail and he advised the designer of the Tay Bridge about wind speeds without taking into account gusting.) However, the laying was in the hands of an engineering genius– Charles Bright – and it was the electrical side which was to cause the worst problems.

Faraday had indeed predicted signal retardation but he and others like Morse had in mind a model of a submarine cable as a hosepipe which took longer to fill with water (signal) as it got longer. The remedy was thus to use a thin wire (so that less electricity was needed to charge it) and high voltages to push the signal through. Faraday's opinion was shared by Morse and by the electrical adviser Dr Whitehouse.

Thomson's researches had given him a clearer mathematical picture of the problem. The current flow in a telegraph wire in air is governed (approximately) by the wave equation

$$c^2 \frac{\partial^2 \phi}{\partial x^2} = \frac{\partial^2 \phi}{\partial t^2}.$$

As we saw in Chapter 64 a pulse $\phi(x, t) = f(t - c^{-1}x)$ travels at a well defined speed c with no change of shape or magnitude with time. Thus signals can be sent as close together as our transmitter can make them and our receiver distinguish them. (Trained operators could manage up to 40 words a minute in Morse code.)

However in undersea cables (of the type proposed) capacitative effects dominate and current flow is governed (approximately) by the heat equation

$$K \frac{\partial^2 \phi}{\partial x^2} = \frac{\partial \phi}{\partial t}.$$

We discussed this problem in Chapter 62 ending up with a series of remarks on the nature of the solution for a heat pulse. In particular, looking at statement (5) at the end of Chapter 62, we are led to the following prediction.

(5)′ For a very long submarine cable of length x the main effect of any sent electric pulse will last for a time T seconds proportional to x^2 (and inversely proportional to K).

If we try to transmit a further signal during this time T the two signals will be jumbled up when received. We are thus roughly limited to sending one signal every T seconds. The maximum rate of signalling is thus inversely proportional to the square of the length of the cable (and directly proportional to K). Thus in leaping from submarine cables of 30 miles length to cables of length 1500 miles retardation effects

become not 50 times but 2500 times worse. Moreover, contrary to received opinion, increasing the voltage (i.e. increasing the pulse height) would, if anything, make things worse. Finally, since K turns out to be directly proportional to the conductivity, the wire should have as large a diameter as possible.

Whitehouse, whose professional reputation was now involved, denied these conclusions. The shareholders of Field's company were largely British and contained many Glaswegians who voted Thomson onto the board of directors. But even as a director Thomson had no authority over the technical advisers. Moreover the production of the cable was now underway on principles contrary to Thomson's own. Testing the completed cable, he was astonished to find some sections conducted only half as well as others. (This was not due to dishonesty or incompetence, the manufacturers were supplying copper to the then highest commercial standards of purity. The need for really high conductivity had not arisen before the Atlantic cable.) Orders were given for greater care to be taken in future.

Realising that the success of the enterprise would depend on a fast, sensitive detector, Thomson set about to invent one. The problem with an ordinary galvanometer is the high inertia of the needle (which thus requires a large kick to make it swing and a long time to stop swinging.) Thomson came up with the mirror galvanometer in which the pointer is replaced by a beam of light. This marvellous idea so impressed Maxwell as to inspire one of his verses (see Appendix D). The other directors who believed, or hoped, that Whitehouse was right, were polite but not enthusiastic about Thomson's activity on their behalf.

In a first attempt in 1857 the cable snapped after 335 miles had been laid. In 1858 new attempts were foiled by storm and breakages but the company went doggedly on and on 5 August 1858, Europe and America were linked by cable. On 16 August it carried a 99-word message of greeting from Queen Victoria to President Buchanan. But that 99-word message took $16\frac{1}{2}$ hours to get through. In vain Whitehouse tried to get his receiver to work – only Thomson's galvanometer was sensitive enough to interpret the minute and blurred messages coming through. In vain Whitehouse used his two thousand volt induction coils to try to push messages through faster – after four weeks of this treatment the cable gave up the ghost; 2500 tons of cable and £350 000 of capital lay useless on the ocean floor.

66

THE TRANSATLANTIC CABLE II

In 1859 you could buy £1000 of stock in Field's company for £30. The submarine cable to India guaranteed by the British Government failed, costing the taxpayer £800 000. Worldwide eleven thousand miles of undersea cable had been laid and only three thousand miles were operating. And in 1861 civil war broke out in the United States.

Yet Field still continued to try and raise interest in his project travelling back and forth across the Atlantic over 30 times (and being sea sick each time). By 1864 things were moving again. Because of the Civil War most of the capital was raised in Britain, half being put up by the firm which was to make the cable. The cable was redesigned in accordance with Thomson's theories. Even more important, strict quality control was exercised: the copper used was so pure that for the next 50 years 'telegraphist's copper' was the purest available.

Once again the British Government supported the project – the importance of quick communications in controlling an empire was evident to everybody. (Before it failed the old cable had carried an order countermanding troop movements which had saved a seventh of the cost of the cable.) Public interest was intense: when Thomson was delayed at his laboratory, the Glasgow London train would be held until he arrived. (Only Thoreau in America and Arnold in England felt that rapid communication was not needed until people had something worthwhile to communicate.)

The new cable was mechanically much stronger but also heavier. Only one ship was large enough to handle it and that was Brunel's *Great Eastern*. Five times larger than any other existing ship she had been built to steam to Australia without refuelling but bad luck and the opening of the Suez Canal had prevented her profitable use and ruined a succession of owners. A separate company was set up to buy her at a knockdown price and lease her to the cable company in return for a shareholding.

But this time there was a competitor. The Western Union Company had decided to build a cable along the overland route across America, Alaska, the Bering Straits, Siberia and Russia to reach Europe the long way round. The Western Union's views

on competition were simple. 'The public has been slow to learn how disastrous opposition to the Western Union Telegraph Company has been. Competitive lines cannot make a go of it and they might as well quit trying' (quoted in A.F. Harlow *Old Wires and New Waves*, Appleton Century 1936).

The commercial success of the cable would thus depend on the rate at which messages could be transmitted. Thomson had predicted the problems of the first cable by mathematics. On the basis of the same mathematics he now promised the company a rate of eight or even 12 words a minute. Half a million pounds was being staked on the correctness of the solution of a partial differential equation.

On 23 July, 1865 the *Great Eastern* set sail. All went well for nine days and then, when 1250 miles had been laid, the cable parted. For a fortnight the expedition fished for the cable in two miles of water. Three times they caught and lifted it but each time the lifting gear failed. Leaving a buoy to mark the spot the expedition sailed for home.

Thomson and the other engineers were now confident that the cable would do all that was required, since communication with the home base had been perfect until the final break. They also felt sure that they could recover the old cable. The company decided to build and lay a new cable and then go back and complete the old one. By now Field had tried three times to lay his cable and failed each time, yet the public subscribed half of the £600 000 required within two weeks.

This time all went well. Cable laying started on 12 July 1866 and the cable was landed on the morning of the 27th. On the 28th the cable was open for business and earned £1000. A director's meeting of the Western Union ordered all work on their projected line to be stopped. Their crews had laid 300 miles of cable in Alaska, 350 miles in Siberia, 400 miles in the Canadian waste but all this was abandoned at a loss of $3 000 000.

On 1 September after three weeks of effort the old cable was recovered and on 8 September a second perfect cable linked America and Europe. A wave of knighthoods swept over the engineers and directors. Investors who had seen their shares drop to almost nothing now received a return of 16% (in an age when government stock paid $2\frac{1}{2}\%$). Even more importantly the cable manufacturers had established a commanding technical lead – in the next 80 years no less than 90% of all undersea cables were supplied by that one company.

Thomson's knowledge and instruments had contributed greatly to the success of the new industry. In turn the patents which he held made him a wealthy man. He could now indulge his love of the sea from his own yacht. (Though even here he and his friend Helmholtz studied the theory of water waves, '"which", [said] Helmholtz, "he loved to treat as a kind of race between us". When Thomson had to go ashore at Inverary for some hours, as he left he said: "Now, mind, Helmholtz, you're not to work at waves while I'm away" (quoted in S.P. Thompson, p. 614).')

Thomson enjoyed the wealth and fame that his work brought him. (His elevation to the Peerage in 1892 was to leave future generations with the vague impression that there had been three great Victorian scientists called, Thomson, Sir William

Thomson and Lord Kelvin.) But he delighted in technological progress itself and in that fact that '... science has tended very much to accelerate the results, and to give the world the benefit of those results earlier than it could have had them, if left to struggle for them and try for them by repeated efforts and repeated failures unguided by such principles as can be evolved from the abstract investigations of science (speech, on being made a freeman of Glasgow in S.P. Thompson, p. 405).'

67

UNIQUENESS FOR THE
HEAT EQUATION I

When discussing the physical meaning of our solutions for the heat equation we have, more or less tacitly, assumed that they were unique. To what extent is this assumption justified?

Problem. *Suppose* $\phi:\mathbb{R} \times \mathbb{R}^+ \to \mathbb{C}$ *is infinitely differentiable with*

(i) $(\partial\phi/\partial t)(x,t) = K(\partial^2\phi/\partial x^2)(x,t)$ *for all* $x\in\mathbb{R}$, $t>0$ $[K>0]$,
(ii) $\phi(x,t)\to 0$ *as* $t\to 0+$.

Does it follow that $\phi(x,t) = 0$ *for all* $x\in\mathbb{R}$, $t>0$?

Our search for an answer to this problem takes us in an unexpected direction. (There is a fair amount of calculation involved but, so long as she follows the drift of the argument, the reader need not worry too much about the details.)

Lemma 67.1. (i) $x^{-n}\exp(-1/2x^2) \leqslant n^{n/2}$ *for all* $x>0$.
(ii) *Define* $h:\mathbb{R}\to\mathbb{R}$ *by*

$$h(x) = \exp(-1/2x^2) \quad \text{for } x>0,$$
$$h(x) = 0 \quad\quad\quad\;\; \text{for } x\leqslant 0.$$

Then h is infinitely differentiable with

$$h^{(r)}(x) = Q_r(x^{-1})\exp(-1/2x^2) \quad \text{for } x>0,$$
$$h^{(r)}(x) = 0 \quad\quad\quad\quad\quad\quad\quad\;\; \text{for } x\leqslant 0,$$

where
$$Q_r(t) = \sum_{s=0}^{r} a_{rs}t^{r+2s}$$

and
$$|a_{rs}| \leqslant 4^r r^{r-s}, \text{ for } r\geqslant s\geqslant 0, r\geqslant 1, \text{ whilst } a_{00} = 1.$$

(iii) *With h as in* (ii) *we have*

$$|h^{(r)}(x)| \leqslant 2^{5r}r^{3r/2} \quad \text{for all } x\in\mathbb{R}, r\geqslant 1.$$

Remark. This lemma is just a more careful reworking of Example 4.2. Although

338

we give a full proof below the reader may prefer to do it for herself.

Proof. (i) Setting

$$f(x) = x^{-n}\exp(-1/2x^2)$$

we see that

$$f'(x) = x^{-n-1}(-n + x^{-2})\exp(-1/2x^2),$$

so that f' is increasing for x running from 0 to $n^{-\frac{1}{2}}$ and decreasing for x running from $n^{-\frac{1}{2}}$ to ∞. Thus

$$f(x) \leqslant f(n^{-1/2}) \quad \text{for all } x > 0$$

(ii) We proceed inductively. Let $P(r)$ be the proposition that h is r times differentiable with

$$h^{(r)}(x) = Q_r(x^{-1})\exp(-1/2x^2) \quad \text{for } x > 0,$$
$$h^{(r)}(x) = 0 \qquad\qquad\qquad\quad \text{for } x \leqslant 0,$$

where

$$Q_r(t) = \sum_{s=0}^{r} a_{r,s} t^{r+2s}$$

and

$$|a_{r,s}| \leqslant 4^r r^{r-s}.$$

Suppose $P(r)$ is true for some $r \geqslant 1$. If $x < 0$ it is obvious that $h^{(r)}$ is differentiable at x with $h^{(r+1)}(x) = 0$. If $x > 0$ we see that $h^{(r)}$ is differentiable at x with

$$h^{(r+1)}(x) = -x^{-2}Q_r'(x^{-1})\exp(-1/2x^2) + x^{-3}Q_r(x^{-1})\exp(-1/2x^2)$$
$$= Q_{r+1}(x^{-1})\exp(-1/2x^2),$$

where

$$Q_{r+1}(t) = -\sum_{s=0}^{r}(r+2s)a_{r,s}t^{r+1+2s} + \sum_{s=0}^{r} a_{r,s}t^{r+1+2(s+1)} = \sum_{u=0}^{r+1} a_{r+1,u}t^{r+1+2u},$$

with

$$a_{r+1,u} = -(r+2u)a_{r,u} + a_{r,u-1}$$

(taking $a_{r,-1} = a_{r,r+1} = 0$). It follows that

$$|a_{r+1,u}| \leqslant 3(r+1)|a_{r,u}| + |a_{r,u-1}| \leqslant 3(r+1)4^r r^{r-u} + r^{r+1-u} \leqslant 4^{r+1}r^{r+1-u},$$

as required.

Finally if $x = 0$ we see that

$$\frac{h^{(r)}(\eta) - h^{(r)}(0)}{\eta} = 0 \to 0 \quad \text{as} \quad \eta \to 0-,$$

whilst (just as in Example 4.2)

$$\frac{h^{(r)}(\eta) - h^{(r)}(0)}{\eta} = \eta^{-1}Q_r(\eta^{-1})\exp(-\eta^{-2}/2) \to 0 \quad \text{as} \quad \eta \to 0+,$$

since $\exp t^2/2 \to \infty$ as $t \to \infty$ faster than any polynomial. Thus $h^{(r)}$ is differentiable at 0 with $h^{(r+1)}(0) = 0$.

Thus if $P(r)$ is true so is $P(r+1)$. But $P(0)$ is trivially true and a simpler version of the inductive step above enables us to deduce $P(1)$ from $P(0)$, so $P(r)$ is true for all r and the stated result is proved.

(iii) Combining (i) and (ii) we have

$$|h^{(r)}(x)| = \left| \sum_{s=0}^{r} a_{r,s} x^{-(r+2s)} \exp(-1/2x^2) \right| \leqslant \sum_{s=0}^{r} |a_{r,s}| |x^{-(r+2s)} \exp(-1/2x^2)|$$

$$\leqslant \sum_{s=0}^{r} |a_{r,s}| (r+s)^{(r+2s)/2}$$

$$\leqslant \sum_{s=0}^{r} 4^r r^{r-s} (2r)^{(r+2s)/2} \leqslant \sum_{s=0}^{r} 4^{2r} r^{r-s} r^{(r+2s)/2} = 4^{2r} \sum_{s=0}^{r} r^{3r/2}$$

$$\leqslant 2^{5r} r^{3r/2} \quad \text{for all} \quad x > 0, r \geqslant 1,$$

and so $|h^{(r)}(x)| \leqslant 2^{5r} r^{3r/2}$ for all $x \in \mathbb{R}, r \geqslant 1.$ ∎

Lemma 67.2. *With the notation of Lemma 67.1, set $g(x) = h(x-1)h(2-x)$. Then $g:\mathbb{R} \to \mathbb{R}$ is an infinitely differentiable function with*

(a) $g(x) = 0$ *for all $x \in [1, 2]$,*
(b) $g(x) > 0$ *for all $x \in (1, 2)$,*
(c) $|g^{(n)}(x)| \leqslant 2^{6n} n^{3n/2}$ *for all $x \in \mathbb{R}, n \geqslant 1.$*

Proof. Conclusions (a) and (b) are obvious. To check (c) we use Leibnitz's formula

$$(D^n f_1 f_2)(x) = \sum_{r=0}^{n} \binom{n}{r} (D^{n-r} f_1)(x)(D^r f_2)(x)$$

together with Lemma 67.1 to obtain

$$|g^{(n)}(x)| \leqslant \sum_{r=0}^{n} \binom{n}{r} |h^{(n-r)}(x-1)| |h^{(r)}(2-x)|$$

$$\leqslant \sum_{r=0}^{n} \binom{n}{r} 2^{5(n-r)} (n-r)^{3(n-r)/2} 2^{5r} r^{3r/2}$$

$$\leqslant 2^{5n} \sum_{r=0}^{n} \binom{n}{r} n^{3(n-r)/2} n^{3r/2} = 2^{5n} \sum_{r=0}^{n} \binom{n}{r} n^{3n/2}$$

$$= 2^{5n} n^{3n/2} (1+1)^n = 2^{6n} n^{3n/2}$$

for all x.   ∎

The only major addition to the information already obtained in Example 4.2 (and discussed again at the end of Appendix C) are the bounds on the size of the derivatives. We use this information to check that certain series converge.

Lemma 67.3. *Let* $R > 0$ *and an integer* $k \geqslant 0$ *be fixed. Then the infinite sums*

$$\sum_{m=0}^{\infty} \frac{g^{(m+k)}(t)}{(2m+1)!} x^{2m+1} \quad and \quad \sum_{m=0}^{\infty} \frac{g^{(m+k)}(t)}{(2m)!} x^{2m}$$

converge uniformly for all $|x| \leqslant R, t \in \mathbb{R}$.

Proof. The treatment of the two sums is similar so we shall only deal with

$$\sum_{m=0}^{\infty} \frac{g^{(m+k)}(t)}{(2m)!} x^{2m}.$$

Observe first that

$$(2m)! = \prod_{r=1}^{2m} r \geqslant \prod_{2m \geqslant r \geqslant m/r} ! \geqslant \prod_{2m \geqslant r \geqslant m/8} (m/8) \geqslant 8^{-2m} m^{7m/4} = 2^{-6m} m^{7m/4} \text{ for } m \geqslant 8,$$

and so, using Lemma 67.2,

$$\left| \frac{g^{(m+k)}(t)}{(2m)!} \right| \leqslant \frac{2^{6(m+k)}(m+k)^{3(m+k)/2}}{(2m)!} \leqslant \frac{2^{12m}(2m)^{3(m+k)/2}}{2^{-6m} m^{7m/4}}$$

$$\leqslant 2^{21m} m^{3(m+k)/2 - 7m/4} \leqslant 2^{21m} m^{-m/8} \text{ for all } m \geqslant 12(k+1).$$

Thus

$$\left| \frac{g^{(m+k)}(t)}{(2m)!} x^{2m} \right| \leqslant (2^{21} R^2)^m m^{-m/8} = (2^{21} R^2 m^{-1/8})^m$$

for all $m \geqslant 12(k+1)$, $|x| \leqslant R$ and $t \in \mathbb{R}$.

Since $m^{-\frac{1}{8}} \to 0$ as $m \to \infty$ it follows that $2^{21} R^2 m^{-\frac{1}{8}} \leqslant 2^{-1}$ for large m and so there exists a constant A with

$$\left| \frac{g^{(m+k)}(t)}{2m!} x^{2m} \right| \leqslant A2^{-m} \text{ for all } m \geqslant 0, |x| \leqslant R \text{ and } t \in \mathbb{R}.$$

Thus if $M \geqslant N$ and $|x| < R$ we have

$$\left| \sum_{m=N}^{M} \frac{g^{(m)}(t)}{(2m)!} x^{2m} \right| \leqslant A \sum_{m=N}^{M} 2^{-m} \leqslant A2^{-N+1} \to 0 \text{ as } N \to \infty.$$

The principle of uniform convergence now shows that

$$\sum_{m=0}^{\infty} \frac{g^m(t)}{(2m)!} x^{2m}$$

converges uniformly for all $|x| \leqslant R, t \in \mathbb{R}$. ∎

The preliminaries are now complete and we can give the answer to the problem posed at the beginning of the chapter.

Example 67.4 (Tychonov). *Let g be as in Lemma 67.2 and write*

$$\phi(x,t) = \sum_{m=0}^{\infty} \frac{g^{(m)}(t)}{(2m)!} x^{2m}.$$

Then $\phi : \mathbb{R} \times \mathbb{R} \to \mathbb{R}$ is infinitely differentiable and

(a) $(\partial\phi/\partial t)(x,t) = (\partial^2\phi/\partial x^2)(x,t)$ *for all $(x,t) \in \mathbb{R}^2$,*
(b) $\phi(x,t) = 0$ *for all $t \notin [1,2]$, $x \in \mathbb{R}$,*
(c) $\phi(0,t) > 0$ *for all $t \in (1,2)$.*

Proof. By Lemma 67.3, ϕ is well defined and by well known results (e.g. Lemma 53.2) we may differentiate the expression

$$\phi(x,t) = \sum_{m=0}^{\infty} \frac{g^{(m)}(t)}{(2m)!} x^{2m}$$

both with respect to x and with respect to t as often as we like to obtain

$$\frac{\partial^u}{\partial x^u}\frac{\partial^v \phi}{\partial t^v}(x,t) = \sum_{2m \geq u} \frac{g^{(m+v)}(t)}{(2m-u)!} x^{2m-u}.$$

In particular,

(a) $(\partial\phi/\partial t)(x,t) = \sum_{m=0}^{\infty} (g^{(m+1)}(t)/(2m)!)x^{2m} = (\partial^2\phi/\partial x^2)(x,t)$.
Since $g(t) = 0$ for all $t \notin [1,2]$, $g^{(m)}(t) = 0$ for all $t \notin [1,2]$ so
(b) $\phi(x,t) = 0$ for all $t \notin [1,2]$, $x \in \mathbb{R}$.
Finally, we observe that
(c) $\phi(0,t) = g(t) > 0$ for all $t \in (1,2)$.   ∎

A simple modification shows that the solution for the heat flow in a semi-infinite rod is also not unique.

Example 67.4′. *Let $\psi(x,t) = \sum_{m=0}^{\infty} (g^{(m)}(t)/(2m+1)!)x^{2m+1}$. Then $\psi : \mathbb{R} \times \mathbb{R} \to \mathbb{R}$ is infinitely differentiable and*

(a) $(\partial\psi/\partial t)(x,t) = (\partial^2\psi/\partial x^2)(x,t)$ *for all $(x,t) \in \mathbb{R}$,*
(b) $\psi(x,t) = 0$ *for all $t \notin [1,2]$, $x \in \mathbb{R}$,*
(c) $(\partial\psi/\partial x)(0,t) \neq 0$ *and so $\psi(\ ,t) \neq 0$*   *for all $t \in [1,2]$*
(d) $\psi(0,t) = 0$ *for all $t \in \mathbb{R}$.*

Proof. Left as a trivial exercise to the reader.   ∎

(Taking the restriction of ψ to the region $\{(x,t) : x \geq 0, t \geq 0\}$ gives a non-zero solution to the heat equation for a semi-infinite rod whose initial temperature is 0 everywhere and whose end is kept at 0 throughout.)

The new solutions that we have found consist of a great blast of heat from infinity, and a little thought suggests that the possibility of such solutions spring

from the fact that the equation used gives no limiting velocity for the propagation of heat. Thus, provided only it is large enough, a very distant disturbance can produce a rapid change in temperature near the origin. We may suspect that the kind of solution described in Example 67.4 cannot occur if we have an equation like the wave equation where disturbances propagate with a finite velocity, and that even with the heat equation we can exclude such solutions by imposing a condition of slow growth at infinity.

To the applied mathematician Example 67.4 is simply an embarrassment reminding her of the defects of a model which allows an unbounded speed of propagation. To the numerical analyst it is just a mild warning that the heat equation may present problems which the wave equation does not. But the pure mathematician looks at it with the same simple pleasure with which a child looks at a rose which has just been produced from the mouth of a respectable uncle by a passing magician.

68

UNIQUENESS FOR THE HEAT EQUATION II

If the discussion in the preceding chapter has shown the reader that problems of uniqueness are not as trivial as is sometimes claimed, then my main object has already been attained. However, it seems only fair to reassure the reader by showing her that some useful uniqueness results can still be obtained.

Theorem 68.1. *Let $\phi: \mathbb{R} \times \mathbb{R}^+ \to \mathbb{R}$ be twice differentiable with*

(i) *$(\partial\phi/\partial t)(x,t) = K(\partial^2\phi/\partial x^2)(x,t)$ for all $x \in \mathbb{R}$, $t > 0$ $[K > 0]$,*
(ii) *$\phi(x,t) \to 0$ as $t \to 0+$ uniformly for x in any chosen interval $[-X, X]$,*
(iii) *$\phi(x,t) \to 0$ as $|x| \to \infty$ uniformly for t in any chosen interval $[0, T]$.*

Then

$$\phi(x,t) = 0 \quad \text{for all} \quad (x,t) \in \mathbb{R} \times \mathbb{R}^+.$$

Remark. A much stronger result is true with (iii) replaced by
(iii)' *$\phi(x,t)\exp(-Ax^2) \to 0$ as $x \to \infty$, uniformly in t for each interval $[0, T]$ and some real A.*

(The proof is not very hard but would constitute a digression from a digression.)

Our proof of Theorem 68.1 depends on results whose statements and proofs recall those of Lemma 31.1 and Theorem 31.2 dealing with a maximum principle for solutions of Laplace's equation

$$\frac{\partial^2\phi}{\partial x^2} + \frac{\partial^2\phi}{\partial y^2} = 0,$$

and any reader who has read Chapter 31 might like to try and provide the proofs herself. (However it is not necessary to have read Chapter 31 to understand what follows.)

Lemma 68.2. *Let Ω be an open set in $\mathbb{R}^2$ with closure $\bar{\Omega}$ and boundary $\partial\Omega$. Suppose that $\bar{\Omega}$ is bounded and that $\phi: \bar{\Omega} \to \mathbb{R}$ satisfies the following conditions*

344

(i) ϕ_{11} and ϕ_2 exist with $\phi_{11} - K\phi_2 > 0$ at all points of Ω,
(ii) ϕ is continuous on $\bar{\Omega}$.

Then there exists an $(x_0, t_0) \in \partial\Omega$ *with* $\phi(x_0, t_0) \geqslant \phi(x, t)$ *for all* $(x, t) \in \bar{\Omega}$.
Proof. Since ϕ is continuous on the closed bounded set $\bar{\Omega}$, it is bounded and attains its bounds (see Appendix B if necessary). Thus we can find an $(x_0, t_0) \in \bar{\Omega}$ with $\phi(x_0, t_0) \geqslant \phi(x, t)$ for all $(x, t) \in \bar{\Omega}$. If $(x_0, t_0) \in \Omega$, then we can find a $\delta > 0$ such that $(x, t) \in \Omega$ whenever $|x - x_0|, |t - t_0| < \delta$. In particular $\phi(x, t_0)$ considered as a function of x is twice differentiable function on $(x_0 - \delta, x_0 + \delta)$ with a maximum at x_0. Thus $(\partial^2 \phi / \partial x^2)(x_0, t_0) \leqslant 0$ and a similar argument gives $(\partial \phi / \partial t)(x_0, t_0) = 0$, so $\phi_{11}(x_0, t_0) - K\phi_2(x_0, t_0) \leqslant 0$ which contradicts hypothesis (i). By *reductio ad absurdum* $(x_0, y_0) \in \partial\Omega$ and we are done.

If we consider Ω of a special type we can say slightly more.

Lemma 68.3. *Let*

$$\Omega = \{(x, t): x_1 < x < x_2, t_1 < t < t_2\},$$
$$\bar{\Omega} = \{(x, t): x_1 \leqslant x \leqslant x_2, t_1 \leqslant t \leqslant t_2\}$$

and

$$\partial\Omega = L_1 \cup L_2 \cup L_3 \cup L_4$$

where

$$L_1 = \{(x_1, t): t_1 \leqslant t \leqslant t_2\}, \quad L_2 = \{(x, t_1): x_1 < x < x_2\},$$
$$L_3 = \{(x_2, t): t_1 \leqslant t \leqslant t_2\}, \quad L_4 = \{(x, t_2): x_1 < x < x_2\}.$$

Suppose

$$X_1 < x_1 < x_2 < X_2, T_1 < t_1 < t_2 < T_2$$

and

$$\phi:(X_1, X_2) \times (T_1, T_2) \to \mathbb{R}$$

is such that ϕ_{11} *and* ϕ_2 *exist with* $\phi_{11} - K\phi_2 > 0$ *at all points of* $(X_1, X_2) \times (T_1, T_2)$. *Then, if* $K > 0$, *there exists an* $(x_0, t_0) \in L_1 \cup L_2 \cup L_3$ *such that* $\phi(x_0, t_0) \geqslant \phi(x, t)$ *for all* $(x, t) \in \bar{\Omega}$.

Proof. By Lemma 68.2 we know that there exists an $(x_0, t_0) \in \partial\Omega = L_1 \cup L_2 \cup L_3 \cup L_4$ with the required properties so all we need show is that $(x_0, t_0) \notin L_4$. Suppose, therefore, that $(x_0, t_0) \in L_4$ so $t_0 = t_2$. The same argument as in Lemma 68.2 shows that $(\partial^2 \phi / \partial x^2)(x_0, t_2) \leqslant 0$ but now we only know that $\phi(x_0, t_2) \geqslant \phi(x, t)$ for $t_2 \geqslant t \geqslant t_1$ and so we can only deduce that $(\partial \phi / \partial t)(x_0, t_2) \geqslant 0$. However, this by itself is sufficient to contradict the hypothesis that $\phi_{11}(x_0, t_2) - K\phi_2(x_0, t_2) > 0$. ∎

Lemma 68.4. *If we replace the hypothesis* $\phi_{11} - K\phi_2 > 0$ *in Lemma 68.3 by the weaker hypothesis* $\phi_{11} - K\phi_2 \geqslant 0$ *the conclusion still holds.*
Proof. To fix ideas we choose an R such that $(x, t) \in \bar{\Omega}$ implies $|x| \leqslant R$ and set $g(x, t) = x^2$. (However, the argument below will work for any $g: \mathbb{R}^2 \to \mathbb{R}$ which is twice differentiable and has $g_{11} - Kg_2 > 0$ everywhere.)

Since $L_1 \cup L_2 \cup L_3$ is a closed bounded set and ϕ is continuous on $L_1 \cup L_2 \cup L_3$, we can find (Theorem B.3) $(x_0, t_0) \in L_1 \cup L_2 \cup L_3$ with $\phi(x_0, t_0) \geqslant \phi(x, t)$ for all

$(x,t) \in L_1 \cup L_2 \cup L_3$. Pick $\varepsilon > 0$ and set $\psi(x,t) = \phi(x,t) + \varepsilon g(x,t)$. Then ψ satisfies the conditions of Lemma 68.3 (in particular $\psi_{11} - K\psi_2 = \phi_{11} - K\phi_2 + \varepsilon(g_{11} - Kg_2) \geqslant 2\varepsilon > 0$) and so we can find an $(x_1,t_1) \in L_1 \cup L_2 \cup L_3$ with $\psi(x_1,t_1) \geqslant \psi(x,t)$ for all $(x,t) \in \bar{\Omega}$.

But $\phi(x_0,t_0) \geqslant \phi(x_1,t_1)$ by the definition given in the first sentence of the last paragraph, whilst $g(x_1,t_1) \leqslant R^2$ (since $(x_1,t_1) \in \bar{\Omega}$ and so $|x_1| \leqslant R$). Thus (since $g(x,t) \geqslant 0$ for all (x,t)) we have

$$\phi(x_0,t_0) + \varepsilon R^2 \geqslant \psi(x_1,t_1) \geqslant \psi(x,t) \geqslant \phi(x,t),$$

and so $\qquad \phi(x_0,t_0) \geqslant \phi(x,t) - \varepsilon R^2$ for all $(x,t) \in \bar{\Omega}$ and all $\varepsilon > 0$.

Letting $\varepsilon \to 0$ we see that $\phi(x_0,t_0) \geqslant \phi(x,t)$ for all $(x,t) \in \bar{\Omega}$. ∎

The proof of Theorem 68.1 from Lemma 68.4 is now routine. If the reader sees this at once she does not need to read the details that follow. (And if she still does not see this after reading the details she has not fully grasped what is going on.) *Proof of Theorem 68.1.* The first and main step consists in showing that given any $\varepsilon > 0$ and any $(x',t') \in \mathbb{R} \times \mathbb{R}^+$ we can prove $\phi(x',t') \leqslant \varepsilon$. To this end choose any $t_2 > t'$ and observe that by hypothesis (iii) we can find an $X > 0$ such that $|\phi(x,t)| \leqslant \varepsilon$ for all $|x| \geqslant X$ and all $t \in [0,t_2]$. Further, by hypothesis (iii) we can find a t_1 with $t' \geqslant t_1 > 0$ such that $|\phi(x,t)| \leqslant \varepsilon$ for all $|x| \leqslant X$ and all $t_1 > t > 0$. Setting $x_2 = X, x_1 = -X$ and using the notation and results of Lemmas 68.3 and 68.4 we see that $\phi(x,t) \leqslant \varepsilon$ for all $(x,t) \in L_1 \cup L_2 \cup L_3$ and so $\phi(x',t') \leqslant \varepsilon$.

Since ε, x and t are arbitrary, it follows at once that $\phi(x,t) \leqslant 0$ for all $(x,t) \in \mathbb{R} \times \mathbb{R}^+$. But $-\phi$ also satisfies the hypotheses of Theorem 68.1 and so also $-\phi(x,t) \leqslant 0$ for all $(x,t) \in \mathbb{R} \times \mathbb{R}^+$. Thus $\phi(x,t) = 0$ for all $(x,t) \in \mathbb{R} \times \mathbb{R}^+$ and we are done. ∎

In Chapter 64 we drew analogies between the behaviour of solutions of the wave equation

$$\frac{\partial^2 u}{\partial x^2} = \frac{\partial^2 u}{\partial t^2}$$

and the heat equation

$$\frac{\partial^2 u}{\partial x^2} = \frac{\partial u}{\partial t}.$$

In this chapter we illustrated analogies between Laplace's equation

$$\frac{\partial^2 u}{\partial x^2} = -\frac{\partial^2 u}{\partial t^2}$$

and the heat equation (see Chapter 31). In many ways the heat equation lies midway between the wave equation and Laplace's equation and the curious behaviour shown in Chapter 67 could be attributed to this.

69

THE LAW OF ERRORS

The most important heuristic principle in statistics is the so called 'law of errors' which states that the final result of a large number of small independent random errors will be approximately normally distributed. Poincaré in his *Calcul des Probabilités* quotes a remark to the effect that 'everybody believes in the law of errors, the experimenters because they think it is a mathematical theorem, the mathematicians because they think it is an experimental fact'.

Of course the law of errors in its most general and sweeping form cannot be true. Suppose that X is a real random variable obtained through the action of a large number of small independent random errors. Then the same statement is true of X^3 yet X and X^3 cannot both be normal. Many distributions in nature are not normal and Lemma 50.5 shows that any mathematical 'law of errors' must be subject to quite severe restrictions.

None the less a mathematical version exists in the form of the central limit theorem first stated by Laplace in 1812 and proved by Tchebychev in 1887. (Certain details were provided later by his pupils Liapounov and Markov.) We give its weakest form.

Theorem 69.1. *Let* $X_1, X_2, \ldots$ *be identically distributed independent real valued random variables with mean* μ *and variance* σ^2. *Then for each* $a < b$ *we have*

$$Pr\left(a \leqslant \frac{X_1 + X_2 + \cdots + X_n - n\mu}{\sigma n^{\frac{1}{2}}} \leqslant b\right) \to \frac{1}{(2\pi)^{\frac{1}{2}}} \int_a^b \exp(-t^2/2)dt \ \ as \ n \to \infty.$$

Remark. Not all random variables have mean and variance. Once again the reader's attention is drawn to the counter example given in Lemma 50.5.

The following more or less plausible argument is often used to justify Theorem 69.1. (The reader should read or reread the beginning of Chapter 50 as far as the proof of Lemma 50.2 (i).)

Heuristic justification of Theorem 69.1. Formally, we have

$$\mathbb{E}\exp(i\lambda(X_j - \mu)) = \mathbb{E}\left(\sum_{m=0}^{\infty}\left(\frac{i\lambda(X_j - \mu)^m}{m!}\right)\right) = \sum_{m=0}^{\infty}\frac{(i\lambda)^m}{m!}\mathbb{E}(X_j - \mu)^m$$

$$= 1 + i\lambda\mathbb{E}(X_j - \mu) - \frac{\lambda^2}{2}\mathbb{E}(X_j - \mu)^2 + \cdots = 1 - \frac{\lambda^2\sigma^2}{2} + \cdots.$$

On the other hand, since the X_j are independent, it follows that the $\exp(i\lambda(X_j - \mu))$ are independent so that

$$\mathbb{E}\exp\left(i\lambda\sum_{j=1}^{n}(X_j - \mu)\right) = \mathbb{E}\left(\prod_{j=1}^{n}\exp(i\lambda(X_j - \mu))\right)$$

$$= \prod_{j=1}^{n}\mathbb{E}\exp(i\lambda(X_j - \mu)) = (1 - \lambda^2\sigma^2/2 + \cdots)^n.$$

In particular, setting

$$S(n) = (X_1 + X_2 + \cdots + X_n - n\mu)/(\sigma n^{\frac{1}{2}}),$$

we have

$$\mathbb{E}\exp(i\lambda S(n)) = \mathbb{E}\exp\left(i(\lambda/\sigma n^{\frac{1}{2}})\sum_{j=1}^{n}(X_j - \mu)\right)$$

$$= (1 - (\lambda^2/2n) + \cdots)^n \to \exp(-\lambda^2/2) \quad \text{as} \quad n \to \infty, \qquad (*)$$

provided that everything is well enough behaved to allow us to use the fact that $(1 + \alpha/n)^n \to e^\alpha$ as $n \to \infty$. On the other hand, if Y is normally distributed with mean 0 and variance 1 we know from Lemma 50.2 (i) that $\mathbb{E}\exp i\lambda Y = \exp(-\lambda^2/2)$ and so $(*)$ yields

$$\mathbb{E}\exp i\lambda S(n) \to \mathbb{E}\exp i\lambda Y,$$

from which we might hope to deduce that

$$S(n) \to Y \quad \text{as} \quad n \to \infty \qquad (**)$$

in some appropriate sense. In particular, we might hope that Theorem 69.1 is true.

In the next chapter we shall toughen up this heuristic argument to give a rigorous proof. But before reading on, the reader might like to try this for herself. (The weak points of the argument as it stands are $(*)$ and $(**)$.)

70

THE CENTRAL LIMIT THEOREM I

The object of the next two chapters is to prove the famous central limit theorem stated as Theorem 69.1 in the last chapter.

Theorem 70.1′ (The central limit theorem). *Let* $X_1, X_2, \ldots$ *be identically distributed real valued random variables with mean* μ *and variance* σ^2. *Then*

$$Pr\left(a \leqslant \frac{X_1 + X_2 + \cdots + X_n - n\mu}{\sigma n^{\frac{1}{2}}} \leqslant b\right) \to \frac{1}{(2\pi)^{\frac{1}{2}}} \int_a^b \exp(-t^2/2)dt \text{ as } n \to \infty.$$

Since we do not really have the axiomatic basis to give a rigorous account for the general problem, we shall use this chapter to prove a very slightly weaker result.

Theorem 70.1. *Suppose the hypotheses of Theorem 70.1′ hold and in addition* X_1 *has a continuous bounded probability density. Then the conclusion of Theorem 70.1′ holds.*

We shall need Theorem 51.1 which the reader may take as an unproved assertion, an axiom, or a definition according to her personal taste. In Chapter 71 I shall sketch what I hope will be a convincing extension to the general case.

Let us start by obtaining two more manageable but equivalent forms of Theorem 70.1.

Lemma 70.2. *Let* $Y_1, Y_2, \ldots$ *be identically distributed real random variables such that* Y_1 *has continuous bounded probability density* f *and* $\mathbb{E}Y = 0$, $\mathbb{E}Y^2 = 1$. *Then*

$$Pr\left(a \leqslant \frac{Y_1 + Y_2 + \cdots + Y_n}{n^{\frac{1}{2}}} \leqslant b\right) \to \frac{1}{(2\pi)^{\frac{1}{2}}} \int_a^b \exp(-t^2/2)dt$$

Proof of Theorem 70.1 from Lemma 70.2. Set $Y_j = (X_j - \mu)/\sigma$. ∎

Lemma 70.3. *Let* $f : \mathbb{R} \to \mathbb{R}^+$ *be a continuous bounded function such that*

$$\int_{-\infty}^{\infty} f(x)dx = 1, \quad \int_{-\infty}^{\infty} xf(x)dx = 0 \quad and \quad \int_{-\infty}^{\infty} x^2 f(x)dx = 1.$$

Then writing $f_1 = f, f_2 = f * f, \ldots, f_n = f * f_{n-1}$ and $g_n(t) = n^{\frac{1}{2}} f_n(n^{\frac{1}{2}} t)$, we have

$$\int_a^b g_n(t) dt \to \frac{1}{(2\pi)^{\frac{1}{2}}} \int_a^b \exp(-t^2/2) dt.$$

Proof of the equivalence of Lemmas 70.2 and 70.3. If we accept Theorem 51.1 (in any of the guises suggested after the statement of Theorem 70.1) then

$$Pr\left(a \leqslant \frac{Y_1 + Y_2 + \cdots + Y_n}{n^{\frac{1}{2}}} \leqslant b\right) = Pr(an^{\frac{1}{2}} \leqslant Y_1 + Y_2 + \cdots + Y_n \leqslant bn^{\frac{1}{2}})$$

$$= \int_{a\sqrt{n}}^{b\sqrt{n}} f_n(t) dt$$

$$= \int_a^b n^{\frac{1}{2}} f_n(n^{\frac{1}{2}} s) ds$$

(making the substitution $s = n^{-\frac{1}{2}} t$). $\blacksquare$

Now that we have translated the statement of the central limit theorem as a result in probability to a statement of the same theorem as a result in analysis we are in a position to prove it. As we might perhaps expect after reading the heuristic justification of Theorem 69.1 in Chapter 69, the key to the matter turns out to reside in the behaviour of $\hat{f}(\lambda)$ for $|\lambda|$ small.

Lemma 70.4. *If f satisfies the condition of Lemma 70.3 then*

$$\frac{\hat{f}(\lambda) - (1 - \lambda^2/2)}{\lambda^2} \to 0 \text{ as } \lambda \to 0.$$

Proof. Observe that if $|s| \geqslant 1$ then

$$\left| e^{-is} - \left(1 - is - \frac{s^2}{2}\right) \right| \leqslant |e^{-is}| + 1 + |s| + s^2 \leqslant 4s^2,$$

whilst if $|s| \leqslant 1$

$$\left| e^{-is} - \left(1 - is - \frac{s^2}{2}\right) \right| = \left| \sum_{r=3}^{\infty} \frac{(is)^r}{r!} \right| \leqslant \sum_{r=3}^{\infty} \frac{|s|^r}{r!} \leqslant \frac{|s|^3}{3!} \sum_{r=0}^{\infty} \left(\frac{|s|}{4}\right)^r = \frac{|s|^3}{6(1 - |s|/4)} \leqslant |s|^3.$$

Thus

$$\left| e^{-is} - \left(1 - is - \frac{s^2}{2}\right) \right| \leqslant 4s^2 \text{ for all } s,$$

whilst if we choose any $0 < \delta < 1$

$$\left| e^{-is} - \left(1 - is - \frac{s^2}{2}\right) \right| \leqslant \delta s^2 \quad \text{for all } |s| \leqslant \delta.$$

We can now use the fact that

$$\int_{-\infty}^{\infty} f(t)\, dt = 1, \quad \int_{-\infty}^{\infty} t f(t)\, dt = 0$$

and

$$\int_{-\infty}^{\infty} t^2 f(t)\, dt = 1$$

to obtain

$$\left| \frac{\hat{f}(\lambda) - (1 - \lambda^2/2)}{\lambda^2} \right| = \lambda^{-2} \left| \int_{-\infty}^{\infty} f(t) e^{-i\lambda t}\, dt - \int_{-\infty}^{\infty} f(t)\left(1 - i\lambda t - \frac{(\lambda t)^2}{2}\right) dt \right|$$

$$= \lambda^{-2} \left| \int_{-\infty}^{\infty} f(t)(e^{-i\lambda t} - (1 - i\lambda t - (\lambda t)^2/2))\, dt \right|$$

$$\leqslant \lambda^{-2} \int_{-\infty}^{\infty} f(t)|e^{-i\lambda t} - (1 - i\lambda t - (\lambda t)^2/2)|\, dt$$

$$= \lambda^{-2} \int_{|\lambda t| \leqslant \delta} f(t)|e^{-i\lambda t} - (1 - i\lambda t - (\lambda t)^2/2)|\, dt$$

$$\quad + \lambda^{-2} \int_{|\lambda t| > \delta} f(t)|e^{-i\lambda t} - (1 - i\lambda t - (\lambda t)^2/2)|\, dt$$

$$\leqslant \lambda^{-2} \int_{|\lambda t| \leqslant \delta} f(t)\delta(\lambda t)^2\, dt + \lambda^{-2} \int_{|\lambda t| > \delta} f(t)4(\lambda t)^2\, dt$$

$$= \delta \int_{|\lambda t| \leqslant \delta} t^2 f(t)\, dt + 4 \int_{|\lambda t| > \delta} t^2 f(t)\, dt$$

$$\leqslant \delta \int_{-\infty}^{\infty} t^2 f(t)\, dt + 4 \int_{|t| > \delta|\lambda|^{-1}} t^2 f(t)\, dt$$

$$= \delta + 4 \int_{|t| > \delta|\lambda|^{-1}} t^2 f(t)\, dt \to \delta \quad \text{as} \quad \lambda \to 0.$$

Since δ was arbitrary, it follows that

$$\frac{\hat{f}(\lambda) - (1 - \lambda^2)}{\lambda^2} \to 0 \quad \text{as} \quad \lambda \to 0. \qquad \blacksquare$$

Remark. Several of the people who read the manuscript of this book have pointed out the following much shorter proof of Lemma 70.4. Since $|xf(x)| \leqslant x^2 f(x)$ for $|x| \geqslant 1$ we know by the comparison test that

$$\int_{-\infty}^{\infty} |ixf(x)e^{i\lambda x}|\, dx = \int_{-\infty}^{\infty} |xf(x)|\, dx$$

converges. Thus, using Theorem 53.5, we may differentiate under the integral sign

to show that $\hat{f}$ is differentiable and

$$\hat{f}'(\lambda) = \frac{d}{d\lambda} \int_{-\infty}^{\infty} f(x) e^{i\lambda x}\, dx = \int_{-\infty}^{\infty} \frac{\partial}{\partial \lambda}(f(x) e^{i\lambda x})\, dx = i \int_{-\infty}^{\infty} x f(x) e^{i\lambda x}\, dx.$$

Similarly, $\hat{f}'$ is differentiable with

$$\hat{f}''(\lambda) = i \int_{-\infty}^{\infty} \frac{\partial}{\partial \lambda}(x f(x) e^{i\lambda x})\, dx = - \int_{-\infty}^{\infty} x^2 f(x) e^{i\lambda x}\, dx.$$

A primitive version of Taylor's theorem tells us that if g is n times differentiable at a then

$$g(a + h) = g(a) + g'(a)h + \cdots + \frac{g^{(n)}(a)h^n}{n!} + \varepsilon(h)h^n$$

with $\varepsilon(h) \to 0$ as $h \to 0$. (For a proof see Hardy's *Pure Mathematics*, §151, or most standard texts.) In this case we obtain

$$\hat{f}(h) = \hat{f}(0) + \hat{f}'(0)h + \frac{\hat{f}''(0)h^2}{2} + \varepsilon(h)h^2$$

with $\varepsilon(h) \to 0$ as $h \to 0$. But the formulae obtained in the last two sentences of the previous paragraph give us

$$\hat{f}'(0) = i \int_{-\infty}^{\infty} x f(x)\, dx = 0, \hat{f}''(0) = - \int_{-\infty}^{\infty} x^2 f(x)\, dx = -1$$

and we know that $\hat{f}(0) = \int_{-\infty}^{\infty} f(x)\, dx = 1$. Thus

$$\hat{f}(h) = 1 - \frac{h^2}{2} + \varepsilon(h)h^2$$

with $\varepsilon(h) \to 0$ as $h \to 0$, and this is the result of Lemma 70.4.   ∎

Personally I find the longwinded proof more satisfying but the reader may well disagree.

It is a simple but illuminating exercise to use the same methods to show that if $f_0 : \mathbb{R} \to \mathbb{R}$ is a bounded positive continuous function with

$$\int_{-\infty}^{\infty} f_0(x)\, dx = 1, \int_{-\infty}^{\infty} x^2 f_0(x)\, dx = \sigma^2$$

and

$$\int_{-\infty}^{\infty} x^2 f_0(x)\, dx = \sigma^2$$

then

$$\lambda^{-2}(\hat{f}_0(\lambda) - (1 - i\lambda\mu - \lambda^2\sigma^2/2)) \to 0 \quad \text{as } \lambda \to 0.$$

We can now state exactly and prove the step marked (∗) in our heuristic justification of Theorem 69.1.

Lemma 70.5. *If g_n is defined as in Lemma 70.3, then*

(i) $g_n(t) \geqslant 0$ *for all* $t \in \mathbb{R}$,

(ii) $\int_{-\infty}^{\infty} g_n(t) dt = 1$,

(iii) $\hat{g}_n(\zeta) \to \exp(-\zeta^2/2)$ *uniformly on each interval* $[-R, R]$ *as* $n \to \infty$.

Proof. (Note that (i) and (ii) are obvious from the probabilistic point of view.)

(i) Trivial computation.

(ii) Using Theorem 51.7 (ii) and making the substitution $t = n^{-\frac{1}{2}}s$ we obtain

$$1 = \left(\int_{-\infty}^{\infty} f(s) \, ds \right)^n = \int_{-\infty}^{\infty} f_n(s) \, ds = \int_{-\infty}^{\infty} n^{\frac{1}{2}} f_n(n^{\frac{1}{2}}t) \, dt = \int_{-\infty}^{\infty} g_n(t) \, dt,$$

as required.

(iii) We note first that making the substitution $s = n^{\frac{1}{2}}t$ and using Theorem 51.7 (iv) we have

$$\hat{g}_n(\zeta) = \int_{-\infty}^{\infty} g_n(t) e^{-i\zeta t} \, dt = \int_{-\infty}^{\infty} n^{\frac{1}{2}} f_n(n^{\frac{1}{2}}t) e^{-i\zeta t} \, dt$$

$$= \int_{-\infty}^{\infty} f_n(s) e^{-i\zeta s/\sqrt{n}} \, ds = \hat{f}_n(\zeta n^{-\frac{1}{2}}) = (\hat{f}(\zeta n^{-\frac{1}{2}}))^n.$$

Thus

$$|\hat{g}_n(\zeta) - (1 - \zeta^2/2n)^n| = |(\hat{f}(\zeta/\sqrt{n}))^n - (1 - \zeta^2/2n)^n|$$

$$= |\hat{f}(\zeta/\sqrt{n}) - (1 - \zeta^2/2n)| \left| \sum_{k=0}^{n-1} (\hat{f}(\zeta/\sqrt{n}))^k (1 - \zeta^2/2n)^{n-k-1} \right|$$

$$\leqslant |\hat{f}(\zeta/\sqrt{n}) - (1 - \zeta^2/2n)| \sum_{k=0}^{n-1} |\hat{f}(\zeta/\sqrt{n})|^k |1 - \zeta^2/2n|^{n-k-1}.$$

But $|\hat{f}(\mu)| \leqslant \int_{-\infty}^{\infty} |f(t)| dt = 1$, so if $0 < |\zeta| \leqslant R$ and $n \geqslant R^2$ we have

$$|\hat{g}_n(\zeta) - (1 - \zeta^2/2n)^n| \leqslant n |\hat{f}(\zeta/\sqrt{n}) - (1 - \zeta^2/2n)|$$

$$= \zeta^2 \frac{|\hat{f}(\zeta/\sqrt{n}) - (1 - \zeta^2/2n)|}{(\zeta/\sqrt{n})^2} \leqslant R^2 \sup_{0 < |\lambda| \leqslant R/\sqrt{n}} \frac{|\hat{f}(\lambda) - (1 - \lambda^2/2)|}{\lambda^2} \to 0 \quad \text{as } n \to \infty,$$

by Lemma 70.4. It follows that $\hat{g}_n(\zeta) - (1 - \zeta^2/2n)^n \to 0$ uniformly on $[-R, R]$ as $n \to \infty$. But $(1 - \zeta^2/2n)^n \to e^{-\zeta^2/2}$ uniformly on $[-R, R]$ so $\hat{g}_n(\zeta) \to e^{-\zeta^2/2}$ uniformly on $[-R, R]$, as required. ∎

We now turn to the step marked (∗∗) in the heuristic justification of Theorem 69.1. To justify this step we use a method analogous to the one used in Chapter 3 to prove Weyl's equidistribution theorem. The reader may find it helpful to read our proof of Theorem 3.1′ before proceeding.

Proof of Lemma 70.3. If $u : \mathbb{R} \to \mathbb{C}$ is a bounded function which is Riemann integrable on each $[a, b]$, let us write

$$G_n(u) = \int_{-\infty}^{\infty} g_n(t)u(t)\,dt - \frac{1}{(2\pi)^{\frac{1}{2}}} \int_{-\infty}^{\infty} e^{-t^2/2} u(t)\,dt.$$

Our object is to show that $G_n(u) \to 0$ for a large class of functions u.

Step 1. Suppose $k:\mathbb{R} \to \mathbb{C}$ is a continuous function with $k(\lambda) = 0$ for $|\lambda| \geqslant R$. Set $u(t) = \hat{k}(t)$. (Thus $u(t) = \int_{-R}^{R} k(\lambda)e^{-i\lambda t}\,d\lambda$ and u will play a role analogous to that of the trigonometric polynomials $P(t) = \sum_{r=-N}^{N} a_r \exp irt$ in Step 3 of the proof of Theorem 3.1′.)

By Lemma 50.2 (ii) we know that

$$\int_{-\infty}^{\infty} e^{-t^2/2} e^{i\lambda t}\,dt = (2\pi)^{\frac{1}{2}} e^{-\lambda^2/2},$$

so, making a change in the order of integration (to be justified in the next paragraph),

$$G_n(u) = \int_{-\infty}^{\infty} (g_n(t) - (2\pi)^{-\frac{1}{2}}\exp(-t^2/2))u(t)\,dt$$

$$= \int_{-\infty}^{\infty} (g_n(t) - (2\pi)^{-\frac{1}{2}}\exp(-t^2/2))\left(\int_{-\infty}^{\infty} k(\lambda)e^{-i\lambda t}\,d\lambda\right) dt$$

$$= \int_{-\infty}^{\infty} \left(\int_{-\infty}^{\infty} (g_n(t) - (2\pi)^{-\frac{1}{2}}\exp(-t^2/2))k(\lambda)e^{-i\lambda t}\,d\lambda\right) dt$$

$$= \int_{-\infty}^{\infty} \left(\int_{-\infty}^{\infty} (g_n(t) - (2\pi)^{-\frac{1}{2}}\exp(-t^2/2))k(\lambda)e^{-i\lambda t}\,dt\right) d\lambda \qquad (\dagger)$$

$$= \int_{-\infty}^{\infty} k(\lambda)\left(\int_{-\infty}^{\infty} g_n(t)e^{-i\lambda t}\,dt - \frac{1}{(2\pi)^{\frac{1}{2}}}\int_{-\infty}^{\infty}\exp(-t^2/2)e^{-i\lambda t}\,dt\right) d\lambda$$

$$= \int_{-\infty}^{\infty} k(\lambda)(\hat{g}_n(\lambda) - \exp(-\lambda^2/2))\,d\lambda = \int_{-R}^{R} k(\lambda)(\hat{g}_n(\lambda) - \exp(-\lambda^2/2))\,d\lambda,$$

and so, using Lemma 70.5 (iii)

$$|G_n(u)| \leqslant \sup_{|\lambda| \leqslant R} |\hat{g}_n(\lambda) - \exp(-\lambda^2/2)| \int_{-R}^{R} |k(\lambda)|\,d\lambda \to 0 \quad \text{as } n \to \infty.$$

The proof that the order of integration may be interchanged at (†) is routine. If we write $F(t, \lambda) = g_n(t)k(\lambda)e^{i\lambda t}$ then $F:\mathbb{R}^2 \to \mathbb{C}$ is continuous and (using Lemma 70.5 (i) and (ii)) we see that

(A1) $\int_{-\infty}^{\infty} |F(t, \lambda)|\,dt = |k(\lambda)| \int_{-\infty}^{\infty} g_n(t)\,dt = |k(\lambda)|$

is a continuous function of λ, whilst

(A2) $\int_{-\infty}^{\infty} |F(t, \lambda)|\,d\lambda = g_n(t) \int_{-\infty}^{\infty} |k(\lambda)|\,d\lambda$

is a continuous function of t. Further,

(B)
$$\int_{-\infty}^{\infty} \left(\int_{-\infty}^{\infty} |F(t,\lambda)| \, d\lambda \right) dt = \int_{-\infty}^{\infty} g_n(t) \int_{-\infty}^{\infty} |k(\lambda)| \, d\lambda \, dt$$

$$= \int_{-\infty}^{\infty} g_n(t) \, dt \int_{-\infty}^{\infty} |k(\lambda)| \, d\lambda = \int_{-\infty}^{\infty} |k(\lambda)| \, d\lambda$$

converges. Thus by Theorem 48.8 $\int_{-\infty}^{\infty} (\int_{-\infty}^{\infty} F(t,\lambda) d\lambda) \, dt$ and $\int_{-\infty}^{\infty} (\int_{-\infty}^{\infty} F(t,\lambda) \, dt) \, d\lambda$ converge and are equal. An exactly similar argument shows that $\int_{-\infty}^{\infty} (\int_{-\infty}^{\infty} \exp(-\lambda^2/2) k(\lambda) e^{i\lambda t} d\lambda) \, dt$ and $\int_{-\infty}^{\infty} (\int_{-\infty}^{\infty} \exp(-\lambda^2/2) k(\lambda) e^{i\lambda t} dt) \, d\lambda$ converge and are equal.

Step 2. Suppose $u, v: \mathbb{R} \to \mathbb{C}$ are bounded functions which are Riemann integrable on each $[a,b]$ and $|u(t) - v(t)| < \varepsilon$ for all $t \in \mathbb{R}$. Then

$$|G_n(u) - G_n(v)| \leqslant \left| \int_{-\infty}^{\infty} (u(t) - v(t)) g_n(t) \, dt \right| + \left| \frac{1}{(2\pi)^{\frac{1}{2}}} \int_{-\infty}^{\infty} (u(t) - v(t)) \exp(-t^2/2) \, dt \right|$$

$$\leqslant \varepsilon \int_{-\infty}^{\infty} g_n(t) \, dt + \frac{\varepsilon}{(2\pi)^{\frac{1}{2}}} \int_{-\infty}^{\infty} \exp(-t^2/2) \, dt = 2\varepsilon \quad \text{for all } n \geqslant 1.$$

Step 3. Suppose $v: \mathbb{R} \to \mathbb{C}$ is uniformly continuous and bounded and $\varepsilon > 0$. By Theorem 49.3 (ii) we know that, taking

$$\Delta_S(\lambda) = \left(1 - \frac{|\lambda|}{S}\right) \quad \text{for } |\lambda| \leqslant S$$

$$\Delta_S(\lambda) = 0 \quad \text{for } S \leqslant |\lambda|,$$

we have

$$\frac{1}{2\pi} \int_{-\infty}^{\infty} \Delta_S(\lambda) \hat{v}(\lambda) \exp i\lambda t \, d\lambda \to v(t) \quad \text{uniformly as } S \to \infty.$$

In particular, taking R large enough and setting

$$k(\lambda) = (2\pi)^{-1} \Delta_R(\lambda) \hat{v}(\lambda), \quad u(t) = \int_{-\infty}^{\infty} k(\lambda) \exp i\lambda t \, dt,$$

we have

$$|u(t) - v(t)| \leqslant \varepsilon/3 \quad \text{for all } t \in \mathbb{R}.$$

Since k and u satisfy the conditions set out in the first 2 sentences of Step 1 we can now find an n_0 such that $|G_n(u)| \leqslant \varepsilon/3$ for all $n \geqslant n_0$. But by the result of Step 2, $|G_n(v) - G_n(u)| \leqslant 2\varepsilon/3$ and so $|G_n(v)| \leqslant |G_n(u)| + |G_n(v) - G_n(u)| \leqslant \varepsilon$ for all $n \geqslant n_0$. It follows that $G_n(v) \to 0$ as $n \to \infty$.

Step 4. It is an easy matter to find, for each ε with $(b-a)/2 > \varepsilon > 0$, uniformly continuous functions $u_+, u_-: \mathbb{R} \to \mathbb{R}$ such that

(a) $1 \geqslant u_+(t), u_-(t) \geqslant 0$ for all $t \in \mathbb{R}$,

(b)$_1$ $u_+(t) = 1$ for all $t \in [a, b]$,

(b)$_2$ $u_-(t) = 0$ for all $t \notin [a, b]$,

(c)$_1$ $u_+(t) = 0$ for all $t \notin [a - \varepsilon, b + \varepsilon]$,

(c)$_2$ $u_-(t) = 1$ for all $t \in [a + \varepsilon, b - \varepsilon]$.

From (a), (b)$_1$ and (b)$_2$ it follows at once that

$$\int_{-\infty}^{\infty} g_n(t)u_+(t)\,dt \geqslant \int_a^b g_n(t)\,dt \geqslant \int_{-\infty}^{\infty} g_n(t)u_-(t)\,dt.$$

On the other hand u_+ and u_- satisfy the conditions set out in the first sentence of Step 3. Thus there exists an $n_0(\varepsilon)$ such that, whenever $n \geqslant n_0(\varepsilon)$ we have $|G_n(u_+)|, |G_n(u_-)| \leqslant \varepsilon$ and so

$$\frac{1}{(2\pi)^{\frac12}} \int_{-\infty}^{\infty} \exp(-t^2/2)u_+(t)dt + \varepsilon \geqslant \int_a^b g_n(t)\,dt \geqslant \frac{1}{(2\pi)^{\frac12}} \int_{-\infty}^{\infty} \exp(-t^2/2)u_-(t)dt - \varepsilon.$$

Using (a) and (c$_2$) we see that

$$\int_{-\infty}^{\infty} \exp(-t^2/2)u_-(t)dt \geqslant \int_{a+\varepsilon}^{b-\varepsilon} \exp(-t^2/2)\,dt$$

$$\geqslant \int_a^b \exp(-t^2/2)\,dt - \left(\int_{b-\varepsilon}^b + \int_a^{a+\varepsilon} \exp(-t^2/2)\,dt \right) \geqslant \int_a^b \exp(-t^2/2)\,dt - 2\varepsilon.$$

Similarly, using (a) and (c$_1$)

$$\int_a^b \exp(-t^2/2)\,dt + 2\varepsilon \geqslant \int_{-\infty}^{\infty} \exp(-t^2/2)u_+(t)\,dt.$$

Thus (since $(2\pi)^{\frac12} \geqslant 2$)

$$\frac{1}{(2\pi)^{\frac12}} \int_a^b \exp(-t^2/2)\,dt + 2\varepsilon \geqslant \int_a^b g_n(t)\,dt \geqslant \frac{1}{(2\pi)^{\frac12}} \int_a^b \exp(-t^2/2)\,dt - 2\varepsilon$$

for all $n \geqslant n_0(\varepsilon)$. So, since ε was arbitrary,

$$\int_a^b g_n(t)\,dt \to \frac{1}{(2\pi)^{\frac12}} \int_a^b \exp(-t^2/2)\,dt,$$

as stated.   ∎

We have thus proved the central limit theorem for random variables with bounded continuous probability density. The proof is not easy but once the reader has mastered it, and is able to write out a proof for herself, she will have learnt most of what this book tries to teach.

71

THE CENTRAL LIMIT THEOREM II

Although the proof of the central limit theorem given in the last chapter applies, essentially unchanged, to all probability distributions, we lack the technical background required to show this. Instead I shall sketch an alternative approach which also extends the theorem to general distributions. The details are not important and once the reader has grasped the essential idea she should either try to prove the results for herself or simply move on to some other part of the book.

We start with the case when X_1 takes only a finite number of possible values.

Theorem 71.1. *Let* $X_1, X_2, \ldots$ *be identically distributed real valued random variables taking only a finite number of values. Then if* X_1 *has mean* μ *and variance* σ^2

$$Pr\left(a \leqslant \frac{X_1 + X_2 + \cdots + X_n - n\mu}{\sigma n^{\frac{1}{2}}} \leqslant b\right) \to \frac{1}{(2\pi)^{\frac{1}{2}}} \int_a^b \exp(-t^2/2)dt$$

The steps by which we prove this result are outlined in the following lemma.

Lemma 71.2. *Let* $X_1, Y_1, X_2, Y_2, X_3, \ldots$ *be independent real valued random variables. Suppose that the* $X_1, X_2, \ldots$ *are identically distributed with mean 0 and variance 1 and take only a finite number of values whilst the* $Y_1, Y_2, \ldots$ *are all normally distributed with mean 0 and variance 1 also. Take* $\varepsilon > 0$ *and write* $Z_j = X_j + \varepsilon Y_j$. *Then*

(i) $Z_1, Z_2, \ldots$ *are independent identically distributed random variables with bounded continuous probability density and mean 0, variance* $1 + \varepsilon^2$.

(ii) $Pr\left(a \leqslant \dfrac{Z_1 + Z_2 + \cdots + Z_n}{n^{\frac{1}{2}}} \leqslant b\right) \to \dfrac{1}{(2\pi)^{\frac{1}{2}}} \displaystyle\int_{a/\sqrt{(1+\varepsilon^2)}}^{b/\sqrt{(1+\varepsilon^2)}} \exp(-t^2/2)dt$ *as* $n \to \infty$.

(iii) $Pr\left(\left|\dfrac{\varepsilon Y_1 + \varepsilon Y_2 + \cdots + \varepsilon Y_n}{n^{\frac{1}{2}}}\right| \leqslant \delta\right) = \dfrac{1}{(2\pi)^{\frac{1}{2}}} \displaystyle\int_{-\delta/\varepsilon}^{\delta/\varepsilon} \exp(-t^2/2)dt$ *for all* $n \geqslant 1$.

(iv) $Pr\left(a \leqslant \dfrac{X_1 + X_2 + \cdots + X_n}{n^{\frac{1}{2}}} \leqslant b\right) \to \dfrac{1}{(2\pi)^{\frac{1}{2}}} \displaystyle\int_a^b \exp(-t^2/2)dt$ *as* $n \to \infty$.

357

Proof. (i) The key step is the proof that Z_1 has continuous bounded probability density and we begin with this. Write $X = X_1$, $Y = Y_1$ and $Z = Z_1$ and suppose X takes the values $a(1), a(2), \ldots, a(m)$ only, with $Pr(X = a(j)) = p_j$. Then (making use of the substitution $t = \varepsilon s + a(j)$)

$$
\begin{aligned}
Pr(Z \leqslant \lambda) &= \sum_{j=1}^{m} Pr(Z \leqslant \lambda, X = a(j)) = \sum_{j=1}^{m} Pr(Y \leqslant \varepsilon^{-1}(\lambda - a(j)), X = a(j)) \\
&= \sum_{j=1}^{m} p_j Pr(Y \leqslant \varepsilon^{-1}(\lambda - a(j))) = \sum_{j=1}^{m} p_j \frac{1}{(2\pi)^{\frac{1}{2}}} \int_{-\infty}^{(\lambda - a(j))/\varepsilon} \exp(-s^2/2) ds \\
&= \sum_{j=1}^{m} p_j \frac{1}{(2\pi)^{\frac{1}{2}}\varepsilon} \int_{-\infty}^{\lambda} \exp(-(t - a(j))^2/(2\varepsilon^2)) dt = \int_{-\infty}^{\lambda} f(t) dt,
\end{aligned}
$$

where

$$
f(t) = \left(\sum_{j=1}^{m} p_j \exp(-(t - a(j))^2/(2\varepsilon^2))/((2\pi)^{\frac{1}{2}}\varepsilon) \right)
$$

is a bounded continuous function.

The same kind of calculations can be used to show directly that $Z_1, Z_2, \ldots, Z_n$ are indeed independent for each n (and so $Z_1, Z_2, \ldots$ are independent) but the reader will probably be prepared to accept this without doing the computation. Finally we note that

$$
\mathbb{E}Z = \mathbb{E}Z + \varepsilon \mathbb{E}Y = 0
$$

and

$$
\mathbb{E}Z^2 = \mathbb{E}X^2 + 2\varepsilon \mathbb{E}X \mathbb{E}Y + \varepsilon^2 \mathbb{E}Y^2 = 1 + \varepsilon^2.
$$

(ii) Use Theorem 70.1 (the central limit theorem for bounded continuous probability densities).
(iii) Lemma 50.3(i) (combined with the results of Chapter 51 if the reader requires).
(iv) Let $\delta, \varepsilon > 0$. Then using (ii) and (iii)

$$
\begin{aligned}
Pr(a \leqslant &(X_1 + X_2 + \cdots + X_n)/n^{\frac{1}{2}} \leqslant b) \\
&\geqslant Pr(a + \delta \leqslant (Z_1 + Z_2 + \cdots + Z_n)/n^{\frac{1}{2}} \\
&\leqslant b - \delta \text{ and } |(\varepsilon Y_1 + \varepsilon Y_2 + \cdots + \varepsilon Y_n)/n^{\frac{1}{2}}| \leqslant \delta) \\
&\geqslant Pr(a + \delta \leqslant (Z_1 + Z_2 + \cdots + Z_n)/n^{\frac{1}{2}} \leqslant b - \delta) \\
&\quad - Pr(|(\varepsilon Y_1 + \varepsilon Y_2 + \cdots + \varepsilon Y_n)/n^{\frac{1}{2}}| > \delta) \\
&= Pr(a + \delta \leqslant (Z_1 + Z_2 + \cdots + Z_n)/n^{\frac{1}{2}} \leqslant b - \delta) - \frac{1}{(2\pi)^{\frac{1}{2}}} \int_{|t| > \delta/\varepsilon} \exp(-t^2/2) dt \\
&\to \frac{1}{(2\pi)^{\frac{1}{2}}} \int_{(a+\delta)/\sqrt{(1+\varepsilon^2)}}^{(b-\delta)/\sqrt{(1+\varepsilon^2)}} \exp(-t^2/2) dt - \frac{1}{(2\pi)^{\frac{1}{2}}} \int_{|t| > \delta/\varepsilon} \exp(-t^2/2) dt.
\end{aligned}
$$

Thus

$$
\liminf_{n \to \infty} Pr(a \leqslant (X_1 + X_2 + \cdots + X_n)/n^{\frac{1}{2}} \leqslant b)
$$

$$\geqslant \frac{1}{(2\pi)^{\frac{1}{2}}} \int_{(a+\delta)/\sqrt{(1+\varepsilon^2)}}^{(b-\delta)/\sqrt{(1+\varepsilon^2)}} \exp(-t^2/2)dt \; - \frac{1}{(2\pi)^{\frac{1}{2}}} \int_{|t| > \delta/\varepsilon} \exp(-t^2/2)dt \quad \text{for all } \delta, \varepsilon > 0.$$

Letting first $\varepsilon \to 0$ and then $\delta \to 0$ we obtain

$$\liminf_{n \to \infty} Pr(a \leqslant (X_1 + X_2 + \cdots + X_n)/n^{\frac{1}{2}} \leqslant b) \geqslant \frac{1}{(2\pi)^{\frac{1}{2}}} \int_a^b \exp(-t^2/2)dt.$$

Exactly the same method of proof gives

$$\liminf_{n \to \infty} Pr((X_1 + X_2 + \cdots + X_n)/n^{\frac{1}{2}} \notin [a,b]) \geqslant \frac{1}{(2\pi)^{\frac{1}{2}}} \int_{t \notin [a,b]} \exp(-t^2/2)dt,$$

so

$$\limsup_{n \to \infty} Pr(a \leqslant (X_1 + X_2 + \cdots + X_n)/n^{\frac{1}{2}}) \leqslant b)$$

$$= 1 - \liminf_{n \to \infty} Pr(((X_1 + X_2 + \cdots + X_n)/n^{\frac{1}{2}}) \notin [a,b])$$

$$\leqslant 1 - \frac{1}{(2\pi)^{\frac{1}{2}}} \int_{t \notin [a,b]} \exp(-t^2/2)dt = \frac{1}{(2\pi)^{\frac{1}{2}}} \int_a^b \exp(-t^2/2)dt.$$

The stated result follows. ∎

Proof of Theorem 71.1. This follows directly from Lemma 71.1(iv) as in the proof of Theorem 70.1 from Lemma 70.2. ∎

The only point in our proof where we used the special characteristics of X_1 was in the proof of Lemma 71.2 (i). Thus (provided we can accept certain intuitively obvious results about random variables) the full central limit theorem (Theorem 70.1′) will follow once we have proved the following generalisation of the main part of Lemma 71.2 (i).

Lemma 71.3. *Let X and Y be independent real valued random variables. Suppose $\mathbb{E}X$ and $\mathbb{E}X^2$ exist with values 0 and 1 whilst Y is normally distributed with mean 0 and variance 1. Then if $\varepsilon > 0$ and $Z = X + \varepsilon Y$, it follows that Z has bounded continuous probability density and mean 0 variance $1 + \varepsilon^2$.*

Lemma 71.3 follows from a closely related result.

Lemma 71.4. *Let X and Y be independent real valued random variables. Then if X has a mean and variance and Y is normally distributed with mean 0 and variance 1, it follows that $Z = X + Y$ has bounded continuous probability density.*

Sketch proof of Lemma 71.3 from Lemma 71.4. Observe that if $\varepsilon^{-1}X + Y$ has bounded continuous probability density, then so does $X + \varepsilon Y = \varepsilon(\varepsilon^{-1}X + Y)$. The computation of the mean and variance depends on simple intuitively obvious properties of the expectation. We obtain

$$\mathbb{E}(X + \varepsilon Y) = \mathbb{E}X + \varepsilon\mathbb{E}Y = 0,$$

$$\mathbb{E}(X + \varepsilon Y)^2 = \mathbb{E}X^2 + 2\varepsilon\mathbb{E}X\mathbb{E}Y + \varepsilon^2\mathbb{E}Y^2 = 1 + \varepsilon^2. \qquad \blacksquare$$

We have thus reduced our problem to that of proving Lemma 71.4.

Sketch proof of Lemma 71.4. Let $G(s) = Pr(X \leqslant s)$. Then G must be an increasing function of s and so Riemann integrable on each interval $[a, b]$. Further $0 \leqslant G(s) \leqslant 1$ for all $s \in \mathbb{R}$ so G is bounded. Writing $E(s) = (2\pi)^{-\frac{1}{2}}\exp(-s^2/2)$, it is intuitively obvious that

$$Pr(Z \leqslant x) = \frac{1}{(2\pi)^{\frac{1}{2}}}\int_{-\infty}^{\infty} Pr(X \leqslant x - t)\exp(-t^2/2)dt = \int_{-\infty}^{\infty} G(x - t)E(t)dt$$

$$= G * Ex,$$

and the result will follow if we can show that $G * E$ is once differentiable with bounded continuous derivative $G * E'$.

In view of Theorem 53.1 this is extremely plausible and there are many possible proofs. For example, writing $F_N(x) = \int_{-N}^{N} E(x - t)G(t)dt$ we note that using the mean value theorem twice

$$\left| \frac{F_N(x + h) - F_N(x)}{h} - \int_{-N}^{N} E'(x - t)G(t)dt \right|$$

$$= \left| \frac{1}{h}\int_{-N}^{N} (E(x + h - t) - E(x - t) - hE'(x - t))G(t)dt \right|$$

$$= \left| \int_{-N}^{N} (E'(x - t + h\theta_1(x - t, h)) - E'(x - t))G(t)dt \right|$$

$$= \left| h\int_{-N}^{N} E''((x - t) + h\theta_2(x - t, h))G(t)dt \right|$$

$$\leqslant 2N|h|\sup_{s\in\mathbb{R}}|E''(s)|\sup_{t\in\mathbb{R}}|G(t)| = 2N|h|\sup_{s\in\mathbb{R}}|E''(s)| \to 0 \quad \text{as } h \to 0$$

for some

$$\theta_1, \theta_2 : \mathbb{R}^2 \to \mathbb{R} \text{ with } 1 \geqslant \theta_1, \theta_2 \geqslant 0.$$

Thus F_N is differentiable with $F_N'(x) = \int_{-N}^{N} E'(x - t)G(t)dt$. But (leaving the detailed verification here and elsewhere in the proof to the reader), F_N' is continuous (use the mean value theorem as before) and

$$|F_N'(x) - G * E'(x)| = \left| \int_{|t| > N} E'(x - t)G(t)dt \right|$$

$$\leqslant \int_{|t| > N} |E'(x - t)|dt \to 0 \text{ uniformly on each } [a, b]$$

$$\text{as } N \to \infty.$$

Thus using Lemma 53.2, $G*E$ is differentiable with continuous derivative $G*E'$. Since

$$|G*E'(t)| \leqslant \sup_{t\in\mathbb{R}}|G(t)| \int_{-\infty}^{\infty} |E'(t)|\,dt = \int_{-\infty}^{\infty} |E'(t)|\,dt,$$

it follows that $G*E'$ is bounded. ∎

We have thus obtained the central limit theorem in the general form given in Chapter 69 (Theorem 69.1). It is a remarkable theorem in probability. But we should not be misled either, as a non-mathematician might be, by the apparent difficulty of the proof nor, as a mathematician might be, by its generality into elevating it to a law of nature. Any application of probability theory to the real world, like any application of classical mechanics, is subject to experimental test. Sometimes it passes this test but sometimes it does not.

PART V
FURTHER DEVELOPMENTS

72

STABILITY AND CONTROL

The technique of using a windmill to grind corn is a complex one. Not only must the windmill sails be always set to catch the wind but their speed of rotation must be regulated so that the millstones grind at their optimum speed. Originally all these tasks were done by the miller but slowly simple semiautomatic and automatic devices were introduced to take over part of the work.

This change was most marked among British millers of the eighteenth century and it was by them that the most difficult problem, that of keeping the sails revolving at the proper speed, was solved. A centrifugal pendulum was employed to measure the angular velocity of the millstone and, working through additional mechanisms, to adjust the sail setting so as to reduce the sail speed if the angular velocity was too large or increase it if the angular velocity was too small.

This invention was taken up by James Watt and formed one of the chain of ingenious improvements to Newcomen's steam engine to which we referred in Chapter 42. Initially enclosed, to hide the secret from the eyes of Watt's competitors, the rotating centrifugal pendulum was soon proudly displayed as the visible symbol of man's control over steam.

Naturally, attempts were made to produce new or improved versions of 'Watt's governor', but not all were successful. In some cases the introduction of the control system led to worse behaviour than before or even to the dangerous phenomenon of 'hunting' in which the process went into wild oscillation.

The first productive mathematical attack on the problem was made by Maxwell (whose wide ranging investigations included enquiries into electromagnetism, colour vision, the kinetic theory and, though this work was not published, the reason why cats always land on their feet). The passage that follows comes from the introduction to his paper 'On Governors' (paper XXXIV of his *Collected Works*).

··· the motion of a machine with its governor consists in general of a uniform motion combined with a disturbance which may be expressed as the sum of several component motions. These components may be of four different kinds:
1. The disturbance may continually increase.

2. It may continually diminish.

3. It may be an oscillation of continually increasing amplitude.

4. It may be an oscillation of continually decreasing amplitude.

The first and third cases are evidently inconsistent with the stability of the motion; and the second and fourth alone are admissible in a good governor. This condition is mathematically equivalent to the condition that all the possible [i.e. real] roots, and all the possible [i.e. real] parts of the impossible [i.e. complex] roots, of a certain equation shall be negative ....

The actual motions corresponding to these impossible [i.e. complex] roots is not generally taken notice of by the inventors of such machines, who naturally confine their attention to the way in which it is *designed* to act; and this is generally expressed by the possible [i.e. real] root of the equation. If, by altering the adjustments of the machine its governing power is continually increased, there is generally a limit at which the disturbance, instead of subsiding more rapidly, becomes an oscillating and jerking motion increasing in violence till it reaches the limit of action of the governor. This takes place when the possible [i.e. real] part of one of the impossible [i.e. complex] roots becomes positive. The mathematical investigation of the motion may be rendered practically useful by pointing out a remedy for these disturbances.

The key to Maxwell's approach lay in the recognition that even for systems given by non linear differential equations the initial response to small disturbances may often (after neglecting second order terms) be described by linear differential equations for which the stability problem is much easier. In the next few chapters we shall deal with techniques for solving the 'linear stability' problem.

The mechanisms discussed by Maxwell were of the kind where by partially controlling the input we hope to keep the output close to a constant value. A much more complicated problem arises when we wish to keep the output close to some function $f(t)$ which we do not know in advance. Such problems arise in driving a car or firing heavy guns at moving targets.

We now wish the output to respond sluggishly and unwillingly to that part of the input which we cannot regulate but to respond quickly and readily to changes in that part of the input which we can regulate. These two objects of stability and control will usually conflict and the best we can hope for is a compromise.

Nineteenth-century engineers were mainly concerned with the steady running of machinery and so identified control with stability. In keeping with this tradition early gliders and aeroplanes were designed to be extremely stable. Thus if an aircraft was flying safely it would take a great deal to change this. On the other hand once it got into trouble (say after an exceptional gust of wind), the high inherent stability meant that it was very difficult to get out of danger. When this was realised aeroplanes were built with much lower stability.

This trend was accelerated by the First World War. Pilots did not mind machines which required constant care to fly under normal conditions, provided they had high

manoeuverability in dog fights. After the war even civil aircraft were built to be only just stable, or even slightly unstable, because of the degree of control this gave the pilot.

Until the 1960s the amount of instability that could be built into a fighter plane was limited by the abilities and the reaction times of the pilot. However, the development of sensing and control devices now means that the pilot no longer flies the aircraft directly and the fighter may be built so as to be extremely unstable (and so manoeuverable). We can measure in the contrast between the crude automatic mechanisms of the miller of 1770 and the sophisticated electronics that will back tomorrow's fighter pilot both the technical and the moral progress that mankind has made in the last 200 years.

73
INSTABILITY

There are many kinds of mathematical instability. For example the equation

$$\dot{x}(t) = t, \qquad x(0) = 1$$

has a solution $x(t) = t^2/2 + 1$ of polynomial growth, whilst the equation

$$\dot{x}(t) = x(t), \qquad x(0) = 1$$

has a solution $x(t) = e^t$ of exponential growth, and the equation

$$\dot{x}(t) = tx(t), \qquad x(0) = 1$$

has a solution $x(t) = e^{t^2/2}$ which grows faster than exponential as $t \to \infty$.

The direct use of the Fourier transform requires that the transformed function $x(t)$ have $\int_{-\infty}^{\infty} |x(t)|\,dt < \infty$ (for otherwise it is difficult to assign a meaning to $\hat{x}(\lambda)$ $= \int_{-\infty}^{\infty} x(t)e^{-i\lambda t}\,dt$). However, if

(1) *we are only interested in the value of $x(t)$ for $t \geqslant 0$, and*

(2) *$x(t)$ grows no faster than exponential,*

then we can study the problem by means of the Laplace transform as follows.

Let $H(t) = 0$ for $t < 0$, $H(t) = 1$ for $t \geqslant 0$ and consider $X(t) = x(t)H(t)e^{-\alpha t}$ where α is a large real number. Then, since $X(t) = 0$ for $t < 0$ and since the factor $e^{-\alpha t}$ converts exponential growth into exponential decrease (for α large enough), we have $\int_{-\infty}^{\infty} |X(t)|\,dt < \infty$. We can now apply the methods of Fourier transform theory to X and then recover $x(t)$ (at least for $t \geqslant 0$) by means of the formula $x(t) = e^{\alpha t}X(t)[t \geqslant 0]$.

The condition that our function $x(t)$ grow no faster than exponential is not very restricting in practice since experience shows that the growth of instability is usually initially exponential. (To a certain extent this corresponds to the mathematical theorem which states that if

$$\mathbf{x}^{(k)}(t) + A_1\mathbf{x}^{(k-1)}(t) + \cdots + A_k\mathbf{x}(t) = \mathbf{c}(t) \qquad (*)$$

[where $\mathbf{x}$ is a function $\mathbb{R}^n \to \mathbb{R}$, A_j a constant $n \times m$ matrix and $\mathbf{c}$ a continuous function $\mathbb{R}^m \to \mathbb{R}$] then, if $\sup_{1 \leqslant i \leqslant m} |c_i(t)|$ grows at most exponentially fast, so does

368

$\sup\limits_{1\leqslant j\leqslant n}|x_j(t)|$. Although it is easy to prove by induction on k both that (*) has solutions and that they behave as stated, the proof is so boring that I have omitted it.)

What about condition (1) which states that we are only interested in the future behaviour of the function $x(t)$? This does not represent a real constraint, since we can always write $x(t) = x_1(t) + x_2(t)$ with $x_1(t) = 0$ for $t \leqslant 0$ and $x_2(t) = 0$ for $t \geqslant 0$. We can then investigate the behaviour of $x_1(t)$ and $x_2(-t)$ by means of the Laplace transform and so recover both the past and future behaviour of $x(t)$.

It could also be argued that we should only be interested in the future behaviour of unstable systems. Since the argument is instructive, though (as is to be expected of any post hoc argument) not entirely convincing, I shall give it here.

Consider the typical unstable system described by

$$\dot{x} + kx = 0, \qquad x(0) = \varepsilon, \qquad\qquad (\dagger)$$

with $k < 0$. Let us say that the system 'explodes' at the first time T when $|x(T)| = A$. Since the solution of ($\dagger$) is $x(t) = \varepsilon e^{-kt}$ we see that the time to explosion is given by

$$T = -k^{-1}(\log A + \log|\varepsilon|^{-1}).$$

Thus the time T to explosion, although sensitive to changes in k, is insensitive to changes in A and ε^{-1}. For example, if, with immense labour, we manage to reduce ε from 10^{-2} to 10^{-4} and increase A from 10^2 to 10^4 we still only double the time before the explosion.

The same phenomenon is familiar from numerical analysis. Suppose we attempt to calculate e^{-t} by considering it as the solution of

$$\ddot{x} - x = 0,$$

$$x(0) = 1, \qquad \dot{x}(0) = -1.$$

If we use a physical system as an analogue of the mathematical one, then the initial conditions $x(0) = 1$, $\dot{x}(0) = -1$ can no longer be satisfied exactly and we must replace them by $x(0) = 1 + \eta_1$, $\dot{x}(0) = -1 + \eta_2$ where η_1 and η_2 are unknown errors of the order of ε say. The solution now becomes

$$x(t) = \tfrac{1}{2}(\eta_1 + \eta_2)e^t + (1 + \tfrac{1}{2}(\eta_1 - \eta_2))e^{-t}$$

and after a time of the order of $-\log \varepsilon$ the *parasitic solution* $\tfrac{1}{2}(\eta_1 + \eta_2)e^t$ will swamp the solution e^{-t} which we wanted. Moreover, increasing the precision with which we can fix the initial conditions has very little effect. For example, if we increase our accuracy from five to ten figures, we only double the time before our calculations become totally unusable.

Thus we expect unstable systems to 'explode' and we expect the 'explosion' to occur sooner rather than later. If we come into a room and see a pencil standing on its point we do not sit down and calculate its position for the last 24 hours (although such a past history has a mathematical meaning), we look for the man who has just balanced it. Unstable systems do not exist in nature, they have to be constructed.

Thus it is natural when studying unstable systems to use a method which deals only with the future $t \geqslant 0$ and not with the past $t < 0$. (But the reader may recall the philosopher Hegel who had the misfortune to produce a metaphysical proof that no unknown planets existed a couple of years before a new one was discovered.)

So much for the general discussion. In the next chapter we return to the style of mathematical demonstration which Heaviside found 'typical of a lot of work made up by the brain-torturers who write books for young people and college students who are going to be Senior Wranglers, perhaps. Let mathematics be humanised if possible. The best of all proofs is to set out the fact descriptively so that it can be seen to be a fact (*Collected Works*, Vol. III, p. 140)'.

Oliver Heaviside was one of those men who always carry around a red rag on the off chance that they may see a bull to wave it at. 'Lord Kelvin used to call me a nihilist. This was a great mistake, (though I did throw a bomb occasionally, to stimulate an official humbug to learn something about electricity and how to apply it) (Vol. III, p. 479).'

Working in the tradition of Maxwell, Kelvin and Fourier ('The only entertaining mathematical work I ever saw (Vol. II, p. 32)') he developed Kelvin's theory of telegraphy and fought a long but ultimately successful battle with the British Post Office as to the correct way of improving the transmission properties of undersea cables. Working with Perry he showed that Kelvin's calculations of the age of the earth would be completely overthrown if the conductivity of the earth's interior were considerably greater than that of the crust.

He fought an entertaining three sided battle in favour of the vector methods developed by himself and Gibbs and against the quaternionic and Cartesian methods then favoured. His many dislikes included Gothic letters (to which he attributed 'the prevalent short sightedness of the German nation (Vol. I, p. 140)'), Lagrange's equations and the principle of least action ('Truly I have never practiced it myself (except with pots and pans) (Vol. III, p. 175)') and 'mathematicians of the Cambridge or conservatory kind, who look the gift-horse in the mouth and shake their heads with solemn smile (Vol. II, p. 12)'.

Heaviside was largely self taught and learned his calculus from the works of Todhunter, Boole and Fourier with the result that his mathematical work lay in a formal tradition which had been obsolete in Europe since the time of Cauchy and Dirichlet and which was beginning to be abandoned even in England. ('It was a long time before the works of the great continental analysts were understood in England and ... [the papers of British analysts in the years 1840–50] ... show a singular and often entertaining mixture of occasional shrewdness and fundamental incompetence (Hardy *Divergent Series*, Oxford 1949, p. 19).')

Formal manipulation produces useful results only in the hands of masters like Euler and Boole. But by a mixture of common sense, physical intuition and mathematical insight Heaviside produced a powerful formal system, his 'operational calculus', for solving ordinary differential and partial differential equations.

The rising generation of British mathematicians could see the regression to a style

that rendered worthless the work of two previous generations but not the new ideas. 'There was a sort of tradition', one of them told Whittaker, 'that a Fellow of the Royal Society could print almost anything he liked in the Proceedings without being troubled by referees: but when Heaviside had published two papers on his symbolic methods, we felt that a line had to be drawn somewhere, so we put a stop to it (Vol. I, p. XXX)'.

Heaviside was deeply hurt but published elsewhere 'for a wider circle of readers who have fewer prejudices though their mathematical knowledge may be to that of the rigorists as a straw to a hay stack' and his operational methods became so widely and successfully used that their justification represented a major challenge to mathematicians. The Laplace transform (of which he probably would not have approved) and the more modern distribution theory (of which he might have approved) were developed, at least in part, to meet this challenge.

74

THE LAPLACE TRANSFORM

In the last chapter we saw that the study of stability and instability might be approached by considering the class $\mathscr{E}^*$ of functions defined as follows.

Definition 74.1*. *We say that a function* $f:\mathbb{R} \to \mathbb{C}$ *belongs to* $\mathscr{E}^*$ *if*

(i) $|f(t)\exp(-at)| \to 0$ *as* $t \to \infty$ *for some* $a \in \mathbb{R}$,

(ii) $f(t) = 0$ *for all* $t < 0$,

(iii)* f *is Riemann integrable on* $[0, b]$ *for all* $b > 0$.

However, it is easier to consider a somewhat more restricted class $\mathscr{E}$.

Definition 74.1. *We say that* $f \in \mathscr{E}$ *if* $f \in \mathscr{E}^*$ *and, further,*

(iii) f *is continuous except possibly at a finite set of points,*

(iv) $\lim\limits_{t \to x+} f(t)$ *and* $\lim\limits_{t \to x-} f(t)$ *exist everywhere and*

$$f(x) = (\lim_{t \to x+} f(t) + \lim_{t \to x-} f(t))/2.$$

For convenience we write $e_a(t) = \exp(at)$ and make the following subsidiary definition.

Definition 74.2. *We write*

$$\mathscr{E}_a = \{f \in \mathscr{E} : f(t)e_{-a}(t) \to 0 \text{ as } t \to \infty\} \quad [a \in \mathbb{R}].$$

The key observation about $\mathscr{E}_a$ (and so about $\mathscr{E}$) is the following.

Lemma 74.3. *If* $f \in \mathscr{E}_a$ *then* fe_{-b} *is Riemann integrable with*

$$\int_{-\infty}^{\infty} |f(t)e_{-b}(t)| dt < \infty \text{ for all } b > a.$$

Proof. This is trivial. Since $f(t)e_{-a}(t) \to 0$ as $t \to \infty$ it follows that there exists an M with $|f(t)e_{-a}(t)| \leqslant M$ for all t. (Formally, choose an R such that $|f(t)e_{-a}(t)| \leqslant 1$ for

372

all $t \geq R$. Suppose f has discontinuities at $x(1) < x(2) < \cdots < x(n) \leq R$ and set $x(0) = 0$, $x(n + 1) = R$. If $g_j(t) = f(t)$ on $[x(j - 1), x(j))$ and $g_j(x(j)) = \lim\limits_{t \to x(j)} f(t)$ then g is continuous on the closed bounded interval $[x(j - 1), x(j)]$ and so bounded with $|g_j(t)| \leq M_j$ for $t \in [x(j - 1), x(j)]$, say. Now set $M = \max(1, e^{-aR})(\max\limits_{1 \leq j \leq n} M_j + 1)$.)

In particular, if $b > a$, then

$$\int_{-S}^{R} |e_{-b}(t)f(t)| \, dt = \int_{0}^{R} |e_{-b}(t)f(t)| \, dt = \int_{0}^{R} |e_{a-b}(t)| |e_{-a}(t)f(t)| \, dt$$

$$\leq M \int_{0}^{R} |e_{a-b}(t)| \, dt \leq \frac{M}{b-a} \quad \text{for all } R, S \geq 0,$$

so $\int_{-\infty}^{\infty} e_{-b} f$ is absolutely convergent. ∎

We can now follow the suggestion of Chapter 73 and consider the Fourier transform $\mathscr{F}(e_{-b}f)(\zeta)$ of $e_{-b}f$. The result depends not only on ζ but on b and this is indicated by writing

$$\mathscr{F}(e_{-b}f)(\zeta) = (\mathscr{L}f)(b + i\zeta).$$

Formally, we define the *Laplace transform* $\mathscr{L}f$ of f as follows. (Observe that $z = b + i\zeta$.)

Definition 74.4. *If $f \in \mathscr{E}_a$ we write*

$$\mathscr{L}f(z) = \int_{-\infty}^{\infty} \exp(-zt)f(t) \, dt$$

for all $z \in \mathbb{C}$ with $\operatorname{Re} z > a$.

Because of the close connection between the Laplace and the Fourier transform we can at once write down an inversion and a uniqueness result.

Theorem 74.5. (i) *If $f \in \mathscr{E}_a$, then, whenever $b > a$ and $t \in \mathbb{R}$,*

$$f(t) = \lim_{k \to \infty} \frac{1}{2\pi} \int_{-k}^{k} (\mathscr{L}f)(b + i\zeta) \exp(b + i\zeta)t \, d\zeta,$$

(ii) *If f, $g \in \mathscr{E}_c$ and $\mathscr{L}f(z) = \mathscr{L}g(z)$ for all $\operatorname{Re} z > c$ then $f = g$.*

Proof. (i) We use the inversion formula of Chapter 61 (Theorem 61.1) applied to fe_{-b}. This gives

$$f(t) = e_b(t)(f(t)e_{-b}(t)) = e_b(t) \lim_{k \to \infty} \frac{1}{2\pi} \int_{-k}^{k} \mathscr{F}(fe_{-b})(\zeta) \exp(i\zeta t) \, d\zeta$$

$$= \lim_{k \to \infty} \frac{1}{2\pi} \int_{-k}^{k} \mathscr{F}(fe_{-b})(\zeta) \exp((b + i\zeta)t) \, d\zeta = \lim_{k \to \infty} \frac{1}{2\pi} \int_{-k}^{k} (\mathscr{L}f)(b + i\zeta) \exp((b + i\zeta)t) \, d\zeta,$$

as required. (Note the use of part (iv) of Definition 74.1 at the points of discontinuity.)
(ii) Take $b > c$ and apply (i) to obtain

$$f(t) = \lim_{k \to \infty} \frac{1}{2\pi} \int_{-k}^{k} ((\mathscr{L}f)(b + i\zeta)) \exp((b + i\zeta)t)d\zeta$$

$$= \lim_{k \to \infty} \frac{1}{2\pi} \int_{-k}^{k} ((\mathscr{L}g)(b + i\zeta)) \exp((b + i\zeta)t)d\zeta = g(t) \qquad \text{for all } t \in \mathbb{R}. \qquad \blacksquare$$

Remark. If f is continuous we can avoid the use of Chapter 61 and instead use the easier (and more satisfactory) Theorem 60.3(i) to obtain (i) and thus (ii). In any case we shall obtain the uniqueness result (ii) by a different and better method in Theorem 75.10 so the reader who does not wish to work through Chapter 61 need not.

The calculation of Laplace transforms is made easier by the existence of a large collection of easy standard formulae. The next two lemmas contain a selection of these but more can be found in text books on the Laplace transform. In what follows we write

$$e_\lambda(t) = \exp \lambda t \qquad\qquad t \in \mathbb{R},$$

$$\begin{cases} H(t) = 0 & t < 0 \\ H(0) = \tfrac{1}{2} & \\ H(t) = 1 & t > 0, \end{cases}$$

$$\tau_n(t) = t^n H(t) \qquad\qquad t \in \mathbb{R},$$

$$(T_a f)(t) = f(t - a) \qquad\qquad t \in \mathbb{R}.$$

[Here $\lambda \in \mathbb{C}$, n is a positive integer and $a \in \mathbb{R}$].

Thus H is the Heaviside 'step function' (with $H(0)$ chosen to suit the purposes of this chapter) and $T_a f$ is the translation of f by a.

Lemma 74.6. (i) *If* $\lambda, \mu \in \mathbb{C}$, $f \in \mathscr{E}_a$, $g \in \mathscr{E}_b$ *and* $c \geqslant a, b$ *then* $\lambda f + \mu g \in \mathscr{E}_c$ *and*

$$\mathscr{L}(\lambda f + \mu g) = \lambda \mathscr{L}f(z) + \mu \mathscr{L}g(z) \qquad \text{for } \mathrm{Re}\, z > c.$$

(ii) *If* $f \in \mathscr{E}_a$, $\lambda \in \mathbb{C}$ *and* $c \geqslant \mathrm{Re}\, \lambda + a$, *then* $e_\lambda f \in \mathscr{E}_c$ *and*

$$\mathscr{L}(e_\lambda f)(z) = \mathscr{L}f(z - \lambda) \qquad \text{for } \mathrm{Re}\, z > c.$$

(iii) $H \in \mathscr{E}_a$ *for all* $a > 0$ *and*

$$\mathscr{L}H(z) = z^{-1} \qquad \text{for } \mathrm{Re}\, z > 0.$$

(iv) $e_\lambda H \in \mathscr{E}_a$ *for all* $a > \mathrm{Re}\, \lambda$ *and*

$$\mathscr{L}e_\lambda H(z) = (z - \lambda)^{-1} \qquad \text{for } \mathrm{Re}\, z > \mathrm{Re}\, \lambda.$$

(v) $\tau_n \in \mathscr{E}_a$ for all $a > 0$, $n \geqslant 0$ and

$$\mathscr{L}\tau_n(z) = n!z^{-n-1} \qquad \text{for Re } z > 0.$$

(vi) If $f \in \mathscr{E}_a$ and $b \geqslant 0$ then $T_b f \in \mathscr{E}_a$ and

$$(\mathscr{L}T_b)(z) = e_{-b}(z)\mathscr{L}f(z) \quad \text{for Re}\, z > a.$$

Remark. Note that we need condition $b \geqslant 0$ in (vi) since if $b < 0$ we cannot guarantee that $T_b f(t) = 0$ for all $t < 0$.

Proofs. These are all trivial and the reader should be able to do them for herself. We sketch the proofs of (ii), (v) and (vi).

(ii) Clearly

$$e_{-c}(e_\lambda f)(t) = e_{\lambda-c}(t)f(t) \to 0 \text{ as } t \to \infty \text{ for all } \mathrm{Re}(\lambda - c) < -a,$$

i.e. for all $c \geqslant \mathrm{Re}\,\lambda + a$. Further

$$\mathscr{L}(e_\lambda f)(z) = \int_0^\infty \exp(\lambda t)f(t)\exp(-zt)dt = \int_0^\infty f(t)\exp(-(z-\lambda)t)dt$$

$$= \mathscr{L}f(z-\lambda) \qquad [\mathrm{Re}\,z > c].$$

(v) We know that $t^n e^{-\varepsilon t} \to 0$ as $t \to \infty$ for all $\varepsilon > 0$. Further by repeated integration by parts

$$\mathscr{L}(\tau_n) = \int_0^\infty t^n \exp(-zt)dt = \left[-\frac{t^n}{z}\exp(-zt) \right]_0^\infty + \frac{n}{z}\int_0^\infty t^{n-1}\exp(-zt)dt$$

$$= \frac{n}{z}\mathscr{L}(\tau_{n-1}) = \cdots = \frac{n!}{z^n}\int_0^\infty \exp(-zt)dt = \frac{n!}{z^{n+1}}[\exp(-zt)]_0^\infty = \frac{n!}{z^{n+1}}$$

$$[\mathrm{Re}\,z > 0].$$

(vi) Observe that

$$e^{-at}f(t-b) = e^{-ab}(e^{-a(t-b)}f(t-b)) \to 0 \quad \text{as } t \to \infty$$

and that, making the substitution $s = t - b$

$$\mathscr{L}(T_b f)(z) = \int_0^\infty f(t-b)\exp(-zt)dt = \int_{-b}^\infty f(s)\exp(-z(s+b))ds$$

$$= \exp(-zb)\int_0^\infty f(s)\exp(-zs)ds = \exp(-zb)(\mathscr{L}f)(z) \qquad [\mathrm{Re}\,z > a].$$

∎

If f is n times differentiable and its n derivatives not only belong to $\mathscr{E}_a$ but are also continuous everywhere, then repeated integration by parts gives the simple formula

$$\mathscr{L}f^{(n)}(z) = \int_0^\infty f^{(n)}(t)e^{-zt}dt = [f^{(n-1)}(t)e^{-zt}]_0^\infty + z\int_0^\infty f^{(n-1)}(t)e^{-zt}dt$$

$$= z\mathscr{L}f^{(n-1)}(z) = \cdots = z^n \mathscr{L}f(z) \qquad \text{for all } \mathrm{Re}\,z > a.$$

(Notice that we can deduce Lemma 74.6(v) in this way.) If we allow discontinuities, then the formulae become more complicated. However it is very useful to have such a formula in the case when we allow discontinuities at 0 (but nowhere else), since, as we shall see in Example 74.8, this gives us a quick method of solving linear differential equations with given initial conditions.

Lemma 74.7. (i) *Suppose that* $f:\mathbb{R}\to\mathbb{C}$ *is such that* $f(t)=0$ *for* $t<0$ *and* f *is continuously differentiable for* $t>0$. *Suppose that* $f(0+)=\lim\limits_{t\to 0+} f(t)$ *and* $f'(0+)$ $=\lim\limits_{t\to 0+} f'(t)$ *exist and we define* $f(0)$ *to be* $\frac{1}{2}f(0+)$ *and* $f'(0)$ *to be* $\frac{1}{2}f'(0+)$. *If* $f'\in\mathcal{E}$ *then* $f\in\mathcal{E}$. *Moreover if* $f, f'\in\mathcal{E}_a$ *then*

$$(\mathcal{L}f')(z) = z(\mathcal{L}f)(z) - f(0+) \qquad \text{for } \operatorname{Re} z > a.$$

(ii) *Suppose that* $f:\mathbb{R}\to\mathbb{C}$ *is such that* $f(t)=0$ *for* $t<0$ *and* f *is* n *times continuously differentiable for* $t>0$. *Suppose that* $f^{(r)}(0+)=\lim\limits_{t\to 0+} f^{(r)}(t)$ *exists for each integer* $0\leqslant r\leqslant n$ *and that we define* $f^{(r)}(0)$ *to be* $\frac{1}{2}f^{(r)}(0+)$. *If* $f^{(n)}\in\mathcal{E}$ *then* $f\in\mathcal{E}$. *Moreover if* $f, f', f'', \ldots, f^{(n)}\in\mathcal{E}_a$ *then*

$$(\mathcal{L}f^{(n)})(z) = z^n(\mathcal{L}f)(z) - \sum_{r=0}^{n-1} f^{(r)}(0+)z^{n-r-1}.$$

Proof. (i) To see that $f\in\mathcal{E}$ if $f'\in\mathcal{E}$ we argue as follows. If $f'\in\mathcal{E}_a$ then, arguing as in the first part of Lemma 74.3, there exists an M with $|f'(s)e^{-as}|\leqslant M$ for all $s>0$. It follows that $|f(t)|\leqslant|f(0+)|+M\int_0^t e^{as}ds$ and so $e^{-bt}f(t)\to 0$ as $t\to\infty$ whenever $b>\max(a,0)$.

The formula is obtained by integrating by parts as follows

$$(\mathcal{L}f')(z) = \int_0^\infty e^{-tz}f'(t)dt = [e^{-tz}f(t)]_0^\infty + z\int_0^\infty e^{-tz}f(t)dt$$

$$= (0 - f(0+)) + z\mathcal{L}f(z) = z\mathcal{L}f(z) - f(0+) \quad \text{for } \operatorname{Re} z > a.$$

(ii) Induction. ∎

Remark. In the general case when f is n times continuously differentiable except possibly at the distinct points $x(1), x(2),\ldots,x(m)$ the formula becomes

$$(\mathcal{L}f^{(n)})(z) = z^n(\mathcal{L}f)(z) - \sum_{j=1}^{m}\exp(-x(j)z)\sum_{r=0}^{n-1}(f^{(r)}(x(j)+) - f^{(r)}(x(j)-))z^{n-r-1}.$$

The reader is left with the easy task of proving this, and, if she wishes, the tedious task of stating exactly the associated lemma.

Remark 2. The definition of the class $\mathcal{E}$ on which the Laplace transform acts and, sometimes, of the Laplace transform $\mathcal{L}$ itself vary from book. For example we could replace condition (iv) of Definition 74.1 by

(iv)′ f *is continuous except, possibly, at zero,* $\lim\limits_{t\to 0+} f(t)$ *exists and* $f(0)=\lim\limits_{t\to 0+} f(t)$.

If we did this then the conclusion of Lemma 74.7 (ii) would take the simpler form

$$\mathscr{L} f^{(n)}(z) = z^n (\mathscr{L} f)(z) - \sum_{r=0}^{n-1} f^{(r)}(0) z^{n-r-1},$$

but the conclusion of Theorem 74.5 (i) would be a little more complicated and Lemma 74.6 (vi) would only hold for continuous f. However, as with the many definitions of the Fourier transform (see Appendix F), these differences lie entirely on the surface and should not cause any problems.

The Laplace transform furnishes a rapid routine method for solving linear differential equations with given initial conditions.

Example 74.8. *Solve* $\ddot{v}(t) - \dot{v}(t) - 6v(t) = 5e^{3t}$ *for* $t \geq 0$ *subject to the initial conditions* $v(0) = 6$, $\dot{v}(0) = 1$.

Solution. (The reader may prefer to start by trying her usual method of solution in order to compare it with the one suggested.) Let us seek $u \in \mathscr{E}$ such that u', $u'' \in \mathscr{E}$ and

$$\ddot{u}(t) - \dot{u}(t) - 6u(t) = 5e^{3t} H(t) \tag{*}$$

and $\lim_{t \to 0+} u(t) = 6$, $\lim_{t \to 0+} \dot{u}(t) = 1$. (As was stated in Chapter 73 it is easy to show, though we shall not bother, that the only solutions to our equations have, at worst, exponential growth so we shall not actually overlook any solutions by this restriction.)

By Lemma 74.7,

$$(\mathscr{L} \dot{u})(z) = z(\mathscr{L} u)(z) - u(0+) = z(\mathscr{L} u)(z) - 6$$

and $\qquad (\mathscr{L} \ddot{u})(z) = z^2 (\mathscr{L} u)(z) - zu(0+) - u'(0+) = z^2 (\mathscr{L} u)(z) - 6z - 1,$

whilst, by Lemma 74.6 (iv), $\mathscr{L}(e_3 H)(z) = 1/(z-3)$. Thus using Lemma 76.4 (i) and taking the Laplace transform of both sides of (*) we obtain

$$(z^2 - z - 6)(\mathscr{L} u)(z) - 6z + 5 = 5/(z-3)$$

(at least, provided $\operatorname{Re} z > a$, for some $a \in \mathbb{R}$). Thus

$$(\mathscr{L} u)(z) = \frac{5}{(z+2)(z-3)^2} + \frac{6z-5}{(z+2)(z-3)}. \tag{**}$$

(Notice how the initial conditions have been absorbed into (**); this is one of the advantages of the method.)

We now try to write the right hand side of (**) in partial fractions as

$$\frac{5}{(z+2)(z-3)^2} + \frac{6z-5}{(z+2)(z-3)} = \frac{A}{(z+2)} + \frac{B}{(z-3)} + \frac{C}{(z-3)^2}.$$

Multiplying out gives

$$5 + (z-3)(6z-5) = A(z-3)^2 + B(z+2)(z-3) + C(z+2).$$

Substituting $z=3$ yields $C=1$. Substituting $z=-2$ yields $A = \frac{18}{5}$. Equating coefficients of z^2 yields $A + B = 6$ and so $B = \frac{12}{5}$. Thus

$$(\mathscr{L}u)(u) = \frac{18}{5}(z+2)^{-1} + \frac{12}{5}(z-3)^{-1} + (z-3)^{-2}$$

$$= \frac{18}{5}\mathscr{L}(He_{-2}) + \frac{12}{5}\mathscr{L}(He_3) + \mathscr{L}(\tau_1 e_3)$$

$$= \mathscr{L}\left(\frac{18}{5}He_{-2} + \frac{12}{5}He_3 + \tau_1 e_3\right).$$

(Using Lemma 74.6 (iii), (v), (ii) and (i).) Thus by the uniqueness of the Laplace transform (Theorem 74.5 (ii))

$$u(t) = \frac{18}{5}H(t)e_{-2}(t) + \frac{12}{5}H(t)e_3 + \tau_1(t)e_3(t).$$

We have thus obtained a solution

$$v(t) = \frac{18}{5}e^{-2t} + \frac{12}{5}e^{3t} + te^{3t} \quad [t \geqslant 0]$$

to our original equation. ∎

A few worked examples should convince the reader that the Laplace transform furnishes a useful technique for solving linear differential equations. (However, the study of differential equations is no more contained in a table of Laplace transforms than the study of geometry in a set of trigonometric tables.)

75

DEEPER PROPERTIES

The Laplace transform is not just a computationally useful variant of the Fourier transform. It has a specific character of its own which springs from the fact that it is analytic.

Let us write $\tau(t) = t [t \geqslant 0]$, $\tau(t) = 0 [t < 0]$.

Lemma 75.1. *Suppose that $f \in \mathscr{E}_a$. Then $\tau f \in \mathscr{E}_b$ for all $b > a$.*
Proof. Arguing as in the first part of Lemma 74.3 there exists an M with $|f(t)e^{-at}| \leqslant M$ for all $t \geqslant 0$. Thus $|t f(t)e^{-bt}| \leqslant Mt e^{-(b-a)t} \to 0$ as $t \to \infty$ whenever $b > a$. ∎

Theorem 75.2. *If $f \in \mathscr{E}_a$ then $\mathscr{L}f$ is analytic with $(\mathscr{L}f)'(z) = -(\mathscr{L}(\tau f))(z)$ for all $\mathrm{Re}\, z > a$.*

It is almost, but not quite, a proof of Theorem 75.2 to say that e^{-tz} is analytic with derivative $-t e^{-tz}$ so, since $(\mathscr{L}f)(z) = \int_0^\infty e^{-tz} f(t)dt$ is the weighted average of such e^{-tz}, this expression must itself be analytic with derivative the weighted average $\int_0^\infty (-t e^{-tz}) f(t)dt$ of the derivatives.

However, such an argument can only fully satisfy a very shallow or a very deep mathematician. Our proof is based on the last of the following chain of important theorems from complex variable theory. (Throughout, 'contour' will mean 'well behaved contour'.)

Theorem 75.3. (i) *(Cauchy). If Ω is open and $g : \Omega \to \mathbb{C}$ is analytic then $\int_C g(z)\,dz = 0$ for any contour C such that C and its interior lie in Ω,*
 (ii) *(Cauchy's formula). If Ω, g and C are as in (i) then*

$$g^{(r)}(w) = \frac{r!}{2\pi i} \int_C \frac{g(z)}{(z - w)^{r+1}}\,dz \text{ for all } w \text{ inside } C.$$

(iii) *(Morera). Suppose Ω is open and $g : \Omega \to \mathbb{C}$ is continuous. If $\int_C g(z)dz = 0$ for all contours C such that C and its interior lie in Ω then g is analytic.*

(iv) *Suppose Ω is open and $g_n : \Omega \to \mathbb{C}$ is analytic for each $n \geqslant 0$. If $g_n \to g$ uniformly on each closed and bounded (i.e. compact) subject of Ω then g is analytic. Moreover $g_n^{(r)}(w) \to g^{(r)}(w)$ as $n \to \infty$ for each $w \in \Omega$.*

Proof. All four results are in every complex variable textbook and the reader who does not know the theorems of Cauchy and Morera is unlikely to profit greatly from the next three chapters. However, the importance of (iv) is not always stressed so we shall prove it here.

(iv) Observe first that if $w \in \Omega$ then we can find an $\varepsilon > 0$ such that $\{z : |z - w| < 2\varepsilon\} \subset \Omega$. Now $g_n \to g$ uniformly on the closed and bounded set $\{z : |z - w| \leqslant \varepsilon\}$ so, since g_n is continuous on $\{z : |z - w| \leqslant \varepsilon\}$, so is g. Thus g is continuous at each $w \in \Omega$ i.e. continuous on Ω.

Turning to the more interesting part of the proof we note that if C is any contour such that C and its interior lie in Ω, then Cauchy's theorem yields

$$\int_C g_n(z)\,dz = 0.$$

But

$$\left| \int_C g_n(z)\,dz - \int_C g(z)\,dz \right| = \left| \int_C g_n(z) - g(z)\,dz \right|$$

$$\leqslant \text{length } C \times \sup_{z \in C} |g_n(z) - g(z)| \to 0 \quad \text{as } n \to \infty,$$

since, by hypothesis, $g_n \to g$ uniformly on the closed and bounded set C. Thus

$$\int_C g(z)\,dz = 0,$$

and, since C was arbitrary, it follows from Morera's theorem that g is analytic.

To compute $g^{(r)}$ at a point $w \in \Omega$ we first choose an $\varepsilon > 0$ such that $\{z \in \Omega : |z - w| < 2\varepsilon\} \subset \Omega$. If C is the circle $z = w + \varepsilon e^{i\theta}$ described as θ runs from 0 to 2π then, by Cauchy's formula,

$$g_n^{(r)}(w) = \frac{r!}{2\pi i} \int_C \frac{g_n(z)\,dz}{(z - w)^{r+1}} \quad \text{and} \quad g^{(r)}(w) = \frac{1}{2\pi i} \int_C \frac{g(z)\,dz}{(z - w)^{r+1}}.$$

Thus

$$|g_n^{(r)}(w) - g^{(r)}(w)| = \left| \frac{r!}{2\pi i} \int_C \frac{g_n(z) - g(z)\,dz}{(z - w)^{r+1}} \right| \leqslant \frac{r!}{2\pi} \text{ length } C \times \sup_{z \in C} \left| \frac{g_n(z) - g(z)}{(z - w)^{r+1}} \right|$$

$$= \frac{1}{2\pi} 2\pi\varepsilon \sup_{z \in C} \frac{|g_n(z) - g(z)|}{\varepsilon^{r+1}} = \varepsilon^{-r} \sup_{z \in C} |g_n(z) - g(z)| \to 0 \quad \text{as } n \to \infty,$$

as required.   ∎

Beside the deep results give above we shall also use a simple computational fact.

Lemma 75.4. *If* $|z| \leqslant 1$ *then*

$$|e^z - 1 - z| \leqslant |z|^2$$

Proof. We have

$$|e^2 - 1 - z| = \left| \sum_{r=2}^{\infty} z^r/r! \right| \leqslant \sum_{r=2}^{\infty} |z|^r/r!$$

$$\leqslant (|z|^2/2) \sum_{n=0}^{\infty} (|z|/2)^n \leqslant \frac{|z|^2/2}{1 - |z|/2} \leqslant |z|^2 \quad \text{for all } |z| \leqslant 1,$$

as required. ∎

Proof of Theorem 75.2. Let us write

$$G_N(z) = \int_0^N f(t) e^{-zt} \, dt \quad [z \in \mathbb{C}],$$

$$G(z) = (\mathscr{L}f)(z) = \int_0^{\infty} f(t) e^{-zt} \, dt \quad [\operatorname{Re} z > a].$$

We shall show that

(i) G_N is analytic in $\mathbb{C}$ (and so in particular on $\Omega = \{z : \operatorname{Re} z > a\}$).
(ii) $G_N \to G$ uniformly on $\{z : \operatorname{Re} z \geqslant b\}$ for each $b > a$ (and so in particular on every closed bounded subset of Ω). The analyticity of $\mathscr{L}f = G$ will then follow from Theorem 75.3 (iv).

Proof of (i). Let $F_N(z) = -\int_0^N t f(t) e^{-zt} \, dt$. Using Lemma 75.4 we see that if $|hN| < 1$

$$\left| \frac{G_N(z+h) - G_N(z)}{h} - F_N(z) \right| = \left| \int_0^N f(t) \frac{(e^{-(z+h)t} e^{-zt} + hte^{-zt}) \, dt}{h} \right|$$

$$= \left| \frac{1}{h} \int_0^N f(t) e^{-zt} (e^{-ht} - 1 + ht) \, dt \right| \leqslant \frac{1}{|h|} \int_0^N |f(t) e^{-zt}| |e^{-ht} - 1 + ht| \, dt$$

$$\leqslant \frac{1}{|h|} \int_0^N |f(t) e^{-zt}| |t^2 h^2| \, dt = |h| \int_0^N |f(t) t^2 e^{-zt}| \, dt \to 0 \quad \text{as } |h| \to 0.$$

Thus G_N is analytic and $G_N' = F_N$.

Proof of (ii). Since $f(t) e^{-at} \to 0$ as $t \to \infty$ we can find an N_0 such that $|f(t) e^{-at}| \leqslant 1$ for all $t \geqslant N_0$. In particular, if $b > a$ and $N \geqslant N_0$, then, for all z with $\operatorname{Re} z \geqslant b$,

$$|G_N(z) - G(z)| = \left| \int_N^{\infty} f(t) e^{-zt} \, dt \right| \leqslant \int_N^{\infty} |f(t) e^{-zt}| \, dt = \int_N^{\infty} |f(t) e^{-at}| |e^{-(z-a)t}| \, dt$$

$$= \int_N^{\infty} \exp(-\operatorname{Re} z - a)t) \, dt = \frac{\exp(-(\operatorname{Re} z - a)N)}{\operatorname{Re} z - a}$$

$$\leqslant \frac{\exp(-(b-a)N)}{b-a} \to 0 \quad \text{as } N \to \infty.$$

Thus $G_N \to G$ uniformly on $\{z : \operatorname{Re} z \geqslant b\}$.

Computation of the derivative. Finally, we observe that the same style of proof as that given for (ii) shows that

$$F_N(z) = -\int_0^N t f(t) e^{-zt} dt \to -\int_0^\infty t f(t) e^{-zt} dt \quad \text{for all } \operatorname{Re} z > a.$$

Thus $G_N'(z) \to -\mathcal{L}(\tau f)(z)$ and $(\mathcal{L}f)'(z) = G'(z) = -(\mathcal{L}(\tau f))(z)$ for all $\operatorname{Re} z > a.$ ∎

Remark. The proof that G_N is analytic is little more than the statement that the weighted sum of analytic functions is analytic. The difficulty in extending the argument to G lies in the fact that, however small h is, $\exp ht - 1 - ht$ does not remain small for large t. A determined analyst would have no difficulty in overcoming this problem by a direct estimate for the 'tail' (see Chapter 70). However, we have chosen an indirect method resembling that used in proving Theorems 53.1 and 53.3.

The study of the Laplace transform is much aided by the fact that the zeros of a (non-zero) analytic function must be isolated.

Lemma 75.5. *Let $D = \{z : |z - z_0| < \rho\}$ with $\rho > 0$. If $f : D \to \mathbb{C}$ is analytic then either $f = 0$ or there exists an $\varepsilon > 0$ such that $f(z) \neq 0$ for $0 < |z - z_0| < \varepsilon$.*

Proof. We can expand f as Taylor series

$$f(z) = \sum_{r=0}^\infty a_r (z - z_0)^r \quad [|z - z_0| < \rho]$$

round z_0. If $a_r = 0$ for all r then $f = 0$. Otherwise we can find an N (possibly 0) such that $a_r = 0$ for $r < N$ and $a_N \neq 0$. The series $\sum_{r=0}^\infty a_{r-N}(z - z_0)^r$ converges to an analytic function $h(z)$ on D. Since $h(z_0) = a_N \neq 0$ and h is continuous we can find an $\varepsilon > 0$ such that $h(z) \neq 0$ for $|z - z_0| < \varepsilon$ and so

$$f(z) = (z - z_0)^N h(z) \neq 0 \quad \text{for } 0 < |z - z_0| < \varepsilon. \qquad\blacksquare$$

We cannot conclude from Lemma 75.5 that if Ω is open and $f : \Omega \to \mathbb{C}$ is a non-zero analytic function then the zeros of f are isolated. For if we take $\Omega = \{z : |z| < 1\} \cup \{z : |z - 3| < 1\}$ and $f(z) = 1$ for $|z| < 1$, $f(z) = 0$ for $|z - 3| < 1$, then f is non-zero but 3 is not an isolated zero. However this is clearly an 'artificial' counter example which depends on the fact that Ω is composed of two disconnected components $\{z : |z| < 1\}$ and $\{z : |z - 3| < 1\}$. If we restrict ourselves to connected open sets, the desired theorem becomes true. (Formally, an open set Ω is *connected* if there do not exist non-empty open sets Ω_1 and Ω_2 such that $\Omega_1 \cup \Omega_2 = \Omega$ but $\Omega_1 \cap \Omega_2 = \emptyset$. If the reader knows about such things well and good. If not, the concept is not central to the argument.)

Theorem 75.6. (i) *Let Ω be a connected open set. If $f:\Omega \to \mathbb{C}$ is analytic, then either $F = 0$ or for each $w \in \Omega$ we can find an $\varepsilon(w) > 0$ such that $f(z) \neq 0$ for all $0 < |z - w| < \varepsilon(w)$.*

(ii) *Let Ω be a connected open set and $f:\Omega \to \mathbb{C}$ an analytic function. If we can find distinct $z_0, z_1, z_2, \ldots, \in \Omega$ such that $z_n \to z_0$ as $n \to \infty$ and $f(z_j) = 0$ for all $j \geqslant 0$ then $f = 0$.*

Proof. (i) This is proved in most complex variable texts.
(ii) This is a restatement of (i). ∎

We can draw two important conclusions concerning the Laplace transform.

Recall first that a function g is called meromorphic on $\mathbb{C}$ if it is analytic except for poles.

Lemma 75.7. (i) *Suppose g_1 and g_2 are meromorphic on $\mathbb{C}$ and $g_1(z) = g_2(z)$ for all z with $\operatorname{Re} z > a$. Then $g_1 = g_2$.*
(ii) *If $\mathscr{L}f$ has a meromorphic extension to the whole of $\mathbb{C}$ then it is unique.*

Sketch of proof. (i) Consider the meromorphic function $g_1 - g_2$. It follows from the definition of a pole that the set E of poles of $g_1 - g_2$ is discrete (i.e. for each $z \in E$, we can find $\varepsilon(z) > 0$ such that $E \cap \{w : 0 < |w - z| < \varepsilon(z)\} = \varnothing$). It seems fairly obvious (but requires a proof involving more topology than I care to introduce, though not more than is contained in a good first course of analytic topology) that $\Omega = \mathbb{C} \backslash E$ is connected. Further, since E is discrete we can find distinct $z_0, z_1, \ldots, \in \Omega$ with $\operatorname{Re} z_k > a + 1$ and $z_k \to z_0$. Thus we can apply Theorem 75.6 (ii) to obtain $g_1(z) - g_2(z) = 0$ for all $z \in \Omega$ and so by continuity E is, in fact, empty and $g_1 = g_2$.
(ii) If $(\mathscr{L}f)(z) = g_1(z) = g_2(z)$ for $\operatorname{Re} z > a$ then we can apply (i) to show $g_1 = g_2$. ∎

Thus, although $\int_0^\infty H(t)e^{-zt}dt$ fails to converge for $\operatorname{Re} z \leqslant 0$, there is no ambiguity in writing $\mathscr{L}H(z) = 1/z$ for all $z \neq 0$ and referring to $1/z$ as the Laplace transform of the Heaviside function H. Note however that, although most of the functions we meet in every day computation have Laplace transforms which extend to meromorphic functions on $\mathbb{C}$, this is not guaranteed. The next example shows the kind of thing that could happen.

Example 75.8. *Let $T = \{-x + iy : x, y > 0, x, y \text{ rational}\}$. Since T is countable we can write $T = \{z(1), z(2), \ldots\}$ with $z(1) = -x(1) + iy(1)$, $z(2) = -x(2) + iy(2), \ldots$ distinct. Let*
$$f_j(t) = e^{-x(j)t} \sin y(j)t \text{ for } t \geqslant 0, \ f_j(t) = 0 \text{ for } t < 0.$$
Then $\sum_{j=1}^\infty 2^{-j} f_j$ converges uniformly on $\mathbb{R}$ to a continuous function $F \in \mathscr{E}_0$ with
$$\mathscr{L}F(z) = \sum_{j=1}^\infty \frac{2^{-j}}{2i}\left(\frac{1}{z - z(j)} - \frac{1}{z - z(j)^*}\right) \text{ for } \operatorname{Re} z > 0.$$

We note that $\{z : z \in T\} \cup \{z^* : z \in T\}$ is dense in $\{z : \operatorname{Re} z \leqslant 0\}$ so it is hardly likely that $\mathscr{L}F$ can have a meromorphic extension to $\mathbb{C}$.

Proof. Observe that $|f_j(t)| \leq 1$ for all $t \in \mathbb{R}$, so that

$$\left| \sum_{j=n}^{m} 2^{-j} f_j(t) \right| \leq \sum_{j=n}^{m} 2^{-j} \leq 2^{-n+1} \to 0 \quad \text{as } n \to \infty,$$

and, by the general principle of uniform convergence, $\sum_{j=1}^{\infty} 2^{-j} f_j$ tends uniformly to some F. Since each f_j is continuous, so is F. Further, $F(t) = 0$ for $t \leq 0$ and $|F(t)| \leq \sum_{j=1}^{N} 2^{-j} |f_j(t)| + \sum_{j=N+1}^{\infty} 2^{-j} \to 2^{-N}$ as $t \to \infty$, so $F \in \mathscr{E}_0$. Finally, if we fix a z with $x = \operatorname{Re} z > 0$ then

$$\left| \sum_{j=1}^{n} 2^{-j} f_j(t) \exp(-zt) - F(t) \exp(-zt) \right| \leq \sum_{j=n+1}^{\infty} 2^{-j} |f_j(t)| |\exp(-zt)|$$

$$\leq \sum_{j=n+1}^{\infty} 2^{-j} \exp(-xt) = 2^{-n} \exp(-xt).$$

so

$$\left| \sum_{j=1}^{n} 2^{-j} (\mathscr{L} f_j)(z) - \mathscr{L} F(z) \right| = \left| \int_{0}^{\infty} \left(\sum_{j=1}^{n} 2^{-j} f_j(t) \exp(-zt) - F(t) \exp(-zt) \right) dt \right|$$

$$\leq \int_{0}^{\infty} \left| \sum_{j=1}^{n} 2^{-j} f_j(t) \exp(-zt) - F(t) \exp(-zt) \right| dt$$

$$\leq 2^{-n} \int_{0}^{\infty} \exp(-xt) \, dt = \frac{2^{-n}}{x} \to 0 \quad \text{as } n \to \infty.$$

Thus $\mathscr{L} F(z) = \sum_{j=1}^{\infty} 2^{-j} \mathscr{L} f_j(z)$ for all z with $\operatorname{Re} z > 0$. But

$$f_j = (2i)^{-1} (He_{-x(j)+iy(j)} - He_{-x(j)-iy(j)}),$$

so, using Lemma 74.6 (i) and (iv), we obtain the desired result. ∎

The second, obvious but useful, consequence of Theorem 75.6 is the following improvement of the uniqueness theorem for the Laplace transform given in Theorem 74.5.

Theorem 75.9. *Suppose $f \in \mathscr{E}_a$ and we can find distinct $z_0, z_1, z_2, \ldots$ with $\operatorname{Re} z_j > a$, $z_n \to z_0$ as $n \to \infty$ and $\mathscr{L} f(z_j) = 0$ for all $j \geq 0$. Then $f = 0$.*
Proof. By Theorem 75.6 (ii), $\mathscr{L} f(z) = 0$ for all $\operatorname{Re} z > a$, so, by Theorem 74.5 (ii), $f = 0$. ∎

Thus, for example, if we can show that $\mathscr{L} f(x) = 0$ for all real $x > b$, it will follow that $f = 0$. A stronger result than this is given in the remarkable theorem of Lerch.

Theorem 75.10 (Lerch). *Suppose $f \in \mathscr{E}_a$ and we can find $b > a$ and $c > 0$ such that $\mathscr{L} f(b + nc) = 0$ for all $n \geq 0$. Then $f = 0$.*
Remark 1. It is not true that if $g: \mathbb{C} \to \mathbb{C}$ is analytic with $g(b + nc) = 0$ for all $n \geq 0$, then $g = 0$. Consider $g(z) = \sin \pi z$ (with $b = 0$, $c = 1$). However, a condition like this

does place major restrictions on the form of an analytic function. (For example, applying the conformal map $z \mapsto 1/z$ to $z^{-m}g(z)$, we see from Theorem 75.6 and the characterisation of removable singularities that if $g(z_n) = 0$ for some sequence $z_n \to \infty$ and further $|g(z)|$ grows slowly as $z \to \infty$, then $g = 0$.)

Remark 2. The proof given below depends neither on complex variable theory nor on Fourier transform theory. It depends solely on elementary real variable theory and the solution of the Hausdorff moment problem given in Chapter 6 (which the reader is invited to reread before continuing).

Proof. Write $\alpha(t) = \int_0^t e^{-bs} f(s) ds$. Then $\alpha : [0, \infty) \to \mathbb{R}$ is continuous and differentiable, $\alpha(0) = 0$, $\alpha'(t) = e^{-bt} f(t)$ and $\alpha(t) \to \mathscr{L} f(b) = 0$ as $t \to \infty$. Integration by parts gives

$$0 = \mathscr{L} f(b + nc) = \int_0^\infty e^{-(b+nc)t} f(t) dt = \int_0^\infty e^{-nct}(e^{-bt} f(t)) dt = \int_0^\infty e^{-nct} \alpha'(t) dt$$

$$= [e^{-nct}\alpha(t)]_0^\infty + nc \int_0^\infty e^{-nct}\alpha(t) dt = nc \int_0^\infty e^{-nct}\alpha(t) dt \quad \text{for all } n \geq 0.$$

Thus $\int_0^\infty e^{-nct}\alpha(t) dt = 0$ for all $n \geq 1$.

We now observe that, making the substitution $u = e^{-ct}$, we obtain

$$\int_0^\infty e^{-nct}\alpha(t) dt = -c^{-1} \int_0^1 u^{n-1} \beta(u) du,$$

where $\beta(u) = \alpha(-c^{-1}\log u)$ for $0 < u < 1$, $\beta(0) = 0$ and $\beta(1) = 0$. Since α is continuous and $\alpha(0) = 0$ whilst $\alpha(s) \to 0$ as $s \to \infty$, it follows that β is continuous. But we have seen that

$$\int_0^1 \beta(u) u^m du = 0 \qquad \text{for all } m \geq 0.$$

Thus, by Hausdorff's moment theorem (Theorem 6.1), $\beta(u) = 0$ for all $0 \leq u \leq 1$ and so $\alpha(t) = 0$ for all $t \geq 0$.

By the fundamental theorem of the calculus

$$e^{-bx} f(x) = \left[\frac{d}{ds} \int_0^s e^{-bt} f(t) dt \right]_{s=x} = \alpha'(x) = 0$$

at all points x where f, and so $e_{-b}f$, are continuous. Thus $f(x) = 0$ at all points x where f is continuous and so, using part (iv) of Definition 74.1, $f(x) = 0$ everywhere. ∎

(If the reader only knows a weak version of the method of integrating by parts she may prefer to consider $\sum \int_{x(j)}^{x(j+1)} e^{-nct}(e^{-bt} f(t)) dt$ in the first paragraph of the proof but the result will be the same.)

Since our proof of Theorem 75.10 does not depend on Theorem 74.5(i), it provides the much more direct proof of Theorem 74.5 (ii) promised in Chapter 74. It is clear that the content and proof of Lerch's theorem will repay careful study.

76

POLES AND STABILITY

The use of complex variable theory enables us to extend the application of the inversion formula of Theorem 74.5 (i) in a remarkable manner.

Theorem 76.1. *Let $f \in \mathscr{E}_a$. Suppose that $\mathscr{L}f$ can be extended to an analytic function on $\{z : \operatorname{Re} z > d\}$ for some $d < a$. Suppose further we can find b and c with $b > a > c > d$ such that $\mathscr{L}f(x + iy) \to 0$ uniformly for $x \in [c, b]$ as $|y| \to \infty$. Then*

$$f(t) = \lim_{R \to \infty} \frac{1}{2\pi} \int_{-R}^{R} (\mathscr{L}f)(c + i\zeta)\exp(c + i\zeta)t\,d\zeta$$

$$= \exp ct \lim_{R \to \infty} \frac{1}{2\pi} \int_{-R}^{R} (\mathscr{L}f)(c + i\zeta)\exp(i\zeta t)d\zeta.$$

Warning. We have shown that the Laplace transform takes 'fairly nasty' functions f to 'very nice' (analytic) function $\mathscr{L}f$. An inversion formula like the one above (or, indeed, that of Theorem 74.5 (i)), must therefore take nice functions to nasty functions. If the reader reflects, she will see that this must produce severe problems when such a formula is used in numerical analysis. Numerical inversion of a Laplace transform should not therefore be undertaken without expert advice.

Proof of Theorem 76.1. Consider the rectangular contour C shown in Figure 76.1, composed of the straight lines C_1 joining $b - iR$ to $b + iR$, C_2 joining $b + iR$ to $c + iR$, C_3 joining $c + iR$ to $c - iR$ and C_4 joining $c - iR$ to $b - iR$.

Since $\mathscr{L}f(z)$ and so $\exp tz\ \mathscr{L}f(z)$ are analytic in $\{z : \operatorname{Re} z > d\}$, Cauchy's theorem shows that

$$\sum_{j=1}^{4} \int_{C_j} \exp(zt)\mathscr{L}f(z)dz = \int_{C} \exp(zt)\mathscr{L}f(z)dz = 0.$$

Now, by Theorem 74.5 (i)

$$\int_{C_1} \exp(zt)\mathscr{L}f(z)dz = i\int_{-R}^{R} \exp((b + i\zeta)t)\mathscr{L}f(b + i\zeta)d\zeta \to 2\pi i f(t).$$

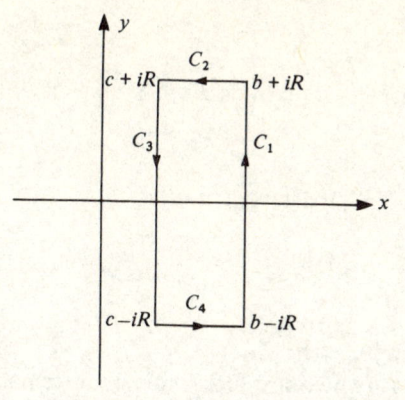

Fig. 76.1. Contour for the inversion of a Laplace transform.

On the other hand,

$$\left| \int_{C_2} \exp zt \, \mathscr{L} f(z) \, dz \right| = \left| i \int_b^c \exp((x+iR)t) \mathscr{L} f(x+iR) \, dx \right|$$

$$\leq (b-c) \exp((|b|+|c|)|t|) \sup_{x \in [c,b]} |\mathscr{L} f(x+iR)|$$

$$\to 0 \quad \text{as } R \to \infty$$

by the hypotheses of the theorem. Similarly,

$$\int_{C_4} \exp(zt) \mathscr{L} f(z) \, dz \to 0,$$

and so,

$$\exp ct \int_{-R}^R \mathscr{L} f(c+i\zeta) \exp(i\zeta t) \, d\zeta = \int_{-R}^R \mathscr{L} f(c+i\zeta) \exp((c+i\zeta)t) \, d\zeta$$

$$= \int_{C_3} i \mathscr{L} f(z) \exp(zt) \, dz = \int_{C_1} + \int_{C_2} + \int_{C_4} (-i) \mathscr{L} f(z) \exp(zt) \, dz$$

$$\to 2\pi f(t) \quad \text{as } R \to \infty. \qquad \blacksquare$$

It is tempting to try to conclude that f cannot grow faster than expected but to do this we must know something about the behaviour of

$$\lim_{R \to \infty} \frac{1}{2\pi} \int_{-R}^R (\mathscr{L}f)(c+i\zeta) \exp(i\zeta t) \, d\zeta \quad \text{as } t \to \infty.$$

Lemma 76.2. *Suppose that the conditions of Theorem 76.1 hold. If we can find K*

and T with

$$\left| \frac{1}{2\pi} \int_{-R}^{R} \mathscr{L}f(c+i\zeta)\exp(i\zeta t)\,d\zeta \right| < K \quad \text{for all } R \geqslant 0, t \geqslant T,$$

then $f \in \mathscr{E}_\gamma$ *for all* $\gamma > c$.

Proof. Using Theorem 76.1 we have, trivially,

$$|f(t)\exp(-\gamma t)| = \left| \exp(c-\gamma)t \lim_{R \to \infty} \frac{1}{2\pi} \int_{-R}^{R} (\mathscr{L}f)(c+i\zeta)\exp i\zeta t\,d\zeta \right|$$

$$\leqslant K \exp(c-\gamma)t \to 0 \quad \text{as} \quad t \to \infty. \qquad \blacksquare$$

The next two lemmas (which are little more than remarks) show that Lemma 76.2 is easy to exploit.

Lemma 76.3. *The additional hypothesis of Lemma 76.2 will certainly hold if* $\int_{-\infty}^{\infty} |\mathscr{L}f(c+i\zeta)|\,d\zeta$ *converges.*

Proof. Trivial. $\qquad \blacksquare$

Lemma 76.4. *Let* $f \in \mathscr{E}_a$. *Suppose that* $\mathscr{L}f$ *can be extended to an analytic function on* $\{z : \operatorname{Re} z > d\}$ *for some* $d < a$. *Suppose further that we can find an* $\eta > 1$ *and* $A, S > 0$ *such that* $|\mathscr{L}f(z)| \leqslant A|z|^{-\eta}$ *for all* $|z| \geqslant S$, $\operatorname{Re} z > d$. *Then* $f \in \mathscr{E}_c$ *for all* $c > d$.

Proof. Observe that $|\mathscr{L}f(x+iy)| \leqslant A|y|^{-\eta}$ for $|y| > S$. Thus $\mathscr{L}f(x+iy) \to 0$ uniformly as $|y| \to \infty$ for all $x > d$ whilst by the comparison test $\int_{-\infty}^{\infty} |\mathscr{L}f(x+iy)|\,dy$ converges for all $x > d$. Thus the hypotheses of Lemma 76.2 are satisfied and we are done. $\qquad \blacksquare$

Lemma 76.4 is interesting in itself but its real importance only appears when we extend it to the case when $\mathscr{L}f$ is allowed to have poles. We make this extension by our usual technique of writing $\mathscr{L}f = \mathscr{L}u_1 + \mathscr{L}u_2 + \cdots + \mathscr{L}u_k + \mathscr{L}g$ where $\mathscr{L}g$ is analytic and each of the u_j are functions whose behaviour we know precisely.

Lemma 76.5. *Suppose* Ω *is an open set and* F *is a meromorphic function having poles at* $\lambda(1)$, $\lambda(2), \ldots, \lambda(k)$ *of order* $n(1), n(2), \ldots, n(k)$. *Then we can find* $A_{jl} \in \mathbb{C}[1 \leqslant l \leqslant n(j), 1 \leqslant j \leqslant k]$ *and an analytic function* $G : \Omega \to \mathbb{C}$ *such that*

$$F(z) = \sum_{j=1}^{k} \sum_{l=1}^{n(j)} A_{jl}(z-\lambda(j))^{-l} + G(z) \quad \text{for all } z \in \Omega \setminus \{\lambda(1), \lambda(2), \ldots, \lambda(k)\}.$$

Proof. Choose $\varepsilon(k) > 0$ such that if $0 \neq |z - \lambda(k)| < \varepsilon(k)$ then $z \in \Omega \setminus \{\lambda(1), \lambda(2), \ldots, \lambda(k-1)\}$. By Laurent's theorem (and the definition of a pole) we can expand $F(z)$ as

$$F(z) = \sum_{l=1}^{n(k)} A_{kl}(z-\lambda(k))^{-l} + \sum_{m=0}^{\infty} B_{km}(z-\lambda(k))^m \quad \text{for } 0 \neq |z-\lambda(k)| < \varepsilon(k).$$

If we define F_{k-1} on $\Omega \setminus \{\lambda(1), \lambda(2), \ldots, \lambda(k-1)\}$ by

$$F_{k-1}(z) = F(z) - \sum_{l=1}^{n(k)} A_{kl}(z - \lambda(k))^{-l} \quad \text{for } z \neq \lambda(k),\ F_{k-1}(\lambda(k)) = B_{k0},$$

then, since $(z - \lambda(k))^{-l}$ is analytic except at $\lambda(k)$, F_{k-1} has the same poles with the same orders as F except possibly at $\lambda(k)$.

But for $|z - \lambda(k)| < \varepsilon(k)$ we know that

$$F_{k-1}(z) = \sum_{m=0}^{\infty} B_{km}(z - \lambda(k))^m,$$

so F_{k-1} is analytic in $\{z : |z - \lambda(k)| < \varepsilon(k)\}$. Thus F_{k-1} is meromorphic on Ω with poles at $\lambda(1), \lambda(2), \ldots, \lambda(k-1)$ of order $n(1), n(2), \ldots, n(k-1)$ and

$$F(z) = \sum_{l=1}^{n(k)} A_{kl}(z - \lambda(k))^{-l} + F_{k-1}(z) \quad \text{for } z \in \Omega \setminus \{\lambda(1), \lambda(2), \ldots, \lambda(k)\}.$$

The result now follows by induction on k. ∎

Lemma 76.6. *Write* $u_{\lambda n}(t) = ((n-1)!)^{-1} H(t) t^{n-1} \exp \lambda t\, [t \in \mathbb{R}]$ *where* $\lambda \in \mathbb{C}$ *and* n *is a strictly positive integer. Then* $u_{\lambda n} \in \mathscr{E}$ *and* $\mathscr{L} u_{\lambda n}(z)$ *can be extended to the function* $(z - \lambda)^{-n}$ *which is analytic for all of* $\mathbb{C}$ *except for a pole at* λ. *Further,*

(i) $\mathscr{L} u_{\lambda n}(x + iy) \to 0$ *uniformly for* $x \in \mathbb{R}$ *as* $|y| \to \infty$,

and

(ii) *If* $c \neq \operatorname{Re} \lambda$ *we can find* K *and* T *such that*

$$\left| \frac{1}{2\pi} \int_{-R}^{R} \mathscr{L} u_{\lambda n}(c + i\zeta) \exp(i\zeta t)\, d\zeta \right| < K \quad \text{for all } R \geqslant 0,\ t \geqslant T.$$

Proof. Using Lemma 74.6 (i), (ii) and (v) we see that $u_{\lambda n} \in \mathscr{E}$ and $\mathscr{L} u_{\lambda n}(z) = (z - \lambda)^{-n}$ for all z with $\operatorname{Re} z > \operatorname{Re} \lambda$. Thus $\mathscr{L} u_{\lambda n}(z)$ extends to the function $(z - \lambda)^{-n}$ on $\mathbb{C} \setminus \{\lambda\}$ as stated. We note that if $|y| \geqslant 2|\lambda|$ then $|((x + iy) - \lambda)^{-n}| \leqslant (|y|/2)^{-n} \to 0$ uniformly for all $x \in \mathbb{R}$ as $|y| \to \infty$ so (i) holds. If $n \geqslant 2$ then it is also clear by comparison of $|(c + i\zeta - \lambda)^{-n}|$ with $|\zeta|^{-2}$ that $\int_{-\infty}^{\infty} |(c + i\zeta - \lambda)^{-n}|\, d\zeta$ converges for $c \neq \operatorname{Re} \lambda$ so condition (ii) holds.

However, in the case $n = 1$ we know that $\int_{-\infty}^{\infty} |(c + i\zeta - \lambda)^{-1}|\, d\zeta$ is divergent so a more delicate argument is needed. Integrating by parts, we obtain

$$\frac{1}{2\pi} \int_{-R}^{R} \mathscr{L} u_{\lambda 0}(c + i\zeta) \exp i\zeta t\, d\zeta = \frac{1}{2\pi} \int_{-R}^{R} \frac{\exp i\zeta t}{c + i\zeta - \lambda}\, d\zeta$$

$$= \frac{1}{2\pi} \left[\frac{\exp i\zeta t}{it(c + i\zeta - \lambda)} \right]_{-R}^{R} + \frac{1}{2\pi} \int_{-R}^{R} \frac{\exp i\zeta t}{t(c + i\zeta - \lambda)^2}\, d\zeta.$$

But

$$\left| \frac{\exp i\zeta t}{(c + i\zeta - \lambda)^2} \right| \leqslant \frac{4}{|\zeta|^2} \quad \text{for } |\zeta| \geqslant 2\lambda,$$

so $\int_{-\infty}^{\infty}|\exp i\zeta t/(c+i\zeta-\lambda)^2|\,d\zeta$ converges to A, say. Thus

$$\left|\frac{1}{2\pi}\int_{-R}^{R}\mathcal{L}u_0(c+i\zeta)\exp i\zeta t\,d\zeta\right|\leqslant\frac{1}{2\pi}\left|\left[\frac{\exp i\zeta t}{it(c+i\zeta-\lambda)}\right]_{-R}^{R}\right|+\frac{A}{2\pi t}$$

$$\leqslant\frac{1}{\pi t}\cdot\frac{1}{|\mathrm{Re}(c-\lambda)|}+\frac{A}{2\pi t}\leqslant1\quad\text{for all }R\geqslant0,\,t\geqslant(A+2|\mathrm{Re}(c-\lambda)|^{-1})/2\pi,$$

and the proof is complete. ∎

Theorem 76.7. *Let* $f\in\mathscr{E}_a$. *Suppose that* $\mathcal{L}f$ *can be extended to a meromorphic function with a finite number of poles on* $\{z:\mathrm{Re}\,z>d\}$ *for some* $d<a$. *Let the poles be at* $\lambda(1),\lambda(2),..,\lambda(k)$ *of order* $n(1),n(2),\ldots,n(k)$ *respectively. Suppose that we can find* $b>a>c>d$ *such that* $\mathrm{Re}\,\lambda(j)>c$ *for* $1\leqslant j\leqslant m$ *and*

(i) $\mathcal{L}f(x+iy)\to0$ *uniformly for* $x\in[b,c]$ *as* $|y|\to\infty$.

(ii) *We can find* K *and* T *such that*

$$\left|\frac{1}{2\pi}\int_{-R}^{R}\mathcal{L}f(c+i\zeta)\exp i\zeta t\,dt\right|\leqslant K\quad\text{for all }R\geqslant0,\,t\geqslant T.$$

Then
$$f(t)=H(t)\sum_{j=1}^{k}\sum_{l=1}^{n(j)}a_{jl}t^{l-1}\exp\lambda(j)t+g(t),$$

where $g\in\mathscr{E}_\gamma$ *for all* $\gamma>c$, $a_{jl}\in\mathbb{C}$ *for all* $1\leqslant l\leqslant n(j)$ *and* $a_{jn(j)}\neq0$ *for all* $1\leqslant j\leqslant k$.

Proof. By Lemma 76.5

$$\mathcal{L}f(z)=\sum_{j=1}^{k}\sum_{l=1}^{n(j)}A_{jl}(z-\lambda(j))^{-l}+G(z),$$

where G is analytic on $\{z:\mathrm{Re}\,z>d\}$. Note that, by the definition of the order of a pole, $A_{jn(j)}\neq0$. Setting

$$u_{\lambda(j)l}(t)=((l-1)!)^{-1}H(t)t^{l-1}\exp\lambda(j)t\quad[t\in\mathbb{R}],$$

we obtain from Lemma 76.6

$$\mathcal{L}f(z)=\sum_{j=1}^{k}\sum_{l=1}^{n(j)}A_{jl}\mathcal{L}u_{\lambda(j)l}(z)+G(z)\quad[\mathrm{Re}\,z>d].$$

The next part of the argument is easy but must be treated carefully. Let us write

$$g(t)=f(t)-\sum_{j=1}^{k}\sum_{l=1}^{n(j)}A_{jl}u_{\lambda(j)l}(t)\quad\text{for all }t\in\mathbb{R}.$$

Since $f\in\mathscr{E}_a$ and
$$u_{\lambda(j)l}\in\mathscr{E}_a\quad[1\leqslant l\leqslant n(j),1\leqslant j\leqslant k]$$

it follows that $g\in\mathscr{E}_a$. Thus $\mathcal{L}g(z)$ exists for $z>\mathrm{Re}\,a$ and we have

$$\mathcal{L}g(z)=\mathcal{L}f(z)-\sum_{j=1}^{k}\sum_{l=1}^{n(j)}A_{jl}\mathcal{L}u_{\lambda(j)l}(z)=G(z)\quad\text{for}\quad z>\mathrm{Re}\,a.$$

Thus $\mathscr{L}g$ can be extended to a function G which is analytic on $\{z:\mathrm{Re}\,z>d\}$. If we can show that g satisfies the conditions of Theorem 76.1 and Lemma 76.2 we shall be done.

Observe first that by condition (i) of our hypotheses and conclusion (i) of Lemma 76.6

$$\mathscr{L}g(x+iy)=G(x+iy)=\mathscr{L}f(x+iy)-\sum_{j=1}^{k}\sum_{l=1}^{n(j)}A_{jl}\mathscr{L}u_{\lambda(j)l}(x+iy)\to 0$$

uniformly for $x\in[b,c]$ as $|y|\to\infty$. Again by Lemma 76.6 we know that we can find K_{jl} and T_{jl} such that

$$\left|\frac{1}{2\pi}\int_{-R}^{R}\mathscr{L}u_{\lambda(j)l}(c+i\zeta)\exp i\zeta t\,d\zeta\right|<K_{jl}\quad\text{for all }R\geqslant 0,\,t\geqslant T_{jl}[1\leqslant l\leqslant n(j),\,1\leqslant j\leqslant k].$$

Thus, taking

$$K_0=K+\sum_{j=1}^{k}\sum_{l=1}^{n(j)}|A_{jl}|K_{jl}$$

and $T_0=\max(T,\max_{j,l}T_{jl})$ we have, using condition (ii) of our hypotheses,

$$\left|\frac{1}{2\pi}\int_{-R}^{R}\mathscr{L}g(c+i\zeta)\exp i\zeta t\,d\zeta\right|\leqslant\left|\frac{1}{2\pi}\int_{-R}^{R}\mathscr{L}f(c+i\zeta)\exp i\zeta t\,d\zeta\right|$$

$$+\sum_{j=1}^{k}\sum_{l=1}^{n(j)}|A_{jl}|\frac{1}{2\pi}\left|\int_{-R}^{R}u_{\lambda(j)l}(c+i\zeta)\exp i\zeta t\,d\zeta\right|\leqslant K_0\quad\text{for all }t\geqslant T_0.$$

Thus, by Lemma 76.2, $g\in\mathscr{E}_\gamma$ for all $\gamma>c$. Setting $a_{jl}=A_{jl}/(l-1)!$ we obtain the complete result. ∎

Thus the behaviour of f as $t\to\infty$ is determined by the right-most poles of its Laplace transform.

Lemma 76.8. *Let f obey the conditions of Theorem 76.7. If $\mathrm{Re}\,\lambda(1)=\mathrm{Re}\,\lambda(2)=\cdots=\mathrm{Re}\,\lambda(p)>\mathrm{Re}\,\lambda(p+1)\geqslant\cdots\geqslant\mathrm{Re}\,\lambda(k)$ and $n(1)=n(2)=\cdots=n(q)>n(q+1)\geqslant\cdots\geqslant n(p)\ [k\geqslant p\geqslant q\geqslant 1]$ then, writing $n=n(1),\ s=\mathrm{Re}\,\lambda(1),\ b_r=a_{rn}$ and $\omega(r)=\mathrm{Im}\,\lambda(r)\ [1\leqslant r\leqslant q]$, we have*

$$\left(t^{n-1}\exp st\sum_{r=1}^{q}b_r(\exp i\omega(r)t)\right)^{-1}f(t)\to 1\quad\text{as }t\to\infty.$$

Thus $f(t)$ behaves like $t^{n-1}\exp st\sum_{r=1}^{q}b_r\exp i\omega(r)t$ as $t\to\infty$.
Proof. Immediate consequence of Theorem 76.7. ∎

In order to show that Lemma 76.8 actually means what it looks as though it means we need a minor lemma.

Lemma 76.9. *Suppose* $b_1, b_2, \ldots, b_p \in \mathbb{C}$ *and* $\omega(1), \omega(2), \ldots, \omega(p) \in \mathbb{R}$. *Then, for any* $T \in \mathbb{R}$,

$$\sup_{t \geqslant T} \left| \sum_{r=1}^{p} b_r \exp i\omega(r)t \right| \geqslant |b_1|$$

and, in particular, if $b_1 \neq 0$,

$$\sum_{r=1}^{p} b_r \exp i\omega(r)t \not\to 0 \quad \text{as } t \to \infty.$$

Proof. (Compare Lemma 8.1.) We observe that

$$\sup_{t \geqslant T} \left| \sum_{r=1}^{p} b_r \exp i\omega(r)t \right| \geqslant \frac{1}{S-T} \int_T^S \left| \sum_{r=1}^{p} b_r \exp i\omega(r)t \right| dt$$

$$\geqslant \frac{1}{S-T} \left| \int_T^S \left(\sum_{r=1}^{p} b_r \exp i\omega(r)t \right) (\exp - i\omega(1)t) dt \right|$$

$$= \frac{1}{S-T} \left| b_1(S-T) + \sum_{r=2}^{p} b_r \left[\frac{\exp i(\omega(r) - \omega(1))t}{i(\omega(r) - \omega(1))} \right]_T^S \right|$$

$$\geqslant |b_1| - \sum_{r=2}^{p} \frac{2|b_r|}{(S-T)(\omega(r) - \omega(1))} \to |b_1| \quad \text{as } S \to \infty. \qquad \blacksquare$$

Lemma 76.10. *Suppose the conditions of Theorem 76.7 are satisfied and* $k \geqslant 1$. *Then*

$$f(t) \to 0 \ \text{ as } \ t \to \infty \qquad \text{if } \max_{k \geqslant j \geqslant 1} \operatorname{Re} \lambda(j) < 0,$$

$$f(t) \not\to 0 \ \text{ as } \ t \to \infty \qquad \text{if } \max_{k \geqslant j \geqslant 1} \operatorname{Re} \lambda(j) = 0,$$

$$\limsup_{t \to \infty} |f(t)| = \infty \qquad \text{if } \max_{k \geqslant j \geqslant 1} \operatorname{Re} \lambda(j) > 0.$$

Proof. Use Lemmas 76.9 and 76.8 remembering that $b_j = a_{jn(j)} \neq 0$. $\qquad \blacksquare$

We specialise to obtain what is, in effect, a stronger version of Lemma 76.4.

Theorem 76.11. *Suppose that* $f \in \mathscr{E}$ *and that* $\mathscr{L}f$ *can be extended to a meromorphic function on* $\Omega = \{z : \operatorname{Re} z > d\}$ *for some* $d < 0$. *Suppose further that* $\mathscr{L}f$ *has only a finite number of poles* $\lambda(1), \lambda(2), \ldots, \lambda(k)$ *say and we can find an* $\eta > 1$ *and* $A, S > 0$ *such that* $|\mathscr{L}f(z)| \leqslant A|z|^{-\eta}$ *for all* $|z| \geqslant S$, $z \in \Omega$. *Then*

$$f(t) \to 0 \ \text{ as } \ t \to \infty \qquad \text{if } \max_{k \geqslant j \geqslant 1} \operatorname{Re} \lambda(j) < 0,$$

$$f(t) \not\to 0 \ \text{ as } \ t \to \infty \qquad \text{if } \max_{k \geqslant j \geqslant 1} \operatorname{Re} \lambda(j) = 0,$$

$$\limsup_{t \to \infty} |f(t)| = \infty \qquad \text{if } \max_{k \geqslant j \geqslant 1} \operatorname{Re} \lambda(j) > 0.$$

Proof. As for Lemma 76.4 replacing the use of Lemma 76.2 by that of Lemma 76.10. ∎

Theorem 76.11 gives us an immediate usable criterion for the stability of many systems. Let us write C_{00}^n for the set of n times continuously differentiable functions $g:\mathbb{R} \to \mathbb{C}$ which are zero outside some interval $[0, T]$ (where T depends on g).

Theorem 76.12. *Suppose that the relation between the 'input function' $g \in \mathcal{E}$ and the 'output function' $f \in \mathcal{E}$ is given by*

$$h(z)(\mathcal{L}f)(z) = (\mathcal{L}g)(z) \quad \text{for Re } z \text{ large.}$$

Suppose that h can be extended to an analytic function on $\{z : \mathrm{Re}\, z > d\}$ for some $d < 0$ and that we can find an integer n with $0 \leqslant n \leqslant 2$ and A, $R > 0$ such that $|h(z)| \geqslant A|z|^{2-n}$ for all $|z| > R$, $\mathrm{Re}\, z > d$. Then $f(t) \to 0$ as $t \to \infty$ for all $g \in C_{00}^n$ if and only if all the zeros of h have strictly negative real part.

In still more picturesque language the system is stable under disturbances lasting a finite time if and only if all the zeros of h lie in the left half plane.

The two minor lemmas that follow will be needed in the proof, but the reader should skip them and only return if she feels the need to dot every i and cross every t.

Lemma 76.13. *For each $\omega \in \mathbb{C}$ there exists a $g \in C_{00}^2$ with $(\mathcal{L}g)(\omega) \neq 0$.*
Proof. Take a non-zero $f \in C_{00}^2$ (e.g. $f(t) = t^3(1-t)^3$ for $0 \leqslant t \leqslant 1$, $f(t) = 0$ otherwise). Since $f \neq 0$ we know that $\mathcal{L}f \neq 0$ (Theorem 75.10 or Theorem 74.5(ii)) and so we can find an $\omega_0 \in \mathbb{C}$ with $\mathcal{L}f(\omega_0) \neq 0$. Let $\lambda = \omega - \omega_0$ and set $g(t) = \exp \lambda t f(t)[t \in \mathbb{R}]$. Then $g \in C_{00}^2$ and by Lemma 74.6 (ii) $\mathcal{L}g(\omega) = \mathcal{L}f(\omega - \lambda) = \mathcal{L}f(\omega_0) \neq 0$. ∎

Lemma 76.14. *There is no loss of generality in Theorem 67.12 if we suppose that h has only a finite number of zeros in $\{z : \mathrm{Re}\, z > d\}$.*
Proof. We shall show that if $d < d' < 0$ then h has only a finite number of zeros in $\{z : \mathrm{Re}\, z > d'\}$. The lemma will then follow on replacing d by d'. Observe that $|h(z)| \geqslant A|z|^{2-n} > 0$ for $|z| > R$, $\mathrm{Re}\, z > d$ so that any zeros of h in $\{z : \mathrm{Re}\, z > d'\}$ must lie in $\{z : \mathrm{Re}\, z > d', |z| \leqslant R\}$.

Suppose, if possible, that h has distinct zeros $z_1, z_2, \ldots$ in $\{z : \mathrm{Re}\, z > d', |z| \leqslant R\}$. Since $\{z : \mathrm{Re}\, z \geqslant d', |z| \leqslant R\}$ is a closed bounded set we can find $z_0 \in \{z : \mathrm{Re}\, z \geqslant d', |z| \leqslant R\}$ and $n(j) \to \infty$ such that $z_{n(j)} \to z_0$ (Lemma B2). Thus by Theorem 75.6(ii), $h = 0$ which is impossible. We have shown that h has only a finite number of zeros in $\{z : \mathrm{Re}\, z > d', |z| \leqslant R\}$ and so in $\{z : \mathrm{Re}\, z > d'\}$, as required. ∎

Proof of Theorem 76.12. By Lemma 76.14 we may assume that h has only a finite

number of zeros in $\{z:\mathrm{Re}\,z>d\}$. Since $g\in\mathscr{E}_d$ (indeed $g\in\mathscr{E}_a$ for all a) $\mathscr{L}g$ is defined and analytic in $\{z:\mathrm{Re}\,z>d\}$ (Theorem 75.2) and so $\mathscr{L}g(z)/h(z)$ is meromorphic there with only a finite number of poles. Thus $\mathscr{L}f$ extends to the meromorphic function $\mathscr{L}g/h$ in $\{z:\mathrm{Re}\,z>d\}$. Further $z^{-n}\mathscr{L}g^{(n)}(z)=\mathscr{L}g(z)$ (Lemma 74.7 or direct integration by parts) whilst, if $g(t)=0$ for $t\notin[0,T]$,

$$|\mathscr{L}g^{(n)}(z)|=\left|\int_0^T g^{(n)}(s)e^{-sz}ds\right|\leqslant T\sup|g^{(n)}(s)|\exp(Td)=B,$$

say, so $|\mathscr{L}f(z)|=|\mathscr{L}g(z)/h(z)|\leqslant BA^{-1}|z|^{-2}$ for all $|z|\geqslant R$, $\mathrm{Re}\,z>d$. Thus Theorem 76.11 applies and $f(t)\to0$ as $t\to\infty$ if and only if all the poles of $\mathscr{L}f=\mathscr{L}g/h$ have strictly negative real part.

But $\mathscr{L}g/h$ can only have a pole in $\{z:\mathrm{Re}\,z>d\}$ if h has a zero, so $f(t)\to0$ as $t\to\infty$ automatically if all the zeros of h have strictly negative real part. On the other hand, if h has a zero at ω, with $\mathrm{Re}\,\omega\geqslant0$ then, by Lemma 76.13, we can find a $g\in C_{00}^n$ such that $\mathscr{L}g(\omega)\neq0$ and so $\mathscr{L}g/h$ has a pole at ω and $f(t)\nrightarrow0$ as $t\to\infty$. ∎

As a simple illustration we have the following lemma (which, however, can be obtained very easily by direct means) illustrating the remarks of Maxwell quoted in Chapter 72.

Lemma 76.15. *Let $n\geqslant2$ and $a_0,a_1,a_2,\ldots,a_n\in\mathbb{C}$ with $a_n\neq0$. Then the solution $f\in\mathscr{E}$ of the system of equations*

$$a_nf^{(n)}(t)+a_{n-1}f^{(n-1)}(t)+\cdots+a_0f(t)=g(t),$$
$$f^{(n)}(0)=f^{(n-1)}(0)=\cdots=f(0)=0$$

obeys the condition $f(t)\to0$ as $t\to\infty$ for all continuous $g:\mathbb{R}\to\mathbb{C}$ with $g(t)=0$ for $t\notin[0,T]$ if and only if all the roots of $a_nz^n+a_{n-1}z^{n-1}+\cdots+a_0=0$ have strictly negative real part.

Proof. By Lemma 74.7,

$$h(z)(\mathscr{L}f)(z)=(\mathscr{L}g)(z),$$

for $\mathrm{Re}\,z$ large, where $h(z)=\sum_{r=0}^n a_rz^r$ satisfies the conditions of Theorem 76.12. ∎

77

A SIMPLE TIME DELAY EQUATION

The reader may well feel that so far I have neither demonstrated any advantages of Laplace transform methods over simpler techniques nor shown any reason for the care with which I approach them. In this chapter I shall give an example where the Laplace transform appears, I hope, to better advantage and in the next chapter an example which demonstrates the need for care in using it.

No one who has used a shower in a student residence can fail to become aware of the problems caused by the fact that the temperature of the water does not respond instantly to the controls. First the water comes out cold so you turn the hot tap on full, only to be scalded ten seconds later. The natural response produces icy cold water after ten seconds of agony. Shivering you then turn the hot tap full on. . . .

The simplest mathematical analogue of such a system is the differential equation

$$F'(s) + KF(s - \eta) = G(s),$$

where $\eta > 0$ and $F(s) = G(s) = 0$ for $s \leqslant 0$. I suggest that before reading any further the reader tries to study the stability of the system without using complex variable methods (i.e. methods which use Cauchy's theorem).

We commence our treatment by making the substitution $s = t\eta$ and writing $f(t) = \eta^{-1}F(\eta t)$, $g(t) = G(\eta t)$, $k = K\eta$. Our equation then becomes

with
$$\left. \begin{array}{l} f'(t) + kf(t-1) = g(t), \\ f(t) = g(t) = 0 \text{ for } t \leqslant 0. \end{array} \right\} \tag{*}$$

It is easy to check that (if g is continuous) (*) has a unique solution and that (if g is bounded) this solution lies in $\mathcal{E}$.

Lemma 77.1. (i) *Let* $g : \mathbb{R} \to \mathbb{C}$ *be a continuous function with* $g(t) = 0$ *for* $t \leqslant 0$. *Then the equation* $f'(t) + kf(t-1) = g(t)$ *has a unique solution such that* $f(t) = 0$ *for* $t \leqslant 0$.
(ii) *If, further, there exists an* M *such that* $|g(t)| \leqslant M$ *for all* $t \in \mathbb{R}$ *then* $f \in \mathcal{E}$.

Proof. (i) We proceed step by step. Suppose that the equation

$$f'(t) + kf(t-1) = g(t), f(t) = 0 \quad \text{for} \quad t \leqslant 0 \tag{*}$$

395

has a unique solution f_n on $(-\infty, n]$. We know from the fundamental theorem of the calculus that the equation

$$f'(t) = g(t) - kf_n(t-1), \ f(n) = f_n(n)$$

has a unique solution

$$F_{n+1}(t) = \int_n^t (g(s) - kf_n(s-1))ds + f_n(n)$$

on the interval $[n, n+1]$. Thus, setting $f_{n+1}(t) = f_n(t)$ for $t \leqslant n$, $f_{n+1}(t) = F_{n+1}(t)$ for $n \leqslant t \leqslant n+1$, we see that (*) has the unique solution f_{n+1} on $(-\infty, n+1]$.

But, trivially, (*) has the unique solution $f_0 = 0$ on $(-\infty, 0]$, so setting $f(t) = f_n(t)$ for $t \leqslant n$, we see that (*) has a unique solution on $\mathbb{R}$.

(ii) As in (i) we proceed step by step. Set $A = |k|e^{-1} + M$. Trivially $|f(t)| = 0 \leqslant e^{At}$ for all $t \leqslant 0$. Suppose we know that $|f(t)| \leqslant e^{At}$ for all $t \leqslant n$. Then

$$|f'(s)| = |g(s) - kf(s-1)| \leqslant |g(s)| + |k||f(s-1)|$$
$$\leqslant M + |k|e^{A(s-1)} \leqslant Me^{As} + |k|e^{-1}e^{As} = Ae^{As} \text{ for all } n \leqslant s \leqslant n+1,$$

and so

$$|f(t) - f(n)| \leqslant \int_n^t |f'(s)| ds \leqslant \int_n^t Ae^{As} ds = e^{At} - e^{An}.$$

It follows that

$$|f(t)| \leqslant |f(n)| + e^{At} - e^{An} \leqslant e^{At} \text{ for all } n \leqslant t \leqslant n+1,$$

and so $|f(t)| \leqslant e^{At}$ for all $t \leqslant n+1$. The full result follows by induction. ∎

Thus if g is a bounded continuous function and $f \in \mathscr{E}$ then (since $g \in \mathscr{E}$ automatically) we can apply the Laplace transform to (*) and obtain (using Lemma 74.6 (vi) and Lemma 74.7 (i))

$$\mathscr{L}g(z) = (\mathscr{L}f')(z) + k(\mathscr{L}\tau_1 f)(z) = (z + ke^{-z})\mathscr{L}f(z) \quad \text{for Re } z \text{ large.}$$

We can thus apply Theorem 76.12.

Lemma 77.2. *Consider the equation*

$$f'(t) + kf(t-1) = g(t)[t \in \mathbb{R}], f(t) = 0 \text{ for } t \leqslant 0 \qquad (*)$$

where g is a once continuously differentiable function with $g(t) = 0$ for $t \leqslant 0$ and $g(t) = 0$ for t large. Then $f(t) \to 0$ as $t \to \infty$ for all such g if and only if all the solutions of $z + ke^{-z} = 0$ have strictly negative real part.

Proof. We have $\mathscr{L}g(z) = h(z)\mathscr{L}f(z)$ where $h(z) = z + ke^{-z}$. Taking $d = -1$, $n = 1$, $A = \frac{1}{2}$ and $R = 2(|k| + 1)e$ we see that the conditions of Theorem 76.12 hold. ∎

The stability of (*) (at least in the restricted sense of Lemma 77.2) thus depends only on the location of the roots of

$$z + ke^{-z} = 0. \qquad (**)$$

For simplicity we restrict ourselves to the case when k is real. If $k \leqslant 0$ then, as we

would expect, (**) has a real non negative root and (*) is unstable. (Observe that if $u:\mathbb{R} \to \mathbb{R}$ is defined by $u(x) = x + ke^{-x}$ then $u(0) \leqslant 0$ but $u(x) \to \infty$ as $x \to \infty$ so there exists an $x_0 \in \mathbb{R}$ with $x_0 \geqslant 0$ and $u(x_0) = 0$.)

In the interesting case when $k > 0$ (i.e. when $f'(t)$ has the opposite sign to $f(t-1)$) the real roots (if any) are negative and we must locate the complex roots (if any). To do this we make use of the powerful *principle of the argument*.

Theorem 77.3. *Suppose $h:\mathbb{C} \to \mathbb{C}$ is analytic and that h has no zeros on a contour C. Then the number of zeros of h within C (multiple zeros being counted multiply) is equal to $(2\pi)^{-1}$ times the change in the argument of h round the contour.*
Proof. There is an unfortunate tendency for modern elementary complex variable texts to try and pretend that this theorem does not exist. None the less, it is true and useful, even if its proof does require a little care. We sketch a (somewhat artificial) proof in Appendix E. A more direct proof is given in Hille's *Analytic Function Theory* (Vol. 1, p. 253). ∎

How can we use the principle of the argument to locate those zeros of (**) with positive real part? We make two trivial observations (throughout, $k > 0$).

Lemma 77.4. (i) *If $\operatorname{Re} z \geqslant 0$ and $|z| > k$ then $z + ke^{-z} \neq 0$.*
(ii) *The equation $z + ke^{-z} = 0$ has a root $z = iy$ [with $y \in \mathbb{R}$ and $y > 0$] if and only if $k \in \{2m\pi + \pi/2 : m \geqslant 0, m \in \mathbb{Z}\}$ and a root $z = iy$ [with $y \in \mathbb{R}, y < 0$] if and only if $k \in \{2m\pi - \pi/2 : m \geqslant 0, m \in \mathbb{Z}\}$.*

Proof. (i) Observe that $|z + ke^{-z}| \geqslant |z| - k|e^{-z}| > |z| - k$ whenever $\operatorname{Re} z > 0$.
(ii) If $iy + ke^{-iy} = 0$, then, considering real parts, we obtain $k \cos y = 0$ and so $y = n\pi + \pi/2$ for some $n \in \mathbb{Z}$. If $y > 0$ then $n \geqslant 1$. Taking imaginary parts we get $(n\pi + \pi/2) + (-1)^{n+1}k = 0$ so, since $k > 0, n$ is even and $y = 2r\pi + \pi/2$ $[r \in \mathbb{Z}^+]$, $k = 2r\pi + \pi/2$. The result for $y < 0$ is obtained by a similar argument or by taking conjugates.

Consider now the contour $C(R)$ shown in Figure 77.1, composed of the straight

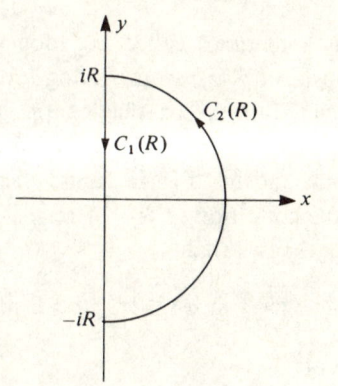

Fig. 77.1. Contour $C(R)$ considered in Lemma 77.5.

line $C_1(R)$ from iR to $-iR$ and the semicircle $C_2(R)$ described by $R\exp i\theta$ as θ runs from $-\pi/2$ to $\pi/2$.

Lemma 77.5. *Suppose $R > k > 0$ and $k \not\equiv \pi/2 \bmod \pi$. If $h(z) = z + ke^{-z}$ then the number of zeros of h with positive real part is equal to $(2\pi)^{-1}$ times the change in argument of $h(z)$ as z runs round the contour $C(R)$.*
Proof. By Lemma 77.4 (ii), h has no zeros on $C(R)$ and by Lemma 77.4 (i) all the zeros of h with positive real part lie within $C(R)$. Applying the principle of the argument (Theorem 77.3) we obtain the result desired. ∎

Usually we cannot compute the change in argument of a function round a contour directly. However the following easy observation is often sufficient to enable us to compute it in a simple manner.

Lemma 77.6. (i) *Let $\theta_1, \theta_2 : [a,b] \to \mathbb{R}$ and $r_1, r_2 : [a,b] \to [0,\infty)$ be continuous. Define $f_1, f_2 : [a,b] \to \mathbb{C}$ by $f_j(t) = r_j(t) \exp i\theta_j(t)$ for $t \in [a,b]$. Then, if*

$$|f_1(t) - f_2(t)| < |f_1(t)| \sin \phi$$

for some $0 < \phi \leqslant \pi/2$ and all $t \in [a,b]$, it follows that

$$|(\theta_2(b) - \theta_2(a)) - (\theta_1(b) - \theta_1(a))| < 2\phi.$$

(ii) *Let C be a well behaved curve in $\mathbb{C}$. Then, if $g_1 g_2 : \mathbb{C} \to \mathbb{C}$ are analytic and*

$$|g_1(z) - g_2(z)| \leqslant |g_1(z)| \sin \phi$$

for some $0 < \phi < \pi/2$ and all $z \in C$, it follows that

$$|[\arg g_2(z)]_C - [\arg g_1(z)]_C| \leqslant 2\phi.$$

Remark 1. If C is a closed contour then $[\arg g_j(z)]_C$ must be a multiple of $2\pi i$. Since $2\phi < \pi$ it follows that $[\arg g_1(z)]_C = [\arg g_2(z)]_C$ and, using Theorem 77.3, we see that g_1 and g_2 have the same number of zeros within C. This is Rouché's theorem.
Remark 2. Lemma 77.6 is sometimes called the 'dog walking lemma'. It states in effect that if a man (position $g_1(z)$) walks a dog (position $g_2(z)$) round a tree (at the origin) then, provided he keeps the length $|g_1(z) - g_2(z)|$ of the leash shorter than his distance $|g_1(z)|$ from the tree, the dog cannot get tangled up round the tree (i.e. the man and his dog go round the tree the same number of times). The criticality of the condition $|g_1(z) - g_2(z)| < |g_1(z)|$ is obvious.
Proof. (i) Simple geometry shows that

$$\{1 + w : |w| < \sin \phi\} \subseteq \{re^{i\theta} : r > 0, |\theta| < \phi\}$$

(just look at Figure 77.2).

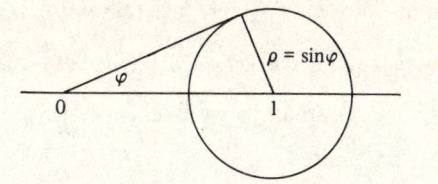

Fig. 77.2. 'Dog walking' with man at 1 and tree at 0.

Thus, since

$$|-1 + f_2(t)/f_1(t)| = |f_1(t) - f_2(t)|/|f_1(t)| < \sin \phi,$$

it follows that

$$f_2(t)/f_1(t) = 1 + (-1 + f_2(t)/f_1(t)) \in \{re^{i\theta} : r > 0, |\theta| < \phi\},$$

and so

$$r_2(t)r_1(t)^{-1} \exp i(\theta_2(t) - \theta_1(t)) \in \{re^{i\theta} : r > 0, |\theta| < \phi\}.$$

whence

$$\theta_2(t) - \theta_1(t) \in \bigcup_{m=-\infty}^{\infty} A_m,$$

where

$$A_m = \{\theta + 2m\pi : |\theta| < \phi\}.$$

But if $\theta_2(a) - \theta_1(a) \in A_{m(0)}$ then $\theta_2(b) - \theta_1(b) \in A_{m(0)}$. (For if $\theta_2(b) - \theta_1(b) \in A_{m(1)}$ with $m(1) \neq m(0)$ then, since the intermediate value theorem shows that $\theta_2(t) - \theta_1(t)$ takes all possible values between $\theta_2(a) - \theta_1(a)$ and $\theta_2(b) - \theta_1(b)$, we must be able to find an $s \in [a, b]$ with $\theta_2(s) - \theta_1(s) \equiv \pi \bmod 2\pi$, and so with $\theta_2(s) - \theta_1(s) \notin \bigcup_{m=-\infty}^{\infty} A_m$ contradicting the result just obtained.) Thus, since

$$\theta_2(a) - \theta_1(a), \quad \theta_2(b) - \theta_1(b) \in A_{m(0)},$$

it follows that

$$|(\theta_2(b) - \theta_2(a)) - (\theta_1(b) - \theta_1(a))| = |(\theta_2(b) - \theta_1(b)) - (\theta_2(a) - \theta_1(a))| < 2\phi.$$

(ii) This is just a special case of (i). ∎

Lemma 77.7. *Suppose $k > 0$, $k \not\equiv \pi/2 \bmod \pi$ and let $h(z) = z + ke^{-z}$. Let $\psi(R)$ be the change in argument of $h(ix)$ as x runs from $-R$ to R. Then*

$$(2\pi)^{-1}(\pi - \psi(R)) \to N \quad as \quad R \to \infty,$$

where N is the number of zeros of h with positive real parts.
Proof. By Lemma 77.5

$$2\pi N = [\arg h]_{C(R)} = [\arg h]_{C_1(R)} + [\arg h]_{C_2(R)} = -\psi(R) + [\arg h]_{C_2(R)},$$

provided only that $R > k$.

We examine $[\arg h]_{C_2(R)}$ using Lemma 77.6. If $\pi/2 > \phi > 0$ then, provided only that $R > k/\sin \phi$, we have

$$|z| \sin \phi = R \sin \phi > k \geqslant |ke^{-z}| \quad \text{for all} \quad z \in C_2(R),$$

and so applying Lemma 77.6 with $g_1(z) = z$, $g_2(z) = h(z)$

$$|\pi - [\arg h]_{C_2(R)}| = |[\arg g_1]_{C_2(R)} - [\arg g_2]_{C_2(R)}| < 2\phi.$$

Thus, since we may take ϕ as small as we like, $[\arg h]_{C_2(R)} \to \pi$ as $R \to \infty$ and the lemma follows. ∎

We have thus reduced the problem of the stability of the system

$$\left. \begin{array}{l} f'(t) + kf(t-1) = g(t), \\ f(t) = g(t) = 0 \text{ for } t \leqslant 0 \end{array} \right\} \qquad (*)$$

with

to calculating the change in argument of $h(ix) = k\cos x + i(x - k\sin x)$ as x runs from $-R$ to R for large R.

But it is easy to graph $p(x) = k^{-1}h(ix)$ consisting as it does of the vector sum of the linear path $ik^{-1}x$ and the clockwise circular path $\cos x - i\sin x$.

For $0 < k < 1$ we get a sinuous curve which develops cusps when $k = 1$ which become loops for $1 < k < \pi/2$. For all these cases $\arg h(ix) = \arg p(x)$ runs from $-\pi/2$ to $\pi/2$ as x runs from $-\infty$ to ∞ and so $\psi(R) \to \pi$ as $R \to \infty$ (see Figure 77.3(a)).

As k increases so does the size of the loops until when $k = \pi/2$ the neck of one

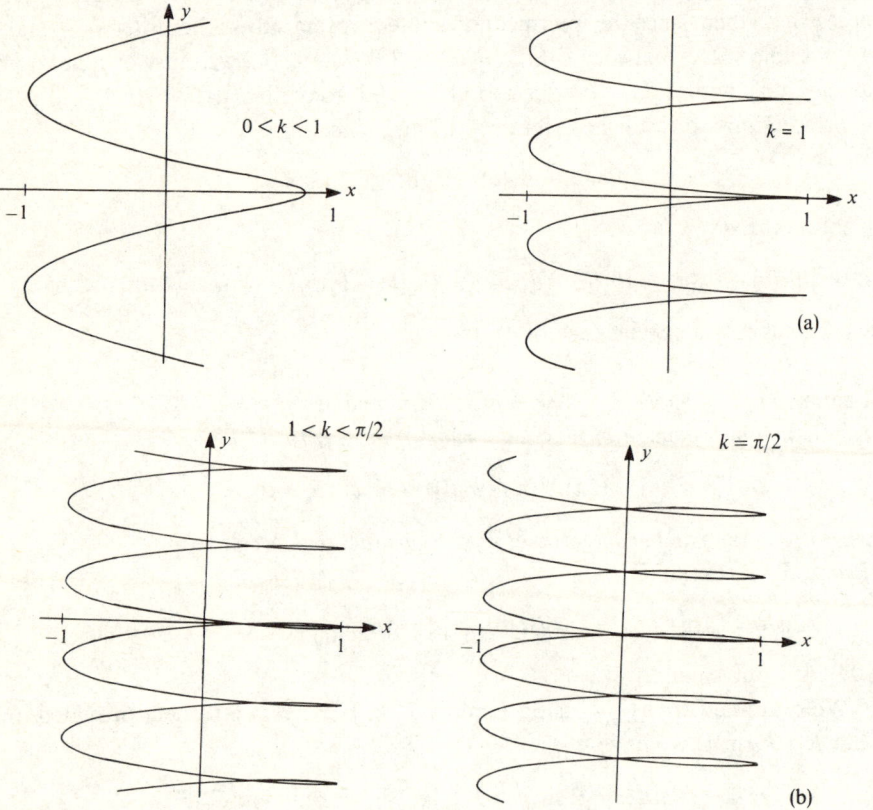

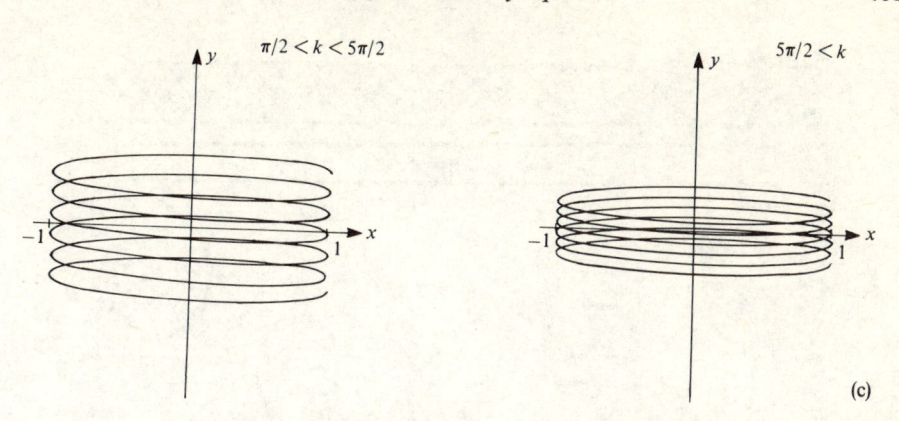

Fig. 77.3. Graphs of $p(x) = k^{-1}h(ix)$ for various values of k.

of the loops occurs at 0 (see Figure 77.3(b)). For $\pi/2 < k < 5\pi/2$ this loop now encloses 0 completely and $\arg h(ix) = \arg p(x)$ now runs from $-\pi/2$ to $-\pi/2$ as x runs from $-\infty$ to ∞. Thus $\psi(R) \to -3\pi$ as $R \to \infty$ for $\pi/2 < k < 5\pi/2$ (see Figure 77.3(c)).

Finally, for $k > 5\pi/2$ we see that 0 is enclosed by many loops and $-\lim_{R \to \infty} \psi(R)$ is correspondingly large. (In fact, as the reader may wish to check, $\lim_{R \to \infty} \psi(R) = -(2M+1)\pi$ for $2(M-1)\pi + \pi/2 < k < 2M\pi + \pi/2$.)

Combining the information just obtained with the results of Lemmas 77.2, Lemma 77.4 (ii) and Lemma 77.7 we obtain the following stability condition.

Theorem 77.8. *Consider the equation*

$$f'(t) + kf(t-1) = g(t)[t \in \mathbb{R}], f(t) = 0 \quad \text{for } t \leqslant 0, \tag{*}$$

where g is a once continuous differentiable function with $g(t) = 0$ for $t \leqslant 0$ and $g(t) = 0$ for t large. Then $f(t) \to 0$ as $t \to \infty$ for all such g if and only if $0 < k < \pi/2$.

Proof. Provided the reader is satisfied with the treatment of $\lim_{R \to \infty} \psi(R)$ given above the proof is complete. If not, then I leave her to supply the extra tedious (and in my view, unnecessary) detail. ∎

Returning to our original problem

$$F'(s) + KF(s - \eta) = G(s)$$

with $F(s) = G(s) = 0$ for $s \leqslant 0$ and with $\eta > 0$, we see that we have stability (in the sense of Theorem 77.8) if and only if $0 < K\eta < \pi/2$. Thus, however small the time delay η, instability will occur for sufficiently large K.

No physical system responds instantly to changes in input though 'under normal operating conditions' the time delay may appear negligible. The results of this chapter as summarised in the previous paragraph suggest that as operating

"I CAN REMEMBER WHEN THIS TRAIN WAS SO HORRIBLY SLOW THAT YOU COULD EAT YOUR MEAL IN COMFORT."

Fig. 77.4. Oscillations due to time delay.

conditions become more extreme even a small time delay may lead to an instability whose existence may not even have been suspected by those '... who naturally confine their attention to the way in which [a system] is designed to act'.

An interesting example occurs in the high speed running of trains. Even in the 1930s the *Great Western Railway* ran a crack train (the *Cheltenham Flyer*) at speeds close to 80 mph but constant and unpleasant vibration meant that the journey was not for the nervous. The smooth running of its successor's 125 mph train stems from a redesign of the wheel surface which in turn owes much to the understanding that even the equations governing the interaction wheel and track involve time delays (Figure 77.4).

78

AN EXCEPTION TO A RULE

Chapter 77 has shown, I hope, that the study of the poles and zeros of Laplace transforms can shed light on interesting stability problems. However, the method used is based on theorems in Chapter 76 which require that the Laplace transforms considered must satisfy certain conditions as $|z| \to \infty$. Can this annoying extra requirement be dropped?

Consider the system of equations (for $t > 0$)

$$\left. \begin{array}{l} f'(t) = tg(t), \\ g'(t) = -tf(t), \\ f(0+) = 0, \, g(0+) = 1. \end{array} \right\} \tag{*}$$

with

Let us seek solutions $f, g \in \mathscr{E}$. By Lemma 75.1 we know that $\tau f, \tau g \in \mathscr{E}$ (where $\tau(t) = t$ for $t \geqslant 0$, $\tau(t) = 0$ for $t < 0$) so $f', g' \in \mathscr{E}$. Thus, using Theorem 75.2 and Lemma 74.7, we have

$$\mathscr{L}(\tau f)(z) = -(\mathscr{L}f)'(z), \; \mathscr{L}(\tau g)(z) = -(\mathscr{L}g)'(z),$$
$$\mathscr{L}(f')(z) = -f(0+) + z(\mathscr{L}f)(z) = z(\mathscr{L}f)(z),$$

and
$$\mathscr{L}(g')(z) = -g(0+) + z(\mathscr{L}g)(z) = -1 + z(\mathscr{L}g)(z).$$

Thus, taking Laplace transforms, (*) becomes

$$\left. \begin{array}{l} z(\mathscr{L}f)(z) = -(\mathscr{L}g)'(z), \\ -1 + z(\mathscr{L}g)(z) = (\mathscr{L}f)'(z). \end{array} \right\} \tag{**}$$

The solution of (**) is probably well within the reader's power and she is invited to attempt it before studying the details below.

Adding and subtracting i times the second equation from the first, we obtain the equivalent system

$$-i + iz(\mathscr{L}g)(z) + z(\mathscr{L}f)(z) = i(\mathscr{L}f)'(z) - (\mathscr{L}g)'(z),$$
$$i - iz(\mathscr{L}g)(z) + z(\mathscr{L}f)(z) = -i(\mathscr{L}f)'(z) - (\mathscr{L}g)'(z),$$

which after rearrangement becomes

$$-i + iz(\mathscr{L}(g - if))(z) + (\mathscr{L}(g - if))'(z) = 0,$$
$$i - iz(\mathscr{L}(g + if))(z) + (\mathscr{L}(g + if))'(z) = 0.$$

Setting $\alpha = \mathscr{L}(g - if)$, $\beta = \mathscr{L}(g + if)$, we have reduced our equations to the simple form

$$-i + iz\alpha(z) + \alpha'(z) = 0,$$
$$i - iz\beta(z) + \beta'(z) = 0.$$

Analogy with the solution of such equations in the real case suggests that we multiply the first equation by $\exp(iz^2/2)$ and the second by $\exp(-iz^2/2)$ to obtain

$$-i\exp(iz^2/2) + \frac{d}{dz}((\exp iz^2/2)\alpha(z)) = 0,$$

$$i\exp(-iz^2/2) + \frac{d}{dz}((\exp - iz^2/2)\beta(z)) = 0.$$

The next stage is also inspired by analogy but now requires a simple preparatory lemma.

Lemma 78.1. *Write $C(z)$ for the straight line path from 0 to z. If $l:\mathbb{C}\to\mathbb{C}$ is analytic everywhere then so is the function L defined by*

$$L(z) = \int_{C(z)} l(w)dw.$$

Moreover, $L'(z) = l(z)$.

Proof. This is simple and standard. Write $C(z_1, z_2)$ for the straight line path from z_1 to z_2. Then by Cauchy's theorem

$$L(z) + \int_{C(z,z+h)} l(w)dw - L(z + h) = \int_{C(0,z)} + \int_{C(z,z+h)} + \int_{C(z+h,0)} l(w)dw = 0,$$

and so

$$\left|\frac{L(z + h) - L(z)}{h} - l(z)\right| = \frac{1}{|h|}\left|\int_{C(z,z+h)} l(w)dw - hl(z)\right|$$

$$= \frac{1}{|h|}\left|\int_{C(z,z+h)} (l(w) - l(z))dw\right|$$

$$\leqslant \sup_{w\in C(z,z+h)} |l(w) - l(z)| \to 0 \quad \text{as } h\to 0. \quad\blacksquare$$

Thus in particular if we write

$$E_1(z) = \int_{C(z)} \exp(iw^2/2)dw,$$

$$E_2(z) = \int_{C(z)} \exp(-iw^2/2)dw,$$

we know that E_1, E_2 are everywhere analytic with $E_1'(z) = \exp(iz^2/2)$, $E_2'(z) = \exp(-iz^2/2)$. Our equations for α and β take the simple form

$$\frac{d}{dz}\left(\exp(iz^2/2)\alpha(z) - i\int_{C(z)} \exp(iw^2/2)dw\right) = 0,$$

$$\frac{d}{dz}\left(\exp(-iz^2/2)\beta(z) + i\int_{C(z)} \exp(-iw^2/2)dw\right) = 0.$$

We can now draw the threads together.

Lemma 78.2. *Suppose* $f, g \in \mathscr{E}_a$ *and satisfy the system of equations* (*) *for* $t > 0$. *Then, writing* $\alpha = \mathscr{L}g - i\mathscr{L}f$, $\beta = \mathscr{L}g + i\mathscr{L}f$, *there exist constants* A *and* B *such that*

$$\alpha(z) = A\exp(-iz^2/2) + i\exp(-iz^2/2)\int_{C(z)} \exp(iw^2/2)dw,$$

$$\beta(z) = B\exp(iz^2/2) - i\exp(iz^2/2)\int_{C(z)} \exp(-iw^2/2)dw,$$

for all z *with* $\operatorname{Re} z > a$.

Proof. If $h: \{z: \operatorname{Re} z > a\}$ is analytic and $h'(z) = 0$ for all z with $\operatorname{Re} z > a$ it follows that h is constant. Thus, using the computations above, there exist constants A and B such that

$$\exp(iz^2/2)\alpha(z) - i\int_{C(z)} \exp(iw^2/2)dw = A,$$

$$\exp(-iz^2/2)\beta(z) + i\int_{C(z)} \exp(-iw^2/2)dw = B,$$

and we have the result stated. ∎

Lemma 78.3. *Suppose* $f, g \in \mathscr{E}_a$ *and satisfy the system of equations* (*) *for* $t > 0$. *Then* $\mathscr{L}f$ *and* $\mathscr{L}g$ *may be extended to functions analytic everywhere in* $\mathbb{C}$.

Proof. Set

$$\tilde{\alpha}(z) = A\exp(-iz^2/2) + i\exp(-iz^2/2)\int_{C(z)} \exp(iw^2/2)dw,$$

$$\tilde{\beta}(z) = B\exp(iz^2/2) - i\exp(iz^2/2)\int_{C(z)} \exp(-iw^2/2)dw,$$

for all $z \in \mathbb{C}$. Then $\tilde{\alpha}, \tilde{\beta}$ are analytic (Lemma 78.1) and (referring to Lemma 78.2), $\tilde{\alpha}(z) = \alpha(z)$, $\tilde{\beta}(z) = \beta(z)$ for all z with $\operatorname{Re} z > a$. It follows that $(\tilde{\beta} - \tilde{\alpha})/2i$ and $(\beta + \alpha)/2$ are extensions of $\mathscr{L}f$ and $\mathscr{L}g$ which are analytic everywhere in $\mathbb{C}$. ∎

Can we therefore deduce that $f(t)$, $g(t) \to 0$ as $t \to \infty$? The next lemma shows that we cannot.

Lemma 78.4. *Set* $f(t) = \sin(t^2/2)$, $g(t) = \cos(t^2/2)$ *for* $t > 0$, $f(0) = 0$, $g(0) = 1/2$ *and* $f(t) = g(t) = 0$ *for* $t < 0$. *Then* $f, g \in \mathscr{E}_a$ *for all* $a > 0$ *and satisfy the system of equations* (*) *for* $t \geqslant 0$.
Proof. Direct computation. ∎

The following Lemma simply embroiders the moral.

Lemma 78.5. (i) *If* $f(t) = \sin t^2/2$ *for* $t \geqslant 0$, $f(t) = 0$ *for* $t < 0$ *then* $f \in \mathscr{E}$ *and* $\mathscr{L}f$ *can be extended to a function* $F: \mathbb{C} \to \mathbb{C}$ *which is everywhere analytic yet* $f(t) \nrightarrow 0$ *as* $t \to \infty$.
(ii) *If* $f_1(t) = t \sin t^2/2$ *for* $t \geqslant 0$, $f_1(t) = 0$ *for* $t < 0$ *then* $f_1 \in \mathscr{E}$ *and* $\mathscr{L}f_1$ *can be extended to a function* $F_1: \mathbb{C} \to \mathbb{C}$ *which is everywhere analytic yet* $f_1(t)$ *oscillates unboundedly as* $t \to \infty$.

Proof. (i) Combine Lemmas 78.3 and 78.4.
(ii) By Lemma 75.1, $f_1 \in \mathscr{E}$. By Theorem 75.2, $\mathscr{L}f_1 = -(\mathscr{L}f)'$. Setting $F_1 = -F'$ we have the required results. ∎

We thus see that any attempt to find properties of an $h \in \mathscr{E}$ by studying the behaviour of its Laplace transform $\mathscr{L}h$ requires a (more or less) explicit study of the behaviour of $\mathscr{L}h(z)$ for large $|z|$.

It is customary to begin courses in mathematical engineering by explaining that the lecturer would never trust his life to an aeroplane whose behaviour depended on properties of the Lebesgue integral. It might, perhaps, be just as foolhardy to fly in an aeroplane designed by an engineer who believed that cookbook application of the Laplace transform revealed all that was to be known about its stability.

My reluctance to fly in such a machine would be based not on the counter example in this chapter, interesting though it is, but on the sentiments expressed by Heaviside (*Collected Works*, Vol. II, p. 4). 'The practice of eliminating the physics by reducing a problem to a purely mathematical exercise should be avoided as much as possible. The physics should be carried on right through, to give life and reality to the problem, and to obtain the great assistance which the physics gives to the mathematics.' Here surely, rather than in the problems highlighted in this chapter, lies the real reason for caution in using the Laplace transform.

79

MANY DIMENSIONS

It is remarkably easy to generalise the theory of Fourier series on $\mathbb{T}$ and Fourier integrals on $\mathbb{R}$ to cover Fourier series on $\mathbb{T}^m$ and Fourier integrals on $\mathbb{R}^m$. To see how this is done, I invite the reader to rewrite certain of the previous chapters with appropriate modifications.

In Chapter 1, let us write $\mathbf{r} \cdot \mathbf{t} = r_1 t_1 + r_2 t_2 + \cdots + r_m t_m$, where $\mathbf{r} = (r_1, r_2, \ldots, r_m) \in \mathbb{Z}^m$ and $\mathbf{t} = (t_1, t_2, \ldots, t_m) \in \mathbb{T}^m$. We then define $\hat{f}$ by

(i)
$$\hat{f}(\mathbf{r}) = \left(\frac{1}{2\pi}\right)^m \int_0^{2\pi} \int_0^{2\pi} \cdots \int_0^{2\pi} f(\mathbf{t}) \exp(-i\mathbf{r} \cdot \mathbf{t}) dt_1 dt_2 \ldots dt_m$$

$$= \left(\frac{1}{2\pi}\right)^m \int_{\mathbb{T}^m} f(\mathbf{t}) \exp i\mathbf{r} \cdot \mathbf{t} \, dV(\mathbf{t}),$$

and $S_n(f, \mathbf{t})$ by

(ii)
$$S_n(f, \mathbf{t}) = \sum_{\max|r_j| \leqslant n} \hat{f}(\mathbf{r}) \exp i\mathbf{r} \cdot \mathbf{t},$$

whilst $\sigma_n(f, \mathbf{t})$ is given by

(iii)
$$\sigma_n(f, \mathbf{t}) = \sum_{\max|r_j| \leqslant n} \left(\prod_{k=1}^{m} \frac{(n+1-|r_k|)}{n+1}\right) \hat{f}(\mathbf{r}) \exp i\mathbf{r} \cdot \mathbf{t}.$$

If we now write

(iv)
$$K_n(\mathbf{t}) = K_n(t_1) K_n(t_2) \ldots K_n(t_m)$$

the reader should have no difficulty in paralleling the work of Chapter 2 to obtain results like the following.

Theorem 2.3*. (i) *If* $f : \mathbb{T}^m \to \mathbb{C}$ *is Riemann integrable then if* f *is continuous at* $\mathbf{t}$
$$\sigma_n(f, \mathbf{t}) \to f(\mathbf{t}) \quad \text{as } n \to \infty.$$

(ii) *If* $f : \mathbb{T}^m \to \mathbb{C}$ *is continuous then*

$$\sigma_n(f, \ ) \to f \text{ uniformly on } \mathbb{T}^m.$$

407

Nor should it be difficult for her to use the methods of Chapter 3 to obtain the multidimensional version of Weyl's theorem.

Theorem 3.1*. *Suppose* $\gamma_1, \gamma_2, \ldots, \gamma_m$ *are real numbers such that the only solution of*

$$q_1\gamma_1 + q_2\gamma_2 + \cdots + q_m\gamma_m = q,$$

with $q, q_1, q_2, \ldots, q_m \in \mathbb{Z}$ *is* $q = q_1 = q_2 = \cdots = q_m = 0.$

Then if $0 \leqslant a_j \leqslant b_j \leqslant 1$ *for all* $1 \leqslant j \leqslant m$, *we have*

$$n^{-1}\operatorname{card}\{1 \leqslant r \leqslant n : a_j \leqslant \langle r\gamma_j \rangle \leqslant b_j \quad \text{for all } 1 \leqslant j \leqslant m\} \to \prod_{j=1}^{m} (b_j - a_j).$$

This result generalises considerably an important theorem of Kronecker.

Theorem 79.1 (Kronecker). *If* $\gamma_1, \gamma_2, \ldots, \gamma_m$ *satisfy the condition of Theorem 3.1* and* $0 \leqslant a_j < b_j \leqslant 1$ *for all* $1 \leqslant j \leqslant m$ *then the system of inequalities*

$$a_j \leqslant \langle r\gamma_j \rangle \leqslant b_j \quad [1 \leqslant j \leqslant m]$$

always has an integer solution with $r > 0$.
Proof. Observe that $\prod_{j=1}^{m}(b_j - a_j) > 0$ and so by Theorem 3.1*

$$\operatorname{card}\{1 \leqslant r \leqslant n : a_j \leqslant \langle r\gamma_j \rangle \leqslant b_j \quad \text{for all } 1 \leqslant j \leqslant m\} \to \infty.$$

The system of inequalities

$$a_j \leqslant \langle r\gamma_j \rangle \leqslant b_j \quad [1 \leqslant j \leqslant m]$$

thus possesses an infinity of solutions and so, in particular, at least one solution.   ∎

Remark 1. The reader should check that the conditions imposed on $\gamma_1, \gamma_2, \ldots, \gamma_m$ in Theorem 3.1* and Theorem 79.1 are not only sufficient but also necessary for the truth of the stated conclusions.
Remark 2. Kronecker's theorem has attracted several proofs based on different principles. (Four including the one above, are given in Chapter XXIII of Hardy and Wright's *Introduction To The Theory of Numbers*.) Anyone who can find such a proof independently can feel well pleased with herself.

When we turn to the discussion of pointwise convergence, we find that the simple results remain simple but the later, complicated results become even more complicated. For example, the statement and proof of the basic Theorem 9.1 go through unchanged.

Theorem 9.1*. *Let* $f : \mathbb{T}^m \to \mathbb{C}$ *be continuous. Then if* $\sum_{\mathbf{r} \in \mathbb{Z}^m} |\hat{f}(\mathbf{r})|$ *converges it follows that*

$$S_n(f, \mathbf{t}) \to f(\mathbf{t}) \text{ uniformly on } \mathbb{T}^m \text{ as } n \to \infty.$$

This can be combined with the following (simplified) version of Lemma 9.5.

Lemma 9.5*. *Suppose* $f : \mathbb{T}^m \to \mathbb{C}$ *is* n *times continuously differentiable. Then there exists a constant* M *such that*

$$|\hat{f}(\mathbf{r})| \leqslant M \|\mathbf{r}\|^{-n} \quad \text{for all} \quad \mathbf{r} \neq 0.$$

Proof. Consider each integer j with $1 \leqslant j \leqslant m$ in turn. Since $\partial^n f / \partial x^n$ is continuous on $\mathbb{T}^m$, it is bounded (Theorem B.3) and we can find an M_j with $|(\partial^n f / \partial x_j^n)(\mathbf{t})| \leqslant M_j$ for all $\mathbf{t} \in \mathbb{T}^m$. Thus if $r_j \in \mathbb{Z}$ and $r_j \neq 0$ we have (Lemma 9.5)

$$\left| \frac{1}{2\pi} \int_{\mathbb{T}} f(t_1, t_2, \ldots, t_j, \ldots, t_m) \exp(-it_j r_j) dt_j \right| \leqslant \frac{M_j}{|r_j|^n}$$

and so

$$\left| \left(\frac{1}{2\pi} \right)^m \int_{\mathbb{T}} \int_{\mathbb{T}} \cdots \int_{\mathbb{T}} f(t_1, \ldots, t_m) \exp(-i \Sigma t_k r_k) dt_1 \ldots dt_m \right|$$

$$\leqslant \left(\frac{1}{2\pi} \right)^{m-1} \int_{\mathbb{T}} \cdots \int_{\mathbb{T}} \frac{M_j}{|r_j|^n} dt_1 \ldots dt_{j-1} dt_{j+1} \ldots dt_m = \frac{M_j}{|r_j|^n}.$$

It follows that

$$|\hat{f}(\mathbf{r})| \leqslant \frac{M_j}{|r_j|^n} \text{ for all } j \text{ with } r_j \neq 0,$$

and so, if $\mathbf{r} \neq 0$,

$$|\hat{f}(\mathbf{r})| \leqslant (\max_{1 \leqslant j \leqslant m} M_j) \min \{|r_j|^{-n} : r_j \neq 0\}$$

$$= (\max_{1 \leqslant j \leqslant m} M_j)(\max_{1 \leqslant j \leqslant m} |r_j|)^{-n} \leqslant \left(\frac{m}{\|\mathbf{r}\|^2} \right)^{n/2} \max_{1 \leqslant j \leqslant m} M_j.$$

Setting $M = m^{n/2} \max_{1 \leqslant j \leqslant m} M_j$ we have the required result. ∎

In this connection we make the following observations.

Lemma 79.2. (i) *If* $a : \mathbb{Z}^m \to \mathbb{C}$ *satisfies the condition* $|a(\mathbf{r})| \leqslant M \|\mathbf{r}\|^{-m-\alpha}$ *for all* $\mathbf{r} \neq 0$, *some* $\alpha > 0$ *and some* M, *then*

$$\sum_{-N}^{N} \sum_{-N}^{N} \cdots \sum_{-N}^{N} |a(r_1, r_2, \ldots, r_m)| \text{ converges as } N \to \infty.$$

(ii) *If* $a(\mathbf{r}) = \|\mathbf{r}\|^{-m}$ *for* $\mathbf{r} \neq 0$ *then*

$$\sum_{-N}^{N} \sum_{-N}^{N} \cdots \sum_{-N}^{N} a(r_1, r_2, \ldots, r_m) \text{ diverges as } N \to \infty.$$

Proof. (i) Since $\|\mathbf{r}\| \geqslant 1$ for all $\mathbf{r} \neq 0, \mathbf{r} \in \mathbb{Z}$, we have, using the inequality of arithmetic and geometric means,

$$(m+1)\|\mathbf{r}\|^2 \geqslant m + \|\mathbf{r}\|^2 = \sum_{j=1}^{m}(1+r_j^2) \geqslant m\left(\prod_{j=1}^{m}(1+r_j^2)\right)^{1/m} \quad \text{for all } \mathbf{r} \neq 0.$$

Thus writing $K = M(m/(m+1))^{-(m+\alpha)/2}$ and $\varepsilon = \alpha/m$ we have

$$|a(r_1, r_2, \ldots, r_m)| \leqslant K \prod_{j=1}^{m}(1+r_j^2)^{-(1+\varepsilon)/2}.$$

It follows that

$$\sum_{-N}^{N}\sum_{-N}^{N}\cdots\sum_{-N}^{N}|a(r_1,r_2,\ldots,r_m)|$$

$$\leqslant K\sum_{-N}^{N}\sum_{-N}^{N}\cdots\sum_{-N}^{N}\prod_{j=1}^{m}(1+r_j^2)^{-(1+\varepsilon)/2} = K\prod_{j=1}^{m}\sum_{r_j=-N}^{N}(1+r_j^2)^{-(1+\varepsilon)/2}$$

$$= K\left(\sum_{r=-N}^{N}(1+r^2)^{-(1+\varepsilon)/2}\right)^m \leqslant K\left(1+2\sum_{n=1}^{\infty}n^{-1-\varepsilon}\right)^m,$$

and, since $\sum_{n=1}^{\infty}n^{-1-\varepsilon}$ converges, we are done.

(ii) Let $\Gamma(k) = \{\mathbf{r} \in \mathbb{Z}^m : 2^{k-1} < \max|r_j| \leqslant 2^k\}$. Thus $\Gamma(k)$ corresponds to a 'hollowed out cube' shown in Figure 79.1. Then $\Gamma(k)$ has at least $2^{m(k-1)}$ points and at each point $\mathbf{r}$ of $\Gamma(k)$, $a(\mathbf{r}) \geqslant (m2^{2k})^{-m/2} = m^{-m/2}2^{-km}$. Thus if $N(q) = 2^q$

$$\sum_{-N(q)}^{N(q)}\sum_{-N(q)}^{N(q)}\cdots\sum_{-N(q)}^{N(q)}|a(r_1,r_2,\ldots,r_m)| \geqslant \sum_{k=1}^{q}\sum_{\Gamma(k)}a(\mathbf{r})$$

$$\geqslant \sum_{k=1}^{q}2^{m(k-1)}m^{-m/2}2^{-km} = m^{-m/2}\sum_{k=1}^{q}2^{-m} = qm^{-m/2}2^{-m} \to \infty \text{ as } q \to \infty. \quad \blacksquare$$

Combining the results above we obtain a generalisation of Theorem 9.6.

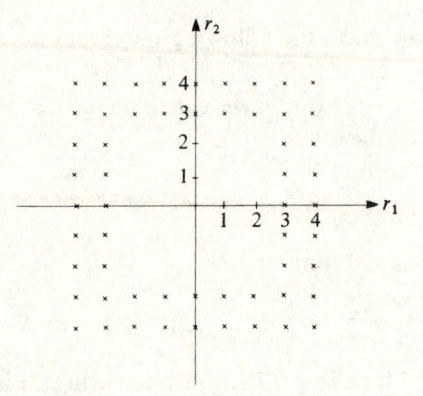

Fig. 79.1. Points of $\Gamma(2)$ when $m = 2$.

Theorem 9.6*. *If $f: \mathbb{T}^m \to \mathbb{C}$ is $m + 1$ times continuously differentiable, then $S_n(f,\ ) \to f$ uniformly.*

However the smooth process of generalisation breaks down when we come to Chapters 16 and 17 dealing with the behaviour of Fourier sums at (simple) discontinuities. Of course, it is not hard to find a restricted class of functions for which the results and proofs go over word for word; but we do not want to obtain any old generalisation, but one which has some chance of being useful. Thus in practice we should be prepared to deal with discontinuities along curves which are not straight lines and functions such as the one defined by

$$f(x, y) = x \qquad \text{for } 0 < x < y < \pi,$$
$$f(x, y) = -y \quad \text{for } 0 < y < x < \pi,$$
$$f(x, y) = 0 \qquad \text{otherwise.}$$

(We leave it to the unambitious reader to guess and to the ambitious reader to find the behaviour of $S_N(f,\ )$ near (π, π).) It thus seems reasonable not to attempt the construction of a general theory but to be prepared to deal 'by hand' with particular cases if and when they occur.

Returning to the path of easy generalisation the reader will readily see that the work on inner products and mean square convergence in Chapters 32–34 goes over to the many dimensional case without problems. For example, she will be able to prove the following improvement on Theorem 9.6*.

Therorem 33.7*. *Suppose $f: \mathbb{T}^m \to \mathbb{C}$ is m times continuously differentiable then*

$$\sum_{-N}^{N} \sum_{-N}^{N} \cdots \sum_{-N}^{N} |\hat{f}(r_1, r_2, \ldots, r_m)|\ converges\ as\ N \to \infty$$

and so, in particular

$$S_n(f,\ ) \to f\ uniformly\ as\ N \to \infty.$$

In the same way we can generalise the notion of the Fourier transform $\hat{f}: \mathbb{R} \to \mathbb{R}$ of a (suitable) function $f: \mathbb{R} \to \mathbb{R}$ to yield the definition of the Fourier transform $\hat{f}: \mathbb{R}^m \to \mathbb{R}$ of a (suitable) function $f: \mathbb{R}^m \to \mathbb{R}$ as follows.

Suppose $f: \mathbb{R}^m \to \mathbb{R}^m$ is continuous and

$$\int_{-\infty}^{\infty} \int_{-\infty}^{\infty} \cdots \int_{-\infty}^{\infty} |f(t_1, t_2, \ldots, t_m)|\,dt_1\,dt_2 \ldots dt_m\ \text{converges.}$$

We say that $f \in L^1 \cap C$ and write

$$\hat{f}(\zeta_1, \zeta_2, \ldots, \zeta_m) =$$

$$\int_{-\infty}^{\infty} \int_{-\infty}^{\infty} \cdots \int_{-\infty}^{\infty} f(t_1, t_2, \ldots, t_n) \exp(-i(\zeta_1 t_1 + \zeta_2 t_2 + \cdots + \zeta_m t_m))\,dt_1\,dt_2 \ldots dt_m$$

or, more briefly,

$$\hat{f}(\zeta) = \int_{\mathbb{R}^m} f(\mathbf{t}) \exp(-i\zeta \cdot \mathbf{t})\,dV(\mathbf{t}),$$

where $\zeta \cdot \mathbf{t} = \sum_{j=1}^{m} \zeta_j t_j$. The definition of convolution follows the same style, so that

$$f * g(\mathbf{t}) = \int_{\mathbb{R}^m} f(\mathbf{t} - \mathbf{s}) g(\mathbf{s}) dV(\mathbf{s})$$

where, for example, $f, g \in L^1 \cap C$.

We will return to the subject of multidimensional Fourier transforms in the next chapter. For the moment I suggest that the reader takes one or two results in the one dimensional case (e.g. Theorem 51.7 and Theorem 46.6) and generalises them.

Let me end this chapter with a word of warning. I have not considered generalisations of notions like the Laplace transform where our proofs relied on results from the theory of analytic functions $g : \mathbb{C} \to \mathbb{C}$. A theory of analytic functions $g : \mathbb{C}^m \to \mathbb{C}^n$ exists, but involves ideas which are much more subtle than any in this book. (For example if $g : \mathbb{C}^2 \to \mathbb{C}$ is analytic it cannot have any isolated zeros. Instead the zeros occur along curves.)

80

SUMS OF RANDOM VECTORS

In Chapters 69–71 we showed that if $X, X(1), X(2), \ldots$ are identically distributed random variables with $\mathbb{E}X = 0$ and $\mathbb{E}X^2 = 1$ then

$$Pr\left(\frac{X(1) + X(2) + \cdots + X(n)}{n^{\frac{1}{2}}} \in [a,b]\right) \to \frac{1}{(2\pi)^{\frac{1}{2}}} \int_a^b \exp(-t^2/2)dt.$$

What happens if we replace the real valued random variables $X, X(1), X(2), \ldots$ by vectors $\mathbf{X}, \mathbf{X}(1), \mathbf{X}(2), \ldots$ in $\mathbb{R}^m$?

If we attempt to reproduce the heuristic argument given in Chapter 69, with $\mathbb{E}\mathbf{X} = \boldsymbol{\mu} = \mathbf{0}$ and multiplication of real numbers replaced by scalar multiplication, we obtain the following heuristic argument.

Heuristic argument. Formally

$$\mathbb{E}\exp i\boldsymbol{\lambda} \cdot \mathbf{X} = \mathbb{E}\left(\sum_{k=0}^{\infty} \frac{(i\boldsymbol{\lambda} \cdot \mathbf{X})^2}{k!}\right) = \mathbb{E}\left(1 + \frac{i\boldsymbol{\lambda} \cdot \mathbf{X}}{1!} - \frac{(\boldsymbol{\lambda} \cdot \mathbf{X})^2}{2!} + \cdots\right)$$

$$= \mathbb{E}\left(1 + i\sum_{j=1}^{m} \lambda_j X_j - \frac{1}{2}\sum_{j=1}^{m}\sum_{l=1}^{m} \lambda_j X_j \lambda_l X_l + \cdots\right)$$

$$= \mathbb{E}\left(1 + i\sum_{j=1}^{m} \lambda_j X_j - \frac{1}{2}\sum_{j=1}^{m}\sum_{l=1}^{m} \lambda_j \lambda_l X_j X_l + \cdots\right)$$

$$= 1 + i\sum_{j=1}^{m} \lambda_j \mathbb{E}X_j - \frac{1}{2}\sum_{j=1}^{m}\sum_{l=1}^{m} \lambda_j \lambda_l \mathbb{E}(X_j X_l) + \cdots$$

$$= 1 - \frac{1}{2}\sum_{j=1}^{m}\sum_{l=1}^{m} \lambda_j \lambda_l \mathbb{E}(X_j X_l) + \cdots.$$

If we make the further simplifying assumption that $\mathbb{E}(X_j X_l) = 0$ for $j \neq l$, $\mathbb{E}(X_j^2) = 1$ for $1 \leqslant j \leqslant m$ we obtain the simpler expression

$$\mathbb{E}\exp i\boldsymbol{\lambda} \cdot \mathbf{X} = 1 - \tfrac{1}{2}\boldsymbol{\lambda} \cdot \boldsymbol{\lambda} + \cdots.$$

413

Following the argument of Chapter 69 and setting

$$\mathbf{S}(n) = (\mathbf{X}(1) + \mathbf{X}(2) + \cdots + \mathbf{X}(n))/n^{\frac{1}{2}},$$

we obtain (with luck),

$$\mathbb{E} \exp i\boldsymbol{\lambda} \cdot \mathbf{S}(n) = (1 - \boldsymbol{\lambda} \cdot \boldsymbol{\lambda}/2n + \cdots)^n \to \exp(-\boldsymbol{\lambda} \cdot \boldsymbol{\lambda}/2).$$

Now a simple calculation gives the following result.

Lemma 50.2*. *Let* $g: \mathbb{R}^m \to \mathbb{R}$ *be given by* $g(\mathbf{t}) = (1/2\pi)^{m/2} \exp(-\mathbf{t} \cdot \mathbf{t}/2)$. *Then* $\hat{g}(\boldsymbol{\lambda})$
$= \exp(-\boldsymbol{\lambda} \cdot \boldsymbol{\lambda}/2)$.
Proof. Using Lemma 50.2

$$\hat{g}(\boldsymbol{\lambda}) = \int_{-\infty}^{\infty} \int_{-\infty}^{\infty} \cdots \int_{-\infty}^{\infty} \left(\frac{1}{2\pi}\right)^{m/2} \exp\left(-\sum_{j=1}^{m} t_j t_j/2\right) \exp\left(-i\sum_{j=1}^{m} \lambda_j t_j\right) dt_1 dt_2 \dots dt_m$$

$$= \int_{-\infty}^{\infty} \int_{-\infty}^{\infty} \cdots \int_{-\infty}^{\infty} \prod_{j=1}^{m} \frac{1}{(2\pi)^{\frac{1}{2}}} \exp(-t_j^2/2) \exp(-it_j \lambda_j) dt_1 dt_2 \dots dt_m$$

$$= \prod_{j=1}^{m} \frac{1}{(2\pi)^{\frac{1}{2}}} \int_{-\infty}^{\infty} \exp(-t_j^2/2) \exp(-it_j \lambda_j) dt_j$$

$$= \prod_{j=1}^{m} \exp(-\lambda_j^2/2) = \exp(-\boldsymbol{\lambda} \cdot \boldsymbol{\lambda}/2). \qquad \blacksquare$$

Thus if $\mathbf{Y}$ is a random variable taking values in $\mathbb{R}^m$ with probability density g (observe that $g(\mathbf{t}) \geqslant 0$ and $\int g(\mathbf{t}) d\mathbf{t} = \hat{g}(\mathbf{0}) = 1$) we have (with luck)

$$\mathbb{E} \exp i\boldsymbol{\lambda} \cdot \mathbf{S}(n) \to \mathbb{E} \exp i\boldsymbol{\lambda} \cdot \mathbf{Y}$$

and $\mathbf{S}(n) \to \mathbf{Y}$ in some sense.

We are thus in possession of a conjecture and a suggested method of proof.

Conjecture 80.1. *If* $\mathbf{X}, \mathbf{X}(1), \mathbf{X}(2), \dots$ *are identically distributed* $\mathbb{R}^m$ *valued random variables with* $\mathbb{E}\mathbf{X} = 0$, $\mathbb{E}X_j X_l = 0$ *for* $j \neq l$, $\mathbb{E}X_j^2 = 1$ *then*

$$Pr\left(\frac{\mathbf{X}(1) + \mathbf{X}(2) + \cdots + \mathbf{X}(n)}{n^{\frac{1}{2}}} \in I\right) \to \left(\frac{1}{2\pi}\right)^{m/2} \int_I \exp(-\mathbf{t} \cdot \mathbf{t}/2) dV(\mathbf{t}),$$

where $\quad I = [a_1, b_1] \times [a_2, b_2] \times \cdots \times [a_m, b_m].$

Exactly as in Chapter 70 the reader will now be able to prove the following version of the conjecture.

Lemma 70.3*. *Let* $f: \mathbb{R}^m \to \mathbb{R}^+$ *be a continuous bounded function such that*

$$\int_{\mathbb{R}^m} f(\mathbf{t}) d\mathbf{t} = 1, \int_{\mathbb{R}^m} f(\mathbf{t}) t_j d\mathbf{t} = 0 \quad for \quad 1 \leqslant j \leqslant m,$$

$$\int_{\mathbb{R}^m} f(\mathbf{t}) t_j t_l dV(\mathbf{t}) = 0 \quad for \quad 1 \leqslant j, l \leqslant m, l \neq j,$$

and
$$\int_{\mathbb{R}^m} f(\mathbf{t}) t_j^2 dV(\mathbf{t}) = 1 \quad for \quad 1 \leqslant j \leqslant m.$$

Then writing $f_1 = f, f_2 = f * f, \ldots, f_n = f * f_{n-1}$ *and* $g_n(\mathbf{t}) = n^{\frac{1}{2}} f_n(n^{\frac{1}{2}} \mathbf{t})$ *we have*

$$\int_I g_n(\mathbf{t}) dt \to \left(\frac{1}{2\pi}\right)^{m/2} \int_I \exp(-\mathbf{t} \cdot \mathbf{t}/2) dV(\mathbf{t}) \quad as \ n \to \infty \ for \ all$$

$$I = [a_1, b_1] \times [a_2, b_2] \times \cdots \times [a_m, b_m].$$

Equivalently, we have the following probabilistic version.

Theorem 70.1*. *Suppose the hypotheses of Conjecture 80.1 hold and in addition X has bounded continuous probability density. Then the conclusions of the conjecture hold.*

The discussion of the next few chapters centres round a version of conjecture 80.1 corresponding to Theorem 71.1 rather than Theorem 70.1.

Theorem 71.1*. *Let* $\mathbf{X}, \mathbf{X}(1), \mathbf{X}(2), \ldots$ *be identically distributed* $\mathbb{R}^m$ *valued random variables taking only a finite number of values. Then if*

$$\mathbb{E}\mathbf{X} = \mathbf{0} \ and \ \mathbb{E}(X_j X_l) = 0 \ for \ j \neq l \ and \ \mathbb{E}(X_j^2) = 1 \ for \ all \ 1 \leqslant j \leqslant m,$$

$$Pr\left(\frac{\mathbf{X}(1) + \mathbf{X}(2) + \cdots + \mathbf{X}(n)}{n^{\frac{1}{2}}} \in I\right) \to \left(\frac{1}{2\pi}\right)^{m/2} \int_I \exp(-\mathbf{t} \cdot \mathbf{t}/2) dV(\mathbf{t}).$$

Proof. As with the previous results this is so close to the one dimensional version that it is left to the reader. ∎

We are particularly interested in the distance $\|\mathbf{S}(n)\|$ of the vector $\mathbf{S}(n) = (\mathbf{X}(1) + \mathbf{X}(2) + \cdots + \mathbf{X}(n))/n^{\frac{1}{2}}$ from the origin.

Lemma 80.2. *Under the conditions of Theorem 71.1* or of Theorem 70.1**

$$Pr(a \leqslant \|\mathbf{S}(n)\|^2 < b) \to F_m(b) - F_m(a) \quad [b \geqslant a \geqslant 0],$$

where
$$F_m(u) = \left(\frac{1}{2\pi}\right)^{m/2} \int_{\|\mathbf{t}\|^2 < u} \exp(-\|\mathbf{t}\|^2/2) dV(\mathbf{t}) \quad [u \geqslant 0].$$

Remark. We consider $\|\mathbf{S}(n)\|^2$ rather than $\|\mathbf{S}(n)\|$ to fit in with normal statistical practice. The function F_m is called 'the distribution function of the χ_m^2 statistic'.

Proof. Let $J(u) = \{\mathbf{t}: \|\mathbf{t}\|^2 < u\}$. It is easy to deduce from the conclusions of Theorem

71.1* (or 70.1*) that

$$Pr\left(\frac{\mathbf{X}(1)+\mathbf{X}(2)+\cdots+\mathbf{X}(n)}{n^{\frac{1}{2}}}\in J(u)\right)\to\left(\frac{1}{2\pi}\right)^{m/2}\int_{J(u)}\exp(-\|\mathbf{t}\|^2/2)dV(\mathbf{t})$$

(see Step 4 of the proof of Lemma 70.3 if necessary). But $\|\mathbf{S}(n)\|^2<u$ if and only if $(\mathbf{X}(1)+\mathbf{X}(2)+\cdots+\mathbf{X}(n))/n^{\frac{1}{2}}\in J(u)$ so $Pr(\|\mathbf{S}(n)\|^2<u)\to F_m(u)$ and the full result follows. ■

The condition $\mathbb{E}(X_jX_l)=0$ for $l\neq j$ appears very restrictive. This is not the case, for reasons which we now sketch. Our discussion is based on the fact that if λ is an arbitrary vector in $\mathbb{R}^n$ then

$$\sum_{j=1}^m\sum_{l=1}^m\mathbb{E}(X_jX_l)\lambda_j\lambda_l=\mathbb{E}((\lambda\cdot\mathbf{X})^2)\geqslant0.$$

Thus, if we write $A_{jl}=\mathbb{E}X_jX_l$, we see that, in the language of linear algebra, $\mathbf{A}=(A_{jl})$ is a symmetric (since $A_{jl}=\mathbb{E}(X_jX_l)=\mathbb{E}(X_lX_j)=A_{lj}$) positive definite matrix.

Now, geometrically, we know that if

$$\sum_{j=1}^m\sum_{l=1}^m a_{jl}x_jx_l=1$$

is the equation of an ellipsoid Λ with respect to a system S of orthogonal coordinates, then, by rotating the coordinate axes so that they lie along the axes of symmetry of the ellipsoid, we can obtain a new system S' of orthogonal coordinates such that Λ has the equation

$$\sum_{j=1}^m a'_jx'^2_j=1$$

in the new system. (Thus if $\mathbf{x}$ has coordinates $(x_1,x_2,\ldots,x_n)$ with respect to S and coordinates $(x'_1,x'_2,\ldots,x'_n)$ with respect to S',

$$\sum_{j=1}^m\sum_{l=1}^m a_{jl}x_jx_l=1\Leftrightarrow\sum_{j=1}^m a'_jx'^2_j=1.)$$

The corresponding algebraic result (which we quote) says that we can find a rotation of our old system S of orthogonal axes to a new system S' such that if λ has coordinates $(\lambda'_1,\lambda'_2,\ldots,\lambda'_m)$ in the new system then (writing $A_{jl}=\mathbb{E}X_jX_l$ as before)

$$\sum_{j=1}^m\sum_{l=1}^m A_{jl}\lambda_j\lambda_l=\sum_{j=1}^m A'_j\lambda'^2_j$$

for some $A'_1,A'_2,\ldots,A'_m\geqslant0$. Since rotation of axes leaves inner products unaltered, it follows that, if $\mathbf{X}$ has coordinates $(X'_1,X'_2,\ldots,X'_m)$ with respect to the new

system S', then

$$\sum_{j=1}^{m} A'_j \lambda_j'^2 = \sum_{j=1}^{m} \sum_{l=1}^{m} A_{jl} \lambda_j \lambda_l = \mathbb{E}(\lambda \cdot \mathbf{X})^2 = \mathbb{E}(\lambda' \cdot \mathbf{X}')^2$$

$$= \sum_{j=1}^{m} \sum_{l=1}^{m} \mathbb{E}(X'_j X'_l) \lambda_j' \lambda_l' \quad \text{for all} \quad \lambda,$$

and so $\mathbb{E}(X'_j X'_l) = 0$ for $j \neq l$, $\mathbb{E}(X_j'^2) = A'_j$ for $1 \leqslant j \leqslant n$. A coordinate rotation has removed the problem.

If all the A'_j are non-zero we now simply replace X'_j by $X'_j / A_j'^{\frac{1}{2}}$ to return to the conditions of Theorems 71.1* or 70.1*. If $A'_m = 0$, say, then this means that $\mathbf{X}$ and so $\mathbf{S}(n)$ is constrained to lie in the $m - 1$ dimensional subspace $\Gamma = \{\mathbf{x} : x'_m = 0\}$. In general, if $A'_m = A'_{m-1} = \cdots = A'_{p+1} = 0$ then $\mathbf{X}$ and $\mathbf{S}(n)$ lie in a p dimensional subspace Γ_0 and, replacing X'_j by $X'_j / A_j'^{\frac{1}{2}} [1 \leqslant j \leqslant p]$, we obtain the conditions of Theorems 71.1* or 70.1* with $\mathbb{R}^m$ replaced by Γ_0 (and m replaced by p).

81

A CHI SQUARED TEST

My pocket calculator has a function 'Ran' which, it is claimed, produces a random integer between 0 and 999, each integer value being equally likely. Leaving aside philosophical problems about the nature of randomness, how should I test whether the performance of the calculator is satisfactory? One way is to press the 'Ran' button 100 times and note the number n_j of times the result falls in the class

$$A_j = \{r : 100(j-1) \leqslant r \leqslant 100j - 1\} \quad [1 \leqslant j \leqslant 10].$$

I can clearly decide in advance that if I get the result of Table 81.1 (i)

Table 81.1 *(i)*

	A_1	A_2	A_3	A_4	A_5	A_6	A_7	A_8	A_9	A_{10}
n_j	15	3	12	2	17	3	28	2	8	10

then it is unlikely that the machine is working properly. Equally I can decide that the result of Table 81.1 (ii)

Table 81.1 *(ii)*

	A_1	A_2	A_3	A_4	A_5	A_6	A_7	A_8	A_9	A_{10}
n_j	10	9	11	10	10	10	10	10	9	11

would be 'too good to be true'. But how can I decide whether the less extreme case of Table 81.1 (iii)

Table 81.1 *(iii)*

	A_1	A_2	A_3	A_4	A_5	A_6	A_7	A_8	A_9	A_{10}
n_j	7	8	11	12	10	8	9	12	12	11

should be taken as a worrying sign or not?

Problem 81.1. *In a sequence of trials the result of each trial falls into one of m disjoint classes $A_1, A_2, \ldots, A_m$. It is claimed that the trials are independent and the probability of the result of any one trial lying in A_j is $p_j > 0$. In an experiment involving n trials it turns out that $N(j)$ trials give rise to a result in $A_j[1 \leqslant j \leqslant m]$. Is this result a good reason for rejecting the hypothesis?*

Observe that we cannot expect a unique answer to this problem and that our judgement of what represents a good answer may depend on what specific version of the problem we have in mind. However the following criteria may be considered appropriate.

(i) *The method of solution should be easy to apply.*

(ii) *The method should not ignore certain pieces of data.*

(Thus the solution 'reject the hypothesis if $|N(1) - p_1 n_1| \geqslant An^{\frac{1}{2}}(p_1(1 - p_1))^{\frac{1}{2}}$' would not be satisfactory.)

Although probability and its relation to the real world had been a subject of thought and research for such men as Laplace and Legendre for over a century, the first person to pose this problem and to obtain a satisfactory solution was Karl Pearson 'the father of modern statistics'. The character of this remarkable man is so well encapsulated in a paragraph of H.M. Walker (*International Encyclopaedia of Statistics*, Macmillan 1978) that I cannot resist its quotation.

[Pearson was asked]... what was the first thing he could remember. He recalled that it was sitting in a high chair and sucking his thumb. Someone told him to stop sucking it and added that unless he did so, the thumb would wither away. He put his two thumbs together and looked at them a long time. 'They look alike to me', he said to himself. 'I can't see that the thumb I suck is any smaller than the other. I wonder if she could be lying to me'. Here in this simple anecdote we have rejection of constituted authority, appeal to empirical evidence, faith in his own interpretation of the meaning of observed data, and, finally, imputation of moral obliquity to a person whose judgement differed from his own.

Let us return to Problem 81.1. If we wish to use the results of the preceding chapter it is natural to consider vectors $\mathbf{X}(1), \mathbf{X}(2), \ldots, \mathbf{X}(n)$ in $\mathbb{R}^m$ given by

$$\mathbf{X}(k) = (\alpha_1(Y_1(k) - \mu_1), \alpha_2(Y_2(k) - \mu_2), \ldots, \alpha_m(Y_m(k) - \mu_k)),$$

where $Y_j(k) = 1$ if the result of the kth trial lies in A_j and $Y_j(k) = 0$ otherwise. We observe that, if as before, we set

$$\mathbf{S}(n) = (\mathbf{X}(1) + \mathbf{X}(2) + \cdots + \mathbf{X}(n))/n^{\frac{1}{2}},$$

then, also,

$$\mathbf{S}(n) = n^{-\frac{1}{2}}(\alpha_1(N(1) - n\mu_1), \alpha_2(N(2) - n\mu_2), \ldots, \alpha_m(N(m) - n\mu_m)).$$

Thus $\mathbf{S}(n)$ can be computed from $N(1), N(2), \ldots, N(m)$ and vice versa (at least if $\alpha_1, \alpha_2, \ldots, \alpha_m > 0$).

How do we choose the constants $\alpha_1, \alpha_2, \ldots, \alpha_m$ and $\mu_1, \mu_2, \ldots, \mu_m$? Almost certainly, we want $\mathbb{E}X(k) = 0$ and so we must take $\mu_j = \mathbb{E}Y_j(k) = p_j$. The choice of $\alpha_1, \alpha_2, \ldots, \alpha_m$ does not affect the answer but does affect the amount of algebraic manipulation involved in obtaining it. More specifically, whatever choice of $\alpha_1, \alpha_2, \ldots, \alpha_m > 0$ we make, the method described in the last four paragraphs of Chapter 80 will work, but only at the cost of increasingly unpleasant algebra. However, careful study of the cases $m = 2$ and $m = 3$ should reveal the 'correct' choice of $\alpha_1, \alpha_2, \ldots, \alpha_m$. In fact if we take $\alpha_j = p_j^{-\frac{1}{2}}$ the algebra becomes practically transparent.

Lemma 81.2. *Let* $Y_1, Y_2, \ldots, Y_m$ *be random variables taking values 0 or 1 and such that*

$$Pr(Y_j = 1, Y_l = 0 \text{ for } l \neq j) = p_j,$$

where $p_1, p_2, \ldots, p_m > 0$ *and* $\sum_{j=1}^{m} p_j = 1$. *Set*

$$\mathbf{X} = (X_1, X_2, \ldots, X_m)$$
$$= (p_1^{-\frac{1}{2}}(Y_1 - p_1), p_2^{-\frac{1}{2}}(Y_2 - p_2), \ldots, p_m^{-\frac{1}{2}}(Y_m - p_m)).$$

Then

(i) $\mathbb{E}X_j = 0$.

(ii) $\mathbb{E}X_j^2 = 1 - p_j$.

(iii) $\mathbb{E}X_j X_l = -p_j^{\frac{1}{2}} p_l^{\frac{1}{2}}$ $\quad [j \neq l]$,

(iv) $\mathbf{X} \in \{\mathbf{x} : p_1^{\frac{1}{2}} x_1 + p_2^{\frac{1}{2}} x_2 + \cdots + p_m^{\frac{1}{2}} x_m = 0\}$.

Proof. (i) $\mathbb{E}X_j = p_j^{-\frac{1}{2}}(p_j(1 - p_j) + (1 - p_j)(-p_j)) = 0$.

(ii) $\mathbb{E}X_j^2 = p_j^{-1}(p_j(1 - p_j)^2 + (1 - p_j)p_j^2) = 1 - p_j$.

(iii) $\mathbb{E}X_j X_l = p_j^{-\frac{1}{2}} p_l^{-\frac{1}{2}}(p_j(1 - p_j)(-p_l) + p_l(-p_j)(1 - p_l) + (1 - p_j - p_l)(-p_j)(-p_l))$

$\qquad = p_j^{-\frac{1}{2}} p_l^{-\frac{1}{2}}(p_j p_l(-(2 - p_j - p_l) + (1 - p_j - p_l))) = -p_j^{\frac{1}{2}} p_l^{\frac{1}{2}}$.

(iv) Immediate. ∎

We see that $\mathbb{E}X_j X_l \neq 0$ and so the conditions of Theorem 71.1* (and so of Lemma 80.2) do not hold. Indeed it is clear that the required conditions cannot hold since $\mathbf{X}$ and so $\mathbf{S}(n)$ is restricted to lie in the $m - 1$ dimensional subspace $\Gamma = \{\mathbf{x} : p_1^{\frac{1}{2}} x_1 + p_2^{\frac{1}{2}} x_2 + \cdots + p_m^{\frac{1}{2}} x_m = 0\}$. Thus the discussion at the end of Chapter 80 (which the reader is invited to reread before proceeding further) suggests the use of a new coordinate system S' with the first $m - 1$ coordinate axes lying in Γ and the mth coordinate axis perpendicular to Γ. We would expect this to be followed by some heavy algebra and the choice of yet another coordinate system S'' (with the same mth coordinate axis). However, the 'correct' choice of α_j above eliminates all this and the required result falls out at once.

Lemma 81.3. *Let* $\mathbf{X}$ *be chosen as in Lemma 81.2. Choose a new system* S' *of Cartesian coordinates such that*

(a) S' *is a rotation of the old system of Lemma 81.2,*

(b) *the mth coordinate axis is perpendicular to* $\Gamma = \{\mathbf{x}:p_1^{\frac{1}{2}}x_1 + p_2^{\frac{1}{2}}x_2 + \cdots + p_m^{\frac{1}{2}}x_m = 0,$
where $\mathbf{x} = (x_1, x_2, \ldots, x_m)$ *in the old coordinate axes*}.

Then, in the new system S', $\mathbf{X} = (X_1', X_2', \ldots, X_{m-1}', 0)$, where the X_j' are real valued variables with

(i) $\mathbb{E}X_j' = 0 \quad [1 \leqslant j \leqslant m-1]$.
(ii) $\mathbb{E}X_j'^2 = 1 \quad [1 \leqslant j \leqslant m-1]$.
(iii) $\mathbb{E}X_j'X_l' = 0 \quad [1 \leqslant j, l \leqslant m-1, j \neq l]$.

Proof. Let $\mathbf{X} = (X_1', X_2', \ldots, X_m')$. Since $\mathbf{X} \in \Gamma$, condition (b) automatically gives $X_m' = 0$. Now let $\boldsymbol{\lambda}$ be a vector with coordinates $(\lambda_1, \lambda_2, \ldots, \lambda_m)$ in the old system and $(\lambda_1', \lambda_2', \ldots, \lambda_m')$ in the new. Observing that rotation of axes leaves lengths (and so scalar products) unchanged we have

$$\sum_{j=1}^m \sum_{l=1}^m \lambda_j' \lambda_l' \mathbb{E}(X_j' X_l')$$

$$= \mathbb{E}\left(\left(\sum_{j=1}^m \lambda_j' X_j'\right)\left(\sum_{l=1}^m \lambda_l' X_l'\right)\right) = \mathbb{E}((\boldsymbol{\lambda} \cdot \mathbf{X})^2) = \sum_{j=1}^m \sum_{l=1}^m \lambda_j \lambda_l \mathbb{E}(X_j X_l)$$

$$= \sum_{j=1}^m (1 - p_j)\lambda_j^2 - \sum\sum_{j \neq l} p_j^{\frac{1}{2}} p_l^{\frac{1}{2}} \lambda_j \lambda_l = \sum_{j=1}^m \lambda_j^2 - \sum_{j=1}^m \sum_{l=1}^m p_j^{\frac{1}{2}} p_l^{\frac{1}{2}} \lambda_j \lambda_l$$

$$= \|\boldsymbol{\lambda}\|^2 - \left(\sum_{j=1}^m p_j^{\frac{1}{2}} \lambda_j\right)^2 = \|\boldsymbol{\lambda}\|^2 - \lambda_m'^2 = \sum_{j=1}^m \lambda_j'^2 - \lambda_m'^2 = \sum_{j=1}^{m-1} \lambda_j'^2,$$

since, by (b) and (a), $\lambda_m' = \sum_{j=1}^m p_j^{\frac{1}{2}} \lambda_j$. Since $\boldsymbol{\lambda}$ was arbitrary we have, on comparing coefficients,

$$\mathbb{E}X_j'^2 = 1 \quad [1 \leqslant j \leqslant m-1],$$
$$\mathbb{E}X_j'X_l' = 0 \quad [1 \leqslant j, l \leqslant m-1, j \neq l].$$

The fact that $\mathbb{E}X_j' = 0$ follows automatically from the fact that $\mathbb{E}\mathbf{X} = 0$. ∎

We are now in a position to use Lemma 80.2.

Theorem 81.4. *In a sequence of trials the result of each trial falls into one of m disjoint classes* $A_1, A_2, \ldots, A_m$. *The trials are independent and the probability of the result of any one trial lying in* A_j *is* p_j *where* $p_j > 0 [1 \leqslant j \leqslant m]$. *If in a sequence of n trials* $N(j, n)$ *trials give rise to a result in* A_j *we write*

$$s(n)^2 = \sum_{k=1}^m \frac{(N(k, n) - p_k n)^2}{p_k n}.$$

Under these conditions

$$Pr(a \leqslant s(n)^2 < b) \to F_{m-1}(b) - F_{m-1}(a) \quad \text{as } n \to \infty$$

where $F_{m-1}(u)$ is the $m-1$ dimensional integral

$$F_{m-1}(u) = \left(\frac{1}{2\pi}\right)^{m-1} \underbrace{\int\int \cdots \int}_{t_1^2 + t_2^2 + \cdots + t_{m-1}^2 < u}$$
$$\times \exp\left(-(t_1^2 + t_2^2 + \cdots + t_{m-1}^2)/2\right) dt_1 \, dt_2 \ldots dt_{m-1}$$
$$= \left(\frac{1}{2\pi}\right)^{m-1} \int_{\|\mathbf{t}\|^2 < u} \exp\left(-\|\mathbf{t}\|^2/2\right) dV(\mathbf{t}) \quad [u \geqslant 0].$$

Proof. As above, take $Y_j(k) = 1$ if the result of the kth trial lies in A_j, $Y_j(k) = 0$ otherwise and set

$$\mathbf{X}(k) = (p_1^{-\frac{1}{2}}(Y_1(k) - p_1), p_2^{-\frac{1}{2}}(Y_2(k) - p_2), \ldots, p_m^{-\frac{1}{2}}(Y_m(k) - p_m)).$$

Clearly, $\mathbf{X}(1), \mathbf{X}(2), \ldots$ are independent random vectors each with the same distribution as the random vector $\mathbf{X}$ of Lemma 81.2.

Now, choosing the axes of Lemma 81.3, we see that $\mathbf{X}$ is a random vector in the $m-1$ dimensional subspace Γ_k with coordinates in that subspace given by $\mathbf{X} = (X_1', X_2', \ldots, X_{m-1}')$ where $\mathbb{E}\mathbf{X} = 0$, $\mathbb{E}X_j'X_l' = 0$ for $j \neq l$ and $\mathbb{E}X_j'^2 = 1$ for all $1 \leqslant j \leqslant m-1$. Since $\mathbf{X}$ takes only a finite number of values we can apply Theorem 71.1* and Lemma 81.1 with $\mathbb{R}^m$ replaced by Γ and m replaced by $m-1$.

In particular, if $\mathbf{S}(n) = (\mathbf{X}(1) + \mathbf{X}(2) + \cdots + \mathbf{X}(n))/n^{\frac{1}{2}}$ we see that, if $b \geqslant a \geqslant 0$,

$$Pr(a \leqslant \|\mathbf{S}(n)\|^2 < b) \to F_{m-1}(b) - F_{m-1}(a) \quad \text{as } n \to \infty.$$

But

$$\|S(n)\|^2 = \frac{1}{n} \sum_{j=1}^{m} \left(\sum_{k=1}^{n} X_j(k)\right)^2 = \frac{1}{n} \sum_{j=1}^{m} \left(\sum_{k=1}^{n} \left(\frac{Y_j(k) - p_j}{p_j^{\frac{1}{2}}}\right)\right)^2$$

$$= \sum_{j=1}^{m} \frac{1}{np_j} (N(j,n) - np_j)^2 = s(n)^2,$$

so we are done. ∎

How does Theorem 81.4 help resolve Problem 81.1?

Possible solution of Problem 81.1 (A chi squared test). Suppose we wish to test the hypothesis advanced in Problem 81.1. We choose in advance a small number $\varepsilon > 0$ (often $\varepsilon = 0{\cdot}05$ or $\varepsilon = 0{\cdot}01$ but we need not be bound by tradition) and numbers $0 \leqslant a < b$ (sometimes $a = 0$ or, with obvious conventions, $b = \infty$) such that

$$1 - (F_{m-1}(b) - F_{m-1}(a)) \approx \varepsilon.$$

We then perform n trials where n is sufficiently large to ensure

$$1 - Pr(a \leqslant s(n)^2 < b) \text{ very close to } 1 - (F_{m-1}(b) - F_{m-1}(a)),$$

i.e. sufficiently large to ensure

$$1 - Pr(a \leqslant s(n)^2 < b) \text{ very close to } \varepsilon.$$

We then look at the actual value $s_0(n)$ of $s(n)$ for this particular sequence of trials.

If $s_0(n) < a$ or $s_0(n) > b$ then we are faced with an outcome which was unlikely (i.e. had probability less than about ε) under the initial hypothesis and we would be inclined to reject the hypothesis. The smaller ε is, the more confident we would be in rejecting the hypothesis.

Remark 1. Our description is not complete since we have not shown how large n must be as a function of $p_1, p_2, \ldots, p_m$ and ε in order to be 'sufficiently large'. (Although

$$Pr(a \leqslant s(n)^2 < b) \to F_{m-1}(b) - F_{m-1}(a)$$

whatever the (non zero) values of $p_1, p_2, \ldots, p_m$ the rate of convergence is not uniform and depends on $p_1, p_2, \ldots, p_m$.) However, further thought and numerical computation gives, as a rule of thumb, that, for $\varepsilon = 0.05$, it suffices to have $np_j \geqslant 5$ for each $m \geqslant j \geqslant 1$.

Remark 2. We reject the hypothesis if $s_0(n)^2$ is either too large or too small. In the first case we reject it because the result is not good enough to support the hypothesis. In the second case we reject it because it is too good.

Remark 3. The function F_m is tabulated as the 'χ_m^2 distribution' or the 'χ^2 distribution with m degrees of freedom'. The work of tabulation is much simplified by the observation that, converting to spherical polar coordinates,

$$F_m(u) = \alpha_m \int_0^u r^m \exp(-r^2/2)\, dr,$$

where α_m is the 'surface area' of the m dimensional unit sphere. Since integration by parts gives (for $m \geqslant 2$)

$$\int_0^u r^m \exp(-r^2/2)\, dr = [-r^{m-1} \exp(-r^2/2)]_0^u + (m-1) \int_0^u r^{m-2} \exp(-r^2/2)\, dr$$

$$= (m-1) \int_0^u r^{m-2} \exp(-r^2/2)\, dr - u^{m-1} \exp(-u^2/2),$$

it will be seen that the work of constructing these particular tables is not outside the range of one man and an old fashioned hand calculator. (The reader will recognise that we have graphed the density F_1' in Figure 14.1.)

Now let me apply the χ^2 test to my pocket calculator with $m = 10$ and $A_1, A_2, \ldots, A_{10}$ as described in the first paragraph. Since I wish to test for good behaviour I take $p_1 = p_2 = \cdots = p_{10} = \frac{1}{10}$. Consulting a book of statistical tables (in this case Biometrika) I find that

$$F_9(2.70) \approx 0.025,$$
$$F_9(19.0) \approx 0.975,$$

and so I take $a = 2.7$, $b = 19$.

I now press the 'Ran' button and after 100 goes I obtain the result of Table 81.1 (iv).

Table 81.1 (iv)

	A$_1$	A$_2$	A$_3$	A$_4$	A$_5$	A$_6$	A$_7$	A$_8$	A$_9$	A$_{10}$
n_j	12	6	8	8	14	9	6	15	13	9

Now

$$s_0^2(100) = \sum_{j=1}^{10} \frac{(n_j - 10)^2}{10} = 9 \cdot 6 \in [a, b),$$

so I cannot criticise the random number generator of my calculator by using its performance in this test.

The knowledgeable reader will have observed that I have not considered the more general chi squared test in which parameters are estimated from the data (as occurs, for example, when we consider 2 × 2 contingency tables). My excuse is that although no new mathematical *tools* are needed, the new probabilistic *ideas* involved are not trivial.

As evidence for this claim I can point out that in Pearson's original paper, after dealing correctly with the problem considered here, he produces an approach to the general case which is both convincing and wrong. For the first 20 years of its existence the Pearson chi squared test must have been wrongly applied in most of the cases when it was used. The correct method in the general case was found by Fisher (and is now taught by rote to 16-year olds). Where Pearson's intuition failed him, we should do well to be cautious.

82

HALDANE ON FRAUD

J.B.S. Haldane was one of those intellectual aristocratic radicals who have enlivened British life and thought for the last two hundred years. He pursued three distinct but simultaneous careers as an experimental physiologist (like his father), as a mathematical geneticist and statistician (he, Fisher and Wright were the first people to give extensive mathematical treatments of Darwinian evolution from the Mendelian viewpoint) and as a scientific journalist. One of the high spots of his stormy public life was his dismissal from Cambridge at the instigation of the 'Sex Viri' ('Six Men', a kind of disciplinary committee) on moral grounds (he had been a co-respondent in a divorce suit). He fought back and gained reinstatement (and the 'Six Men' added one to their number to become the 'Septem Viri'). (On a more peaceful level he wrote a curious children's book *My Friend Mr Leaky* which I remember as one of my childhood favourites.) Here is an article which he wrote for *Eureka* (the Cambridge undergraduate mathematical journal) in 1941. It was entitled '*The Faking of Genetical Results*'.

My father published a number of papers on blood analysis. In the proofs of one of them the following sentence, or something very like it, occurred: 'Unless the blood is very thoroughly faked, it will be found that duplicate determinations rarely agree.' Every biochemist will sympathise with this opinion. I may add that the verb 'to lake', when applied to blood, means to break up the corpuscles so that it becomes transparent.

In genetical work also, duplicates rarely agree unless they are faked. Thus I may mate two brother black mice, both sons of a black father and a white mother, with two white sisters, and one will beget 10 black and 15 white young; the other 15 black and 10 white. To the ingenuous biologist this appears to be a bad agreement. A mathematician will tell him that where the same ratio of black to white is expected in each family, so large a discrepancy (though how best to compare discrepancies is not obvious) will occur in about 26 per cent of all cases. If the mathematician is a rigorist he will say the same thing a little more accurately in a great many more words.

A biologist who has no mathematical knowledge, and, what is vastly more

serious, no scientific honour, will be tempted to fake his results, and say that he got 12 black and 13 white in one family, and 13 black and 12 white in the other. The temptation is generally more subtle. In one of a number of families where equality is expected he gets 19 black and 6 white mice. It looks much more like a ratio of 3 black to 1 white. How is he to explain it? Wasn't that the cage whose door once seemed to be insecurely fastened? Perhaps the female got out for a while or some other mouse got in. Anyway he had better reject the family. The total gives a better fit to expectation if he does so, by the way. Our poor friend has forgotten the binomial theorem. A study of the expansion of $(1 + x/2)^{25}$ would have shown him that as bad a fit or worse would be obtained with a probability of 122753×2^{-23}, or $\cdot 0146$. There is nothing at all surprising in getting one family as aberrant as this in a set of 20. But he is now on a slippery slope.

He gets his Ph.D. He wants a fellowship and time is short. But he has been reading *Nature* and noticed two letters* to that journal of which I was joint author, in which I might appear to have hinted at faking by my genetical colleagues. Thoroughly alarmed, he goes to a venal mathematician. Cambridge is full of mathematicians who have been so corrupted by quantum mechanics that they use series which are clearly divergent, and not even proved to be summable. Interrupting such a one in the midst of an orgy of Bhabha and benzedrine, our villain asks for a treatise on faking. 'I am trying to reconcile Milne, Born and Dirac, not to mention some facts which don't seem to agree with any of them, or with Eddington', replies the debauchee, 'and I feel discontinuous in every interval; but here goes'.

I suppose you know the hypothesis you want to prove. It wouldn't be a bad thing to grow a few mice or flies or parrots or cucumbers or whatever you're supposed to be working on, to see if your hypothesis is anywhere near the facts. Suppose in a given series of families you expect to get four classes of hedgehogs or whatnot with frequencies p_1, p_2, p_3, p_4, and your total is S, I shouldn't advise you to say you just got Sp_1, Sp_2, Sp_3 and Sp_4, or even the nearest whole number. Here is what you'd better do. Say you got A_1, A_2, A_3 and A_4, and evaluate

$$\chi^2 = \frac{(A_1 - Sp_1)^2}{Sp_1} + \frac{(A_2 - Sp_2)^2}{Sp_2} + \cdots.$$

Your χ^2 has three degrees of freedom. That is to say you can say you got A_1 red, A_2 green and A_3 blue hedgehogs. But you will then have to say you got $S - A_1 - A_2 - A_3$ purple ones. Hence the expected value of χ^2 is 3, and its standard error $\sqrt{6}$; so choose your A's so as to give a χ^2 anywhere between 1 and 6. This is called faking of the first order. It isn't really necessary. You might have $p_1 = \frac{9}{16}$, $p_2 = p_3 = \frac{3}{16}$, $p_4 = \frac{1}{16}$ and $A_1 = 9$, $A_2 = A_3 = 3$, $A_4 = 1$. The

*U. Philip and J.B.S. Haldane (1939). *Nature*, 143, p. 334. Hans Grüneberg and J.B.S. Haldane (1940). *Nature*, 145, p. 704.

probability of getting this is

$$\frac{16!3^{24}}{9!(3!)^2 1!16^{16}},$$

which is only just under ·04. However, it looks better not to get the exact numbers expected, and if you do it on a population of hundreds or thousands you may be caught out.

Your second order faking is the same sort of thing. Supposing your total is made up of n families, and you say the rth consisted of $a_{r1}, a_{r2}, a_{r3}, a_{r4}$ members of the four classes, s_r in all, you take

$$\frac{(a_{r1} - s_r p_1)^2}{s_r p_1} + \frac{(a_{r2} - s_r p_2)^2}{s_r p_2} + \cdots$$

and sum for all values of r. Your total ought to be somewhere near $3n$. The standard error is $\sqrt{6n}$, and it's better to be too high than too low. A chap called Moewus in Berlin who counted different types of algae (or so he said), got such a magnificent agreement between observed and theoretical results, that if every member of the human race had repeated his work once a month for 10^{12} years, they might expect as good a fit on one occasion (though not with great confidence). So Moewus certainly hadn't done any second order faking. Of course I don't suggest that he did any faking at all. He may have run into one of those theoretically possible miracles, like the monkey typing out the text of Hamlet by mere luck. But I shouldn't have a miracle like that in your fellowship dissertation.

There is also a third order faking. The $4n$ different components of χ^2 should be distributed round their mean in the proper way. That is to say, not merely their mean, but their mean square, cube and so on, should be near the expected values (but not too near). But I shouldn't worry too much about the higher orders. The only examiner who is likely to spot that you haven't done them is Haldane, and he'll probably be interned as a Red before you send your thesis in. Of course you might get R.A. Fisher, which would be quite as bad. So if you are worried about it you'd better come back and see me later'.

Man is an orderly animal. He finds it very hard to imitate the disorder of nature. In fact the situation is the exact opposite of what the reader of Paley's *Evidences* might expect. But the problem is an interesting one, because it raises in a sharp and concrete way the question of what is meant by randomness, a question which, I believe, has not been fully worked out. The number of independent numerical criteria of randomness which can be applied increases with the number of observations, but much more slowly, perhaps as its logarithm. The criteria now in use have been developed to search for excessive irregularity, that is to say, unduly bad fit between observation and hypothesis. It does not follow that they are so well adapted to a search for an unduly good fit. Here, I believe, is a real problem for students of probability. Its solution might lead

to a better set of axioms for that very far from rigorous but none the less fascinating branch of mathematics.

The piece above calls forth three or four remarks by way of commentary.

Remark 1. Haldane was not interned. During the war he worked for the Admiralty on physiological problems associated with deep sea diving, conducting hair raising experiments on himself, his wife and his colleagues.

Remark 2. The last sentence of Haldane's article was out of date even when written and may reflect the rather insular and 'antitheoretical' attitude of English statisticians before 1945. Kolmogorov's axiomatisation of 1933 showed that all of classical probability theory could be put on a rigorous basis, and laid the foundation for the magnificent edifice which is modern probability theory. Of course, the problem of how and why probability theory applies to the real world remains, but this is now seen to be a philosophical rather than a mathematical question.

Remark 3. The reluctance of the scientific community to accept the possibility of fraud is illustrated by the fact that Moewus was still cited in the literature (and even spoken of as a possible Nobel prize winner) until 1953. However, no one else ever succeeded in repeating his experiments.

Remark 4. Unfortunately the statistical war against fraud is now over and the cheaters have won. The kind of tests proposed by Haldane depended on the fact that 'higher order faking' required a great deal of computational work. The invention and accessibility of the computer means that the computational work involved has ceased to be a problem for the dishonest scientist. In the physical and biological sciences the possibility that others will attempt to replicate experiments may act as a sufficient deterrent but in purely statistical subjects like sociology and experimental psychology the problems raised by potential fraud have still to be faced.

83

AN EXAMPLE OF OUTSTANDING STATISTICAL TREATMENT I

In the next three chapters we shall be looking at the work of Sir Cyril Burt, honorary fellow of Jesus College, Oxford, fellow of the British Academy and the first psychologist to be knighted. We shall use the description of his work given by his pupil, Professor Eysenck, a man who, to quote the brief biography in one of his books, 'advocates the highest degree of rigour in the design of psychological experiments and is very critical of much loose thinking current under the guise of psychology'.

In 1981 it was possible to buy two books by Eysenck entitled *Race, Intelligence and Education* (Temple Smith, London 1971) and *The Inequality of Man* (Temple Smith, London 1973) in which he describes his views on the heritability of intelligence. (I shall refer to them as [R] and [I].) Both books contain the following impressive table (Table 83.1), taken from the work of Burt and showing 'the actual distribution of IQ's found in his population for fathers in the six occupational groups he uses and

Table 83.1. *Eysenck's presentation of Burt's data*

IQ	Professional		Clerical III	Skilled IV	Semi-skilled V	Unskilled VI	Total
	Higher I	Lower II					
50–60						1	1
60–70					5	18	23
70–80				2	15	52	69
80–90			1	11	31	117	160
90–100			8	51	135	53	247
100–110			16	101	120	11	248
110–120		2	56	78	17	9	162
120–130		13	38	14	2		67
130–140	2	15	3	1			21
140 +	1	1					2
Total	3	31	122	258	325	261	1000
Mean IQ	139.7	130.6	115.9	108.2	97.8	84.9	100

Table 2. Distribution of intelligence according to occupational class: adults.

IQ	Professional		Clerical	Skilled	Semi-skilled	Unskilled	Total
	Higher I	Lower II	III	IV	V	VI	
50–60					1	1	2
60–70				1	6	15	22
70–80			3	12	23	32	70
80–90		1	8	33	55	62	159
90–100		2	21	53	99	75	250
100–110	1	6	31	70	85	54	247
110–120		12	35	59	38	16	160
120–130	1	8	18	22	13	6	68
130–140	1	2	6	7	5		21
140 +				1			1
Total	3	31	122	258	325	261	1000
Mean IQ	120.8	114.7	107.8	104.6	98.9	92.9	100

Table 3. Distribution of intelligence according to occupational class: children.

The data in these two tables were collected by Sir Cyril Burt on some 40,000 adults and children and have been reduced to a base of 1,000. (This means that the total of 3 for the higher professional category actually refers to 120 fathers.)

(The table and commentary are reproduced directly from [R].)

for children of fathers in the group' ([I] page 12). The tables are copied from the sixth reimpression of [R].

Professor Eysenck draws heavily on Burt's work because 'of the outstanding quality and statistical treatment in his studies' ([I] *p.* 120).

Burt's tables illustrate not only the fact that occupational class is very closely connected to IQ but also the phenomenon of regression towards the mean, that is the tendency of the IQ's of children to be closer to the mean of population than those of their fathers. Figure 83.1 (redrawn from [I] p. 92) shows this very clearly.

Remark. The reader who has not met 'regression towards the mean' before may find this astonishing. In fact it is a rather general statistical phenomenon which occurs when we pair the results of experiments. To take a simple example let us throw a pair of dice and record the scores X_i and Y_i of the first and second dice. Let us write $A_r = \{i : X_i = r\}$ for the set of throws in which $X_i = r$. What can we say about the expected value of Y_i given that $i \in A_r$ (i.e. $X_i = r$)? Since X_i and Y_i are independent, this will be $3\frac{1}{2}$ whatever the value of r. In other words there will be a tendency for the Y_i to be closer to the mean score than their associated X_i (see Figure 83.2).

In the same way there is a tendency for the X_i to be closer to the mean score than their associated Y_i. In a more interesting context, not only are the heights of sons closer to the mean of the population than their fathers (so that tall fathers tend to have shorter sons) but also the heights of fathers are closer to the mean of the population than their sons (so that tall sons tend to have shorter fathers).

The value of these tables, and Burt's work in general, was questioned by Kamin in

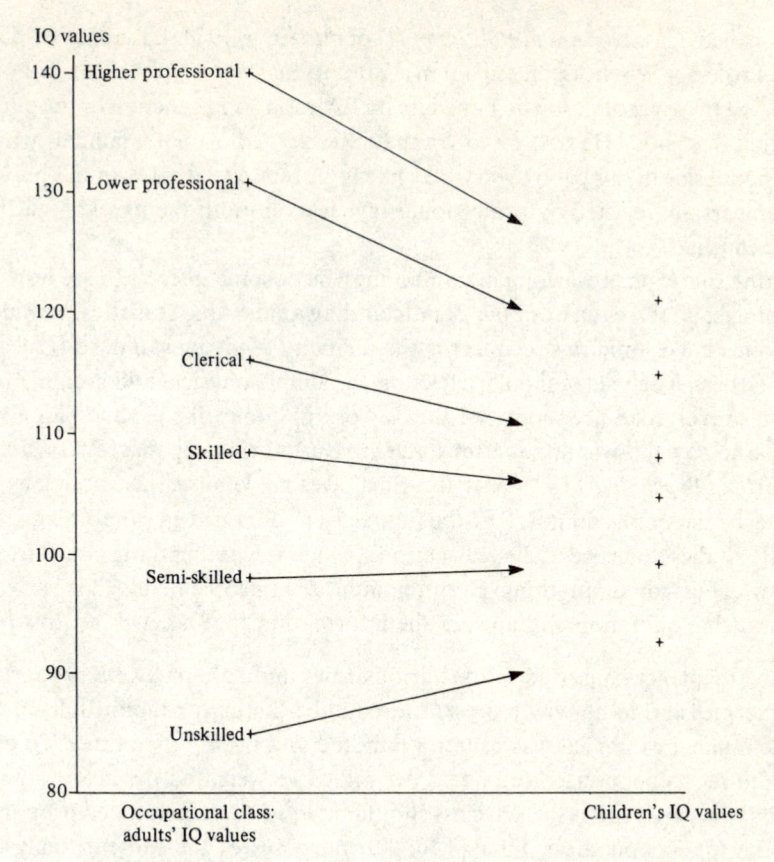

Fig. 83.1. Regression to the mean in Burt's data.

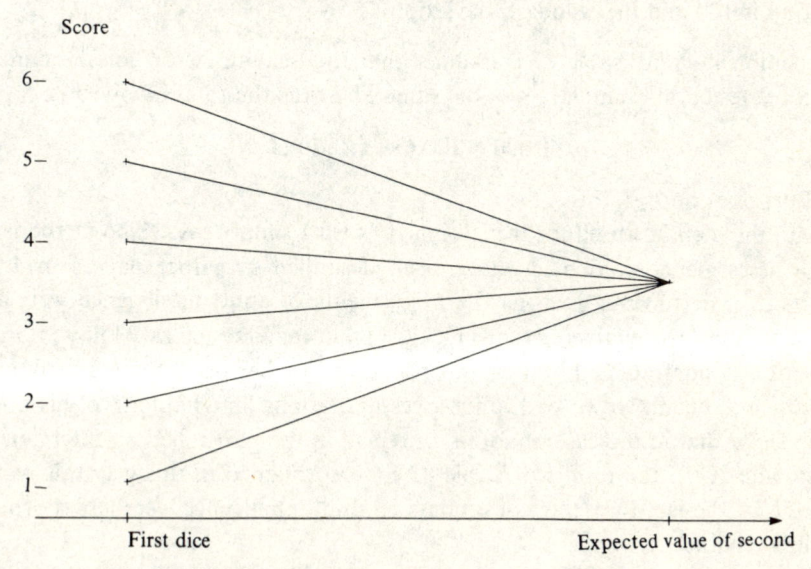

Fig. 83.2. Regression to the mean in dice throwing.

a book called *The Science and Politics of IQ* published in 1974 (Lawrence Erlbaum). But as Professor Eysenck pointed out in a letter to *Encounter* (July 1976): 'Kamin, of course, is a rat psychologist who boned up on IQ testing and genetics for the purpose of writing this book.' He goes on to say that: 'Readers who are not familiar with the professional side of this controversy may like to be reminded that Kamin's book has been universally rejected by professional reviewers in both the psychological and genetic journals.'

Bearing this explicit warning in mind it may be of some interest to see how a rat psychologist discusses an example 'of outstanding quality and statistical technique'. One point that Kamin raises concerns the difficulty of giving standard IQ tests to 40 000 fathers. The reader should reflect on the simple physical and administrative problems involved. Since reports of Burt's work in books like [I] and [R] give no details as to how this was done we must refer to Burt's original paper (*British Journal of Statistical Psychology*, **14** (1961)). But Burt gives no details either though he does indicate, as the books do not, that the figures were obtained by combining several separate studies done over a 50-year period. In fact as Kamin points out, Burt gives very few details about anything. Even the number of people involved is left vague. The following quotation contains all the information that he gives on this point.

> The frequencies inserted in the various rows and columns were proportional frequencies and in no way represent the number actually examined: from Class 1 the number examined was nearer a hundred and twenty than three. To obtain the figures to be inserted (numbers per mille) we weighted the actual numbers so that the proportions in each class should be equal to the estimated proportions for the total population. Finally, for the purposes of the present analysis we have rescaled our assessments of intelligence, so that the mean of the whole group is 100 and the standard deviation 15.

The commentary to Table 83.1 assumes that the 'scaling factor' for the number of people tested in each class was the same and thus the total involved is

$$40\,000 = 1000 \times \tfrac{120}{3} \text{ adults,}$$

but Burt does not say this.

As to the methods used for measuring IQ's Burt simply says: 'The methods by which assessments were made have been described in earlier papers or LCC reports.... For obvious reasons the assessments of adult intelligence were less thorough and less reliable.' According to Kamin (and, so far as I know, no one has seriously contradicted him on this point) the 'earlier papers' are also thin on experimental detail. However Kamin gives quotations (in Chapter 3 of his book) which show that in the case of adults Burt had in the past relied on interviewing to estimate IQ. If the results of Table 83.1 were obtained in this way this would account for the fact that 'the assessments of adult intelligence were less thorough and less reliable'.

As the reader presumably knows, the problem raised by such methods is not

simply one of inaccuracy (indeed an experienced psychologist like Burt might be quite good at this sort of guesswork) but the possibility of unconscious systematic bias (for example there might be a tendency to underestimate the IQ of unskilled workers). A similar problem for the estimates of involving the children springs from the fact that, as Kamin documents, Burt was not always careful to distinguish between crude IQ scores and 'adjusted assessments'.

Thus this particular example of 'outstanding quality and statistical treatment' breaks the following rules on the reporting of experimental data

(1) *Always give details of the experimental procedure (so that other people can redo the experiment or, in any case, form their own opinion on the magnitude of possible errors).*
(2) *As far as possible give the data in untreated form (so that other people may do their own data analysis).*
(3) *When transforming data explain the method used (so that other people may decide for themselves whether your method is appropriate).*

By failing to warn the reader of these problems those authors who used Burt's data to buttress their arguments broke a fourth rule (whose consistent application is, however, beyond the moral strength of most of us, including the present author).

(4) *Do not attempt to conceal the weaknesses of your arguments or to ignore the strong points of your opponents.*

One of the things which strikes the reader of the *Origin of Species* is the honesty with which Darwin discusses the difficulties faced by his theory and it is perhaps this honesty which explains the final persuasiveness of his arguments. In his auto-biography Darwin explained that: 'I had during many years followed a golden rule, namely that whenever a published fact, a new observation or thought came across me, which was opposed to my general results, to make a memorandum of it without fail and at once; for I had found by experience that such facts and thoughts were far more apt to escape from memory than favourable ones.'

84

AN EXAMPLE OF OUTSTANDING STATISTICAL TREATMENT II

Kamin's criticism of Sir Cyril Burt's research did not stop with the results described in the last chapter. One method of attempting to disentangle the effects of heredity and environment is to investigate identical (and so genetically identical) twins who by some accident have been brought up in different households (and so have different environments). Unfortunately even if identical twins are separated there might still be a tendency for them to be brought up in similar homes.

One answer to this problem is indicated in the next quotation.

'That this criticism is without value can be shown by looking at the actual homes of the separated identical twins studied by Sir Cyril Burt. He classified parents' or foster parents' occupation on a six point scale, ranging from (1) higher professional... to (6) unskilled. The correlation between the socio-economic status of the home in which the one twin was brought up correlated 0.03 with that in which the other twin was brought up!'   [R, p. 67]

The paper of Burt's referred to is indeed remarkable. There are in the literature reports of 122 pairs of identical twins reared apart. Of these 53 come from Burt's 1966 paper which is the only one to give the socio-economic status of the homes involved.

The sceptical reader may ask what sort of measure of socio-economic status is worth quoting to two significant figures. She may also wonder *why* there was apparently no tendency whatsoever to place identical twins in similar homes. Kamin raised these objections but his main objection was much simpler and more devastating. He reread some of Burt's earlier papers and came up with the following table.

Date of paper	1955	1958	1966
Number of identical twins reared apart	21	over 30	53
Correlation of IQ for such twins	0.771	0.771	0.771
Number of identical twins reared together	83	not given	95
Correlation of IQ for such twins	0.944	0.944	0.944

Thus although the number of identical twins reared apart more than doubled between 1955 and 1966 the IQ correlation remained unchanged to the third decimal place.

Figures like these are unlikely to be the result of actual experiment, but Burt's friends and colleagues loyally insisted that they must be the result of error rather than fraud. Professor Hernstein, who like Professor Eysenck, had made extensive and uncritical use of Burt's work, stated that 'Burt was a towering figure in 20th century psychology. I think it is a crime to cast such doubt over a man's career' (*Science*, **194**, p. 196). Professor Vernon was of the opinion that 'Burt's inconsistencies do not justify rejecting the enormous corpus of his contributions to mental measurement, though it is of course unfortunate that we cannot tell how frequently or where other errors occurred' (p. 172, P.E. Vernon *Intelligence, Heredity and Environment*, Freeman, San Francisco, 1978. Note, however, the retraction printed at the beginning of the book.)

Further fuel was added to the flames when a journalist named Gillie published an article suggesting that Burt had invented two of his collaborators Howard and Conway (so that, for example, a series of papers by 'Burt and Howard' were actually by Burt alone). Burt's defenders reacted indignantly to what Eysenck, in an article in *Encounter* (January 1977), called this 'unjustified and completely irresponsible accusation of fraud' but, although new evidence could still turn up, it would seem (as of 1986) that Gillie was justified.

In view of the suspicion surrounding Burt's work A.M. and A.D.B. Clarke in England and Dorfman in America reanalysed the figures discussed in the previous chapter looking for indications of irregularity. Since the next chapter is based on Dorfman's account of their findings (for details and some discussion see *Science*, **194**, 916–9; **201**, 1177–86; **204**, 242–6; **205**, 1204; and **206**, 144–6) the reader should bear in mind a warning given by Professor Eysenck to readers of the *Daily Telegraph*. 'There are a number of errors in Professor Dorfman's work which make it much less valuable than it could have been. There are many things that he has just misunderstood.' In a different vein she might like to re-examine Table 83.1 and Figure 83.2 to see if there is anything in them which strikes her as odd.

AN EXAMPLE OF OUTSTANDING
STATISTICAL TREATMENT III

It is known that when IQ tests are administered to a large number of children the results are distributed approximately (but not exactly) in a normal distribution. Dorfman suggests that the distributions reported by Burt are much closer to the normal distribution than they ought to be. If we follow standard statistical procedure and ignore Burt's splitting of the category 130 + into 130–140 and 140 + and his splitting of the category 70 − into 60–70 and 50–60 then my calculations (which the reader is invited to verify; it is, unfortunately, very easy to make numerical mistakes in this kind of work) give the following results.

Table 85.1. *Predicted and actual figures in Burt's data*

IQ range	70 −	70–80	80–90	90–100	100–10	110–20	120–30	130 +
Predicted	22.7	68.5	161.3	247.5	247.5	161.3	68.5	22.7
Parents	24	69	160	247	248	162	67	23
Children	24	70	159	250	247	160	68	22

Here the predicted value $V(I)$ for a given range I is the expected number of times we would obtain a result in I if we performed 1000 independent trials using a normally distributed random variable mean 100 and variance 15. Thus if X is normally distributed mean 100 and variance 15 whilst Y is normally distributed mean 0 and variance 1

$$V(I) = 1000 Pr(X \in [a, b])$$
$$= 1000 Pr((X - 100)/(15)^{\frac{1}{2}} \in [(a - 100)/(15)^{\frac{1}{2}}, (b - 100)/(15)^{\frac{1}{2}}])$$
$$= 1000 Pr(Y \in [(a - 100)/(15)^{\frac{1}{2}}, (b - 100)/(15)^{\frac{1}{2}}]),$$

so $V(I)$ may be calculated from standard tables.

Remark. One of the readers of the manuscript felt that my 'standard statistical procedure' of merging the categories 50–60 and 60–70 into 70− and the similar treatment for 130 + was, in fact, a 'biased subjective fiddle'. For the reasons given

in the paragraph numbered (2) below I disagree but the reader may wish to redo the computations without grouping.

The values given by Burt are certainly very close to the predicted values (particularly for the parents). Are they too close? This is clearly a case for trying out the methods of Chapter 81 but, even before we begin, we run into the problem that Burt's figures are proportions and not the actual results. We must therefore do our calculations for a collection of $1000N$ parents and discuss what happens if N is indeed 40, or if N is in fact much closer to 1 (as more suspicious commentators like Kamin have suggested).

We now rewrite the data of Table 85.1 to form Table 85.2 and bring it into line with the notation of Chapter 81

Table 85.2. *Predicted and actual figures in the notation of Chapter 81*

A_j	A_1	A_2	A_3	A_4	A_5	A_6	A_7	A_8
p_j	0.0227	0.0685	0.1613	0.2475	0.2475	0.1613	0.0685	0.0227
n_j	24N	69N	160N	247N	248N	162N	67N	23N
n'_j	24N	70N	159N	250N	247N	160N	68N	22N

and look at

$$s_p(N)^2 = \sum_{k=1}^{8} \frac{(n_k - 1000Np_k)^2}{1000Np_k},$$

$$s_c(N)^2 = \sum_{k=1}^{8} \frac{(n'_k - 1000Np_k)^2}{1000Np_k}.$$

Suppose now we conducted the following idealised experiment. We take a characteristic which is exactly normally distributed with mean 100 and variance 15 and measure its value for $1000N$ people writing n_k (or n'_k) for the number of cases found in the range A_k. For this idealised experiment the results of Chapter 81 predict that $s_p(N)^2$ and $s_c(N)^2$ will be distributed as χ^2 distributions with 7 degrees of freedom.

My calculations (which again the reader is invited to check) give

$$s_p(N)^2 = 0.13N$$

and
$$s_c(N)^2 = 0.20N.$$

Consulting tables we see that a χ^2 random variable with 7 degrees of freedom will take a value less than 1 with probability less than 0.01 and a value greater than 1 and less than 16 with probability about 0.97. Thus if we performed our ideal experiment the results for both parents and children would be rather unlikely for $N \leqslant 5$ whilst if $N = 40$ they would cause no surprise.

There are two important ways in which our idealised experiment differs from the experiment described by Burt.

(1) Burt rescaled his results to have mean 100 and variance 15 *before* assigning the classes A_k. Thus his n_k will tend to be closer to $1000 N p_k$ than ours and his $s_p(N)^2$ will tend to be smaller.

(2) Although, in general, a χ^2 test becomes more sensitive as the size of the population studied increases, this is not the case here because Burt has rounded his figures. Thus a change of one unit in the table corresponds to a change in the assignment of N people and may cause the value of $s(N)^2$ to change sharply. (Consider, for example, the effect on $s_c(N)^2$ of changing n_1 to $25N$ and n_8 to $21N$ in the case when $N = 40$.)

Since the methods of Chapter 81 do not apply directly we must use our own judgement in estimating the importance of these two problems. My own view is that when $N \leqslant 5$ the fact that $s_p(N)^2$ and $s_c(N)^2$ are simultaneously so small remains suspicious even taking (1) into account. (But the reader should take time to think and come to her own conclusion.) On the other hand if $N = 40$ the values of $s_c(N)^2$ and $s_p(N)^2$ seem so reasonable (on the assumption that IQ is actually normally distributed) that (2) loses most of its force. If $5 < N \leqslant 20$ then (1) and (2) give us additional reasons for suspending judgement.

Thus if we use tests based on the assumption that Burt was dealing with an exact normal distribution we have cause for suspicion only if N is of the order of 5 or less and not if N is of the order of 40 (as the commentary to Table 83.1 states).

However as Dorfman points out measurements of human characteristics such as height, weight and IQ (or more properly, the results of a given IQ test) do not reveal an exactly normal distribution, and a set of figures which would not be suspicious if drawn from a normal population may become so if drawn from a population which is not exactly normal. Is Burt's data too close to normal, given that IQ is not quite normally distributed? Earlier we saw that the values $s_p(N)^2/N = 0.13$ and $s_c(N)^2/N = 0.20$ computed for Burt's data would be about the size we would expect if $N = 40$ and the sampling was from a normal distribution. Might they still be much lower than we would expect from the actual not quite normal distribution?

Here, clearly, pure thought will not give the answer. Instead we must look at the results of other surveys and compare them with Burt's. What Dorfman did was to take the results of several large scale surveys of height, weight and IQ, standardise them to have mean 100 and variance 15 and then look at $s(M)^2/M$ where

$$s(M)^2 = \sum_{k=1}^{8} \frac{(m_k - 1000 M q_k)^2}{1000 M q_k},$$

$1000M$ is the number of people involved, m_k the number of (rescaled) measurements falling in the range B_k chosen as close to Burt's A_k as possible and q_k is the probability of a single measurement falling in B_k on the assumption that we are dealing with a normal distribution.

Remark 1. We use $s(M)^2/M$ as a measure of dissimilarity from normal rather than $s(M)^2$ since if we simultaneously multiply M and all the m_k by some scalar λ (i.e.

we change the population size M but not the proportions m_k/M) we wish our measure of dissimilarity to remain unchanged.

Remark 2. It is important that the reader appreciates that the choice of $s(M)^2/M$ as a measure of dissimilarity is an ad hoc one which she is free to reject. Naturally, I hope that she will accept it temporarily for the sake of argument but when she has finished the chapter she should return to this and think again before making a final decision.

Dorfman shows that Burt's values of $s(N)^2/N$ are much lower than any others in his collection of surveys. For example a survey of the IQ of Scottish school children gave $s(M)^2 = 2085$ for the sample of 35 809 boys (so that $s(M)^2/M = 58.23$ in this case) and $s(M)^2 = 1608$ for the sample of 34 996 girls (so that $s(M)^2/M = 45.95$). The values of $s(M)^2/M$ are over 200 times larger than Burt's.

Of course, since we do not know what tests Burt used and since in any case his tested population (London adults and children) was different we are not comparing exact like with exact like. However, the surveys cited by Dorfman (the reader must consult his article in *Science,* **201**, 1177–86 for details) indicate that Burt's s values are two orders of magnitude lower than we would expect for a survey of IQ or weight and one order of magnitude lower than one would expect for a survey of height. We must remember also that, according to Burt himself, 'the assessments of adult intelligence were less thorough and less reliable' so that we would expect Burt's s_p to be large rather than small. With this additional evidence I feel bound to accept Dorfman's claim that the marginal totals do not represent the 'actual distribution of IQ's found in the population' but have been fixed by Burt himself. (But the reader need not feel bound to follow me. Such things must always be a matter of opinion rather than proof.)

What follows if we accept Dorfman's claim? We already know, since Burt says so, that 'the proportions in each class' have been fixed so as to equal the 'estimated proportions in the population'. (Dorfman suggests that the proportions are taken from a 1926 publication of Spielman and Burt.) Thus if Burt actually started from real data he has adjusted it so as to fit preassigned marginal distributions. This cannot be done by the method he claims to have used: 'We weighed the actual numbers so that the proportions in each class should be equal to the estimated proportions for the total population. Finally for the purposes of the present analysis we have rescaled our assessments of intelligence so that the mean of the whole group is 100 and the standard deviation 15.' (Observe that if we first weigh our data to fit one set of marginals and *then* reweigh our new figures to agree with the other set we lose the agreement with the first set.)

There do exist methods for fitting data to preassigned marginals but they are computationally tedious and bear little resemblance to Burt's description of his procedure. (Moreover, any method which calls for data to be fitted to a large number of constraints is exceedingly dangerous. To quote Fisher, '... the statistician is not elucidating but falsifying the facts who rearranges them to give an artificial

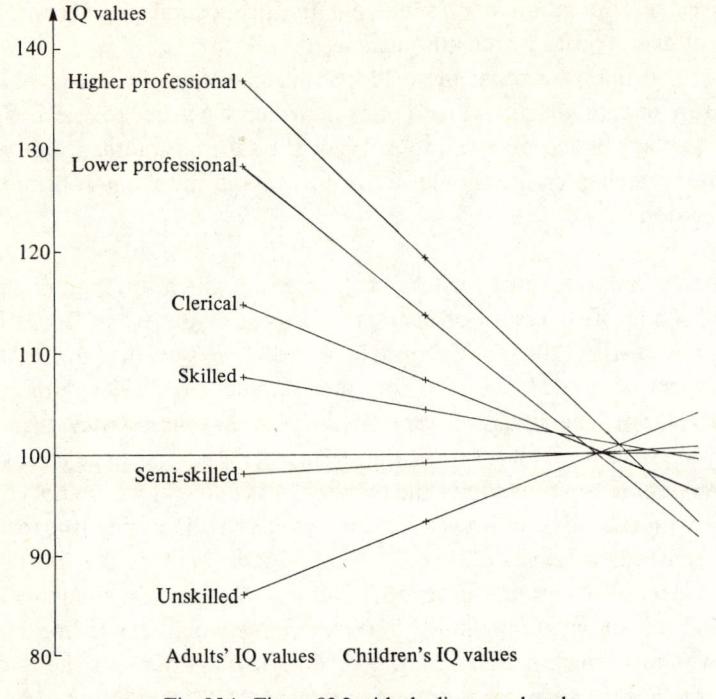

Fig. 85.1. Figure 83.2 with the lines produced.

appearance of regularity' (p. 41 *The Design of Experiments*, Oliver and Boyd, Edinburgh 1935).) If we were faced with an isolated problem like this in the work of an otherwise highly reputable scientist we would certainly give him the benefit of the doubt but this is not an isolated problem and Burt's twin data show that he is not beyond suspicion. There seems to be a strong case for supposing that the tables of Burt given in Figure 83.1 do not represent actual data.

Further evidence of this view is contained in the diagram given as Figure 83.2. If we redraw it and extend the lines of the diagram they come remarkably close to intersecting as shown in Figure 85.1.

In fact, as the Clarkes seem to have been the first to realise, the mean IQ's given by Burt fit very closely to the simple formula

$$\mu_c = \tfrac{1}{2}(\mu_p + 100),$$

Table 85.3. *Burt's figures compared to the Clarkes' formula*

Class	I	II	III	IV	V	VI
$\frac{1}{2}(\mu_p + 100)$	119.85	115.3	107.95	104.1	98.9	92.45
μ_c	120.8	114.7	107.8	104.6	98.9	92.6
$\frac{1}{2}(\mu_p + 100) - \mu_c$	−0.95	0.6	0.15	−0.5	0	−0.15

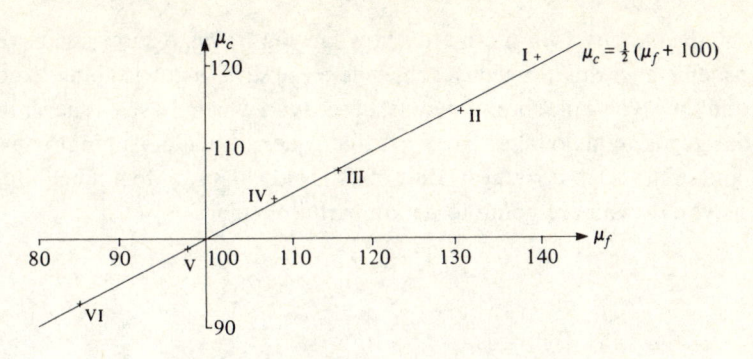

Fig. 85.2. Graphical representation of Table 85.3

where μ_p is the mean IQ of the fathers in a given class and μ_c that of the children. The closeness of the fit is shown in Figure 85.1 and Table 85.4.

Now if we were investigating a physical law given by the Clarkes' formula and we had an excellent experimental setup we might hope for an agreement as good as that shown in Figure 85.2. But it is very unlikely that the relation between the mean IQ's of father and son satisfy such a simple law so accurately. (Dorfman does however quote a paper by Conway, a pseudonym, according to Gillie, of Burt, which makes use of such a law.) Once again we have strong grounds for suspecting the reality of the data.

My object in this discussion has not been to paint a stunning picture of scientific truth triumphing over falsehood. Even if Burt was guilty of fraud (and there are still honest and reputable men who have looked at the evidence and come to the verdict 'not proven') he was no satanic villain but a lonely and, inspite of the honours heaped upon him, disappointed man who could not bear not to have the last word in scientific controversy. The reader who wishes for a sympathetic and, it is generally agreed, unbiased view of Burt and the controversies surrounding him should consult L. Hearnshaw's biography (*Cyril Burt, Psychologist*, Hodder and Stoughton 1981).

I do not wish to leave the reader with the impression that the chief problem in evaluating data is fraud. In my view, unconcious bias and overoptimistic treatment of sources of error represent far more serious problems. And these faults are normally due not to lack of honesty but to lack of introspection. Nor do I wish to give the reader the impression that statistical methods are a strong weapon against fraud. For the reasons outlined at the end of Chapter 82 I believe that the determined cheat can now produce fraudulent data which, from the statistical point of view, are indistinguishable from the real thing.

Rather I have tried to show the reader that 'batallions of figures, like batallions of men are not always as strong as is supposed'. I have tried to show that the real world and the real people who inhabit it are much queerer than statistical text books suggest. My moral is best expressed in the father's advice to his son as recorded in Runyon's *The Idyll of Miss Sarah Brown*.

'Son', the old guy says, 'no matter how far you travel, or how smart you get, always remember this: Someday, somewhere', he says, 'a guy is going to come to you, and show you a nice brand-new set of cards on which the seal is never broken, and this guy is going to offer to bet you that a jack of spades will jump out of the deck and squirt cider in your ear. But son', the old guy says, 'do not bet him, for as sure as you do you are going to get an earful of cider'.

86

WILL A RANDOM WALK RETURN?

It is pleasant, perhaps surprising and certainly useful, that so much of Fourier theory generalises so easily from one to many dimensions. However life would be very dull if the study of $\mathbb{R}^m$ consisted only of rewriting theorems about $\mathbb{R}$. Fortunately, from this point of view, raising the dimension does introduce new and interesting problems. As a simple example we give a result due to Pólya.

Theorem 86.1. *Consider a particle performing a random walk on the lattice* $\mathbb{Z}^m$ *according to the following rules.*

(i) *If at time* $t = n$ *the particle is at* $\mathbf{k} = (k_1, k_2, \ldots, k_m) \in \mathbb{Z}^m$ *then the probability* $p(\mathbf{k}, \mathbf{k}')$ *that it will be at* $\mathbf{k}' = (k_1', k_2', \ldots, k_m') \in \mathbb{Z}^m$ *at time* $t = n + 1$ *is given by*

$$p(\mathbf{k}, \mathbf{k}') = (2m)^{-1} \quad if \ \sum_{j=1}^{n} |k_j - k_j'| = 1,$$

$$p(\mathbf{k}, \mathbf{k}') = 0 \quad otherwise.$$

(ii) *The steps are probabilistically independent.*
(iii) *At time* $t = 0$ *the particle is at* $\mathbf{0}$.
Then
(a) *Pr (particle returns to* $\mathbf{0}$ *at some time* $t > 0) = 1$ *if* $m \leqslant 2$.
(b) *Pr (particle returns to* $\mathbf{0}$ *at some time* $t > 0) < 1$ *if* $m \geqslant 3$.
Remark. The reader should have no difficulty in deducing that under these circumstances:

(i) If $m \leqslant 2$ then, with probability 1, the particle visits each point of $\mathbb{Z}^m$ infinitely often.
(ii) If $m \geqslant 3$ then, with probability 1, the distance of the particle from the origin tends to infinity as $t \to \infty$.

Our proof of Theorem 86.1 makes use of simple related probabilistic results.

Lemma 86.2. (i) *Let* q_n *be the probability that the particle retuns to the origin at*

443

least n times and let u_n be the probability that the particle returns to the origin exactly n times. Write $q = q_1$. Then for all $n \geqslant 0$

$$q_n = q^n,$$
$$u_n = q^n(1 - q).$$

(ii) (a) *If $q = 1$ then with probability 1 the particle returns to the origin infinitely often.*

    (b) *If $q < 1$ then with probability 1 the particle returns to the origin only finitely often.*

(iii) *Let p_k be the probability that the particle returns to the origin at time k. Then*

    (a) *If $\sum_{k=1}^{\infty} p_k$ diverges,*

        $q = Pr$ *(particle returns to the origin at some time $t > 0$)* $= 1$

    (b) *If $\sum_{k=1}^{\infty} p_k$ converges*

        $q = Pr$ *(particle returns to the origin at some time $t > 0$)* < 1.

Proof. (i) Writing out the argument in detail we obtain

$$
\begin{aligned}
q_{n+1} &= Pr \text{ (return to origin at least } n+1 \text{ times)} \\
&= Pr \text{ (return at least } n+1 \text{ times} \,|\, \text{return at least } n \text{ times)} \\
&\quad \times Pr \text{ (return at least } n \text{ times)} \\
&= Pr \text{ (return at least 1 time} \,|\, \text{return at least 0 times)} \, q_n \\
&= Pr \text{ (return at least 1 time)} \, q_n = q q_n.
\end{aligned}
$$

Since $q_0 = 1$ it follows by induction that $q_n = q^n \,[n \geqslant 0]$. Again

$$
\begin{aligned}
u_n &= Pr \text{ (exactly } n \text{ returns)} \\
&= Pr \text{ (at least } n \text{ returns)} - Pr \text{ (at least } n+1 \text{ returns)} \\
&= q_n - q_{n+1} = q^n(1 - q).
\end{aligned}
$$

(ii) Observe that

$$Pr \text{ (infinite number of returns)}$$

$$= 1 - Pr \text{ (finite number of returns)} = 1 - \sum_{n=0}^{\infty} u_n,$$

which takes the value 1 or 0 according as $q = 1$ or $q < 1$.

(iii) (a) If $q < 1$ then by (ii) we know that with probability 1 the number Z of returns to the origin is finite and by (i) that

$$\mathbb{E}Z = \sum_{n=0}^{\infty} n u_n = \sum_{n=0}^{\infty} n q^n (1 - q) = \frac{(1-q)}{(1-q)^2} = \frac{1}{1-q} < \infty.$$

But, if Z_j is the random variable with value 1 if the particle returns to **0** when $t = j$ and value 0 otherwise, we have

$$Z = \sum_{j=1}^{\infty} Z_j$$

and so,

$$\mathbb{E}Z = \sum_{j=1}^{\infty} \mathbb{E}Z_j = \sum_{j=1}^{\infty} p_j,$$

so (if $q < 1$), $\sum_{j=1}^{\infty} p_j$ converges.

(b) On the other hand, if $q = 1$, then, writing $Z(k) = k$ if the particle makes k or more returns to the origin, $Z(k) = r$ if the particle makes $r \leqslant k$ returns to the origin, we know that $Z(k) = k$ with probability 1 and $Z(k) \leqslant \sum_{j=1}^{\infty} Z_j$. Thus

$$k = \mathbb{E}Z(k) \leqslant \sum_{j=1}^{\infty} \mathbb{E}Z_j = \sum_{j=1}^{\infty} p_j,$$

and so (if $q = 1$), $\sum_{j=1}^{\infty} p_j$ diverges. ∎

With these rather trivial matters out of the way, we turn to the proof of the main theorem.

Proof of Theorem 86.1. Step 1. Write $\mathbf{k}(j)$ for the position of the particle at time j. Lemma 84.2 suggests that we try to find out whether $\sum_{j=1}^{\infty} p_j$ diverges or not and in particular that we should start by attempting to compute $p_j = Pr(\mathbf{k}(j) = \mathbf{0})$. It turns out (as we might expect) to be just as easy to compute $Pr(\mathbf{k}(j) = \mathbf{r})$ for all $\mathbf{r} \in \mathbb{Z}^m$.

The key idea is to consider the function $f_j: \mathbb{T}^m \to \mathbb{C}$ given by

$$f_j(\mathbf{x}) = \sum_{\mathbf{r} \in \mathbb{Z}^m} Pr(\mathbf{k}(j) = \mathbf{r}) \exp(i\mathbf{r} \cdot \mathbf{x}).$$

(Thus f_j is a trigonometric polynomial with $\hat{f}_j(\mathbf{r}) = Pr(\mathbf{k}(j) = \mathbf{r})$.) By a calculation which echoes several earlier ones we see that

$$f_j(\mathbf{x}) = \mathbb{E}(\exp i\mathbf{k}(j) \cdot \mathbf{x}) = \mathbb{E}\left(\exp i \sum_{q=1}^{j} \mathbf{e}(q) \cdot \mathbf{x} \right),$$

where $\mathbf{e}(q) = \mathbf{k}(q) - \mathbf{k}(q-1)$ (i.e. $\mathbf{e}(q)$ is the qth step). Since the $\mathbf{e}(q)$ are independent (condition (ii)) and identically distributed, so are the $\exp i\mathbf{e}(q) \cdot \mathbf{x}$ and so

$$f_j(\mathbf{x}) = \mathbb{E}\left(\exp i \sum_{q=1}^{j} \mathbf{e}(q) \cdot \mathbf{x} \right) = \mathbb{E}\left(\prod_{q=1}^{j} \exp i\mathbf{e}(q) \cdot \mathbf{x} \right)$$

$$= \prod_{q=1}^{j} \mathbb{E}(\exp i\mathbf{e}(q) \cdot \mathbf{x}) = (\mathbb{E} \exp i\mathbf{e}(1) \cdot \mathbf{x})^j = f_1(\mathbf{x})^j.$$

Thus

$$Pr(\mathbf{k}(j) = \mathbf{r}) = \hat{f}_j(\mathbf{r}) = \left(\frac{1}{2\pi} \right)^m \int_{\mathbb{T}^m} f_j(\mathbf{x}) \exp(-i\mathbf{r} \cdot \mathbf{x}) dV(\mathbf{x})$$

$$= \left(\frac{1}{2\pi} \right)^m \int_{\mathbb{T}^m} (f_1(\mathbf{x}))^j \exp(-i\mathbf{r} \cdot \mathbf{x}) dV(\mathbf{x}),$$

and, in particular,

$$p_j = Pr(\mathbf{k}(j) = 0) = \left(\frac{1}{2\pi}\right)^m \int_{\mathbb{T}^m} (f_1(\mathbf{x}))^j dV(\mathbf{x}).$$

Step 2. There is no theorem of analysis which would enable us to write

$$\sum_{j=0}^{\infty} p_j = \left(\frac{1}{2\pi}\right)^m \int_{\mathbb{T}^m} \sum_{j=0}^{\infty} (f_1(\mathbf{x}))^j dV(\mathbf{x}) = \left(\frac{1}{2\pi}\right)^m \int_{\mathbb{T}^m} \frac{1}{1 - f_1(\mathbf{x})} dV(\mathbf{x})$$

without requiring further information about f_1. However, we do know that $|f_1(\mathbf{x})| = |\mathbb{E} \exp i e(1) \cdot \mathbf{x}| \leqslant \mathbb{E} |\exp i e(1) \cdot \mathbf{x}| = \mathbb{E} 1 = 1$. Thus if we fix $0 < \rho < 1$, we see that $|(\rho f_1(\mathbf{x}))^m| \leqslant \rho^m$ and

$$\sum_{j=0}^{N} \rho^j (f_1(\mathbf{x}))^j \to \frac{1}{1 - \rho f_1(\mathbf{x})} \quad \text{uniformly as } N \to \infty.$$

It follows that we may integrate term by term to obtain

$$\sum_{j=0}^{\infty} \rho^j p_j = \sum_{j=0}^{\infty} \left(\frac{1}{2\pi}\right)^m \int_{\mathbb{T}^m} \rho^j (f_1(\mathbf{x}))^j dV(\mathbf{x}) = \left(\frac{1}{2\pi}\right)^m \int_{\mathbb{T}^m} \frac{1}{1 - \rho f_1(\mathbf{x})} dV(\mathbf{x}).$$

Since $p_j \geqslant 0$, it is easy to check that $\sum_{j=0}^{\infty} p_j$ diverges if and only if $\sum_{j=0}^{\infty} \rho^j p_j \to \infty$ as $\rho \to 1-$. Thus, by Lemma 86.2,
(a) Pr(particle returns to origin) $= 1$ if

$$\int_{\mathbb{T}^m} \frac{1}{1 - \rho f_1(\mathbf{x})} dV(\mathbf{x}) \to \infty \quad \text{as } \rho \to 1-$$

(b) Pr (particle returns to origin) < 1 if

$$\int_{\mathbb{T}^m} \frac{1}{1 - \rho f_1(\mathbf{x})} dV(\mathbf{x}) \quad \text{is bounded as } \rho \to 1-.$$

Step 3. So far we have avoided calculating $f_1(\mathbf{x})$. (Thus our argument so far applies to all random walks satisfying the obvious conditions (ii) and (iii) of the theorem.) A simple computation gives

$$f_1(\mathbf{x}) = \mathbb{E} \exp (i e(1) \cdot \mathbf{x}) = \sum_{r=1}^{m} (2m^{-1} \exp i x_r + 2m^{-1} \exp - i x_r) = m^{-1} \sum_{r=1}^{m} \cos x_r.$$

If $|x_r| \geqslant 10^{-2}$ then $\cos x_r \leqslant 1 - 10^{-5}$ (using a very rough bound) and so, if we write $\|\mathbf{x}\| = \sqrt{(\sum_{r=1}^{m} x_r^2)}$ and

$$A = \{\mathbf{x} \in \mathbb{T}^m : \|\mathbf{x}\| < 10^{-2} \quad \text{for all } 1 \leqslant r \leqslant m\},$$

we know that $f_1(\mathbf{x}) \leqslant 1 - m^{-1} 10^{-5}$ for all $\mathbf{x} \notin A$ and so

$$\int_{\mathbf{x} \notin A} \frac{1}{1 - \rho f_1(\mathbf{x})} dV(\mathbf{x}) \leqslant \int_{\mathbf{x} \notin A} \frac{1}{m^{-1} 10^{-5}} dV(\mathbf{x}) \leqslant (2\pi)^m m 10^5 \quad \text{for all } \rho < 1.$$

Thus
$$\int_{\mathbb{T}^m} \frac{1}{1-\rho f_1(\mathbf{x})} dV(\mathbf{x}) \to \infty,$$

or not, according as $\int_A 1/(1-\rho f_1(\mathbf{x})) dV(\mathbf{x})$ does.

On the other hand, if $\mathbf{x} \in A$ then for each $1 \leqslant r \leqslant m$ we know that $|x_r| \leqslant 10^{-2}$ and so

$$|1 - \cos x_r - |x_r|^2/2| = \left| \sum_{n=2}^{\infty} (-|x_r|^2)^n/(2n)! \right| \leqslant |x_r|^2 \sum_{n=2}^{\infty} |x_r|^{2n-2}/(2n)!$$

$$\leqslant |x_r|^2 \sum_{n=1}^{\infty} 10^{-4n} \leqslant |x_r|^2/4.$$

Thus
$$|x_r|^2 \geqslant 1 - \cos x_r \geqslant |x_r|^2/4,$$

and so, summing, $\|\mathbf{x}\|^2 = \sum_{r=1}^{m} |x_r|^2 \geqslant m - \sum_{r=1}^{m} \cos x_r \geqslant \|\mathbf{x}\|^2/4,$

whence $\quad m^{-1}\|\mathbf{x}\|^2 \geqslant 1 - f_1(\mathbf{x}) \geqslant m^{-1}\|\mathbf{x}\|^2/4 \quad$ for all $\mathbf{x} \in A.$

Thus if $1 > \rho > \frac{1}{2}$ the trivial equality $1 - \rho f_1(\mathbf{x}) = (1-\rho) + \rho(1 - f_1(\mathbf{x}))$ yields

$$(1-\rho) + m^{-1}\|\mathbf{x}\|^2 \geqslant 1 - \rho f_1(\mathbf{x}) \geqslant (1-\rho) + (8m)^{-1}\|\mathbf{x}\|^2,$$

and thus

$$\int_A \frac{1}{(1-\rho) + (8m)^{-1}\|\mathbf{x}\|^2} dV(\mathbf{x}) \geqslant \int_A \frac{1}{1-\rho f_1(\mathbf{x})} dV(\mathbf{x}) \geqslant \int_A \frac{1}{(1-\rho) + m^{-1}\|\mathbf{x}\|^2} dV(\mathbf{x}).$$

Converting to spherical polars we know that

$$\int_A \frac{1}{(1-\rho) + C\|\mathbf{x}\|^2} dV(\mathbf{x}) = \alpha_m \int_0^\tau \frac{r^{m-1}}{(1-\rho) + Cr^2} dr,$$

where α_m is the surface area of the unit sphere in $\mathbb{R}^m$ and $\tau = 10^{-2}.$

But (if $C \neq 0$) it is easy to check that

$$\int_0^\tau \frac{r^{m-1}}{(1-\rho) + Cr^2} dr \text{ diverges as } \rho \to 1- \text{ if and only if}$$

$$\int_\delta^\tau \frac{r^{m-1}}{r^2} dr = \int_\delta^\tau r^{m-3} dr \text{ diverges as } \delta \to 0+,$$

and that $\quad \int_0^\tau r^{m-3} dr$ diverges if and only if $m=1$ or $m=2.$

Thus $\quad \int_A \frac{1}{1-\rho f_1(\mathbf{x})} dV(\mathbf{x}) \to \infty$ as $\rho \to 1-$ if and only if $m=1$ or $m=2,$

and so the particle returns with probability 1 if and only if the dimension m is 1 or 2. ∎

The argument is clearly a very general one.

Theorem 86.1′. *Consider a particle performing a random walk on the lattice* $\mathbb{Z}^m$ *according to the following rules.*

(i) *If at time n the particle is at* $\mathbf{k} = (k_1, k_2, \ldots, k_n)$ *then the probability* $p(\mathbf{k}, \mathbf{k}')$ *that it will be at* $\mathbf{k}' = (k_1', k_2', \ldots, k_m') \in \mathbb{Z}^m$ *at time* $n+1$ *is given by* $p(\mathbf{k}, \mathbf{k}') = P(\mathbf{k} - \mathbf{k}')$ *where*

(A) $P(\mathbf{e}) = 0$ *for* $\|\mathbf{e}\| \geqslant R$.

(B) $\sum P(\mathbf{e})\mathbf{e} = \mathbf{0}$.

(C) *We cannot find* $\boldsymbol{\lambda} = (\lambda_1, \lambda_2, \ldots, \lambda_m) \in \mathbb{R}^m$ *with* $\boldsymbol{\lambda} \neq 0$ *such that, writing* $\Gamma = \{\mathbf{e} \in \mathbb{Z}^m : \sum \lambda_j e_j = 0\}$, *we have* $\sum_{\mathbf{e} \in \Gamma} P(\mathbf{e}) = 1$.

(ii) *The steps are probabilistically independent.*

(iii) *At time* $t = 0$ *the particle is at* $\mathbf{0}$.

Then

(a) *Pr (particle returns at* $\mathbf{0}$ *at some time* $t > 0$) $= 1$ *if* $m \leqslant 2$.

(b) *Pr (particle returns to* $\mathbf{0}$ *at some time* $t > 0$) < 1 *if* $m \geqslant 3$.

Remark. Condition (A) says that the particle only makes bounded jumps, condition (B) says that its 'average drift' is zero and condition (C) says that the motion of the particle is not confined to a lower dimensional subspace. Conditions (B) and (C) are natural ones. The reader may like to see how far she can weaken condition (A).
Proof. I shall just sketch the proof leaving the (considerable) detail to the reader if she wishes to fill it in. We concentrate on proving (a). Clearly, the arguments of Theorem 86.1 go through word for word until the calculation of f_1 in Step 3. Observe that

$$\exp i\mathbf{e} \cdot \mathbf{x} = 1 + i\mathbf{e} \cdot \mathbf{x} - (\mathbf{e} \cdot \mathbf{x})^2/2 + \varepsilon(\mathbf{e}, \mathbf{x})\|\mathbf{x}\|^2,$$

where $\varepsilon(\mathbf{e}, \mathbf{x}) \to 0$ as $\|\mathbf{x}\| \to 0$. Thus, using (B),

$$f_1(\mathbf{x}) = \sum P(\mathbf{e})\exp(i\mathbf{e} \cdot \mathbf{x}) = 1 + i\left(\sum P(\mathbf{e})\mathbf{e}\right) \cdot \mathbf{x} - \sum P(\mathbf{e})(\mathbf{e} \cdot \mathbf{x})^2/2 + \eta(\mathbf{x})\|\mathbf{x}\|^2$$

$$= 1 - \sum P(\mathbf{e})(\mathbf{e} \cdot \mathbf{x})^2/2 + \eta(\mathbf{x})\|\mathbf{x}\|^2 = 1 - Q(\mathbf{x}) + \eta(\mathbf{x})\|\mathbf{x}\|^2,$$

where $\eta(\mathbf{x}) \to 0$ as $\mathbf{x} \to \mathbf{0}$ and $Q(\mathbf{x}) = \sum q_{rs}x_r x_s$.

We observe that $Q(\mathbf{x}) \geqslant 0$ for all $\mathbf{x}$ and that $Q(\mathbf{x}) > 0$ unless $\mathbf{e} \cdot \mathbf{x} = 0$ for all e with $P(\mathbf{e}) \neq 0$. Using (C), we deduce that $Q(\mathbf{x}) > 0$ for all $\mathbf{x} \neq \mathbf{0}$ (i.e. Q is strictly positive definite). It follows by the kind of argument used at the end of Chapter 80 (we can always fit one sphere inside an ellipsoid and another outside) or by simple analytic arguments (since Q is bounded on the closed bounded set $\{\mathbf{x}: \|\mathbf{x}\| = 1\}$, it attains its upper and lower bounds at $\mathbf{x}_1, \mathbf{x}_2$ say; write $A = Q(\mathbf{x}_1)$, $B = Q(\mathbf{x}_2)$) that we can find $A, B > 0$ such that $A\|\mathbf{x}\|^2 \geqslant Q(\mathbf{x}) \geqslant B\|\mathbf{x}\|^2$ for all $\mathbf{x}$.

It follows that $1 - f_1(\mathbf{x})$ behaves like $\|\mathbf{x}\|^2$ as $\mathbf{x} \to \mathbf{0}$ and thus $\int 1/(1 - f_1(\mathbf{x}))dV(\mathbf{x})$ diverges if $m = 1$ or $m = 2$. The result (a) follows as in the proof of Theorem 86.1. At first glance the proof of (b) should also run as in Theorem 86.1 but there is a minor annoying hiccup.

Clearly, the behaviour of $\int 1/(1 - \rho f_1(\mathbf{x}))dV(\mathbf{x})$ as $\rho \to 1-$ (at least as far as

possible divergence goes) is governed by the behaviour of $f_1(\mathbf{x})$ near those points where $f_1(\mathbf{x}) = 1$. Now $f_1(0) = 1$ always and in the example given as Theorem 86.1, $f_1(\mathbf{x}) \neq 1$ for $\mathbf{x} \neq 0$. However in certain cases $f(\mathbf{x})$ may take the value 1 at other points (for example if $m = 4$ and $P(\mathbf{e}) = 2^{-4}$ if $|e_j| = 1[1 \leqslant j \leqslant 4]$, $P(\mathbf{e}) = 0$ otherwise then $f(\mathbf{x}) = \cos x_1 \cos x_2 \cos x_3 \cos x_4$ and $f(\pi, \pi, \pi, \pi) = 1$).

We get round this difficulty by observing that $f(\mathbf{x}) = 1$ if and only if $\mathbf{e} \cdot \mathbf{x} \equiv 0 \bmod 2\pi$ for all $\mathbf{e}$ with $P(\mathbf{e}) \neq 0$. Condition (C) tells us that the system of equations $\mathbf{e} \cdot \mathbf{x} = 0$ for all $\mathbf{e}$ with $P(\mathbf{e}) \neq 0$ has no non zero solutions and so the system of equations of the previous sentence can only have a finite number of solutions $\mathbf{y}_1$, $\mathbf{y}_2, \ldots, \mathbf{y}_p \in \mathbb{T}^m$, say. We then show that $1 - f(\mathbf{x})$ behaves like $\|\mathbf{x} - \mathbf{y}_q\|^2$ near $\mathbf{y}_q$ for each $1 \leqslant q \leqslant p$ and argue as before. ∎

In this chapter we have only dealt with the interesting case in which the 'average drift' is zero. However in Chapter 88 we shall also refer to the case in which there is drift. In order to provide some background, we give an example in which the drift is non zero.

Lemma 86.3. *Consider a particle performing a random walk on $\mathbb{Z}$ according to the following rules.*

(i) *If at time $t = n$ the particle is at k the probability that it will be at $k + 1$ at time $t = n + 1$ is p, the probability it will be at $k - 1$ at time $t = n + 1$ is q and the probability that it will be at k at time $t = n + 1$ is $1 - p - q$.*
(ii) *The steps are probabilistically independent.*
(iii) *At time $t = 0$ the particle is at 0.*

Then if $p \neq q$ and $p + q \neq 0$

(a) *Pr (particle returns to 0 at some time $t > 0$) < 1*
and
(b) *Pr (particle returns to 0 infinitely often) $= 0$.*

Proof. (This is included for completeness. The reader may use any excuse to omit it.) By Lemma 86.2 consequence (b) follows from (a). We may thus concentrate on proving (a), and since the cases $p = 0$ or $q = 0$ are trivial we may assume $p, q \neq 0$ and $p \neq q$.

Let $u_N(r)$ be the probability that a particle starting at $0 < r < N$ hits 0 before it hits N and set $u_N(0) = 1$, $u_N(N) = 0$. Then

$$u_N(k) = p u_N(k + 1) + (1 - p - q) u_N(k) + q u_N(k - 1) \quad [1 \leqslant k \leqslant N - 1]$$

i.e.
$$0 = p u_N(k + 1) - (p + q) u_N(k) + q u_N(k - 1),$$

and, solving this difference equation, we obtain

$$u_N(k) = A\alpha^k + B\beta^k \quad [0 \leqslant k \leqslant N]$$

with α and β the roots of

$$0 = pt^2 - (p + q)t + q,$$

i.e. $$0 = (t - 1)(pt - q).$$

Thus $$u_N(k) = A + B(q/p)^k \quad [0 \leqslant k \leqslant N]$$

for some constants A and B.

Since $u_N(0) = 1$ and $u_N(N) = 0$, it follows that $A + B = 1$ and

$$u_N(k) = 1 - \frac{1 - (q/p)^k}{1 - (q/p)^N}.$$

We conclude that, if $k \geqslant 1$

Pr(particle returns from k to 0)

$$= \lim_{N \to \infty} Pr \text{(particle returns from } k \text{ to 0 before it hits } N) = \lim_{N \to \infty} u_N(k)$$

$$= \lim_{N \to \infty} 1 - \frac{1 - (q/p)^k}{1 - (q/p)^N} = \begin{cases} (q/p)^k & \text{if } p > q, \\ 1 & \text{if } p < q. \end{cases}$$

Further, by symmetry, we see that, if $k \leqslant -1$,

$$Pr \text{(particle returns from } k \text{ to 0)} = \begin{cases} 1 & \text{if } p > q, \\ (p/q)^k & \text{if } p < q. \end{cases}$$

It follows that

Pr (particle starting from zero returns at some $t > 0$)

$= Pr$ (at first go moves right) Pr (particle returns from 1 to 0)

$\quad + Pr$ (at first go moves left) Pr (particle returns from -1 to 0)

$\quad + Pr$ (at first go stays at 0)

$$= \begin{cases} p(q/p) + q + (1 - p - q) & \text{if } p > q, \\ p + q(p/q) + (1 - p - q) & \text{if } q > p, \end{cases} = \begin{cases} 1 + (q - p) & \text{if } p > q, \\ 1 + (p - q) & \text{if } q > p, \end{cases} < 1,$$

as required.   ■

(The method of proof of Theorem 86.1 will also work as the reader may verify if she wishes. However, she is warned that the treatment of the limit of $(2\pi)^{-1} \int_T 1/(1 - \rho f_1(x)) dx$ requires careful thought.)

Lemma 86.3 and results like it are the mathematical expression of the practical observation that: 'All horse players die broke.' In an unfair game the weaker player may break even a few times but eventually he will go into the red and stay there. 'If I wanted to gamble', said the multimillionaire Paul Getty, 'I would buy a casino'.

87

WILL A BROWNIAN MOTION RETURN?

Before proceeding further, the reader is invited to reread Chapters 13 and 14. In these chapters we discussed a possible model for Brownian motion as a 'limit' (in a hand waving sense) of random walks on a grid of step length h and step time τ where $\tau^{-1}h^2$ was kept fixed and h (and thus τ) were allowed to tend to zero.

For simplicity the discussion was confined to two dimensions but a glance suffices to show that we could have considered n dimensions. Let us restate our conclusions given as 'Plausible Result 14.3' in this slightly more general context.

Plausible Result 87.1. *There is a family of random continuous paths* $\mathbf{Z}:\mathbb{R} \to \mathbb{R}^n$ *with the following properties.*

(i)
$$Pr\left((\mathbf{Z}(t+\delta t) - \mathbf{Z}(t))/\delta t^{\frac{1}{2}} \in \prod_{r=1}^{n} [a_r, b_r] \right)$$

$$= \left(\frac{1}{2\pi} \right)^{n/2} \int_{a_1}^{b_1} \int_{a_2}^{b_2} \cdots \int_{a_n}^{b_n} \exp\left(-\sum_{r=1}^{n} x_r^2/2 \right) dx_1 dx_2 \ldots dx_n,$$

for all $\delta t > 0$, $b_r > a_r [n \geqslant r \geqslant 1]$ *and* t, *independent of previous history.*

(ii) $Pr(\mathbf{Z}:\mathbb{R} \to \mathbb{R}^n$ *is nowhere differentiable)* $= 1$.

(iii) *Let* Ω *be a (reasonably well behaved) set and* $\partial\Omega$ *its boundary. Let* $G:\partial\Omega \to \mathbb{R}$ *be a continuous function. Then if we write* $\phi(\mathbf{x})$ *for the expectation of* $G(\mathbf{Z}(t))$ *given that* $\mathbf{Z}(0) = \mathbf{x} \in \Omega$ *and* t *is the first* $s > 0$ *such that* $\mathbf{Z}(s) \in \partial\Omega$, *it follows that*

$$\nabla^2 \phi = 0 \text{ and } \phi(\mathbf{y}) = G(\mathbf{y}) \text{ for all } \mathbf{y} \in \partial\Omega.$$

In the last chapter we saw that under quite general conditions a random walk on a lattice will return to its starting point with probability 1 if the dimension is 1 or 2 (Theorem 86.1'). It is easy to see from arguments of the type used in the proof of Lemma 86.2 that this means that (with probability 1) a random walk (in dimension 1 or 2) visits each grid point infinitely often. Can we conclude that (a) our Brownian motion in the plane will visit each point of $\mathbb{R}^2$ infinitely often (with probability 1), or (b) only that it will pass arbitrarily near each point of $\mathbb{R}^2$ (with probability 1), or (c) is nothing of the sort true? Unless our probabilistic intuition has completely

451

misled us we may exclude possibility (c), but it is not obvious (to the author) whether we should guess (a) or (b).

In fact (b) is the case and we may show this by means of the following neat argument. (Here, as usual, $\|\mathbf{x}\| = \sqrt{(\sum_{r=1}^{n} x_r^2)}$.)

Plausible Lemma 87.2. *Consider the system of Plausible Result 87.1. Suppose $a < \|\mathbf{x}\| < b$ and write $p_{ab}(\mathbf{x})$ for the probability that a path $\mathbf{Z}:\mathbb{R} \to \mathbb{R}^n$ with $\mathbf{Z}(0) = \mathbf{x}$ hits the circle $\|\mathbf{y}\| = a$ before it hits the circle $\|\mathbf{y}\| = b$. (In other words $p_{ab}(\mathbf{x})$ is the probability that*

$$\inf\{s > 0 : \|\mathbf{Z}(s)\| \leqslant a\} < \inf\{s > 0 : \|\mathbf{Z}(s)\| \geqslant b\}.)$$

Then

(i) *If $n = 1$, $p_{ab}(\mathbf{x}) = (b - \|\mathbf{x}\|)/(b - a)$.*
(ii) *If $n = 2$, $p_{ab}(\mathbf{x}) = (\log b - \log \|\mathbf{x}\|)/(\log b - \log a)$.*
(iii) *If $n \geqslant 3$, $p_{ab}(\mathbf{x}) = (\|\mathbf{x}\|^{2-n} - b^{2-n})/(a^{2-n} - b^{2-n})$.*

Proof. We use the result and notation of Result 87.1 (iii). Let $\Omega = \{\mathbf{y} \in \mathbb{R}^n : a \leqslant \|\mathbf{y}\| \leqslant b\}$. Define G by $G(\mathbf{y}) = 1$ if $\|\mathbf{y}\| = a$, $G(\mathbf{y}) = 0$ if $\|\mathbf{y}\| = b$. The expectation of $G(\mathbf{Z}(t))$, where t is the first time that $\mathbf{Z}(s)$ hits the boundary, is $p_{ab}(\mathbf{x})1 + (1 - p_{ab}(\mathbf{x}))0 = p_{ab}(\mathbf{x})$, so $p_{ab}(\mathbf{x}) = \phi(\mathbf{x})$ where

$$\nabla^2 \phi = 0 \quad \text{in } \Omega,$$

and $\quad\quad\quad\quad \phi(\mathbf{x}) = 1 \quad \text{if } \|\mathbf{x}\| = a, \phi(\mathbf{x}) = 0 \quad \text{if } \|\mathbf{x}\| = b.$

We know (by Theorem 31.3) that the equation and boundary conditions just given admit no more than one solution, so if we can find that solution we shall have found $p_{ab}(\mathbf{x})$.

Most of my readers will be able to write down such a solution at sight. If not, they can observe that we must clearly look for radially symmetric solutions with $\phi(\mathbf{x}) = \psi(r)[r = \|x\|]$. Our equation and boundary conditions become

$$\frac{1}{r^{n-1}}\frac{d}{dr}r^{n-1}\frac{d\psi}{dr} = 0, \psi(a) = 1, \psi(b) = 0,$$

giving $\quad\quad\quad\quad \dfrac{d}{dr}r^{n-1}\dfrac{d\psi}{dr} = 0,$

whence $\quad\quad\quad\quad \dfrac{d\psi}{dr} = \dfrac{A}{r^{n-1}}$

and $\quad\quad\quad\quad \psi(r) = B + A\displaystyle\int_a^r \frac{1}{s^{n-1}}ds$

for constants A and B chosen to fit the boundary conditions. The solution is thus $\phi(\mathbf{x}) = \psi(\|\mathbf{x}\|)$ where, for $a \leqslant r \leqslant b$,

(i) $\psi(r) = (b - r)/(b - a)$ $\quad\quad\quad\quad\quad$ if $n = 1$,

(ii) $\psi(r) = (\log b - \log r)/(\log b - \log a)$ if $n = 2$,

(iii) $\psi(r) = (r^{2-n} - b^{2-n})/(a^{2-n} - b^{2-n})$ if $n \geqslant 3$.

Since $p_{ab}(\mathbf{x}) = \phi(\mathbf{x})$ the stated result follows. ∎

All we need to do now is interpret the meaning of the result above.

Plausible Lemma 87.3. *Let* $\mathbf{x} \in \mathbb{R}^n$, $\mathbf{x} \neq \mathbf{0}$ *and let us consider paths with* $\mathbf{Z}(0) = \mathbf{x}$,

(i) *If* $n \leqslant 2$ *then*

$$Pr\{\|\mathbf{Z}(t)\| \leqslant \varepsilon \text{ for some } t \geqslant 0\} = 1 \text{ for all } \varepsilon > 0.$$

(ii) *If* $n \geqslant 3$ *then*

$$Pr\{\|\mathbf{Z}(t)\| \leqslant \varepsilon \text{ for some } t \geqslant 0\} \to 0 \text{ as } \varepsilon \to 0+$$

(iii) *If* $n \geqslant 2$ *then*

$$Pr\{\mathbf{Z}(t) = \mathbf{0} \text{ for some } t \geqslant 0\} = 0.$$

This result is completed by one with a slightly different proof.

Plausible Lemma 87.3. (iv) *If* $n = 1$ *then*

$$Pr\{\mathbf{Z}(t) = \mathbf{0} \text{ for some } t \geqslant 0\} = 1.$$

Sketch proofs. (i) Observe that, by the definition of $p_{\varepsilon b}$

$$Pr\{\|\mathbf{Z}(t)\| < \varepsilon \text{ for some } t \geqslant 0\} \geqslant p_{\varepsilon b}$$

for all $b > \|\mathbf{x}\|$.

Allowing $b \to \infty$ we see that (if $n = 1$ or $n = 2$) $p_{\varepsilon b} \to 1$ and so

$$Pr\{\|\mathbf{Z}(t)\| \leqslant \varepsilon \text{ for some } t \geqslant 0\} = 1.$$

(ii) Although we do not give a formal proof it seems clear that

$$p_{ab} \to Pr\{\|\mathbf{Z}(t)\| \leqslant a \text{ for some } t \geqslant 0\} \text{ as } b \to \infty.$$

Thus if $n \geqslant 3$

$$Pr\{\|\mathbf{Z}(t)\| \leqslant \varepsilon \text{ for some } t \geqslant 0\} = \lim_{b \to \infty} p_{\varepsilon b}$$

$$= \|\mathbf{x}\|^{2-n}/\varepsilon^{2-n} = \varepsilon^{n-2}/\|\mathbf{x}\|^{n-2} < 1.$$

(iii) Let us write p_{0b} for the probability that $\mathbf{Z}(s)$ hits $\mathbf{0}$ before it hits the circle $\|\mathbf{r}\| = b$. (That is to say

$$p_{0b} = Pr\{\text{There exists a } t > 0 \text{ with } \mathbf{Z}(t) = \mathbf{0} \text{ and } \|\mathbf{Z}(s)\| < b \text{ for all } 0 \leqslant s \leqslant b\}).$$

Plainly, $p_{ab} \geqslant p_{0b}$ for all $b > a > 0$. But if we fix b and let $a \to 0+$ we see that (for

$n \geqslant 2)p_{ab} \to 0$. Thus $p_{0b} = 0$ for all b and (again without formal proof) it seems clear that this implies

$$Pr\{\text{There exists a } t > 0 \text{ with } \mathbf{Z}(t) = \mathbf{0}\} = \lim_{b \to \infty} p_{0b} = 0.$$

(iv) It is possible to construct a proof using the fact that, if $\mathbf{Z}(t)$ is to the right of $\mathbf{0}$ and $\mathbf{Z}(s)$ to the left of $\mathbf{0}$, then $\mathbf{Z}(u) = \mathbf{0}$ for some u between t and s. However, it is simpler (though less illuminating) to compute $q_b(x)$ the probability that a path commencing at x passes through b before it passes through $0[0 \leqslant x \leqslant b]$. Just as in Lemma 87.2 we see that $q_b(x) = x/b$ and so $q_b(x) \to 0$ as $b \to \infty$ for x fixed. A similar result holds if $b \leqslant x \leqslant 0$ and we deduce the result required.   ∎

We may restate Lemma 87.3 in an apparently more general manner.

Plausible Lemma 87.4. *Let* $\mathbf{x}, \mathbf{y} \in \mathbb{R}^n$, $\mathbf{x} \neq \mathbf{y}$ *and let* $t_1 \geqslant 0$. *The following results hold for paths with* $\mathbf{Z}(t_1) = \mathbf{x}$.

(i) *If* $n \leqslant 2$ *then*

$$Pr\{\|\mathbf{Z}(t) - \mathbf{y}\| \leqslant \varepsilon \text{ for some } t \geqslant t_1\} = 1 \text{ for all } \varepsilon > 0.$$

(ii) *If* $n \geqslant 3$ *then*

$$Pr\{\|\mathbf{Z}(t) - \mathbf{y}\| \leqslant \varepsilon \text{ for some } t \geqslant t_1\} \to 0 \text{ as } \varepsilon \to 0 +.$$

(iii) *If* $n \geqslant 2$ *then*

$$Pr\{\mathbf{Z}(t) = \mathbf{y} \text{ for some } t \geqslant t_1\} = 0.$$

(iv) *If* $n = 1$ *then*

$$Pr\{\mathbf{Z}(t) = \mathbf{y} \text{ for some } t \geqslant t_1\} = 1.$$

Sketch proof. Observe that if $\mathbf{Z}(t)$ is Brownian so is $\mathbf{Z}(t + t_1) - \mathbf{y}$ and apply Lemma 87.3.   ∎

We thus see that there are fundamental differences between Brownian motion in one dimension (where, with probability 1, a path passes through any given point), two dimensions (where, with probability 1, a path does not hit a given point but will pass arbitrarily close to it) and three or more dimensions (where, with probability 1, a path will always remain some distance away from a given point). In many ways the two dimensional case is the most interesting and we return to its exclusive study in the next three chapters.

88

ANALYTIC MAPS OF
BROWNIAN MOTION

<hr/>

In the last chapter we saw one reason why (mathematical) Brownian motion is particularly interesting for $\mathbb{R}^2$. Another reason is given by the well known identification of $\mathbb{R}^2$ with $\mathbb{C}$ (where (x, y) corresponds to $x + iy$). We can therefore talk about Brownian motion on $\mathbb{C}$ with Plausible Result 14.3 (i) taking the following form.

Plausible Result 88.1. *There is a family of random continuous maps $Z:\mathbb{R} \to \mathbb{C}$ such that*

$$Pr\{(Z(t + \delta t) - Z(t))/\delta t^{\frac{1}{2}} \in \{x + iy : x \in [a,b], y \in [c,d]\}\}$$
$$= \frac{1}{2\pi} \int_a^b \int_c^d \exp(-(x^2 + y^2)/2)dxdy,$$

independent of previous history.

The following question was posed and answered by Paul Lévy.

Question 88.2. *If $f:\mathbb{C} \to \mathbb{C}$ is a non constant analytic function, what can we say about the family of paths $f(Z)$?*

Let us say that the family of paths described in Plausible Result 88.1 is in standard Brownian motion. If we have such a family Z and we consider the family $Z(kt)$ (k real $k > 0$) we say that the new family is in Brownian motion with rate k. We make the following observations.

Lemma 88.3. *Consider a family Γ of random maps Z which is in standard Brownian motion. Consider the new family Γ_f of random maps $f(Z)$ where f is a function from $\mathbb{C}$ to $\mathbb{C}$.*

(i) *If $f(z) = z + w_0$ then Γ_f is again in standard Brownian motion.*
(ii) *If $f(z) = e^{i\theta}z$ (θ real) then Γ_f is again in standard Brownian motion.*
(iii) *If $f(z) = kz$ (k real, $k > 0$) then Γ_f is in Brownian motion with rate k^2.*
(iv) *If $f(z) = \lambda z + z_0 (\lambda \neq 0)$ then Γ_f is in Brownian motion with rate $|\lambda|^2$.*

Remark. These results should be evident from the discussion of Chapter 14. If not, the reader should reread Chapter 14 until they become evident.

Proof. (i) Observe that

$$Pr\{(Z(t + \delta t) + w_0) - (Z(t) + w_0))/\delta t^{\frac{1}{2}} \in \{x + iy : x \in [a,b], y \in [c,d]\}\}$$
$$= Pr\{(Z(t + \delta t) - Z(t))/\delta t^{\frac{1}{2}} \in \{x + iy : x \in [a,b], y \in [c,d]\}\}.$$

(ii) Let Ψ be any reasonable set in $\mathbb{C}$. Then

$$Pr\{(Z(t + \delta t) - Z(t))/\delta t^{\frac{1}{2}} \in \Psi\}$$

$$= \frac{1}{2\pi} \int\int_{x + iy \in \Psi} \exp(-(x^2 + y^2)/2) dx dy = \frac{1}{2\pi} \int\int_{z \in \Psi} \exp(-|z|^2/2) dA(z).$$

Thus writing $\Psi(0) = \{x + iy : x \in [a,b], y \in [c,d]\}$ and $\Psi(\theta) = \{e^{-i\theta} z : z \in \Psi(0)\}$ we have

$$Pr\{(e^{i\theta} Z(t + \delta t) - e^{i\theta} Z(t))/\delta t^{\frac{1}{2}} \in \Psi(0)\} = Pr\{(Z(t + \delta t) - Z(t))/\delta t^{\frac{1}{2}} \in \Psi(\theta)\}$$

$$= \frac{1}{2\pi} \int\int_{z \in \Psi(\theta)} \exp(-|z|^2/2) dA(z) = \frac{1}{2\pi} \int\int_{z \in \Psi(0)} \exp(-|z|^2/2) dA(z),$$

as required. (The reader may consider that we have taken excessive trouble to prove the obvious fact that Brownian motion is rotation invariant. The author tends to agree.)

(iii) The tautologous equality

$$Pr\{(Z(t + \delta t) - Z(t))/\delta t^{\frac{1}{2}} \in \{x + iy : x \in [a,b], y \in [c,d]\}\}$$
$$= Pr\{(Z(s + \delta s) - Z(s))/\delta s^{\frac{1}{2}} \in \{x + iy : x \in [a,b], y \in [c,d]\}\}$$

yields the slightly less obvious equality

$$Pr\{(Z(t + \delta t) - Z(t))/\delta t^{\frac{1}{2}} \in \{x + iy : x \in [k^{-1}a, k^{-1}b], y \in [k^{-1}c, k^{-1}d]\}\}$$
$$= Pr\{(Z(k^2 t + k^2 \delta t) - Z(k^2 t))/(k\delta t^{\frac{1}{2}}) \in \{x + iy : x \in [k^{-1}a, k^{-1}b],$$
$$y \in [k^{-1}c, k^{-1}d]\}\},$$

whence

$$Pr\{(kZ(t + \delta t) - kZ(t))/\delta t^{\frac{1}{2}} \in \{x + iy : x \in [a,b], y \in [c,d]\}\}$$
$$= Pr\{(Z(k^2(t + \delta t)) - Z(k^2 t))/\delta t^{\frac{1}{2}} \in \{x + iy : x \in [a,b], y \in [c,d]\}\},$$

and the result follows. (This simply confirms the observation of Chapter 14 that the time scale of Brownian motion varies with the square of the length scale.)

(iv) Let $\lambda = |\lambda| e^{i\theta}$ and set $f_1(z) = z + \lambda^{-1} w_0$, $f_2(z) = e^{i\theta} z$, $f_3(z) = |\lambda| z$. Since $f(z) = f_3(f_2(f_1(z)))$ the required result follows. ∎

Plausible Lemma 88.4. (i) *Let Γ be a family of random maps Z in standard Brownian motion. Consider the new family Γ_f of random maps $f(Z)$. If f is locally of the form $f(z) \approx w_0 + \lambda z$ then Γ_f is locally in Brownian motion with rate $|\lambda|^2 [\lambda \neq 0]$.*

(ii) *If* $f:\mathbb{C} \to \mathbb{C}$ *is analytic then locally*

$$f(z) \approx f(z_0) + (z - z_0)f'(z_0) = (f(z_0) - z_0 f'(z_0)) + f'(z_0)z.$$

Remarks. The vagueness of the formulation of (i) prevents anything but a vague and qualified concurrence on the reader's part. Part (ii) is just the definition of analyticity and may be sharpened using Taylor's theorem to give $|f(z) - (f(z_0) + (z - z_0)f'(z_0))| \leqslant A|z - z_0|^2$ for all $|z - z_0| < \delta$ and some A and $\delta > 0$. (The approximation given is thus reassuringly good.)

Plausible Lemma 88.5′. *Let Γ be a family of random maps Z in standard Brownian motion. If $f:\mathbb{C} \to \mathbb{C}$ is analytic and non constant then, ignoring an event of probability 0, the family Γ_f of random maps is locally in Brownian motion with rate $|f'(Z)|^2$.*
Plausible Argument. We combine the two parts of Plausible Lemma 88.4. This cannot be done for any Z such that $f'(Z(t)) = 0$ for some time $t \in \mathbb{R}$ i.e. for those Z which pass through a zero of f'. But the zeros of f' are isolated (and in particular there are only a finite number in any disc surrounding 0, see e.g. Lemma 75.6 (i)) and the probability that a particular Z passes through a particular point is 0 (Plausible Lemma 87.4 (iii)). Thus we conclude that the probability that a particular Z passes through a zero of f' is 0. Hence the event $f'(Z(t)) = 0$ for some t has probability zero and, ignoring this event, we have the result. ∎

Our next remark bears on the peculiar status of events with probability zero. Consider the 3 random variables X_1, X_2, X_3 defined as follows.

(i) X_1 takes any value in $[0, 1]$ and

$$Pr(a \leqslant X_1 \leqslant b) = b - a \quad [0 \leqslant b \leqslant a \leqslant 1].$$

(ii) X_2 takes any value in $[0, 1] \backslash \{1/2\}$ and

$$Pr(a \leqslant X_2 \leqslant b) = b - a \quad [0 \leqslant b \leqslant a \leqslant 1].$$

(iii) X_3 takes any value in $[0, 1] \cup \{2\}$ and

$$Pr(a \leqslant X_3 \leqslant b) = b - a \quad [0 \leqslant b \leqslant a \leqslant 1],$$
$$Pr(X_3 = 2) = 0.$$

Now X_1, X_2 and X_3 are different objects (for example, X_3 can take the value 2 but X_1 and X_2 cannot) but from the point of view of probability theory they behave in exactly the same way (although the event $X_3 = 2$ is possible it has probability 0). We therefore usually identify X_1, X_2 and X_3. In the same way, if we have two probabilistic models which give the same probabilities to the same events, we usually identify them although certain events which are impossible in one system are possible (though, of course, with probability 0) in the other. With this convention (which we have already adopted implicitly in using Plausible Result 88.1 as the definition of Brownian motion) we may drop the words 'ignoring an event of probability 0' from the statement of Plausible Lemma 88.5′.

Plausible Lemma 88.5. *Let* Γ *be a family of random maps* Z *in standard Brownian motion. If* $f : \mathbb{C} \to \mathbb{C}$ *is analytic and non constant then the family* Γ_f *of random maps* $f(Z)$ *is locally in Brownian motion with rate* $|f'(Z)|^2$.

Suppose now we let a particle move in standard Brownian motion so that its position at time t is $Z(t)$. At the same time we film the motion of a particle at $f(Z(t))$ but instead of filming at a constant rate we let the camera run faster or slower so that the number of frames per second is proportional to $|f'(Z(t))|^2$ at time t. Plausible Lemma 87.5 says, in effect, that if we now show a collection of such films at a constant number of frames per second the result will be indistinguishable from a collection of films of particles in (ordinary) Brownian motion at a constant rate.

Remark. The suspicious reader may notice that there are two ways in which our strategy of filming with a variable speed camera may go wrong. In the first the camera runs so fast that we use up an infinite amount of film in a finite time (this would happen, for example, if the number of frames per second were proportional to $(1-t)^{-1}$ for $0 \leqslant t < 1$). In the second the camera runs so slowly that we only use a finite amount of film in an infinite time (this would happen, for example, if the number of frames per second was proportional to $e^{-|t|}$).

Fortunately the first possibility is excluded by the fact that $Z(t)$ and thus $|f'(Z(t))|$ are continuous and so bounded on any closed interval. The second possibility could occur, but only with probability 0. To see this, choose a $z_0 \in \mathbb{C}$ with $f'(z_0) \neq 0$. By Plausible Lemma 87.4 we know that (with probability 1) $Z(t)$ will pass close to z_0 infinitely often and so, we would expect the total length of time for which $Z(t)$ is close to z_0 to be infinite (with probability 1). Thus we would have $|f'(Z(t))| \geqslant \frac{1}{2}|f'(z_0)|$ for an infinite amount of time, and, since the camera runs at a rate proportional to $|f'(Z(t))|^2$ we use an infinite amount of film (with probability 1). Thus by excluding an event of probability 0, as we did in the transition between Plausible Lemmas 88.5′ and 88.5 we may ensure that our filming does not run into the kind of trouble described above.

We have shown that although the family of paths $f(Z(t))$ do not, in general, describe Brownian motion we can, by retiming them individually, obtain a family of paths which do describe Brownian motion. Let us call a family of paths which can be so retimed a family of Brownian paths.

Plausible Theorem 88.6 (Paul Lévy). *Let* Γ *be a family of random maps* Z *in standard Brownian motion. If* $f : \mathbb{C} \to \mathbb{C}$ *is analytic and non constant then the family* Γ_f *of random maps* $f(Z)$ *is a family of Brownian paths.*

Why is Lévy's observation so important? Simply because a family of Brownian paths is obtained by (individually) retiming Brownian motion and any property of Brownian motion which is unaffected by retiming will remain true for Brownian paths. For example Plausible Result 14.3 (iii) takes the following form.

Plausible Result 88.7. *Consider a family of Brownian paths Z. Let $\phi(z)[z\in\Omega]$ be the expectation of $G(Z(t))$ given that $Z(0) = z$ and t is the first $s > 0$ such that $Z(s)\in\partial\Omega$. Then $\nabla^2\phi = 0$ and $\phi(w) = G(w)$ for all $w\in\partial\Omega$.*

Combined with Lévy's theorem this gives us another way of seeing why the conformal mapping enables us to obtain a new solution of Laplace's equation from an old one.

Another result which carries over is Plausible Lemma 87.4 (i).

Plausible Lemma 88.8. *Consider a family of Brownian paths $W:\mathbb{R}\to\mathbb{C}$. Then for each $w_0\in\mathbb{C}$ and $\varepsilon > 0$*

$$Pr\{|W(t) - w_0| < \varepsilon \text{ for some } t\} = 1.$$

Combining this with Lévy's theorem (Plausible Theorem 88.6) we obtain the following result.

Plausible Lemma 88.9. *Consider a family of Brownian paths $Z:\mathbb{R}\to\mathbb{C}$. Then for each $w_0\in\mathbb{C}$ and $\varepsilon > 0$*

$$Pr\{|f(Z(t)) - w_0| < \varepsilon \text{ for some } t\} = 1.$$

Now the fact that something has probability 1 does not mean that it happens *every time*. (If X is a random variable uniformly distributed on $[0,1]$, the event $\{X \neq \frac{1}{2}\}$ has probability 1 but it is not certain that $X \neq \frac{1}{2}$. The event $\{X = \frac{1}{2}\}$ has probability 0 but is not impossible.) However, if an event has probability 1 it certainly happens *sometimes*. Thus the event described in Plausible Lemma 88.9 sometimes occurs.

Plausible Lemma 88.10. *If $f:\mathbb{C}\to\mathbb{C}$ is a non constant analytic function then given $w_0\in\mathbb{C}$ and $\varepsilon > 0$ we can find a $u:\mathbb{R}\to\mathbb{C}$ such that*

$$|f(u(t)) - w_0| < \varepsilon \text{ for some } t\in\mathbb{R}.$$

Proof. Take a Z for which the result of Plausible Lemma 88.9 is true and set $u = Z$. ∎

But, if we look carefully, we see that Plausible Lemma 88.10 is simply a complicated way of stating the following theorem.

Theorem 88.11 (Casorati–Weierstrass). *If $f:\mathbb{C}\to\mathbb{C}$ is a non constant analytic function then given $w_0\in\mathbb{C}$ there exists a $z_0\in\mathbb{R}$ such that*

$$|f(z_0) - w_0| < \varepsilon.$$

We have thus recovered the classical theorem that the range of a non constant analytic function is dense in $\mathbb{C}$. (The reader will probably have met the slightly

harder version of this theorem which states that in the neighbourhood of an isolated essential singularity a meromorphic function takes values arbitrarily close to any assigned complex number. Theorem 88.11 could thus be called the 'little' theorem of Casorati–Weierstrass.) In the next two chapters we shall show how a deeper study of the topological properties of Brownian motion (and so of Brownian paths) leads to the marvellous 'little' theorem of Picard.

As a preparation I suggest that the reader consider the arguments of the present chapter for the case $f(z) = \exp z$ considering in particular the behaviour of $f(Z(t))$ when $\operatorname{Re} Z(t)$ is large and negative.

89

WILL A BROWNIAN
MOTION TANGLE?

Suppose we take a piece of board with an origin marked at 0 and a drawing pin stuck in it at A say. We throw a piece of string on the board both of whose ends fall near 0. We fix both ends at 0 and then try to draw the string in to a bundle round 0. This may be possible (see Figure 89.1(a)) or impossible because the string always catches on A (Figure 89.1(b)). In the second case we say that the string is tangled round A.

The result that follows may be stated precisely and proved by using the ideas of algebraic topology.

Plausible Lemma 89.1. *If $f:[a,b] \to \mathbb{R}^2$ is a continuous function with $f(a) = f(b) = 0$ and $f(t) \neq x_0$ for all $t \in [a,b]$ then the point $f(t)$ moves n times clockwise round x_0 as t runs from a to b where n is an integer. The curve is tangled round x_0 if and only if $n \neq 0$.*

The reader should convince herself that this lemma is plausible before continuing.

Does a Brownian motion in $\mathbb{R}^2$ starting at Q tangle itself round a given point x_0 or does it remain untangled? To make the question more precise let us consider a specific Brownian motion $y(t)$ with $y(0) = 0$. Since Brownian motion in $\mathbb{R}^2$ fails to pass through any given point with probability 1 we may assume that $y(t) \neq x_0$ for all t. Consider a small $\delta > 0$ and a very small $\varepsilon > 0$ and times $s_1 < t_1 < \cdots$ defined

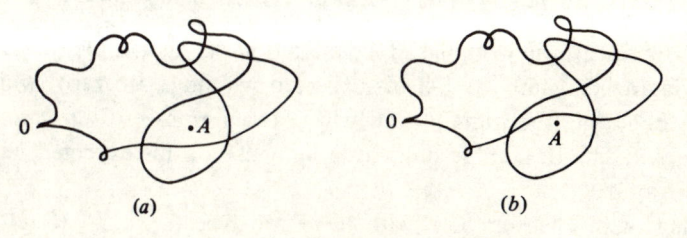

(a)   (b)

Fig. 89.1. Will the string catch?

461

Fig. 89.2. Path of Brownian motion between times s_j and t_j.

by $t_0 = 0$

$$s_{j+1} = \inf\{t \geqslant t_j : \|\mathbf{y}(t)\| = \delta\},$$
$$t_j = \inf\{t \geqslant s_j : \|\mathbf{y}(t)\| = \varepsilon\}.$$

Thus in some sense s_j marks the jth time $\mathbf{y}$ leaves the immediate vicinity of $\mathbf{0}$ and t_j the jth time that $\mathbf{y}$ returns (see Figure 89.2).

We can now state our question more precisely. Let us say that the Brownian motion $\mathbf{y}(t)$ starting with $\mathbf{y}(t) = 0$ is tangled at t_j if the curve $\mathbf{z}_j : [0, t_j + 1] \to \mathbb{R}$ defined by

$$\mathbf{z}_j(t) = \mathbf{y}(t) \qquad \text{for } 0 \leqslant t \leqslant t_j,$$
$$\mathbf{z}_j(t_j + s) = (1 - s)\mathbf{y}(t_j) \quad \text{for } 0 \leqslant s \leqslant 1,$$

is tangled. What can we say about the tangledness of $\mathbf{y}$ at $t_1, t_2, t_3, \ldots$?

Observe first that, in the time interval $[t_j, t_{j+1}]$, $\mathbf{y}(t)$ will move (approximately) N_j times round $\mathbf{x}_0$ where N_j is an integer, and that the curve $\mathbf{z}_j(t)$ will move (exactly) $\sum_{k=1}^{j} N_k$ times round $\mathbf{x}_0$ as t runs from 0 to $t_j + 1$. Thus $\mathbf{y}$ will be tangled at t_j if and only if $\sum_{k=1}^{j} N_k \neq 0$. The following facts are (more or less) clear.

(A) $P(N_k = r) = P(N_k = -r)$ by symmetry,
(B) $P(N_k = 0) < 1$.

(Let $\mathbf{x}_0 = (x_0, 0)$ to fix ideas. It is clear that, with positive probability, $\mathbf{y}(t)$ will cross the line segment $\{(x, 0) : x > x_0\}$ in the interval $[t_k, t_{k+1}]$ and that of the paths that do cross at least half will return in such a way that $N_k \neq 0$. Thus $P(N_k = 0) < 1$.)

(C) N_j and N_{j+1} are 'nearly independent' if ε is sufficiently small.

Thus if N_j represents the winnings of a gambler on his jth gamble we can say that he is playing a fair game (by (A) his expected winnings are zero) which is not trivial (by (B) he will from time to time win or lose a non-zero sum) and in which his winnings at the jth gamble have little influence on his expected winnings at the $j + 1$th gamble.

By analogy with Theorem 86.1 (with $m = 1$) we expect $\sum_{k=1}^{j} N_k$ to return to the value 0 infinitely often. However, common sense or a more delicate analysis (such

as the magnificent treatment in Chapter 3, Vol. 1 of the Second Edition of Feller *An Introduction to Probability Theory And Its Applications*) suggests that for most of the time (i.e. most values of j) $\sum_{k=1}^{j} N_k$ will be non-zero. Using Lemma 89.1 we now translate this statement back to one on Brownian motion.

Plausible Theorem 89.2. *With probability* 1, $\mathbf{y}(t)$ *is not tangled at* t_j *for infinitely many values of* j. *However, with probability* 1 *also,*

$$M^{-1} \times (\text{number of values of } 1 \leqslant j \leqslant M \text{ with } \mathbf{y} \text{ untangled at } t_j) \to 0 \text{ as } M \to \infty.$$

Thus, with probability 1, Brownian motion is tangled most of the time but very occasionally untangles.

What happens if we provide another drawing pin B in addition to A? There are now many more ways in which the string can tangle. For example Figure 89.3 below shows a case in which the string is not tangled round A or round B but none the less we can not draw the string into a bundle round 0.

A little thought shows (but does not prove) that the process of tangling can still be described quite simply. Consider a path that starts and finishes near 0. It may be described as a succession of actions of the form 'wind once round A clockwise', 'wind once round B anticlockwise' and so on. Writing a for 'wind once round A anticlockwise', a^{-1} for 'wind once round A clockwise', b for 'wind once round B anticlockwise' and b^{-1} for 'wind once round B clockwise', we may describe a tangle by the sequence a, a^{-1}, b, b, a^{-1} (wind anticlockwise round A, then clockwise round A, then anticlockwise round B, then anticlockwise round B again, then clockwise round A) or even more briefly as $aa^{-1}bba^{-1}$.

The notation suggests writing $aa^{-1}bba^{-1} = b^2a^{-1}$ and the reader should have no difficulty in convincing herself that the tangle $aa^{-1}bba^{-1}$ of Figure 89.4 (i) can be distorted to the tangle b^2a^{-1} of Figure 89.4 (ii) (twice anticlockwise round B and then clockwise round A) without the string catching on A or B and without moving the two ends at 0.

Plausible Lemma 89.3. *If we write the state of the string as* $\varepsilon_1 \varepsilon_2 \dots \varepsilon_n$ *with* $\varepsilon_j = a, a^{-1}$, b *or* b^{-1} *then the string is tangled if and only if the expression* $\varepsilon_1 \varepsilon_2 \dots \varepsilon_n$ *does not cancel in the obvious way to the identity.*

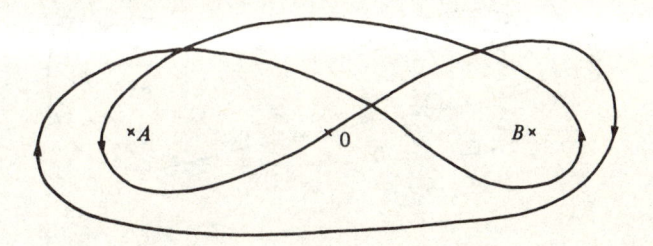

Fig. 89.3. String tangled round two points jointly but not separately.

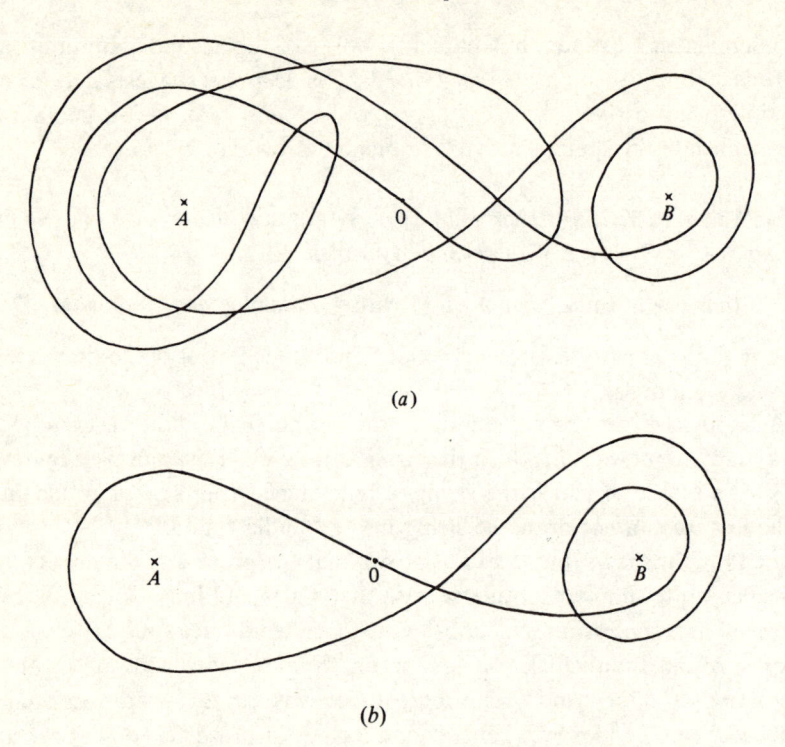

(a)

(b)

Fig. 89.4.(a) The tangle $aa^{-1}bba^{-1}$. (b) The tangle b^2a^{-1}.

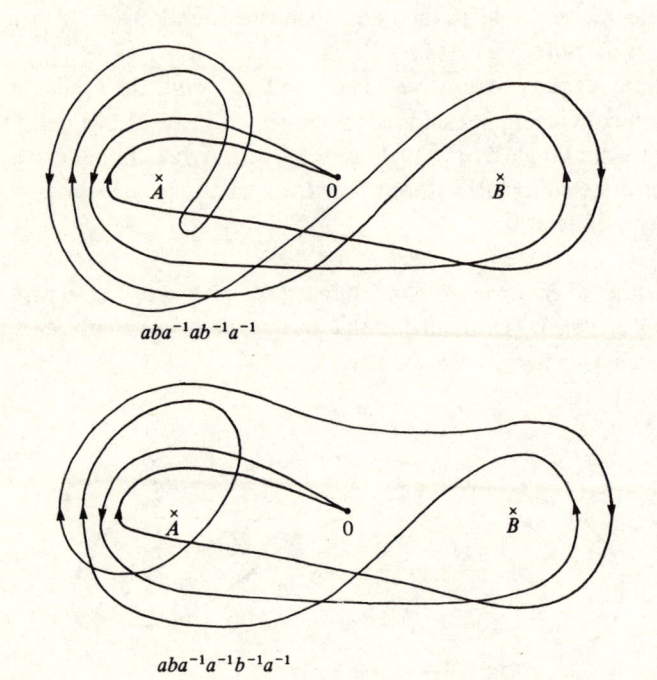

$aba^{-1}ab^{-1}a^{-1}$

$aba^{-1}a^{-1}b^{-1}a^{-1}$

Fig. 89.5. The non-tangle $aba^{-1}ab^{-1}a^{-1}$ and the tangle $aba^{-1}a^{-1}b^{-1}a^{-1}$.

Thus $aba^{-1}ab^{-1}a^{-1}$ does not give a tangle but $aba^{-1}a^{-1}b^{-1}a^{-1}$ does (Figure 89.5).

Just as the description of a rational number can have different forms e.g. $\frac{6}{12} = \frac{7}{14} = \frac{-3}{-6}, \ldots$, etc. so the description of our string can have different forms: e.g. $abbaa^{-1}b^{-1}b$, $abbaa^{-1}bb^{-1}$, $aa^{-1}abb, \ldots$, etc. But just as a rational number can be expressed uniquely in its lowest terms, e.g. $\frac{6}{12} = \frac{1}{2}$, so the description of our string has a shortest representation e.g. $abbaa^{-1}b^{-1}b = abb$. We say that our string is more or less tangled according as the shortest representation is more or less long.

Lemma 89.4. *If x has shortest representation $\varepsilon_1\varepsilon_2\ldots\varepsilon_n$ with $n \geqslant 1$ and y has shortest representation $\varepsilon_{n+1}\varepsilon_{n+2}\ldots\varepsilon_{n+m}$ with $m \geqslant 1$ then xy has shortest representation $\varepsilon_1\varepsilon_2\ldots\varepsilon_{n+m}$ unless $\varepsilon_n = \varepsilon_{n+1}^{-1}$.*

Let us now return to Brownian motion. We have seen that Brownian motion tangles most of the time with only one point. What happens with two? For simplicity we place A and B symmetrically at $\mathbf{x}_0$ and $-\mathbf{x}_0$. We define s_j, t_j and $\mathbf{z}_j$ as before and ask about the tangledness of y at $t_1, t_2, t_3, \ldots$.

In the case of just one point the path between t_j and t_{j+1} was almost as likely to help unwind the tangle as to wind it. But now Lemma 89.4 shows that for every kind of path that starts unwinding the tangle we have three which tangle it further. More precisely if the path from 0 to t_j is described by $\varepsilon(1)\varepsilon(2)\ldots\varepsilon(n)$ (in its shortest form) than the path from t_j to t_{j+1} will be more tangled unless the path from t_j to t_{j+1} reduces to the identity (in which case there is no change) or the path from t_j to t_{j+1} is written in its shortest form as $\varepsilon(n+1)\varepsilon(n+2)\ldots\varepsilon(n+m)$ with $\varepsilon(n+1) = \varepsilon(n)^{-1}$ (and even in this case it could still become more tangled).

Thus anyone who hopes that the string will untangle is, in effect, playing against loaded dice. And, as we saw illustrated at the end of Chapter 87, anyone playing against loaded dice should expect, even if they start with a few runs of good luck, eventually to go into debt and stay there.

Plausible Theorem 89.5′. *Let us replace the single point of Theorem 89.2 by two points. Then, with probability 1, a Brownian motion is such that for some j_0 (depending on $\mathbf{y}$), $\mathbf{y}(t)$ is tangled at t_j for all $j \geqslant j_0$.*

For the purposes of the next chapter we recast this plausible theorem in a more exact form. (We also allow the Brownian motion to start at a point $\mathbf{x}_0$ near $\mathbf{0}$. This slight change, which, obviously, does not affect the arguments used above will be needed in the next chapter.)

Plausible Theorem 89.5. *Let $\varepsilon > 0$ be small compared to 1 and let $\|\mathbf{x}_0\| < \varepsilon$. Then, with probability 1, a Brownian motion $\mathbf{y}:[0,\infty) \to \mathbb{R}^2$ starting at $\mathbf{x}_0$ will have an associated S with the following property. Suppose $T > S$ and $|\mathbf{y}(T)| \leqslant \varepsilon$. Then any*

path $\mathbf{w}:[0, T+1] \to \mathbb{R}^2$ *with*

$$\mathbf{w}(t) = \mathbf{y}(t) \qquad \text{for } 0 \leqslant t \leqslant T,$$
$$\mathbf{w}(T+1) = \mathbf{x}_0,$$
$$\|\mathbf{w}(t)\| \leqslant \varepsilon \qquad \text{for } T \leqslant t \leqslant T+1,$$

is tangled round the set $\{-1, 1\}$.

Remark. Note that S depends on $\mathbf{w}$. Some Brownian motions tangle quickly, some slowly but they all (with probability 1) tangle in the end.

90

LA FAMILLE PICARD
VA A MONTE CARLO

In Chapter 88 we saw how the study of analytic maps of Brownian motion lead to the 'little' theorem of Casorati and Weierstrass.

Theorem 90.1. *If $f:\mathbb{C} \to \mathbb{C}$ is analytic and non-constant then the range of f is dense in $\mathbb{C}$.*

In this chapter we show how the additional ideas of Chapter 89 lead to the deeper 'little' theorem of Picard.

Theorem 90.2. *If $f:\mathbb{C} \to \mathbb{C}$ is analytic and non-constant then the range of f omits at most one point of $\mathbb{C}$.*

In other words we can choose $w_0 \in \mathbb{C}$ such that for each $w \neq w_0$ the equation $f(z) = w$ has a solution. If $f(z) = \exp z$ then we already know that the result is true with $w_0 = 0$. If f is a polynomial then the fact that every non-constant polynomial has a root shows that the range of f is $\mathbb{C}$. But Picard's theorem says that *any* arbitrarily chosen analytic function such as $f(z) = z \sin z + \exp(z^2 + \exp z)$ has range $\mathbb{C}$ or $\mathbb{C} \backslash \{w_0\}$. This theorem is one of the glories of nineteenth-century mathematics, a jewel 'five words long that on the stretched forefinger of all time sparkles forever'. And the discovery by Burgess Davis that this result could be proved using Brownian motion was a triumph for modern probability theory.

Let us sketch Davis' remarkable proof. We need some simple preparations. Suppose f is a non-constant analytic function whose range omits z_1 and z_2. By replacing f by $\lambda(f - \mu)$ if necessary, we may suppose $z_1 = 1$ and $z_2 = -1$ (take $\mu = (z_1 + z_2)/2$, $\lambda = 2/(z_1 - z_2)$). Choose $\varepsilon > 0$ to be small. By the theorem of Casorati and Weierstrass (Theorem 89.1) we can find z_0 such that $|f(z_0)| < \varepsilon/2$. By the continuity of f we can find a $\delta > 0$ such that $|z - z_0| < \delta$ implies $|f(z) - f(z_0)| < \varepsilon/2$ and so $|f(z)| < \varepsilon$.

Just as in Chapter 88, we now identify $\mathbb{R}^2$ and $\mathbb{C}$ and consider a particle describing Brownian motion $z(t)$ starting with $z(0) = z_0$. In Chapter 88 we saw that $f(z(t))$

describes a Brownian path (that is to say a Brownian motion whose rate may vary, i.e. whose local clock may run fast or slow). But Theorem 88.5 tells us that a Brownian motion tangles round two points and so a Brownian path tangles round two points (with probability 1).

Plausible Lemma 90.3. *With probability 1, there exists an S such that if $T > S$ and $|f(z(t))| \leqslant \varepsilon$ then any path $w\colon [0, T+1] \to \mathbb{C}$ with*

$$
\begin{aligned}
w(t) &= f(z(t)) &&\text{for } 0 \leqslant t \leqslant T, \\
w(T+1) &= z_0, \\
|w(t)| &\leqslant \varepsilon &&\text{for } T \leqslant t \leqslant T+1,
\end{aligned}
$$

is tangled round the set $\{-1, 1\}$.

Proof. This is a straight translation of Plausible Theorem 89.5. ∎

But we also know that (with probability 1) the Brownian motion $z(t)$ passes infinitely often arbitrarily close to its starting point z_0.

Plausible Lemma 90.4. *With probability 1, there exists a $T > S$ such that*

$$
|z(T) - z_0| \leqslant \delta.
$$

Proof. Choose z_1 with $|z_1 - z_0| \leqslant \delta/2$, $z_1 \neq z_0$ and use Plausible Lemma 87.4 (i) to show that (with probability 1) there exists a $T > S$ with $|z(T) - z_1| \leqslant \delta/2$. ∎

We now draw these two results together.

Lemma 90.5. *If the events described in Plausible Lemmas 90.3 and 90.4 occur then defining $u\colon [0, T+1] \to \mathbb{C}$ by*

$$
\begin{aligned}
u(t) &= z(t) &&\text{for } 0 \leqslant t \leqslant T, \\
u(T+s) &= s z_0 + (1-s)z(T) &&\text{for } 0 \leqslant s \leqslant 1,
\end{aligned}
$$

we know that $f \circ u(t) = f(u(t))$ defines a path $f \circ u\colon [0, T+1] \to \mathbb{C}$ which tangles round $\{-1, 1\}$.

Proof. Observe that

$$
|u(T+s) - z_0| = (1-s)|z(T) - z_0| < \delta,
$$

so $|f(u(T+s)) - f(z_0)| \leqslant \varepsilon/2$ for $0 \leqslant s \leqslant 1$ and so $|f \circ u(t)| \leqslant \varepsilon$ for $T \leqslant t \leqslant T+1$. The conditions of Plausible Lemma 90.3 thus apply. ∎

As we remarked in Chapter 88, the fact that something has probability 1 does not mean that it happens *every time* but if an event has probability 1 it certainly happens *some times*. Thus the events described in Plausible Lemmas 90.3 and 90.4 sometimes occur.

Plausible Lemma 90.6. *There is a continuous map* $u:[0, T+1] \to \mathbb{C}$ *such that*

$$u(0) = u(T+1) = z_0$$

and $f \circ u$ *tangles round* $\{-1, 1\}$.
Proof. Use Lemmas 90.3, 90.4 and 90.5. ∎

At first sight the result just obtained looks harmless enough but it leads directly into a contradiction.

Lemma 90.7. *Suppose* $g:\mathbb{C} \to \mathbb{C}$ *is continuous and the range of* g *does not contain* -1 *and* 1. *Then if* $v:[0, T+1] \to \mathbb{C}$ *is a continuous map such that*

$$v(0) = v(T+1) = z_0,$$

then $g \circ v$ *does not tangle round* $\{-1, 1\}$.
Proof. For $0 \leqslant s \leqslant 1$ set $G_s(t) = g((1-s)v(t))$ $[0 \leqslant t \leqslant T+1]$. Then, if we take G_s to represent a loop of string at time s, we see before us a method of continuously crumpling the loop $G_0 = g \circ v$ into a point $G_1 = z_0$. Since the range of g does not contain 1 or -1, none of the loops G_s pass through 1 or -1 and so the loop $G_0 = g \circ v$ cannot have been tangled. ∎

Since the only assumption we made about f to obtain Lemma 90.6 was that its range omitted two distinct values, it follows that this assumption is impossible and Picard's theorem (Theorem 90.2) must be true.
Remark. At first sight it might seem that the arguments above would also give a contradiction if f omitted only one value 1, say, since Plausible Theorem 89.2 tells us that $f(z(t))$ spends most of its time tangled round 1. But it is also true that $z(t)$ spends most of its time far away from z_0. As the reader may check explicitly in the case $f(z) = \exp z + 1$, $z_0 = 0$, the times t that $z(t)$ is close to z_0 correspond exactly to the times when $f \circ z$ is untangled. There is thus no contradiction.

In Chapter 14 and Chapters 87–90 I have attempted to give an intuitive account of the mathematical ideas connected with Brownian motion. Some of the gaps in my arguments will be obvious to the reader, but she is warned that here as elsewhere in mathematics it is the obvious gaps which are easiest to fill. (For example the motion of 'untangled' is easily formalised as 'homotopic to a point' and Lemma 89.3 then becomes a fairly elementary result of algebraic topology.) It is the problems which lurk on the periphery of our field of vision and which are felt rather than seen which are most difficult to identify and resolve. In the case of mathematical Brownian motion these subtle problems appear to be most easily identified and resolved within the framework of measure theory and then only by extremely careful argument. But prizes like the proof of Burgess Davis make such work worthwhile.

PART VI
OTHER DIRECTIONS

91

THE FUTURE OF MATHEMATICS
VIEWED FROM 1800

The discovery of the calculus was followed by a century of brilliant exploitation led by men like Euler and Lagrange. But towards the end of that century it became clear that several problems of physics and pure mathematics were resisting all attacks. In addition the lack of any clear and agreed foundation for the calculus led to paradoxes and obscurities which might be ignored but could not be resolved.

In a much quoted letter to d'Alembert in 1781, Lagrange wrote

It seems to me that the mine is already almost too deep, and unless we discover new seams we shall sooner or later have to abandon it. Today Physics and Chemistry offer more brilliant and more easily exploited riches; and it seems that the taste of the century has turned entirely in that direction. It is not impossible that the mathematical positions in the Academies will one day become what the University chairs in Arabic are now.

(p. 386, Vol. 13 of Lagrange's *Collected Works*).

These sentiments were echoed by Delambre in his summary for the Academy of Sciences of the state of the mathematical sciences ('Rapport Historique sur les Progrès des Sciences Mathématiques Depuis 1789'). Speaking of the calculus he described the general view of such discoveries. 'Prepared by the work of many centuries, when they are sufficiently ripe to be plucked by genius, their first effect is to excite admiration and their second to impose on following generations immense labours whose glory will never equal their difficulty.'

It would be difficult (he continues) and perhaps rash to try and analyse the chances which the future offers for the progress of mathematics; in almost all its branches we are halted by insurmountable difficulties and only improvements of detail seem to remain for us. All movements which cannot be reduced to small oscillations about a stable point subject to simple laws, all solutions which we are not given initially to a first approximation, all these things seem beyond our grasp...

After remarking that the fundamental principles of hydrodynamics still eluded

mathematicians, he goes on: – 'All these difficulties seem to show that the power of our methods is almost exhausted, just as the power of ordinary algebra was in the time of Leibnitz and Newton and that we need some new method to enable us to study transcendental functions and the equations which contain them.'

Delambre was mathematical secretary to the Academy of Sciences and his report appeared in 1810. Three years previously the prefect of Isère had submitted to the Academy a memoir on the conduction of heat which was received without enthusiasm. Looking back, we now see this memoir as heralding the surge of new mathematical methods and results which were to mark the new century.

92

WHO WAS FOURIER? I

(The contents of the next two chapters are based on the excellent biography by Herivel *Joseph Fourier The Man and the Physicist*, Oxford 1975.)

Joseph Fourier was born in 1768 in Auxerre, the ninth child of a master tailor. Although the death of his father left him an orphan at the age of ten, his intelligence gained him a free place at the local Benedictine school. At the end of a brilliant school career he applied to enter the artillery only to be informed that such a profession was only open to those of noble blood and was closed to him 'even if he were a second Newton'.

Fourier began to prepare to enter the Benedictine teaching order but, whatever his plans may have been, the course of his life was violently altered by the outbreak of the French Revolution. 'As the natural ideas of equality developed, it was possible to conceive the sublime hope of establishing among us a free government exempt from kings and priests and to free from this double yoke the long usurped soil of Europe. I readily became enamoured of this cause, in my opinion the greatest and the most beautiful which any nation has undertaken.'

However, the Revolution was soon threatened by problems of its own making. The collapse of royal authority and the effects of revolutionary zeal and political misjudgement created administrative chaos whilst at the same time driving France into war against most of Europe. A succession of military defeats aggravated by treachery, economic problems and internal revolution put the Republic in danger and brought Robespierre's party to power. In the next two years France was reorganised for war and central power reasserted. Both at a national and a local level, educated men were needed to fill the administrative vacuum caused by the revolution. Fourier, who was now teaching at his old school, became first an active member of the revolutionary committee of Auxerre and then its president.

The situation of the new Republic called for ruthless measures which the government, conscious of its own revolutionary virtues, was well prepared to take. Treachery was fought by a political terror in which opponents both to the left and right were executed and, as the definition of treachery was extended, it became clear that no one was safe. Fourier himself was arrested, released and then

rearrested. A deputation from Auxerre which, with considerable courage, went to Paris to plead his case, was told – 'Yes, he speaks well, but we no longer have any need of musical patriots.' Only the fall of Robespierre saved Fourier's head.

However Fourier's release did not mark the end of his troubles. As coup d'état followed coup d'état, and the revolution swung erratically to the right he would remain a marked man. No one had been executed in Auxerre but Fourier had been an agent of the terror there. His arrest was on a charge of Hébertism and the Hébertists were to the left of Robespierre. The word 'terrorist' then, like 'Trotskyist' now, denoted a defeated yet feared opponent.

Luckily an opportunity to leave Auxerre now presented itself. A new college (the Ecole Normale) was being set up in Paris to help train teachers and Fourier was nominated as a pupil by a neighbouring district. Fourier could now study under men like Lagrange, Monge and Laplace and escape his terrorist past. Fourier's talents were soon noted, but the college was not successful and its closure was followed by further problems for Fourier.

'We shudder when we think that the pupils of the Ecole Normale were chosen under the reign of Robespierre and his protégés. It is only too true that Balme and Fourier, pupils of the department of Yonne have long professed the atrocious principles and infernal maxims of the tyrants. Nevertheless they prepare to become teachers of our children. Is it not to vomit their poison in the bosom of innocence (From an address to the National Convention, quoted by Herivel).' Fourier was again arrested, released, rearrested and finally, following yet another political swing, released to become a teacher at the new Ecole Polytechnique.

Here Fourier remained for three years. That his talent was recognised is shown by the fact that he succeeded Lagrange in the Chair of Analysis and Mechanics. This quiet interlude was ended by a government order to join the invasion of Egypt. Ostensibly intended to liberate Egypt from the Turks and to threaten the British position in India, the expedition may have been seen by the government as a way of keeping a troublesome general as far away as possible and by the general (Napoleon) as the first step toward becoming Emperor of the East. Fourier was one of a group of scientists and intellectuals intended to form part of the immense cultural benefits that France was to bestow on Egypt.

Unfortunately the effects of stunning French victories on land were negated by Nelson's destruction of the French invasion fleet. Learning of serious military and domestic problems in France, Napoleon deserted his army and rushed home to 'save France'. A rapid and successful coup d'état enabled him to read the dispatches in which General Kléber (his successor in Egypt) explained exactly what he thought of this behaviour.

Both before and after Napoleon's departure, Fourier occupied several important administrative and political posts in Egypt. When the French expedition finally surrendered in 1801 and Fourier was repatriated, Napoleon offered him the post of Prefect of the Department of the Isère centred round Grenoble. (France had been divided into 83 Departments and each Prefect governed his Department on

behalf of the central government.) Although he could have continued as Professor at the Polytechnique, Fourier accepted the offer. Herivel suggests that Egypt had given him a taste for administration and that he hoped to rise higher. Herivel also suggests that Fourier's close association with Kléber after Napoleon's departure accounts for the fact that these hopes were not fulfilled.

Fourier seems to have been a popular and efficient Prefect. His greatest achievement during his 14 years of office was by reconciling the conflicting interests of some forty communities to enable the swamps of Bourgoin to be drained. The draining of twenty thousand acres of swamps resulted in major economic and health benefits and was achieved during a period more noted for grandiose paper plans than for concrete achievements. Fourier's other administrative memorial was a new road across the Alps (now Route N91).

Apart from his prefectorial duties Fourier helped organise the *Description of Egypt*. This work written by the intellectuals attached to the Egyptian expedition did much to inspire European interest in Egypt and was thus one of the two permanent results of that expedition. (The other was the discovery of the Rosetta Stone, a trilingual inscription which was to provide the key to the deciphering of hieroglyphics.) Fourier's main contribution was the general introduction – a survey of Egyptian history up to modern times. (An Egyptologist with whom I discussed this described the introduction as a masterpiece and a turning point in the subject. He was surprised to hear that Fourier also had a reputation as a mathematician.) On a personal level he encouraged Champollion, a linguistic infant prodigy, to take up Egyptology and used his position as prefect to preserve his protegé from conscription. It was Champollion who eventually deciphered the Rosetta Stone.

Napoleon's domestic policy carried to its natural conclusion a political compromise which had been developing since the fall of Robespierre. The old powers of the church and the nobility were to be reconciled with the new power of the bourgeoisie and an administration loyal to the state alone. Henceforward men were to be equal before the law but the state renounced any attempt to achieve economic equality, or indeed any form of economic control. Fourier who had risen through his own talent but who 'could give lessons in theology to bishops and politeness to the pre-1790 parliamentarians' both symbolised and in his position as Prefect aided the new compromise.

93

WHO WAS FOURIER? II

'Yesterday was my 21st birthday, at that age Newton and Pascal had [already] acquired many claims to immortality.' Fourteen years after he wrote the postscript above, Fourier was prefect of Isère but had still no claim to the immortality he had craved as a young man. The work on the zeros of algebraic polynomials which he had pursued since his time at Auxerre still gains him a footnote in algebra textbooks – but a footnote is not immortality. His lecturing while at the Ecole Polytechnique had been much praised – but a good lecture is the most ephemeral of triumphs.

However, in 1804 he took up the subject of the propagation of heat. The field of Newtonian mechanics had already been worked over by several masters but the physics of heat, light, electricity and magnetism had still to be brought under the rule of mathematics. The choice of one of these fields thus requires no explanation. The particular choice of heat may have been connected with an obsessional need for warmth which Fourier acquired in Egypt.

In three remarkable years Fourier found the fundamental equations for heat conduction, developed new methods to solve them, applied these methods in a wide variety of cases and produced experimental evidence to support his solutions. At the end of 1807 he had submitted a memoir containing this work to the Academy. A commission consisting of Lagrange, Laplace, Monge and Lacroix was set up to examine it.

Instead of the enthusiastic acceptance that Fourier may have hoped for, he ran into two major criticisms. On the mathematical side neither Laplace nor Lagrange could accept the validity of his use of Fourier series. For example Laplace could not believe that $\cos x$ could possibly be expressed using a sine series. On the physical side Laplace and his pupils had already attacked the problem of heat conduction from a different angle. Unable to accept that Fourier's approach was superior, Laplace, Biot and Poisson attacked his derivation of the equations of conduction as lacking in rigour.

In 1811 the Academy gave Fourier a second chance to promote his theory by offering its grand prize in mathematics for work on the theory of heat conduction.

Fourier resubmitted his earlier essay together with some further results. (The most important addition was the introduction of the Fourier transform and its use, in the manner of Chapter 55, in studying the cooling of infinite solids.) However, although Fourier gained the prize, the accompanying report made it clear that Lagrange and Laplace had not withdrawn their objections.

The attitude of Lagrange and Laplace has been severely criticised by writers of modern texts on mathematical methods as (to quote one of them) 'typical in its demonstration of the difference between pure and applied mathematicians'. Since Laplace was the greatest mathematical physicist of the age and the author of the *Traité de Mécanique Céleste*, whilst Lagrange (who admittedly often worked in pure mathematics) was the author of the equally influential *Mécanique Analytique* this seems a curious judgement.

In fact Laplace and, particularly, Lagrange, did not simply doubt the rigour but also the truth of Fourier's assertions concerning expansions in trigonometric series. They may also have doubted the utility of such expansions. Why should Fourier's expansion in trigonometric series be superior to an expansion in continued fractions, or infinite products, or formal power series or any of the hundred and one pretty tricks of the eighteenth century mathematicians? With hindsight we can see (or at least persuade ourselves that we can see) why Fourier series are much better adapted than Taylor series to a whole class of physical problems. Laplace and Lagrange could not see into the future and their doubts are surely more a tribute to the originality of Fourier's methods than a reproach to mathematicians who Fourier greatly respected (and, in Lagrange's case, admired).

Soon new political worries were added to Fourier's academic ones (the Academy had published neither his first nor his second memoir leaving his results open to plagiarism or rediscovery). Napoleon had continued the aggressive foreign policy of the revolution with results depicted in vast battle pieces by a generation of French painters and, on a smaller scale, by Goya. So long as French forces were victorious the cost of the policy would be borne by the defeated powers and not by France but from 1812 onwards the tide of war turned. In 1814 enemy troops entered Paris, Napoleon was forced to abdicate and Louis XVIII installed as a constitutional monarch.

The new king understood, even if some of his followers did not, that the Napoleonic domestic compromise would have to form the basis of his rule. Nor could all of France's experienced administrators be replaced by the few men who had stayed loyal to him in exile. Fourier remained in Grenoble as prefect of the Isère.

However the last act of the Napoleonic adventure was still to be played. The new government was short of money and made the mistake of trying to economise on the pensions of Napoleon's old armies, and, worse, the pension of Napoleon himself. Tired of his role as a ruler of the little kingdom of Elba, Napoleon returned to France and began a march towards Paris. Troops sent out to arrest him went over to him without firing a shot. Grenoble was the first large town on Napoleon's line of march. Fourier's attempts to organise resistance proved fruitless and he

and the commanding general left Grenoble by one gate as Napoleon entered by another.

Whilst at Grenoble, Fourier had acted as a loyal and indeed zealous servant of the King. But the continuing collapse of the Royalist position seems to have persuaded him to throw in his lot with Napoleon and five days later we find him as Napoleon's prefect of the Rhône.

Twenty years before when Fourier was president of the revolutionary committee of Auxerre, the republic had lacked everything from experienced soldiers to gunpowder; but hope and enthusiasm had saved France. Now there was material and organisation but only Napoleon's promise of a new regime that was to be at once revolutionary, liberal and imperial. Passive and active resistance increased and Fourier himself was dismissed for failure to comply with orders. Napoleon was defeated at Waterloo and the King returned.

Fourier was now without a job and in deep disfavour with the new regime. The pension to which his years of public service would normally have entitled him was refused on technical grounds. An internal minute concerning his case sums up the government's attitude. 'The prefect of Grenoble at the time of the arrival of Bonaparte should not be surprised not to have a pension.'

The low point of Fourier's fortunes did not last long. The prefect of Paris was an old pupil and friend and had gained sufficient credit with the new government to be able to appoint him as Director of the Statistical Bureau of the Seine. (There are demographers who only know of Fourier as an important figure in the development of French governmental statistics.) In 1816 he was elected to the Academy of Sciences but the King refused to confirm his election. In 1817 the Academy tried again and this time his election was allowed. The physical and mathematical results of his theory of heat began to gain general acceptance. In 1822 his prize essay was at last published and he was elected permanent mathematical secretary of the Academy.

Fourier was now able to enter upon the role of grand old man encouraging younger talent (Liouville, Sturm, Dirichlet, Navier) and attaining that curious summit of French distinction, election to the Académie Française. He did not completely stop mathematical work and some of his later work concerns problems in what is now called linear programming. (Darboux, the nineteenth-century editor of his collected work clearly wondered why such a great mind should be interested in such out of the way puzzles.)

'There was at the Academy of Sciences', wrote Victor Hugo, seeking a contrast with the growing fame of the socialist Charles Fourier, 'a celebrated Fourier whom posterity has forgotten'. But posterity has not forgotten Joseph Fourier the Egyptologist, mathematician, physicist and public servant.

94

WHY DO WE COMPUTE?

Oxford stories lose little in the telling. Titchmarsh wrote of Hardy that: 'I worked on the theory of Fourier integrals under his guidance for a good many years before I discovered that this theory has applications in applied mathematics, if the solution of certain differential equations can be called "applied"' (*JLMS* **25**, p. 85, 1950).

Changes in the undergraduate curriculum have ensured that few pure mathematicians today can enjoy such a sheltered upbringing. None the less there must be several who, like myself, have been greatly surprised by the information that large amounts of computer time are spent by crystallographers actually computing Fourier transforms numerically. Yet although the technical details are complex it is not hard to see why Fourier transforms should be so intimately linked with crystallography and optics in general.

Consider an incident wave of light (Figure 94.1) given by (the real part of)

$$\phi_{inc} = \phi \exp i(\mathbf{k}_0 \cdot \mathbf{r} - \omega t)$$

falling on a particle at the origin. In the simplest case we may suppose that this

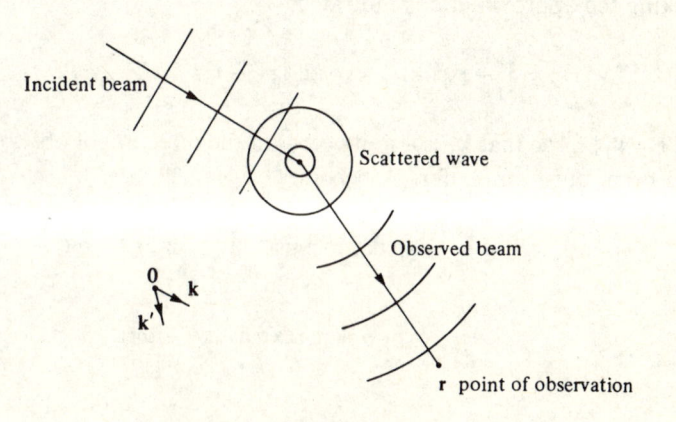

Fig. 94.1. Incident and scattered wave.

481

gives rise to a scattered spherical wave (Figure 94.1) given by (the real part of)

$$\phi_{sc} = \frac{\phi A}{\|\mathbf{r}\|} \exp i(l\|\mathbf{r}\| - \omega t),$$

where A is some constant associated with the particle.

If the particle is placed at $\mathbf{r}'$ then, writing $\mathbf{s} = \mathbf{r} - \mathbf{r}'$, we have

$$\phi_{inc} = (\phi \exp i\mathbf{k}_0 \cdot \mathbf{r}') \exp i(\mathbf{k}_0 \cdot \mathbf{s} - \omega t)$$

and we obtain a new scattered spherical wave

$$\phi'_{sc} = \frac{\phi A}{\|\mathbf{s}\|} \exp i\mathbf{k}_0 \cdot \mathbf{r}' \exp i(l\|\mathbf{s}\| - \omega t)$$

$$= \frac{\phi A}{\|\mathbf{r} - \mathbf{r}'\|} \exp i\mathbf{k}_0 \cdot \mathbf{r}' \exp i(l\|\mathbf{r} - \mathbf{r}'\| - \omega t).$$

Suppose now that $\|\mathbf{r}\|$ is very large compared with $\|\mathbf{r}'\|$. In an obvious notation

$$\|\mathbf{r} - \mathbf{r}'\| = \sqrt{(\|\mathbf{r} - \mathbf{r}'\|^2)} = \sqrt{\left(\sum_{j=1}^{3} (x_j - x'_j)^2 \right)}$$

$$= \sqrt{\left(\sum_{j=1}^{3} x_j^2 - 2\sum_{j=1}^{3} x_j x'_j + \sum_{j=1}^{3} x'^2_j \right)}$$

$$= \|\mathbf{r}\| \sqrt{((1 - 2\mathbf{r} \cdot \mathbf{r}'/\|\mathbf{r}\|^2 + \|\mathbf{r}'\|^2/\|\mathbf{r}\|^2)}$$

$$= \|\mathbf{r}\| (1 - \mathbf{r} \cdot \mathbf{r}'/\|\mathbf{r}\|^2 + O(\|\mathbf{r}\|^{-2}))$$

$$= \|\mathbf{r}\| - \mathbf{r} \cdot \mathbf{r}'/\|\mathbf{r}\| + O(\|\mathbf{r}\|^{-1}).$$

(Here, as usual, $O(\|\mathbf{r}\|^{-1})$ indicates an error of the order of magnitude $\|\mathbf{r}\|^{-1}$.) Similarly,

$$\|\mathbf{r} - \mathbf{r}'\|^{-1} = \|\mathbf{r}\|^{-1} + O(\|\mathbf{r}\|^{-2}),$$

and, making the approximations indicated,

$$\phi'_{sc} \approx \frac{\phi A}{\|\mathbf{r}\|} \exp i\mathbf{k}_0 \cdot \mathbf{r}' \exp i(l(\|\mathbf{r}\| - \mathbf{r} \cdot \mathbf{r}'/\|\mathbf{r}\|) - \omega t).$$

If we set $\mathbf{r} = \mathbf{k}_1 \|\mathbf{r}\|$ so that $\mathbf{k}_1$ is a unit vector in the direction of observation of the scattered beam our formula for ϕ'_{sc} becomes

$$\phi'_{sc}(\mathbf{r}, t) \approx \frac{\phi A}{\|\mathbf{r}\|} \exp i\mathbf{k}_0 \cdot \mathbf{r}' \exp i(l(\|\mathbf{r}\| - \mathbf{k}_1 \cdot \mathbf{r}') - \omega t)$$

$$= \frac{\phi A}{\|\mathbf{r}\|} \exp(- i\mathbf{k} \cdot \mathbf{r}') \exp i(l\|\mathbf{r}\| - \omega t),$$

with $\mathbf{k} = l\mathbf{k}_1 - \mathbf{k}_0$

Now suppose we have a collection of particles at $\mathbf{r}_1, \mathbf{r}_2, \ldots, \mathbf{r}_n$ with associated 'A

values' of A_1, A_2,...,A_n. Then (assuming that secondary scattering may be neglected)

$$\phi_{sc}(\mathbf{r}, t) \approx \sum_{m=1}^{n} \frac{\phi A_m}{\|\mathbf{r}\|} \exp(-i\mathbf{k} \cdot \mathbf{r}_m) \exp i(l\|\mathbf{r}\| - \omega t)$$

$$\approx \frac{\phi}{\|\mathbf{r}\|} \exp i(l\|\mathbf{r}\| - \omega t) \sum_{m=1}^{n} A_m \exp(-i\mathbf{k} \cdot \mathbf{r}_m).$$

If we replace the sum over a large number n of particles by an integral and ignore second order terms in the usual way our formula becomes

$$\phi_{sc}(\mathbf{r}, t) = \frac{\phi}{\|\mathbf{r}\|} \exp i(l\|\mathbf{r}\| - \omega t) \int A(\mathbf{x}) \exp(-i\mathbf{k} \cdot \mathbf{x}) dV(\mathbf{x})$$

$$= \frac{\phi}{\|\mathbf{r}\|} \exp i(l\|\mathbf{r}\| - \omega t)\hat{A}(\mathbf{k}).$$

In general the crystallographer can only measure the amplitude $|\phi_{sc}(\mathbf{r}, t)|$ of the wave $\phi_{sc}(\mathbf{r}, t)$ and thus (knowing $\mathbf{r}$ and ϕ) the value of $|\hat{A}(\mathbf{k})|$. The basic problem of crystallography is thus 'knowing $|\hat{A}(\mathbf{k})|$ for various values of $\mathbf{k}$ to find A'. Since $A(-\mathbf{x}) = 2\pi\hat{\hat{A}}(\mathbf{x})$ it would appear that we need to know the value of $\hat{A}(\mathbf{k})$ (rather than $|\hat{A}(\mathbf{k})|$) at all points $\mathbf{k}$ (rather than a selection) in order to find A from $\hat{A}$. In fact the crystallographer starts with other information (or guesses) about A and uses this additional information to help find A.

We shall pursue the topic no further, partly because it would take us too far afield, but mainly because the author has exhausted his very limited knowledge of the subject. I hope, however, that the reader has seen enough to show her that, even if Fourier transforms had not existed elsewhere in physics, the wave nature of light would have forced the optician and the crystallographer to invent them. I hope also to have given a plausible explanation of why certain sciences require the actual numerical computation of Fourier transforms on an industrial scale.

In the next chapter we shall consider a special problem whose solution involves the same ideas. Since the details are more specific their effect may be more convincing.

Postscript

A few months before I saw the proofs for this book, the 1985 Nobel Prize in Chemistry was awarded to Hauptman and Karle for developing a new method of finding A from measurements of $|\hat{A}(\mathbf{k})|$. The method is highly mathematical and makes use, among other things, of Weyl's equidistribution theorem combined with the central limit theorem and work of Toeplitz on Fourier series of non-negative functions.

95

THE DIAMETER OF STARS

We are used to thinking of stars as being so far away as to act as point sources of light. The fact that they appear to us as 'twinkling' patches of light is due to atmospheric effects. But, surprising as it may seem to a layman, the nearest stars are sufficiently close that, if it were not for the effects of the atmosphere, a good photograph using a good telescope would show them as tiny discs. Since observations of the nearest stars at six-monthly intervals (i.e. using a diameter of the earth's orbit as a surveyor's base line) enable astronomers to measure the distance of these stars, knowledge of the apparent diameter (i.e. the diameters of the discs on the photographic plate) would then enable us to calculate the true diameters of the nearest stars.

However, the blurring due to atmospheric effects is much greater than the apparent diameter we wish to observe. How can we get round this problem? Soon we will be able to use the 'big science' method and spend our way out of trouble by putting our telescope in orbit above the atmosphere. A more elegant (and considerably cheaper) solution has been found by Labeyrie.

Suppose we photograph a point source at time t and suppose that, without atmospheric effects, it would appear at $\mathbf{0}$ on our photographic plate. Owing to atmospheric effects we obtain a picture whose 'brightness' at a point $\mathbf{x}$ on the plate is $\lambda K_t(\mathbf{x})$ (where λ is the 'brightness' of our original point source). Suppose now that we have a second point source which has 'brightness' λ' and would (in the absence of atmospheric effects) appear as a point at $\mathbf{y}'$ on the plate. If $\|\mathbf{y}'\|$ is very small (i.e. the second point source is very close to the first in the sky) we would expect light coming from it to be affected in the same way as for the first and so we expect a picture with 'brightness' $\lambda' K_t(\mathbf{x} - \mathbf{y}')$.

More generally, if we had point sources of 'brightness' $\lambda_1, \lambda_2, \ldots, \lambda_n$ whose images would (in the absence of atmospheric effects) lie at $\mathbf{y}_1, \mathbf{y}_2, \ldots, \mathbf{y}_n$ with $\|\mathbf{y}_1\|$, $\|\mathbf{y}_2\|, \ldots, \|\mathbf{y}_n\|$ very small we would expect an image of the form

$$\sum_{j=1}^{n} \lambda_j K_t(\mathbf{x} - \mathbf{y}_j).$$

Replacing the sum by an integral in the usual way we would expect that a 'true image' $f(\mathbf{x})$ would be replaced by an 'actual image'

$$\iint f(\mathbf{y})K_t(\mathbf{x}-\mathbf{y})\,dA(\mathbf{y})$$

(provided $f(\mathbf{y})=0$ except when $\|\mathbf{y}\|$ is very small). In other words, if in the absence of atmospheric effects, we would get an image f, then in the presence of atmospheric effects, we will get an image $f*K_t$ where K_t represents the blurring effect of the atmosphere.

The reader should note that K_t is not fixed but (since it is due to random atmospheric effects) itself varies randomly in time. Thus our problem appears to be: 'Extract f from $f*K_t$ where K_t is an unknown random function' and so appears to be insoluble. The first idea that comes to mind is probably that of 'averaging' the image received at various times $t(1)$, $t(2),\ldots,t(n)$, say, to obtain

$$n^{-1}\sum_{j=1}^{n}f*K_{t(j)}=f*\sum_{j=1}^{n}n^{-1}K_{t(j)}=f*\tilde{K},$$

where $\tilde{K}=n^{-1}\sum_{j=1}^{n}K_{t(j)}$ is the 'average blur'. But we know little more about the 'average blur' $\tilde{K}$ than we know about its random constituents.

Labeyrie's idea is to take the Fourier transform of our image $\phi_t=f*K_t$. Since convolution becomes multiplication under Fourier transforms, $\hat{\phi}_t=\hat{f}\hat{K}_t$ and so the zeros of $\hat{\phi}_t$ are the zeros of $\hat{f}$ and of $\hat{K}_t$. Of course for one photograph we cannot distinguish the zeros of $\hat{f}$ and those of $\hat{K}_t$. But since K_t is random, so are the zeros of $\hat{K}_t$! Thus if we take a sequence $\hat{\phi}_{t(1)}, \hat{\phi}_{t(2)},\ldots,\hat{\phi}_{t(n)}$ the only zeros of the $\hat{\phi}_{t(j)}$ common to all the j will be those of $\hat{f}$, the other zeros being the random ones of each random $\hat{K}_{t(j)}$. In particular, if we superimpose pictures of the $|\hat{\phi}_{t(j)}|$ to form

$$\psi(\zeta)=\sum_{j=1}^{n}|\hat{\phi}_{t(j)}(\zeta)|=\sum_{j=1}^{n}|\hat{f}(\zeta)||\hat{K}_{t(j)}(\zeta)|$$

the zeros of $\hat{f}(\zeta)$ will stand out clearly as the zeros of $\psi(\zeta)$.

Of course the zeros of $\hat{f}$ do not suffice to determine f in general any more than the zeros of a function g determine g. (Observe, for example, that $\hat{f}$ and $(f*f)^{\wedge}$ have the same zeros.) However, we are not dealing with an arbitrary function f but with the 'true' image of a star which we expect to be a uniform disc. In other words, we expect

$$f(\mathbf{x})=\lambda D(\varepsilon^{-1}(\mathbf{x}-\mathbf{x}_0)),$$

where
$$D(\mathbf{x})=1 \quad \text{for } |\mathbf{x}|\leqslant 1,$$
$$D(\mathbf{x})=0 \quad \text{for } |\mathbf{x}|\geqslant 1,$$

ε is the radius of the image, $\mathbf{x}_0$ its centre and λ some positive real number. We know neither $\mathbf{x}_0$ nor ε and we wish to find ε.

The reader already knows the formulae which enable us to obtain $\hat{f}$ in terms of $\hat{D}$ but it can do no harm to rederive them. In effect, making the substitutions $\mathbf{y} = \mathbf{x} - \mathbf{x}_0$ and $\mathbf{w} = \varepsilon^{-1}\mathbf{y}$ we obtain

$$\hat{f}(\zeta) = \iint_{\mathbb{R}^2} \lambda D(\varepsilon^{-1}(\mathbf{x} - \mathbf{x}_0)) \exp(-i\zeta\cdot\mathbf{x})\,dA(\mathbf{x})$$

$$= \iint_{\mathbb{R}^2} \lambda D(\varepsilon^{-1}\mathbf{y}) \exp(-i\zeta\cdot(\mathbf{y} + \mathbf{x}_0))\,dA(\mathbf{y})$$

$$= \lambda \exp(-i\zeta\cdot\mathbf{x}_0) \iint_{\mathbb{R}^2} D(\varepsilon^{-1}\mathbf{y}) \exp(-i\zeta\cdot\mathbf{y})\,dA(\mathbf{y})$$

$$= \lambda\varepsilon^2 \exp(-i\zeta\cdot\mathbf{x}_0) \iint_{\mathbb{R}^2} D(\mathbf{w}) \exp(-i\varepsilon\zeta\cdot\mathbf{w})\,dA(\mathbf{w})$$

$$= \lambda\varepsilon^2 \exp(-i\zeta\cdot\mathbf{x}_0)\hat{D}(\varepsilon\zeta).$$

Thus the zeros of $\hat{f}$ are given by $\hat{D}(\varepsilon\zeta) = 0$.

Since D is radially symmetric, so is $\hat{D}$. Thus the zeros of $\hat{D}$ will appear in rings and so the zeros of $\hat{f}$ will appear in rings whose spacing is inversely proportional to the required radius ε. (Thus the further apart the rings, the smaller the radius of the 'true' image.)

In order to compute ε we need to know the location of the rings of zeros of $\hat{D}$. Since we could compute $\hat{D}$ numerically in the same way as we will have to compute $\hat{\phi}_{t(j)}$ this presents no problems. The reader may, however, be interested to see an explicit formula for $\hat{D}$ in terms of Bessel functions. (This expression may also quieten the fears of any reader who has observed that, whilst the argument above shows that the zeros, if any, of $\hat{D}$ form circular rings, it does not show that $\hat{D}$ actually has any zeros.)

Lemma 95.1. *Suppose* $D:\mathbb{R}^2 \to \mathbb{R}$ *is given by*

$$D(\mathbf{x}) = 1 \quad for \; \|\mathbf{x}\| \leqslant 1,$$
$$D(\mathbf{x}) = 0 \quad otherwise.$$

Then $\hat{D}(\zeta) = 2\pi J_1(\|\zeta\|)/\|\zeta\|$ *where* J_1 *is the first Bessel function.*

Remark. Our proof makes use of the relations

$$\int_0^{2\pi} \exp(-ia\cos\theta)\,d\theta = 2\pi J_0(a), \tag{1}$$

$$\int_0^x y^n J_{n-1}(y)\,dy = x^n J_n(x), \tag{2}$$

which may be found proved in most standard mathematical methods texts.

Proof. By definition

$$\hat{D}(\zeta) = \int\int D(\mathbf{x}) \exp - i\boldsymbol{\zeta}\cdot\mathbf{x}\, dA(\mathbf{x}).$$

Choosing polar coordinates (r, θ) in such a way that $\boldsymbol{\zeta}$ has coordinates $(\zeta, 0)$ we see that

$$\hat{D}(\zeta) = \int_0^1 \int_0^{2\pi} r \exp - (i\zeta r \cos \theta)\, d\theta\, dr.$$

Making use of relation (1) above and making the substitution $s = \zeta r$ we obtain

$$\hat{D}(\zeta) = 2\pi \int_0^1 r J_0(\zeta r)\, dr = \frac{2\pi}{\zeta^2} \int_0^\zeta s J_0(s)\, ds.$$

Thus by relation (2)

$$\hat{D}(\zeta) = 2\pi J_1(\zeta)/\zeta,$$

which is the required result. (The case $\zeta = 0$ must be considered separately but here, trivially, $\hat{D}(0) = 2\pi$.) ∎

Just as in the previous chapter we have sketched a method which requires the numerical computation of several Fourier transforms and the reconstitution of a function f from a limited knowledge of $\hat{f}$ (in this case the location of its zeros) and other information about the form of f (in this case that f is a (scalar multiple of) the characteristic (indicator) function of a disc). The method of Labeyrie's gives results consistent with Michelson's interferometer results but because of its simplicity has been applied to many more stars. (Over 30 have now had their diameters measured in this way.)

96

WHAT DO WE COMPUTE?

Between the land of the pure mathematician where the formula is the solution and the real world of simple men and stubborn machines lies the twilight domain of the theoretical numerical analyst. In the next few chapters we shall consider some of the problems which arise when we try to compute Fourier transforms numerically. But at the same time we will make assumptions which cannot be true in the real world. (For example, we shall assume that our 'computer' stores numbers with no loss in accuracy.) If this procedure makes the reader unhappy she should remember that a simple partial solution may be more illuminating than a labyrinthine complete one.

We shall confine our attention to the one dimensional case. In the two dimensional case, if $f : \mathbb{R}^2 \to \mathbb{C}$ is given, then

$$\hat{f}(\zeta) = \iint f(\mathbf{x}) \exp - i\zeta \cdot \mathbf{x} \, dA(\mathbf{x})$$

$$= \int_{-\infty}^{\infty} \int_{-\infty}^{\infty} f(x_1, x_2) \exp(-i(\zeta_1 x_1 + \zeta_2 x_2)) \, dx_1 \, dx_2$$

$$= \int_{-\infty}^{\infty} \left(\int_{-\infty}^{\infty} f(x_1, x_2) \exp(-i\zeta_1 x_1) dx_1 \right) \exp(-i\zeta_2 x_2) dx_2$$

$$= \int_{-\infty}^{\infty} g(\zeta_1, x_2) \exp(-i\zeta_2 x_2) \, dx_2,$$

where
$$g(\zeta_1, x_2) = \int_{-\infty}^{\infty} f(x_1, x_2) \exp(-i\zeta_1 x_1) \, dx_1.$$

Thus the computation of the two dimensional Fourier transform reduces to the computation of one dimensional Fourier transforms. (But remember the general rule that if it takes N operations to do something in one dimension, it takes N^2 operations in two dimensions.) Similar considerations apply in three and higher dimensions.

In numerical analysis we are not given the value of a function $f:\mathbb{R}\to\mathbb{C}$ everywhere. In particular, we will have no knowledge of the values of f outside some interval $[-R,S]$. We must therefore assume that $\int_S^\infty |f(x)|\,dx$ and $\int_{-\infty}^{-R}|f(x)|\,dx$ are sufficiently small to ensure the goodness of the approximation

$$\hat{f}(\zeta) \approx \int_{-R}^{S} f(x)\exp-i\zeta x\,dx.$$

Even within the interval $[-R,S]$ we will know the value of f only at a finite number of points $x_0, x_1, \ldots, x_{N-1}$. To make life simple (and our final result elegant) we shall assume that the $x_0, x_1, \ldots, x_{N-1}$ are equally spaced with $x_j = -R + j(R+S)/N$, say. The natural approximation now consists in replacing the integral by a sum

$$\hat{f}(\zeta) \approx (R+S)N^{-1} \sum_{j=0}^{N-1} f(x_j)\exp(-i\zeta x_j)$$

$$\approx K_\zeta N^{-1} \sum_{j=0}^{N-1} f(-R+j(R+S)/N)\exp(-i\zeta j(R+S)/N),$$

with $K_\zeta = (R+S)\exp i\zeta R$.

Just as we can only sample f at a finite number of points, so we can only calculate $\hat{f}$ at a finite number of points. It turns out that a good choice of sample points is given by taking $\zeta = m\eta$ where $\eta = 2\pi/(R+S)$ and m is an integer. We then get

$$\hat{f}(m\eta) = K_\zeta N^{-1} \sum_{j=0}^{N-1} f(-R+j(R+S)/N)\exp(-2\pi i j m/N)$$

$$= K_\zeta N^{-1} \sum_{j=0}^{N-1} f(-R+j(R+S)/N)\omega^{-mj},$$

where $\omega = \exp(2\pi i/N)$ is an Nth root of unity.

Setting $a_j = f(-R+j(R+S)/N)$, we see that the problem of computing the Fourier transforms of f numerically has been reduced to the problem of computing

$$F(m) = N^{-1} \sum_{j=0}^{N-1} a_j \omega^{-mj}$$

for different values of m.

We do not, of course, claim that $K_\zeta F(m)$ can be a good approximation to $\hat{f}(m\eta)$ for all m. There are two evident and complementary reasons for this. Firstly we observe that F is periodic with $F(m+N) = F(m)$ whilst $\hat{f}(m\eta)$ is not (indeed, by the Riemann–Lebesgue Lemma (Theorem 52.11), $\hat{f}(m\eta)\to 0$ as $m\to\infty$). Secondly, we note that

$$\hat{f}(m\eta) = \int_{-\infty}^{\infty} f(x)\exp(-im\eta x)\,dx$$

involves 'integrating f against an exponential with wave length $\lambda = 2\pi/|m\eta|$'. Thus, if we only sample f at points $x_1, \ldots, x_N$ a distance $\delta = (R + S)/N$ apart, we cannot hope to estimate $\hat{f}$ unless δ is quite small compared with the wave length λ.

Thus we expect $K_\zeta^{-1} F(m)$ to be a good approximation to $\hat{f}(m)$ only when $|m|/N$ is quite small. However, in the next chapter we shall see that it may be worthwhile to compute F for all m. (Since F is periodic this amounts to computing $F(m)$ for a range like $-N/2 < m \leqslant N/2$ or $0 \leqslant m \leqslant N - 1$.) One important reason for undertaking this extra computation is given in the following result.

Theorem 96.1. *If we write*

$$F(m) = N^{-1} \sum_{j=0}^{N-1} f(x_j)\omega^{-mj},$$

then the values of $F(0)$, $F(1), \ldots, F(N-1)$ completely determine the values of $f(x_0)$, $f(x_1), \ldots, f(x_{N-1})$.
Proof. See Theorem 97.6. ∎

Other reasons will emerge in the course of the proof of this result.

97

FOURIER ANALYSIS ON
THE ROOTS OF UNITY

In Chapter 93 we saw that, under certain assumptions, the numerical evaluation of Fourier transforms reduced to the evaluation of sums of the form

$$F(m) = N^{-1} \sum_{j=0}^{N-1} a_j \omega^{-jm},$$

where $a_0, a_1, \ldots, a_{N-1} \in \mathbb{C}$, $\omega = \exp 2\pi i/N$ and m is an integer.

In this chapter we shall see that such sums can be treated in a manner analogous to Fourier coefficients on $\mathbb{T}$ by considering them as Fourier coefficients on the group G of Nth roots of unity. Initially our treatment will parallel that given in Chapter 32 (which the reader should quickly reread) but the fact that G is finite will mean that there are no problems about convergence and the treatment rapidly simplifies.

Getting down to detail, let us write $G = \{1, \omega, \omega^2, \ldots, \omega^{N-1}\}$ and $C(G)$ for the set of functions $f: G \to \mathbb{C}$. (There are no problems about continuity since G is finite.) We write $(f, g) = N^{-1} \sum_{j=0}^{N-1} f(\omega^j) g(\omega^j)^*$ where z^* denotes the complex conjugate of z.

Lemma 97.1 (*cf. Lemma 32.1*). *If $f, g, h \in C(G)$, then*

(i) $(f, g) = (g, f)^*,$

(ii) $(\lambda f + \mu g, h) = \lambda(f, h) + \mu(g, h),$

(iii) (f, f) *is real and* $(f, f) \geqslant 0,$

(iv) *If* $(f, f) = 0$ *then* $f = 0.$

Proof. Trivial. For example,

$$(f, f) = N^{-1} \sum_{j=0}^{N-1} f(\omega^j) f(\omega^j)^* = N^{-1} \sum_{j=0}^{N-1} |f(\omega^j)|^2,$$

so (f, f) is real, $(f, f) \geqslant 0$ and if $(f, f) = 0$ then $|f(\omega^j)|^2 = 0$ and $f(\omega^j) = 0$ for all $0 \leqslant j \leqslant N - 1$, i.e. $f = 0$. ∎

Lemma 97.2 (The Cauchy, Schwartz, Buniakowski inequality). *If $f, g \in C(G)$ then* $|(f, g)|^2 \leqslant (f, f)(g, g)$ *with equality if and only if* $\lambda f + \mu g = 0$ *for some* λ, $\mu \in \mathbb{C}$ *not both zero.*

Proof. Exactly as for Lemma 32.2.   ∎

Defining $\| f \|_2 = (f, f)^{\frac{1}{2}}$ we obtain the analogue of Lemma 32.3.

Lemma 97.3. *If $f, g \in C(G)$ and $\lambda \in \mathbb{C}$ then*

(i) $\| \lambda f \|_2 = |\lambda| \| f \|_2$,

(ii) $\| f \|_2 \geqslant 0$ *with equality if and only if* $f = 0$,

(iii) *(Triangle inequality)* $\| f \|_2 + \| g \|_2 \geqslant \| f + g \|_2$.

Proof. Exactly as for Lemma 32.3.   ∎

Now let us define $e_m : G \to \mathbb{C}$ by $e_m(\omega^j) = \omega^{mj}$.

Lemma 97.4. (i) $(e_m, e_m) = 1$,

(ii) $(e_n, e_m) = 0$ *for* $n \not\equiv m \bmod N$.

Proof. (i) By definition

$$(e_m, e_m) = N^{-1} \sum_{j=0}^{N-1} e_m(\omega^j) e_m(\omega^j)^* = N^{-1} \sum_{j=0}^{N-1} 1 = 1.$$

(ii) By definition

$$(e_m, e_n) = N^{-1} \sum_{j=0}^{N-1} e_m(\omega^j) e_n(\omega^j)^* = N^{-1} \sum_{j=0}^{N-1} \omega^{(m-n)j}$$

$$= N^{-1}(\omega^{(m-n)N} - 1)/(\omega^{m-n} - 1) = N^{-1}(1 - 1)/(\omega^{m-n} - 1)$$

$$= 0, \text{ whenever } m \not\equiv n \bmod N.$$ ∎

The way is now clear to obtain an analogue of Theorem 32.5.

Lemma 97.5. *If* $g_0 = \sum_{j=0}^{N-1} (f, e_j) e_j$ *and* $g = \sum_{j=0}^{N-1} \lambda_j e_j$ *then* $\| f \|_2^2 \geqslant \sum_{j=0}^{N-1} |(f, e_j)|^2$ *and*

$$\| f - g \|_2 \geqslant \| f - g_0 \|_2 = \sqrt{\left(\| f \|_2^2 - \sum_{j=0}^{N-1} |(f, e_j)|^2 \right)},$$

with equality if and only if $\lambda_j = (f, e_j) [0 \leqslant j \leqslant N - 1]$.

Proof. Just as for Theorem 32.5.   ∎

It is thus in keeping with our other notations to write

$$\hat{f}(e_j) = (f, e_j) = N^{-1} \sum_{j=0}^{N-1} f(\omega^j) \omega^{-mj}.$$

The connection with the sum given in the first paragraph emerges if we consider a function f with $f(\omega^j) = a_j$ since then

$$F(m) = N^{-1} \sum_{j=0}^{N-1} a_j \omega^{-mj} = N^{-1} \sum_{j=0}^{N-1} f(\omega_j) e_m(\omega^j)^* = \hat{f}(m).$$

Thus, in our new formalism, the evaluation of sums $N^{-1}\sum_{j=0}^{N-1} a_j \omega^{-mj}$ becomes the evaluation of Fourier coefficients of an $f \in C(G)$.

Lemma 32.5 tells us that $S(f) = \sum_{j=0}^{N-1} \hat{f}(e_j) e_j$ is the best mean square approximation to f among sums of the form $\sum_{j=0}^{N-1} \lambda_j e_j$. Since we are dealing with a finite set G of points, we may well hope to have $S(f) = f$ and this is indeed the case.

Theorem 97.6. *If $f \in C(G)$ then $f = \sum_{j=0}^{N-1} \hat{f}(e_j) e_j$.*
(It is, of course, clear that if $f = \sum_{j=0}^{N-1} \lambda_j e_j$ then $\hat{f}(e_k) = \sum_{j=0}^{N-1} \lambda_j(e_j, e_k) = \lambda_k$.)

Theorem 97.6 is easily proved by direct computation. None the less we shall give two indirect proofs. The first is included to tie in our results with those of elementary linear algebra and will not be referred to again in this chapter.

Proof of Theorem 97.6 by linear algebra. Observe that $C(G)$ is a vector space over $\mathbb{C}$ of dimension N (since it has a basis $g_0, g_1, \ldots, g_{N-1}$ with $g_j(\omega^j) = 1$, $g_j(\omega^k) = 0$ if $\omega^j \neq \omega^k$). The functions $e_0, e_1, \ldots, e_{N-1}$ are orthonormal and so linearly independent (since, if

$$\sum_{j=0}^{N-1} \lambda_j e_j = 0,$$

we have

$$\lambda_k = \sum_{j=0}^{N-1} \lambda_j (e_j, e_k) = \left(\sum_{j=0}^{N-1} \lambda_j e_j, e_k \right) = 0 \quad \text{for each } 0 \leqslant k \leqslant N-1).$$

Any set of N linearly independent vectors in a space of dimension N form a basis and so form a spanning set. Thus, if $f \in C(G)$, we have $f = \sum_{j=0}^{N-1} \lambda_j e_j$ for some choice of $\lambda_0, \lambda_1, \ldots, \lambda_{N-1}$. Automatically,

$$\hat{f}(e_k) = \left(\sum_{j=0}^{N-1} \lambda_j e_j, e_k \right) = \lambda_k,$$

so we are done. ∎

For our second proof we use the model set out in Chapter 2 and used repeatedly since then. (The reader may also like to compare and contrast the first part of Chapter 18 concerning the possible divergence at a point of Fourier sums on $\mathbb{T}$.) *Proof of Theorem 97.6 in the style of Chapter 2.* Suppose $f \in C(G)$. Then writing

$$S(f, \omega^r) = \sum_{j=0}^{N-1} \hat{f}(e_j) e_j(\omega^r),$$

we have

$$S(f,\omega^r) = \sum_{j=0}^{N-1} \left(N^{-1} \sum_{k=0}^{N-1} f(\omega^k)e_j(\omega^k)^* \right) e_j(\omega^r)$$

$$= N^{-1} \sum_{k=0}^{N-1} \left(\sum_{j=0}^{N-1} f(\omega^k)e_j(\omega^k)^* e_j(\omega^r) \right)$$

$$= N^{-1} \sum_{k=0}^{N-1} f(\omega^k) \sum_{j=0}^{N-1} e_j(\omega^k)^* e_j(\omega^r)$$

$$= N^{-1} \sum_{k=0}^{N-1} f(\omega^k) \sum_{j=0}^{N-1} e_j(\omega^{r-k})$$

$$= N^{-1} \sum_{k=0}^{N-1} f(\omega^k)D(\omega^{r-k}),$$

where $D(\omega^l) = \sum_{j=0}^{N-1} e_j(\omega^l)$. Noting that $\omega^{k+N} = \omega^k$ we see that the substitution $l = r - k$ gives

$$S(f,\omega^r) = N^{-1} \sum_{l=0}^{N-1} f(\omega^{r-l})D(\omega^l).$$

The structure of D is particularly simple. If $1 \leqslant l \leqslant N - 1$,

$$D(\omega^l) = \sum_{j=0}^{N-1} e_j(\omega^l) = \sum_{j=0}^{N-1} \omega^{jl} = 0.$$

If $l = 0$

$$D(\omega^0) = \sum_{j=0}^{N-1} e_j(\omega^0) = \sum_{j=0}^{N-1} 1 = N.$$

Thus

$$S(f,\omega^r) = N^{-1} \sum_{l=0}^{N-1} f(\omega^{r-l})D(\omega^l) = N^{-1}Nf(\omega^r) = f(\omega^r) \quad \text{for all } r$$

and

$$S(f, \ ) = f(\ ),$$

as stated. ∎

We now see why Theorem 96.1 is true.

Proof of Theorem 96.1. Define $g \in C(G)$ by $g(\omega^j) = f(x_j)$. Then, in the notation of Theorem 97.6, $\hat{g}(e_m) = F(m)$ and so

$$f(x_j) = g(\omega^j) = \sum_{m=0}^{N-1} \hat{g}(e_m)e_m(\omega^j) = \sum_{m=0}^{N-1} F(m)\omega^{jm}.$$

Thus the values of $F(0), F(1), \ldots, F(N-1)$ completely determine the values of $f(x_0), f(x_1), \ldots, f(x_{N-1})$. ∎

In Chapters 51 and 52 (and more particularly in Theorem 51.5 and Theorem 52.6) we saw that proofs like our second proof of Theorem 96.1 could be expressed

in a general form by introducing the notion of convolution. Since the expression corresponding to the formula $\sigma_n(f) = f * K_n$ of Chapter 2 and $S_n(f) = f * D_n$ of Chapter 18 is here

$$S(f, \omega^r) = N^{-1} \sum_{l=0}^{N-1} f(\omega^{r-l}) D(\omega^l),$$

our definition of convolution springs to the eye.

Definition 97.7. *If $f, g \in C(G)$ we write*

$$f * g(\omega^r) = N^{-1} \sum_{l=0}^{N-1} f(\omega^{r-l}) g(\omega^l).$$

Convolution on $C(G)$ behaves in a similar way to convolution on $C(\mathbb{T})$ and $(L^1 \cap C)(\mathbb{R})$. However the proofs (and, sometimes, the statements of the results) are simpler since we are not dealing with intervals over $\mathbb{T}$ or $\mathbb{R}$ but with finite sums. (In particular, the interchange of two finite sums presents no theoretical difficulty!) For example the following lemma parallels Theorem 52.2 and Lemma 52.3.

Lemma 97.8. *Suppose that $f, g, h \in C(G)$ and $\lambda \in \mathbb{C}$. Then*

(i) $N^{-1} \sum_{r=0}^{N-1} f * g(\omega^r) = (N^{-1} \sum_{r=0}^{N-1} f(\omega^r))(N^{-1} \sum_{r=0}^{N-1} g(\omega^r))$.

(ii) $N^{-1} \sum_{r=0}^{N-1} |f * g(\omega^r)| \leqslant (N^{-1} \sum_{r=0}^{N-1} |f(\omega^r)|)(N^{-1} \sum_{r=0}^{N-1} |g(\omega^r)|)$.

(iii) $(f * g)^\wedge(e_j) = \hat{f}(e_j) \hat{g}(e_j)$ for all $0 \leqslant j \leqslant N - 1$.

(iv) $f * g = g * f$.

(v) $f * (g * h) = (f * g) * h$.

(vi) $f * (g + h) = f * g + f * h$.

(vii) $(\lambda f) * g = \lambda (f * g)$.

Proof. We prove (iii) and leave it to the reader to prove a selection of the rest.

(iii)

$$(f * g)^\wedge(e_j) = N^{-1} \sum_{m=0}^{N-1} (f * g)(\omega^m) e_j(\omega^m)^*$$

$$= N^{-2} \sum_{m=0}^{N-1} \sum_{r=0}^{N-1} f(\omega^{m-r}) g(\omega^r) e_j(\omega^m)^*$$

$$= N^{-2} \sum_{m=0}^{N-1} \sum_{r=0}^{N-1} f(\omega^{m-r}) e_j(\omega^{m-r})^* g(\omega^r) e_j(\omega^r)^*$$

$$= N^{-1} \sum_{r=0}^{N-1} \left(N^{-1} \sum_{m=0}^{N-1} f(\omega^{m-r}) e_j(\omega^{m-r})^* \right) g(\omega^r) e_j(\omega^r)^*$$

$$= N^{-1} \sum_{r=0}^{N-1} \hat{f}(e_j) g(\omega^r) e_j(\omega^r)^* = \hat{f}(e_j) \hat{g}(e_j),$$

as required. ∎

In contrast to the case of convolution over $C(\mathbb{T})$ and over $C \cap L^1(\mathbb{R})$, convolution over $C(G)$ does have an identity (as we have already seen).

Lemma 97.9. (i) *If $D(\omega^r) = 0$ for $1 \leqslant r \leqslant N$ and $D(1) = N$ then*

$$f * D = f \quad \text{for all } f \in C(G).$$

Proof. Direct computation.   ∎

Thus in algebraic language $(C(G), +, *)$ is a commutative ring with identity D. However it is not (unless $N = 1$) a field since it possesses 'divisors of zero'.

Lemma 97.9. (ii) *Let $N \geqslant 2$. Then $e_0, e_1 \neq 0$ yet*

$$e_0 * e_1 = 0.$$

Proof. Let $f = e_0 * e_1$. Then

$$\hat{f}(e_j) = \hat{e}_0(e_j)\hat{e}_1(e_j) = \begin{cases} 1.0 = 0 & \text{for } j = 0, \\ 0.1 = 0 & \text{for } j = 1, \\ 0.0 = 0 & \text{for } N - 1 \geqslant j \geqslant 2. \end{cases}$$

Thus $f = 0$. (Alternatively we could use direct computation.)   ∎

The reader may well feel that results obtained so easily cannot be very valuable. But the combinatorial study of finite systems depends on the recognition of patterns and patterns which have proved so useful in the infinite case ought to be of some use in the finite case.

98

HOW DO WE COMPUTE?

Until recently the numerical computation of Fourier transforms was considered a thing to be avoided whenever possible. Users, like crystallographers, who insisted on such computation could tie up the largest computer systems for hours on end.

It is easy to see why this should be the case. Roughly speaking the time taken by a numerical calculation on a computer is proportional to the number of operations involved. To compute $\hat{f}$ at one point ζ by means of the formula

$$\hat{f}(\zeta) \approx N^{-1} \sum_{j=0}^{N-1} f(x_j) \exp(-i\zeta x_j)$$

involves at least $N + 1$ multiplications and $N - 1$ additions. Thus, it would appear, if the user starting from the value of f at N points $x_0, x_2, \ldots, x_{N-1}$, demands to know the value of $\hat{f}$ at N points $\zeta_0, \zeta_1, \ldots, \zeta_{N-1}$ then we shall need at least $2N^2$ multiplications and additions. Since, in practice, users demanded $N = 10^3$, this made the computation of Fourier transforms a very slow business. (The introduction of a new generation of faster machines was followed, in accordance with well understood principles, by the users raising N to 10^4 so that the method sketched above demanded a hundredfold increase in the number of operations.)

The situation was transformed by the following simple observation.

Theorem 98.1. *Let G_N be the group of Nth roots of unity and let G_{2N} be the group of $2N$th roots of unity. Suppose that, given $\omega_N = \exp 2\pi i/N$, we can compute all the Fourier coefficients of any $F \in C(G')$ using no more than M multiplications and additions. Then, given $\omega_{2N} = \exp i\pi/N$, we can compute all the Fourier coefficients of any $f \in C(G_{2N})$ using no more than $2M + 8N$ additions and multiplications.*

Proof. We need a little notation. If $f \in C(G_{2N})$ is given we define $F_a, F_b \in C(G_N)$ by

$$F_a(\omega_N^r) = f(\omega_{2N}^{2r}),$$
$$F_b(\omega_N^r) = f(\omega_{2N}^{2r+1}).$$

We further define $E_j \in C(G_N)$ by

$$E_j(\omega_N^r) = \omega_N^{jr}$$

and $e_k \in C(G_{2N})$ by

$$e_k(\omega_{2N}^s) = \omega_{2N}^{sk}.$$

Note that $E_{j+N} = E_j$.

We can now give the details of the computation.

Step 1. We compute ω_{2N}^s for $0 \leqslant s < 2N$. This requires at most $2N$ multiplications. (A real life machine might find complex multiplication harder to handle than real multiplication but in our idealised discussion we ignore this.)

Step 2. We now know $\omega_N = \omega_{2N}^2$. Thus by hypothesis we can compute all the Fourier coefficients of F_a and F_b using at most $2M$ additions and multiplications.

Step 3. We now observe (and this is the key point) that

$$\hat{f}(e_k) = (2N)^{-1} \sum_{r=0}^{2N-1} f(\omega_{2N}^r) e_k(\omega_{2N}^r)$$

$$= 2^{-1}\left(N^{-1} \sum_{u=0}^{N-1} f(\omega_{2N}^{2u}) e_k(\omega_{2N}^{2u}) + N^{-1} \sum_{v=0}^{N-1} f(\omega_{2N}^{2v+1}) e_k(\omega_{2N}^{2v+1}) \right)$$

$$= 2^{-1}\left(N^{-1} \sum_{u=0}^{N-1} F_a(\omega_N^u) E_k(\omega_N^u) + N^{-1} \sum_{v=0}^{N-1} F_b(\omega_N^v) E_k(\omega_N^v) \omega_{2N}^k \right)$$

$$= 2^{-1}(\hat{F}_a(E_k) + \omega_{2N}^k \hat{F}_b(E_k)).$$

Thus, for each $0 \leqslant k < 2N - 1$, $\hat{f}(e_k)$ can now be computed using at most one addition and two multiplications. We can now compute all the Fourier coefficients of f using at most $2N \times 3 = 6N$ additions and multiplications.

Summary

We have computed all the Fourier coefficients of f using at most $2M + 8N$ additions and multiplications.∎

The way is now clear for a simple induction.

Theorem 98.2. *Let $N = 2^n$ and let G_N be the group of Nth roots of unity. Then given $\omega_N = \exp 2\pi i/N$ we can compute all the Fourier coefficients of any $F \in C(G_N)$ using no more than $n2^{n+2} = 4N \log_2 N$ additions and multiplications.*

Proof. Let $M(n)$ be the number of additions and multiplications required to compute all the Fourier coefficients of any $F \in C(G_{2^n})$ once given $\omega_{2^n} = \exp 2^{-n+1}\pi i$. If $n = 1$ we have

$$\hat{F}(e_0) = 2^{-1}(F(1) + F(-1)),$$
$$\hat{F}(e_1) = 2^{-1}(F(1) + (-1)F(-1)),$$

so, certainly, $M(1) \leqslant 5 \leqslant 1.2^{1+2}$. Further, if $M(n) \leqslant n2^{n+2}$, we know from Theorem 98.1 that $M(n+1) \leqslant 2M(n) + 8.2^n$ and so

$$M(n+1) \leqslant 2n2^{n+2} + 8.2^n = (n+1)2^{n+3}.$$

The stated result follows by induction.∎

The method implicitly proposed in the proofs is a simple one and should be easy to implement. (The details will vary according to the system used. Even the author does not believe that the best way of obtaining ω^r is to multiply ω by itself r times.) If $N = 2^{10} = 1024$ then it uses at most 40.2^{10} operations to compute all the Fourier coefficients of an $F \in C(G_N)$ whereas the simple minded approach uses at least $2.2^{20} = 2048.2^{10}$ operations and promises to take about 50 times as long.

If $N = 2^{20} = 1048576$ and we have a computing machine which performs a million operations a second then the new method uses at most 80.2^{20} operations and will compute all the Fourier coefficients of $C(G_N)$ in under two minutes. The old method will use at least 2.2^{40} operations and take at least 23 days. More importantly, in the time the new method uses to calculate all its results, the old method could only compute 40 values of the formula

$$\hat{f}(\zeta) \approx K_\zeta N^{-1} \sum_{j=0}^{N-1} f(x_j) \exp(-i\zeta x_j).$$

The method outlined in this chapter is known as the Fast Fourier Transform (or FFT) and its publication by Cooley and Tukey revolutionised the numerical computation of Fourier transforms. The practicality of a method like that of Chapter 95 (measurement of the diameters of stars) depends on the availability of cheap (i.e. fast) numerical transforms.

Once the method was established it became clear that it had a long and interesting prehistory going back as far as Gauss. But until the advent of computing machines it was a solution looking for a problem. Mathematics, perhaps more than any other human endeavour, is littered with ideas born out of their time. (Consider, for example, Fourier's work on linear programming, a subject whose utility also depends on the existence of computers, or Wilbraham's discovery of the Gibbs phenomenon.)

We conclude with the simple remark that the formula for f in terms of $\hat{f}$ given by Theorem 97.6,

$$f(\omega^r) = \sum_{j=0}^{N-1} \hat{f}(e_j) e_j(\omega^r) = \sum_{j=0}^{N-1} \hat{f}(e_j) \omega^{rj}$$

has (apart from a factor of N^{-1}) essentially the same form as the definition of $\hat{f}$ in terms of f

$$\hat{f}(e_r) = N^{-1} \sum_{j=0}^{N-1} f(\omega^j) e_r(\omega^j)^* = N^{-1} \sum_{j=0}^{N-1} f(\omega^j) \omega^{-rj}.$$

Thus it should take roughly the same number of operations to compute f from $\hat{f}$ as it takes to compute $\hat{f}$ from f.

Theorem 98.2′. *Let $N = 2^n$ and let G_N be the group of Nth roots of unity. Then given $\omega_N = \exp 2\pi i / N$ and $\hat{F}$ we can compute all the values of $F \in C(G_N)$ using no more than $n2^{n+2} = 4N \log_2 N$ additions and multiplications.*

Proof. Left as a recommended exercise for the reader. ∎

99

HOW FAST CAN WE MULTIPLY?

In Chapter 95 we discussed the problem of computing the Fourier coefficients of an $f \in C(G_N)$ using the fewest possible additions and multiplications. A similar problem (though of much less practical importance) occurs when we wish our computer to perform algebraic manipulations.

Problem 99.1. *Given the coefficients $a_0, a_1, \ldots, a_p \in \mathbb{C}$ and $b_0, b_1, \ldots, b_q \in \mathbb{C}$ of two polynomials*

$$P(t) = \sum_{r=0}^{p} a_r t^r, \quad Q(t) = \sum_{s=0}^{q} b_s t^s,$$

how many operations are required to find the coefficients $c_0, c_1, \ldots, c_{p+q}$ of their product

$$P(t)Q(t) = \sum_{u=0}^{p+q} c_u t^u?$$

We restate the problem in a more concrete form.

Problem 99.1′. *Given $a_0, a_1, \ldots, a_p \in \mathbb{C}$ and $b_0, b_1, \ldots, b_q \in \mathbb{C}$ how many operations are required to find $c_0, c_1, \ldots, c_{p+q}$ where*

$$c_u = \sum_{r=0}^{u} a_{u-r} b_r \quad [0 \leqslant u \leqslant p+q].$$

A simple minded approach working directly from the formula $c_u = \sum_{r=0}^{u} a_{u-r} b_r$ requires $(p+1)(q+1)$ multiplications and $(p+1)(q+1) - (p+q+1)$ additions. Since in most practical applications we would expect $p, q \leqslant 100$ we could well be satisfied with the direct method. However it is interesting to ask if we could do better than this in the case when p and q are large. The reader might like to consider this problem for herself before proceeding further.

In fact we can make a considerable improvement. The key lies in the observation

that the formula

$$c_u = \sum_{r=0}^{u} a_{u-r} b_r \quad [0 \leqslant u \leqslant p+q]$$

has a distinct flavour of convolution.

Lemma 99.2. *Let G be the group of Nth roots of unity and let $\omega = \exp 2\pi i/N$. Suppose $N \geqslant p+q+1$ and $f, g \in C(G)$ are defined by*

$$\begin{aligned}
f(\omega^r) &= a_r & [0 \leqslant r \leqslant p], \\
f(\omega^r) &= 0 & [p+1 \leqslant r \leqslant N-1], \\
g(\omega^s) &= b_s & [0 \leqslant s \leqslant q], \\
g(\omega^s) &= 0 & [q+1 \leqslant s \leqslant N-1].
\end{aligned}$$

Then writing $c_u = \sum_{r=0}^{u} a_{u-r} b_r [0 \leqslant u \leqslant p+q]$ we obtain

$$\begin{aligned}
f*g(\omega^u) &= c_u & [0 \leqslant u \leqslant p+q], \\
f*g(\omega^u) &= 0 & [p+q+1 \leqslant u \leqslant N-1].
\end{aligned}$$

At first glance we seem simply to have exchanged one notation for another. But we have just seen how to compute Fourier coefficients very rapidly and we know that the effect of taking Fourier coefficients is to convert the complicated operation of convolution into the simple operation of (pointwise) multiplication.

Theorem 99.3. *Suppose $2^n = N \geqslant p+q+1$. Then given $\omega = \exp 2\pi i/N$ we can compute all the coefficients c_u of the product $\sum_{u=0}^{p+q} c_u t^u$ of two polynomials $\sum_{r=0}^{p} a_r t^r$ and $\sum_{s=0}^{q} b_s t^s$ using no more than $2^n + 3n2^{n+2} = N + 12N \log_2 N$ additions and multiplications.*

Proof. We use the notation of Lemma 99.2 and Chapter 98. The calculation runs as follows.

Step 1. Compute all the Fourier coefficients of f and g. By Theorem 98.2 this requires at most $8N \log_2 N$ additions and multiplications.

Step 2. Compute all the Fourier coefficients of $f*g$ using the formula

$$(f*g)^\wedge(e_j) = \hat{f}(e_j)\hat{g}(e_j) \quad [0 \leqslant j \leqslant N-1],$$

established in Lemma 97.8 (iii). This requires exactly N multiplications.

Step 3. Compute the values of $f*g$ using the values of $(f*g)^\wedge$ found in Step 2. By Theorem 98.2' this can be done in at most $4N \log_2 N$ additions and multiplications.

Summary

Since $c_u = f*g(\omega^u) \; [0 \leqslant u \leqslant p+q]$ we have obtained the coefficients required using at most $8N \log_2 N + N + 4N \log_2 N = N + 12N \log_2 N$ additions and multiplications. ∎

If p and q are about the same size the number of operations required by our new method increases as $p \log p$ whereas that required by the direct method increases as p^2. However, the advantage of the bound $N + 12N \log_2 N$ over $2(p+1)(q+1) - (p+q+1)$ only becomes evident for values of p substantially larger than 100.

We conclude this chapter with a brief glance at a problem which is slightly more concerned with the actual 'hardware' of the computer than those we have so far considered. As the reader knows the underlying machinery of a computer only handles integers whose values lie between 0 and $Q - 1 = 2^r - 1$. A larger integer would be stored as a sequence $a_0, a_1, \ldots, a_p$ of integers with $0 \leqslant a_j \leqslant Q - 1$ and $a = \sum_{r=0}^{p} a_r Q^r$. The machine can only add, multiply and carry integers lying between 0 and $Q - 1$.

If we wish to multiply two very large integers $a = \sum_{r=0}^{p} a_r Q^r$ and $b = \sum_{s=0}^{q} b_s Q^s$ and obtain an exact answer then we are faced with a problem strongly reminiscent of Problem 99.1. (Note, however that, as for Problem 99.1, a 'good' solution only becomes important for values of p and q much larger than those normally met with.) Analogy with Problem 99.1 suggests splitting the problem into two steps.

Step 1. Compute and store

$$c_u = \sum_{r=0}^{u} a_{u-r} b_r \quad [0 \leqslant u \leqslant p + q].$$

(Observe that $0 \leqslant c_u < pQ^2$ but that, in general, c_u will require several stores.)
Step 2. Compute and store

$$\sum_{u=0}^{p+q} c_u Q^u.$$

Fast multiplication using the procedure above depends on being able to compute the c_u rapidly since Step 2 itself can be done rapidly. We can compute the c_u directly as Fourier coefficients but, although this already gives substantial savings, the method is inelegant since complex numbers like ω require special treatment at this basic level. By using instead a kind of Fourier transform on the integers mod $2^n - 1$ Strassen and Schönhage obtained a method of multiplying $\sum_{r=0}^{N} a_r Q^r$ and $\sum_{s=0}^{N} b_s Q^s$ in at most $AN \log N \log \log N$ operations for some constant A.

100

WHAT MAKES A GOOD CODE?

The techniques of modern 'digital' communication have made familiar the idea that any piece of information, whether a written message, a photograph or even a sound can be transmitted in the form of a number. From this point of view a secret code consists of a finite subset U of the positive integers $\mathbb{Z}^+$ (the possible messages), a finite subset V of $\mathbb{Z}^+$ (the possible coded messages) together with a function $T:U \to V$ (the encoding function) and a function $S:V \to U$ (the decoding function) such that $ST:U \to U$ is the identity.

Remark. In fact even this definition fails to cover all possibilities since, for example, we could suppose T 'multivalued' with the value $T(u)$ being chosen at random from a set $Q(u) \subseteq Q$ such that $v \in Q(u)$ implies $S(v) = u$. But we must start somewhere.

As a simple example let us take $U = V = \{n : 0 \leqslant n \leqslant N - 1\}$ and define $T:U \to V$ and $S:V \to U$ by the relations $T(u) \equiv u + M \bmod N$, $S(v) \equiv v - M \bmod N$. We consider the problem faced by 'opponents' who wish to decipher messages written using this code.

In general we must assume that our opponents know or guess the method of coding that we use. For the sake of illustration we may suppose that their information includes the value of N but not, at least initially, the value of M. Thus if we choose M at random and only use the code once, it is unbreakable, since trial decodes $S'_r(v) \equiv v - r \bmod N$ allowing r to run from 0 to $N - 1$ will give all possible messages without any indication of which to choose. (This is the principle of 'one time codes'.)

Suppose, however, that we use our code repeatedly to send messages $u_1, u_2, \ldots, u_n$ and that our opponents have intercepted the corresponding coded messages $v_1, v_2, \ldots, v_n$. Suppose, further, that they know that $u_1 \in U_1$, $u_2 \in U_2, \ldots, u_n \in U_n$. (Here U_j may simply be 'the subset of U corresponding to English words'. Or they may know that the elements of U_j contain certain proper names or set phrases. During the Second World War the British attacked minor targets like navigation buoys just to elicit coded messages containing known sequences of words for their code breakers to work on.) Then, eventually, they may expect a situation in which only one value of $0 \leqslant r \leqslant N - 1$ will give trial decodes $S'_r(v_j) \equiv v_j - r \bmod N$ with

503

$S'_r(v_j) \in U_j$. (Write $U_j - v_j = \{m \in U : m \equiv r' - v_j \bmod N$ for some $r' \in U_j\}$, then, in the example given, this will occur when $\bigcap_{j=1}^n (U_j - v_j)$ contains only one member which will be the required r.) They then know that $S(v) \equiv v - r \bmod N$ and have cracked the code.

More generally we may have a family Σ of pairs (T, S) of encoding and decoding functions. We select $(T_0, S_0) \in \Sigma$ and use this system to send messages $u_1 \in U_1$, $u_2 \in U_2, \ldots, u_n \in U_n$. Our opponents know $\Sigma, U_1, U_2, \ldots, U_n$ and $v_1 = T_0(u_1)$, $v_2 = T_0(u_2), \ldots, v_n = T_0(u_n)$. What properties should Σ have to make life as hard as possible for our opponents and as easy as possible for ourselves?

Note first that we wish to use our code. Thus if it takes a year to encode a message and a year to decode it, our coding system will find few users. We thus have our first criterion.

(1A) *Coding and decoding must be quick.*

The speed of coding depends on how it is done but nowadays we would expect to use a computer. The speed with which a computer calculates a function depends on the speed with which it performs elementary operations (such as addition, multiplication and moving data from one store to another). Our criterion can now be stated more precisely.

(1B) *For each $(T, S) \in \Sigma$ and each $u \in U$, $v \in V$ the computation of $T(u)$ must take no more than N_1 operations and that of $S(v)$ no more than N_2 operations.*

The numbers N_1 and N_2 depend on the use of the code and the technology available.

Let us turn now to the problems facing our opponent. Here the criterion is simple but vague.

(2A) *It should be very hard indeed to find S even knowing $\Sigma, U_1, U_2, \ldots, U_n, v_1$, $v_2, \ldots, v_n$.*

How hard depends of course on $U_1, U_2, \ldots, U_n$, but history shows that if n is large, eventually the opponent will come into possession of a U_j which is very small, or, indeed, consists of one point. (For example an embassy may transmit a known newspaper article or political speech using the code.)

Knowing this we must make our demand more precise by making the U_j one point sets $U_j = \{u_j\}$.

(2B) *It should be very hard indeed to find S even when n is large and Σ, u_1, $u_2, \ldots, u_n \in U$, $v_1 = T(u_1), v_2 = T(u_2), \ldots, v_n = T(u_n)$ are known.*

Notice that this criterion emphasises the futility of the code used as an example. Once u_1 and $v_1 \equiv u_1 + M \bmod N$ are known, the value of M is known and the code is broken.

The use of machines to code and decode has led inevitably to the use of machines to break codes. As in other fields this has led to some loss of romance particularly in Britain where a long succession of chess players, experts in early Church history,

beautiful women spies and assorted eccentrics have served their country well. (Wallis owed his appointment as Savilian professor of geometry at Oxford to his success as a code breaker for the parliamentary side in the English Civil War. Some of his mathematical techniques are believed to have a cryptological inspiration.) The use of machines also means that criterion (2B) can be made more precise.

(2C) *Even when n is large and $\Sigma, u_1, u_2, \ldots, u_n \in U$ and $v_1 = T(u_1), v_2 = T(u_2), \ldots, v_n = T(u_n)$ are known, the number of operations required to find S should be at least N_3.*

It is an open secret that some of the fastest and largest computers in existence are to be found in government 'Security Agencies' so N_3 must be very large indeed.

As it stands, condition (2C) is strong but still not as precise as we would wish. For example in some cases the choice of a particular $u_1, u_2, \ldots, u_n$ may make the code easy to break. We can avoid this difficulty by strengthening the condition.

(2D) *If Σ is known then, even given a chosen n and a chosen sequence $u_1, u_2, \ldots, u_n \in U$ together with $v_1 = T(u_1), v_2 = T(u_2), \ldots, v_n = T(u_n)$, the number of operations required to find S should be at least N_3.*

The problem of finding a 'good' code is thus reduced to reconciling (1B) and (2D) for N_3 much larger than N_1 or N_2. The problem is now precise but gives no indication of where, if anywhere, a solution might be found. However, Diffie and Hellman showed how by making the problem still more difficult the area of search could be narrowed to manageable proportions. Specifically, they proposed replacing condition (2D) by a new condition.

(2E) *If Σ and T are known the number of operations required to find S should be at least N_3.*

Observe that (2E) includes (2D) since our opponents can now compute $v_1 = T(u_1)$, $v_2 = T(u_2), \ldots$ themselves starting from whatever $u_1, u_2, \ldots$ they want. Notice also that we need no longer keep both S and T secret but only S.

Using this clue Rivest, Shamir and Adleman proposed a system Σ of codes which would be good if the following plausible conjecture was true.

Conjecture 97.1. *Let $N(d)$ be the number of operations required to factorise an arbitrary integer of size about 2^d. Then $d^{-m}N(d) \to \infty$ as $d \to \infty$ for all m.*

In the next two chapters we will sketch a simple version of their idea.

101

A LITTLE GROUP THEORY

In previous chapters we looked at some of the uses of the multiplicative group of the nth roots of unity. Under the name of the cyclic group of order n, it furnishes the simplest example of an Abelian group whilst under the name of the additive group of the integers modulo n, it forms a simple but very useful tool in number theory. In this chapter we introduce another kind of Abelian group which also plays an important role in number theory.

Let $\mathbb{Z}_n$ be the set of integers mod n. (More formally $\mathbb{Z}_n$ is the collection of equivalence classes $[a] = \{b : b \equiv a \bmod n\}$.) Recall that b and c are said to be coprime if they have no common factors. We make an obvious remark.

Lemma 101.1. (i) *If a and n are coprime so are $a + nm$ and n.*
Proof. If q divides n and $a + nm$ then q divides n and $a = a + nm - nm$. ∎

Thus we can define

$$G = \{[a] \in \mathbb{Z}_n : a \text{ and } n \text{ are coprime}\}$$

without ambiguity.

In the same vein we have the following results.

Lemma 101.1. (ii) *If $[a_1] = [a_2]$ and $[b_1] = [b_2]$ then $[a_1 b_1] = [a_2 b_2]$.*
(iii) *If $[a], [b] \in G$ then $[ab] \in G$.*

Proof. (ii) Since $a_1 - a_2, b_1 - b_2 \equiv 0 \bmod n$ we have

$$a_1 b_1 - a_2 b_2 = (a_1 - a_2)b_1 + a_2(b_1 - b_2) \equiv 0 \bmod n,$$

as required.
(iii) We give two proofs based on two distinct fundamental ideas.

Proof A. (By uniqueness of factorisation.) If $[ab] \notin G$ then ab and n have a common factor $r \geqslant 2$. If p is a prime factor of r then p divides ab and so by the uniqueness of factorisation p divides at least one of a and b. Let us suppose p divides a. Since p divides n, it follows that $[a] \notin G$.

Proof B. (By Euclid's algorithm.) Since $[a], [b] \in G$, it follows by Euclid's algorithm that we can find integers r, s, u, v with

$$ra + sn = 1, ub + vn = 1.$$

Thus

$$1 = (ra + sn)(ub + vn) = (ru)ab + (sub + rav + svn)n,$$

and so ab and n can have no common factors, i.e. $[ab] \in G$. ∎

We can now give an unambiguous definition of multiplication $\times$ on G by

$$[a] \times [b] = [ab],$$

which is closed in the sense that if $[a], [b] \in G$ then $[a] \times [b] \in G$. As usual we write $[a][b] = [a] \times [b]$. We wish to show that $(G, \times)$ is an Abelian group. All the conditions except the existence of an inverse are trivial.

Lemma 101.1. *If* $[a], [b], [c] \in G$ *then*

(iv) $[a]([b][c]) = ([a][b])[c]$ *(Associative law).*
(v) $[a][b] = [b][a]$ *(Commutative law).*
(vi) $[1][a] = [a]$ *(Existence of unit).*

Proof. (iv)   Evidently $[a]([b][c]) = [a]([bc]) = [a(bc)] = [(ab)c] = [ab][c] = ([a][b])[c]$. (v) and (vi) are even more trivial. ∎

The existence of an inverse is a more subtle matter.

Lemma 101.1. (vi) *If* $[a] \in G$ *we can find* $[A] \in G$ *such that* $[a][A] = [1]$.
Proof. We again give two proofs.
Proof A. (By uniqueness of factorisation.) We prove first that if $[b], [c] \in G$ and $[b][a] = [c][a]$ then $[b] = [c]$. For if $[b][a] = [c][a]$ then by definition $[ba] = [ca]$, $ba \equiv ca$ and $(b - c)a \equiv 0 \bmod n$. Thus n divides $(b - c)a$ so, since a and n have no common factor, n divides $(b - c)$ i.e. $b \equiv c \bmod n$ and $[b] = [c]$.

Now let the distinct elements of G be $[a_1] = [1], [a_2], [a_3], \dots, [a_N]$. By the first paragraph $[a_1][a], [a_2][a], [a_3][a], \dots, [a_N][a]$ are also distinct and so must be the elements of G under new names. In particular for some $1 \leqslant k \leqslant N$ we have $[a_k][a] = [1]$ and writing $A = a_k$ we obtain the result stated.
Proof B. (By Euclid's algorithm.) Since a and n are coprime, we can find integers A and N such that $Aa + Nn = 1$. The expression just given shows that A and n are coprime so that $[A] \in G$ and $[A][a] = [1]$ as required. ∎

We shall call the group G whch we have obtained the *group of units modulo* n. The number of elements of G is equal to the number of integers a with $1 \leqslant a \leqslant n - 1$ which are coprime to n. We denote this function by $\phi(n)$ and call it Euler's ϕ function. Since G is a group, we may apply Lagrange's theorem.

Theorem 101.2'. *If* $[a] \in G$ *then* $[a]^{\phi(n)} = [1]$.

Proof. The order of any element of a group divides the order of the group. ∎

Remark. Since G is Abelian, there is an alternative proof which does not use Lagrange's theorem. Using the results and notation of Proof A of Lemma 101.1 (vi) we know that, if $[a_1], [a_2], \ldots, [a_N]$ are the distinct elements of G, so are $[a][a_1], [a][a_2], \ldots, [a][a_N]$. Thus

$$[a_1][a_2]\ldots[a_N] = ([a][a_1])([a][a_2])\ldots([a][a_N]),$$

i.e. $$([a_1][a_2]\ldots[a_N]) = [a]^N([a_1][a_2]\ldots[a_N]),$$

and so $[1] = [a]^N$. Since $N = \phi(n)$ we have the result.

Translating we obtain a theorem of Euler.

Theorem 101.2. *If* a *and* n *are coprime then* $a^{\phi(n)} \equiv 1 \bmod n$.

Proof. Immediate from Theorem 101.2'. ∎

If p is prime then $\phi(p) = p - 1$ and we obtain Fermat's theorem.

Theorem 101.3. *If* p *is prime and* p *does not divide* a *then* $a^{p-1} \equiv 1 \bmod p$.

Proof. Immediate from Theorem 101.2. ∎

Euler's original proof is given in Chapter VI of Hardy and Wright.

We conclude this easy chapter with an example of a group of units modulo n which is not cyclic.

Example 101.4. *Let* G *be the group of units modulo* 8. *Then writing* $x = [3]$, $y = [5]$ *we have*

$$G = \{e, x, y, xy\} \text{ with } xy = yx, x^2 = y^2 = e,$$

(i.e. $G = C_2 \times C_2$ *the Klein* 4 *group*).

Proof. Direct verification. ∎

Observe that whilst Euler's theorem yields $a^4 \equiv 1 \bmod 8$ whenever a and 8 are coprime we have, in fact, the stronger result $a^2 \equiv 1 \bmod 8$ since all elements of G have order 1 or 2.

102

A GOOD CODE?

In Chapter 100 we discussed the properties which might be demanded from a family Σ of secret codes. In this chapter we describe the code invented by Rivest, Shamir and Adleman.

We start with two very large primes p and q. Write $N = pq$. In the notation of Chapter 100 we take

$$U = V = \{u \in \mathbb{Z} : 0 \leqslant u \leqslant N - 1\}.$$

We choose a coprime to $(p - 1)(q - 1)$ and define $T : U \to V$ by

$$T(u) \equiv u^a \bmod N.$$

How can we recover u from $T(u)$ knowing p and q?

Lemma 102.1. *If p and q are prime then $\phi(pq) = (p - 1)(q - 1)$.*
Proof. Any integer with a factor in common with pq must be divisible by p and/or q. Thus the only integers m with $1 \leqslant m \leqslant pq - 1$ having a factor in common with pq are $p, 2p, \ldots, (q - 1)p$ and $q, 2q, \ldots, (p - 1)q$. Since these are all distinct, we know that there are exactly $(p - 1) + (q - 1)$ integers m with $1 \leqslant m \leqslant pq - 1$ having a factor in common with pq. Hence

$$\phi(pq) = (pq - 1) - ((q - 1) + (p - 1)) = (p - 1)(q - 1). \qquad \blacksquare$$

By Theorem 101.2 it follows that

$$u^{(p-1)(q-1)} \equiv 1 \bmod N,$$

and so

$$u^t \equiv u \bmod N$$

whenever $t \equiv 1 \bmod (p - 1)(q - 1)$. Since a was chosen to be coprime to $(p - 1)(q - 1)$ we can find, using Euclid's algorithm, integers b_1 and b_2 such that

$$ab_1 + (p - 1)(q - 1)b_2 = 1.$$

Let b be the integer defined by the conditions

$$b \equiv b_1 \bmod (p-1)(q-1) \quad \text{and} \quad 0 \leqslant b < (p-1)(q-1).$$

Then $ab \equiv 1 \bmod (p-1)(q-1)$ and so defining $S\colon V \to U$ by

$$S(v) \stackrel{*}{\equiv} v^b \bmod N,$$

we have $ST(u) \equiv u^{ab} \equiv u \bmod N$ and $ST(u) = u$, as required.

Does this code satisfy the criterion of easy coding and decoding as formalised in Condition (1B) of Chapter 99? By definition

$$T(u) \equiv u^a \bmod N.$$

Suppose $N < 2^d$. By using relations of the form

$$(u^2)^2 \equiv u^4, (u^4)^2 \equiv u^8, \ldots, (u^{d-2})^2 \equiv u^{d-1} \bmod N,$$

we can work out $T(u)$ using no more than $2d - 2$ multiplications modulo N. Since N is large we will need to use multiple precision arithmetic so we must expect the number of elementary operations involved in each multiplication modulo N to rise as d^2 if we go about it the obvious way and as $d \log d \log \log d$ if we use the fast multiplication techniques mentioned in Chapter 99. Thus, in the first case, the total number of operations will rise as $d \cdot d^2 = (\log_2 N)^3$ and in the second as $(\log_2 N)^2 \log_2 \log_2 N \log_2 \log_2 \log_2 N$. At the moment values of N of the order of 2^{500} are easily practicable.

Does the code satisfy the criterion for being hard to crack formalised in Condition (2E) of Chapter 100?

Problem 102.2'. *Given N which is known to be the product of two large prime numbers and an integer a which is known to be coprime to N, roughly how many operations are required to find a function $S\colon \{r\colon 0 \leqslant r \leqslant N-1\} \to \{r\colon 0 \leqslant r \leqslant N-1\}$ such that*

$$S(u^a) = u \quad \text{for all } u?$$

Working modulo N throughout we see that if $u^{ab} \equiv u$ for all $0 \leqslant u \leqslant n-1$ then $S(u) = S(u^{ab}) = S((u^b)^a) = u^b$ so, by choosing suitable trial u it is easy starting from S to find ab with $u^{ab} \equiv u \bmod N$ for all u. Our problem can thus be restated.

Problem 102.2. *Given N which is known to be the product of two large prime numbers and an integer a which is known to be coprime to $\phi(N)$, roughly how many operations are required to find an integer b such that*

$$u^{ab} \equiv u \bmod N \quad \text{for all } u?$$

We now make three plausible conjectures.

Conjecture 102.3(A). *Provided N and a have no special features then knowledge of b will enable us to find the factors of N very rapidly.*

Remark. Observe, for example, that if $a = 1$ or, less trivially, if $p = 2^r + 1, q = 2^s + 1$

then N and a have special features which render conjecture 102.3(A) false without some restriction.

Conjecture 102.3(B). *Given N which is known to be the product of two large prime numbers and an integer a coprime to $\phi(N)$ the number of operations required to factor N is a function $\psi(N)$ such that $N^{-r}\psi(N) \to \infty$ as $N \to \infty$ for all r.*

Conjecture 102.3(C). *Given N with no special features the number of operations required to factor N is a function $\psi(N)$ such that $N^{-r}\psi(N) \to \infty$ as $N \to \infty$ for all r.*

Conjectures 102.3(A) and (B) together imply that our code is very hard to crack for large N. It seems very likely that, provided the notion of 'special feature' is carefully defined, Conjecture 102.2(A) is true. If Conjecture 102.3(C) is true we would expect the related Conjecture 102.3(B) to be true. However, the evidence for Conjecture 102.3(C) is simply the negative one that in 300 years of investigation by very able mathematicians no fast factoring method has been found. We are left with a suggestive idea but no proof.

Remark. The reader may wonder how, if the factorisation problem is so hard, we can find sufficient large primes to make our code work. The reason lies in the existence of fast tests for what might be called 'engineer's primes'. More precisely given $\varepsilon > 0$ and a large d there exist tests involving about d^3 operations such that if a random integer $2^d \leqslant N \leqslant 2^{d+1}$ passes all the tests then the probability that N is prime is greater than $1 - \varepsilon$. In practice, we would choose a very small ε and a random selection of large integers until we found two 'engineer's primes' p and q. We would then use these 'engineer's primes' to construct our code.

If Conjectures 102.3(B) and (A) are true then, as we have observed, Condition (2E) of Chapter 100 holds.

(2E) *Even if T is known the number of operations required to find S is very large.*

Thus by broadcasting knowledge of T but keeping S secret we can ensure that many can write but few can read.

The symmetry of T and S for the given code ensures that the mirror condition also holds.

(2E') *Even if S is known the number of operations required to find T is very large.*

Thus by keeping T secret but broadcasting S we can ensure that many can read but few can write. This property is particularly valuable if we wish to produce documents like credit cards or identity papers which may have to be verified (read) by many people but which must be protected against forgery (unauthorised writing).

'A science', wrote G.H. Hardy, 'is said to be useful if its development tends to accentuate the existing inequalities in the distribution of wealth, or more directly promotes the destruction of human life.' With credit cards and identity papers a small part of number theory has entered the sphere of the useful sciences.

In *A Mathematician's Apology* (Chapter 21) Hardy clarified his position.

If the theory of numbers could be employed for any practical and obviously honourable purpose, if it could be turned directly to the furtherance of human happiness or the relief of human suffering, as physiology or even chemistry can, then surely neither Gauss nor any other mathematician would have been so foolish as to decry or regret such applications. But science works for evil as well as for good (and particularly of course, in time of war); and both Gauss and lesser mathematicians may be justified in rejoicing that there is one science at any rate, and that their own, whose very remoteness from ordinary human activities should keep it gentle and clean.

There is loss as well as gain for the mathematical community when research into the factorisation of large numbers could become important enough to be classified as a state secret.

Postscript

Although the problem of factorisation has always been of interest to mathematicians the glamour which now surrounds the subject has lead to a great increase in the number and intensity of attacks on it. Thus as this chapter passes through its proof stage I can add the news that efficient prime testing algorithms now remove the need to use 'engineer's primes' and that Lenstra has discovered a new factorising algorithm based on ideas from Algebraic Geometry which works much faster than anything known previously on a large class of integers. However the coding scheme described above is still secure.

Postscript 1989

The factorisation of 100-digit numbers will soon be routine. This represents spectacular progress compared with what was believed possible twenty years ago but will not worry commercial users who will just switch to 200-digit numbers.

103

A LITTLE MORE
GROUP THEORY

When we studied Fourier series on $\mathbb{T}$, we used exponentials $e_n:\mathbb{T}\to\mathbb{C}$ given by $e_n(t)=\exp int$. When we studied Fourier transforms on $\mathbb{R}$, we used exponentials $e_\lambda:\mathbb{R}\to\mathbb{C}$ given by $e_\lambda(x)=\exp i\lambda x$. When we generalised to several dimensions, we used exponentials $e_n:\mathbb{T}^m\to\mathbb{C}$ with $e_n(\mathbf{t})=\exp i\mathbf{n\cdot t}$ and $e_\lambda:\mathbb{R}^m\to\mathbb{C}$ given by $e_\lambda(x)=\exp i\lambda\cdot\mathbf{x}$. Finally in Chapter 94 when we studied the group $G=\{1,\omega,\omega^2,\dots,\omega^N\}$ of Nth roots of unity we used an analogous function $e_n:G\to\mathbb{C}$ given by $e_n(\omega^r)=\omega^{rn}$.

In each case we obtained the same kind of 'Fourier theory' with the same kind of theorems and proofs. What are the fundamental properties which unite all these cases and can we hope to find other cases to which the same set of ideas will be applicable? It is very rare that we can find a complete answer to such a question but even a very partial response may indicate an interesting field of enquiry. In this instance we note that in each of the cases given in the first paragraph we have an Abelian group G, with unit ι say, and a map $e:G\to\mathbb{C}$ such that

(a) $e(x+y)=e(x)e(y)$,
(b) $e(\iota)=1$.

What can we say of such functions, at least in the case when G is finite but, for the moment, not necessarily Abelian?

Lemma 103.1. *Let G be a finite group and $e:G\to\mathbb{C}$ a function such that*

(a) $e(xy)=e(x)e(y)$,
(b) $e(\iota)=1$.

Then $|e(x)|=1$ for all $x\in G$.
Proof. Let G have order N. By Lagrange's theorem $x^N=\iota$ so $e(x)^N=e(x^N)=e(\iota)=1$ and $e(x)$ is an Nth root of unity. ∎

Since $S=\{\lambda\in\mathbb{C}:|\lambda|=1\}$ is an Abelian group under the usual definition of multiplication, Lemma 103.1 can be restated.

Lemma 103.1'. *Under the conditions of Lemma 103.1, e is a homomorphism from G to S.*

Proof. This is a simple restatement of Lemma 103.1.   ∎

Remark. The group S is, of course, isomorphic to $\mathbb{T}$ but we shall not use this fact.

Just as in Chapter 97 we write $C(G)$ for the collection of functions $f:G \to \mathbb{C}$ and, if G has N elements,

$$(f, h) = N^{-1} \sum_{x \in G} f(x)h(x)^* \quad [f, h \in C(G)].$$

The reader should see at a glance that Lemmas 97.1, 97.2 and 97.3 of Chapter 97 (which state, in effect, that $C(G)$ is an inner product space) are true. We now wish to study the homomorphisms $e:G \to S$ which we hope will be good analogues of the exponentials.

We start with two lemmas which deal with results which we took for granted when we treated our earlier exponentials. Write $\hat{G}$ for the set of homomorphisms $e:G \to S$. If $e_1, e_2 \in \hat{G}$ define the pointwise product $e_1 e_2$ by $e_1 e_2(g) = e_1(g)e_2(g) [g \in G]$.

Lemma 103.2. *The set $\hat{G}$ is an Abelian group under pointwise multiplication. More precisely,*

(i) *If $e_1, e_2 \in \hat{G}$ then $e_1 e_2 \in \hat{G}$.*
(ii) *If $e_1, e_2, e_3 \in \hat{G}$ then $(e_1 e_2)e_3 = e_1(e_2 e_3)$.*
(iii) *If $e_1, e_2 \in \hat{G}$ then $e_1 e_2 = e_2 e_1$.*
(iv) *Let $e_0(g) = 1$ for all $g \in G$. Then $e_0 \in \hat{G}$ and $e_1 e_0 = e_1$ for all $e_1 \in \hat{G}$.*
(v) *If $e_1 \in \hat{G}$ then $e_1^* \in G$ and $e_1 e_1^* = e_0$.*

Proof. (i) If $g_1, g_2 \in G$ then

$$e_1 e_2(g_1 g_2) = e_1(g_1 g_2)e_2(g_1 g_2) = (e_1(g_1)e_1(g_2))(e_2(g_1)e_2(g_2))$$
$$= (e_1(g_1)e_2(g_1))(e_1(g_2)e_2(g_2)) = e_1 e_2(g_1)e_1 e_2(g_2),$$

using the fact that S is an Abelian group.

(ii), (iii), (iv) left to the reader.

(v) If $g_1, g_2 \in G$ then

$$e_1^*(g_1 g_2) = (e_1(g_1 g_2))^* = (e_1(g_1)e_2(g_2))^* = e(g_1)^* e_1(g_2)^* = e_1^*(g_1)e_1^*(g_2).$$

Further, if $g \in G$ then

$$e_1 e_1^*(g) = e_1(g)e_1(g)^* = |e_1(g)|^2 = 1.$$ ∎

Let us always write e_0 for the element of $\hat{G}$ given by $e_0(g) = 1 [g \in \hat{G}]$.

Lemma 103.3. *If $e \in \hat{G}$ and $e \neq e_0$ then $\sum_{g \in G} e(g) = 0$.*

Proof. Since $e \neq e_0$ we can find an $x \in G$ with $e(x) \neq 1$. Since G is a group xg ranges

over G as g ranges over G and so

$$\sum_{g \in G} e(g) = \sum_{g \in G} e(xg) = \sum_{g \in G} e(x)e(g) = e(x) \sum_{g \in G} e(g).$$

Thus $(1 - e(x)) \sum_{g \in G} e(g) = 0$ and so (since $e(x) \neq 1$) $\sum_{g \in G} e(g) = 0$. ∎

We can now prove the result corresponding to Lemma 97.4.

Lemma 103.4. *If* $e_1, e_2 \in \hat{G}$ *then*

(i) $(e_1, e_1) = 1$,
(ii) $(e_1, e_2) = 0$ *if* $e_1 \neq e_2$.

Proof. (i) By definition

$$(e_1, e_1) = N^{-1} \sum_{g \in G} e_1(g)e_1(g)^* = N^{-1} \sum_{g \in G} 1 = 1.$$

(ii) Using Lemma 103.2 we observe that $e_1 e_2^* \in \hat{G}$ and $e_1 e_2^* \neq e_0$ so, using Lemma 103.3

$$(e_1, e_2) = N^{-1} \sum_{g \in G} e_1(g)e_2(g)^* = N^{-1} \sum_{g \in G} e_1 e_2^*(g) = 0.$$ ∎

As in Chapter 97 we have an analogue of Theorem 32.5.

Lemma 103.5. *If* $h_0 = \sum_{e \in \hat{G}} (f, e)e$ *and* $h = \sum_{e \in \hat{G}} \lambda_e e$ *then*

$$\|f\|_2^2 \geqslant \sum_{e \in \hat{G}} |(f, e)|^2$$

and
$$\|f - h\|_2 \geqslant \|f - h_0\|_2 \geqslant \sqrt{\left(\|f\|_2^2 - \sum_{e \in \hat{G}} |(f, e)|^2 \right)}$$

with equality if and only if $\lambda_e = (f, e) [e \in \hat{G}]$.
Remark. The sum $\sum_{e \in \hat{G}}$ is finite because $\hat{G}$ is finite. (Observe that $e(g)$ is an Nth root of unity for each $g \in G$.)
Proof. Just as for Theorem 32.5. ∎

It is thus in keeping with our other notations to write

$$\hat{f}(e) = (f, e) = N^{-1} \sum_{g \in G} f(g)e(g)^*,$$

but when we now try to prove the analogue of Theorem 97.6 the smooth flow of generalisation comes to an abrupt halt. We gave two proofs of Theorem 97.6, one by the methods of linear algebra, and one by the methods of Chapter 2. Let us see what the methods yield in the more general case.

First we try linear algebra. (The result and its proof are intended to tie in our results with those of elementary linear algebra and will not be used again.)

Theorem 103.6. (i) *The group $\hat{G}$ has at most N elements.*
(ii) *The formula*

$$f = \sum_{e \in \hat{G}} \hat{f}(e)e$$

holds for all $f \in C(G)$ if and only if $\hat{G}$ has exactly N elements.
Proof. (i) We observe, just as in the first proof of Theorem 97.6, that $C(G)$ is a vector space over $\mathbb{C}$ of dimension N. Since the elements of $\hat{G}$ are orthonormal, they are linearly independent and so there can exist at most N of them.
(ii) A formula of the form

$$f = \sum_{e \in \hat{G}} a_e(f)e \quad [a_e(f) \in \mathbb{C}]$$

holds for all $f \in C(G)$ if and only if $\hat{G}$ spans $C(G)$. But the elements of $\hat{G}$ are linearly independent so such formulae can hold for all $f \in C(G)$ if and only if $\hat{G}$ has exactly N elements. By Lemma 103.5

$$f = \sum_{e \in \hat{G}} a_e(f)e$$

implies $a_e(f) = \hat{f}(e)$ so we are done.   ∎

Before trying the method of Chapter 2 we need a result echoing Lemma 103.3.

Lemma 103.7. *If $x \in G$ and there exists an $e_x \in \hat{G}$ with $e_x(x) \neq 1$ then $\sum_{e \in \hat{G}} e(x) = 0$.*
Proof. Since $\hat{G}$ is a group $e_x e$ ranges over $\hat{G}$ as e ranges over $\hat{G}$ and so

$$\sum_{e \in \hat{G}} e(x) = \sum_{e \in \hat{G}} (e_x e)(x) = \sum_{e \in \hat{G}} e_x(x)e(x) = e_x(x) \sum_{e \in \hat{G}} e(x).$$

Thus $(1 - e_x(x)) \sum_{e \in \hat{G}} e(x) = 0$ and so (since $e_x(x) \neq 1$) $\sum_{e \in \hat{G}} e(x) = 0$.   ∎

We now try the method of Chapter 2.

Theorem 103.8. *The following two statements are equivalent.*
(i) *For each $x \in G$ with $x \neq \iota$ there exists an $e_x \in \hat{G}$ with $e_x(x) \neq 1$.*
(ii) *If $f \in C(G)$ then*

$$f = \sum_{e \in \hat{G}} \hat{f}(e)e.$$

Moreover, if (i) holds then $\hat{G}$ has exactly N elements.
Proof. We show first that (ii) implies (i) and then that (i) implies (ii).

(ii) implies (i)

Choose $f \in C(G)$ with $f(x) = 0$, $f(\iota) = 1$. If $e(x) = 1$ for all $e \in \hat{G}$ then

$$0 = f(x) = \sum_{e \in \hat{G}} \hat{f}(e)e(x) = \sum_{e \in \hat{G}} \hat{f}(e) = \sum_{e \in \hat{G}} \hat{f}(e)e(\iota) = f(\iota) = 1$$

which is absurd.

(i) *implies* (ii)

As in Chapter 97 assume that $f \in C(G)$. Then writing $S(f, x) = \sum_{e \in \hat{G}} \hat{f}(e)e(x)$ we have

$$S(f, x) = \sum_{e \in \hat{G}} \left(N^{-1} \sum_{g \in G} f(g)e(g)^* \right) e(x) = N^{-1} \sum_{g \in G} \left(\sum_{e \in \hat{G}} f(g)e(g)^*e(x) \right)$$

$$= N^{-1} \sum_{g \in G} f(g) \sum_{e \in \hat{G}} e(g)^* e(x) = N^{-1} \sum_{g \in G} f(g) \sum_{e \in \hat{G}} e(g^{-1}x)$$

$$= N^{-1} \sum_{g \in G} f(g) D(g^{-1}x),$$

where $D(y) = \sum_{e \in \hat{G}} e(y)$. Noting that, since G is a group, $g^{-1}x$ ranges through G as g ranges through G we see that the substitution $y = g^{-1}x$ gives

$$S(f, x) = N^{-1} \sum_{y \in G} f(xy^{-1}) D(y).$$

The structure of D is particularly simple. Using hypothesis (i) and Lemma 103.7 we see that if $y \neq \iota$

$$D(y) = \sum_{e \in \hat{G}} e(g) = 0.$$

On the other hand if $y = \iota$ we obtain

$$D(\iota) = \sum_{e \in \hat{G}} e(\iota) = \sum_{e \in \hat{G}} 1 = M,$$

where M is the order of $\hat{G}$. Thus

$$S(f, x) = N^{-1} \sum_{y \in G} f(xy^{-1}) D(y) = N^{-1} M f(x) \quad \text{for all } x \in G.$$

Now if $f = e_0$, the formula just becomes

$$e_0(x) = N^{-1} M e_0(x),$$

so $1 = N^{-1}M$ and $\hat{G}$ has exactly N elements as stated in the last part of the theorem. We thus have

$$S(f, x) = N^{-1} N f(x) = f(x) \quad \text{for all } x \in G,$$

and this is the desired result (ii). ∎

Theorems 103.6 and 103.8 both state that the desired relation $f = \sum_{e \in \hat{G}} \hat{f}(e)e$ will hold for all $f \in C(G)$ only if $\hat{G}$ is large enough, i.e. only if there are enough homomorphisms $e: G \to S$. Let us see if this assumption is justified in the case of S_3 the smallest non-Abelian group.

Lemma 103.9. *Let S_3 be the group generated by a, b with $a^2 = b^3 = \iota$ and $ab^2 = ba$. Then S_3 has six elements but $\hat{S}_3$ has only 2.*

Proof. $e: S_3 \to S$ be a homomorphism. Then $e(a)e(b)^2 = e(ab^2) = e(ba) = e(b)e(a) =$

$e(a)e(b)$ so $e(b) = 1$. Since $e(a)^2 = e(a^2) = e(\iota) = 1$ we have $e(a) \in \{-1, 1\}$. The only possible homomorphisms $e : S_3 \to S$ are thus e_0 and e_1 defined by $e_0(a^r b^s) = 1$ and $e_1(a^r b^s) = (-1)^r [r = 0, 1; s = 0, 1, 2]$. Both e_0 and e_1 are readily seen to be homomorphisms.   ∎

This counterexample shows that any generalisation of Fourier analysis to non commutative groups must be more daring than anything proposed here. (Such a generalisation was found by Frobenius but that is another story.) In the next chapter we shall show that in the case of finite Abelian groups G the group $\hat{G}$ is large enough and our generalisation will work smoothly.

104

FOURIER ANALYSIS ON
FINITE ABELIAN GROUPS

The main object of this chapter is to show that if G is a finite Abelian group then $\hat{G}$ contains enough elements to enable us to do Fourier analysis. More specifically we wish to prove the following result.

Theorem 104.1. *Let G be a finite Abelian group with identity ι. Then for each $x \in G$ with $x \neq \iota$ there exists an $e_x \in \hat{G}$ with $e_x(x) \neq 1$.*

Together with Theorem 103.8 this gives the desired result.

Theorem 104.2. *Let G be a finite Abelian group. Then G and $\hat{G}$ have the same order, and if $f \in C(G)$,*

$$f = \sum_{e \in \hat{G}} \hat{f}(e)e.$$

Proof. Use Theorems 103.8 and 104.1. ∎

The key step in the proof of Theorem 104.1 is given in the following lemma.

Lemma 104.3. (i) *Let G be a finite Abelian group and H a subgroup of G. Then given any $a \in G$ and any homomorphism $e : H \to S$ we can find a homomorphism $\tilde{e} : K \to S$, where K is the subgroup generated by a and H, such that $\tilde{e}(h) = e(h)$ for all $h \in H$.*
Proof. Let r be the least integer with $r \geq 1$ and $a^r \in H$. (If N is the order of G, $a^N = e \in H$ so such an integer does exist. Note that if $a \in H$ then $r = 1$.) We claim that every element k of K can be written in one and only one way as $k = a^s h$ with $0 \leq s \leq r - 1$ and $h \in H$.

To show the existence of such a representation we observe that (since G is Abelian) every $k \in K$ can be written as $k = a^t h_1$ with $h_1 \in H$. Let s be the integer with $0 \leq s \leq r - 1$ and $s \equiv t \bmod r$. Then $t - s = mr$ for some integer m so $a^{t-s} = (a^r)^m \in H$ and $k = a^s h$ where $h = a^{t-s} h_1 \in H$. To show uniqueness, suppose $a^{s(1)} h_1 = a^{s(2)} h_2$ with $0 \leq s(1) \leq s(2) \leq r - 1$ and $h_1, h_2 \in H$. Then $0 \leq s(2) - s(1) \leq r - 1$ and

519

$a^{s(2)-s(1)} = h_1 h_2^{-1} \in H$ so (by the definition of r) $s(2) - s(1) = 0$ giving $s(1) = s(2)$ and $h_1 = h_2$.

Choose $\omega \in S$ so that $\omega^r = e(a^r)$ (for example, if $e(a^r) = \exp i\theta$, we could take $\omega = \exp(i\theta/r)$) and set $\tilde{e}(a^s h) = \omega^s e(h)$ for all $0 \leqslant s \leqslant r - 1$, $h \in H$. By the previous paragraph $\tilde{e}:K \to S$ is a well defined function. It is clear that $\tilde{e}(h) = e(h)$ for all $h \in H$ so all that remains to prove is that $\tilde{e}$ is a homomorphism. But if $0 \leqslant s(1), s(2) < r - 1$ and $h_1, h_2 \in H$ then, either $s(1) + s(2) \leqslant r - 1$ and

$$\tilde{e}(a^{s(1)}h_1 a^{s(2)}h_2) = \tilde{e}(a^{s(1)+s(2)}h_1 h_2) = \omega^{s(1)+s(2)}e(h_1 h_2)$$
$$= \omega^{s(1)+s(2)}e(h_1)e(h_2) = \omega^{s(1)}e(h_1)\omega^{s(2)}e(h_2) = \tilde{e}(a^{s(1)}h_1)\tilde{e}(a^{s(2)}h_2),$$

or

$$r \leqslant s(1) + s(2) \leqslant 2r - 2,$$

and

$$\tilde{e}(a^{s(1)}h_1 a^{s(2)}h_2) = \tilde{e}(a^{s(1)+s(2)-r}a^r h_1 h_2)$$
$$= \omega^{s(1)+s(2)-r}e(a^r h_1 h_2) = \omega^{s(1)+s(2)-r}e(a^r)e(h_1)e(h_2)$$
$$= \omega^{s(1)+s(2)-r}\omega^r e(h_1)e(h_2) = \omega^{s(1)+s(2)}e(h_1)e(h_2) = \tilde{e}(a^{s(1)}h_1)\tilde{e}(a^{s(2)}h_2).$$

Thus $\tilde{e}$ is indeed a homomorphism $\tilde{e}:K \to S$. ∎

Since Lemma 104.3 (i) is the key to what follows, the reader should not proceed until she can prove it for herself and recognises how easy the proof is. It is clear what the next steps must be.

Lemma 104.3. (ii) *Let G be a finite Abelian group and H a subgroup of G. Then given any homomorphism $e:H \to S$ we can find a homomorphism $\tilde{e}:G \to S$ such that $\tilde{e}(h) = e(h)$ for all $h \in H$.*

Proof. Choose $a_1, a_2, \ldots, a_n \in G$ such that, writing $K_1 = H$ and K_{j+1} for the subgroup generated by K_j and $a_j[1 \leqslant j \leqslant n]$, we have $K_{n+1} = G$. (We could, for example, choose $a_1, a_2, \ldots, a_n$ to be all the elements of G.) Write $e_1 = e$. By Lemma 104.3 (i) we can find inductively homomorphisms $e_j:K_j \to S$ in such a way that $e_{j+1}(k) = e_j(k)$ for all $k \in K_j$. Setting $\tilde{e} = e_{n+1}$ we have the desired result. ∎

We can now prove Theorem 104.1 in a very slightly stronger form which will come in useful later on.

Theorem 104.1'. *Let x be an element of order n is a finite Abelian group G. Then there exists an $e_x \in \hat{G}$ such that $e_x(x^r) = \exp 2\pi i r/n$ for $0 \leqslant r \leqslant n - 1$.*

Proof. Let $H = \{\iota, x, x^2, \ldots, x^{n-1}\}$ be the subgroup of G generated by x. Let $\omega = \exp 2\pi i/n$ and set $e(x^r) = \omega^r [0 \leqslant r \leqslant n - 1]$. It is easy to see that $e:H \to S$ is a homomorphism. By Lemma 104.3(ii) we can find a homomorphism $\tilde{e}:G \to S$ such that $\tilde{e}(h) = e(h)$ for all $h \in H$. Setting $e_x = \tilde{e}$, we see that $e_x \in \hat{G}$ and $e_x(x^r) = e(x^r) = \omega^r$, so we are done. ∎

Applying the result above to the groups discussed in Chapter 101 we obtain the following immediate consequences.

Theorem 104.4. *Let G be the group of units modulo n. Then there are exactly $\phi(n)$ distinct maps $e : G \to \mathbb{C}$ such that*

(i) *$e([m])$ is a root of unity for each $[m] \in G$,*
(ii) *$e([m_1][m_2]) = e([m_1])e([m_2])$ for all $[m_1]$, $[m_2] \in G$.*

Write $\hat{G}$ for this set of maps. Then if f is a map from G to $\mathbb{C}$ we have

$$f([m]) = \sum_{e \in \hat{G}} \hat{f}(e)e([m]) \quad \text{for all } [m] \in G$$

where
$$\hat{f}(e) = \phi(n)^{-1} \sum_{[m] \in G} f([m])e([m])^*.$$

Proof. Apply Theorem 104.2 and Lemma 103.1′. ∎

Theorem 104.4 is all that we shall need in the remainder of the book. In the remainder of the chapter we digress to consider the problem of what the elements $e \in \hat{G}$ actually look like.

Our investigation involves looking at the structure of G. We begin with two routine lemmas.

Lemma 104.5. *Let G be an Abelian group.*

(i) *If G contains an element x of order r and an element y of order s with r and s coprime then xy has order rs.*
(ii) *If G contains an element x of order r and an element y of order s then G contains an element z whose order is the lowest common multiple of r and s.*

Remark. Note that, in general, xy need not have order equal to the lowest common multiple of the order of x and y. In $C_2 = \{1, -1\}$ the (multiplicative) group of two elements, $x = y = -1$ has order 2 but $xy = 1$ has order 1.
Proof. (i) Suppose $(xy)^a = \iota$. Then $y^{ar} = \iota^a y^{ar} = x^{ar} y^{ar} = (xy)^{ar} = \iota$ so $ar \equiv 0 \bmod s$. Since r and s are coprime, it follows that $a \equiv 0 \bmod s$. Similarly, $a \equiv 0 \bmod r$ so, since r and s are coprime, $a \equiv 0 \bmod rs$.
(ii) Factorising, we can find primes $p(1), p(2), \ldots, p(l), q(1), q(2), \ldots, q(k)$ and positive integers $N(i) \geqslant n(i)$ $[1 \leqslant i \leqslant l]$, $M(j) \geqslant m(j)$ $[1 \leqslant j \leqslant k]$ such that

$$r = \prod_{i=1}^{l} p(i)^{N(i)} \prod_{j=1}^{k} q(j)^{m(j)}, \quad s = \prod_{i=1}^{l} p(i)^{n(i)} \prod_{j=1}^{k} q(j)^{M(j)}.$$

Set $r_1 = \prod_{i=1}^{l} p(i)^{N(i)}$, $r_2 = \prod_{j=1}^{k} q(j)^{m(j)}$, $s_1 = \prod_{i=1}^{l} p(i)^{n(i)}$, $s_2 = \prod_{j=1}^{k} q(j)^{M(j)}$. Then x^{r_2} has order r_1, y^{s_1} has order s_2 and r_1, s_2 are coprime. Thus, by (i), $z = x^{y_2}y^{s_2}$ has order $r_1 s_2$. Since, by inspection $r_1 s_2$ is the lowest common multiple of r and s we are done. ∎

Lemma 104.6. *Let G be a finite Abelian group. If n is the largest order of any element of G then the order of any element of G divides n.*

Proof. Let x be an element of G of maximal order and let y be any element of G. If y has order m then by Lemma 104.5 (ii) there exists an element z of order N the lowest common multiple of m and n. But by the definition of n we have $N \leqslant n$. Thus $N = n$ and m divides n. ∎

The proof of the next Lemma is simple but not routine, since it makes use of Theorem 104.1′.

Lemma 104.7. *Let G be a finite Abelian group and let x be an element of G of largest order, n say. Then we can find a subgroup K of G such that each $g \in G$ can be written uniquely in the form $g = x^r k$ with $0 \leqslant r \leqslant n - 1$ and $k \in K$. The order of every element of K divides n.*

Proof. By Theorem 104.6′ we can find an $e \in \hat{G}$ such that $e(x^m) = \exp 2\pi i m/n$. Let $K = e^{-1}(\{1\}) = \{g \in G : e(g) = 1\}$. Since e is a homomorphism, K is a subgroup of G (the kernel of e). If $g \in G$ then, by Lemma 104.6, $g^n = \iota$ so $e(g)^n = e(g^n) = e(\iota) = 1$. Thus $e(g)$ is an nth root of 1 i.e. $e(g) = \exp 2\pi i r/n$ for some $0 \leqslant r \leqslant n - 1$. Set $k = x^{-r}g$. Then $g = x^r k$ and $e(k) = e(x^r)^{-1}e(g) = 1$ so $k \in K$. Conversely, if $k_1, k_2 \in K$ and $D \leqslant r(1)$, $r(2) \leqslant n - 1$ then the equality $x^{r(1)}k_1 = x^{r(2)}k_2$ implies

$$\exp 2\pi i r(1)/n = e(x^{r(1)}) = e(x^{r(1)})e(k_1) = e(x^{r(1)}k_1) = e(x^{r(2)}k_2) = \exp 2\pi i r(2)/n,$$

so that $r(1) = r(2)$ and $k_1 = k_2$. Thus each element $g \in G$ can be written in one and only one way as $g = x^r k$ with $0 \leqslant r \leqslant n - 1$ and $k \in K$. That the order of every element of K divides n is obvious, since (as we have already noted) the order of every element of G divides n. ∎

From Lemma 104.7 it is a simple step to a complete description of the structure of a finite Abelian group.

Theorem 104.8. *If G is a finite Abelian group with more than one element then we can find positive integers $n(1) \geqslant n(2) \geqslant \cdots \geqslant n(m) \geqslant 2$ such that $n(j+1)$ divides $n(j)[1 \leqslant j \leqslant m - 1]$ and elements $x_1, x_2, \ldots, x_m \in G$ such that x_j has order $n(j)$ $[1 \leqslant j \leqslant m]$ with the property that each $x \in G$ can be written in one and only one way as*

$$x = x_1^{r(1)}x_2^{r(2)}\ldots x_m^{r(m)},$$

with $0 \leqslant r(j) \leqslant n(j) - 1$ $[1 \leqslant j \leqslant m]$.

Proof. We use induction on Lemma 104.7. ∎

It is not hard to see that the sequence $n(1), n(2), \ldots, n(m)$ characterises the finite Abelian group G up to isomorphism. (If two finite Abelian groups have the same sequence they are obviously isomorphic. If one finite Abelian group G has the associated sequence $n(1), n(2), \ldots, n(k), n(k+1), \ldots, n(m)$ and another group G' has the associated sequence $n(1), n(2), \ldots, n(k), n'(k+1), \ldots, n'(m')$ with $n(k+1) > n'(k+1)$ then G and G' have a different number of elements with order a multiple

of $n(k+1)$ so G and G' are not isomorphic. If one group G has the associated sequence $n(1)$, $n(2), \ldots, n(m)$ and another G' has the associated sequence $n(1)$, $n(2), \ldots, n(m')$ with $m > m'$ then G and G' have different orders and cannot be isomorphic. Thus each associated sequence defines a unique Abelian group up to isomorphism.) The algebraists thus call Theorem 104.8 *the structure theorem for finite Abelian groups*. It has a clear analogy with the basis theorem for finite dimensional vector spaces.

Once we have characterised G in the form of Theorem 104.8 it is easy to characterise $\hat{G}$.

Theorem 104.9. *If G is given as in Theorem 104.8 then $\hat{G}$ consists of the functions* $e_{\omega(1)\omega(2)\cdots\omega(m)}$ *given by*

$$e_{\omega(1)\omega(2)\ldots\omega(m)}(x_1^{r(1)} x_2^{r(2)} \ldots x_m^{r(m)}) = \omega(1)^{r(1)} \omega(2)^{r(2)} \ldots \omega(m)^{r(m)},$$

where $\omega(j)$ is an $n(j)$th root of unity $[1 \leqslant j \leqslant m]$.

Proof. By inspection $e_{\omega(1)\omega(2)\ldots\omega(m)} \in \hat{G}$ and $e_{\omega(1)\omega(2)\ldots\omega(m)} = e_{\omega'(1)\omega'(2)\ldots\omega'(m)}$ if and only if $\omega(1) = \omega'(1)$, $\omega(2) = \omega'(2), \ldots, \omega(m) = \omega'(m)$. Thus we have $n(1)n(2)\ldots n(m)$ distinct $e_{\omega(1)\omega(2)\ldots\omega(m)}$. But G has $n(1)n(2)\ldots n(m)$ elements and $\hat{G}$ has the same number of elements as G, so the elements of $\hat{G}$ are precisely the $e_{\omega(1)\omega(2)\ldots\omega(m)}$, as required. ∎

Remark. If G is the cyclic group of Chapter 97 generated by $\omega = \exp 2\pi i / N$, then $\hat{G}$ consists of the functions e_{ω^n} defined by $e_{\omega^n}(\omega^r) = \omega^{nr}$. Writing $e_n = e_{\omega^n}$ we recover the 'exponentials' of Chapter 97.7.

Example 104.10. *Let G be the group of units modulo 8. Then $\hat{G}$ has four elements* $e_{1,1}, e_{1,-1}, e_{-1,1}, e_{-1,-1}$ *whose values are tabulated below.*

	[1]	[3]	[5]	[7]
$e_{1,1}$	1	1	1	1
$e_{1,-1}$	1	1	-1	-1
$e_{-1,1}$	1	-1	1	-1
$e_{-1,-1}$	1	-1	-1	1

Proof. Take $x_1 = [3]$, $x_2 = [5]$ and apply Theorem 104.9. ∎

Remark. The reader may well have observed that Theorem 104.1 and so Theorem 104.2 are easy deductions from the structure theorem for finite Abelian groups (Theorem 104.8) and may ask why we did not start from there. For answer let her take four or five algebra textbooks at random and look at the proof of the

structure theorem that they give. In most of them she will find that the author has given way to the algebraist's habit of only proving simple results as corollaries of hard general theorems. Rather than leave the reader with the impression that the study of $\hat{G}$ for a finite Abelian group G depended on the 'structure theorem for finitely generated modules' or some equally impressive piece of machinery, it seemed better to start from scratch.

105

A FORMULA OF EULER

In *Mathematician's Miscellany* (Methuen, 1953) Littlewood gives a brief autobiographical sketch. In it he recalls that '...the Euler formula $\sum n^{-s} = \prod (1 - p^{-s})^{-1}$... was introduced to us at school, as a joke (rightly enough, and in excellent taste)'.

For the rest of the book p will denote a prime so that, for example, $\sum_{p<N}$ means the sum over primes less than N.

Lemma 105.1. *If $s > 0$*

$$\prod_{p<N} (1 - p^{-s})^{-1} = \sum_{n \in Q(N)} n^{-s},$$

where $Q(N)$ is the set of integers $n \geq 1$ all of whose prime factors are less than N.

Proof. Since $0 < p^{-s} < 1$

$$(1 - p^{-s})^{-1} = 1 + p^{-s} + p^{-2s} + \cdots.$$

But convergent series of positive terms can be multiplied term by term. Thus, if the primes less than N are $p_1, p_2, \ldots, p_k$, we have

$$\prod_{p<N} (1 - p^{-s}) = \prod_{j=1}^{k} (1 - p_j^{-s}) = \prod_{j=1}^{k} (1 + p_j^{-s} + p_j^{-2s} + \cdots)$$

$$= \sum_{r(1)=0}^{\infty} \sum_{r(2)=0}^{\infty} \cdots \sum_{r(k)=0}^{\infty} (p_1^{r(1)} p_2^{r(2)} \ldots p_k^{r(k)})^{-s}.$$

Now, by the uniqueness theorem for prime factorisation, each element q of $Q(N)$ can be written in one and only one way as $q = p_1^{r(1)} p_2^{r(2)} \ldots p_k^{r(k)}$ with $r(j) \geq 0$ an integer. Thus

$$\prod_{p<N} (1 - p^{-s})^{-1} = \sum_{n \in Q(N)} n^{-s},$$

as stated. ∎

Theorem 105.2 (Euler's formula). *If $s > 1$ then*

$$\prod_{p<N} (1 - p^{-s})^{-1} \to \sum_{n=1}^{\infty} n^{-s} \quad as \ N \to \infty.$$

525

Proof. Observe that

$$\sum_{n=1}^{\infty} n^{-s} \geqslant \sum_{n \in Q(N)} n^{-s} \geqslant \sum_{n=1}^{N-1} n^{-s}$$

and

$$\sum_{n=1}^{N-1} n^{-s} \to \sum_{n=1}^{\infty} n^{-s} \quad \text{as } N \to \infty.$$

Thus

$$\sum_{n \in Q(N)} n^{-s} \to \sum_{n=1}^{\infty} n^{-s},$$

and so

$$\prod_{p<N} (1 - p^{-s})^{-1} \to \sum_{n=1}^{\infty} n^{-s} \quad \text{as } N \to \infty. \qquad \blacksquare$$

Writing $\prod_p (1-p^{-s})^{-1}$ for the limit of $\prod_{p<N}(1-p^{-s})^{-1}$ we obtain Euler's formula in its usual statement

$$\prod_p (1 - p^{-s})^{-1} = \sum_{n=1}^{\infty} n^{-s} \quad [s > 1].$$

As our method of proof shows, Euler's formula may be considered as the analytic expression of the uniqueness theorem for prime factorisation in number theory.

Given any product it is natural to take logarithms. Formally, Euler's formula yields

$$\log \sum_{n=1}^{\infty} n^{-s} = \sum_p - \log(1 - p^{-s}).$$

But, to the first order in x, $\log(1-x) = -x$ for x small. Using this approximation we obtain (at least formally)

$$\log \sum_{n=1}^{\infty} n^{-s} \approx \sum_p p^{-s},$$

giving rise to the hope that $\log \sum_{n=1}^{\infty} n^{-s}$ and $\sum p^{-s}$ will behave similarly.

In the next lemma we shall see that this hope is justified.

Lemma 105.3. (i) *If $|x| \leqslant \frac{1}{2}$ then $|\log(1-x) + x| \leqslant x^2$.*

(ii) *If $s > 1$ then*

$$\left| \sum_{p<N} \log(1-p^{-s}) + \sum_{p<N} p^{-s} \right| \leqslant \sum_{n=1}^{\infty} n^{-2}.$$

(iii) *If $s > 1$ then $\sum p^{-s}$ converges and*

$$\left| \log \sum_{n=1}^{\infty} n^{-s} - \sum_p p^{-s} \right| \leqslant \sum_{n=1}^{\infty} n^{-2}.$$

Proof. (i) Using the Taylor expansion (valid for $|x| < 1$),

$$|\log(1-x)+x| = \left| -\sum_{r=2}^{\infty} x^r/r \right| \leqslant x^2 \sum_{r=0}^{\infty} |x|^r/2$$

$$\leqslant x^2 \sum_{r=0}^{\infty} 2^{-r-1} = x^2 \quad \text{for all } |x| < \tfrac{1}{2}.$$

(ii) Since $p \geqslant 2$ we have $p^{-s} < \tfrac{1}{2}$ and so

$$\left| \sum_{p<N} \log(1-p^{-s}) + \sum_{p<N} p^{-s} \right| \leqslant \sum_{p<N} |\log(1-p^{-s}) + p^{-s}|$$

$$\leqslant \sum_{p<N} p^{-2s} \leqslant \sum_{p<N} p^{-2} \leqslant \sum_{n=1}^{N} n^{-2} \leqslant \sum_{n=1}^{\infty} n^{-2}.$$

(iii) Since $\sum_{p<N} p^{-s} \leqslant \sum_{n<N} n^{-s}$ and $\sum_{n=1}^{N} n^{-s}$ converges it follows (since $p^{-s} > 0$) that $\sum p^{-s}$ converges. But, using Euler's formula, we have

$$\sum_{p<N} \log(1-p^{-s}) = -\log \prod_{p<N} (1-p^{-s})^{-1} \to -\log \sum_{n=1}^{\infty} n^{-s},$$

so using (ii) we obtain

$$\left| -\log \sum_{n=1}^{\infty} n^{-s} + \sum_{p} p^{-s} \right| \leqslant \sum_{n=1}^{\infty} n^{-2},$$

and the result. ∎

Thus $\log \sum_{n=1}^{\infty} n^{-s}$ and $\sum p^{-s}$ behave similarly as $s \to 1$ from above.

Theorem 105.4. (i) $\sum p^{-s} \to \infty$ as $s \to 1$ *from above.*
(ii) *The sum* $\sum p^{-1}$ *diverges.*

Proof. Since $\sum_{n=1}^{\infty} n^{-1}$ diverges, $\sum_{n=1}^{\infty} n^{-s} \to \infty$ as $s \to 1$ from above. (Observe, for example, that $\liminf_{s \to 1+} \sum_{n=1}^{\infty} n^{-s} \geqslant \liminf_{s \to 1+} \sum_{n=1}^{N} n^{-s} = \sum_{n=1}^{N} n^{-1}$ for any $N \geqslant 1$.) Thus $\log \sum_{n=1}^{\infty} n^{-s} \to \infty$ as $s \to 1$ from above and so, by Lemma 105.3 (iii), $\sum_p p^{-s} \to \infty$ as $s \to 1$ from above. Since $p^{-1} > p^{-s}$ for each $s > 1$, it follows that $\sum_{p<N} p^{-1}$ diverges as $N \to \infty$. ∎

Since (as the reader may readily check) we have nowhere assumed in this chapter that the set of primes is infinite, we have a new proof of a theorem of Euclid.

Theorem 105.5 (Euclid). *There are an infinite number of primes.*
Proof (Euler). Since $\sum_{p<N} 1/p$ diverges as $N \to \infty$, there must be an infinite number of primes. ∎

However, if all we wanted to do was to prove Euclid's result using Euler's idea, we could have stopped after Lemma 105.1.

Proof of Theorem 105.5 using Lemma 105.1 alone. If there were only a finite number of primes then we could find an N larger than any prime. Automatically $Q(N)$ would consist of all integers $n \geqslant 1$ and so setting $s = 1$ we obtain

$$\prod_{p < N} (1 - p^{-1})^{-1} = \sum_{n=1}^{\infty} n^{-1}.$$

But the left hand side of this equation is a well defined real number while the right hand side diverges. Theorem 105.5 follows by *reductio ad absurdum.* ∎

There are two reasons for proceeding the way we did. The first is to provide a model to be followed in the next chapter. The second is that Theorem 105.4 is much stronger than Theorem 105.5. To say that $\sum_{k \in K} k^{-1}$ diverges is a much stronger statement than to say that K is infinite. The set $L = \{n^2 : n \geqslant 1\}$ of squares is infinite but $\sum_{l \in L} l^{-1}$ is convergent.

As a simple example of what we mean consider the following consequence of Theorem 105.4.

Lemma 105.6. *Let $\pi(x)$ be the number of primes not exceeding x. Then for each $\varepsilon > 0$*

$$\limsup_{x \to \infty} x^{-1} \pi(x) (\log x)^{1+\varepsilon} = \infty.$$

(Thus $\pi(x) \geqslant x/(\log x)^{1+\varepsilon}$ infinitely often.)
Proof. Suppose not. Then we can find an $\varepsilon > 0$ and a constant A such that $\pi(x) \leqslant Ax/(\log x)^{1+\varepsilon}$ for all $x \geqslant 1$. Thus

$$\sum_{2^{n-1} \leqslant p < 2^n} p^{-1} \leqslant 2^{-(n-1)} \sum_{2^{n-1} \leqslant p < 2^n} 1 \leqslant 2^{-n+1} \sum_{p < 2^n} 1 \leqslant 2^{-n+1} \pi(2^n)$$
$$\leqslant 2^{-n+1} A 2^n / (n \log 2)^{1+\varepsilon} \leqslant B n^{-(1+\varepsilon)}$$

for some constant B and all $n \geqslant 1$. Since $\sum_{n=1}^{\infty} n^{-(1+\varepsilon)}$ converges, so does $\sum_{n=1}^{\infty} \sum_{2^{n-1} \leqslant p < 2^n} p^{-1} = \sum p^{-1}$. The required result follows by *reductio ad absurdum.* ∎

A more striking, though ultimately less fruitful, application of Euler's formula is the following.

Slogan 105.7′. *The probability that two positive integers chosen at random are not coprime is $6/\pi^2$.*

Heuristic Argument

Let the first integer be K and the second L. Let A_p be the event that the prime p does not divide both K and L, B_p the event that p divides K and C_p the event that p divides L. Since B_p and C_p are independent

$$Pr(A_p) = 1 - Pr(B_p \cap C_p) = 1 - Pr(B_p) Pr(C_p).$$

But exactly one integer in every p is divisible by p so $Pr(B_p) = Pr(C_p) = p^{-1}$ and $Pr(A_p) = (1 - p^{-2})$.

Now let $p(1),\ p(2), \ldots, p(n)$ be the primes less than N. The events $A_{p(1)}$, $A_{p(2)}, \ldots, A_{p(n)}$ are independent and so

$Pr(K$ and L have a common prime factor less than $N)$

$$= Pr\left(\bigcap_{k=1}^{n} A_{p(k)}\right) = \prod_{k=1}^{r} Pr(A_{p(k)}) = \prod_{k=1}^{r} (1 - p(k)^{-2}) = \prod_{p<N} (1 - p^{-2}).$$

Euler's formula (Theorem 105.2) tells us that

$$\prod_{p<N} (1 - p^{-2}) = \left(\prod_{p<N} (1 - p^{-2})^{-1}\right)^{-1} \to \left(\sum_{n=1}^{\infty} n^{-2}\right)^{-1} \quad \text{as } N \to \infty,$$

whilst Lemma 33.6 (iv) says that $\sum_{n=1}^{\infty} n^{-2} = \pi^2/6$. Thus
$Pr(K$ and L have a common prime factor less than $N) \to 6/\pi^2$ as $N \to \infty$ and so

$$Pr(K \text{ and } L \text{ have a common factor}) = 6/\pi^2.$$

The arguments just given constitute a (rather amateurish) three card trick. Until we specify a probability distribution on the positive integers, Slogan 105.7′ remains meaningless but, once we specify such a distribution, supporting arguments like 'the events $A_{p(1)}, A_{p(2)}, \ldots, A_{p(n)}$ are independent' become false. None the less they suggest a true statement and a rigorous proof. Let us write $|A|$ for the number of elements in a set A.

Theorem 105.7. *If*

$$A(N) = \{(n, m): 1 \leqslant n, m \leqslant N \text{ and } m \text{ and } n \text{ are not coprime}\}$$

then

$$N^{-2}|A(N)| \to 6/\pi^2 \quad \text{as } N \to \infty.$$

Proof. The proof falls into two distinct parts. In the first part we calculate the exact probability that two randomly chosen integers between 1 and N are coprime. In the second we show that this exact expression is close to the value suggested by the heuristic argument.

Step 1. (We use the notation $p|n$ to mean p divides n.) If $p, q, r, \ldots$ are distinct primes let us write

$$D_p(N) = \{(n, m): 1 \leqslant n, m \leqslant N, p|n, p|m\},$$
$$D_{pq}(N) = \{(n, m): 1 \leqslant n, m \leqslant N, p, q|n, p, q|m\}$$
$$= \{(n, m): 1 \leqslant n, m \leqslant N, pq|n, pq|m\},$$
$$D_{pqr}(N) = \{(n, m): 1 \leqslant n, m \leqslant N, p, q, r|n, p, q, r|m\}$$
$$= \{(n, m): 1 \leqslant n, m \leqslant N, pqr|n, pqr|m\}, \ldots,$$

and so on. We observe that

$$|D_p(N)| = [N/p]^2, |D_{pq}(N)| = [N/pq]^2, |D_{pqr}(N)| = [N/pqr]^2, \ldots,$$

and so on. Thus using the inclusion exclusion principle we have

$$A(N) = \sum_{p<N} |D_p(N)| - \sum_{p,q<N} |D_{pq}(N)| + \sum_{p,q,r<N} |D_{pqr}(N)| - \cdots,$$

and thus

$$N^{-2}A(N) = N^{-2}\left(\sum_{p<N} [N/p]^2 - \sum_{p,q<N} [N/pq]^2 + \sum_{p,q,r<N} [N/pqr]^2 - \cdots \right).$$

(Here $\sum_{p,q,r<N}$ means the sum over triples of distinct primes less than N and so on. Notice that, since there are only finitely many primes less than N, the expression for $N^{-2}A(N)$ contains only finitely many terms.)

Step 2. Fix an integer M for the moment and let $N \geqslant M$. We set

$$\alpha(N,M) = N^{-2}\left(\sum_{p<M} [N/p]^2 - \sum_{p,q<M} [N/pq]^2 + \sum_{p,q,r<M} [N/pqr] - \cdots \right)$$

and observe that

$$|N^{-2}A(N) - \alpha(N,M)|$$

$$= \left| N^{-2}\left(\sum_{p<N} [N/p]^2 - \sum_{p,q<N} [N/pq]^2 + \sum_{p,q,r<N} [N/pqr]^2 - \cdots \right) \right.$$

$$\left. - N^{-2}\left(\sum_{p<N} [N/p]^2 - \sum_{p,q<N} [N/pq]^2 + \sum_{p,q,r<N} [N/pqr]^2 - \cdots \right) \right|$$

$$\leqslant N^{-2} \sum_{u \in C(N,M)} [N/u]^2,$$

where $C(N,M)$ is a finite subset of the positive integers which does not contain any u with $1 \leqslant u \leqslant M-1$. It follows that

$$|N^{-2}A(N) - \alpha(N,M)| \leqslant N^{-2} \sum_{u \in C(N,M)} (N/u)^2 \leqslant \sum_{u \in C(N,M)} (1/u)^2$$

$$\leqslant \sum_{u=M}^{\infty} (1/u)^2 \leqslant \sum_{u=M}^{\infty} (1/u(u-1))$$

$$= \sum_{u=M}^{\infty} ((u-1)^{-1} - u^{-1}) = (M-1)^{-1}.$$

But as $N \to \infty$ with M fixed

$$\alpha(N,M) = \sum_{p<M} N^{-2}[N/p]^2 - \sum_{p,q<M} N^{-2}[N/pq]^2 + \sum_{p,q,r<M} N^{-2}[N/pqr]^2 - \cdots$$

$$\to \sum_{p<M} (1/p)^2 - \sum_{p,q<M} (1/pq)^2 + \sum_{p,q,r<M} (1/pqr)^2 - \cdots = \prod_{p<M} (1 - p^{-2}),$$

and we know from Theorem 105.2 and Lemma 33.6 (iv) that as $M \to \infty$

$$\prod_{p<M} (1 - p^{-2}) = \left(\prod_{p<M} (1 - p^{-2})^{-1} \right)^{-1} \to \left(\sum_{n=1}^{\infty} n^{-2} \right)^{-1} = 6/\pi^2.$$

Thus given any $\varepsilon > 0$ we can find an $M(\varepsilon)$ such that $(M(\varepsilon) - 1)^{-1} < \varepsilon/3$ and $|\prod_{p<M(\varepsilon)}(1 - p^{-2}) - 6/\pi^2| < \varepsilon/3$. We now choose an $N(\varepsilon)$ such that $|N^{-2}\alpha(N) - \alpha(N, M(\varepsilon))| < \varepsilon/3$ for all $N \geqslant N(\varepsilon)$ and obtain

$$|N^{-2}\alpha(N) - 6/\pi^2| \leqslant |N^{-2}\alpha(N) - \alpha(N, M(\varepsilon))| + |\alpha(N, M(\varepsilon)) - \prod_{p<M(\varepsilon)} (1 - p^{-2})|$$

$$+ |\prod_{p<M(\varepsilon)}(1 - p^{-2}) - 6/\pi^2| < \varepsilon/3 + \varepsilon/3 + \varepsilon/3 = \varepsilon \quad \text{for all } N \geqslant N(\varepsilon).$$

Since ε was arbitrary we are done. ∎

The formula of Euler thus enables us to deduce results in number theory (the infinity of primes, the proportion of coprime pairs) from results in analysis (the divergence of $\sum n^{-1}$, the value of $\sum n^{-2}$). 'A joke...and in excellent taste.'

106

AN IDEA OF DIRICHLET

'The careful study of the natural sequence of primes leads one to conjecture several properties whose truth may be made to appear arbitrarily likely by repeated numerical testing, but where the discovery of a truly rigorous proof is attended with the greatest difficulty. One of the most noteworthy results of this kind is obtained when one considers the distribution of the remainders of the primes after division by a fixed divisor'.

With these words Dirichlet begins his attack on a fifty year old conjecture of Legendre.

Conjecture 106.1. *If a and d are coprime positive integers then the arithmetical progression a, a + d, a + 2d, . . . contains an infinity of primes.*
(Thus if, for example, we take $a = 3$, $d = 8$ then the conjecture states that there are an infinity of primes of the form $8b + 3$. The condition that a and d be coprime is clearly necessary since each $a + nd$ is divisible by the highest common factor of a and d.)

(Fourier seems to have thought such researches a waste of talent. There is a well known letter from Jacobi to Legendre (Jacobi, *Collected Works*, Vol. I, p. 454) in which Jacobi complains that Poisson had repeated in a report a remark of Fourier where he 'reproaches Abel and myself for not having given priority to our research in the theory of heat conduction'. 'It is true' continues Jacobi, 'that Fourier was of the opinion that the chief end of mathematics was the public good and the explanation of natural phenomena; but a philosopher such as he was should have known that the only goal of science is the honour of the human spirit and in this respect a question in the theory of numbers is as valuable as a problem in physics.' Fourier's opinion is, however, still shared by many people, some of whom have not done quite as much as Fourier for 'the public good and the explanation of natural phenomena'.)

Dirichlet models his proof of the conjecture on Euler's proof of the infinity of primes. Writing $\sum_{p \equiv a}$ for the sum over all primes congruent to $a \bmod d$, he sets out to prove that $\sum_{p \equiv a} p^{-1}$ diverges by studying the behaviour of $\sum_{p \equiv a} p^{-s}$ as $s \to 1$ from above. The expression $\sum_{p \equiv a} p^{-s}$ is difficult to deal with and we would wish to replace

it by an expression of the form $\sum_p g(p)$ where we sum over all primes. This is easily done, since $\sum_{p \equiv a} p^{-s} = \sum_p \delta_a(p) p^{-s}$, where $\delta_a(r) = 1$ if $r \equiv a$ mod d and $\delta_a(r) = 0$ otherwise. But unless we can find a useful form for δ_a we are no further advanced.

The way out is indicated by Theorem 104.4.

Lemma 106.2. *Let G be the group of units modulo d and let $f_a : G \to \mathbb{C}$ be defined by $f_a([a]) = 1, f_a([b]) = 0$ otherwise. Then*

$$f_a([m]) = \sum_{e \in \hat{G}} \phi(d)^{-1} e([a])^* e([m]).$$

Proof. Apply Theorem 104.4 to f_a. ∎

The result is easily transferred from G to $\mathbb{Z}$.

Lemma 106.3. *We adopt the notation of Lemma 106.2. If $g \in C(G)$ define $\tilde{g} : \mathbb{Z} \to \mathbb{C}$ by*

$$\tilde{g}(m) = g([m]) \quad \text{for } m \text{ and } d \text{ coprime,}$$
$$\tilde{g}(m) = 0 \qquad \text{otherwise.}$$

Then (for a and d coprime)

$$\delta_a(m) = \tilde{f}_a(m) = \phi(d)^{-1} \sum_{e \in \hat{G}} \tilde{e}(a)^* \tilde{e}(m) \quad \text{for all } m \in \mathbb{Z}.$$

Proof. Direct from Lemma 106.2. ∎

We are thus led to the following formal definition.

Definition 106.4. *We call $\chi : \mathbb{Z} \to \mathbb{C}$ a character (modulo d), if there exists an $e \in \hat{G}$ (where G is the group of units modulo d) such that*

$$\chi(m) = e([m]) \quad \text{for } m \text{ and } d \text{ coprime,}$$
$$\chi(m) = 0 \qquad \text{otherwise.}$$

We write χ_0 for the character given by

$$\chi_0(m) = 1 \quad \text{for } m \text{ and } d \text{ coprime,}$$
$$\chi_0(m) = 0 \quad \text{otherwise,}$$

and call χ_0 the principal character.

By convention $\sum_\chi$ will mean the sum over all the characters.

Lemma 106.5. (i) *If χ is a character*

$$\chi(m_1 m_2) = \chi(m_1) \chi(m_2),$$

(ii) $\delta_a(m) = \phi(d)^{-1} \sum_\chi \chi(a)^* \chi(m)$ *(for a and d coprime).*
Proof. (i) Immediate.

(ii) This is Lemma 106.3. ∎

We thus have

$$\sum_{p \equiv a} p^{-s} = \sum_p \delta_a(p)p^{-s} = \sum_p \left(\phi(d)^{-1} \sum_\chi \chi(a)^* \chi(p)p^{-s} \right)$$

$$= \phi(d)^{-1} \sum_\chi \chi(a)^* \sum_p \chi(p)p^{-s}.$$

Hence the behaviour of $\sum_{r \equiv a} p^{-s}$ depends on that of $\sum_p \chi(p)p^{-s}$ for each χ.

Lemma 106.6. *The sum* $\sum_{p \equiv a} p^{-1}$ *diverges (and there are thus an infinite number of primes* $p \equiv a \bmod d$) *if* $\sum_p \chi(p)p^{-s}$ *remains bounded as* $s \to 1$ *from above for each* $\chi \neq \chi_0$.
Proof. Observe first that since $\sum_p p^{-s}$ is convergent for $s > 1$ (by comparison with $\sum n^{-s}$) and since $|\chi(p)| \leq 1$ the sum $\sum_p \chi(p)p^{-s}$ is absolutely convergent for $s > 1$. The manipulations above are thus justified and we have indeed,

$$\sum_{p \equiv a} p^{-s} = \phi(d)^{-1} \sum_\chi \chi(a)^* \sum_p \chi(p)p^{-s}$$

$$= \phi(d)^{-1} \sum_p \chi_0(p)p^{-s} + \phi(d)^{-1} \sum_{\chi \neq \chi_0} \chi(a)^* \sum_p \chi(p)p^{-s}.$$

But (unless p divides d) any prime p is automatically coprime to d and $\chi_0(p) = 1$. Thus $\sum_p \chi_0(p)p^{-s} = \sum_{p \nmid d} p^{-s}$ and so by the results of the previous chapter $\sum_p \chi_0(p)p^{-s} \to \infty$ as $s \to 1 +$. If $\sum_p \chi(p)p^{-s}$ remains bounded for each $\chi \neq \chi_0$ it will follow that $\sum_{p \equiv a} p^{-s} \to \infty$ as $s \to 1 +$ and so, since $p^{-1} > p^{-s}$ for $s > 1$, that $\sum_{p \equiv a} p^{-1}$ diverges.   ∎

Since $\sum n^{-1}$ and so $\sum p^{-1}$ diverges slowly, it appears quite likely that there will be sufficient cancellation (after all $\sum_{r=nd+k+1}^{(n+1)d+k} \chi(r) = 0$) to ensure that $\sum \chi(p)p^{-s}$ remains bounded as $s \to 1 +$ for $\chi \neq \chi_0$. But how can we prove this? The key lies in the observation that there exists an 'Euler formula' for characters.

Lemma 106.7. (i) *If* $s > 0$ *then, with the notation of Lemma* 105.1,

$$\prod_{p < N} (1 - \chi(p)p^{-s})^{-1} = \sum_{n \in Q(N)} \chi(n)n^{-s}.$$

(ii) (*The Euler–Dirichlet formula*) *If* $s > 1$ *then*

$$\prod_{p < N} (1 - \chi(p)p^{-s})^{-1} \to \sum_{n=1}^\infty \chi(n)n^{-s} \quad as \ N \to \infty.$$

Proof. (i) Since $|\chi(p)p^{-s}| < 1$, the sum $\sum_{r=0}^\infty (\chi(p)p^{-s})^r$ is absolutely convergent and we may multiply term by term to get (in the notation of Lemma 105.1)

$$\prod_{p < N} (1 - \chi(p)p^{-s})^{-1} = \prod_{j=1}^k (1 - \chi(p_j)p_j^{-s})^{-1}$$

$$= \prod_{j=1}^{k} (1 + \chi(p_j)p_j^{-s} + \chi(p_j)^2 p_j^{-2s} + \cdots)$$

$$= \sum_{r(1)=0}^{\infty} \sum_{r(2)=0}^{\infty} \cdots \sum_{r(k)=0}^{\infty} \chi(p_1)^{r(1)} \chi(p_2)^{r(2)} \cdots \chi(p_k)^{r(k)} (p_1^{r(1)} p_2^{r(2)} \cdots p_k^{r(k)})^{-s}$$

$$= \sum_{r(1)=0}^{\infty} \sum_{r(2)=0}^{\infty} \cdots \sum_{r(k)=0}^{\infty} \chi(p_1^{r(1)} p_2^{r(2)} \cdots p_k^{r(k)}) (p_1^{r(1)} p_2^{r(2)} \cdots p_k^{r(k)})^{-s}$$

$$= \sum_{n \in Q(N)} \chi(n)n^{-s},$$

using the multiplicative property of χ (Lemma 106.5(i)) and, as before, the uniqueness of prime factorisation.

(ii) Since $|\chi(n)n^{-s}| \leqslant n^{-s}$, the sum $\sum_{n=1}^{\infty} \chi(n)n^{-s}$ is absolutely convergent and

$$\left| \sum_{n \in Q(N)} \chi(n)n^{-s} - \sum_{n=1}^{\infty} \chi(n)n^{-s} \right| \leqslant \sum_{n \notin Q(N)} n^{-s} \leqslant \sum_{n \geqslant N} n^{-s} \to 0 \quad \text{as } N \to \infty.$$

Thus

$$\prod_{p < N} (1 - \chi(p)p^{-s})^{-1} \to \sum_{n=1}^{\infty} \chi(n)n^{-s},$$

as required. ∎

We have now generalised Lemma 105.1 and Theorem 105.2 (Euler's formula). Can we do the same for Lemma 105.3? In Lemma 105.3 we dealt with the (real valued) logarithm of the real positive number $(1 - p^s)$. Now we have to deal with the logarithm of a number $(1 - \chi(p)p^{-s})$ which may not be real. As the reader knows there is no way of defining a unique logarithmic function $\mathbb{C} \backslash \{0\} \to \mathbb{C}$ which will not lead to trouble in some circumstances. However, if we restrict ourselves to a 'cut' complex plane with the negative real axis removed and define $lg : \mathbb{C} \backslash \{x : x \text{ real } x \leqslant 0\} \to \mathbb{C}$ by $lg(re^{i\theta}) = \log r + i\theta \; [-\pi < \theta < \pi]$, then lg is well defined and $lg(1 + z) = \sum_{r=1}^{\infty} (-z)^r / r$ for all $|z| < 1$.

Lemma 106.8. (i) *If* $|z| \leqslant \frac{1}{2}$ *then* $|lg(1 - z) + z| \leqslant |z|^2$.

(ii) *If* $s > 1$ *then*

$$\left| \sum_{p < N} lg(1 - \chi(p)p^{-s}) + \sum_{p < N} \chi(p)p^{-s} \right| \leqslant \sum_{n=1}^{\infty} n^{-2}.$$

(iii) *If* $s > 1$ *then* $\sum lg(1 - \chi(p)p^{-s})$ *and* $\sum \chi(p)p^{-s}$ *converge (uniformly for* $s > 1 + \delta$ *$[\delta > 0]$) and*

$$\left| \sum_{p} lg(1 - \chi(p)p^{-s}) + \sum_{p} \chi(p)p^{-s} \right| \leqslant \sum_{n=1}^{\infty} n^{-2}.$$

Proof. (i) Using the Taylor expansion mentioned above

$$|lg(1-z)+z| = \left|-\sum_{r=2}^{\infty} z^r/r\right| \leqslant |z|^2 \sum_{r=0}^{\infty} |z|^r/2$$

$$\leqslant |z|^2 \sum_{r=0}^{\infty} 2^{-r-1} = |z|^2 \quad \text{for all } |z| < \tfrac{1}{2}.$$

(ii) The proof follows that of Lemma 105.3 (ii) exactly.

(iii) Since $|\chi(p)p^{-s}| \leqslant p^{-s}$ and $\sum p^{-s}$ converges (Lemma 105.3 (iii)) it follows that $\sum \chi(p)p^{-s}$ converges absolutely for all $s > 1$. Further, using (i), we obtain

$$|lg(1 - \chi(p)p^{-s})| \leqslant |\chi(p)p^{-s}| + |\chi(p)p^{-s}|^2 \leqslant 2p^{-1-\delta},$$

so $\sum lg(1 - \chi(p)p^{-s})$ also converges absolutely for all $s > 1$ and uniformly for $s > 1 + \delta$. The inequality

$$\left|\sum_p lg(1 - \chi(p)p^{-s}) - \sum_p \chi(p)p^{-s}\right| \leqslant \sum_{n=1}^{\infty} n^{-2}$$

follows from (ii). ∎

Unfortunately, the fact that we are dealing with the logarithms of complex numbers means that there is absolutely no reason to suppose that $\sum_p lg(1 - \chi(p)p^{-s})$ equals $lg(\prod_p(1 - \chi(p)p^{-s}))$. The example $z_1 = z_2 = \cdots = z_8 = \exp \pi i/4$ shows that the relation $\sum lg z_j \neq lg \prod z_j$ may fail even for finite products (in the case given

$$\sum_{j=1}^{8} lg z_j = \sum_{j=1}^{8} \pi i/4 = 2\pi i \neq 0 = lg\, 1 = lg \prod_{j=1}^{8} z_j).$$

For infinite products we may expect even worse behaviour.

Example 106.9. *If we define* $f_n : \mathbb{R} \to \mathbb{C}$ *by*

$$\begin{aligned}
f_n(s) &= 1 & &\text{for } |s-1| \leqslant 2^{-n}, \\
f_n(s) &= \exp(\pi i 2^{n-1}(2^{-n} - |s-1|)) & &\text{for } 2^{-n} \geqslant |s-1| \geqslant 2^{-n-1}, \\
f_n(s) &= \exp \pi i/4 & &\text{for } 2^{-n-1} \geqslant |s-1|,
\end{aligned}$$

then

$$\sum_{n=1}^{\infty} lg\, f_n(s) \text{ converges for all } s \neq 1$$

and

$$\left|\prod_{n=1}^{N} f_n(s)\right| = 1 \quad \text{for all } N \text{ and all } s,$$

yet

$$\left|\sum_{n=1}^{\infty} lg\, f_n(s)\right| \to \infty \quad \text{as } s \to 1.$$

Proof. Simple verification based on the observation that, if $2^{-M} \geqslant |s-1| \geqslant 2^{-M-1}$ and $N \geqslant M$, then

$$\sum_{n=1}^{N} lg\, f_n(s) = \sum_{n=1}^{M} lg\, f_n(s) = \pi i((M-1)/4 + 2^{M-1}(2^{-M} - |s-1|)). \quad ∎$$

In the next chapter we shall see that the problem just considered is a technical rather than a fundamental difficulty. We conclude this chapter by showing that we are already in a position to obtain some encouraging results in two cases where all the characters are real and so the problem does not arise.

Lemma 106.10. (i) *There are an infinity of primes of the form $3n + 1$ and an infinity of primes of the form $3n + 2$.*

(ii) *There are an infinity of primes of the form $4n + 1$ and an infinity of primes of the form $4n + 3$.*

Proof. (i) The group G of units modulo 3 consists of two elements $[1]$ and $[2]$. Thus $\hat{G}$ contains two elements e_0 and e_1 which are easily seen to be given by the relations $e_0([1]) = e_0([2]) = 1$ and $e_1([1]) = 1$, $e_1([2]) = -1$. These elements e_0 and e_1 in turn give rise to the principal character χ_0 and a character χ_1 given by

$$\chi_1(3n) = 0, \quad \chi_1(3n + 1) = 1, \quad \chi_1(3n + 2) = -1 \, [n \in \mathbb{Z}].$$

We observe that

$$\sum_{n=1}^{\infty} \chi_1(n) n^{-s} = 1 - 2^{-s} + 4^{-s} - 5^{-s} + 6^{-s} - \cdots = \sum_{m=0}^{\infty} (-1)^m v_m(s),$$

where $v_{2n}(s) = (3n + 1)^{-s}$ and $v_{2n+1}(s) = (3n + 2)^{-s}$.

But we know how to deal with alternating sequences of decreasing modulus. Since $v_m(s) \geqslant v_{m+1}(s)$ for all $s > 0$, we have, for all $s \geqslant \delta$,

$$\left| \sum_{m=N}^{M} (-1)^m v_m(s) \right| \leqslant v_N(s) \leqslant N^{-s} \leqslant N^{-\delta} \to 0 \quad \text{as } N \to \infty.$$

Thus by the general principle of uniform convergence, $\sum_{m=0}^{\infty} (-1)^m v_m(s)$ converges uniformly to a limit $L(s, \chi_1)$, say, on each set $\{s : s > \delta\}$ $[\delta > 0]$. Since v_m is continuous on $\{s : s > \delta\}$, it follows that $L(\ , \chi_1)$ is. Thus $\sum_{n=1}^{\infty} \chi_1(n) n^{-s}$ converges to a continuous function $L(s, \chi_1)$ for all $s > 0$. We note that

$$\left| \sum_{m=1}^{N} (-1)^m v_m(s) \right| \leqslant v_1(s) \leqslant \tfrac{1}{2},$$

and so
$$1 + \sum_{m=1}^{N} (-1)^m v_m(s) \geqslant \tfrac{1}{2} \quad \text{for all } s \geqslant 1 \text{ and all } N \geqslant 1.$$

Thus
$$L(s, \chi_1) \geqslant \tfrac{1}{2} \quad \text{for all } s \geqslant 1.$$

Since $(1 - \chi_1(p)p^{-s})$ and $L(s, \chi_1)$ are real, we can take (real) logarithms of both sides of the Euler Dirichlet formula (Lemma 106.7 (ii)) to obtain

$$\sum_{p < N} \log(1 - \chi_1(p)p^{-s})^{-1} \to \log L(s, \chi_1),$$

whence, since $\log(1 - \chi_1(p)p^{-s})^{-1} = lg(1 - \chi_1(p)p^{-s})$, Lemma 106.8, gives

$$\left| \log L(s, \chi_1) - \sum_{p} \chi_1(p)p^{-s} \right| \leqslant \sum_{n=1}^{\infty} n^{-2} \quad \text{for all } s > 1.$$

Other directions

But $L(\ ,\chi_1)$ is continuous at 1 so $L(s,\chi_1) \to L(1,\chi_1)$ as $s \to 1$ from above. Since $L(1,\chi_1) \geqslant \frac{1}{2}$, it follows in particular that $L(1,\chi_1) \neq 0$ and so $\log L(s,\chi_1)$ tends to a finite limit $\log L(1,\chi_1)$. The relation

$$|\log L(s,\chi_1) - \sum_p \chi_1(p)p^{-s}| \leqslant \sum_{n=1}^{\infty} n^{-2}$$

now tells us that $\sum_p \chi_1(p)p^{-s}$ remains bounded as $s \to 1$ from above. By Lemma 106.6 applied with $a = 1$ and $a = 2$, it follows that there exist an infinity of primes of the form $3n + 1$ and $3n + 2$.

(ii) Left as a very strongly recommended exercise for the reader. ∎

The proof above depended on the fact that $L(\ ,\chi_1)$ is continuous at 1 (so that $\lim_{s \to 1+} L(s,\chi_1)$ exists) and that $L(1,\chi_1) \neq 0$ so that $\lim_{s \to 1+} L(s,\chi_1) \neq 0$ and $\lim_{s \to 1+} \log L(s,\chi_1)$ exists. These are the key points in the general proof also.

107

PRIMES IN SOME
ARITHMETICAL PROGRESSIONS

It is not hard to gain more information about the sum $\sum_{n=1}^{\infty} \chi(n) n^{-s}$ by making use of simple criteria introduced by Abel and Dirichlet.

Lemma 107.1. (i) *(Abel) Let* $a_1, a_2, \ldots, a_n$ *be a decreasing sequence of positive terms and let* $b_1, b_2, \ldots$ *be a sequence of complex numbers such that* $|\sum_{r=1}^{m} b_r| \leq S$ *for all* $1 \leq m \leq n$. *Then*

$$\left| \sum_{r=1}^{n} a_r b_r \right| \leq S a_1.$$

(ii) *(Dirichlet) Let* $I \subseteq \mathbb{R}$ *and let* $a_1, a_2, \ldots$ *be a sequence of functions* $a_j : I \to \mathbb{R}$ *such that* $a_j(s) \geq a_{j+1}(s) \geq 0$ *for all* $s \in I$ *and* $j \geq 1$. *Suppose further that* $a_j(s) \leq \alpha_j$ *for all* $s \in I$ *and* $j \geq 1$ *and that* $b_1, b_2, \ldots$ *is a sequence of complex numbers such that* $|\sum_{r=l}^{m} b_r| \leq S$ *for all* $m \geq l \geq 1$. *Then if* $\alpha_j \to 0$

$$\sum_{r=1}^{\infty} b_r a_r(s) \text{ converges uniformly}$$

and
$$\left| \sum_{r=1}^{\infty} b_r a_r(s) \right| \leq S \alpha_l \quad \text{for all } s \in I.$$

Proof: (i) Write $S_0 = 0$, $S_l = \sum_{r=1}^{l} b_r [1 \leq l \leq n]$ and $a_{n+1} = 0$. By partial summation (i.e. Abel summation)

$$\sum_{r=1}^{n} a_r b_r = \sum_{r=1}^{n} a_r(S_r - S_{r-1}) = \sum_{r=1}^{n} (a_r - a_{r+1}) S_r,$$

and so, since $a_r - a_{r+1} \geq 0$ and we have set $a_{n+1} = 0$,

$$\left| \sum_{r=1}^{n} a_r b_r \right| \leq \sum_{r=1}^{n} |(a_r - a_{r+1}) S_r| = \sum_{r=1}^{n} (a_r - a_{r+1}) |S_r|$$

$$\leq \sum_{r=1}^{n} (a_r - a_{r+1}) S = (a_1 - a_{n+1}) S = a_1 S.$$

(ii) By part (i)

$$\left|\sum_{r=l}^{m} a_r(s)b_r\right| \leqslant a_l(s)S \leqslant \alpha_l S \to 0 \quad \text{as } l \to \infty \text{ for all } s.$$

Thus by the general principle of uniform convergence $\sum_{r=1}^{\infty} a_r(s)b_r$ converges uniformly on I. Since $|\sum_{r=1}^{m} a_r(s)b_r| \leqslant \alpha_l S$ for all m it follows at once that $|\sum_{r=1}^{\infty} a_r(s)| \leqslant \alpha_l S$ for all $s \in I$. ■

We supplement this with a trivial remark.

Lemma 107.2. *If* $\chi \neq \chi_0$ *then* $|\sum_{r=1}^{m} \chi(r)| \leqslant \phi(d)/2$.
Proof. Choose N so that $l + Nd \leqslant m \leqslant l + (N+1)d - 1$. Then, by Lemma 103.3 and the definition of χ, $\sum_{r=1}^{l+Nd-1} \chi(r) = 0$, $\sum_{r=1}^{l+(N+1)d-1} \chi(r) = 0$ and

$$\sum_{r=l+Nd}^{l+(N+1)d-1} |\chi(r)| = \phi(d) \text{ so that}$$

$$2\left|\sum_{r=l}^{m} \chi(r)\right| = \left|\sum_{r=l}^{m} \chi(r) - \sum_{r=l}^{l+Nd-1} \chi(r)\right| + \left|\sum_{r=l}^{l+N(d+1)-1} \chi(r) - \sum_{r=l}^{m} \chi(r)\right|$$

$$= \left|\sum_{r=l+Nd}^{m} \chi(r)\right| + \left|\sum_{r=m+1}^{l+N(d+1)-1} \chi(r)\right|$$

$$\leqslant \sum_{r=l+Nd}^{l+N(d+1)-1} |\chi(r)| = \phi(d), \text{ and so } \left|\sum_{r=l}^{m} \chi(r)\right| \leqslant \phi(d)/2.$$

■

(The proof is unimportant but if the reader does not find the result obvious she should think until it is.)

We can now obtain some useful results concerning $\sum_{n=1}^{\infty} \chi(n)n^{-s}$.

Lemma 107.3. *Suppose* $\chi \neq \chi_0$. *Then*

(i) $\sum_{n=1}^{\infty} \chi(n)n^{-s}$ *converges to a continuous function* $L(s, \chi)$ *for all* $s > 0$.
(ii) $\sum_{n=1}^{\infty} \chi(n)n^{-s} \log n$ *converges uniformly to a continuous function for* $s > \delta$ *whenever* $\delta > 0$.
(iii) $L(s, \chi)$ *is continuously differentiable on* $s > 0$.

Proof. (i) By Lemma 107.1 (ii) with $I = \{s : s > \delta\}$, $a_r = r^{-s}$ and $b_r = \chi(r)$ we know that $\sum_{n=1}^{\infty} \chi(n)n^{-s}$ converges uniformly to $L(s, \chi)$, say, for all $s > \delta$. Since $\chi(n)n^{-s}$ is continuous, so is $L(s, \chi)$. Since δ is arbitrary, $\sum_{n=1}^{\infty} \chi(n)n^{-s}$ converges for all $s > 0$ and its limit $L(s, \chi)$ is continuous.

(ii) If $s > 0$ then $n^{-s} \log n$ is a decreasing sequence tending to 0. The arguments of (i) thus apply without change.

(iii) Set $f_m(s) = \sum_{n=1}^{m} \chi(n)n^{-s}$. Then for all $0 < a < b$ we know that f_m is a differen-

tiable function on $[a, b]$ with continuous derivative $f'_m(s) = -\sum_{n=1}^{m} \chi(n)n^{-s}\log n$. But, by (i) and (ii), $f_m(s) \to L(s, \chi)$ and $f'_m(s) \to g(s) = -\sum_{n=1}^{\infty} \chi(n)n^{-s}\log n$ uniformly on $[a, b]$. It follows that $L(s, \chi)$ is differentiable on $[a, b]$ with continuous derivative $L'(s, \chi) = -\sum_{n=1}^{\infty} \chi(n)n^{-s}\log n$. Since a and b were arbitrary, it follows that $L(s, \chi)$ is continuously differentiable for all $s > 0$. ∎

Remark. In Chapter 108 we shall see that $L(s, \chi)$ is even better behaved than this result suggests.

Next we show that not only is $L(s, \chi)$ continuous on $s > 0$ but it has no zeros in the region $s > 1$.

Lemma 107.4. (i) $\exp(-\sum_p lg(1 - \chi(p)p^{-s})) = L(s, \chi)$ *for* $s > 1$.
(ii) $L(s, \chi) \neq 0$ *for* $s > 1$.

Proof. (i) We saw in Lemma 106.8 (iii) that $\sum_{p<N} lg(1 - \chi(p)p^{-s})$ converges to a finite limit $\sum_p lg(1 - \chi(p)p^{-s})$ as $N \to \infty$ for all $s > 1$. Thus by the continuity of the exponential function

$$\prod_{p<N} (1 - \chi(p)p^{-s})^{-1} = \prod_{p<N} \exp(-lg(1 - \chi(p)p^{-s}))$$

$$= \exp(-\sum_{p<N} lg(1 - \chi(p)p^{-s}))$$

$$\to \exp(-\sum_p lg(1 - \chi(p)p^{-s})) \text{ as } N \to \infty.$$

But the Euler Dirichlet formula tells us that (again for $s > 1$)

$$\prod_{p<N} (1 - \chi(p)p^{-s})^{-1} \to L(s, \chi),$$

so
$$L(s, \chi) = \exp(-\sum_p lg(1 - \chi(p)p^{-s})),$$

as required.

(ii) Since the exponential function never takes the value 0, the fact that $L(s, \chi) \neq 0$ (for $s > 1$) follows directly from (i). ∎

It follows that we can define a logarithm of $L(s, \chi)$ (at least for $s > 1$).

Lemma 107.5. *Choose A such that* $\exp A = L(2, \chi)$. *Then writing*

$$\log L(s, \chi) = A + \int_2^s \frac{L'(t, \chi)dt}{L(t, \chi)} \quad [s > 1],$$

we see that $\log L(s, \chi)$ *is a continuously differentiable function on $s > 1$ such that*

(i) $\exp \log L(s, \chi) = L(s, \chi)$ $[s > 1]$.
(ii) *If $L(1, \chi) \neq 0$ then $\log L(s, \chi)$ tends to the finite limit* $A + \int_2^1 (L'(t, \chi)/L(t, \chi))dt$ *as $s \to 1$ from above.*

Remark. If we did not know that $L(t, \chi) \neq 0$ for $t > 1$ we would not (in general) be able to state that $L'(t, \chi)/L(t, \chi)$ was a well defined continuous function for $t > 1$ and so that the integral existed.

Proof. By the fundamental theorem of the calculus $\log L(s, \chi)$ is differentiable with derivative $L'(s, \chi)/L(s, \chi)$. Thus (noting again that $L(s, \chi) \neq 0$ for $s > 1$)

$$\frac{d}{ds}\left(\frac{1}{L(s, \chi)} \exp(\log L(s, \chi))\right)$$

$$= -\frac{L'(s, \chi)}{L(s, \chi)^2} \exp(\log L(s, \chi)) + \frac{L'(s, \chi)}{L(s, \chi)^2} \exp(\log L(s, \chi)) = 0 \quad [s > 1],$$

and so (by the mean value theorem)

$$\frac{1}{L(s, \chi)} \exp(\log L(s, \chi)) = \frac{1}{L(2, \chi)} \exp(\log L(2, \chi))$$

$$= \frac{1}{L(2, \chi)} \exp A = 1,$$

whence $\exp(\log L(s, \chi)) = L(s, \chi)$, as required.

If $L(1, \chi) \neq 0$ then $L'(t, \chi)/L(t, \chi)$ is continuous on $[1, 2]$. Part (ii) is thus trival. ∎

We can now establish a relation between $\log L(s, \chi)$ as defined in Lemma 107.5 and $-\sum_p lg(1 - p^{-s})$ and so get round the problem of the non uniqueness of the complex logarithm.

Lemma 107.6. (i) *There is a fixed integer M such that*

$$\log L(s, \chi) + \sum_p lg(1 - \chi(p)p^{-s}) = 2\pi Mi \quad \textit{for all } s > 1.$$

(ii) *If* $\log L(s, \chi)$ *remains bounded as* $s \to 1$ *from above then so does* $\sum \chi(p)p^{-s}$.
(iii) *If* $L(1, \chi) \neq 0$ *then* $\sum \chi(p)p^{-s}$ *remains bounded as* $s \to 1$ *from above* $[\chi \neq \chi_0]$.

Proof. (i) Write

$$f(s) = \log L(s, \chi) + \sum_p lg(1 - \chi(p)p^{-s}).$$

Since $\sum_p lg(1 - \chi(p)p^{-s})$ converges uniformly for $s > 1 + \delta$, $\sum_p lg(1 - \chi(p)p^{-s})$ is a continuous function for $s > 1 + \delta [\delta > 0]$ and so for $s > 1$. Thus $f(s)$ is continuous for $s > 1$. But, by Lemma 107.4 (i) and Lemma 107.5,

$$\exp \log L(\chi, s) = L(\chi, s) = \exp - \sum_p lg(1 - \chi(p))p^{-s},$$

so that $\exp f(s) = 1$ for all $s > 1$ and so $(2\pi i)^{-1} f(s) \in \mathbb{Z}$ for each $s > 1$. Also by the

intermediate value theorem an integer valued continuous function on $\{s : s > 1\}$ is constant. Thus $(2\pi i)^{-1} f(s) = M$ for all $s > 1$ and some M.

(ii) Using Lemma 106.8 (iii) and the result just proved we obtain

$$|\log L(s, \chi) - \sum_p \chi(p) p^{-s}| \leqslant 2\pi M + \sum_{n=1}^{\infty} n^{-2} \quad \text{for all } s > 1,$$

whence the result.

(iii) Just apply Lemma 107.5 (ii). ∎

We have arrived at the end of the first part of Dirichlet's proof of Legendre's conjecture.

Theorem 107.7. *If $L(1, \chi) \neq 0$ for all $\chi \neq \chi_0$ then there are an infinite number of primes of the form $a + nd$.*

Proof. Combine Lemma 106.6 and Lemma 107.6 (iii). ∎

The reader will see that this chapter has been mainly concerned with technical problems. If I had assumed that the reader was more familiar with certain parts of analysis like infinite products and, especially, complex logarithms, then the existence of a function $\log L(1, \chi)$ with the properties described in Lemmas 107.5 and 107.6 could have been established much more quickly. However, the problems, though technical, are real and the effect of greater knowledge is not to make them vanish but simply easier to understand and overcome. (In Dirichlet's original paper he also takes the technical details slowly, and even finds time to remind the reader that, although the infinite sum

$$1 - \tfrac{1}{2} + \tfrac{1}{3} - \tfrac{1}{4} + \tfrac{1}{5} - \tfrac{1}{6} + \cdots$$

and its rearrangement

$$1 + \tfrac{1}{3} - \tfrac{1}{2} + \tfrac{1}{5} + \tfrac{1}{7} - \tfrac{1}{4} \cdots$$

both converge, they converge to different limits.)

Let us use Theorem 107.7 in a couple of examples.

Lemma 107.8. (i) *There are an infinity of primes of the form $5n + a$ for $a = 1, 2, 3, 4$.*
(ii) *There are an infinity of primes of the form $8n + a$ for $a = 1, 3, 5, 7$.*

Proof. (i) The group G of units mod 5 is cyclic generated by the element $[2]$. (Observe that $[2]^2 = [4]$, $[2]^3 = [3]$ and $G = \{[1], [2], [3], [4]\}$.) Thus (by Theorem 104.9 or inspection) $\hat{G}$ is a cyclic group $\{e_0, e_1, e_2, e_3\}$ with $e_2 = e_1^2, e_3 = e_1^3$ and

$$e_1([2]) = i, e_1([3]) = e_1([2]^3) = e_1([2])^3 = -i, \; e_1([4]) = e_1([2])^2 = -1.$$

By a simple adhoc argument (or by using Theorem 104.9) we find the four characters $\chi_0, \chi_1, \chi_2, \chi_3$ whose values are tabulated below.

	$5n+1$	$5n+2$	$5n+3$	$5n+4$
χ_0	1	1	1	1
χ_1	1	i	-1	$-i$
χ_2	1	-1	1	-1
χ_3	1	$-i$	-1	i

To see that $\sum_{n=1}^{\infty} \chi_2(n)n^{-1} \neq 0$, observe that by Lemma 107.1 (ii) or inspection

$$\left| \sum_{n=4}^{\infty} \chi_2(n)n^{-1} \right| \leqslant \tfrac{2}{4} = \tfrac{1}{2},$$

whilst

$$\sum_{n=1}^{3} \chi_2(n)n^{-1} = 1 - \tfrac{1}{2} + \tfrac{1}{3} = \tfrac{5}{6}.$$

Thus

$$\left| \sum_{n=1}^{\infty} \chi_2(n)n^{-1} \right| \geqslant \tfrac{5}{6} - \tfrac{1}{2} = \tfrac{1}{3},$$

so $\sum_{n=1}^{\infty} \chi_2(n)n^{-1} \neq 0$.

To see that $\sum_{n=1}^{\infty} \chi_1(n)n^{-1}$ and $\sum_{n=1}^{\infty} \chi_3(n)n^{-1}$ are non-zero, observe that

$$\left(\sum_{n=1}^{\infty} \chi_3(n)n^{-1} \right)^* = \sum_{n=1}^{\infty} \chi_3(n)^* n^{-1} = \sum_{n=1}^{\infty} \chi_1(n)n^{-1},$$

so if one sum is non-zero so is the other. It thus suffices to show that

$$2^{-1} \left(\sum_{n=1}^{\infty} \chi_1(n)n^{-1} + \sum_{n=1}^{\infty} \chi_3(n)n^{-1} \right) = \sum_{n=1}^{\infty} 2^{-1}(\chi_1(n) + \chi_3(n))n^{-1}$$

is. But

$$\sum_{n=1}^{\infty} 2^{-1}(\chi_1(n) + \chi_3(n))n^{-1} = 1 - \tfrac{1}{3} + \tfrac{1}{6} - \tfrac{1}{8} + \cdots = 1 - (\tfrac{1}{3} - \tfrac{1}{6} + \tfrac{1}{8} - \cdots)$$

$$\geqslant 1 - 1/3 > 0,$$

so $\sum \chi_1(n)n^{-1}$ and $\sum \chi_3(n)n^{-1}$ are non-zero. Thus $\sum_{n=1}^{\infty} \chi(n)n^{-1} \neq 0$ for all $\chi \neq \chi_0$ and the desired result follows from Theorem 107.7.

(ii) We found $\hat{G}$ in this case in Example 104.10 so we may read off the four characters χ_0, χ_1, χ_2, χ_3, whose values are tabulated below.

	$8n+1$	$8n+3$	$8n+5$	$8n+7$
χ_0	1	1	1	1
χ_1	1	1	-1	-1
χ_2	1	-1	1	-1
χ_3	1	-1	-1	1

In this case it is particularly easy to show that $\sum_{n=1}^{\infty} \chi(n)n^{-1} \neq 0$ for $\chi \neq \chi_0$, since (by Lemma 107.1 (ii) or inspection) $|\sum_{n=3}^{\infty} \chi(n)n^{-1}| \leqslant \frac{2}{3}$, whilst $\sum_{n=1}^{2} \chi(n)n^{-1} = 1$ so $\sum_{n=1}^{\infty} \chi(n)n^{-1} \geqslant 1 - \frac{2}{3} = \frac{1}{3}$ and $\sum_{n=1}^{\infty} \chi(n)n^{-1} \neq 0$ for $\chi \neq \chi_0$. The desired result now follows from Theorem 107.7. ∎

We thus have what an algebraic topologist might call a 'theorem proving machine'. Given a and d we can compute the group G of units modulo d, find $\hat{G}$ and so the appropriate characters. By computing sufficiently long finite sums $\sum_{n=1}^{N} \chi(n)n^{-1}$ we can expect to show that $\sum_{n=1}^{\infty} \chi(n)n^{-1} \neq 0$ for each $\chi \neq \chi_0$ and so use Theorem 107.7 to verify Legendre's conjecture in the given case. But it is not clear how we can hope to prove that there exist no exceptional values of d for which $L(1,\chi) = 0$ for some $\chi \neq \chi_0$ and for which our attempted proof will fail.

108

EXTENSION FROM REAL
TO COMPLEX VARIABLE

We have reduced the proof of Legendre's conjecture to showing that $L(1,\chi) \neq 0$ for all $\chi \neq \chi_0$. It turns out to be easy to prove this provided that χ takes complex (i.e. non real) values but hard if χ only takes the values 1 and -1. Several proofs are now available in this second case but none are both short and illuminating. Since every book, however full of asides, must come to an end I have chosen a short proof due to de la Vallée Poussin which depends on complex variable theory. (A real variable proof due to Mertens is given in Chapter 14 of Rademacher's *Lectures on Elementary Number Theory*, Blaisdell, New York 1966.)

Euler's proof of the existence of infinity of primes depends only on the behaviour of $\sum n^{-s}$ when $s = 1$ (see the proof of Theorem 105.5 using Lemma 105.1). Dirichlet's ideas require the study of $\sum n^{-s}$ for $s > 1$ and $\sum \chi(n)n^{-s}$ for $s \geqslant 1$. Riemann showed that more progress could be made by studying $\sum n^{-s}$ and $\sum \chi(n)n^{-s}$ as functions of a complex variable s. In this chapter we discuss in a leisurely way how $\sum \chi(n)n^{-s}$ can be extended to a meromorphic function on $\{s \in \mathbb{C} : \operatorname{Re} s > 0\}$.

Lemma 108.1. *Suppose that $B : \mathbb{R} \to \mathbb{R}$ is bounded on $\mathbb{R}$ (i.e. $|B(t)| \leqslant M$ for all t and some M) and B is Riemann integrable on every bounded interval. Then $\int_1^\infty B(x)x^{-s-1}\,dx$ is a well defined analytic function of s in the region of $\mathbb{C}$ given by $\operatorname{Re} s > 0$.*

Proof. We give two proofs. The first refers back to the chapters on the Laplace transform. The second is solely for those who have not read the appropriate chapters.

First proof. Write $f(t) = B(e^t)$ for $t > 0$, $f(t) = 0$ otherwise. Then $f \in \mathscr{E}^*$ where $\mathscr{E}^*$ is the set defined in Definition 74.1*. (In the chapters on the Laplace transform we considered a slightly more restricted class $\mathscr{E}$ of functions but the reader will easily check that the definitions and results we shall need all extend to the more general case.) Making the substitution $x = e^t$, we obtain

$$\int_1^\infty B(x)x^{-s-1}\,dx = \int_0^\infty f(t)e^{-st}\,dt = (\mathscr{L}f)(s),$$

where $\mathscr{L}f(s)$ is defined for $\operatorname{Re} s > 0$. Theorem 75.2 now tells us that $\mathscr{L}f$ is analytic for $\operatorname{Re} s > 0$ and the result follows.

Second proof. (We just outline the main steps leaving the details to the reader. If she has looked at Chapter 75 she will recognise a simple reworking of the proof of Theorem 75.2.)

Step 1. $|\eta^{-1}(x^{-z-\eta-1} - x^{-z-1} + \eta \log x \, x^{z-1})| \to 0$ as $\eta \to 0$ uniformly for $1 \leqslant x \leqslant N$ and z fixed.

Step 2. Write

$$G_N(z) = \int_1^N B(x) x^{-z-1} dx \text{ and } H_N(z) = - \int_1^N B(x) \log x \, x^{-z-1} dx.$$

By the result of Step 1

$$|\eta^{-1}(G_N(z+\eta) - G_N(z) - \eta H_N(z))| \to 0 \quad \text{as } \eta \to 0.$$

Thus G_N is analytic on $\mathbb{C}$.

Step 3. Since $G_N(z) \to \int_1^\infty B(x) x^{-z-1} dx$ uniformly in the region $\operatorname{Re} z > \delta$ and the uniform limit of analytic functions is itself analytic (Theorem 75.3 (iv)), we see that $\int_1^\infty B(x) x^{-z-1} dx$ is analytic in the region $\operatorname{Re} z > \delta$ for all $\delta > 0$. The full result follows. ∎

It turns out that if $\chi \neq \chi_0$ then $\sum_{n=1}^\infty \chi(n) n^{-s}$ can be written in the form suggested above and that $\sum_{n=1}^\infty \chi_0(n) n^{-s}$ can be written in a related form.

Lemma 108.2. (i) *If we write $S(x) = \sum_{1 \leqslant m \leqslant x} \chi(m)$ then $S : \mathbb{R} \to \mathbb{R}$ is bounded on $\mathbb{R}$ and Riemann integrable on every bounded interval. Further, $\sum_{n=1}^\infty \chi(n) n^{-s}$ converges to $s \int_0^\infty S(x) x^{-s-1} dx$ whenever $\operatorname{Re} s > 0$.*

(ii) *If we write $S_0(x) = 0$ for $x < 0$ and $S_0(x) = \sum_{1 \leqslant m \leqslant x} \chi_0(m) - \phi(d) x / d$ for $x \geqslant 0$ then $S_0 : \mathbb{R} \to \mathbb{R}$ is bounded on $\mathbb{R}$ and Riemann integrable on every bounded interval. Further,*

$$\sum \chi_0(n) n^{-s} \text{ converges to } s \int_1^\infty S_0(x)^{-s-1} dx + \frac{\phi(d) s}{d(s-1)}$$

whenever $\operatorname{Re} s > 1$.

Proof. (i) Since S is a simple step function, it is automatically Riemann integrable on bounded intervals. That S is bounded follows from Lemma 104.2. To establish the main result we use partial summation (as in the proof of Lemma 104.1 (i)) to see that

$$\sum_{n=1}^N \chi(n) n^{-s} = \sum_{n=1}^N (S(n) - S(n-1)) n^{-s}$$

$$= \sum_{n=1}^N S(n)(n^{-s} - (n+1)^{-s}) + S(N)(N+1)^{-s}$$

$$= \sum_{n=1}^{N} s \int_{n}^{n+1} S(n)x^{-s-1}dx + S(N)(N+1)^{-s}$$

$$= s \sum_{n=1}^{N} \int_{n}^{n+1} S(x)x^{-s-1}dx + S(N)(N+1)^{-s}$$

$$= s \int_{1}^{N+1} S(x)x^{-s-1}dx + S(N)(N+1)^{-s}$$

$$\rightarrow s \int_{1}^{\infty} S(x)x^{-s-1}dx \quad \text{as } N \rightarrow \infty.$$

(ii) Since S_0 is the sum of a simple step function and a linear function, S_0 is automatically Riemann integrable on bounded intervals. That S_0 is bounded follows from the observation that $S_0(Nd) = 0$ and so for any $Nd \leqslant x \leqslant (N+1)d$ we have $|S_0(x)| \leqslant \phi(d)$. To establish the main result we consider $S(x) = \sum_{1 \leqslant m \leqslant x} \chi_0(m)$ as before, and obtain, by partial summation

$$\sum_{n=1}^{N} \chi_0(n)n^{-s} = s \int_{1}^{N+1} S(x)x^{-s-1}dx + S(N)(N+1)^{-s}$$

$$= s \int_{1}^{N+1} (S_0(x) + \phi(d)x/d)x^{-s-1}dx$$

$$\quad + (S_0(N) + \phi(d)N/d)(N+1)^{-s}$$

$$= s \int_{1}^{N+1} S_0(x)x^{-s-1}dx + \frac{s\phi(d)}{d} \int_{1}^{N+1} x^{-s}dx$$

$$\quad + S_0(N)(N+1)^{-s} + \frac{\phi(d)}{d} \frac{N}{(N+1)^s}$$

$$= s \int_{1}^{N+1} S_0(x)x^{-s-1}dx + \frac{s\phi(d)}{d} \frac{(N+1)^{-s+1}-1}{1-s}$$

$$\quad + S_0(N)(N+1)^{-s} + \frac{\phi(d)}{d} \frac{N}{(N+1)^s}$$

$$\rightarrow s \int_{1}^{\infty} S_0(x)x^{-s-1}dx + \frac{\phi(d)s}{d(s-1)} \quad \text{as } N \rightarrow \infty. \qquad \blacksquare$$

We can now deduce the results we need.

Theorem 108.3. (i) *If $\chi \neq \chi_0$ then $\sum_{n=1}^{\infty} \chi(n)n^{-s}$ converges to an analytic function on $\{s \in \mathbb{C} : \operatorname{Re} s > 0\}$.*

(ii) *If, with the notation of Lemma 108.2 (ii), we write*

$$f_0(s) = s \int_{1}^{\infty} S_0(x)x^{-s-1}dx,$$

then $f_0:\{s\in\mathbb{C}:\operatorname{Re}s>0\}\to\mathbb{C}$ *is a well defined analytic function such that*

$$\sum_{n=1}^{\infty}\chi_0(n)n^{-s}\ \text{converges to}\ f_0(s)+\frac{\phi(d)s}{d(s-1)}\quad\text{on}\ \{s\in\mathbb{C}:\operatorname{Re}s>1\}.$$

Proof. Combine Lemmas 108.2 and 108.1. ∎

Notice that, whilst $\sum_{n=1}^{\infty}\chi(n)n^{-s}$ converges for $\operatorname{Re}s>0$ when $\chi\neq\chi_0$, it is not true that $\sum_{n=1}^{\infty}\chi_0(n)n^{-s}$ converges for $1>\operatorname{Re}s>0$. (If s is real and $1>s>0$ then

$$\sum_{n=1}^{Nd}\chi_0(n)n^{-s}\geqslant\sum_{m=1}^{N}\phi(d)(md)^{-s}=\phi(d)d^{-s}\sum_{m=1}^{N}m^{-1}\to\infty\ \text{as}\ N\to\infty.$$

The reader should check why the proof of Lemma 108.2 (ii) breaks down for $1>\operatorname{Re}s>0$ and why the proof of Lemma 108.2 (i) does not.)

The situation is not novel. If the reader has looked at Chapter 75 (on the properties of the Laplace transform considered as an analytic function) she will recall several cases where the integral defining the Laplace transform $\mathscr{L}f(z)=\int_{-\infty}^{\infty}f(x)e^{-zx}dx$ only converges on part of $\mathbb{C}$ but the function $\mathscr{L}f$ itself can be continued as a meromorphic function over a larger domain. (For example if $H(x)=1$ for $x\geqslant0$, $H(x)=0$ for $x<0$ then $\int_{-\infty}^{\infty}H(x)e^{-zx}dx$ only converges for $\operatorname{Re}z>0$ but the function z^{-1} to which it converges for $\operatorname{Re}z>0$ is defined as a meromorphic function on $\mathbb{C}$.) She will also recall that under normal circumstances the extension can be shown to be unique (Lemma 75.7).

Lemma 108.4. *Let* $\Omega=\{s:\operatorname{Re}s>0\}$ *and let* $f_1,f_2:\Omega\to\mathbb{C}$ *be analytic functions such that* $f_1(s)=f_2(s)$ *for* $\operatorname{Re}s>1$. *Then* $f_1(s)=f_2(s)$ *for* $\operatorname{Re}s>0$.
Proof. Let $g=f_1-f_2$. Then g is an analytic function on the (connected) set Ω and so (by Theorem 75.6 (i)) either g is identically zero or all its zeros are isolated. But $g(s)=0$ for $\operatorname{Re}s>1$ so g has zeros which are not isolated. Thus $g(s)=0$ for all s with $\operatorname{Re}s>0$ and we are done. ∎

We can now define $L(s,\chi)$ for all χ and all s with $\operatorname{Re}s>0$.

Definition 108.5. (i) *If* $\chi\neq\chi_0$ *then* $L(s,\chi)=\sum_{n=1}^{\infty}\chi(n)n^{-s}$ *for all* s *with* $\operatorname{Re}s>0$.
(ii) *If* $\operatorname{Re}s>0$ *we write*

$$L(s,\chi_0)=f_0(s)+\frac{s\phi(d)}{d(s-1)},$$

where $f_0:\{s:\operatorname{Re}s>0\}\to\mathbb{C}$ *is the unique analytic function with* $f_0(s)=\sum_{n=1}^{\infty}\chi_0(n)n^{-s}-s\phi(d)\,(d(s-1))^{-1}$ *for* $\operatorname{Re}s>1$.

(The existence of f_0 is guaranteed by Theorem 108.3 (ii), the uniqueness by Lemma 108.4.)

The following lemma restates results already obtained in this chapter in such a

way as to link our new functions $L(s, \chi)$ defined for $\mathrm{Re}\, s > 0$ with the old functions $L(s, \chi)$ of the previous chapters which were only defined for s real and $s > 0$ (if $\chi \neq \chi_0$) or s real and $s > 1$ (if $\chi = \chi_0$).

Lemma 108.6. (i) *If $\chi \neq \chi_0$ then $L(s, \chi)$ is analytic for $\mathrm{Re}\, s > 0$. If s is real and $s > 0$ then $L(s, \chi) = \sum_{n=1}^{\infty} \chi(n) n^{-s}$.*
(ii) *The function $L(s, \chi_0)$ is analytic for $\mathrm{Re}\, s > 0$ except for a simple pole at $s = 1$ with residue $\phi(d)/d$. If s is real and $s > 1$ then $L(s, \chi_0) = \sum_{n=1}^{\infty} \chi_0(n) n^{-s}$.*

Proof. This lemma is essentially a slightly restricted form of Theorem 108.3. ∎

Many of the results which we proved in the two previous chapters concerning the behaviour of $L(s, \chi)$ for s real and $s > 1$ carry over to the case when $s \in \mathbb{C}$ and $\mathrm{Re}\, s > 1$. In the next chapter we shall need, in particular, the fact that $L(s, \chi_0) \neq 0$ for $\mathrm{Re}\, s > 1$. We give a sequence of results which includes this fact.

Lemma 108.7. *If $\mathrm{Re}\, s > 1$ then*

(i) (*The Euler–Dirichlet formula*)

$$\prod_{p < N} (1 - \chi(p) p^{-s})^{-1} \to \sum_{n=1}^{\infty} \chi(n) n^{-s} = L(s, \chi),$$

(ii) $\exp(-\sum_p lg(1 - \chi(p) p^{-s})) = L(s, \chi)$,
(iii) $L(s, \chi) \neq 0$.

Proof. (i) This follows the proof of Lemma 106.7 (i) and (ii) word for word except that inequalities of the form $|\chi(n) n^{-s}| \leq n^{-s}$ must be replaced by $|\chi(n) n^{-s}| \leq n^{-\mathrm{Re}\, s}$.
(ii) This follows the proof of Lemma 107.4 (i) word for word.
(iii) This follows the proof of Lemma 107.4 (ii) word for word. ∎

Remark. There are several important functions in number theory with the property that $\psi(uv) = \psi(u)\psi(v)$ $[u, v \geq 1]$ and for which we have an Euler–Dirichlet type formula $\prod_p (1 - \psi(p) p^{-s}) = \sum_{n=1}^{\infty} \psi(n) n^{-s}$. We are therefore led to study sums of the form $\sum_{n=1}^{\infty} a_n n^{-s}$. If $|a_n| \leq K$ for all n then, writing $S(x) = \sum_{n < x} a_n n^{-s}$, partial summation yields

$$\sum_{n=1}^{\infty} a_n n^{-s} = s \int_1^{\infty} S(x) x^{-s-1} dx \quad \text{for } \mathrm{Re}\, s > 1,$$

and the substitution $x = e^t$ yields $\sum_{n=1}^{\infty} a_n n^{-s} = s(\mathscr{L}g)(s)$, where $g(t) = S(e^t)$ for $t \geq 0$, $g(t) = 0$ for $t < 0$, and the integral defining $\mathscr{L}g$ converges for $\mathrm{Re}\, s > 1$. If the a_n are periodic (i.e. $a_{n+d} = a_n$ for each n) then we can do better and obtain

$$\sum_{n=1}^{\infty} a_n n^{-s} = s(\mathscr{L}h)(s) + As/(s-1) \quad \text{for } \mathrm{Re}\, s > 1,$$

where the integral defining $\mathscr{L}h$ converges for $\mathrm{Re}\, s > 0$. (The reader is invited to fill in

the details. She may also like to compute h and A when $a_n = 1$ and we consider $\zeta(s) = \sum_{n=1}^{\infty} n^{-s}$.)

The author wonders whether the Laplace transform will keep its place in the standard mathematical methods course for very much longer. If it is only used as a quick mechanical method for solving linear differential equations then this cuts across today's tendency to leave mechanical things to machines. (Admittedly, machines are not much good at symbol manipulation but then neither are human beings.) However for the reasons given in the previous paragraph the Laplace transform will survive as a tool in pure mathematics.

109

PRIMES IN GENERAL
ARITHMETICAL PROGRESSIONS

We now return to our attempt to show $L(\chi, 1) \neq 0$ for all $\chi \neq \chi_0$. Our starting point is a lemma along the lines of Lemma 107.4.

Lemma 109.1. (i) *If s is real and $s > 1$ then*

$$\prod_{\chi} L(s, \chi) = \exp\left(-\sum_p \sum_{\chi} lg(1 - \chi(p)p^{-s})\right).$$

(ii) *If s is real and $s > 1$ then $\prod_{\chi} L(s, \chi)$ is real and*

$$\prod_{\chi} L(s, \chi) \geq 1.$$

(iii) $\prod_{\chi} L(s, \chi) \nrightarrow 0$ *as $s \to 1$.*

Proof. (i) Using Lemma 107.4(i), we have

$$\prod_{\chi} L(s, \chi) = \prod_{\chi} \exp\left(-\sum_p lg(1 - \chi(p)p^{-s})\right) = \exp\left(-\sum_{\chi}\sum_p lg(1 - \chi(p)p^{-s})\right)$$

$$= \exp\left(-\sum_p \sum_{\chi} lg(1 - \chi(p)p^{-s})\right).$$

(There are no analytic difficulties since the sum $\sum_{\chi}$ and product $\prod_{\chi}$ are finite.) (ii) Since we have defined lg in such a way that

$$lg(1 + z) = -\sum_{r=1}^{\infty} (-z)^r/r \quad \text{for } |z| < 1$$

(see Lemma 106.8 and the preceding discussion), it follows that

$$-\sum_{\chi} lg(1 - \chi(p)p^{-s}) = \sum_{\chi}\sum_{r=1}^{\infty} \chi(p^r)p^{-rs}r^{-1}$$

$$= \sum_{r=1}^{\infty}\sum_{\chi} \chi(p^r)p^{-rs}r^{-1} = \sum_{r=1}^{\infty} (rp^{rs})^{-1}\sum_{\chi}\chi(p^r) = \sum_{r=1}^{\infty} (rp^{rs})^{-1}\phi(d)\delta(p^r),$$

where $\delta(m) = 1$ if $m \equiv 0 \bmod d$, $\delta(m) = 0$ otherwise. In particular (and this is all

552

that we need), it follows that $-\sum_\chi lg(1 - \chi(p)p^{-s})$ is real and positive (i.e. greater than or equal to zero). Thus $-\sum_p \sum_\chi lg(1 - \chi(p)p^{-s})$ is real and positive and so $\exp(-\sum_p \sum_\chi lg(1 - \chi(p)p^{-s}))$ is real and greater than or equal to 1. The equality of (i) completes the proof.

(iii) This is an immediate consequence of (ii). ∎

The work of Chapter 108 in extending the definitions of the Dirichlet L functions $L(s, \chi)$ to the right half of the complex plane now bears fruit.

Lemma 109.2. *There can be at most one character $\chi \neq \chi_0$ with $L(1, \chi) = 0$.*

Proof. If $\chi \neq \chi_0$ then we know that $L(s, \chi)$ is analytic for $\operatorname{Re} s > 0$ (Lemma 108.6 (i)) and is not identically zero (Lemma 107.4 (i)). Thus there exists an integer $n(\chi) \geq 0$ and an analytic function $h_\chi : \{s \in \mathbb{C} : \operatorname{Re} s > 0\} \to \mathbb{C}$ such that $h_\chi(1) \neq 0$ and $L(s, \chi) = (s - 1)^{n(\chi)} h_\chi(s)$. (If $n(\chi) \geq 1$, we say that $L(\ , \chi)$ has a zero of order $n(\chi)$ at 1.) On the other hand we know that $L(s, \chi_0)$ is analytic for $\operatorname{Re} s > 0$ except for a simple pole at 1 (Lemma 108.6 (ii)). Thus there exists an analytic function $h_{\chi_0} : \{s \in \mathbb{C} : \operatorname{Re} s > 0\} \to \mathbb{C}$ such that $h_{\chi_0}(1) \neq 0$ and $L(s, \chi) = (s - 1)^{-1} h_{\chi_0}(s)$.

Thus, writing $H(s) = \prod_\chi h_\chi(s)$ and $N = \sum_{\chi \neq \chi_0} n(\chi) - 1$, we see that $H : \{s \in \mathbb{C} : \operatorname{Re} s > 0\} \to \mathbb{C}$ is analytic (and so continuous) with $H(1) \neq 0$ whilst

$$\prod_\chi L(\chi, s) = (s - 1)^N H(s) \quad \text{for } \operatorname{Re} s > 0.$$

If $N \geq 1$ then, automatically,

$$\prod_\chi L(\chi, s) = (s - 1)^N H(s) \to 0 H(1) = 0 \quad \text{as } s \to 1.$$

Referring back to Lemma 109.1 (iii), we conclude that $N \leq 0$ and so $\sum_{\chi \neq \chi_0} n(\chi) \leq 1$. Thus $n(\chi) = 0$ for all $\chi \neq \chi_0$ with at most one exception and this gives the result we want. ∎

The fact that at most one $L(1, \chi)$ can be zero shows that certain $L(1, \chi)$ cannot be.

Lemma 109.3. *If χ is a character which takes non real values then $L(1, \chi) \neq 0$.*

Proof. If χ is a character then, fairly obviously, so is χ^* (where by definition $\chi^*(n) = \chi(n)^*$ the complex conjugate of $\chi(n)$). Moreover $L(1, \chi^*) = L(1, \chi)^*$ so if $L(1, \chi) = 0$ it follows that $L(1, \chi^*) = 0$. If χ only takes real values, then $\chi = \chi^*$ and there is no contradiction, but, if χ takes values which are not real, $\chi \neq \chi^*$ and this contradicts the conclusion of Lemma 109.2. ∎

The proof of Legendre's conjecture is thus reduced to showing that $L(1, \chi) \neq 0$ when $\chi \neq \chi_0$ takes only the values 0, 1 and -1. This point gave Dirichlet much difficulty and is still the least transparent part of the proof. As I said earlier, we shall follow de la Vallée Poussin's proof which takes the ideas of Lemmas 109.2 and 109.3 much further.

Lemma 109.4. *Suppose* $\chi \neq \chi_0$ *takes only the values* 0, 1 *and* -1. *Set* $\psi(s) = L(s, \chi)L(s, \chi_0)/L(2s, \chi_0)$. *Then*

  (i) *ψ is well defined and analytic for* $\mathrm{Re}\, s > \frac{1}{2}$ *except, possibly, for a simple pole at 1.*
  (ii) *If* $L(1, \chi) = 0$ *then 1 is a removable singularity (and ψ may be defined at 1 so as to make ψ analytic everywhere in* $\mathrm{Re}\, s > \frac{1}{2}$).
  (iii) *In any case,* $\psi(s) \to 0$ *as* $s \to \frac{1}{2}$ *from the right through real values of s.*

Proof. (i) We know that $L(s, \chi_0)$ is analytic for $\mathrm{Re}\, s > 1$ (Lemma 108.6(i)) and has no zeros in this region. Thus $L(2s, \chi_0)$ is analytic with no zeros in $\mathrm{Re}\, s > \frac{1}{2}$ and so $1/L(2s, \chi_0)$ is analytic in $\mathrm{Re}\, s > \frac{1}{2}$. Since $L(s, \chi)$ is analytic for $\mathrm{Re}\, s > 0$ and $L(s, \chi_0)$ is analytic except for a simple pole at $s = 1$, it follows that $\psi(s) = L(s, \chi)L(s, \chi_0)/L(2s, \chi_0)$ is analytic for $\mathrm{Re}\, s > \frac{1}{2}$ except, possibly, for a simple pole at 1.
  (ii) This follows easily from the same sort of considerations as arise in the proof of Lemma 109.2.
  (iii) Since $L(s, \chi)$ and $L(s, \chi_0)$ are analytic and so continuous in a neighbourhood of $s = \frac{1}{2}$, whilst $L(2s, \chi_0) \to \infty$ it follows that $\psi(s) \to 0$ as $s \to \frac{1}{2}$ through the reals from the right. ∎

    We now show that the fact that $\psi(s) \to 0$ as $s \to \frac{1}{2}$ from the right is incompatible with the possibility that 1 is a removable singularity. We start by expressing $\psi(s)$ in the form $\sum a_n n^{-s}$ for $\mathrm{Re}\, s > 1$.

Lemma 109.5. *Let ψ be defined as in Lemma* 109.4. *If* $\mathrm{Re}\, s > 1$ *then*

  (i)
$$\prod_{\chi(p) = 1, p < N} \left(\frac{1 + p^{-s}}{1 - p^{-s}} \right) \to \psi(s) \quad \text{as } N \to \infty.$$

  (ii) *There exist subsets* $Q(1)$ *and* $Q(2)$ *of the positive integers such that*

$$\prod_{\chi(p) = 1, p < N} (1 + p^{-s}) \to \sum_{n \in Q(1)} n^{-s}$$

and
$$\prod_{\chi(p) = 1, p < N} (1 - p^{-s})^{-1} \to \sum_{n \in Q(2)} n^{-s} \quad \text{as } N \to \infty.$$

  (iii) $\psi(s) = \sum_{n=1}^{\infty} a_n n^{-s}$ *where the a_j are real, $a_1 = 1$ and $a_j \geqslant 0$ for all $j \geqslant 1$.*

Proof. (i) By the Euler–Dirichlet formula

$$\prod_{p < N} (1 - \chi(p)p^{-s})^{-1} \to L(s, \chi),$$

$$\prod_{p < N} (1 - \chi_0(p)p^{-s})^{-1} \to L(s, \chi_0),$$

$$\prod_{p < N} (1 - \chi_0(p)p^{-2s})^{-1} \to L(2s, \chi_0),$$

and so

$$\prod_{p < N} \frac{(1 - \chi_0(p)p^{-2s})}{(1 - \chi_0(p)p^{-s})(1 - \chi(p)p^{-s})} \to \psi(s) \quad \text{as } N \to \infty.$$

But if $\chi(p) = 0$ then $\chi_0(p) = 0$ whilst if $\chi(p) = -1$ then $\chi_0(p) = 1$ and so in either case

$$\frac{(1 - \chi_0(p)p^{-2s})}{(1 - \chi_0(p)p^{-s})(1 - \chi(p)p^{-s})} = 1.$$

In the remaining possible case when $\chi(p) = 1$ we have $\chi_0(p) = 1$ again and

$$\frac{(1 - \chi_0(p)p^{-2s})}{(1 - \chi_0(p)p^{-s})(1 - \chi(p)p^{-s})} = \frac{1 + p^{-s}}{1 - p^{-s}}.$$

The formula now follows.

(ii) Use the method of Lemma 106.7 (as modified in the proof of Lemma 108.7(i)). Clearly $Q(2)$ is the set of all strictly positive integers n whose prime factors p satisfy $\chi(p) = 1$ whilst $Q(1)$ is the subset of $Q(2)$ consisting of those n which have no factors of the form p^k with $k \geqslant 2$.

(iii) By comparison with the sum $\sum_{n=1}^{\infty} n^{-\mathrm{Re}\,s}$ both sums $\sum_{n \in Q(1)} n^{-s}$ and $\sum_{n \in Q(2)} n^{-s}$ are absolutely convergent. They can be multiplied term by term to give

$$\psi(s) = \lim_{n \to \infty} \prod_{\chi(p)=1, p < N} \left(\frac{1 + p^{-s}}{1 - p^{-s}} \right) = \sum_{n \in Q(1)} n^{-s} \sum_{m \in Q(2)} m^{-s} = \sum_{n=1}^{\infty} a_n n^{-s},$$

where a_j is real, $a_1 = 1$, and $a_j \geqslant 0$ for each $j \geqslant 1$. ∎

The importance of Lemma 109.5 is that it gives us information about the Taylor expansion of ψ about a point.

Lemma 109.6. (i) *If* $\mathrm{Re}\,s > 1$ *then*

$$\psi^{(m)}(s) = \sum_{n=1}^{\infty} a_n (-\log n)^m n^{-s},$$

where the a_n *are as in Lemma 109.5.*

(ii) *If* s_0 *is real and* $s_0 > 1$ *then* $(-1)^m \psi^{(m)}(s_0) > 0$.

(iii) *If* ψ *is analytic for* $|s - s_0| < \rho$ *then*

$$\psi(s) = \sum_{m=0}^{\infty} \psi^{(m)}(s_0)(s - s_0)^m / m!$$

(iv) *If* ψ *is analytic for* $\mathrm{Re}\,s > \frac{1}{2}$ *then* $\psi(s) \nrightarrow 0$ *as* $s \to \frac{1}{2}$ *through real values from the right.*

Proof. (i) Let $\psi_N(s) = \sum_{n=1}^{N} a_n n^{-s}$ (so ψ_N is analytic). Then, since the a_n are real and positive, we know that

$$|\psi(s) - \psi_N(s)| \leqslant \sum_{n=N+1}^{\infty} a_n n^{-1-\delta} \to 0 \quad \text{as } N \to \infty,$$

and so

$$\psi_N(s) \to \psi(s) \text{ uniformly for } \mathrm{Re}\,s > 1 + \delta \; [\delta > 0].$$

By Theorem 75.3 (iv) (on the uniform limit of analytic functions) $\psi_N^{(m)}(s) \to \psi^{(m)}(s)$ for each s with $\operatorname{Re} s > 1 + \delta [\delta > 0]$. Thus

$$\psi^{(m)}(s) = \lim_{N \to \infty} \psi_N^{(m)}(s) = \lim_{N \to \infty} \sum_{n=1}^{N} a_n (-\log n)^m n^{-s} = \sum_{n=1}^{\infty} a_n (-\log n)^{mn^{-s}}$$

whenever $\operatorname{Re} s > 1$.

(ii) Immediate.

(iii) Taylor's theorem. (This is a point where we use complex variable techniques to the fullest. As we pointed out in Chapter 4, Example 4.2, there is no corresponding real variable result.)

(iv) Fix some $s_0 > 1$. By (iii) with $\rho = s_0 - \frac{1}{2}$, and (ii) we have

$$\psi(s) = \sum_{m=0}^{\infty} \psi^{(m)}(s_0)(s - s_0)^m / m! \geqslant \psi(s_0) \geqslant a_1 = 1$$

for all s real with $\frac{1}{2} < s < s_0$. Thus $\psi(s) \nrightarrow 0$ as $s \to \frac{1}{2}$ through real values with $s > \frac{1}{2}$. ∎

Lemma 109.7. *If $\chi \neq \chi_0$ takes only the values 0, 1 and -1 then $L(1, \chi) \neq 0$.*
Proof. By Lemma 109.4 (ii), Lemma 109.6 (iv) and Lemma 109.4 (iii) the alternative would lead to a contradiction. ∎

Combining Lemmas 109.3 and 109.7 we see that $L(1, \chi) \neq 0$ for all $\chi \neq \chi_0$. Referring to Theorem 107.7, we see that we can now replace the conjecture of Legendre by the theorem of Dirichlet.

Theorem 109.8. *If a and d are coprime positive integers then the arithmetical progression $a, a + d, a + 2d, \ldots$ contains an infinity of primes.*

'Dirichlet', wrote Jacobi to their mutual patron Humboldt, 'has founded a new branch of mathematics which uses the infinite series, introduced by Fourier for the study of heat, to discover properties of prime numbers' (quoted in *Von Fermat bis Minkowski*, W. Schalan, H. Opolka, Springer, 1980). Looking at Dirichlet's theorem and its proof we may echo Keats' lines on Chapman's translation of Homer.

> Much have I travelled in the realms of gold
> And many goodly states and kingdoms seen;
> Round many western isles have I been
> Which bards in fealty to Apollo hold.
> Oft of one wide expanse had I been told
> That deep-brow'd Homer ruled as his demesne;
> Yet did I never breathe its pure serene
> Till I heard Chapman speak out loud and bold:
> Then I felt like some watcher of the skies

When a new planet swims into his ken;
Or like stout Cortez when with eagle eyes
He star'd at the Pacific – and all his men
Look'd at each other with wild surmise
Silent, upon a peak in Darien.

110

A WORD FROM OUR FOUNDER

(The following passage is translated from the introduction to Fourier's *Théorie Analytique de la Chaleur*.)

The equations of heat conduction like those of sound or small oscillations for liquids belong to one of the most recently discovered branches of science which it is most important to extend. Having established these differential equations we must obtain their solution which involves passing from a general to a particular solution obeying all the boundary conditions. This difficult investigation requires a special calculus based on new theorems.... The method which results leaves nothing vague or undetermined in the solutions; it leads us to the point where we can obtain numerical values, a necessary property of any such method, without which we obtain nothing but useless formal transformations.

These same theorems which enable us to solve the heat equation have immediate applications to questions in general analysis and dynamics whose solutions have long been sought.

The profound study of nature is the most fruitful source of mathematical discovery. Not only does this study, by proposing a fixed goal for our research, have the advantage of excluding vague questions and calculations without result. It is also a sure way of shaping Analysis itself, and discovering which parts it is most important to know and must always remain part of the subject. These fundamental principles are those which are found in all natural phenomena.

We see, for example, that the same expression which the geometers have studied abstractly and which, from this point of view belongs to pure mathematics also describes the motion of light through the atmosphere, determines the laws governing heat conduction in solids and is involved in all the main questions of probability theory.

The analytic equations, unknown to the classical geometers, which Descartes was the first to introduce for the study of curves and surfaces do not only appear in the study of geometrical figures and rational mechanics, they cover all general systems. There can not be a language more universal or more simple, more free

from error and obscurity, that is to say more worthy of expressing the relationship between natural phenomena.

From this point of view, mathematical analysis is as extensive as nature herself. It defines all observable relationships, measures time, space, force, temperature. This difficult science grows slowly but once ground has been gained it is never relinquished. It grows, without ceasing, in size and strength, amid the constant error and confusion of the human spirit.

Its main property is clarity; analysis has no means of expressing confused notions. Analysis brings together the most disparate phenomena and discovers the hidden analogies which unify them. If the material escapes us like air or light because of its extreme fineness, if the bodies are placed far from us in the immensity of space, if man desires to know the aspect of the heavens for times separated by many centuries, if the effects of gravity and temperature occur in the interior of the solid earth at depths which will remain for ever inaccessible, yet mathematical analysis can still grasp the laws governing the phenomena. Analysis makes them actual and measurable and seems to be a faculty of human reason meant to compensate for the brevity of life and the imperfection of our senses. And which is still more remarkable it follows the same path when applied to all phenomena and interprets them all in the same language as if to attest the unity and simplicity of the design of the universe and to make still more evident that unchanging order which presides over all natural laws.

APPENDIX A:
THE CIRCLE $\mathbb{T}$

There are several ways of constructing the circle $\mathbb{T}$. One is to take the interval $[0, 2\pi]$ bend it round and solder its ends together (i.e. identify 0 and 2π) as in Figure A.1. The *analytic* structure of the circle is then clear. A function $f:\mathbb{T} \to \mathbb{C}$ is obtained from a function $\tilde{f}:[0, 2\pi] \to \mathbb{C}$ with $\tilde{f}(0) = \tilde{f}(2\pi)$, f will be continuous if $\tilde{f}$ is and so on.

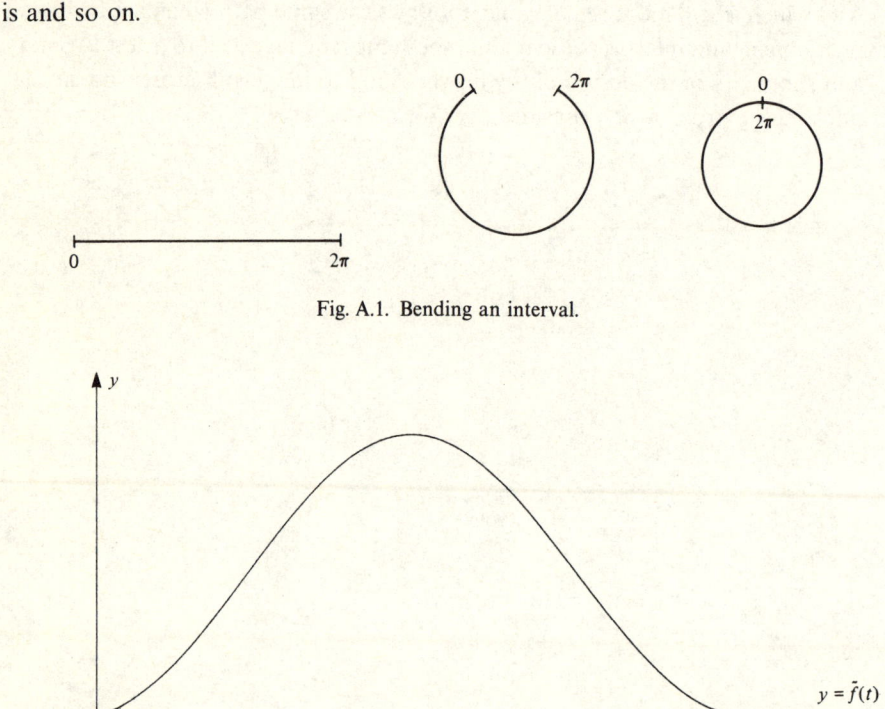

Fig. A.1. Bending an interval.

Fig. A.2. Endpoint behaviour.

560

This method is unsatisfactory in that it assigns a special place to 0 and 2π. Thus for example $f:\mathbb{T}\to\mathbb{C}$ is differentiable if and only if $\bar{f}$ is differentiable on $[0,2\pi]$, $\bar{f}(0)=\bar{f}(2\pi)$ and, in addition, the right derivative of $\bar{f}$ at 0 equals the left derivative of $\bar{f}$ at 2π as in Figure A.2.

A more satisfactory method is to take the real line $\mathbb{R}$ and wind it into a circle by identifying $\theta,\ \theta+2\pi,\ \theta+4\pi,\dots,\theta-2\pi,\ \theta-4\pi,\dots$ and so on, as in Figure A.3. Thus every point of $\mathbb{T}$ will now have several names, to wit, $\theta+2n\pi$ for all $n\in\mathbb{Z}$. We now view a function $f:\mathbb{T}\to\mathbb{C}$ as being obtained from a function $\tilde{f}:\mathbb{R}\to\mathbb{C}$ which is periodic with period 2π. We can then say that f is differentiable at the point of $\mathbb{T}$ called after θ if and only if $\tilde{f}$ is differentiable at θ (Figure A.4).

This method also allows us to do a limited amount of arithmetic on $\mathbb{T}$. We say that the sum of two points with names θ_1 and θ_2 is the point with name $\theta_1+\theta_2$. If n is an integer then we say that n times a point with name θ is the point with name $n\theta$.

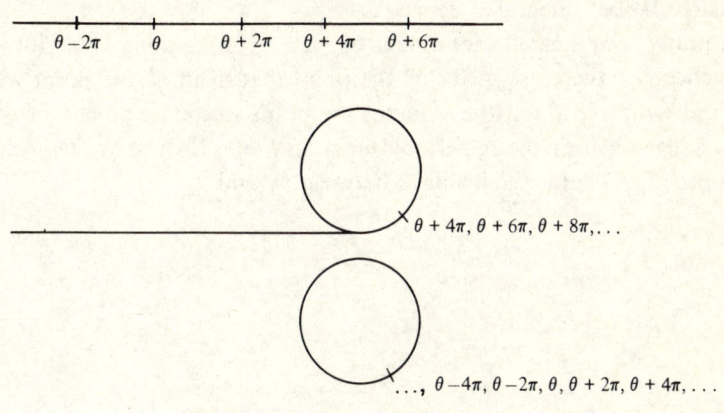

Fig. A.3. Rolling up the real line.

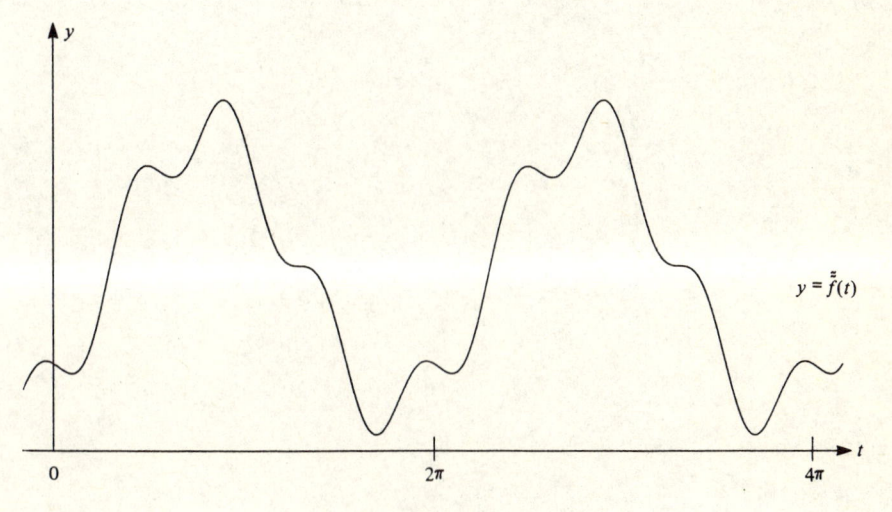

Fig. A.4. Unrolling a function on $\mathbb{T}$.

It is essential in definitions of the type given above to make sure that they are independent of the choice of name for the point. For example we cannot define $\frac{1}{2}$ times a point with name θ to be $\theta/2$ since the point with name θ also has name $\theta + 2\pi$ but $(\theta + 2\pi)/2 = \theta/2 + \pi$ is not another name of the point with name $\theta/2$. The reader should construct a similar counter example to show that we cannot define the product of two points in the obvious way.

The distance between two points whose names are θ_1 and θ_2 is defined as $\min\{|\theta_1 - \theta_2 + 2n\pi| : n \in \mathbb{Z}\}$. We shall denote by $[\theta_1, \theta_2]$ the set of points with names θ where $\theta_1 \leqslant \theta \leqslant \theta_2 [\theta_1 \leqslant \theta_2]$. If $0 \leqslant \theta_2 - \theta_1 \leqslant 2\pi$, then we write

$$\int_{[\theta_1, \theta_2]} f(x)\, dx = \int_{\theta_1}^{\theta_2} f(x)\, dx = \int_{\theta_1}^{\theta_2} \tilde{f}(x)\, dx$$

(provided the last integral is defined).

Naturally such a careful treatment of such a simple thing is terribly restrictive in practice. We therefore speak of 'the point θ' instead of 'the point whose name is θ' and write $\exp i\theta$ for 'the value at the point whose name is θ of the function $f : \mathbb{T} \to \mathbb{C}$. derived from the 2π periodic function $\tilde{f} : \mathbb{R} \to \mathbb{C}$ given by $\tilde{f}(t) = \exp it \, [t \in \mathbb{R}]$'. We write $|\theta_1 - \theta_2|$ for the distance between θ_1 and θ_2.

APPENDIX B:
CONTINUOUS FUNCTIONS ON
CLOSED BOUNDED SETS

In this appendix we extend the main elementary results on continuous functions on closed bounded intervals in $\mathbb{R}$ to cover continuous functions on closed bounded sets in $\mathbb{R}^m$. The reader who already knows these extensions will find nothing new and should not bother to read any further.

If $x \in \mathbb{R}^m$ we write $x = (x(1), x(2), \ldots, x(m))$, where $x(j)$ is the jth coordinate of x. We use the usual Euclidean metric $|x - y| = \sqrt{(\sum_{j=1}^m (x(j) - y(j))^2)}$ and write $x_r \to x$ if $|x_r - x| \to 0$ as $r \to \infty$. The key lemma is the following:

Lemma B.1. *Suppose $x_1, x_2, \ldots$ are such that $|x_n(j)| \leqslant M$ for all $1 \leqslant j \leqslant m, n \geqslant 1$. Then we can find $n(1) < n(2) < \cdots$ and $x \in \mathbb{R}^m$ such that $x_{n(r)} \to x$ as $r \to \infty$.*
Proof. Since $|x_n(1)| \leqslant M$, we know from the one dimensional case that there exists a subsequence $n(1, 1) < n(1, 2) < \cdots$ and an $x(1) \in \mathbb{R}$ such that $x_{n(1, r)}(1) \to x(1)$. But $|x_{n(1, r)}(2)| \leqslant M$ for all r and so we can find a subsequence $n(2, 1) < n(2, 2) < \cdots$ of the $n(1, r)$ and an $x(2) \in \mathbb{R}$ such that $x_{n(2, r)}(2) \to x(2)$. Continuing inductively, we can find for each $2 \leqslant j \leqslant m$ a subsequence $n(j, 1) < n(j, 2) < \cdots$ of the sequence $n(j - 1, r)$ and an $x(j) \in \mathbb{R}$ such that $x_{n(j, r)}(j) \to x(j)$.

Set $n(r) = n(m, r)$. Then, since $n(r) = n(m, r)$ is a subsequence of the sequence $n(j, r)$ for each $1 \leqslant j \leqslant m$, we have $x_{n(r)}(j) \to x(j) [1 \leqslant j \leqslant m]$ and so $x_{n(r)} \to x$. ∎

As an immediate consequence we obtain the following corollary.

Lemma B.2. *Let K be a closed bounded set in $\mathbb{R}^m$. Then given a sequence $x_n \in K$ we can find an $x \in K$ and a subsequence $x_{n(r)} \to x$ as $r \to \infty$.*
Proof. Since K is bounded, it follows by definition that there exists an M with $|y(j)| \leqslant M$ for all $1 \leqslant j \leqslant m$ and $y \in K$. Thus $|x_n(j)| \leqslant M$ for all $1 \leqslant j \leqslant m, n \geqslant 1$ and by Lemma B.1 we can find $n(1) < n(2) < \cdots$ and $x \in \mathbb{R}^m$ with $x_{n(r)} \to x$. But $x_{n(r)} \in K$ and K is closed so $x \in K$. ∎

We can now prove the results we require.

Theorem B.3. *Let K be a closed non-empty bounded set in $\mathbb{R}^m$ and $f : K \to \mathbb{R}$ a*

continuous function. Then

(i) *There exists an M with $|f(y)| \leqslant M$ for all $y \in K$.*

(ii) *There exist $y_1, y_2 \in K$ with $f(y_1) \geqslant f(y) \geqslant f(y_2)$ for all $y \in K$.*

(In other words f is bounded and attains its bounds.)

Proof. (i) Suppose that f is unbounded in K. Then we can find $x_n \in K$ with $|f(x_n)| \geqslant n$. By Lemma B2 we can find an $x \in K$ and a subsequence $x_{n(r)} \to x$. Since f is continuous, $f(x_{n(r)}) \to f(x)$. But $|f(x_{n(r)})| \geqslant n(r) \to \infty$ as $r \to \infty$ and the result follows by contradiction.

(ii) Since $\{f(y) : y \in K\}$ is a non-empty bounded set in $\mathbb{R}$, it possesses a supremum μ say. Thus we can find $x_n \in K$ with $f(x_n) \geqslant \mu - 1/n$. By Lemma B.2 we can find a $y_1 \in K$ and a subsequence $x_{n(r)} \to y_1$. Since f is continuous, $f(x_{n(r)}) \to f(y_1)$ and so $f(y_1) \geqslant \mu$. But $y_1 \in K$ and so $\mu \geqslant f(y_1)$ whence $\mu = f(y_1)$ and f attains its supremum. The existence of a suitable y_2 follows in the same way. ∎

A trivial modification gives the next result.

Theorem B.4. *Let K be a non-empty closed bounded set in $\mathbb{R}^m$ and $g : K \to \mathbb{C}$ a continuous function. Then there exists a $y_1 \in K$ with $|g(y_1)| \geqslant |g(y)|$ for all $y \in K$.*

Proof. Set $f(x) = |g(x)|$ and apply Theorem B.3. ∎

Our final result shows that a continuous function on a closed bounded set is uniformly continuous.

Theorem B.5. *Let K be a closed bounded set in $\mathbb{R}^m$ and $f : K \to \mathbb{C}$ a continuous function. Then given any $\varepsilon > 0$ we can find a $\delta(\varepsilon) > 0$ such that whenever $x, y \in K$ and $|x - y| < \delta$ we have $|f(x) - f(y)| < \varepsilon$.*

Proof. Suppose not. Then we can find an $\varepsilon > 0$ and $x_n, y_n \in K$ such that $|x_n - y_n| < 1/n$ and yet $|f(x_n) - f(y_n)| \geqslant \varepsilon$. By Lemma B.2 we can find an $x \in K$ and a subsequence $x_{n(r)} \to x$. We observe that

$$|y_{n(r)} - x| \leqslant |y_{n(r)} - x_{n(r)}| + |x_{n(r)} - x| \to 0,$$

and so $y_{n(r)} \to x$ as $r \to \infty$, also. Since f is continuous, we deduce that

$$\varepsilon \leqslant |f(x_{n(r)}) - f(y_{n(r)})| \to |f(x) - f(x)| = 0,$$

so $\varepsilon \leqslant 0$ and the contradiction proves the theorem. ∎

APPENDIX C:
WEAKENING HYPOTHESES

Throughout this book and particularly in the second half dealing with functions on $\mathbb{R}$ rather than $\mathbb{T}$, we have used rather strong hypotheses to simplify our proofs. However, even with the simple tools at our disposal, it is possible to extend our results in a more or less routine manner.

As an example we show how the continuity condition of Theorem 51.7 may be dropped to give the following result.

Theorem C.1. *Suppose f and g are bounded Riemann integrable functions with $\int_{-\infty}^{\infty} |f(y)| \, dy, \int_{-\infty}^{\infty} |g(y)| \, dy < \infty$. Then*

  (i) *$f * g \in L^1 \cap C$ and $f * g$ is bounded,*
 (ii) *$\int_{-\infty}^{\infty} f * g(y) \, dy = \int_{-\infty}^{\infty} f(y) \, dy \int_{-\infty}^{\infty} g(y) \, dy$,*
(iii) *$\int_{-\infty}^{\infty} |f * g(y)| \, dy \leqslant \int_{-\infty}^{\infty} |f(y)| \, dy \int_{-\infty}^{\infty} |g(y)| \, dy$,*
(iv) *$(f * g)^{\wedge}(\zeta) = \hat{f}(\zeta)\hat{g}(\zeta)$ for all $\zeta \in \mathbb{R}$.*

Our proof depends on the following plausible fact.

Lemma C.2. *If $f: \mathbb{R} \to \mathbb{C}$ is a Riemann integrable function with $\int_{-\infty}^{\infty} |f(y)| \, dy < \infty$ then given $\varepsilon > 0$ we can find a bounded function $f_0 \in L^1 \cap C$ with $\int_{-\infty}^{\infty} |f(y) - f_0(y)| \, dy < \varepsilon$.*

Lemma C.2 follows in the usual way from a slightly simpler result.

Lemma C.3. *If $f: \mathbb{R} \to \mathbb{R}$ is a Riemann integrable function with $\int_{-\infty}^{\infty} |f(y)| \, dy < \infty$ then given $\varepsilon > 0$ we can find a bounded function $f_0 \in L^1 \cap C$ with $\int_{-\infty}^{\infty} |f(y) - f_0(y)| \, dy < \varepsilon$.*
Proof of Lemma C.2 from Lemma C.3. Let f be as in Lemma C.2. Setting $u = \operatorname{Re} f$, $v = \operatorname{Im} f$ we see that u and v satisfy the hypotheses of Lemma C.3 and so we can find bounded functions $u_0, v_0 \in L^1 \cap C$ with

$$\int_{-\infty}^{\infty} |u(y) - u_0(y)| \, dy, \int_{-\infty}^{\infty} |v(y) - v_0(y)| \, dy < \varepsilon/2.$$

Setting $f_0 = u_0 + iv_0$, we have the desired result.   ∎

Proof of Lemma C.3. This divides into several easy parts.
Step 1. Since $\int_{-\infty}^{\infty} |f(y)| \, dy < \infty$ we can find an R such that

$$\int_{|y| > R} |f(y)| < \varepsilon/3.$$

If we write

$$f_1(y) = f(y) \text{ for } |y| \leqslant R, \, f_1(y) = 0 \quad \text{otherwise,}$$

this result may be rewritten as $\int_{-\infty}^{\infty} |f_1(y) - f(y)| \, dy < \varepsilon/3$.
Step 2. By the definition of the Riemann integral we can find $-R = x_0 < x_1 < x_2 < \cdots < x_{n+1} = R$ and $m_0, m_1, \ldots, m_n \in \mathbb{R}$ such that

(i) $m_j \geqslant f(x)$ for $x \in [x_j, x_{j+1}]$ $[0 \leqslant j \leqslant n]$,

(ii) $\int_{-R}^{R} f(x) \, dx + \varepsilon/3 \geqslant \sum_{j=0}^{n} m_j(x_{j+1} - x_j)$.

Thus writing

$$f_2(y) = m_j \quad \text{for} \quad y \in [x_j, x_{j+1}), 0 \leqslant j \leqslant n, f_2(y) = 0 \quad \text{otherwise,}$$

we have

$$\int_{-\infty}^{\infty} |f_2(y) - f_1(y)| \, dy = \int_{-R}^{R} (f_2(y) - f_1(y)) \, dy \leqslant \varepsilon/3.$$

Step 3. Choose a $\delta > 0$ so small that $2\delta < (x_{j+1} - x_j)$ and $24(n+1)|m_j|\delta < \varepsilon$ for all $0 \leqslant j \leqslant n$. Then set $m_{-1} = m_{n+1} = 0$ and write

$$f_0(x) = 0 \quad \text{if } x \leqslant x_0 - \delta \quad \text{or} \quad x_{n+1} + \delta < x,$$
$$f_0(x) = m_j \quad \text{if } x_j + \delta < x \leqslant x_{j+1} - \delta \quad [0 \leqslant j \leqslant n],$$
$$f_0(x_j + s\delta) = ((1 + s)m_j + (1 - s)m_{j-1})/2 \quad \text{if} \quad -1 < s \leqslant 1 \quad [0 \leqslant j \leqslant n+1].$$

We observe that f_0 is continuous and that

$$\int_{-\infty}^{\infty} |f_0(y) - f_2(y)| \, dy = \sum_{j=0}^{n} \int_{x_j - \delta}^{x_j + \delta} |f_0(y) - f_2(y)| \, dy$$

$$\leqslant \sum_{j=0}^{n} 2\delta(|m_{j-1}| + |m_j|) < \sum_{j=0}^{n} \frac{\varepsilon}{6(n+1)} < \frac{\varepsilon}{3}.$$

Thus combining the formulae at the end of Steps 1, 2 and 3, we have

$$\int_{-\infty}^{\infty} |f_0(y) - f(y)| \, dy < \varepsilon.$$ ∎

Remark. The proof above may be illustrated by the sequence of curves in Figure C.1. If the reader does not consider the proof to be trivial she has not understood what is going on.

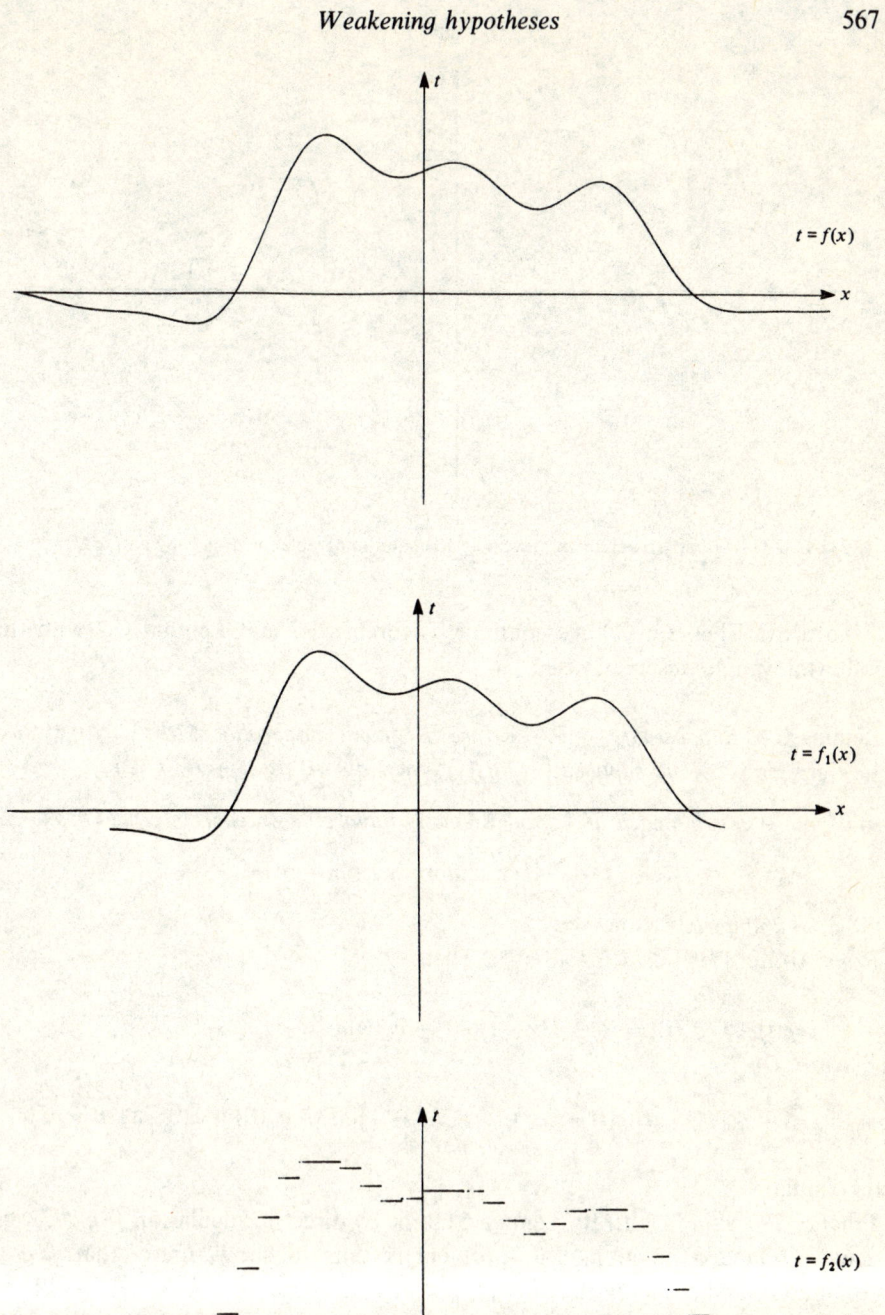

$t = f(x)$

$t = f_1(x)$

$t = f_2(x)$

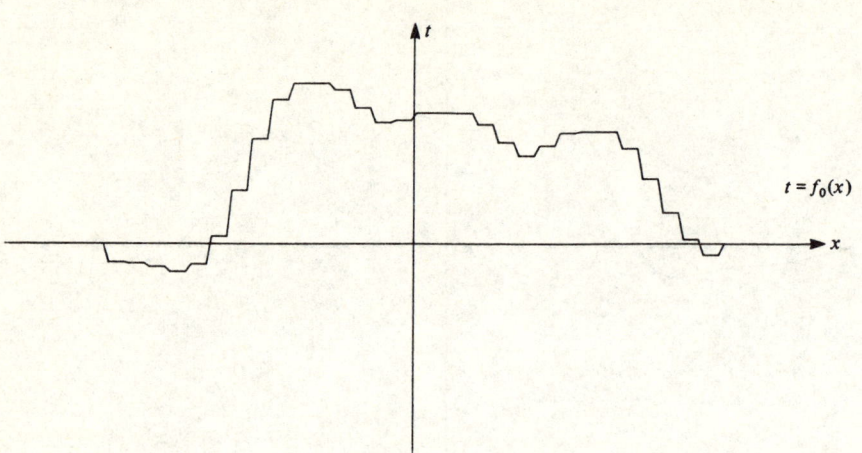

Fig. C.1. How to approximate an absolutely Riemann integrable function f by an $f_0 \in L^1 \cap C$.

To prove Theorem C.1 we combine Theorem 51.7 and Lemma C.2 with the following simple observations.

Lemma C.4. *Suppose* g_m, *$g : \mathbb{R} \to \mathbb{C}$ are Riemann integrable with* $\int_{-\infty}^{\infty} |g_m(y)| \, dy$, $\int_{-\infty}^{\infty} |g(y)| \, dy < \infty$ *and suppose* $\int_{-\infty}^{\infty} |g_m(y) - g(y)| \, dy \to 0$ *as* $m \to \infty$. *Then*

(i) *If* $f : \mathbb{R} \to \mathbb{C}$ *is bounded on* $\mathbb{R}$ *and Riemann integrable on each interval* $[a, b]$

$$f * g_m \to f * g \text{ uniformly as } m \to \infty.$$

(ii) *$\hat{g}_m \to \hat{g}$ uniformly as* $m \to \infty$.

Proof. (i) Let $|f(s)| \leqslant M$ for all $s \in \mathbb{R}$. Then

$$|f * g(t) - f * g_m(t)| = \left| \int_{-\infty}^{\infty} f(x - t)(g(t) - g_m(t)) \, dt \right|$$

$$\leqslant \int_{-\infty}^{\infty} |f(x - t)(g(t) - g_m(t))| \, dt \leqslant M \int_{-\infty}^{\infty} |g(t) - g_m(t)| \, dt \to 0 \quad \text{as } m \to \infty.$$

(ii) Similar. ∎

Proof of Theorem C.1. (i) By Lemma 51.3, or by direct recalculation $f * g$ is a well defined bounded function. Our problem is thus to show, firstly that $f * g$ is continuous, and secondly that $\int_{-\infty}^{\infty} |f * g(t)| \, dt$ converges. To do this we shall need both parts (i) and (ii) of Theorem 51.7.

Observe first that, by Lemma C.3, we can find bounded functions $f_m, g_m \in L^1 \cap C$ with

$$\int_{-\infty}^{\infty} |f(y) - f_m(y)| \, dy, \int_{-\infty}^{\infty} |g(y) - g_m(y)| \, dy \to 0 \quad \text{as } m \to \infty.$$

By Lemma 51.3 (i) and (iii) we have

(a)$_{nm}$ f_n*g_m continuous and bounded,

(b)$_{nm}$
$$\int_{-R}^{R} |f_n*g_m(y)|\,dy \leqslant \int_{-\infty}^{\infty} |f_n*g_m(y)|\,dy$$

$$\leqslant \int_{-\infty}^{\infty} |f_n(y)|\,dy \int_{-\infty}^{\infty} |g_m(y)|\,dy \quad \text{for all } R>0.$$

But keeping n fixed and letting $m\to\infty$, we know from Lemma C.4 that $f_n*g_m \to f_n*g$ uniformly so, since the uniform limit of continuous function is continuous,

(a)$_n$ f_n*g is continuous.

Also if h_m is continuous and $h_m \to h$ uniformly on a bounded interval $[a,b]$ then

$$\int_a^b h_m(x)\,dx \to \int_a^b h(x)\,dx.$$

Thus
$$\int_{-R}^{R} |f_n*g_m(y)|\,dy \to \int_{-R}^{R} |f_n*g(y)|\,dy,$$

and since

$$\left| \int_{-\infty}^{\infty} |g_m(y)|\,dy - \int_{-\infty}^{\infty} |g(y)|\,dy \right| \leqslant \int_{-\infty}^{\infty} \big||g_m(y)|-|g(y)|\big|\,dy$$

$$\leqslant \int_{-\infty}^{\infty} |g_m(y)-g(y)|\,dy \to 0 \quad \text{as } m\to\infty,$$

it follows that

(b)$_n$ $\int_{-R}^{R}|f_n*g(y)|\,dy \leqslant \int_{-\infty}^{\infty}|f_n(y)|\,dy \int_{-\infty}^{\infty}|g(y)|\,dy.$

Letting $n\to\infty$ and repeating the arguments, we get

(a) $f*g$ continuous,
(b) $\int_{-R}^{R}|f*g(y)|\,dy \leqslant \int_{-\infty}^{\infty}|f(y)|\,dy \int_{-\infty}^{\infty}|g(y)|\,dy,$

for all R. Since $|f*g(y)|\geqslant 0$ for all y, this means that $\int_{-\infty}^{\infty}|f*g(y)|\,dy$ converges with

$$\int_{-\infty}^{\infty}|f*g(y)|\,dy \leqslant \int_{-\infty}^{\infty}|f(y)|\,dy \int_{-\infty}^{\infty}|g(y)|\,dy.$$

Thus $f*g\in L^1\cap C$.
(ii) This follows from (iv) on putting $\zeta=0$, so we shall not prove it separately.
(iii) Proved as the last but one sentence of the proof of (i).
(iv) Using (iii) we have

$$\int_{-\infty}^{\infty}|f_n*g_m(y)-f_n*g(y)|\,dy = \int_{-\infty}^{\infty}|f_n*(g-g_m)(y)|\,dy$$

$$\leqslant \int_{-\infty}^{\infty}|f_n(y)|\,dy \int_{-\infty}^{\infty}|g(y)-g_m(y)|\,dy \to 0,$$

so, using Theorem 51.7 (iv) and Lemma C.4 (ii)

$$\hat{f}_n(\zeta)\hat{g}_m(\zeta) = (f_n * g_m)^\wedge(\zeta) \to (f_n * g)^\wedge(\zeta) \quad \text{as } m \to \infty.$$

But by Lemma C.4 (ii) again

$$\hat{g}_m(\zeta) \to \hat{g}(\zeta) \quad \text{as } m \to \infty,$$

so

$$\hat{f}_n(\zeta)\hat{g}(\zeta) = (f_n * g)^\wedge(\zeta).$$

Repeating the same arguments with $n \to \infty$ we get

$$\int_{-\infty}^{\infty} |f_n * g(y) - f * g(y)| \, dy \to 0,$$

$$\hat{f}_n(\zeta)\hat{g}(\zeta) = (f_n * g)^\wedge(\zeta) \to (f * g)^\wedge(\zeta), \hat{f}_n(\zeta) \to \hat{f}(\zeta)$$

and

$$(f * g)^\wedge(\zeta) = \hat{f}(\zeta)\hat{g}(\zeta),$$

as required.   ■

Practically all the results of this book can be generalised in this manner. If the reader feels unhappy because this has not been done, she may tackle the task herself as an extended exercise.

Unfortunately the boundedness restriction in Theorem C.1 cannot be dropped within the context of Riemann integration as the next two examples make clear.

Example C.5. *Given* $x \in \mathbb{R}$ *and* $R > 0$ *we can find an* $f \in L^1 \cap C$ *such that*

(i) $f(t) \geqslant 0$ *for all* $t \in \mathbb{R}$,

(ii) $\int_{-\infty}^{\infty} f(t) \, dt \leqslant 1$,

(iii) $\int_{-\infty}^{\infty} f(x - t) f(t) \, dt$ *diverges*,

(iv) $f(t) = 0$ *for all* $|t| \leqslant R$.

Example C.6. *We can find an* $f \in L^1 \cap C$ *such that*

(i) $f(t) \geqslant 0$ *for all* $t \in \mathbb{R}$,

(ii) $\int_{-\infty}^{\infty} f(t) \, dt \leqslant 1$,

(iii) $\int_{-\infty}^{\infty} f(x - t) f(t) \, dt$ *diverges for all* $x \in \mathbb{Q}$.

(Informally f is a positive continuous function with $\int_{-\infty}^{\infty} f(t) \, dt$ convergent and $f * f$ divergent at all rational points.)

Construction of Example C.5. See Figure C.2. Let N be an integer with $N \geqslant |x| + R + 1$. For each integer $n \geqslant N$ we choose a $K_n \geqslant 0$ and a δ_n with $\frac{1}{2} \geqslant \delta_n \geqslant 0$ and define f as follows

$$\begin{aligned}
f(t) &= K_n \delta_n^{-1}(\delta_n - |n - t|) && \text{for } |n - t| < \delta_n, n \geqslant N, \\
f(t) &= K_n \delta_n^{-1}(\delta_n - |(x - n) - t|) && \text{for } |(x - n) - t| < \delta_n, n \geqslant N, \\
f(t) &= 0 && \text{otherwise.}
\end{aligned}$$

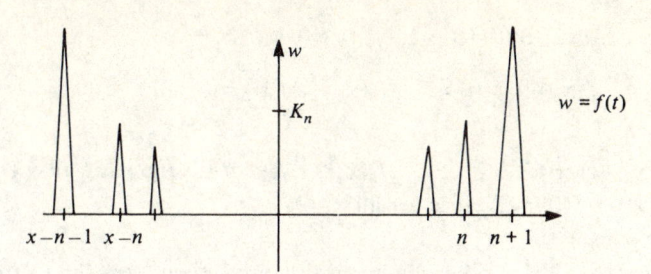

Fig. C.2. An unpleasant function.

Then, provided $\sum_{n=N}^{\infty} K_n \delta_n$ converges, we have

$$\int_{-\infty}^{\infty} f(t)\, dt = 2 \sum_{n=N}^{\infty} \int_{n-\delta_n}^{n+\delta_n} f(t)\, dt = 2 \sum_{n=N}^{\infty} K_n \delta_n.$$

On the other hand,

$$\int_{-\infty}^{\infty} f(x-t)f(t)\, dt \geqslant \sum_{n=N}^{\infty} \int_{n-\delta_n}^{n+\delta_n} f(x-t)f(t)\, dt = \sum_{n=N}^{\infty} \int_{n-\delta_n}^{n+\delta_n} f(t)^2\, dt$$

$$= \sum_{n=N}^{\infty} 2K_n^2 \delta_n^{-2} \int_0^{\delta_n} s^2\, ds = \frac{2}{3} \sum_{n=N}^{\infty} K_n (K_n \delta_n).$$

Thus if we take, for example, $K_n \delta_n = 2^{-n-2}$ and $K_n = 2^{n+2}$ (so that $\delta_n = 2^{-2(n+2)}$) we have a function of the required type. ∎

Construction of Example C.6. Since $\mathbb{Q}$ is countable, we can write $\mathbb{Q} = \{x_1, x_2, x_3, \ldots\}$. Using Example C.5 we can construct $f_n \in L^1 \cap C$ such that

(i)$_n$ $f_n(t) \geqslant 0$ for all $t \in \mathbb{R}$,
(ii)$_n$ $\int_{-\infty}^{\infty} f_n(t)\, dt \leqslant 1$,
(iii)$_n$ $\int_{-\infty}^{\infty} f_n(x_n - t)f(t)\, dt$ diverges,
(iv)$_n$ $f_n(t) = 0$ for all $|t| \leqslant n$.

Condition (iv)$_n$ tells us that for each t all but a finite number of terms in $\sum_{n=1}^{\infty} 2^{-n} f_n(t)$ are zero. Thus $\sum_{n=1}^{\infty} 2^{-n} f_n(t)$ converges to $f(t)$ say at each $t \in \mathbb{R}$.
Further $f(t) = \sum_{n=1}^{m} 2^{-n} f_n(t)$ for $t \in [-m, m]$ so f is continuous on $[-m, m]$ and

(i)$'_m$ $f(t) \geqslant 0$ for all $t \in [-m, m]$,
(ii)$'_m$

$$\int_{-m}^{m} f(t)\, dt = \sum_{n=1}^{m} 2^{-n} \int_{-m}^{m} f_n(t)\, dt \leqslant \sum_{n=1}^{m} 2^{-n} \int_{-\infty}^{\infty} f_n(t)\, dt \leqslant \sum_{n=1}^{m} 2^{-n} < 1.$$

Since this is true for all $m \geqslant 1$, we deduce that f is continuous on $\mathbb{R}$ and

(i) $f(t) \geqslant 0$ for all $t \in \mathbb{R}$,
(ii) $\int_{-\infty}^{\infty} f(t)\, dt \leqslant 1$,

so, indeed, $f \in L^1 \cap C$. But $f(t) \geq 2^{-n} f_n(t) \geq 0$, so

$$f(x_n - t)f(t) \geq 2^{-2n} f_n(x_n - t)f_n(t) \quad \text{for all } t,$$

and

(iii)′ $\int_{-\infty}^{\infty} f(x_n - t)f(t)dt \geq 2^{-2n}\int_{-\infty}^{\infty} f_n(x_n - t)f_n(t)dt$ diverges for all $n \geq 1$, i.e.

(iii) $\int_{-\infty}^{\infty} f(x - t)f(t)dt$ diverges for all $x \in \mathbb{Q}$.　∎

We remark that if the Lebesgue integral is used, then even the $f * f$ of Example C.6 can be considered as a well defined Lebesgue integrable function and Theorem C.1 continues to hold in the following form.

Theorem C.1′. *Suppose* f *and* g *are Lebesgue integrable functions with* $\int_{-\infty}^{\infty}|f(y)|dy, \int_{-\infty}^{\infty}|g(y)|dy < \infty$. *Then*

(i) $f * g$ *is a well defined Lebesgue integrable function and* $\int_{-\infty}^{\infty}|f * g(y)|dy < \infty$,

(ii) $\int_{-\infty}^{\infty} f * g(y)dy = \int_{-\infty}^{\infty} f(y)dy \int_{-\infty}^{\infty} g(y)dy$,

(iii) $\int_{-\infty}^{\infty}|f * g(y)|dy \leq \int_{-\infty}^{\infty}|f(y)|dy \int_{-\infty}^{\infty}|g(y)|dy$,

(iv) $(f * g)^{\wedge}(\zeta) = \hat{f}(\zeta)\hat{g}(\zeta)$.

Proof. This should form part of any respectable course on Lebesgue measure which includes Fubini's theorem. Alternatively the result may be obtained in the manner of this appendix without using Fubini's theorem by using Egorov's theorem (in a form corresponding to Lemma C.2) and the monotone convergence theorem (prove the result first for $f, g \geq 0$ by considering f_m, g_m with $f_m(x) = f(x)$ for $|x| \leq m, f(x) \leq m$ and $f_m(x) = 0$ otherwise).　∎

The increased elegance of results like Theorem C.1′ is an important reason for using the Lebesgue rather than the Riemann integral. As Hardy wrote in 1922: 'No account of the theory of Fourier's series can possibly satisfy the imagination if it takes no account of the ideas of Lebesgue; the loss of elegance and of simplicity of statement is overwhelming (p. 437, Vol. VII, *Collected Works*).' In this book by a judicious selection of topics (the reader may have noticed that Part III dealing with orthogonal functions for $C(\mathbb{T})$ has no echo in our discussion of Fourier transforms) and by evading certain problems (such as the treatment of double integrals of discontinuous functions) I have managed to avoid using Lebesgue measure but measure theory seems to be indispensible in the further development of Fourier analysis. Mathematicians find it easier to understand and enjoy ideas which are clever rather than subtle. Measure theory is subtle rather than clever and so requires hard work to master. However no subject shows better that in Lebesgue's own words.... 'a generalisation made not for the empty pleasure of generalising but to solve existing problems, is always fruitful (p. 374, *Collected Works*, Vol. II).'

We end this appendix by looking in a rather different direction towards an improvement of Lemma C.2 which is sometimes useful.

We observe that in the proof of Lemma C.3 $f_0 = f_2 * k$ where $k(x) = \delta^{-1}/2$ for $-\delta < x \leqslant \delta$, $k(x) = 0$ otherwise. The reader is expressly invited to show that if $K: \mathbb{R} \to \mathbb{R}$ is any Riemann integrable function with

(i) $K(s) \geqslant 0$ for all $s \in \mathbb{R}$,
(ii) $K(s) = 0$ for $|s| > \delta$,
(iii) $\int_{-\infty}^{\infty} K(s)\,ds = 1$,

then, writing $F_0 = f_2 * K$, we have

(iv) $F_0(s) = 0$ for $|s| > R + \delta$,
(v) $\int_{-\infty}^{\infty} |F_0(s) - f_2(s)|\,ds < \varepsilon/3$,

and so

(v)' $\int_{-\infty}^{\infty} |F_0(s) - f(s)|\,ds < \varepsilon$.

But the results of Chapters 52 and 53 show that, if K is n times differentiable, so is $f_2 * K$, so by a suitable choice of K we can make F_0 as many times differentiable as we want. Armed with this insight we can go still further. Define $h: \mathbb{R} \to \mathbb{R}$ by $h(x) = \exp(-1/x^2)$ for $x > 0$ and $h(x) = 0$ for $x \leqslant 0$. The arguments of Example 4.2 show that h is infinitely differentiable and so, setting $k(x) = h(x - \delta)h(x + \delta)$, we have

(i)' $k(s) \geqslant 0$ for all $s \in \mathbb{R}$,
(ii)' $k(s) = 0$ for $|s| > \delta$,
(iii)' $\int_{-\infty}^{\infty} k(s)\,ds > 0$,

and k infinitely differentiable. Setting $A = \int_{-\infty}^{\infty} k(s)\,ds$ and $K = A^{-1}k$, we obtain $f_0 = f_2 * K$ as an infinitely differentiable function.

In effect we have proved the following improvement of Lemma C.2.

Lemma C.7. *If $f: \mathbb{R} \to \mathbb{C}$ is a Riemann integrable function with $\int_{-\infty}^{\infty} |f(y)|\,dy < \infty$ then, given $\varepsilon > 0$, we can find an infinitely differentiable function f_0 and an $R > 0$ such that $f_0(x) = 0$ for $|x| > R$ and $\int_{-\infty}^{\infty} |f(y) - f_0(y)|\,dy < \varepsilon$.*
Proof. This follows the sketch above and is recommended to the reader as an exercise. ∎

As an example of the use of this stronger version of Lemma C.3 we give yet another proof of yet another version of the Riemann–Lebesgue Lemma (see e.g. Theorem 52.11).

Theorem C.8. *If $f: \mathbb{R} \to \mathbb{C}$ is a Riemann integrable function with $\int_{-\infty}^{\infty} |f(y)|\,dy < \infty$ then $\hat{f}(\zeta) \to 0$ as $|\zeta| \to \infty$.*
Proof. Given $\varepsilon > 0$, we can choose f_0 as in Lemma C.7. Since $f(y) = f'(y) = 0$ for $|y| > R$, we have

$$\hat{f}_0(\zeta) = \int_{-\infty}^{\infty} f_0(y) \exp - i\zeta y \, dy$$

$$= \left[f_0(y) \frac{\exp - i\zeta y}{-i\zeta} \right]_{-\infty}^{\infty} + \frac{1}{i\zeta} \int_{-\infty}^{\infty} f_0'(y) \exp - i\zeta y \, dy$$

$$= \frac{1}{i\zeta} \int_{-R}^{R} f_0'(y) \exp - i\zeta y \, dy.$$

Thus $$|\hat{f}_0(\zeta)| \leqslant \frac{1}{|\zeta|} \int_{-R}^{R} |f'(y)| dy \to 0 \quad \text{as } |\zeta| \to \infty,$$

and, in particular, we can find an $S > 0$ with $|\hat{f}_0(\zeta)| < \varepsilon$ for all $|\zeta| > S$. It follows that

$$|\hat{f}(\zeta)| \leqslant |\hat{f}(\zeta) - \hat{f}_0(\zeta)| + |\hat{f}_0(\zeta)|$$

$$\leqslant \int_{-\infty}^{\infty} |f(y) - f_0(y)| dy + \varepsilon \leqslant 2\varepsilon \quad \text{for all } |\zeta| > S,$$

and we are done.   ∎

APPENDIX D:
ODE TO A GALVANOMETER

R.V. Jones (*Bulletin of the IMA*, February 1975) tells the story of how Maxwell invited Kelvin to look at one of his optical experiments.

When Kelvin looked through the eyepiece he saw what was undoubtedly the phenomenon that Maxwell had described, but in addition there was the image of a little man dancing about in the field of view. Maxwell had achieved this by the addition of a zoetrope, a device in which he was much interested and for which he had drawn the animated diagrams. Kelvin could not help asking: 'What is the little man there for?' 'Have another look, Thomson,' said Maxwell 'and you should see'. Kelvin had another look but was no wiser. 'Tell me Maxwell', he said impatiently, 'what *is* he there for?...' 'Just for fun, Thomson'.

Maxwell's sense of fun was also expressed in typical Victorian light verse and parody. Here is the first of two poems entitled *Lectures to Women on Physical Science*.

      PLACE – A small alcove with dark curtains.
            The Class consists of one member.

      SUBJECT – Thomson's Mirror Galvanometer.

The lamp-light falls on blackened walls,
   And streams through narrow perforations,
The long beam trails o'er pasteboard scales,
   With slow-decaying oscillations.
Flow, current, flow, set the quick light-spot flying,
Flow current, answer light-spot, flashing, quivering, dying,
   O look! how queer! how thin and clear,
     And thinner, clearer, sharper growing
   The gliding fire! with central wire,
     The fine degrees distinctly showing.
Swing, magnet, swing, advancing and receding,

Swing magnet! Answer dearest, What's your final reading?
   O love! you fail to read the scale
      Correct to tenths of a division.
   To mirror heaven those eyes were given,
      And not for methods of precision.
Break contact, break, set the free light-spot flying;
Break contact, rest thee, magnet, swinging, creeping, dying.

APPENDIX E:

THE PRINCIPLE OF THE ARGUMENT

Having claimed in Chapter 77 that the principle of the argument is fairly easy to prove, I feel under some obligation to include a proof. However the effect of treating this result in isolation is to produce an air of artificiality out of keeping with its true nature.

We start with a simple observation.

Theorem E.1. *Suppose* $h:\mathbb{C} \to \mathbb{C}$ *is analytic and that* h *has no zeros on a contour* C. *Then the number of zeros* h *within* C (*multiple zeros being counted multiply*) *is equal to* $(2\pi i)^{-1} \int_c (h'(z)/h(z)) \, dz$.

Sketch of proof. Let h have zeros $z_1, z_2, \ldots, z_m$ of orders $n(1)$, $n(2), \ldots, n(m)$, say within C so h'/h is analytic within C except at $z_1, z_2, \ldots, z_n$. Now by the definition of $n(j)$ we can find a $\delta(j) > 0$ and an f_j analytic and non zero in $\{z : |z - z_j| < \delta(j)\}$ such that $h(z) = (z - z_j)^{n(j)} f_j(z)$ for all z with $|z - z_j| < \delta(j)$. Thus

$$\frac{h'(z)}{h(z)} = \frac{n(j)(z - z_j)^{n(j)-1} f_j(z) + (z - z_j)^{n(j)} f_j'(z)}{(z - z_j)^{n(j)} f_j(z)}$$

$$= \frac{n(j)}{(z - z_j)} + \frac{f_j'(z)}{f_j(z)} \quad \text{for } |z - z_j| < \delta(j)$$

and h'/h has a pole with residue $n(j)$ at z_j.

Since h'/h is continuous within and in a neighbourhood of C (except at z_1, $z_2, \ldots, z_n$), Cauchy's theorem now gives

$$\int_C \frac{h'(z)}{h(z)} \, dz = 2\pi i \sum \text{residue} = 2\pi i \sum_{j=1}^m n(j),$$

which is the stated result. ∎

In an old fashioned elementary text, Theorem E.1 would be followed by the formal manipulation

$$2\pi i \sum_{j=1}^{m} n(j) = \int_c \frac{h'(z)}{h(z)} \, dz = \int_c \frac{d}{dz} (\log h(z)) \, dz = [\log h(z)]_c = [\log|h(z)| + i \arg h(z)]_c$$

$$= [\log|h(z)|]_c + i[\arg h(z)]_c = 0 + i[\arg h(z)]_c = i[\arg h(z)]_c,$$

where $[f(z)]_c$ denotes the change in value of f as z runs round the contour and must therefore be zero when f is the single valued $\log|h|$ but need not be zero when f is the multivalued $\arg h$. It follows that $2\pi \sum_{j=1}^{m} n(j) = [\arg h(z)]_c$.

This presentation with its sudden appeal to multivaluedness smacks too much of a pea and thimbles game to be entirely satisfactory. However, it does form the basis of the more extended proof which follows.

Lemma E.2. *Let* $\gamma:[a,b] \to \mathbb{C}$ *be once continuously differentiable and let* $h:\mathbb{C} \to \mathbb{C}$ *be analytic with no zeros on* $\{\gamma(t):t \in [a,b]\}$. *Then if* $h(\gamma(a)) = |h(\gamma(a))| \exp i\theta_a [\theta_a \text{ real}]$ *there exists one and only one continuous function* $\theta:[a,b] \to \mathbb{R}$ *such that* $\theta(a) = \theta_a$ *and* $h(\gamma(t)) = |h(\gamma(t))| \exp i\theta(t)$. *This function is given by*

$$\theta(t) = \theta_a + \text{Im} \int_0^t \frac{h'(\gamma(s))}{h(\gamma(s))} \gamma'(s) \, ds.$$

Proof (Existence). Let $L(t) = \int_a^t \frac{h'(\gamma(s))}{h(\gamma(s))} \gamma'(s) \, ds \quad [t \in [a,b]]$.

Then by the fundamental theorem of the calculus

$$L'(t) = \frac{h'(\gamma(t))}{h(\gamma(t))} \gamma'(t)$$

and so

$$\frac{d}{dt} (h(\gamma(t)) \exp(-L(t))) = \gamma'(t) h'(\gamma(t)) \exp(-L(t))$$

$$- h(\gamma(t)) L'(t) \exp(-L(t)) = 0,$$

whence

$$h(\gamma(t)) \exp(-L(t)) = h(\gamma(a)) \exp(-L(a)) = h(\gamma(a)) \quad \text{for all } t \in [a,b].$$

But

$$h(\gamma(a)) = \exp(\log|h(\gamma(a))| + i\theta_a),$$

so

$$h(\gamma(t)) = \exp(\text{Im } L(t) + \log|h(\gamma(a))| + i\theta_a)$$

$$= |h(\gamma(t))| \exp(i(\theta_a + \text{Im } L(t))) \quad \text{for all } t \in [a,b].$$

Since L is continuous the result follows on setting $\theta(t) = \theta_a + \text{Im } L(t)$.

Uniqueness

Suppose $\phi:[a,b]\to\mathbb{R}$ is a continuous function such that $\phi(a)=\theta_a$ and $h(\gamma(t))=|h(\gamma(t))|\exp i\phi(t)$. Then $\phi(t)-\theta(t)\equiv 0 \bmod 2\pi$ for each $t\in[a,b]$. It follows that $\psi=(2\pi)^{-1}(\phi-\theta)$ is an integer valued continuous function on $[a,b]$ and so constant. (Suppose $\psi(t_1)\neq\psi(t_2)$. Then choose a non integral valued real number τ between $\psi(t_1)$ and $\psi(t_2)$. By the intermediate value theorem there exists a t_0 between t_1 and t_2 with $\psi(t_0)=\tau$ which contradicts the fact that ψ is integer valued.) Thus $\theta(t)-\phi(t)=\theta(a)-\phi(a)=0$ for all $t\in[a,b]$. $\blacksquare$

Lemma E.3. *The result of Lemma E.2 holds if γ is piecewise continuously differentiable.*

Proof. Apply Lemma E.2 to each piece. $\blacksquare$

Theorem E.4. *Let C be a contour given by the piecewise once continuously differentiable function $\gamma:[a,b]\to\mathbb{C}$ with $\gamma(a)=\gamma(b)$. Suppose that $h:\mathbb{C}\to\mathbb{C}$ is analytic and has no zeros on C. Then given $\theta_a\in\mathbb{R}$ with $h(\gamma(a))=|h(\gamma(a))|\exp i\theta_a$ there exists a unique continuous function $\theta:[a,b]\to\mathbb{R}$ such that $\theta(a)=\theta_a$ and $h(\gamma(t))=|h(\gamma(t))|\exp i\theta(t)$. The number of zeros of h within C (multiple zeros being counted multiply) is equal to $(2\pi)^{-1}(\theta(b)-\theta(a))$.*

Proof. Everything except the last sentence is a repeat of Lemma E.3. The last sentence follows from Theorem E.1 and the observation that, by definition,

$$\frac{1}{2\pi i}\int \frac{h'(z)}{h(z)}\,dz = \frac{1}{2\pi i}\int_a^b \frac{h'(\gamma(s))}{h(\gamma(s))}\gamma'(s)\,ds.$$

Thus

$$\text{number of zeros} = \frac{1}{2\pi i}\int_a^b \frac{h'(\gamma(s))}{h(\gamma(s))}\gamma'(s)\,ds,$$

and, taking the real part of both sides,

$$\text{number of zeros} = \frac{1}{2\pi}\,\text{Im}\int_a^b \frac{h'(\gamma(s))}{h(\gamma(s))}\gamma'(s)\,ds = \frac{1}{2\pi}(\theta(b)-\theta(a)). \qquad \blacksquare$$

Theorem 77.3 may be considered as a less precise (but more comprehensible) restatement of Theorem E.4.

APPENDIX F:

CHASE THE CONSTANT

Let $f:\mathbb{R}\to\mathbb{R}$ be a suitably well behaved function. If we set

$$\mathscr{F}f(\lambda) = \frac{1}{A}\int_{-\infty}^{\infty} f(t)e^{iB\lambda t}\,dt,$$

then, as the reader should check,

$$\mathscr{F}(\mathscr{F}f)(t) = \frac{2\pi}{A^2|B|}f(-t),$$

$$(\mathscr{F}f')(\lambda) = -iB(\mathscr{F}f)(\lambda),$$

$$(\mathscr{F}f)'(\lambda) = iB(\mathscr{F}h)(\lambda) \quad \text{(with } h(t) = tf(t)\text{)},$$

and
$$\mathscr{F}(f*g)(\lambda) = A(\mathscr{F}f)(\lambda)(\mathscr{F}g)(\lambda).$$

These formulae suggest no less than six 'natural' choices of A and B in defining the Fourier transform (viz. $A = (2\pi)^{\frac{1}{2}}$, $B = \pm 1$; $A = 1$, $B = \pm 2\pi$; $A = 1$, $B = \pm 1$) and all of these will be found in practice. However it seems to me that the majority of authors use either $A = 1$, $B = 1$ or (as I have done in this book) $A = 1$, $B = -1$ and that there is no particular reason to go against the majority in this case.

APPENDIX G:
ARE SHARE PRICES
IN BROWNIAN MOTION?

Most of the figures in this book have been computer generated by David Jackson. I should like to thank him for helping 'to give artistic verisimilitude to an otherwise bald and unconvincing narrative'. Since my book has given rise to his diagrams it is not unreasonable that one of his diagrams should have given rise to this last appendix.

Let us rework Chapter 14 in one rather than two dimensions. We now consider a particle which moves from the point nh on $\mathbb{R}$ to the point $(n + 1)h$ with probability $\frac{1}{2}$

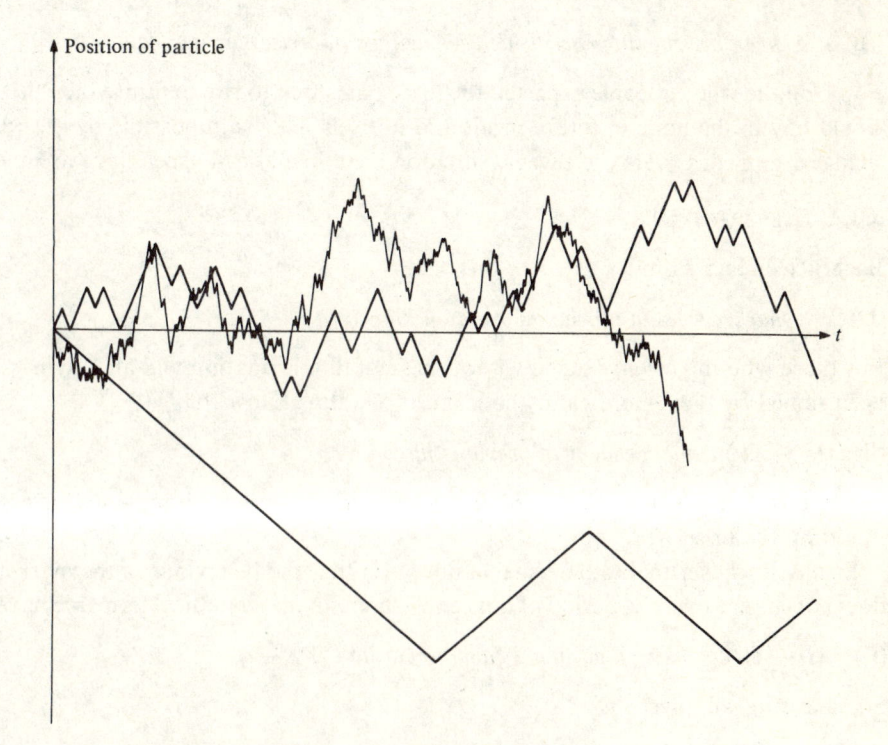

Fig. G.1. Scaled random walks with successively shorter step lengths.

and to the point $(n-1)h$ with probability $\frac{1}{2}$, each step taking a time τ and being independent of previous history. Just as in Chapter 14 we let the step length h decrease towards 0 keeping $\tau^{-1}h^2$ fixed. David Jackson has generated some typical paths for various values of h and these are shown in Figure G.1.

If the reader looks at these graphs she cannot, I think, fail to be reminded of the construction of a nowhere differentiable function illustrated in Figure 11.1. But she may also be reminded of the graphs of stockmarket prices which appear in the financial press.

Five years before Einstein's paper on Brownian motion which led to the work of Perrin described in Chapter 13, a pupil of Poincaré called Bachelier wrote a thesis on the behaviour of stock market prices. His initial argument may, I think, be summarised as follows.

Let $X(t)$ be the price of a certain government security (or a ton of a certain commodity or a share in some company) at a time t, and $X(s)$ the price at some later time $s > t$. What can be said about the random variable $X(s) - X(t)$? The following propositions seem very plausible.

(A) *The price $X(t)$ is one at which as many people wish to buy as to sell.*

For if more people wished to buy than to sell the owners could (and therefore would) demand a higher price. (A similar argument covers the reverse situation.) From this it follows

(B) *At a given instant the market believes neither in a rise nor in a fall.*

For if, on the whole, people expected the price of a stock to rise in future then they would buy in the hope of future profit and not sell. This is impossible by (A). (A similar argument covers the reverse situation.) Mathematically speaking we have

(C) $\mathbb{E}(X(s) - X(t)) = 0$.

Bachelier further assumes

(D) *The market acts on all the information known to it.*

For those who might have acted without using full information will already have been ruined by their more canny colleagues. We thus know that

(E) $X(s) - X(t)$ *is independent of previous history.*

For by (D) the market has extracted all possible information from previous history in setting the price $X(t)$.

Finally Bachelier makes the usual assumption that the behaviour of the market does not change over the period of time considered. In mathematical terms we have

(F) $X(s) - X(t)$ *is distributed like a random variable $Y(s - t)$.*

Summarising we have

Bachelier's principle (first version). *In an (ideal) stock-market the price function X is a random variable such that $X(s) - X(t)$ is distributed like a certain random variable $Y(s - t)$ independent of previous history $[s > t]$. Further $\mathbb{E}Y(t) = 0$ for all $t > 0$.*

Since $X(s + t) = (X(s + t) - X(t)) + (X(t) - X(0))$ it follows that the sum of two independent variables distributed like $Y(s)$ and like $Y(t)$ must be distributed like $Y(s + t)$. Bachelier observes that

(*)   *Y(t) is normal with mean 0 variance $\sigma^2 t$*

is a solution for this problem but fails to exclude the possibility that other solutions exist. For example the argument of Lemma 50.3 (ii) shows (as the reader should check) that

(**)   *Y(t) is Cauchy with density $t/\pi(t^2 + x^2)$*

is another possible solution (at least if we replace the condition '$\mathbb{E}Y(t) = 0$' by the condition '$Y(t)$ symmetric').

Bachelier backed up his solution (*) by an appeal to de Moivre's theorem on the lines that we followed in Chapter 14. Here his intuition did not fail him for, many years later, Lévy proved a very deep mathematical version of the 'Law of Errors' (Chapter 69) and used it to obtain the following theorem.

Theorem (Lévy). *If X satisfies Bachelier's conditions and if, further, X is continuous (with probability 1) then Y(t) is normal with mean 0 and variance $\sigma^2 t$.*

Thus if we add the following natural condition

(G) *The share price X(t) varies continuously*

to our postulates for an (ideal) stock-market we obtain

Bachelier's principle (final version). *In an (ideal) stock-market the price function X is a random variable such that*

$$Pr\left(\frac{X(s) - X(t)}{\sigma(s - t)^{\frac{1}{2}}} \in [a, b]\right) = \left(\frac{1}{2\pi}\right)^{\frac{1}{2}} \int_a^b \exp\left(-\frac{x^2}{2}\right) dx,$$

independent of previous history.

In other words, in an (ideal) stock-market share prices will be in Brownian motion.

Bachelier goes on to compare his theoretical predictions with the actual behaviour of the market in French Government bonds. The agreement is excellent. He also solves what we would now call the problem of first crossing times (what is the probability distribution of T_a the first time t that $X(t) = a$). For the modern reader Bachelier's thesis is full of remarkable anticipations of later work. But (and

here I speak as the former secretary of a mathematics degree committee) considered as a doctoral dissertation in mathematics it looks very odd indeed. It must have required both insight and courage on the part of the examiners to pass it. As it was they gave it a 'mention honorable' rather than the 'mention très honorable' given to theses of the first rank and though he eventually became professor at Besançon even this minor distinction was a long time in coming. By the time the mathematical community rediscovered his thesis Wiener, Lévy and Kolmogorov had gone far beyond the mathematical results contained in it. (However Kolmogorov specifically acknowledged a debt to Bachelier.)

Economists do not read obscure theses in mathematics but in 1953 the eminent statistician M.G. Kendall wrote a long paper studying the behaviour of various financial markets over a long time. (This paper is also notable as one of the first to use a computer to analyse large amounts of data.) He was looking for statistical regularities but he did not find them. Instead:

> it seems that the change in price from one week to the next is practically independent of the change from that week to the week after. This alone is enough to show that it is impossible to predict the price from week to week from the series itself. And if the series really is wandering, any systematic movements such as trends or cycles which may be 'observed' in such series are illusory. The series looks like a 'wandering' one, almost as if once a week the Demon of Chance drew a random number from a symmetrical population of fixed dispersion and added it to the current price to determine the next week's price. And this, we may recall, is not the behaviour in some small backwater market. The data derive from the Chicago wheat market over a period of fifty years.

(Kendall's paper and Bachelier's thesis are reprinted in *The Random Character of Stock Market Prices*, P.H. Cootner, ed., MIT, 1964.)

'It may be', he speculated, 'that the motion is genuinely random and that what looks like a purposive movement over a long period is merely a kind of economic Brownian motion.' Stimulated by Kendall's data economists have now discovered Bachelier. His principle has been tested repeatedly for many markets. As one might expect the agreement is not perfect but it is good. Share prices in the real world do perform (approximate) Brownian motion.

INDEX

Abel
  convergence test, 539–40
  reproached by Fourier, 532
Adleman, secret codes, 505, 509
Airy, failures of imagination, 30, 333
analytic function
  Cauchy Riemann equations, 121
  importance in number theory, 546–56
  key theorems, 379–80, 382–3, 397–9
  Laplace transform is analytic, 379–82
  real part harmonic, 122–3
  (and vice versa), 124, 126–9, 136–8
Arnold, catastrophe theory, 112
author
  clad in a little brief authority, 584
  clad in nothing, 395

Bachelier, economic Brownian motion,
  582–4
Bell, invention of telephone, 25
Bessel function, 174, 486–7
Brown, discoverer of Brownian motion, 43
Brownian motion
  economic, 581–4
  mathematical (*see* separate entry)
  physical, 43–5
Brownian path
  defined, 458
  Lévy's theorem, 458
  properties, 459
  used, 459–60, 467–9
Burgess Davis, (*see* Davis, Burgess)
Burt, 429–41

Cambridge
  difficulty of exams, in old days, 270–1
  mathematicians, corrupted by quantum
    mechanics, 426

mathematicians, Heaviside's sour view
  of, 370
probity of teachers, 272
Carleson, convergence theorem, 75
Casorati Weierstrass little theorem, proof
  via Brownian motion, 459
catastrophe theory, 111–12
Cauchy
  distribution, 246–52, 583
  inversion formula, independent
    discovery of, 295
  Liouville's theorem, 175
central limit theorem
  counter example, 251–2
  general form, 357–61
  multidimensional, 413–7
  proof, 347–56
  used to construct Brownian motion, 50
  warnings, 347, 361
character
  defined, 533
  principal, 533
characteristic function, another name for
  Fourier transform q.v., 245
Chebychev (*see* Tchebychev)
chi squared
  distribution, 415, 423
  test, 418–24, 436–8
Clarke, A.M. and A.D.B., on Burt, 435,
  440–1
codebreaking
  English Civil War, 504–5
  World War II, 92, 503
code making
  general criteria, 503–5
  proposal of Rivest, Shamir and
    Adleman, 509–11
convolution on the circle
  defined, 259

Final Remark

Indexing is the author's last task. Apart from reflecting on the probable appearance in the near future of a 'Journal of Fuzzy Indexing Theory' all that I can now do is to thank readers in advance for all corrections and suggestions for improvements they may care to send me.